Behavioral Genetics of the Mouse
Volume I

Genetics of Behavioral Phenotypes

The first volume in the new **Cambridge Handbooks in Behavioral Genetics**, *Behavioral Genetics of the Mouse* provides baseline information on normal behaviors, essential in both the design of experiments using genetically modified or pharmacologically treated animals, and in the interpretation and analyses of the results obtained. Offering a comprehensive overview of the genetics of naturally occurring variation in mouse behavior; from perception and spontaneous behaviors, such as exploration, aggression, social interactions, and motor behaviors, to reinforced behaviors such as the different types of learning. Also included are numerous examples of potential experimental problems, which will aid and guide researchers trying to troubleshoot their own studies. A lasting reference, the thorough and comprehensive reviews offer an easy entrance into the extensive literature in this field, and will prove invaluable to students and specialists alike.

Wim E. Crusio is a Research Director at the Centre National de la Recherche Scientifique (CNRS). With over 35 years' experience in mouse behavioral neurogenetics and the analysis of mouse behavior, his recent interests cover mouse models of depression, autism, and Fragile X syndrome.

Frans Sluyter is an expert in (wild) house mouse behavior and neurogenetics at the Bioscience Project, Wakefield, MA, USA. His interests range from aggression, addiction, depression, and (stress) coping strategies to gene expression profiles and consciousness.

Robert T. Gerlai has held numerous positions in academia, biotechnology, and the pharmaceutical research industry and is currently a Professor in the Department of Psychology at the University of Toronto. He studies biological and genetic mechanisms of behavior using reverse and forward genetic and psychopharmacological methods in mice and zebrafish.

Susanna Pietropaolo is a Researcher at the Centre National de la Recherche Scientifique (CNRS). She is an expert in the behavioral analysis of the laboratory mouse, with a special interest in interspecific social behaviors. Her recent research focuses on mouse models of social dysfunction, including autism and Fragile X syndrome.

Cambridge Handbooks in Behavioral Genetics

Series editor

Wim E. Crusio

Behavioral Genetics of the Mouse Volume I

Genetics of Behavioral Phenotypes

Edited by
Wim E. Crusio
University of Bordeaux and CNRS, Talence, France
Frans Sluyter
BioScience Project, Wakefield, MA, USA
Robert T. Gerlai
University of Toronto at Mississauga, Mississauga, Ontario, Canada
Susanna Pietropaolo
University of Bordeaux and CNRS, Talence, France

CAMBRIDGE
UNIVERSITY PRESS

CAMBRIDGE UNIVERSITY PRESS
Cambridge, New York, Melbourne, Madrid, Cape Town,
Singapore, São Paulo, Delhi, Mexico City

Cambridge University Press
The Edinburg Building, Cambridge CB2 8RU, UK

Published in the United States of America by Cambridge University Press, New York

www.cambridge.org
Information on this title: www.cambridge.org/9781107034815

First published 2013

Printed and bound by CPI Group (UK) Ltd, Croydon CR0 4YY

A catalogue record for this publication is available from the British Library

Library of Congress Cataloguing in Publication data
Behavioral genetics of the mouse.
volumes cm. – (Cambridge handbooks in behavioural genetics)
ISBN 978-1-107-03481-5 (hardback)
1. Mice – Genetics. 2. Behavior genetics. I. Crusio, W. E., editor of compilation.
QH432.B44 2013
599.35135 – dc23 2012030953

ISBN 978-1-107-03481-5 Hardback

Dedications

To my mother, Annie Crusio-Jordans
Wim E. Crusio

To my son Sam
Frans Sluyter

I would like to thank Julia, my wife, and Flora and Mark, my children, who supported me throughout this book project with their love and happiness.
Robert T. Gerlai

Contents

Contributors

Nazia M. Alam
Department of Physiology and Biophysics,
Weill Cornell Medical College,
Burke Medical Research Institute,
White Plains, New York, USA

Enrico Alleva
Section of Behavioral Neuroscience,
Department of Cell Biology and Neurosciences,
Istituto Superiore di Sanita,
Roma, Italy

Hiroyuki Arakawa
Department of Psychology,
University of New York,
Binghampton, New York, USA

Robert H. Benno
Department of Biology,
William Paterson University,
Wayne, New Jersey, USA

Fred G. Biddle
Department of Medical Genetics,
Alberta Children's Hospital Research Institute for Child and Maternal Health,
Faculty of Medicine,
Department of Biological Sciences,
Faculty of Science,
University of Calgary,
Calgary, Alberta, Canada

D. Caroline Blanchard
Pacific Biosciences Research Center and
Department of Genetics and Molecular Biology,
John A. Burns School of Medicine,
University of Hawaii,
Hawaii, USA

Robert J. Blanchard
Pacific Biosciences Research Center and
Department of Psychology,
University of Hawaii,
Hawaii, USA

Richard J. Bodnar
Neuropsychology Doctoral Sub-program and
Department of Psychology,
Queens College,
City University of New York,
New York, USA

John D. Boughter, Jr.
Department of Anatomy and Neurobiology,
University of Tennessee Health Center,
Memphis, Tennessee, USA

Igor Branchi
Section of Behavioral Neuroscience,
Department of Cell Biology and Neurosciences,
Istituto Superiore di Sanita,
Roma, Italy

Richard E. Brown
Department of Psychology,
Dalhousie University,
Halifax, Nova Scotia, Canada

Abel Bult-Ito
Behavioral and Evolutionary Neuroscience Laboratory,
Department of Biology and Wildlife,
University of Alaska Fairbanks,
Fairbanks, Alaska, USA

Jonathan M. Cachat
Department of Pharmacology,
Tulane University Medical School,
New Orleans, Louisiana, USA

Peter R. Canavello
Department of Pharmacology,
Tulane University Medical School,
New Orleans, Louisiana, USA

Francesca Cirulli
Section of Behavioral Neuroscience,
Department of Cell Biology and Neurosciences,
Istituto Superiore di Sanita,
Roma, Italy

Giovanni Colacicco
Institute of Anatomy,
University of Zürich,
Zürich, Switzerland

John C. Crabbe
Department of Behavioral Neuroscience,
Oregon Health and Science University; and
Portland Alcohol Research Center,
Department of Veterans Affairs Medical Center,
Portland, Oregon, USA

Jacqueline N. Crawley
Laboratory of Behavioral Neuroscience,
Intramural Research Program,
National Institute of Mental Health,
Bethesda, Maryland, USA

Wim E. Crusio
Institut de Neurosciences Cognitives et
Intégratives d'Aquitaine,
Université de Bordeaux, and CNRS,
Talence, France

Sietse F. de Boer
Department of Behavioral Physiology,
Biological Center,
University of Groningen,
Haren, The Netherlands

Ekrem Dere
Institute of Experimental Psychology,
Department of Physiological Psychology,
Heinrich–Heine University of Düsseldorf,
Düsseldorf, Germany

Brenda A. Eales
Department of Medical Genetics,
Alberta Children's Hospital Research Institute
for Child and Maternal Health,
Faculty of Medicine,
Department of Biological Sciences,
Faculty of Science,
University of Calgary,
Calgary, Alberta, Canada

Robert T. Gerlai
Department of Psychology,
University of Toronto at Mississauga,
Mississauga, Ontario, Canada

Howard K. Gershenfeld
Department of Psychiatry,
University of Texas Southwestern Medical Center,
Dallas, Texas, USA

Thomas J. Gould
Department of Psychology,
Temple University,
Philadelphia, Pennsylvania, USA

Martin E. Hahn
Emeritus,
Department of Biology,
William Paterson University,
Wayne, New Jersey, USA

Peter C. Hart
Department of Pharmacology,
Tulane University Medical School,
New Orleans, Louisiana, USA

Andrew Holmes
Section on Behavioral Science and Genetics,
Laboratory for Integrative Neuroscience,
National Institute on Alcoholism and Alcohol Abuse,
National Institutes of Health,
Rockville, Maryland, USA

Joseph P. Huston
Institute of Experimental Psychology,
Department of Physiological Psychology,
Heinrich–Heine University of Düsseldorf,
Düsseldorf, Germany

Allan V. Kalueff
Department of Pharmacology,
Tulane University Medical School,
New Orleans, Louisiana, USA

Benjamin Kest
Neuropsychology Doctoral Sub-program and
Department of Psychology,
College of Staten Island,
City University of New York,
New York, USA

Robert Lalonde
Université de Rouen,
Faculté de Médecine et de Pharmacie,
Rouen, Cedex, France and CHUM/St-Luc,
Unité de Recherche en Sciences Neurologiques,
Montréal, Québec, Canada

Sarah R. Lewis-Levy
Neuropsychology Doctoral Sub-program and
Department of Psychology,
Queens College,

City University of New York,
New York, USA

Hans-Peter Lipp
Institute of Anatomy,
University of Zürich,
Switzerland

Sheree F. Logue
Wyeth Research, Princeton, New Jersey, USA

Stephen C. Maxson
Department of Psychology,
University of Connecticut,
Storrs, Connecticut, USA

Jeffrey S. Mogil
Department of Psychology and Centre for Research on Pain,
McGill University,
Montreal, Québec, Canada

Douglas A. Monks
Department of Psychology,
University of Toronto at Mississauga,
Mississauga, Ontario, Canada

Dennis L. Murphy
Department of Pharmacology,
Tulane University Medical School,
New Orleans, Louisiana, USA

Lee Niel
Department of Population Medicine,
Ontario Veterinary College,
University of Guelph,
Guelph, Ontario, Canada

Timothy P. O'Leary
Department of Psychology,
Dalhousie University,
Halifax, Nova Scotia, Canada

Susanna Pietropaolo
Institut de Neurosciences Cognitives et
Intégratives d'Aquitaine,
Université de Bordeaux,
Talence, France

Peter K.D. Pilz
Institut für Neurobiologie,
Universität Tübingen,
Tübingen, Germany

Claudia F. Plappert
Institut für Neurobiologie,
Universität Tübingen,
Tübingen, Germany

Bernard Possidente
Department of Biology,
Skidmore College,
Saratoga Springs,
New York, USA

Glen T. Prusky
Department of Physiology and Biophysics,
Weill Cornell Medical College,
Burke Medical Research Institute,
White Plains, New York, USA

Laura Ricceri
Istituto Superiore di Sanità,
Section of Neurotoxicology
and Neuroendocrinology,
Department of Cell Biology and Neuroscience,
Rome, Italy

Heather Schellinck
Department of Psychology,
Dalhousie University,
Halifax, Nova Scotia, Canada

Herbert Schwegler
Institut für Anatomie,
Otto von Guericke Universität,
Magdeburg, Germany

Burton Slotnick
Department of Psychology,
University of South Florida,
Tampa, Florida, USA

Frans Sluyter
Bioscience Project,
Wakefield, MA, USA

Shad B. Smith
Department of Psychology and Centre for
Research on Pain,
McGill University,
Montreal, Québec, Canada

Catherine Strazielle
Université Henri Poincaré,
Nancy I, Laboratoire de Pathologie Moléculaire et Cellulaire
des Nutriments,
and Service de Microscopie Electronique,
Faculté de Médecine,
Vandoeuvre-les-Nancy, France

Douglas Wahlsten
Department of Psychology,
University of Alberta,
Edmonton, Alberta, Canada

Hans Welzl
Division of Neuroanatomy and Behavior,
Institute of Anatomy,
University of Zürich,
Zürich, Switzerland

James F. Willott
Department of Psychology,
University of South Florida, Tampa, Florida, USA
and The Jackson Laboratory,
Bar Harbour, Maine, USA

David P. Wolfer
Institute of Anatomy,
University of Zürich; Institute of Human Movement Sciences and Sport, ETH Zürich;
Zürich Center of Integrative Human Physiology,
University of Zürich,
Zürich, Switzerland

Armin Zlomuzica
Institute of Experimental Psychology,
Department of Physiological Psychology,
Heinrich–Heine University of Düsseldorf,
Düsseldorf, Germany

Preface

From the earliest beginnings at the start of the twentieth century, behavior geneticists have been interested in explaining individual differences in behavior: why do not all individuals within a given species display the same levels of aggressiveness, intelligence, curiosity, and such? The answer, of course, is that not all individuals share the same environment and that there can be quite large genetic differences between them. Together, these environmental and genetic variations, and their interplay, cause individuals to differ from each other.

Of this genetic variation, only little is directly due to *de novo* mutations (which in most cases lead to pathological variation). Hence, if one is interested in the causation of *naturally-occurring* variation in behavior (i.e., which has not been weeded out by natural selection and hence represents adaptively acceptable phenotypical characteristics), one should concentrate on the genetic variation that already exists in populations. Induced mutations may be more useful in uncovering the physiological regulation of behavior and may allow the construction of valuable animal models of human neuropsychiatric disorders. Of course, in some species, such as humans and great apes, ethical and practical considerations prevent the study of mutations other than spontaneous ones.

The present Handbook is intended to give an overview of the current state-of-the-art in behavior genetics of the mouse. It consists of three volumes and is part of a larger series of *Cambridge Handbooks in Behavior Genetics*. This first volume covers our current knowledge about the genetic underpinnings of naturally-occurring, non-pathological individual differences in behavior. Mouse models for human neuropsychiatric disorders, including those based on induced mutations, will be the subject of the next volume in this series. The third volume will present detailed protocols for the different tests of mouse behavior described in the first two volumes.

While editing this book, we were aided by many colleagues who acted as anonymous reviewers of one or more of the chapters. We extend our warmest thanks to them. For manifold reasons, the gestation time of this book has been much longer than originally foreseen. We therefore thank the many authors not only for their excellent contributions, but also for their patience and continued faith in us and this project.

Chapter 1

Behavior genetics

Where do we come from and where are we going?

Wim E. Crusio and Robert T. Gerlai

Behavior genetics

Genetics is the science that studies the nature and action of genes, their transmission from parents to offspring, and their allele frequencies within populations. As Muller (1922) formulated it, "the question as to what the general principle of gene construction is ... is the most fundamental question of genetics." This question has largely been answered and at present there is a huge amount of literature about the chemical structure of the genetic material, DNA, as well as vast databases containing the complete genomic sequences of an increasing list of organisms. Although (rather important) questions still remain, in principle at least the functioning of DNA is understood. This even led Caspari (1979) to conclude that "Genetics is in many respects a dead science." Obviously, many researchers living in the current exciting times where ever more detailed genetic information and increasingly sophisticated genetic tools become available probably disagree with this, but whether or not genetics has transcended from being a scientific endeavor in its own right to becoming an applied technology, the fact remains that it represents an elaborate and integrated system of well-established facts and theory that can be used by other fields of the life sciences, such as behavioral science. The traits studied by genetics are referred to as phenotypes (Crusio, 2002). Behavioral traits are one class of phenotypes and this choice of subject matter characterizes behavior genetics.

Although many consider behavior genetics to be a young field, one could regard Francis Galton (1822–1911) as the first behavior geneticist. However, one of the more explicit signs of the birth of behavior genetics as a separate scientific field was Hall's seminal chapter on psychogenetics in Stevens' *Handbook of Experimental Psychology* (1951). The starting point of behavior genetics is nevertheless more conveniently placed with the publication of Fuller and Thompson's book *Behavior Genetics* (1960), which indicated in the words of Lindzey et al. (1971) "a fully developed self-awareness of an important new specialty." In this chapter we will first discuss the focus of study of behavior genetics, behavioral phenotypes, followed by an exploration of the goals of the field and ending with a brief appraisal of how far we have come in realizing these goals.

Behavioral phenotypes

Fuller and Wimer (1973) divided phenotypes into two rather broad categories: somatophenes and psychophenes. This classification was extended by Fuller (1979b) and can briefly be outlined as follows. Firstly, we have somatophenes. These are characteristics such as body size and shape, pigmentation, etc. They are defined therefore by *structural* criteria. These somatophenes may be divided further into chemophenes, as for instance type of hemoglobin, and morphenes, such as body shape. Secondly, we have behavioral phenotypes, which sometimes also are referred to as psychophenes. These are measured directly or indirectly from behavior and are therefore defined by *process* rather than by structure. A further subdivision of psychophenes leads to the recognition of ostensible and inferred psychophenes. The former are based on the occurrence, frequency, and intensity of an objectively defined (behavioral) act. Inferred psychophenes are more general attributes or states of an organism such as anxiety levels and emotionality. The third and last category of phenotypes is formed by the syndromes. These are groups of psychophenes, usually occurring together with somatophenes. Some well-known examples include Down's syndrome, schizophrenia, and Fragile X syndrome.

Psychophenes and syndromes are the subjects of behavior genetics. Of these, inferred psychophenes and syndromes are usually the most difficult to define or interpret. The syndrome schizophrenia, for instance, is not always easy to delineate from related syndromes such as bipolar disorder (DSM-IV-TR, 2000; Kraemer et al., 2007), which of course can raise doubt about the justification of a patient's inclusion in or exclusion from an experimental group. Similarly, inferred psychophenes often give rise to interpretational difficulties, e.g., the concepts of "emotionality" (Fuller and Thompson, 1978) or "anxiety" (Stanford, 2007). Therefore, it is often safer to reduce the inferred psychophenes to the ostensible psychophenes on which they are based. Unfortunately, with the current emphasis on "translational" research, nowadays this is done only rarely and analyses abound of constructs such as "behavioral despair," "anxiety," etc. that are not always well defined (or even defined only

Behavioral Genetics of the Mouse: Volume I. *Genetics of Behavioral Phenotypes*, eds. Wim E. Crusio, Frans Sluyter, Robert T. Gerlai, and Susanna Pietropaolo. Published by Cambridge University Press.

operationally, hence the frequent discrepancies between results of different tests that are supposed to measure the same construct). On a more positive note, as stated by Fuller (1979a), "Perhaps this is . . . the very core of psychology. Here is where the action is. (But) . . . behavior geneticists should be aware of its inferential status and of possible arbitrary judgment in the definition of its members."

In selecting the appropriate psychophene for a behavior genetic study, several aspects should be kept in mind. The chosen phenotype should be one that can be measured reliably (and for economical reasons also with relative ease) and be of interest for ethological, psychological, and/or evolutionary theory (or as a model of a neuropsychiatric phenotype, the subject of Volume II of this Handbook). Furthermore, there are important advantages to choosing a measure of some elements of behavior that are relevant to the organism under natural conditions (Gerlai, 1999; Gerlai and Clayton, 1999) and, of course, at least some genetic variance (be it natural or induced) for the phenotype of choice should be present in the population studied (Fuller, 1979c; Henderson, 1979).

Aims and purposes of behavior genetics

The aims and purposes of behavior genetics have been formulated many times and by many different persons (mostly in the past: it appears that nowadays researchers are less inclined to spend time reflecting on the how and why of their chosen field of investigation). To start, Hall (1951) formulated four objectives of what he termed "psychogenetics": "(1) to discover whether a given behavior pattern is transmitted from generation to generation, (2) to determine the number and nature of the genetic factors involved in the trait, (3) to locate the gene or genes on the chromosomes, and (4) to determine the manner in which the genes act to produce the trait" (statements that still sound astonishingly modern, as do, in fact, large parts of the rest of Hall's chapter). Twenty years later, Thiessen (1972) posed eight questions for what had now become commonly called "behavior genetics": "(1) Is the observed behavior influenced by variations in genotype? (2) What proportions of the measured variability of a trait are the result of genetic and environmental factors? (3) Given a clear-cut genetic effect, how many genes are operating? (4) What is the frequency with which the gene appears within a population or a species group? (5) How is the gene modified by changes in the course of development or by environmental contingencies? (6) What structure and physiological processes intervene between the genetic constitution of an organism and the ultimate expression of behavior? (7) Does the trait have adaptive significance (that is, reproductive fitness), and is it subject to natural and artificial selection pressures? (8) What are the phylogenetic relationships of the behavior with related species?" Dewsbury (1978) condensed this to six very similar questions. Fuller and Thompson (1978) further reduced this to three basic questions: whether the psychophene is transmitted genetically, how the genes are distributed in space and time, and how the genes produce their behavioral effects. Finally, these questions were reduced to just two fundamental problems by van Abeelen (1979) who saw the goal of behavior genetics in the analysis of the phylogenetic as well as the phenogenetic causes of the psychophenes studied.

Following this, we can say that the ultimate aims of behavior genetics are twofold. The first aim concerns the investigation of the physiological substrates of psychophenes and the role of the environment therein (the phenogenetic aspect of the causation of behavior). At this point the profound influence of the environment on most of an organism's behavior must be stressed again. Not only are environmental effects one of the major sources of non-genetic variation, but genotype × environment interactions are also very important (Wahlsten et al., 2003). Therefore, the environmental contribution towards a psychophene is one of the major concerns of behavior genetics, too. The second aim of behavior genetics lies in analyzing the role of psychophenes in individual fitness, which of course includes the evolutionary history of the chosen behavior (the phylogenetic aspect of the causation of behavior).

Regardless of which one of these two aims was being addressed, genetics originally inquired about individual differences, that is, how differences between individuals come about either in a gene–physiological or in an evolutionary sense. Nowadays, the stress is often much more on the gene–physiological aspect, frequently completely ignoring individual variation: the question has become how genes lead to the expression of a certain psychophene, regardless of the question whether this psychophene is variable within the population or not.

However, the analysis of naturally-occurring genetic variation has important merits. First of all, the genetic differences represent physiologically relevant variation. The differences are obviously not so dramatic as to jeopardize viability and, more importantly, the reproductive fitness of the animal, and they represent conditions that have enabled the animal to survive successfully. Another advantage of analyzing naturally-occurring genetic variability is that by doing so one may be able to link the different variants to certain ecological conditions and/or explain the natural selection forces, i.e., the evolutionary past, which shaped the behavior in question.

Of course, genes that influence phenotypes with a high fitness component, i.e., those characteristics that are crucial from an evolutionary perspective, should show limited or no variability due to the strong selection pressures exerted on them (Broadhurst and Jinks, 1974). However, non-variable loci can nowadays also be studied using modern recombinant DNA technologies that allow the introduction of artificial novel mutations, approaches called reverse (or targeted) mutagenesis and forward (or random) mutagenesis.

How far have we come?

In the late 1960s and early 1970s many scientists still needed to be convinced that heredity could, in fact, influence behavioral differences between individuals. Proving that this was the case was hampered by the fact that molecular–genetic techniques enabling the localization of genes were still rudimentary or non-existent. However, by the late 1980s and the early 1990s, almost all serious scientists had come to accept the importance of genetics for understanding interindividual variation in behavior (Plomin et al., 2003). As may be expected, this realization occurred earlier in animal genetics than in human genetics. However, once this realization had taken root, it became possible to expand behavior–genetic research beyond the simple calculation of heritability (h^2, the proportion of phenotypical variance in a population that can be attributed to heredity). This was an important advance, since "heritability analysis" in itself is not very interesting or useful, apart from the questions whether h^2 differs significantly from 0 (showing that significant genetic effects are present) or from 1 (demonstrating that significant environmental influences are present). In parallel with the enormous advances in molecular genetics, it has now become increasingly feasible to investigate the *mechanisms* underlying interindividual differences and identify and analyze the underlying genes, the subject matter of the current volume. In addition, going beyond classical methods such as genetic selection and random mutagenesis, it is now possible to modify genes in animals in a targeted manner by inserting foreign genetic material into the genome in such a way that it gets expressed or by "knocking out" specific genes. The latter methods have become invaluable tools in investigating the gene–physiological bases of behavior and, in addition, have made it possible to create pertinent models for single-gene disorders, the subject matter of Volume II of this Handbook.

Where are we going?

Predicting the future course of a field of scientific endeavor is always a risky undertaking at best and futile at worst. It is difficult to impossible to know for certain what new methods and techniques will become available even in the near future and, even when they are already in existence, it is often nearly impossible to correctly predict their impact. A case in point is the enormous growth of molecular–genetic methods. In the late 1980s and early 1990s, many predicted that it would now become relatively easy to localize genes for complex characters such as behavior. The Human Genome Project was touted to lead to cures for many devastating disorders within years of the completion of the sequencing of the human genome. Of course, by now we know that this has not been the case. As so often, reality has shown that things are more complicated and more difficult than we thought (or hoped) in our initial enthusiasm, and preciously few genes have been identified for any behavior, be it in mice, humans, or other organisms. Indeed, understanding how genes influence brain function and behavior is a much more complicated endeavor than originally forecast. The danger is that by creating such false expectations, we risk losing our scientific credibility with society at large and, in consequence, with policy makers. History up till now has taught us to be careful in our future expectations.

The above notwithstanding, we feel that some predictions can be made with a certain level of confidence. The arrival of new genetic tools, such as the expanded set of BXD Recombinant Inbred Strains (Peirce et al., 2004) or the Collaborative Cross (Churchill et al., 2004), may finally allow us to identify some of the genes responsible for the myriad of quantitative trait loci (QTLs) that have been localized in the past two decades. Similarly, after many years of churning out huge streams of gene-expression data, DNA microarray technology along with advances in bioinformatics is now slowly morphing into a new field called systems genetics (Schughart and SYSGENET consortium, 2010) with promising new tools to enhance our understanding of how the manifold interactions between genes cause interindividual variation. The increasing sophistication with which cell-type-restricted and temporally controlled gene targeting may be performed appears also promising, as do novel technologies including the utilization of miRNAs or RNA interference. These and many other methods will undoubtedly allow behavior geneticists to arm themselves with increasingly precise and controlled genetic tools. Obviously, the end of history has not yet been reached in behavior genetics.

References

Broadhurst, P.L. and Jinks, J.L. (1974) What genetical architecture can tell us about the natural selection of behavioural traits. In van Abeelen, J.H.F. (ed.), *The Genetics of Behaviour*. North-Holland, Amsterdam, The Netherlands, pp. 43–63.

Caspari, E. (1979) The goals and future of behavior genetics. In Royce, J.R. and Mos, L.P. (eds.), *Theoretical Advances in Behavior Genetics, NATO Advanced Study Institutes Series D: Behavioral and Social Sciences*. Sijthoff and Noordhoff, Alphen aan den Rijn, The Netherlands, pp. 661–679.

Churchill, G.A., Airey, D.C., Allayee, H., Angel, J.M., Attie, A.D., Beatty, J., et al. (2004) The Collaborative Cross, a community resource for the genetic analysis of complex traits. *Nat Genet* 36: 1133–1137.

Crusio, W.E. (2002) My mouse has no phenotype. *Genes Brain Behav* 1: 71.

Dewsbury, D.A. (1978) *Comparative Animal Behavior*. McGraw-Hill, New York.

DSM-IV-TR (2000) *Diagnostic and Statistical Manual of Mental Disorders, Fourth Edition Text Revision (DSM-IV-TR)*. American Psychiatric Association, Washington, DC.

Fuller, J.L. (1979a) Comment. In Royce, J.R. and Mos, L.P. (eds.), *Theoretical Advances in Behavior Genetics, NATO Advanced Study Institutes Series D: Behavioral and Social Sciences*. Sijthoff and Noordhoff, Alphen aan den Rijn, The Netherlands, p. 290.

Fuller, J.L. (1979b) The taxonomy of psychophenes. In Royce, J.R. and Mos, L.P. (eds.), *Theoretical Advances in Behavior Genetics, NATO Advanced Study Institutes Series D: Behavioral and Social Sciences*. Sijthoff and Noordhoff, Alphen aan den Rijn, The Netherlands, pp. 483–504.

Fuller, J.L. (1979c) Reply to comments. In Royce, J.R. and Mos, L.P. (eds.), *Theoretical Advances in Behavior*

Genetics, NATO Advanced Study Institutes Series D: Behavioral and Social Sciences. Sijthoff and Noordhoff, Alphen aan den Rijn, The Netherlands, pp. 511–513.

Fuller, J.L. and Thompson, W.R. (1960) *Behavior Genetics*, 1st edn. John Wiley and Sons, New York.

Fuller, J.L. and Thompson, W.R. (1978) *Foundations of Behavior Genetics.* C.V. Mosby Company, St. Louis, MO, USA.

Fuller, J.L. and Wimer, R.E. (1973) Behavior genetics. In Dewsbury, D.A. and Rethlingshafer, D.A. (eds.), *Comparative Psychology. A Modern Survey.* McGraw-Hill, New York, pp. 197–237.

Gerlai, R. (1999) Ethological approaches in behavioral neurogenetic research. In Crusio, W.E. and Gerlai, R.T. (eds.), *Handbook of Molecular-Genetic Techniques for Brain and Behavior Research, Techniques in the Behavioral and Neural Sciences*, Vol. 13. Elsevier, Amsterdam, pp. 605–613.

Gerlai, R. and Clayton, N.S. (1999) Analysing hippocampal function in transgenic mice: an ethological perspective. *Trends Neurosci* 22: 47–51.

Hall, C.S. (1951) The genetics of behavior. In Stevens, S.S. (ed.), *Handbook of Experimental Psychology*, 1st edn. John Wiley and Sons, New York, pp. 304–329.

Henderson, N.D. (1979) Adaptive significance of animal behavior. In Royce, J.R. and Mos, L.P. (eds.), *Theoretical Advances in Behavior Genetics, NATO Advanced Study Institutes Series D: Behavioral and Social Sciences.* Sijthoff and Noordhoff, Alphen aan den Rijn, The Netherlands, pp. 243–287.

Kraemer, H.C., Shrout, P.E., and Rubio-Stipec, M. (2007) Developing the *Diagnostic and Statistical Manual V*: what will "statistical" mean in DSM-V? *Soc Psychiatry Psychiatr Epidemiol* 42: 259–267.

Lindzey, G., Loehlin, J., Manosevitz, M., and Thiessen, D. (1971) Behavioral genetics. *Ann Rev Psychol* 22: 39–94.

Muller, H.J. (1922) Variation due to change in the individual gene. *Am Nat* 56: 32–49.

Peirce, J.L., Lu, L., Gu, J., Silver, L.M., and Williams, R.W. (2004) A new set of BXD recombinant inbred lines from advanced intercross populations in mice. *BMC Genet* 5: 7.

Plomin, R., DeFries, J.C., Craig, I.W., and McGuffin, P. (2003) Behavioral genetics. In Plomin, R., DeFries, J.C., Craig, I.W. and McGuffin, P. (eds.), *Behavioral Genetics in the Postgenomic Era.* American Psychological Association, Washington, DC, pp. 3–15.

Schughart, K. and SYSGENET consortium (2010) SYSGENET: a meeting report from a new European network for systems genetics. *Mamm Genome* 21: 331–336.

Stanford, S.C. (2007) The open field test: reinventing the wheel. *J Psychopharmacol* 21: 134–135.

Thiessen, D.D. (1972) *Gene Organization and Behavior.* Random House, New York.

Van Abeelen, J.H.F. (1979) Ethology and the genetic foundations of animal behavior. In Royce, J.R. and Mos, L.P. (eds.), *Theoretical Advances in Behavior Genetics, NATO Advanced Study Institutes Series D: Behavioral and Social Sciences.* Sijthoff and Noordhoff, Alphen aan den Rijn, The Netherlands, pp. 101–112.

Wahlsten, D., Metten, P., Phillips, T.J., Boehm, S.L., 2nd, Burkhart-Kasch, S., Dorow, J., et al. (2003) Different data from different labs: lessons from studies of gene–environment interaction. *J Neurobiol* 54: 283–311.

Chapter 2

Natural neurobiology and behavior of the mouse

Relevance for behavioral studies in the laboratory

Hans-Peter Lipp and David P. Wolfer

Summary

The house mouse (*Mus musculus*) has its origins presumably in Asia. Among many other small rodents, it represents an extremely flexible and adaptive species. Commensalism with human civilization and agriculture resulted in a worldwide distribution. While biology and behavior in the laboratory are well-documented, ecological–behavioral studies in natural or naturalistic environments are comparatively rare.

The relevance of the natural organization of behavior for laboratory tests appears to depend on an intracerebral hierarchy of sensory abilities and related behavioral processing crucial for survival: defensive fear-related behaviors, exploration and foraging strategies, olfactory communication and reproductive behavior, behavioral flexibility, and, lowest in the hierarchy, cognitive processing and complex memory. The relative lack of higher-order associative cortex in the house mouse also implies that the mouse hippocampus and prefronto-limbic cortex remain as the main associative structures, yet predominantly orchestrating ethologically relevant processes. Thus, experimental and genetic manipulations of the mouse brain for behavioral analysis need to consider its evolutionary adaptations and constraints.

These ideas shall be illustrated with some examples of outdoor studies in mice.

Introduction

The house mouse (*Mus musculus* ssp.) represents, as humans, rats, and sparrows, a recent evolutionary success story (Bonhomme et al., 1984; Bronson, 1984). From its origin in the Indian subcontinent some 500 000 years ago, this species ramified into the ancient Middle East from where it spread all over the world, following humans to almost every place except the arctic regions (Boursot et al., 1993). The reasons for its ability to follow humans (commensalism) and for its remarkable capacity to adapt to a large variety of habitats not shared with humans remain largely unknown (Berry and Bronson, 1992; Frynta et al., 2005). Its domestication, initially by fanciers, and much later on by scientific institutions, makes it the most frequently used laboratory animal at present.

While the mouse remains probably the best-investigated species with respect to genetics, cell biology, and physiology, knowledge about its brain and behavior is comparatively rudimentary, despite the many reports of the behavior of genetically modified animals. Much of that knowledge is inferred from rats: studies elucidating brain-behavior mechanisms in mice themselves are not abundant. Part of this problem is the paucity of ethological studies in naturalistic and semi-naturalistic environments, a prerequisite for meaningful interpretation of phenotypic changes in transgenic mouse models (Gerlai and Clayton, 1999). Thus far, behavioral studies of wild mice in naturalistic environments are rare, and appear to be of little interest as evident by the neglect of the beautiful monography by Crowcroft (1966). Most of them have focused on reproductive biology (Bronson, 1979; Drickamer et al., 1999, 2000), others on habitat structure (Pennycuik et al., 1987; Plesner-Jensen et al., 2003), on behavioral mechanisms underlying fluctuations of population densities (Van Oortmerssen, 1971), and on effects of predator pressure on populations (Arthur et al., 2005). Even less frequent are studies on learning abilities and behavioral traits of normal and experimentally manipulated laboratory mice and strains in naturalistic environments (Blanchard and Blanchard, 2003; Dudek et al., 1983; Glickman and Morrison, 1969). Clearly, the necessity of such approaches has been recognized by many behavioral scientists, but observing mice in naturalistic environments has been technically difficult and tedious.

Natural constraints for behavioral phenotyping in genetically modified mice

A casual survey of the many papers describing behavioral phenotypes of genetically modified mice reveals a conundrum: the majority of them describe altered hippocampus-dependent behaviors irrespective of whether the mutation was specific for hippocampal neurons or ubiquitously expressed in the brain. Quite often, hippocampus-dependency is also taken as a synonym for cognitive changes, particularly so if the modified gene is considered as being important for cellular processes underlying memory formation, and the mouse line is intended to serve as a model of human psychopathology. On the other

Behavioral Genetics of the Mouse: Volume I. *Genetics of Behavioral Phenotypes*, eds. Wim E. Crusio, Frans Sluyter, Robert T. Gerlai, and Susanna Pietropaolo. Published by Cambridge University Press.

hand, closer inspection of such reports reveals that the observed changes not only include popular hippocampus-dependent tasks, such as the water maze, the radial maze, and contextual fear conditioning, but also "exploratory" inspectional behavior observed in object recognition tasks and spontaneous alternation in T-mazes. Thus, while not evident from reading a single paper, a comprehensive view conveys the impression that most genetic manipulations result in behavioral manifestations linked in one way or the other to hippocampal function. Another puzzling finding is that when mice are tested for social behavior, in particular intermale aggression, there appears to be a disproportionably high number of targeted mutations enhancing or reducing aggression with no obvious relation to brain structures known to regulate murine aggression (Maxson and Canastar, 2003; Miczek et al., 2001). Similarly, many transgenic mouse lines show up- or down-regulation of open field activity.

In order to understand this, it is necessary to recall the natural constraints of phenotypic expression of mutations, as well as of lesions and pharmacological manipulations. These constraints include: (1) the species-specific (ethological) behaviors that a mouse employs to cope with natural or test situations in the laboratory; (2) the species-specific neural output pathways for translating brain processes into behavior; (3) the role of the hippocampus and limbic system in orchestrating species-specific behavior; and (4) the evolutionary pressure modeling and ordering different neural circuits according to their relevance for survival and biological fitness of *Mus musculus*. The following paragraphs shall try to sketch these points.

Ethology

It is undisputed that the behavioral repertoire of the house mouse ought to be considered when interpreting behavioral changes following experimental manipulation of the brain. Despite the efforts of ethologically oriented behavioral scientists such as Bolles (1970), Blanchard and Blanchard (1988), and, for genetically modified mice, Gerlai and Clayton (1999), their caveats are too often not considered. For example, it is well known that the defensive response of mice to natural and learned threatening stimuli includes freezing, flight, and risk assessment, and that mice can shift, depending on the situation and stimulus, rather flexibly from one of these behaviors to another. Nonetheless, most studies employing contextual fear conditioning (during which a mouse is placed in an environment where it previously received signaled punishment) use only the time spent freezing as an equivalent of a memory trace. While the presence of freezing indicates qualitatively a memory trace, quantitative variation of this freezing response may equally indicate either a decay in memory or else a shift in response pattern as the mouse tries locate the threat or to get out of that environment after some time. Thus, changes in very different brain mechanisms may result in similar reduction of freezing.

Neural output

A subtler point is how brain processes translate into behavioral events. To recall a basic feature: regardless of the species-specific adaptation of the brain for sensory and cognitive abilities, the endpoint of all cerebral processing in vertebrates is contractions of striate and smooth muscles. The former manifest as species-specific changes of body parts or movements across space, the latter as changes in the autonomous nervous system, being mostly invisible, less frequently also apparent to an observer as piloerection, changes in skin color, or secretion of pheromones, to name a few. These two output pathways are found in all vertebrate species, the neural programs for activation and inhibition being located in the reticular formation of the rostral brainstem. The midbrain controls species-specific behavioral events in form of a stop-and-go principle, summating facilitatory and inhibitory signals according to locally stored memory, exteroceptive (visual, somatosensory, auditory, and olfactory) and interoceptive (hormonal and gustatory/chemical) inputs. Hormonal information from the hypothalamic receptors and olfactory input reaches the midbrain through a chain of reciprocally interconnected structures and axons, enabling a coarse evaluation of rewarding properties (go signals) along the laterally running medial forebrain bundle, while aversive/alarming signals are conveyed via more medially running fibers to the central gray and brainstem nuclei of the autonomous nervous system. Taken together, the ultimate behavioral control is governed by stop-and-go signals originating in the midbrain, but there are important species differences in the modulation of these processes, depending on the degree of forebrain development (encephalization). This is most evident by considering the neuroanatomical differences between mice and men.

Hippocampus of mice and men

The main difference between the mouse (or rodent brain in general) and the primate brain is that the limbic output from prefrontal cortex, hippocampus, and amygdala acts more directly on subcortical motor systems activating the midbrain stop-and-go system than in primates. This is due to two connective properties. For one, in rodents, a considerable portion of the limbic output from prefrontal cortex, hippocampus, and amygdala terminates in dorsal and ventral basal ganglia whose output is directed chiefly to the rostral brainstem, inhibiting or activating ongoing species-specific motor acts. In primates, the output of the basal ganglia converges preferentially on the motor thalamus and thus on the primary motor cortex, whereas output from limbic basal ganglia reaches the midbrain and the intralaminar thalamus, which in turn may control neocortical processing by ascending systems (Lipp and Wolfer, 1998). Thus, motor activity in humans reflects eventually the neural activity of the entire neocortex, whereas motor activity in rodents reflects primarily limbic processing acting on midbrain structures. Consequently, large portions of the rodent neocortex can be

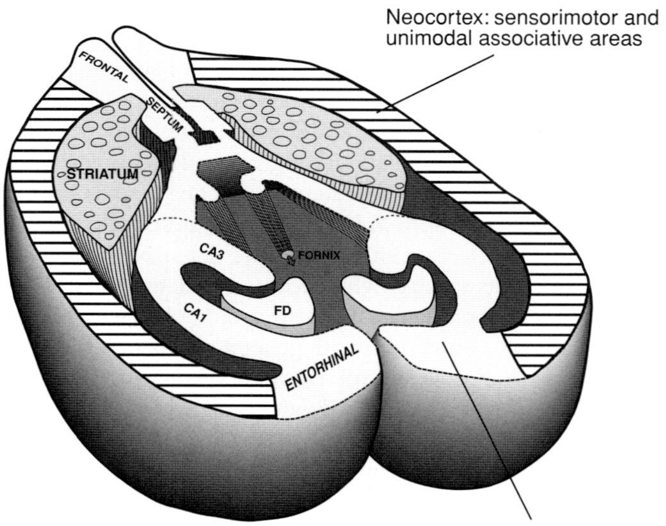

Figure 2.1 Schematic view of a horizontal section through the mouse brain. Hatched areas denote neocortex devoted to sensory, motor, and modality-specific analysis. Note that the hippocampal formation and the (small) prefrontal areas remain as the main higher-order associative cortex.

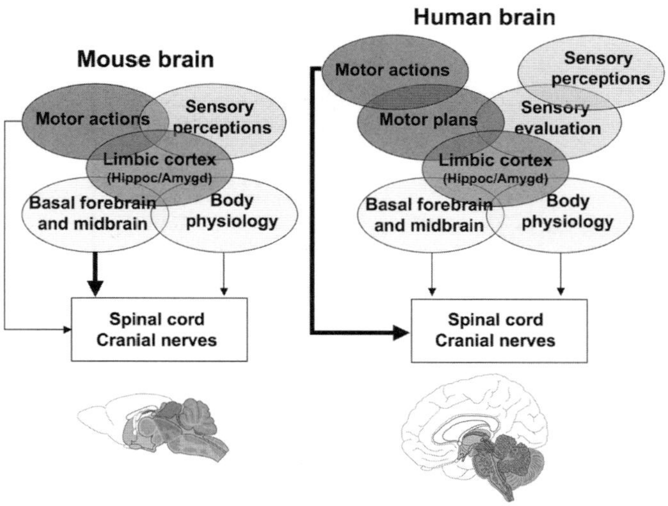

Figure 2.2 Organization of behavioral output pathways in mouse and human brain, and the central role of the limbic cortex which, in the mouse, directs its outputs preferentially towards the midbrain. Thus, most processing in the forebrain results in immediate motor reactions. In humans, large portions of the behavioral output system are shifted towards the motor cortex, the limbic system acting on the neocortex, both through mesencephalic and thalamic feedback loops and reciprocal connections with higher-order associative cortex. This causes iterative processing, resulting in constant adaptation and thus less abrupt changes of ongoing motor activities. In terms of inputs, the mouse hippocampus receives motor and sensory information, without much preprocessing through higher-order associative areas as observed in the human brain.

removed without disabling gross motor activity (Figure 2.1) (Huston and Borbely, 1973).

The second connective difference is the relative lack of polymodal (higher-order) associative cortex in rodents, most clearly seen in mice in which most of the neocortex comprises primary sensory areas and unimodal association cortex (Figure 2.2). In essence, the hippocampus is their main associative cortex, linking both pre-processed sensory and motor information. It is thus likely to be (variably) involved in any kind of complex learning. In humans, the hippocampal formation connects primarily with polymodal motor (executive) or polymodal sensory neocortex (Figure 2.2). Hence, lesions or malfunctions of the human hippocampal formation and proximally connected limbic areas manifest themselves as deficits in memory or cognition but have little impact on ongoing motor behavior except for verbal communication. On the other hand, malfunctions of the mouse hippocampus and associated limbic structures will result in *both* impaired orchestration of species-specific motor responses in the midbrain, evident as hyper-reactivity plus movement stereotypies, and impaired spatial abilities. This is because the only substrate for finely matching directed movement with sensory information has been disabled. One may note that this view would predict the presence of place cells in the rodent hippocampus but the relative paucity of such cells in the hippocampus of monkeys or humans.

Taken together, a certain amount of "hippocampal" behavioral impairments in mice is likely to occur in many transgenic mouse models showing a behavioral phenotype. They may be equally observed after specific inactivation of hippocampal substructures, after ubiquitous impairment of neuronal function in the forebrain, or even in mutations sparing the hippocampus. This happens because most efferent fibers of the mouse forebrain are targeting rather directly the same stop-and-go system in the midbrain, and because the mouse hippocampus interacts primarily with modality-specific parts of the neocortex. Thus, these behavioral signs cannot be taken as an indicator of cognitive malfunction or memory impairments in the human sense, but are a relatively fine indicator of neuronal malfunction within but also outside the hippocampus. In practice, employing "hippocampal" tests for behavioral phenotyping of genetically modified mice is useful for screening but of limited value for testing psychological concepts.

Ecological constraints

The frequently observed up- and down-regulation of putative "non-hippocampal" behaviors such as aggression or open field activity indicates another interpretation problem. If observed in a given mutant line, there is often a penchant to attribute this to malfunction of a particular neural subsystem regulating that behavior. However, such interpretations neglect the fact that most subsystems regulating species-specific behaviors (via the midbrain stop-and-go system) interact homeostatically, particularly those competing antagonistically for a motor output requiring approach, avoidance or immobility. This is most evident in hypothalamic brain stimulation studies capable of activating simultaneously rewarding and aversive neural subsystems (Lipp, 1978, 1979). In terms of genetic manipulations,

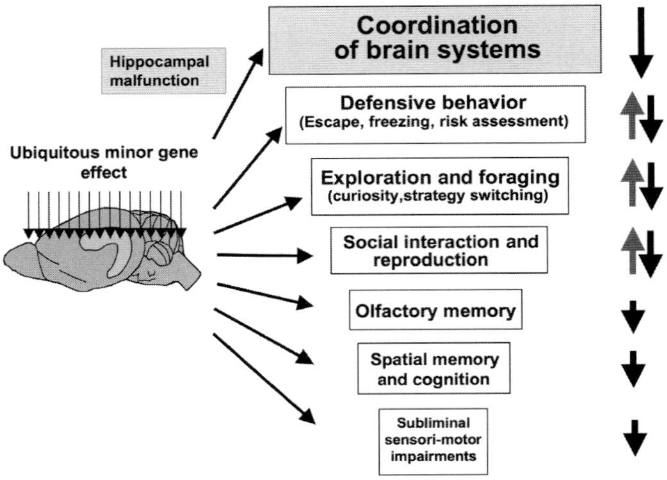

Figure 2.3 Hierarchy of expected behavioral phenotypes of murine mutations that have ubiquitous action in the central nervous system, or entail extensive pleiotropy, arrows to the right indicating up- or down-regulation. "Hippocampus-dependent" behaviors are expected to occur most often, alone or in combination, because of the size of hippocampus, its role as largest associative brain region, and because of many other brain regions mimicking hippocampal function by jointly acting on the midbrain stop-and-go system. The other potential effects of ubiquitous mutations on specific behavioral phenotypes have a likelihood decreasing in parallel with their ecological and functional significance. Thus, the next likely candidates for phenotypical up- or down-regulation are changes in the balance between defensive and exploratory behavioral tendencies, possibly also behavioral flexibility versus rigidity, followed by up- or down-regulation of aggressive behavior and social interactions. Because of the importance of smells for the daily life and reproduction of mice, unspecific mutations affecting concomitantly the olfactory systems have also an increased likelihood of phenotypical manifestation in olfactory-dependent behaviors, albeit less likely as gain-of-function. Specific effects on spatial memory and learning may occur but confounds with neural systems co-mediating the midbrain stop-and-go system are to be expected. Finally, there might be many subtle sensori-motor deficits yet difficult to observe at the behavioral level.

this implies that up- or down-regulation of particular behaviors may often reflect an altered homeostatic balance between neural subsystems rather than alteration of a given neural subsystem. Obviously, behavioral observation alone cannot discriminate between the two possibilities.

Yet, it is reasonable to assume that natural selection is carefully tuning the balance between such systems according to species, ecological niche and even individual propensities within a population. This is indicated by the very rapid effects of natural selection in both mutant mice and mice carrying natural genetic variability (see below), and also by the observations that many targeted mutations entail, somewhat unpredictably, up- or down-regulation of behavioral traits seemingly unrelated to the targeted mutation. In mice, it would seem that there is a hierarchy of such processes according to the importance for survival and biological fitness.

This shall be exemplified by assuming a targeted mutation with ubiquitous but minor effects on neuronal functioning across the entire forebrain (Figure 2.3). Because of its relatively large size, a certain degree of hippocampal malfunction is likely to occur. This will be preferentially reflected in malcoordination of spatial behavior but also in shifted balances between antagonistic systems governing species-specific behavior. In addition, the balance between ecologically important behaviors is likely to be tuned additionally by non-hippocampal systems. For mice as a small and highly predated species, the most important behavioral system is the one regulating the reactivity to external stimuli (many of them potentially threatening) and the selection of antagonistic defensive behaviors; that is, immobility versus flight, because this determines life or death. Depending on the local situation, either behavior can be appropriate. Thus, mutations with general effects may increase or decrease the propensity for running versus freezing (and might so be mistaken as up- or down-regulation of memory in contextual fear conditioning).

A second class of antagonistic behavior is the propensity of exploring and foraging necessary to locate food, which, however, bears an increased risk of predation, and must thus be subject to a carefully tuned check-and-balance system resulting either in more curious or more fearful animals. Again, a shifted balance may be mistaken as increased genuine curiosity or fear, respectively.

While less important for daily survival, social interaction and reproduction are of paramount importance for a short-living social species. Particularly in male mice, many minor genetic disturbances do have the potential to alter the balance between attack and flight, thereby affecting social status. This might explain why so many mouse mutants appear hypo- or hyperaggressive.

On the other hand, the functional relevance of memory processing may be of lesser importance for a species with a lifespan that, under natural conditions, rarely exceeds 6 months. Given that male mice distribute daily up to 40 mg of major urinary proteins (Beynon and Hurst, 2004; Hurst and Beynon, 2004), one would expect that olfactory memory mechanisms are by far the most important ones for these species. Likewise, olfactory processing is critical for the survival of pups.

Surprisingly from a psychologist's point of view, yet unsurprising for ethologists, spatial memory and cognition do not appear to be of tremendous importance for a species preferring to move in a well-known, spatially confined, and mostly dark environment along olfactory paths

Finally, one might expect that a mutation acting ubiquitously in the central nervous system of a mouse is likely to impair a variety of sensory and motor processes. However, if the mutation does not have a strong effect, a phenotype may be difficult to detect. For example, an overall decrease of 20% in synaptic transmission in all neurons may entail phenotypic changes at the behavioral level. However, measuring concomitant minor impairment in sensory or motor processing would need extensive behavioral and neurophysiological studies to document it.

The next sections describe an approach of how to study the effects of genetic and classic lesions on brain and behavior of mice living in semi-naturalistic environments, and they will illustrate some of the theoretical points made above.

Figure 2.4 Outdoor pens for studying learning and natural selection in laboratory mice in summer (a) and winter (b).

Studying behavior and survival of mice in outdoor settings

Principles

One procedure suitable to test the effects of genetic manipulations on general biological fitness and ecologically relevant behaviors is to release mice into outdoor settings for a limited period during summer and early fall. Provided that spacious shelters are available, fluctuating meteorological conditions in this time are well tolerated by mice, except by some inbred strains (see below in this section). On the other hand, the mice are at risk for aerial predation as soon as they leave the protected shelters and pathways leading to outdoor sites, and they have to face a tremendous change in environment. This offers a convenient opportunity for testing whether behavioral changes observed in the laboratory predict behavior in naturalistic environments, and also for testing whether the mice suffer from unrecognized maladaptive effects of mutated genes.

If the interest is on natural selection rather than short-term adaptation, mice genotyped for modified and wildtype alleles are released in proportions matching Mendelian inheritance (e.g., the founder population includes 25% homozygous mutants, 50% heterozygous, and 25% wildtype mice, resulting in a balanced distribution of 50% wildtype and 50% mutant alleles). The mice are then left with food *ad libitum* and recaptured every year for genotyping, being re-released afterwards. This approach permits observing natural selection effects on brain and behavior, or the elimination or accumulation of targeted mutations in the offspring of the released animals.

The simplest method for assessing the impact of a treatment or of a genetic mutation is re-trapping ear-tagged mice after a defined period, be this after a few weeks, or just every year. A more sophisticated method is animal monitoring by using implantable passive radio-frequency identification (RFID) transponders (Dell'Omo et al., 1998, 2000).

In cooperation with Russian behavioral geneticists at Moscow State University, we had the opportunity to build a field

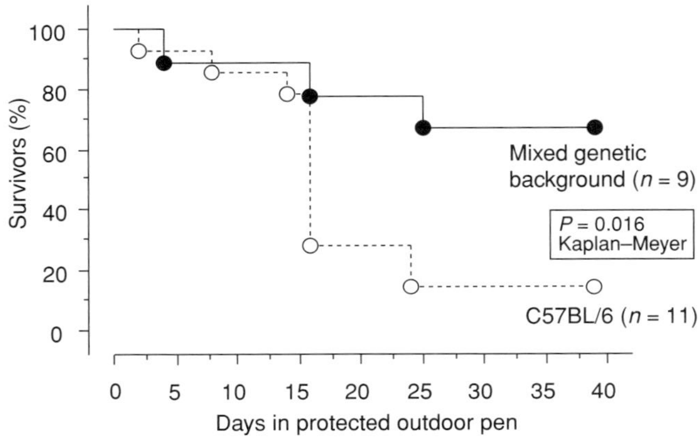

Figure 2.5 Short-term survival of female C57BL/6 and mice of mixed genetic background (random-bred from diallel cross C57BL/6J, C3H/J, NZB/J, and DBA/2J). The curves include both mice with hippocampal lesions and control mice. Presence of mice was monitored by means of subcutaneously implanted microchips.

station for studying the effects of natural selection on artificially mutated genes, brain traits, and associated behavior, and also the effects of genetic manipulations and experimental lesions on behavioral abilities in naturalistic environments. For this purpose, the field station contained several large outdoor pens (Figure 2.4) and a field laboratory permitting local neurohistology of mice. This chapter will review exemplary studies that permitted us to identify main determinants and species-specific constraints of the behavior of the house mouse in naturalistic conditions.

Short-term survival of normal inbred strains

Pilot studies and short-time experiments clearly indicate that hybrid mice with mixed genetic background always adapted easily to summer conditions, while this was not always true for inbred strains. The effects of mixed versus inbred genetic background is evident in Figure 2.5 showing more severe losses of female C57BL/6 mice as compared to random-bred mice,

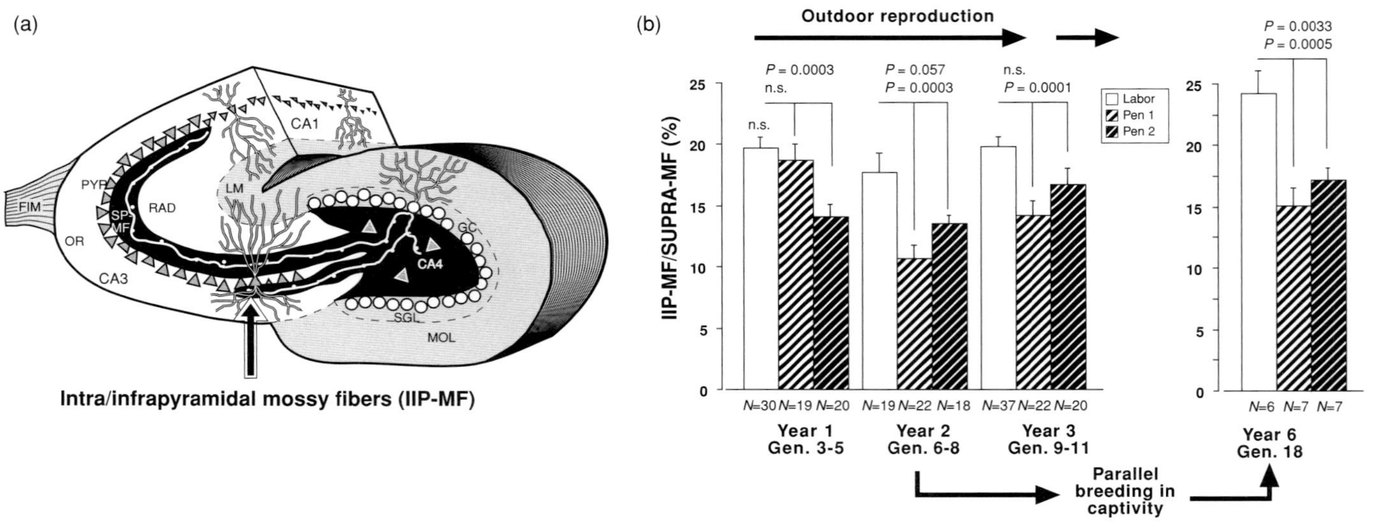

Figure 2.6 Intra/infrapyramidal mossy fiber (IIP-MF) projections and natural selection over 3 years in outdoor pens. (a) Diagram of mouse hippocampus with mossy-fiber projections. (b) Reduction in the extent of the IIP-MF projections as observed during 3 years of living in outdoor pens. The reduction of the IIP-MF in the feralized mice remained after having transferred the mice to mouse facilities and through embryo transfer, which indicates natural selection.

even when the enclosure was additionally protected with a net against aerial predators. These mice had food *ad libitum*, suffered no bad weather, had empty space in two shelters, and no or little social stress. Thus, the perhaps most simple conclusion from these adaptation studies is that inbred strains are likely to carry maladaptive traits not evident in the laboratory. Consequently, our laboratory avoids behavioral testing of mutant mice on an inbred background and prefers testing hybrids according to the Banbury recommendations (Wolfer et al., 2002).

When male and female mice are released together, the number of male mice in unprotected outdoor pens decreased generally faster than the number of females, regardless of genotype. This can be expected from the social structure of mice characterized by dominant males chasing subordinates relentlessly also in naturalistic settings (Crowcroft, 1966; Ely et al., 1976). Thus, dominant males tend to occupy the protected shelters, presumably forcing subordinates to enter unprotected risk areas more frequently. Therefore, the cerebral regulation of intermale aggression is one of the decisive mechanisms in both short-term adaptation studies of individual mice and multi-generation studies on natural selection.

Long-term selection of hippocampal mossy fiber traits and associated behavior

Background

Another line of research in natural selection originated from the discovery that hereditary variation of a hippocampal structural trait in rats and mice appeared to be correlated with learning abilities in laboratory tasks (Schwegler and Lipp, 1983). Hippocampal mossy fibers are the axons of dentate granule cells terminating in defined layers above and below the pyramidal target neurons in hippocampal subregion CA3 (suprapyramidal (SP-MF) and intra/infrapyramidal mossy fibers (IIP-MF) respectively; Figure 2.6a). Somewhat surprisingly, the extent of the IIP-MF projection along the basal dendrites was often correlated with performance in a variety of hippocampus-dependent tasks; for example, negatively with two-way avoidance learning, and positively with radial maze learning and the efficiency of platform reversal learning in the water maze. This had been verified in a long series of studies using strains selectively bred for extremes in behavior, inbred, and random-bred strains of mice, and ontogenetic manipulations of the IIP-MF projection (for reviews see Crusio and Schwegler, 2005; Lipp et al., 2006). In many cases, reduced IIP-MF projections appeared to mimic a mild hippocampal lesion (also known to improve two-way avoidance learning), while extended projections appeared to be associated with a factor reflecting an intact basic hippocampal function necessary for complex (mostly spatial) learning. However, the extent of the IIP-MF appeared also to be correlated with behaviors not considered as hippocampus-dependent, such as strength of paw preference being more pronounced in mice with large IIP-MF projections (Lipp et al., 1996), and reduced attack latencies as observed in intermale aggression in mouse strains with small IIP-MF projections (Guillot et al., 1994; Sluyter et al., 1994). These latter observations provided a hint that the mouse hippocampus might be mediating behavioral mechanisms not necessarily predicted by the human hippocampal lesion syndrome. Our explanatory hypothesis was that structural mossy fiber variations might pre-set individual behavioral reactivity to distracting stimuli of exteroceptive or interoceptive origin. Thus, small IIP-MF would be associated with short attack latencies and superior two-way avoidance learning (requiring immediate motor reaction as operant response), while such high reactivity would be detrimental for most complex learning tasks requiring attention and suppression of inappropriate responses.

However, at that time it was also not clear whether the observed morpho-behavioral correlations in the laboratory were of any biological relevance, since they were mostly observed in inbred strains deemed degenerated by many biologists. Also, as it is often the case in behavioral neuroscience, there were studies not confirming the expected correlations with hippocampus-dependent behavior (Lipp et al., 2006). Finally, ethologists were critical whether the behavioral experiments designed for mice by psychologists, pharmacologists, and neuroscientists were of much relevance for revealing biologically important mechanisms regulating the behavior in mice in naturalistic environments.

Knowing that artificial selection in the laboratory for extremes in hippocampus-dependent behaviors entails morphological changes in the mossy-fiber distribution (Hausheer-Zarmakupi et al., 1996; Schwegler and Lipp, 1983), we decided to test whether and how experimental natural selection would act on the hippocampal mossy-fiber distribution and the genetically associated behavioral traits of mice. Rapid natural selection in either direction would indicate that genetic variations of this trait were of paramount behavioral relevance for regulating behavior under naturalistic conditions, while the direction of the selection towards larger or smaller IIP-MF projections might provide clues for the natural functional role of this brain trait.

Breeding a founder generation and observing natural selection

Four inbred mouse strains, two with extended IIP-MF projections (C3H/J and C57BL/6J) and two strains with scanty IIP-MF projections (NZB/J and DBA/2J), were crossed in order to obtain founder populations of F1 hybrids ensuring that everyone contained equal proportions of genes from the four strains, a procedure known as diallel cross (Crusio et al., 1986). One founder population was bred by random mating in standard mouse facilities at Moscow State University, two populations were released into adjacent outdoor pens of 400 m^2, each one containing dry underground shelters of 4 m^2 filled with hay, branches, and wooden planks. Food *ad libitum* was delivered in the shelters; the amount consumed providing a rough estimate of the actual population in a pen. Starting 1 year after release, all mice in the two pens were caught every summer. From each pen, a sample of 15–20 mice was then randomly drawn for perfusion and histological processing of the hippocampus at the field station, the other mice were released again and left undisturbed for another year. The brains of these mice and from a matching sample bred in captivity were then analyzed morphometrically for the extent of the hippocampal mossy fiber projection as visualized by Timm staining (Schwegler and Lipp, 1983), thereby permitting to monitor the adaptation of the IIP-MF distribution over time.

Figure 2.6b shows an excerpt of this study covering 3 years of outdoor reproduction. In comparison to captivity-bred animals, there was a significant reduction of the IIP-MF projections in both populations already in the 2nd year after release of the founder mice, the difference appearing stabilized in the 3rd year. Other data not shown here revealed that the same difference was also observed in a 4th year (Poletaeva et al., 2001), after which both pen populations crashed with no animals left surviving the 5th winter outdoors. Obviously, these data alone would not prove natural selection as the changes in mossy fiber distribution might also reflect environmental factors modulating this trait epigenetically. However, in the 2nd year, samples of mice from the two pens were transferred back to animal facilities and bred there further. Repeated analysis of their hippocampal mossy fiber distributions in the 15th (bred in Moscow) and 18th generation (bred in Zurich after embryo transfer) revealed significant differences between the lines (Figure 2.6b).

Behavioral studies in Moscow with the naturally selected lines showed increased defecation in the open field, reduced activity in exploratory arenas after repeated exposures, and, interestingly, an improved ability of extrapolating the movements of reward towards a hidden consumption site (Poletaeva et al., 2001). Studies in Zurich showed no parallel selection effects in the water maze and open field (Rissi, 2002), but naturally selected lines showed neophobia towards a novel object in an open field (Ceschi, 2004).

Taken together, this study showed an extremely rapid natural selection effect on both hippocampal circuitry and behavior, and proved that the association between IIP-MF and behavior is of ecological relevance. In addition, it supported the hypothesis of an association between reduced IIP-MF and behavioral reactivity. Obviously, a rapid effect of natural selection was to alter the balance between genetic factors favoring exploration and defensive reactions such as neophobia. The mice had to live in the shelters throughout the year. Intentional exploration of the pen territory (as expected for carriers with expanded IIP-MF projections) bears a high risk of predation by owls, while the social structure in the shelter (a few highly aggressive dominant males and many quickly escaping subordinate males) favors again selection for high behavioral reactivity (either for fight or flight), at least in males. As the lifestyle in these types of pens is not demanding in terms of spatial abilities and complex learning, the lacking selection effects on performance in classic cognitive tests such as the water maze is unsurprising. Most likely, other genetically dependent brain traits supporting the observed phenotype were subject to selection as well. Nonetheless, the speed of genetic adaptation of the IIP-MF projection implies that natural selection in this type of pens acted upon a few major loci only.

Hippocampal malfunction in naturalistic environments

The classic approach to study hippocampal function is its inactivation, by means of physical or, more recently, by genetic lesions. It is evident that this approach does not reveal primarily hippocampal function but rather the functions of the remaining intact brain (O'Keefe and Nadel, 1978). Nonetheless, it permits to identify behaviors and brain functions depending critically on an intact hippocampus. Historically, effects of hippocampal lesions on species-specific naturalistic behaviors, chiefly of rats, were investigated during the period from 1950 to 1975, and were dominated by interpretations conceiving the hippocampus as inhibiting inappropriate behavioral responses. Under the influence of the cognitive map theory of hippocampal function (O'Keefe and Nadel, 1978), the research focus moved to studies on spatial behavior, cognition, and memory.

Hippocampal lesions and naturalistic behavior of rodents

As discussed below, the most important behavior of rodents in their natural environment are defense responses (flight, freezing, aggression) to threats, predators and individuals of the same species. There have been several studies dealing with defense responses of hippocampally lesioned rats in threatening situations, most of them discussed by O'Keefe and Nadel (1978) and Blanchard et al. (1970). In short, they indicate hyperreactivity towards threatening (sometimes also towards novel stimuli), and reduced freezing responses. Social dominance behavior of male rats having undergone neonatal X-ray exposure (severely reducing the number of dentate granule cells) was found to be impaired when facing controls except in competition for receptive females (Wallace et al., 1981). More recently, Deacon and co-workers studied altered naturalistic behaviors in the laboratory after hippocampal lesions in C57BL/6 mice (Deacon et al., 2002; Deacon and Rawlins, 2005). They found a severe deficit in burrowing (removing pebbles or food pellets) from a tube in the home cage; minor deficits in food hoarding and nest building; lacking exploratory behavior; fear of entering tubes; and reduced hyponeophagia. Largely similar findings have been reported for rats (Antonavich et al., 1997; Kim, 1960) and gerbils (Glickman et al., 1970).

Hippocampal lesions and malfunction in naturalistic environments

On the other hand, hippocampal lesion studies in the natural environment of a species are rare. They have been mostly performed in homing pigeons in which such lesions impair but do not prevent long-distance homing thus permitting experimental analysis of the hippocampal deficits that appear to be linked to recognition of relevant landmarks in the vicinity of the loft (Bingman et al., 1995; Gagliardo et al., 2002). In laboratory settings, Glickman and Morrison (1969) observed that hippocampally lesioned mice were preferred victims of an owl sitting in an enclosure housing lesioned and control animals. They attributed this to a hyperactivity of the lesioned mice. Using an ingenious forerunner of the modern RFID technology, Ely et al. (1976) monitored long-term activity and social interactions of male mice with large electrolytic hippocampal lesions in a small arena containing several home cages, feeders and activity wheels, and intact female mice. Their main findings were transient postoperative hyperactivity and a long-term lack of clear social dominance by a particular male as typically observed in control mice.

But what about hippocampal malfunction in outdoor settings? To answer this question, mice with a forebrain-specific deletion of the trkB neurotrophin receptor were studied (Vyssotski et al., 2002). These mice lack trkB receptors not only in the hippocampus but also in most parts of the forebrain differentiating postnatally. Thus, a hippocampal syndrome, aggravated by prefrontal and basal ganglia malfunction, characterizes the behavior of these mutants. In the Morris water maze, they showed extreme perseverative thigmotaxis (wall hugging) that prevented learning the location of an invisible platform, but also escape from the water by swimming towards a visible platform. This left the question whether the behavioral impairment was caused either by deficits in spatial memory and learning, or, alternatively, by loss of behavioral flexibility. Eight *trkB* mutants, 13 wildtype, and 22 heterozygous mice were implanted with RFID transponders and released for 21 days into a large outdoor pen (10 × 10 m). The enclosure contained two shelters and eight computer-controlled feeder boxes delivering food portions for every mouse only during their first visit, thus mimicking a radial maze set-up (Figure 2.7a–c). A net and an electrical fence provided protection against predators.

Every third day, mice received food *ad libitum* inside the shelters, because there was concern whether mice with impaired spatial learning abilities would starve when forced to learn a feeder-patrolling task. However, all mice learned to patrol the boxes correctly within a few days (Figure 2.7d). Interestingly, significant differences emerged during those days with free food available. Wildtype mice remained inside the shelters, while all homozygous mutants continued to patrol the boxes in their habitual way, the heterozygous mutants showing intermediate scores (Figure 2.7e). Thus, this experiment showed that the predominant deficit in the water maze, namely reduced behavioral flexibility in changing and adopting search strategies, re-appeared in other form under naturalistic conditions. In addition, this study revealed a residual capacity for spatial learning not evident in the water maze. Interestingly, these results were predicted by earlier results of radial maze learning by *trkB* mutants, which showed that they were able learn this task but committed more repetitive errors than wildtype mice, heterozygous mice falling in-between (Minichiello et al., 1999).

Conclusions and outlook

Ethological and ecological constraints

The outdoor approach clearly revealed some ethological peculiarities of mouse behavior that should be considered when using genetically defined mouse models for human cognition and psychopathology. The main feature is that almost every event and process in the mouse brain appears to be translated immediately into movements or their suppression. While of

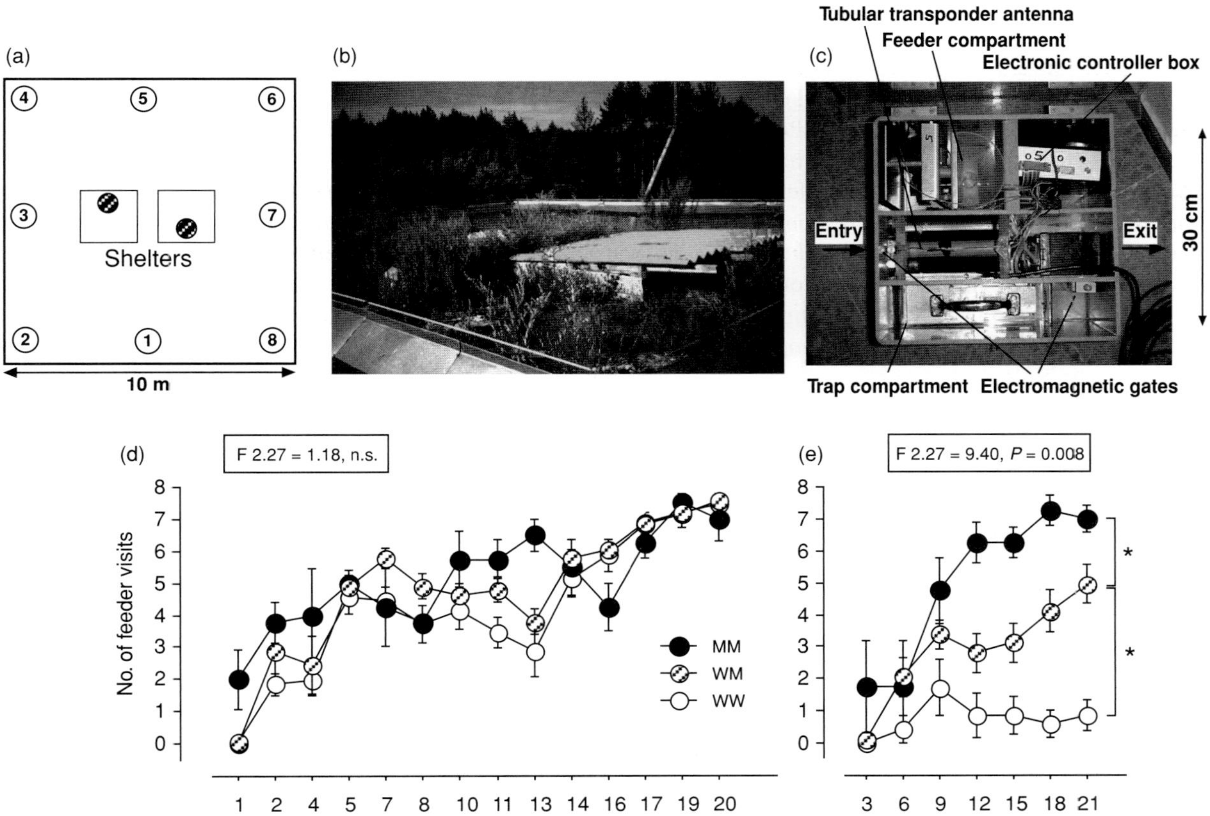

Figure 2.7 Food-site patrolling of mutant mice lacking the neurotrophin receptor trkB in an outdoor pen. (a) Arrangement of feeder sites delivering food only upon the first visit. (b) View of test site. (c) Computer-controlled feeder box delivering selectively food to mice identified by means of implanted microchips. (d) Successful learning of patrolling feeders as observed in mutant and control mice. (e) When food was delivered inside the shelters, without need for patrolling, homozygous mutants continued to patrol whereas wildtype mice switched immediately to the new location of food. Note that heterozygous mutants showed an intermediate impairment in behavioral flexibility. (Modified from Vyssotski et al., 2002.)

obvious necessity for a minuscule and highly predated species, these immediate stop-or-go reactions make it difficult to identify and to differentiate the underlying neural processes, which moreover often appear to act antagonistically. It has been said before that such a stop-or-go dichotomy also characterizes human behavior. There is a difference, however. The larger and thus less predated a species is, the less its need for immediate motor reactions, and the better its suitability for analysis of brain mechanisms by means of experimental psychology. Body size permits leisure and consideration time not available for a mouse, thus limiting the expression of psychopathology in this species.

On the other hand, this limitation is useful, albeit only at second glance. The presence of strongly acting antagonistic neural systems for stop-or-go implies mechanisms maintaining and regulating the balance between such systems. These would represent naturally regulatory sites for shifting behavioral propensities ontogenetically or genetically. This claim is supported chiefly by the very rapid effects of natural selection on the balance between exploratory and neophobic tendencies. It is unlikely that mossy-fiber variations reflect variations of an exploration or fear system, but it is conceivable that the correlated change in behavioral reactivity towards interoceptive or exteroceptive stimuli entails phenotypes that appear either neophobic or careless. One may note that most populations of vertebrate species (including humans) show a bimodal distribution of individuals classified according to such dichotomous behavioral traits, e.g., copers versus non-copers (Sluyter et al., 1995), or high- versus low-sensation seekers (Dellu et al., 1996; Zuckerman, 1990). Likewise, many human psychiatric diseases are characterized by similar dichotomies (Loas, 1996). Thus, searching for genetic control of such balancing regulatory systems or mechanisms may benefit from using mice, whose behavior appears to reflect a permanent battle between incompatible motor tendencies.

Benefits of outdoor testing

This approach of observing brain–behavior relationships of mice in naturalistic settings originated from dissatisfaction with conventional phenotyping in the laboratory. The thorough analysis of several thousand mutant mice, including about 100 mutations, did not yield a clearly recognizable differentiation of behavioral changes according to neural systems involved – clearly the goal of most researchers wishing to analyze genotype-to-phenotype pathways. While observation in naturalistic settings has not achieved this goal so far, it nonetheless has provided important information for behavioral phenotyping of targeted mutations. (1) Outdoor testing revealed that many of the genetically modified mice said to be normal with respect to gross anatomy and behavior appear to suffer from hidden problems not or barely recognizable in the laboratory. (2) It enabled evaluation of the relevance of behavioral tests in the laboratory as compared to the real-life situation of mice. (3) Finally, outdoor testing with natural selection is the only way to recognize whether mutations thought to enhance cognitive abilities of mice do not impair biological fitness.

Concerns about outdoor testing of genetically modified mice

The approach presented here meets the legal requirements of all Western countries and also of Russia. However, it cannot and will not have many followers, simply because the construction and maintenance of outdoor pens requires considerable efforts and resources not available to most laboratories working with genetically modified mice.

On the other hand, it also meets questions and criticisms that vary with cultural and scientific background. The most frequent objection results apparently from an unspecified fear that such mice might escape and spread over the world, in analogy to genetically manipulated plants and crops. However, we firmly believe that outdoor testing in pens with mice carrying targeted deletions or species-specific transgenes does not bear such risks. The reason is simple. Given the worldwide number of mice and their generation cycle, it is likely that deletions and point mutations of all loci in the mouse genome are constantly occurring. One can agree that some of these mutations do convey to mice the ability to spread all over the world, but it would seem that they have occurred in the past, some 500 000 years ago, and have laid the foundation for the omnipresence of this species. Conversely, one should avoid conducting such studies using mice subjected to xenogenetic manipulations that may alter unpredictably the architecture of the mouse genome and its phenotypic manifestations.

Acknowledgments

The reported studies were continually supported by Swiss National Foundation for Scientific Research, by grants for Swiss–Russian institutional partnership SCOPES (IP 51224, IP62645, IP111081), the Russian Foundation for Basic Research, and the NCCR "Neural Plasticity and Repair." We acknowledge the expert help and support of numerous collaborators. Pre-eminent among them are Inga I. Poletaeva, Valentin and Svetlana Pazhetnov, Natasha Bologova, Marina G. Pleskacheva, Alexei L. Vyssotski, Giacomo Dell'Omo, Irmgard Amrein, Robert M. J. Deacon, Rosmarie Lang, and Inger Drescher-Lindh.

References

Antonavich, F.J., Melton, C.S., Wu, P., and Davis, J.N. (1997) Nesting and shredding behavior as an indicator of hippocampal ischemic damage. *Brain Res* 764: 249–252.

Arthur, A.D., Pech, R.P., and Dickman, C.R. (2005) Effects of predation and habitat structure on the population dynamics of house mice in large outdoor enclosures. *Oikos* 108: 562–572.

Berry, R.J. and Bronson, F.H. (1992) Life history and bioeconomy of the house mouse. *Biol Rev Camb Philos Soc* 67: 519–550.

Beynon, R.J. and Hurst, J.L. (2004) Urinary proteins and the modulation of chemical scents in mice and rats. *Peptides* 25: 1553–1563.

Bingman, V.P., Jones, T.-J., Strasser, R., Gagliardo, A., and Ioalè, P. (1995) Homing pigeons, hippocampus and spatial cognition. In Alleva, E., Fasolo, A., Lipp, H.-P., Nadel, L. and Ricceri, L. (eds.), *Behavioural Brain Research in Naturalistic and Semi-Naturalistic Settings: Possibilities and Perspectives.* Kluwer, Dordrecht, The Netherlands, pp. 207–224.

Blanchard, D.C. and Blanchard, R.J. (1988) Ethoexperimental approaches to the biology of emotion. *Annu Rev Psychol* 39: 43–68.

Blanchard, R.J. and Blanchard, D.C. (2003) Bringing natural behaviors into the laboratory: a tribute to Paul MacLean. *Physiol Behav* 79: 515–524.

Blanchard, R.J., Blanchard, D.C., and Fial, R.A. (1970) Hippocampal lesions in rats and their effect on activity, avoidance, and aggression. *J Comp Physiol Psychol* 71: 92–101.

Bolles, R.C. (1970) Species-specific defense reactions and avoidance learning. *Psychol Rev* 77: 32–48.

Bonhomme, F., Catalan, J., Britton-Davidian, J., Chapman, V.M., Moriwaki, K., Nevo, E., et al. (1984) Biochemical diversity and evolution in the genus *Mus. Biocheml Genet* 22: 275–303.

Boursot, P., Auffray, J.-C., Britton-Davidian, J., and Bonhomme, F. (1993) The evolution of house mice. *Annu Rev Ecol Syst* 24: 119–152.

Bronson, F.H. (1979) The reproductive ecology of the house mouse. *Q Rev Biol* 54: 265–299.

Bronson, F.H. (1984) The adaptability of the house mouse. *Sci Am* 250: 116–125.

Ceschi, A. (2004) *Effects of Experimental Natural Selection on Exploratory Activity and Anxiety in Mice: Assessment by Means of Three Behavioral Paradigms.* Institute of Anatomy, University of Zürich, Zürich, p. 35.

Crowcroft, P. (1966) *Mice All Over.* G.T. Foulis and Co., London.

Crusio, W.E., Genthner-Grimm, G., and Schwegler, H. (1986) A quantitative-genetic analysis of hippocampal variation in the mouse. *J Neurogenet* 3: 203–214.

Crusio, W.E. and Schwegler, H. (2005) Learning spatial orientation tasks in the radial-maze and structural variation in the hippocampus in inbred mice. *Behav Brain Funct* 1: 3.

Deacon, R.M., Croucher, A., and Rawlins, J.N. (2002) Hippocampal cytotoxic lesion effects on species-typical behaviours in mice. *Behav Brain Res* 132: 203–213.

Deacon, R.M. and Rawlins, J.N. (2005) Hippocampal lesions, species-typical behaviours and anxiety in mice. *Behav Brain Res* 156: 241–249.

Dell'Omo, G., Ricceri, L., Wolfer, D.P., Poletaeva, I.I., and Lipp, H.-P. (2000) Temporal and spatial adaptation to food restriction in mice under naturalistic conditions. *Behav Brain Res* 115: 1–8.

Dell'Omo, G., Shore, R.F., and Lipp, H.-P. (1998) An automated system based on microchips for monitoring individual activity of wild small mammals. *J Exp Zool* 280: 97–99.

Dellu, F., Piazza, P.V., Mayo, W., Le Moal, M., and Simon, H. (1996) Novelty-seeking in rats – biobehavioral characteristics and possible relationship with the sensation-seeking trait in man. *Neuropsychobiology* 34: 136–145.

Drickamer, L.C., Feldhamer, G.A., Mikesic, D.G., and Holmes, C.M. (1999) Trap-response heterogeneity of house mice (*Mus musculus*) in outdoor enclosures. *J Mammal* 80: 410–420.

Drickamer, L.C., Gowaty, P.A., and Holmes, C.M. (2000) Free female mate choice in house mice affects reproductive success and offspring viability and performance. *Anim Behav* 59: 371–378.

Dudek, B.C., Adams, N., Boice, R., and Abbott, M.E. (1983) Genetic influences on digging behaviors in mice (*Mus musculus*) in laboratory and seminatural settings. *J Comp Psychol* 97: 249–259.

Ely, D.L., Greene, E.G., and Henry, J.P. (1976) Minicomputer monitored social behaviour of mice with hippocampal lesions. *Behav Biol* 16: 1–29.

Frynta, D., Slabová, M., Vachová, H., Volfová, R., and Munclinger, P. (2005) Aggression and commensalism in house mouse: a comparative study across Europe and the Near East. *Aggress Behav* 31: 283–293.

Gagliardo, A., Odetti, F., Ioale, P., Bingman, V.P., Tuttle, S., and Vallortigara, G. (2002) Bilateral participation of the hippocampus in familiar landmark navigation by homing pigeons. *Behav Brain Res* 136: 201–209.

Gerlai, R. and Clayton, N.S. (1999) Analysing hippocampal function in transgenic mice: an ethological perspective. *Trends Neurosci* 22: 47–51.

Glickman, S.E., Higgins, T.J., and Isaacson, R.L. (1970) Some effects of hippocampal lesions on the behavior of the Mongolian gerbil. *Physiol Behav* 5: 931–938.

Glickman, S.E. and Morrison, B.J. (1969) Some behavioral and neural correlates of predation susceptibility in mice. *Comm Behav Biol* 4: 361–367.

Guillot, P.-V., Roubertoux, P.L., and Crusio, W.E. (1994) Hippocampal mossy fiber distributions and intermale aggression in seven inbred mouse strains. *Brain Res* 660: 167–169.

Hausheer-Zarmakupi, Z., Wolfer, D.P., Leisinger-Trigona, M.-C., and Lipp, H.-P. (1996) Selective breeding for extremes in open-field activity of mice entails a differentiation of hippocampal mossy fibers. *Behav Genet* 26: 167–176.

Hurst, J.L. and Beynon, R.J. (2004) Scent wars: the chemobiology of competitive signalling in mice. *Bioessays* 26: 1288–1298.

Huston, J.P. and Borbely, A.A. (1973) Operant conditioning in forebrain ablated rats by use of rewarding brain stimulation. *Brain Res* 50: 467–472.

Kim, C. (1960) Nest building, general activity, and salt preference of rats following hippocampal ablation. *J Comp Physiol Psychol* 53: 11–16.

Lipp, H.-P. (1978) Aggression and flight behaviour of the marmoset monkey *Callithrix jacchus*: an ethogram for brain stimulation studies. *Brain Behav Evol* 15: 241–259.

Lipp, H.-P. (1979) Differential hypothalamic self-stimulation behaviour in Roman high-avoidance and low-avoidance rats. *Brain Res Bull* 4: 553–559.

Lipp, H.-P., Amrein, I., and Wolfer, D.P. (2006) Natural genetic variation of hippocampal structures and behavior – an update. In Jones, B.C. (ed.), *Neurobehavioral Genetics: Methods and Applications, Second Edition*, 2nd edn. CRC Press, Boca Raton, FL, USA.

Lipp, H.-P., Collins, R.L., Hausheer-Zarmakupi, Z., Leisinger-Trigona, M.-C., Crusio, W.E., Nosten-Bertrand, M., et al. (1996) Paw lateralization and intra/infrapyramidal mossy fibers in the hippocampus of the mouse. *Behav Genet* 26: 167–176.

Lipp, H.-P. and Wolfer, D.P. (1998) Genetically modified mice and cognition. *Curr Opin Neurobiol* 8: 272–280.

Loas, G. (1996) Vulnerability to depression: a model centered on anhedonia. *J Affect Disord* 41: 39–53.

Maxson, S.C. and Canastar, A. (2003) Conceptual and methodological issues in the genetics of mouse agonistic behavior. *Horm Behav* 44: 258–262.

Miczek, K.A., Maxson, S.C., Fish, E.W., and Faccidomo, S. (2001) Aggressive behavioral phenotypes in mice. *Behav Brain Res* 125: 167–181.

Minichiello, L., Korte, M., Wolfer, D.P., Kühn, R., Unsicker, K., Cestari, V., et al. (1999) Essential role for TrkB receptors in hippocampus-mediated learning. *Neuron* 24: 401–414.

O'Keefe, J. and Nadel, L. (1978) *The Hippocampus as a Cognitive Map*. Clarendon Press, Oxford.

Pennycuik, P.R., Reisner, A.H., and Westwood, N.H. (1987) Effects of variations in the availability of food and home sites and of culling on populations of house mice housed in out-door pens. *Oikos* 50: 33–41.

Plesner-Jensen, S., Gray, S.J., and Hurst, J.L. (2003) How does habitat structure affect activity and use of space among house mice? *Anim Behav* 66: 239–250.

Poletaeva, I.I., Pleskacheva, M.G., Markina, N.W., Perepiolkina, O.W., Scheffrahn, H., Wolfer, D.P.P., et al. (2001) [Environmental habitat-related pressure: behavioral alterations and morphological changes in the brain of the house mouse] Russian. *Ecologia* 3: 231–236.

Rissi, S. (2002) *Moosfasersystem und Verhalten bei der Hausmaus nach Embryonentransfer*. Institute of Anatomy, University of Zürich, Zürich, p. 33.

Schwegler, H. and Lipp, H.-P. (1983) Hereditary covariations of neuronal circuitry and behavior: correlations between the proportions of hippocampal synaptic fields in the regio inferior and two-way avoidance in mice and rats. *Behav Brain Res* 7: 1–39.

Sluyter, F., Bult, A., Lynch, C.B., van Oortmerssen, G.A., and Koolhaas, J.M. (1995) A comparison between house mouse lines selected for attack latency or nest-building: evidence for a genetic basis of alternative behavioral strategies. *Behav Genet* 25: 247–252.

Sluyter, F., Jamot, L., Van Oortmerssen, G.A., and Crusio, W.E. (1994) Hippocampal mossy fiber distributions in mice selected for aggression. *Brain Res* 646: 145–148.

Van Oortmerssen, G.A. (1971) Biological significance, genetics and evolutionary origin of variability in behaviour within and between inbred strains of mice (*Mus musculus*). A behaviour genetic study. *Behaviour* 37: 1–91.

Vyssotski, A.L., Dell'Omo, G., Poletaeva, II, Vyssotsk, D.L., Minichiello, L., Klein, R., et al. (2002) Long-term monitoring of hippocampus-dependent behavior in naturalistic settings: mutant mice lacking neurotrophin receptor TrkB in the forebrain show spatial learning but impaired behavioral flexibility. *Hippocampus* 12: 27–38.

Wallace, R.B., Graziadei, R., and Werboff, J. (1981) Behavioral correlates of focal hippocampal x-irradiation in rats II. Behavior related to adaptive function in a natural setting. *Exp Brain Res* 43: 207–212.

Wolfer, D.P., Crusio, W.E., and Lipp, H.-P. (2002) Knockout mice: simple solutions to the problems of genetic background and flanking genes. *Trends Neurosci* 25: 336–340.

Zuckerman, M. (1990) The psychophysiology of sensation seeking. *J Pers* 58: 313–345.

Chapter 3

Ethogram of the mouse

Wim E. Crusio, Frans Sluyter, and Robert T. Gerlai

Introduction

Regarding the question how we can record and quantify animal behavior, two main schools exist whose traditional answers have been, and perhaps still are, quite different. Comparative psychologists (mainly North American) argue that one has to design a laboratory experiment in which precise procedural control is implemented, using well-designed hardware and software solutions, so that all possible experimental factors will be controlled, error variation will be minimized, and quantification of behavior is achieved in an unbiased manner, preferably with the use of automated recording techniques. This approach emphasizes cross-species similarities and focuses on common features that may represent the fundamental core of behavioral mechanisms and/or brain function. Ethologists (mainly European) argue that instead of forcing the animal to conform to the rigid boundaries set by the test design, allowing the animal to exhibit its natural spontaneous behavior provides a better characterization of behavioral mechanisms and functional aspects of the brain. Given that each species possesses a broad and complex behavioral repertoire that may manifest in a species-specific and test-situation-dependent manner not easily quantifiable using automated techniques, ethologists often prefer observation-based quantification of behavior. Briefly, the differences between the comparative psychological and ethological approaches concern both experimental control and the method of behavioral quantification. Another main difference between these two schools is the higher importance that ethologists attach to naturalistic approaches, stressing the fact that animals and their behavioral characteristics have evolved and that ecological relevance therefore is important, whereas comparative psychology tends to be "nature-blind." For example, ethologists argue that taking species-specific characteristics (perceptual, motivational, or motor features, for example) into account and designing "nature-aware" experiments according to the unique features of the species under study may actually make it easier to get to the common underlying biological and behavioral mechanisms. The discussion as to which approach is more useful is still ongoing (see, e.g., Gerlai and Clayton, 1999), but a synthesis is not impossible and many examples of this can be found in the present volume.

As indicated above, the differences between these two schools are not only theoretical but also have clear practical implications. Comparative psychologists tend to measure behavioral performance differently than ethologists do. In an active avoidance shuttle box experiment using electric shocks, for example, a comparative psychologist would argue that amount of shuttling activity (frequency of going back and forth between the compartments of a test apparatus) represents motivation, or that the latency to exit the place where shock is to be received is a good measure of acquired memory. The behavioral parameters the comparative psychologist measures are strictly defined by the experimental apparatus and procedure and also are usually quite limited in number. This is a great advantage when it comes to computerized recording of behavior and when high-throughput testing is required. However, sometimes this approach may lead to false results. For example, once we (Gerlai et al., personal observation) found a fish that produced a shuttling frequency count of 600 per minute, a speed that even these agile animals (Paradise fish, *Macropodus opercularis*) could not possibly achieve. It turned out that this particular subject discovered a different strategy than what the experimenter had intended for him. He stayed in the middle of the doorway in between the two compartments of the shuttle box and, with his pectoral fin beating at 10 Hz frequency right in front of the photocell detector, he confused the shock system and as a result the computer never gave the command "electric shock on." The behavior the fish performed was similar to a fear-related posture "creeping" that has been described and well-investigated before (e.g., Gerlai and Csányi, 1994). Another anecdote of such alternative solutions is that of a mouse that assumed a unique body position stepping onto parts of the shock grid and walls of the shuttle box so as to avoid the shock (Herbert Schwegler, Magdeburg, Germany, personal communication), thereby obtaining a "learning score" (i.e., number of shock avoidances) that normally would be interpreted as a failure to learn the task at all. These examples suggest that automated behavioral recording, and perhaps any artificial behavioral paradigm in general may, at least on occasion, turn out not to be able to cope with the complex responses that animals exhibit. The solution suggested by

Behavioral Genetics of the Mouse: Volume I. *Genetics of Behavioral Phenotypes*, eds. Wim E. Crusio, Frans Sluyter, Robert T. Gerlai, and Susanna Pietropaolo. Published by Cambridge University Press.

ethologists is to actually *look* at the subject, record its motor and posture patterns, and thus obtain a rich description of what exactly the animal was doing. In the words of van Abeelen (1965, p. 15): "Most behaviour-genetical studies have restricted themselves to very few aspects of behaviour only. For a further development it seems important that behaviour genetics should to a greater extent adopt the ethological procedures of direct observation and the drawing up of ethograms." This refers to the basic ethological technique of breaking down the seemingly continuous behavior of an animal to a succession of distinct and discrete motor and posture patterns, which together form the *ethogram* (e.g., Huntingford, 1984).

Despite the foregoing, it is important to note that we do not advocate abandoning rigorous experimental control or abandoning computerized and automated behavioral quantification, but we are arguing that combining the two approaches makes sense not only from a theoretical (Gerlai and Clayton, 1999) but also from a practical standpoint. It is also notable that there are now concerted efforts towards the establishment of automated computer image analysis methods that can detect and quantify a large number of motor and posture patterns (for examples see Gerlai, 2002). In sum, the recording of behavioral elements may be an important addition to any behavioral study. In the following pages, we will review the ethogram of the house mouse and subsequently discuss how quantification of the behavioral elements of the mouse may be done using twenty-first century technology.

Ethogram of the mouse

One of the earliest descriptions of the mouse ethogram, possibly *the* earliest, was published almost half a century ago by one of the founders of the field of behavior genetics, Hans van Abeelen (1963), and few have been published since. In keeping with ethological practice, van Abeelen chose *descriptive* names for his behavioral elements (such as "leaning against wall"), rather than the *functional* names used by, for example, Scott (1958), who grouped different behavioral elements into exploratory, ingestive, agonistic, and sexual behaviors. This is an important point that cannot be stressed enough: learning, anxiety, aggression, and such cannot be measured directly but are *inferred* from more simple behavioral measures (such as number of correct responses in a maze-learning task, number of attacks in a resident–intruder test, or amount of time spent on an open arm in a plus-maze test; see Henderson, 1979).

The description of the mouse ethogram below closely follows van Abeelen's (1963, 1965), with some later modifications (Crusio and van Abeelen, 1986), and a few additions (Brown et al., 1999; van Oortmerssen, 1971). The different behavioral elements are grouped according to the situations in which they may be observed. Some very good quality drawings of different elements can be found in van Abeelen's original article (1963). To quantify the behavior using these qualitative descriptions, the frequency and/or duration of the different elements may be measured.

Single mouse in an unfamiliar new environment

Most of the behavioral elements described below can also be observed in familiar environments such as an animal's home cage, although the frequencies with which they occur may differ vastly according to the specific test situation (see Chapter 15).

Locomotor activity: The amount of ambulation displayed by the animal. This can be quantified (as centimeters traveled) by using automatic tracking software or by counting the number of crossings of lines drawn on the floor of the testing apparatus. In general, the tail is ignored in these measures (as the animals seem to do themselves). If line crossings are used as a measure, then a more exact definition (such as: crossing the line with at least three paws) is needed. Locomotor activity is sometimes subdivided according to the zone of the observation cage in which it occurs (i.e., center activity and activity close to the walls, thigmotaxis).

Rearing: The animal stands upright on its hind legs, with the tail usually serving as a support. The forepaws are free from any surface. The animal will sometimes repeat this movement in a rapid succession. Van Abeelen (1963) originally termed this behavior "reconnoitering," but this term has some undesirable functional implications.

Leaning: Very similar to rearing, but here the animal places one or both forepaws against the wall. Often, but not always, this is combined with sniffing of the wall. This behavior is sometimes combined with the previous one, but this is not recommended since it has been shown that these behaviors have different genetic underpinnings (see Chapter 15).

Sniffing: Unless the animal is freezing (see below) sniffing occurs continually and is characterized by (often rapid) movements of the nasal skin. When quantifying this behavior, it is often only counted if the animal is standing still and the behavior is clearly directed to a particular spot.

Jumping: All four legs simultaneously lose contact with the floor. Rarely, an animal will display a rapid succession of jumps, mostly while it is standing in a corner of the observation cage. The movement can be initiated from a crouched, sitting, or upright position (mostly leaning, rarely rearing).

Gnawing: Occasionally animals will gnaw at edges of the floor and walls. As this is often difficult to see, it may be scored only when audible when quantifying the behavior.

Defecation: Often observed especially at the beginning of an observation. It can be quantified simply by counting the number of boluses deposited.

Urination: Can be quantified by the presence or absence of urine at the end of the session, or by counting the number of different urinary spots produced (in the latter case the researcher will often put absorbent paper on the floor of the apparatus).

Freezing: Apart from breathing, the animal is completely motionless including, typically, an absence of movements of the nasal skin (see *sniffing*, above). Van Abeelen (1963) originally called this "staring at the observer," as it often occurs in consequence of a movement made by the observer, but, again, later discarded this usage because of its functional implications. Freezing sometimes occurs without any apparent cause. Obviously, nowadays experimenters will generally score behavior from videotapes, eliminating unwanted observer interference. Freezing can be quantified by measuring bout frequency and duration.

Grooming: This is a quite complex behavior and several components, such as wiping, fur-licking, combing, scratching, and face-cleaning can be distinguished (see Chapter 20). Often, however, the different components are lumped together and the behavior is then quantified by measuring bout frequency and duration. Grooming often occurs when the animal habituates to the testing environment and reduces its exploration.

Tail rattling: This behavior often occurs in a mouse that has suddenly caught sight of some object or the observer (or another mouse, see the next section). The intensity is variable and sometimes only the tip of the tail moves a little.

Sitting upright: The animal sits on its haunches with the forepaws not touching the floor.

Carrying: If food or other materials are present in the observation cage, animals will occasionally carry this, generally to the corner of the cage, which is a favorite eating spot.

Eating: The animal assumes a sitting posture, with its back strongly flexed, while eating.

Digging: If the floor is covered with sawdust or some other material, digging will occur.

Stripping or fraying: The animal draws material (grass, paper) through the mouth using the forepaws while making gnawing movements with its jaws and jerky head movements, splitting the material into smaller parts.

Hauling in: Pulling of nesting material with the forepaws and pulling movements of the mouth towards a specific spot (often the nest).

Arranging or nest building: While sitting in the nest, the animal distributes nesting material by pushing movements of the forepaws.

Two unfamiliar male mice in an unfamiliar environment

Most of the behavioral elements described below can also be observed in familiar environments such as an animal's home cage and between familiar males, although the frequencies with which different behaviors occur may differ vastly according to the exact testing situation (see Chapter 24). In addition, most of the behaviors described above can also occur in this situation and their descriptions will generally not be repeated here.

Tail rattling: As described above. This behavior occurs especially when animals are confronted with each other for the first time. It can be quantified by measuring bout frequency and latency to first occurrence.

Dancing position: The animal is sitting facing the opponent, with the head thrust forward and the ears folded backwards.

Boxing: Drumming on the opponent with the forepaws.

Kicking: A hard kick with the hind-leg, occurs mostly during wrestling.

Anogenital sniffing: One animal (or sometimes both simultaneously) sniffs the anogenital area of the other.

Wrestling: The animals roll over and over, accompanied by biting, mainly in the flanks.

Chasing: One animal chases the other and bites it in the lower back, tail, or hind-legs.

Following: As the previous one, but without biting taking place. Often the ambulation speed of the animals is lower, too.

Fleeing: The opposite of the foregoing, rushing away in an often seemingly random manner, often accompanied by squeaks.

Submissive posture: The mouse rears on its hind legs with the head in an upright position (cf. dancing position), with the forelegs held in front. The animal remains motionless and squeals when touched by the other mouse.

Mounting: See next subsection.

Social grooming: Grooming of the opponent mouse.

Crawling over/under: One animal crawls over or under the other animal.

A male and a female mouse together

This situation is usually studied in the home cage, but most of the behavioral elements described below can also be observed in unfamiliar environments such as an open field, although the frequencies with which different behaviors occur may differ vastly according to the exact testing situation (see Chapter 22). In addition, most of the behaviors described in the previous subsections can also occur in this situation and their descriptions will generally not be repeated here.

Creeping in front of partner: The animal crawls slowly immediately in front of the other mouse, its body sometimes touching the floor.

Touching: Nudging of the partner underneath the head and in the neck with the mouth. This is often followed by social grooming.

Mounting: The male places his forepaws on the female's back and moves his lower abdomen towards her genitals. Mounting sometimes also occurs at the other side, near the head.

Copulation and falling over: Together, the copulating pair falls on their sides and remains in that position for some moments. Ejaculation occurs at this point.

Lying flat: This behavior is only seen in males, especially when the female engages in a lot of parrying (see below). The animal shows a prolonged stretching of the trunk which lasts for about half a minute.

Licking the genitals: This occurs especially after copulation. The female will usually lift a hind-leg while engaging in this behavior.

Parrying: Sidelong movements, made with the forepaws as well as the hind-legs, to keep the male at bay during attempts at approach and mounting.

Adult mouse with pups

This situation generally concerns female mice, although occasionally the interactions of male mice with pups are also studied. As before, most of the behaviors described in the previous subsections can also occur in this situation and their descriptions will generally not be repeated here. The behaviors listed here are based on the observations of Brown et al. (1999).

Sniffing pups: The animal touches the pups with the nose accompanied by the typical movements of the nasal skin.

Licking pups: The animal licks any part of the pup's body.

Resting with pups: The animal is inactive but (disregarding the tail) is in bodily contact with at least one pup.

Resting alone: As the foregoing, but without any contact with any pup.

Crouching over pups: The animal assumes an arched-back position over the pups. This generally is accompanied by nursing (suckling).

Nursing: At least one pup is being suckled.

Retrieving: The animal pushes or carries a pup by taking a part of the pup's body into the mouth. This generally occurs when pups for some reason find themselves outside of the nest.

Examples of the experimental use of the ethogram

One telling example of the advantages that small-grained observation of behavior may offer was the finding of a spontaneous mutation affecting behavior in C57BL/6J (B6) mice (Crusio et al., 1991). By means of direct observation of exploratory behavior in an open field it was established that mice from the B6 strain that had been maintained over many generations in Nijmegen (the Netherlands) by Hans van Abeelen showed increased frequencies of wall-leaning but decreased frequencies of rearing-up when compared to B6 mice obtained from The Jackson Laboratory (Bar Harbor, ME, USA; Crusio et al., 1991). However, had these two behaviors been summed into a "vertical activity" category (as happens almost always when behavior is recorded automatically), no difference between the Nijmegen and Jackson B6 lines would have appeared and the mutation, which also affected hippocampal neuroanatomy and behaviors related to learning and aggression (Jamot et al., 1994; Sluyter et al., 1999), would have gone unnoticed.

Gerlai et al. (1999) investigated the role of EphA tyrosine kinase receptors in learning a fear conditioning task (see Chapter 31) in which the traditional way to assess the strength of fear memory is to measure freezing. The assumption is that a higher amount of freezing exhibited by the animal in a test situation indicates better memory of the electric shock that was delivered there before the test. The concern about how to measure fear (and thus memory) was twofold. First, freezing may not be the only way the animal expresses its fear. Second, perhaps forms of immobility unrelated to freezing and fear may appear and these may be confused by the experimenter with freezing. Thus both false negative (animal shows signs of fear other than freezing and thus experimenter misses fear reaction) and false positive (animal is passive and experimenter falsely quantifies it as freezing) findings may arise. To avoid this, Gerlai et al. (1999) recorded more behaviors, quantifying grooming and long-body posture or stretch attend posture, which has been found to be associated with fear and anxiety (Blanchard et al., 2001). The latter appears in response to mild pain or discomfort when the situation is somewhat ambiguous or after an obvious presence of danger is not apparent any more. Grooming, on the other hand, most often appears in situations where the animal is more habituated and hence, presumably, more relaxed (Gerlai et al., 1999). This behavior could thus be interpreted to correlate negatively with fear but is associated with lack of locomotion. Gerlai et al. (1999) argued that if increase or decrease of freezing is correlated with the above behaviors (positive correlation with long-body and negative correlation with grooming) then these changes may be interpreted as alterations in fear levels and not just as changes in motor performance or activity. Indeed, this is exactly what they have found. Figure 3.1 shows the results obtained with mice treated either with the antagonist agent (EphA5-IgG) that is expected to deactivate the kinases or with the control agent (CD4-IgG) that is expected have no effect. It shows the

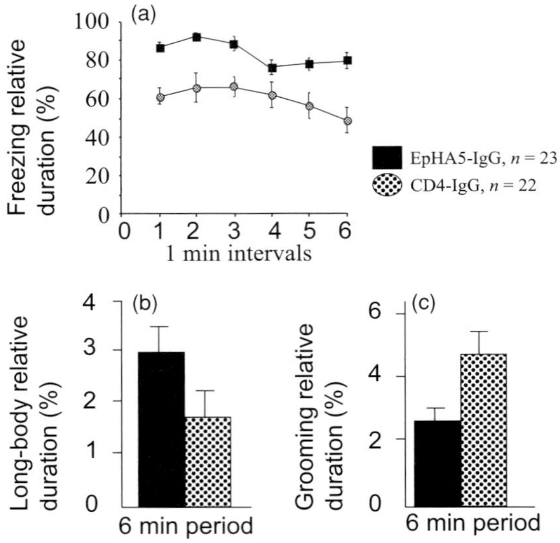

Figure 3.1 Disruption of EphA tyrosine kinase activity leads to correlated behavioral responses in freezing, long-body posture, and grooming in C57BL/6J mice. Behavior was recorded 24 hours after training in the shock chamber. The recording session was 6 minutes long. The previous training included three electric shocks (0.7 mA, 1 s long each) delivered through a metal bar grid on the floor of the shock chamber. Note that during testing, no shocks were delivered and increased fear responses to the shock chamber are regarded as evidence of memory of the shock received 24 hours before. Means ± SEM are shown. Sample sizes (n) are also indicated. Panel (a) shows how much freezing the animals performed in the shock chamber (black squares: EphA5-IgG antagonist-infused mice; gray-hatched circles: CD4-IgG control-infused mice). Panels (b) and (c) show the results for long-body posture and grooming, respectively (black bars: EphA5-IgG antagonist-infused mice; gray-hatched bars: CD4-IgG control-infused mice). Note that the latter two behaviors occurred for a much shorter period of time than freezing did and thus they are quantified for the entire 6 minute recording session as opposed to 1 minute intervals of the session. (Modified from Gerlai et al., 1999.)

behavioral responses to the shock chamber (note that now no electric shocks are administered) where the mice received three electric shocks 24 hours before the test. The results demonstrate a significant reduction of freezing in the antagonist treated mice as compared to the control group. In addition to freezing, the effect can also be seen in long-body posture, which is also reduced, and in grooming, which is increased. Thus changes in the latter two behavioral elements correlate well with changes in freezing and this suggests that the antagonist-treated mice are less afraid of the shock chamber. The point here is that in this particular experiment freezing turned out to be the dominant form of fear reaction and thus appropriately reflected level of fear. But depending on the type of experimental treatment and procedural details of the test, high grooming values (which do not indicate fear) could have been mistaken for freezing by an automated recording method (falsely leading to the conclusion of increased fear) and high amounts of long-body posture (which do reflect increased fear) could have been interpreted as reduction of freezing (leading to the false conclusion of reduced fear).

Future of the ethogram: computer vision and other high-tech applications

With the advent of recombinant DNA technologies and large-scale chemical (random) mutagenesis studies, the number of mouse mutants waiting to be properly assessed (nowadays fashionably called "phenotyping") has dramatically increased. Similarly, there are a large number of pharmaceutical compounds that one could test and, given that the least understood human diseases are those that affect the central nervous system and that these diseases represent the biggest potential markets for drug companies, the need for testing the efficacy of compounds, i.e., their ability to alter brain function and behavior, is enormous. Briefly, high-throughput screening is an absolute must.

However, the traditional ethological method of observation-based quantification of the ethogram requires a live human observer and thus is too labor intensive and time consuming. Latencies to enter a compartment or the number of lever presses, examples of classical comparative animal psychology variables, however, are easy to quantify using automated computerized methods. Is this then the death of ethological measures?

While the problem of automatic computerized quantification of motor and posture patterns is not easy, it is not impossible to solve. Numerous examples already exist suggesting that in the future such applications will be possible. The development of three-dimensional video-input based image analysis systems has started (reviewed in Gerlai, 2002). These software applications, called "computer vision," have already been demonstrated to be able to distinguish complex and dynamic human facial expressions (Pantic and Patras, 2006). Animal behavior software and hardware companies have started to add certain functionalities to their video-tracking systems that now allow the user to distinguish the front end and the rear end of the animal or whether the animal is rearing or moving along or curled up or not.

Although still rudimentary in terms of resolution and precision compared to the human observer, the above computer applications signal the coming of sophisticated methods with which behavior will be quantified with an ever-increasing precision. Eventually all that is needed to be done is to establish a library of force prints (if one uses force transducer technology, e.g., Fitch et al., 2002), or dynamic images (video-tracking and image analysis), for example, and correlate their temporal appearance with the behavioral elements recoded by a human observer. Once a strong correlation is established, the computer can take over and the recording of motor and posture patterns will be possible in a high-throughput manner. It would appear, therefore, that the advent of high-throughput phenotyping will wed the experimental rigor of the comparative psychology approach with the increased sensitivity of the ethological method to detect subtle effects of mutations and other experimental treatments on behavior.

References

Blanchard, R.J., Yang, M., Li, C.I., Gervacio, A., and Blanchard, D.C. (2001) Cue and context conditioning of defensive behaviors to cat odor stimuli. *Neurosci Biobehav Rev* 25: 587–595.

Brown, R.E., Mathieson, W.B., Stapleton, J., and Neumann, P.E. (1999) Maternal behavior in female C57BL/6J and DBA/2J inbred mice. *Physiol Behav* 67: 599–605.

Crusio, W.E., Schwegler, H., and van Abeelen, J.H.F. (1991) Behavioural and neuroanatomical divergence between two sublines of C57BL/6J inbred mice. *Behav Brain Res* 42: 93–97.

Crusio, W.E. and van Abeelen, J.H.F. (1986) The genetic architecture of behavioural responses to novelty in mice. *Heredity* 56: 55–63.

Fitch, T., Adams, B., Chaney, S., and Gerlai, R. (2002) Force transducer-based movement detection in fear conditioning in mice: a comparative analysis. *Hippocampus* 12: 4–17.

Gerlai, R. (2002) Phenomics: fiction or the future? *Trends Neurosci* 25: 506–509.

Gerlai, R. and Clayton, N.S. (1999) Analysing hippocampal function in transgenic mice: an ethological perspective. *Trends Neurosci* 22: 47–51.

Gerlai, R. and Csányi, V. (1994) Artificial bidirectional selection for a species specific behavioural element, staccato movement, in paradise fish (*Macropodus opercularis*). *Anim Behav* 48: 1293–1300.

Gerlai, R., Shinsky, N., Shih, A., Williams, P., Winer, J., Armanini, M., et al. (1999) Regulation of learning by EphA receptors: a protein targeting study. *J Neurosci* 19: 9538–9549.

Henderson, N.D. (1979) Adaptive significance of animal behavior. In Royce, J.R. and Mos, L.P. (eds.), *Theoretical*

Advances in Behavior Genetics, NATO Advanced Study Institutes Series D: Behavioral and Social Sciences. Sijthoff and Noordhoff, Alphen aan den Rijn, The Netherlands, pp. 243–287.

Huntingford, F. (1984) *The Study of Animal Behaviour*. Chapman and Hall, London.

Jamot, L., Bertholet, J.-Y., and Crusio, W.E. (1994) Neuroanatomical divergence between two substrains of C57BL/6J inbred mice entails differential radial-maze learning. *Brain Res* 644: 352–356.

Pantic, M. and Patras, I. (2006) Dynamics of facial expression: recognition of facial actions and their temporal segments from face profile image sequences. *IEEE Trans Syst Man Cybern B Cybern* 36: 433–449.

Scott, J.P. (1958) *Animal Behavior. The College Library of Biological Sciences*. University of Chicago Press, Chicago.

Sluyter, F., Marican, C.C., and Crusio, W.E. (1999) Further phenotypical characterisation of two substrains of C57BL/6J inbred mice differing by a spontaneous single-gene mutation. *Behav Brain Res* 98: 39–43.

van Abeelen, J.H.F. (1963) Mouse mutants studied by means of ethological methods.I. Ethogram. *Genetica* 34: 79–94.

van Abeelen, J.H.F. (1965) *An Ethological Investigation of Single-Gene Differences in Mice*. University of Nijmegen, The Netherlands, p. 79.

van Oortmerssen, G.A. (1971) Biological significance, genetics and evolutionary origin of variability in behaviour within and between inbred strains of mice. *Behaviour* 38: 1–92.

Chapter 4

Replicability and reliability of behavioral tests

Douglas Wahlsten and John C. Crabbe

Introduction

When a behavioral test is repeated, we hope the results will be essentially the same, despite small variations in the data that inevitably occur with repetition. *Replicability* of a test refers to the repetition of an entire experiment with different samples of subjects, done either in the same laboratory separated by a period of time or done in different laboratories. *Reliability* of a test expresses the tendency of the same individual to have the same score on different test occasions. These two concepts are to some extent independent, in that a test found to be highly reliable in one laboratory might nevertheless yield different results in another laboratory if the genotype × laboratory environment interaction is substantial. Likewise, two laboratories might obtain the same results when the test has very low reliability; for example, neither laboratory might observe significant genetic effects.

In this chapter, we illustrate the concepts with data collected on mouse inbred strains in our two laboratories. Brief details of behavioral tests and mouse strains are given in the Appendix. More complete descriptions are available from published articles as well as protocols deposited along with the data in the Mouse Phenome Database (http://phenome.jax.org/).

Replicability

In designed experiments with laboratory animals differing in genotype, the contribution of genotype is often assessed with an analysis of variance (ANOVA). This tests the hypothesis that variability within strains, which is assumed to be environmental in origin, is less than that among strain means, which reflects genetic sources of influence. Consider an experiment where learning to remain on an accelerating rotarod was compared for 21 inbred mouse strains tested simultaneously in two laboratories (Rustay et al., 2003). Figure 4.1a shows that mice tended to remain on the rod longer in Edmonton than Portland when averaged over all strains, so that the ANOVA main effect of laboratory was unquestionably significant ($P < 0.00001$), and the strain main effect was also significant ($P < 0.00001$). The strain by laboratory interaction was clearly significant as well ($P = 0.0005$), yet the pattern of strain differences was fairly similar in the two laboratories. Figure 4.1b shows the linear correlation of $r = 0.66$ between strain means in the two laboratories. For most of the strains, the ranking in the two laboratories was nearly the same. For the strain NOD/LtJ that was farthest from the regression line, its rank was first in Portland and third in Edmonton, a very minor discrepancy. SPRET/Ei was near the bottom in Portland but above the mean of all strains in Edmonton, a more striking difference in results. The significant interaction term and the fact that strain correlation was substantially less than 1.0 indicate that results were different for some but not all strains. Apart from this statement, it is necessary to conduct a careful examination of the data in order to interpret the extent of replicability. We regard the rotarod data as showing good but imperfect replicability of the strain means in the two laboratories. Effects of ethanol on rotarod performance, on the other hand, were virtually the same in the two laboratories; although the strains were differentially impaired by ethanol, the strain sensitivities were similar in the two laboratories (Rustay et al., 2003).

The relative merits of the ANOVA and correlational approaches for assessing reliability are explored in Figure 4.2 for hypothetical data on four strains in two laboratories. Values are given as true, population values, much as one would use in doing a power and sample-size calculation (Bausell and Li, 2002; Cohen, 1988; Murphy and Myors, 2004). Case A expresses complete additivity, where strains are 50 units apart, laboratories differ by 20 units, and the laboratory effect is the same for every strain. The population standard deviation among the four strain means, $\sigma_s = 55.9$, is the strain main effect. Obviously, the correlation of strain means must be 1.0 and there should be no interaction effect in the ANOVA. In each of the other five cases, the strain means across laboratories are the same as in Case A, and laboratory means over the four strains are 175 and 195 for Laboratories 1 and 2, respectively. Therefore, for all other cases, the cell means expected from an additive relation of strain and laboratory are the cell means for Case A.

In the five examples, B–F, two of the four strains have scores the same as in Case A, whereas two other strains show an interaction effect. In Case B, strains S2 and S3 differ greatly in L1 but are the same in L2, such that the cell means deviate by 25 points from those in Case A. This reduces the correlation to

Behavioral Genetics of the Mouse: Volume I. *Genetics of Behavioral Phenotypes*, eds. Wim E. Crusio, Frans Sluyter, Robert T. Gerlai, and Susanna Pietropaolo. Published by Cambridge University Press.

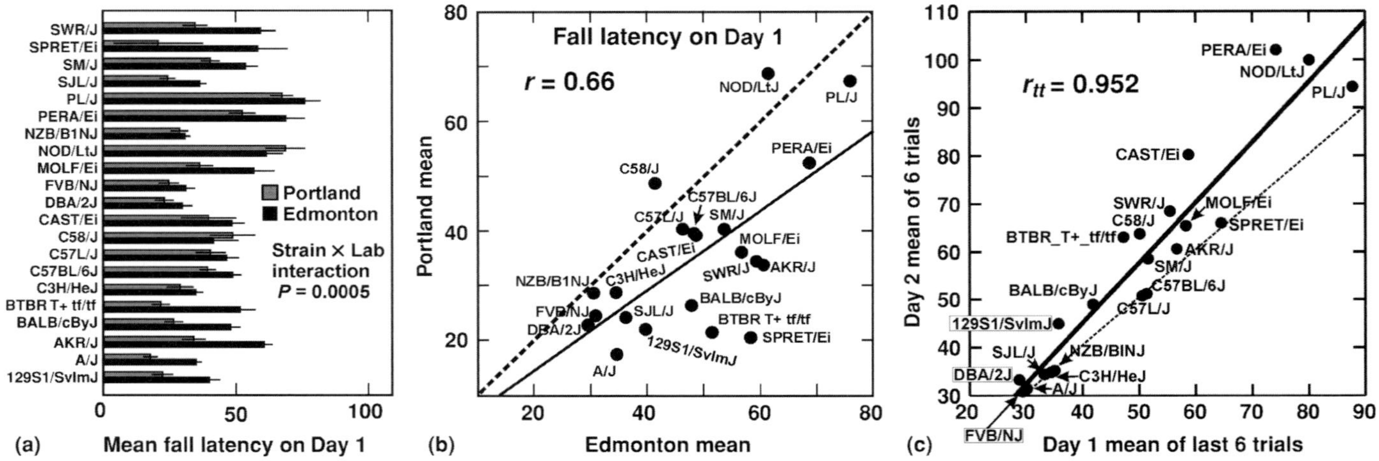

Figure 4.1 (a) Performance on 10 trials on 1 day on the accelerating rotarod tested simultaneously at two sites with 21 inbred strains (data from Rustay et al., 2003). The strain by laboratory analysis of variance (ANOVA) indicated a clearly significant interaction effect in which results were very similar for certain strains (e.g., C57L, NOD, NZB) but quite different for others (BTBR, SPRET). (b) The same strain means in a scatter plot indicated a substantial but imperfect correlation. (c) The correlation of strain means across six trials on 2 successive days was very high indeed, thanks in part to the inclusion of a wide range of strains. Data from Edmonton and Portland are pooled. The dashed line indicates equal scores on the 2 days. Because the solid regression line is above the dashed line, scores were generally higher on the second day.

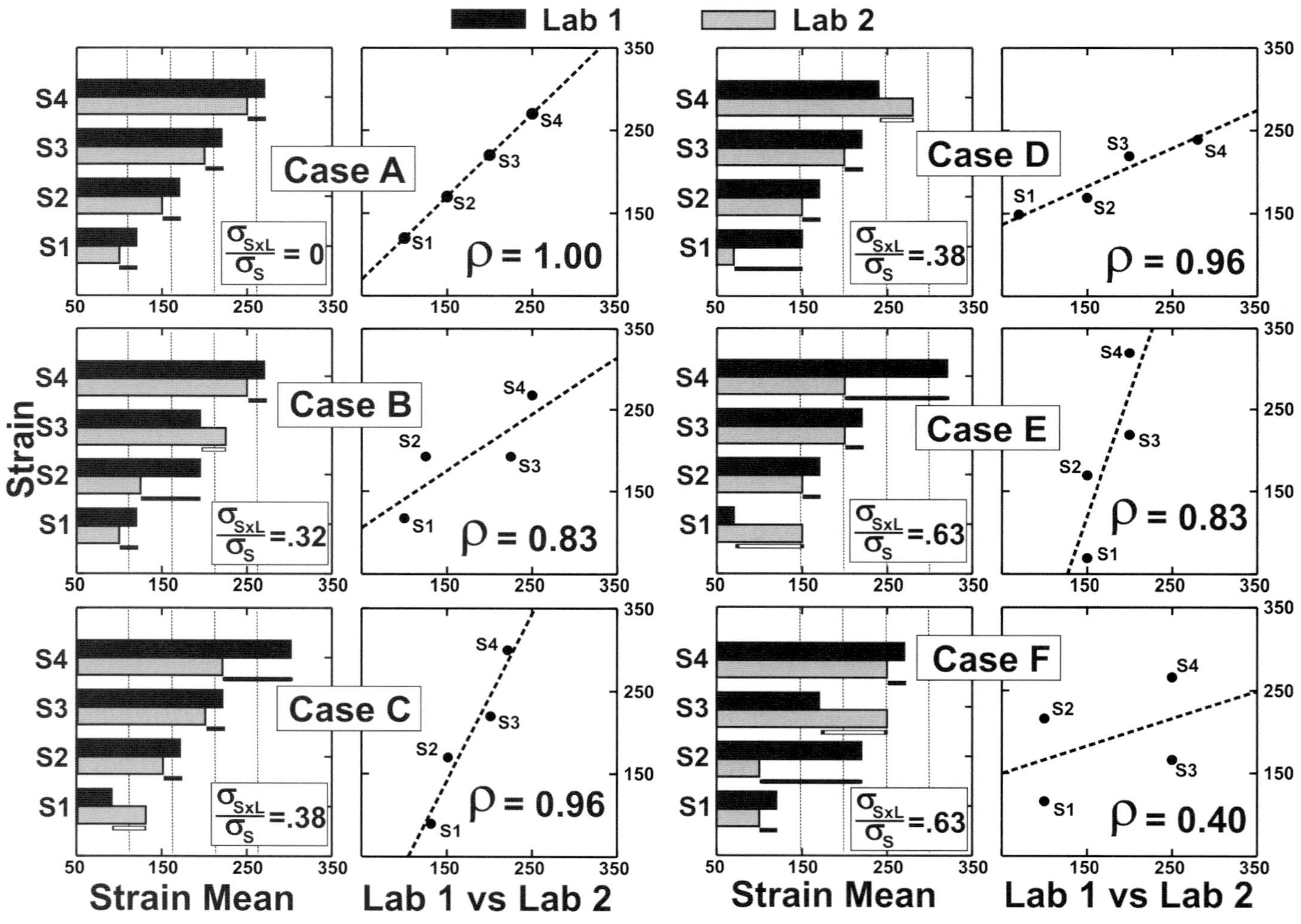

Figure 4.2 Six hypothetical examples of four strains tested in two laboratories, expressed as population means in a bar graph and scatter plots between laboratories. Small bars next to each pair of laboratory means for a strain emphasize the size of the laboratory difference. Black bars indicate Laboratory 2 mean exceeds Laboratory 1, whereas white bars indicate Laboratory 1 mean is greater. The inset for each panel compares the relative effect size from ANOVA due to the interaction versus due to the strain main effect (see text). Case A: Additive effects of strain and laboratory yield a perfect correlation (ρ) of means and no interaction effect ($\sigma_{SxL} = 0$). Cases B–D: When the strain rank orders are nearly the same in both laboratories, the interaction effect is moderate and strain correlations are high. Case E: A large interaction effect is in this case consistent with a high strain correlation. Case F: An interaction effect of the same magnitude as in Case E involves a major laboratory difference in strain rank orders and therefore a diminished correlation.

$\rho = 0.83$, but the similarity of strain means across laboratories is still substantial. To compute the interaction effect, sum the squared deviations of each cell mean in Case B from the expected values under additivity (Case A) and divide by the number eight (number of cells) to find variance of the interaction effect. Take the square root to obtain the population standard deviation of strain by laboratory interaction departures from additivity, which for Case B is $\sigma_{S \times L} = 17.7$. This effect size is considerably lower than the strain main effect of $\sigma_S = 55.9$, but larger than the laboratory effect of $\sigma_L = 10.0$. Because the strain main effect σ_S is the same for all six cases in Figure 4.2, the relative magnitudes of the interaction effect in cases B through F can be judged from the ratio of the value of $\sigma_{S \times L}$ to the strain main effect σ_S. These indicators of effect size are especially useful because, when divided by the standard deviation within groups (σ_W), they determine Cohen's effect size $f = \sigma_S/\sigma_W$ for the main effect and $f = \sigma_{S \times L}/\sigma_W$ for the interaction that are used to find sample size needed to confer adequate power on the test.

In Case C, the highest and lowest strains differ greatly between laboratories but in opposite directions, and their cell deviations from additivity are 30 units. Consequently, the interaction effect increases to $\sigma_{S \times L} = 21.2$. Despite the larger interaction effect, the strain correlation becomes very large ($\rho = 0.96$). In Case D the laboratory effects in strains S1 and S4 are opposite those in Case C but the deviations from additivity are also 30 units, the interaction is also $\sigma_{S \times L} = 21.2$, and the strain correlation is also $\rho = 0.93$. Thus, for these two cases where deviations of four of eight cells from additivity are 30 units, the interaction effects and strain correlations are identical. In both Cases C and D, the strain rank orders are the same in the two laboratories, despite the substantial interaction effects.

The deviation from additivity invariably accounts for the ANOVA interaction effect $\sigma_{S \times L}$, but the strain correlation does not always follow in synchrony. For Case E, the middle two strains have the same laboratory effect as in Cases C and D, but for L1 the strains S1 and S2 are the same and S3 and S4 are the same, whereas in L2 the strains S1 and S2 differ by 100 units, as do strains S3 and S4, resulting in deviations from additivity of 50 units in the four cells and a much larger $\sigma_{S \times L} = 35.4$.

The strain correlation, however, decreases to $\rho = 0.83$, the same as for Case B. Case F also has four cells with deviations from additivity of 50 units and $\sigma_{S \times L} = 35.4$, but the strain correlation decreases radically to $\rho = 0.40$. When the correlation is this low, there are major discrepancies in the strain rank orders in the two laboratories.

The interaction effect in the ANOVA depends on the extent of deviations from additivity but is not influenced by which of the various strains shows the largest difference between laboratories. The strain correlation, on the contrary, depends strongly on the range of strain means as well as which specific strains show the largest difference between laboratories, and the correlation can be augmented or greatly diminished by a strain by laboratory interaction effect. We conclude that both the correlation of strain means and the interaction term in the ANOVA need to be considered in order to evaluate the stability or replicability of genetic experiments across laboratories. First and foremost, the investigator should view a scatter plot of the pattern of means in the two laboratories and interpret the abstract statistics in the light of perceived patterns in the data.

It is clear from these examples, and many other ANOVAs we have conducted, that a small but nevertheless statistically significant interaction between genotype and laboratory does not necessarily imperil the replicability of behavioral test data. On the other hand, a large main effect of genotype in the absence of a significant genotype × laboratory interaction is a dependable sign that results have been replicated. Caution is warranted when drawing conclusions from ANOVAs, because the statistical power of the ANOVA to detect a real interaction tends to be lower than the power to detect main effects (Wahlsten, 1990, 2006), and replicability of results may appear to be misleadingly good when sample sizes are small.

Fixed or random effects

In these examples and most of our research on replicability across laboratories, we have employed a fixed effects model for the ANOVA, whereas others (Kafkafi et al., 2005) advocate use of a mixed model in which genotype is regarded as a random effect. If we were to choose strains more or less randomly from a larger population of available strains (Beck et al., 2000; Petkov et al., 2004), this would be appropriate. One of the great advantages of the Mouse Phenome Database (http://phenome.jax.org/), however, is that researchers in different laboratories agree to focus their efforts on a carefully selected subset of available strains arranged in priority Lists A to D, each having 10 strains (Grubb et al., 2004; Paigen and Eppig, 2000). Top priority was conferred by us on List A strains because of their superior reproduction and lower cost as well as reasonable genetic diversity. By working with a common set of strains, researchers can compute strain correlations for different phenotypes measured in different laboratories. When working with a random effects model, the researcher is allowed to make generalizations about a larger population of genotypes on the basis of a sample. We do not believe generalizations beyond the specific List A strains examined in two specific laboratories are justified. On the contrary, if a different sample of strains, perhaps priority List B, were to be examined, phenotypic strain correlations could be considerably different than for the strains in List A. Neither would we care to generalize our results on cross-laboratory replicability to results that might be obtained when different pairs of laboratories compare their data. Neither, of course, does the replicability of one set of behavioral assays in one set of strains predict anything about the replicability of other tests, or even other variants of the same tests.

Enhancing replicability

There are several simple steps that can enhance the replicability of experiments on heredity and behavior:

1. It is of utmost importance that investigators work with the same genetic material, choosing the identical inbred strains and substrains in a strain survey or placing a targeted mutation onto the same inbred background. When different substrains are employed, comparing two studies can teach something about the generality of results but not genuine replicability of an experiment. There is much to be gained by many laboratories working with the strains nominated by the Mouse Phenome Database (http://phenome.jax.org/).
2. Careful attention to the details of apparatus and test protocol can insure that two laboratories are indeed administering the same test. As studies of task parameters within a laboratory have amply demonstrated, many seemingly minor factors can influence test results (Chesler et al., 2002; Dotson and Spector, 2005; Izidio et al., 2005; Tordoff et al., 2005; Wang et al., 2005).
3. Choosing strains with extreme high and low scores also helps to insure a large strain main effect in both laboratories and a high strain correlation, although this cannot forestall a troublesome but real interaction.
4. Working with substantial numbers of strains, more than 10 at a very minimum, tends to minimize the impact of one strain that may be very sensitive to local conditions. A study might take note of that one susceptible strain while finding that data for most strains are similar.

It might seem that increasing sample size would also increase replicability because it would minimize false positive results that could give the misleading appearance of a significant effect in one laboratory but not another. Nevertheless, when the null hypothesis of no genetic by treatment interaction effect is true, sample size has no bearing on replicability. On the other hand, if there is a genuine strain by laboratory interaction effect, larger samples are more likely to reveal this. Thus, replicability will appear to be greater when smaller samples are used, because neither laboratory would have a very good chance of detecting any but the largest effect. The use of small samples to achieve this dubious end is not recommended.

When investigators seek to assess replicability with an ANOVA to test for interaction, it is important that they choose a sample size that is sufficient to detect the interaction with adequate power. The crucial consideration is what kind of interaction would constitute a threat to replicability. Researchers need to devise a numerical model of results that embodies a noteworthy interaction for the kind of experiment they plan to conduct. The interaction effect is then computed from the deviations of the group means in the model with interaction from group means in a strictly additive model where the main effects have the same size but there is no interaction (Cohen, 1988; Wahlsten, 2006). If they employ the appropriate sample size in the experiment and in fact do not observe a significant interaction in the ANOVA, this does not prove there is no interaction whatsoever, but it is good evidence that any interaction

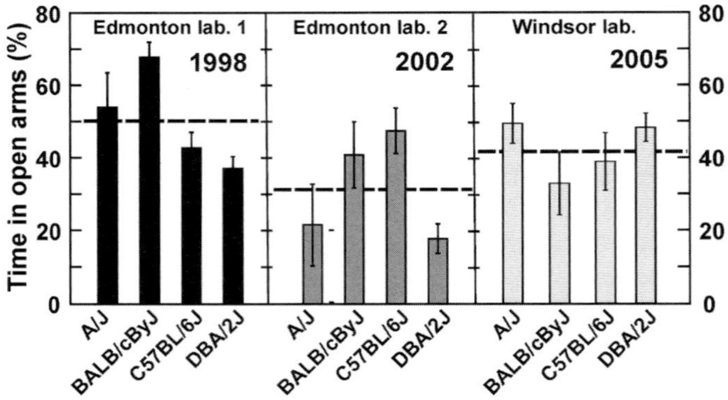

Figure 4.3 Percent time in the open arms of an elevated plus-maze used to assess anxiety-related behavior. The maze and test protocol were the same in all three experiments conducted in the Wahlsten laboratory and have been described previously (Crabbe et al., 1999; Wahlsten et al., 2003). (a) The first laboratory produced remarkably high levels of open-arm exploration. (b) When the laboratory was moved to a centralized animal facility in another wing of the same building, anxiety levels were considerably higher. (c) In the latest laboratory location, results are intermediate. Furthermore, the patterns of strain differences vary across the three laboratories.

is smaller than the kind of interaction that they believe would imperil replicability.

Replicability is not the ultimate good

As we have learned, using identical strains and identical tests cannot always guarantee identical results because laboratory environments usually differ and may exert significant influences on test scores (Crabbe et al., 1999; Wahlsten et al., 2003). While undisputed replicability of test data is generally seen as a good thing, a failure to replicate in another laboratory may reveal important environmental effects that warrant further investigation. A test such as the elevated plus-maze may give substantially different results across laboratories not because it is a bad test but rather because it is very sensitive to anxiety-inducing features of the animal facility. For example, one of us (D. W.) has occupied three different laboratories in the past few years, and in each laboratory the identical elevated plus-mazes were used with the same four inbred strains (Figure 4.3) and the same test protocol. In 1998 at the University of Alberta laboratory, all mice were housed close to the testing room in a quiet colony room dedicated to that study, but in 2002 the operation was moved to a bustling central animal facility perfused with odors from different mammalian species and the mice had to be transported down a long hallway on a cart to reach the testing room. Accordingly, the average amount of exploration of the open arms of the elevated plus-maze was considerably lower in the new location. The laboratory at the Great Lakes Institute in Windsor housed mice in cages (MICE system from Animal Care Systems) that isolated them from extraneous noises and the test area was within an enclosed, compact suite of rooms that was far from any other rodents or cats. Average open-arm exploration in those more recent data was intermediate, between the two previous laboratories for the four strains also used in those studies.

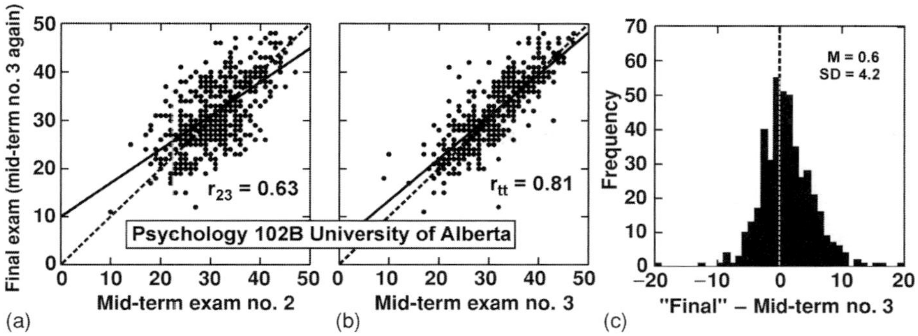

Figure 4.4 Data on two mid-term exams and the "final" exam, which was in fact a repetition of mid-term exam no. 3. (a) Correlation of two mid-term exams with different test items was moderate. (b) Test–retest reliability was $r_{tt} = 0.81$ for mid-term exam no. 3 in this unusual circumstance. (c) Most students showed very little improvement from the first to the second administration of mid-term no. 3. Thanks to Donald Heth for access to these unique data.

A good test of anxiety-related behavior ought to show dramatic variations among laboratories in baseline test scores when laboratory environments vary over a wide range, as we know they do in many instances. In other words, replicability of test results is not the ultimate good in this field. Our goal is to understand the multifarious influences on animal behavior and, for many investigators, to construct valid animal models of problem behaviors in humans. It is well known that many anxiety-related behaviors in humans are strongly influenced by conditions of living. Hence we should not be at all disappointed to find strong environmental influences on animal models of anxiety.

Reliability

Reliability of a test is high when the measure of an individual is nearly the same on different test occasions. Low reliability or unreliability of a test increases the variance of scores within a group of individuals and thereby reduces the statistical power – the ability of the experiment to detect real effects. Reliability can sometimes be increased by administering more trials or longer trials in a behavioral test. If repeated testing is not feasible or is very expensive, there may be other ways to enhance reliability. At the same time, some behaviors fluctuate greatly from moment to moment or day to day, and reliability of tests that aim to measure these will tend to be low.

If the test is repeated on different days, the indicator is *test–retest reliability* (r_{tt}). If the occasions are different items, trials, or minutes within a test session, it is the *internal consistency* of the test. Test–retest reliability is expressed by a simple Pearson correlation, whereas internal consistency can be estimated in several ways. Internal consistency is not reported often in the literature on animal behavior, and its primary value occurs when a test changes the animal so greatly that it cannot readily be repeated, as commonly occurs in the domain of learning and memory or psychopharmacology.

Test–retest reliability

In psychological testing of humans, test–retest reliability is usually assessed in a formal setting when the test is standardized on a carefully selected sample of the population, but the test is seldom administered twice to the same person in routine applications. An exception to this rule occurred at the University of Alberta in 1993 when a clerical error was made in the printing of the final exam for a large introductory psychology class. Although the cover sheet to each exam booklet properly labeled it as a final exam for that section of the course, the following pages were actually an exact reprint of the third mid-term exam that students had taken previously. A small number of astute pupils noticed the striking similarity and brought this to the attention of the professor (not D. W.!) who was proctoring the exam, but many others did not seem to realize what had happened. After a moment of panic, the professor instructed the students, who were soon to leave campus for the summer, to ignore the faux pas and continue answering the multiple-choice questions. As shown in Figure 4.4, the correlation of the scores on the "final" exam was considerably higher with mid-term exam no. 3 than mid-term exam no. 2, and was in fact a good estimate of the reliability of the third mid-term exam in a real-life situation. The high reliability of $r = 0.81$ suggests that the test was a good measuring instrument, but the mean score on the second try was a paltry 0.6 item better after several more weeks of brilliant lectures and earnest study.

Figure 4.5 describes a typical situation in work with mice. Eight strains were assessed with six of each sex. Each animal was given 10 trials on 2 successive days on an accelerating rotarod test of motor coordination. The training is usually done for only 10 trails prior to evaluating deficits caused by ethanol (Rustay et al., 2003), but another 10 trials were given in one study in order to assess test–retest reliability. First, we consider the phenotypic reliability by looking at the individual performance of each mouse. The mice improved greatly over the first 10 trials, but they required additional practice the next day to regain the top running speed achieved earlier (Figure 4.5a). The correlation of latencies to fall on the first trial of each day was close to zero (Figure 4.5b) and average latencies were considerably longer on the second day, which provides solid evidence of learning. The correlation of the average of the two sets of 10 trials (Figure 4.5c) was similar to the psychology exam results (Figure 4.4b). To consider the reliability of the genetic differences, we averaged performance for each strain over all 12 mice tested. Figure 4.5d shows that the genetic contribution to performance increased as we considered more trials to index strain performance. Finally, Figure 4.5e shows that test–retest reliability improved with the inclusion of more trials.

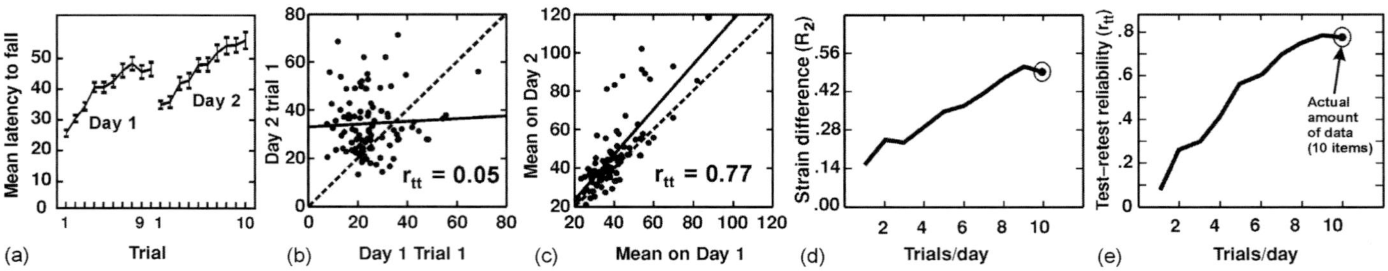

Figure 4.5 Performance on the accelerating rotarod by eight inbred strains of mice, six males and six females each, where data are pooled over strain and sex. All were tested in the Crabbe laboratory in Portland. (a) Average performance improved greatly across 10 trials within a day, while early trials on Day 2 were lower than maximum performance at the end of Day 1. (b) The correlation of fall latencies on Trial 1 on Days 1 and 2 was very low, indicating that the mice had changed considerably through training. (c) Correlation of the means of all 10 trials on the 2 days was very high, indicating good test–retest reliability. (d) After the fact, the magnitude of the strain difference in average fall latency was recomputed for fewer trials per day. The more trials in a day, the larger the strain effect, up to about eight trials per day. (e) Test–retest reliability for the same data as in (d) increases as the number of trials increases.

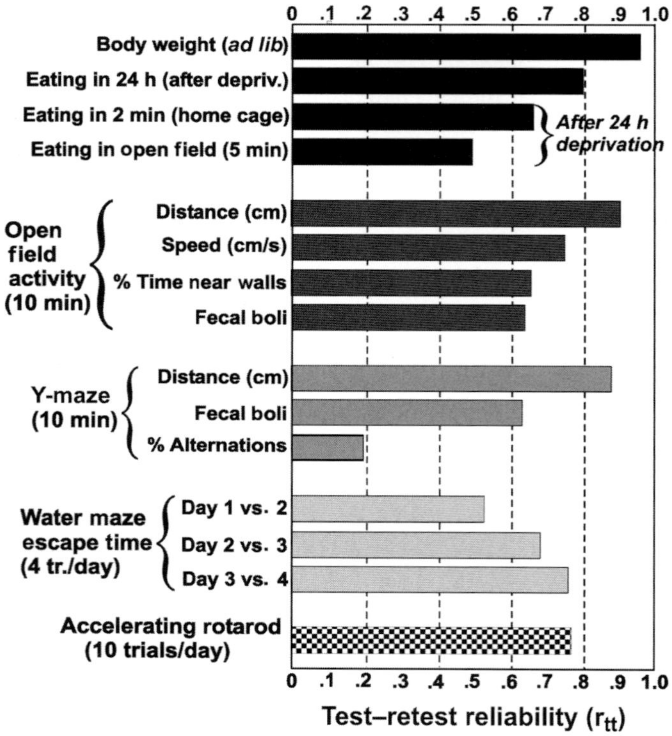

Figure 4.6 Test–retest reliabilities for different measures collected on 2 successive days for the same eight inbred strains (six males, six females in each) in five different test situations in a single laboratory. All but the rotarod data were from Edmonton, while rotarod was done in Portland.

Test–retest reliability is shown for different measures on five kinds of tests in Figure 4.6. For discrete and occasional events such as defecation in the apparatus, reliability tends to be lower than for a continuous behavior such as motor activity, and it is notably low for Y-maze alternation. In our laboratories, the reliabilities of tests that we regard as useful for research are often in the range from $r = 0.7$ to 0.9 when computed for individual scores rather than strain means. If reliability falls below $r = 0.5$, the experimenter is well advised to improve the methods before embarking on a series of studies using that test. These data reveal that reliability varies considerably across different measures of behavior in a single test situation. In this respect, it is hazardous to assert in a general sense that one kind of behavioral test, e.g., open field, is more reliable than another. For a behavior such as distance traveled, data are almost as stable from one day to the next as body weight, whereas reliabilities are considerably lower for time near a wall and defecation.

Critical choice of a subject population

Although reliability is often understood to be a property of a test that involves an apparatus and protocol, it is critically dependent on the population from which the sample of animals is drawn. The wider the range of individual genotypes, the higher will be the potential reliability of the test. When comparing test reliabilities across different laboratories, it is important that the laboratories measure mice from the same strains. For the data in Figure 4.5 and Figure 4.6, 8–12 mice of each of the same eight strains (List A of the Mouse Phenome Project, minus the wild-derived strains) were studied in a single laboratory. The correlations within a single strain were usually lower than for the pooled sample. For example, on the rotarod (Figure 4.5), the average within-strain correlation between means on the 2 days was $r = 0.51$, considerably less than $r = 0.77$ for the entire sample of 96 mice. Thus, one effective way to enhance the reliability of a test is to select a wide genetic and phenotypic range of strains. The balance between genetic variation among strains and non-genetic variation within strains will be influenced by the number of strains and sample size for each strain, and a systematic comparison of reliabilities for different kinds of tests should use the same number of strains and mice per strain for each test.

In research using targeted mutations where a mutation is back-crossed onto a standard strain background, perhaps C57BL/6J, it would be most relevant to assess test reliability using just that one strain, in which case reliability would be based entirely on non-genetic individual differences. A large sample of mice should then be studied to insure a good estimate of the correlation. For the rotarod example in Figure 4.5, the test–retest correlation for 12 C57BL/6J mice was a

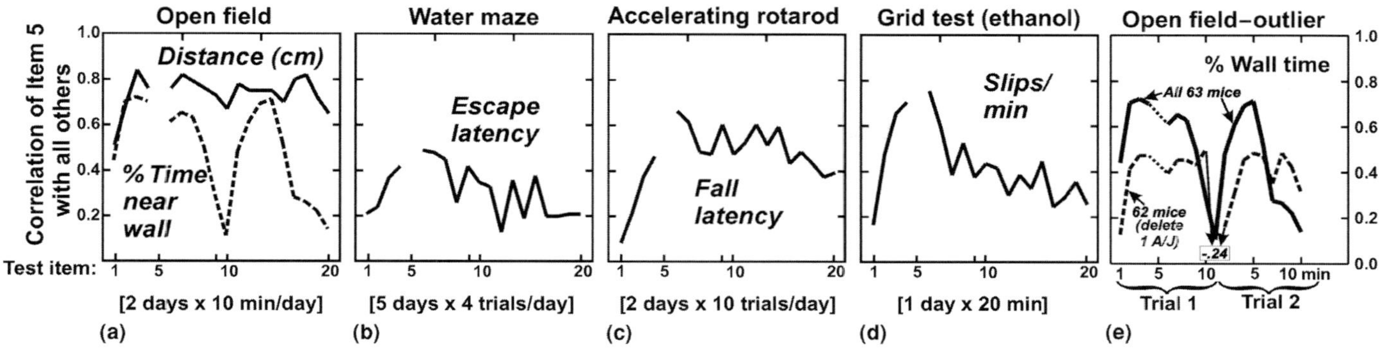

Figure 4.7 Four tests, each with 20 items, showing the correlation of item no. 5 with all other items. Data in each test are from eight inbred strains with four to six of each sex. (a) Distance in the open field shows a fairly stable correlation after the first few minutes, whereas wall hugging changes greatly during a trial. (b) For escape latency in version two of a four-arm water maze (data from Wahlsten et al., 2005), correlations are generally lower than for open field distance and are lower with trials early and later in training. (c) Fall latency on the accelerating rotarod shows intermediate levels of correlation and low similarity with initial trials. (d) Slips in a grid test under the influence of ethanol show marked changes in correlation with different phases of acquisition. (e) For percent time near walls in the open field, correlations were considerably increased by a single A/J mouse with extreme scores.

respectable $r = 0.79$, but it was only 0.4 for DBA/2J and 0.12 for FVB/NJ.

Generally speaking, test–retest reliability should be estimated using the same range of genotypes, the same test procedures, and the same measures as will be used in a later study of a substantive question in behavioral or neural biology. The only major difference should be that a second test is administered in the initial study done to assess reliability. If the researcher is primarily interested in correlations among strain means, then it would be informative to compute reliability for strain means (Figure 4.1c and Figure 4.5e), provided the sample size is about the same as in a future study. Working with strain means greatly enhances the contribution of genetic variation at the expense of within-strain variation, and in our study with 21 strains the reliability for means of rotarod performance was very high indeed.

Boosting reliability by using more trials or longer trials

When the dependent variable in an experiment is the mean of performance across time or trials, reliability depends strongly on trial length or number of trials. Increasing the amount of data collected for each individual can boost reliability. Likewise, a preliminary study with a long training procedure may reveal that good results are possible with fewer trials, thereby making the study more efficient. Figure 4.5d and e present data in which all animals received 10 training trials on the rotarod on 2 successive days (Figure 4.5a). Data were then truncated post hoc to create shorter sessions and reliability was recomputed. Reliability improved dramatically by increasing number of trials; however, no further improvement in reliability or the magnitude of strain differences was seen after eight trials. This approach cannot be the sole criterion for deciding the number of training trials, because a few extra trials at the end of a session may improve performance on early trials the next day. Nevertheless, it is far easier on the budget than running separate groups of animals with different numbers of daily trials.

Internal consistency for dynamic characteristics

When a behavioral test cannot be readily repeated for an individual, for example in a complex learning task, it may be helpful to assess internal consistency of the test. An elaborate methodology has been devised for estimating internal consistency of tests with humans (Traub, 1994), but this is based on assumptions that have dubious validity for laboratory animals. Classical test theory in human psychometrics usually requires that all test items assess the same construct, e.g., intelligence, to some degree, and that items are carefully arranged to make average scores on the first few and last few items about the same. Furthermore, taking the test is not supposed to change the construct appreciably, because all subjects are presumed to have long experience with the general test-taking situation. In work with laboratory animals, on the other hand, we prefer to work with experimentally naïve subjects and evaluate reactions to an entirely novel situation. Behavioral processes that are most active early in a trial or a series of trials are often very different from those invoked after substantial amounts of experience with a test. Consequently, the correlation of scores on a trial midway through training tends to be highest with adjacent trials and lower, sometimes much lower, with scores earlier and later in a test (Figure 4.7). The analysis of internal consistency in laboratory animals is perhaps better done with the aid of generic statistical methods rather than those from human psychometrics. Alternatives to classical test theory address some of the shortcomings of the usual psychometric approach (Shavelson and Webb, 1991), but advantages over generic multivariate methods are not compelling.

Most animal behaviors change over the duration of a single trial or across several trials that are administered in the same way. Reliability in this kind of situation expresses the extent to which animals that score high early in testing continue to rank highly throughout the test. Repeated measures ANOVA can be applied to assess internal consistency of performance when there is systematic change within an individual by including a within-subjects term to represent

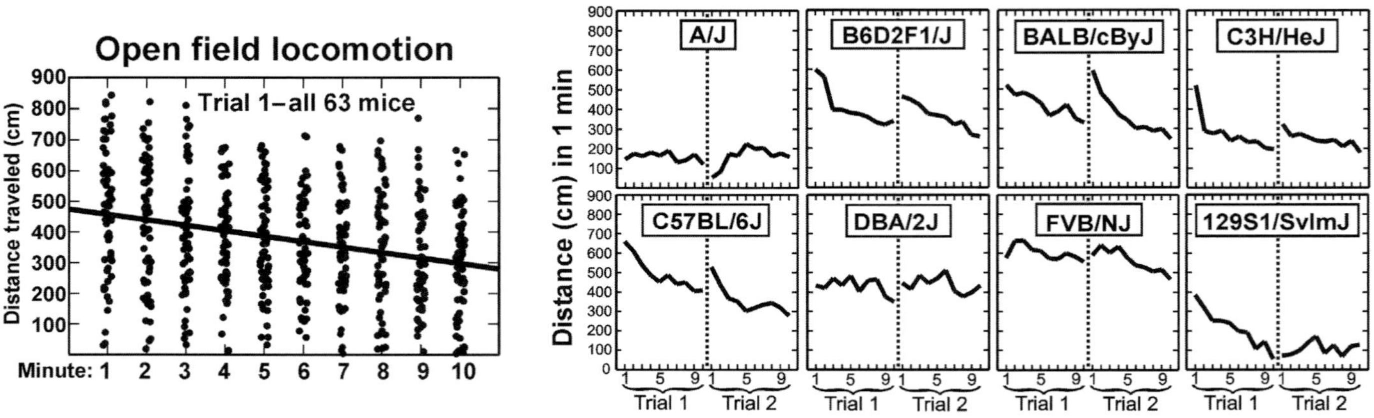

(a) (b)

Figure 4.8 (a) Distributions of distance traveled in an open field in each minute for 63 mice from eight genetic groups. Data are jittered slightly to reveal overlapping points. Internal consistency of the test (Table 4.1) represents the proportion of variance attributable to differences between the 63 mice after the overall linear decline in distance is removed from the data. (b) Distance in an open field in each minute on two trials on successive days, showing different patterns of change within and between days for seven inbred strains and one F1 hybrid that contributed data in Figure 4.8a.

trials or time periods within a trial. The experimental design entails several (n) individuals measured on k occasions. The n individuals constitute a sample from some larger population and therefore are a random factor in the statistical model. In this case the order in which measurements are made is meaningful, and trial or time period constitutes a fixed factor in the analysis, making this equivalent to the mixed Model III as discussed by Winer (1991). The model for a score of individual i in period j is $Y_{ij} = \mu + \pi_i + \tau_j + (\pi\tau)_{ij} + \varepsilon_{ij}$. In classical test theory, π_i is viewed as the underlying true phenotypic value, but it is more plausibly regarded as the hypothetical mean value of a characteristic over very many measurement occasions. The exception term ε_{ij} is the unsystematic deviation on occasion j. In classical test theory, ε_{ij} is regarded as measurement "error," whereas in work with laboratory animals the fluctuations in behavior from time to time are usually quite real and do not arise from any shortcoming of the measuring device. It should be emphasized that no claim is made here about the genetic or environmental origins of the value π_i for any individual, although ε_{ij} must of course be entirely non-genetic. τ_j is the trial or time effect, and an interaction term is included to express possible differences between individuals in the slope of the change over periods. In the mixed model for a single factor repeated measures ANOVA, there is no way to distinguish the subject × time period interaction from the term representing non-systematic fluctuations, and the residual term includes any individual differences in slopes. Hence, the partition of variation is SStotal = SSb.I + SStime + SSresiduals, where SSb.I is the between-individuals component.

When n individuals are measured in k time periods, the analysis estimates internal consistency of performance. Reliability is indicated by the effect size indicator in a random effects ANOVA model, the *intraclass correlation* for individuals ρ_I, which expresses the proportion of total variance that is attributable to average differences among individuals, while the remainder $1 - \rho_I$ represents unsystematic fluctuations between measurements. This method removes from consideration the variance arising from systematic changes across time periods. The intraclass correlation for individuals is derived from the effect size indicator 2, using the unbiased estimator obtained from equation (4.1):

$$\hat{\theta} = \frac{MS_{b.I} - mMS_{residual}}{kmMS_{residual}} \text{ where } m = \frac{(n-1)(k-1)}{[(n-1)(k-1)-2]} \quad (4.1)$$

The internal consistency or reliability of the mean of scores across all trials or all time periods is $\hat{\rho}_k = k\hat{\theta}/(1 + k\hat{\theta})$. When $k > 2$, the value is nearly identical to Chronbach's coefficient alpha that is prevalent in human psychometrics.

An illustration is provided by data on open field activity of eight strains of mice, where distance traveled during each of 10 samples, 1 minute long, was measured with video-tracking. On average, activity declined gradually for all 63 animals (Figure 4.8a). The term for time period in the repeated measures ANOVA removed the average pace of change from the calculation of consistency. For the activity data, internal consistency was very high. For all 63 mice combined, distance in 1 minute was expressed by the linear equation $Y = 475.3$ cm -17.75cm/min*Minute. This linear trend was removed from each of 10 scores for each mouse in Figure 4.9a by subtracting the actual distance in each minute from (475.3–17.75*Minute). Consistency is then equivalent to a one-way ANOVA where the between "group" factor is the 63 mice and the within "group" variance is among the 10 scores for each mouse. Data are rank ordered by mean score in Figure 4.9a to make it easier to visualize the differences between and within mice. The standard deviation within a mouse ($S = 97$ cm) was far less than the standard deviation for all 630 scores combined ($S = 178$ cm), which indicates that consistency is very high. Figure 4.9b shows the alternation index in a Y-maze, where consistency was much lower

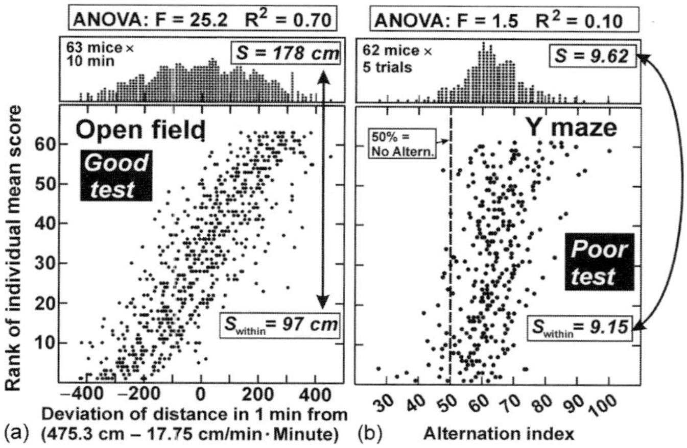

Figure 4.9 Scores for individual mice after the linear trend over minutes or trials was first removed using a regression equation. The top frequency distributions show all scores combined over all mice, and the values are plotted below for each individual, where individuals are ranked according to mean score. The ANOVA expresses the proportion of total variance that is attributable to the differences among individual mice. (a) Open field test with distance measured in each of 10 minutes. Variability within an individual was far less than variability for the pooled data, which indicates high consistency of individual performance. (b) Y-maze alternation index with low consistency, showing substantial overlap of scores for most mice. Random arm entry would yield an alternation score of 50%, and most mice were above this value on most trials.

and removing the linear trend over trials had little impact on variance within a mouse. For the activity data, internal consistency calculated using equation (4.1) as intraclass correlation was very high. Note that the consistency of open field distance estimated from the repeated measures ANOVA as $\rho_1 = 0.71$ for a single measure in Table 4.1 is very close to the multiple R^2 value from the one-way ANOVA on individuals after the linear trend is removed (Figure 4.9a). We note that if the temporal trend in a data set were curvilinear, the same approach could be taken by employing an appropriate higher-order best-fit equation to model and remove that trend.

Table 4.1 presents internal consistencies for several measures on different tests and, where feasible, compares them with test–retest reliabilities. Internal consistency of performance within a single test session is generally higher than test–retest reliability for different sessions separated by 24 hours or more.

Limitations of a single indicator of reliability and the need for thorough analysis of data

While an internal consistency index of $p_k = 0.96$ suggests that distance traveled in the open field is highly consistent, this one quantity conceals as much as it conveys. The profiles of change in behavior across time or trials are strongly dependent on the genotype of the animal (Figure 4.8b). The DBA/2J strain shows little sign of habituation either within or between days, BALB/cByJ shows pronounced habituation within a session but little change over days, C57BL/6J declines both within and between sessions, while 129S1/SvImJ habituates rapidly during the first session and then remains low thereafter. The complexity of the patterns is indicated by a significant strain by day by trial interaction term in a repeated measures ANOVA ($F = 2.22$, $df = 63/495$, $P < 0.000001$). A more intricate pattern is evident from regression lines for individual mice in each strain (Figure 4.10a), where the multiple R^2 value indicates the proportion of variance among distance measures in each minute that is attributable to the linear change over time. For some individuals, this R^2 value is remarkably high, whereas other mice show little or no change across time. Within a strain there is notable heterogeneity as well.

Multiple regression can be adapted to capture the full extent of variation across time on two trials on successive days both between and within strains. The model shown in Figure 4.10b utilizes "centered" variables that allow a good test of the slope by trial interaction because the statistical independence or tolerance of all three predictors is 1.0. There are two trials, and centered trial number (cTrial) is −0.5 and 0.5, whereas centered minute (cMinute) in a 10 minute trial ranges from −4.5 to 4.5. The equation is estimated for each individual. The graph plots the standardized regression coefficients for the difference between days and the slope within a day on the x and y axes, and an asterisk shows where the slope by day interaction is significant. This way of expressing the results conveys quite effectively and efficiently the full range of individual differences within and between strains of mice. The terse summation of internal consistency by a single intraclass correlation $\rho_k = 0.96$ for the same set of data suggests to the reader that differences between individuals are very large, which is substantially true, and that every mouse is highly consistent across time, which is not true. Furthermore, it provides not the slightest hint of the dramatic and interesting interactions with genotype that are obvious in Figures 4.8 and 4.10.

Outliers can inflate consistency scores

Reliability analysis can be highly misleading unless the investigator conducts a careful examination of the data. One of the most effective ways to spot trouble is to scrutinize a scatter plot of individual scores to locate radically exceptional cases. Reliability can be strongly influenced by one extreme animal that happens to be consistently extreme on successive days. Figure 4.7e shows the impact of just one A/J mouse on internal consistency of time spent near the walls of an open field. The trial always began when the mouse was placed in the center of the chamber. On both days, that mouse remained immobile in the center for about 5 minutes and then walked slowly to the wall, where it remained until the end of the trial. The data for this one animal greatly elevated the interindividual correlations across minutes. Both test–retest reliability and internal consistency are very sensitive to outliers in the data. Because bad data can make a test appear good, caution is warranted when interpreting published reliability coefficients.

Table 4.1 Reliability analysis for behaviors measured in *k* time periods or trials.[a]

				1 measure	Mean of *k* measures		
Task	Measure	*k* items	Strains (*n*)	$\hat{P}_1$	$\hat{P}_k$	Chronbach's coefficient α	r_{tt}
Open field	Distance (cm)	10 min	8(63)	0.707	0.960	0.960	0.906
Open field	% Time near walls	10 min	8(62)	0.367	0.853	0.853	0.679
Y-maze	No. arms entered	5 trials	9(62)	0.636	0.897	0.897	0.71
Y-maze	Alternation index (%)	5 trials	9(62)	0.182	0.526	0.513	0.27
Four-arm water maze, version 2	Escape latency	10 trials (of 20 total)	8(64)	0.386	0.863	0.823	nm
Four-arm water maze, version 2	No. arms entered	10 trials (of 20 total)	8(64)	0.077	0.456	0.456	nm
Accelerating rotarod	Latency to fall	10 trials	9(108)	0.392	0.866	0.866	0.774
Grid test	Horizontal activity	10 min	9(102)	0.576	0.931	0.932	nm
Grid test	Slips/horizontal activity	10 min	9(102)	0.121	0.580	0.581	nm

[a] Data for open field, Y-maze, and water maze were collected in Edmonton, while rotarod and grid test results are from Portland. The four-arm water maze is described in detail by Wahlsten et al. (2005). The eight strains were 129S1/SvImJ, A/J, BALB/cByJ, BTBR T+ tf/J, C3H/HeJ, C57BL/6J, DBA/2J, and FVB/NJ. In several studies the B6D2F1/J hybrid was also included; nm indicates the measure was not meaningful because extensive learning made it impossible to repeat the test for an individual.

Conclusions

In the literature on behavioral and neural genetics, repeatability of an experiment is sometimes said to express the reliability of results. Here we propose a more formal distinction between replicability or repeatability of an entire experiment versus reliability or consistency of individual performances across a series of trials or time within a study. By adopting this terminology, investigators can convey their ideas with greater clarity and perhaps become aware of new patterns in their data. They ought to assess reliability of a test before embarking on an elaborate experiment, thereby avoiding failure because of an insensitive behavioral assay.

The methods presented here for assessing replicability and reliability of behavior test data rely entirely on familiar, generic statistical methods such as Pearson correlation, multiple regression, and ANOVA. Courses on these topics are available at all universities, and several excellent computer programs (e.g., SAS, SYSTAT, SPSS, S+, R) are available to do the calculations and plot the results. The flexibility of these methods offers an important advantage over classical psychometric test theory. In the twentieth century, behavior genetics was often done by biological psychologists who themselves had been trained in psychometrics, whereas the field is increasingly populated by researchers with backgrounds in molecular biology. The common language of generic statistical methods offers distinct advantages in such a diverse field.

Acknowledgments

Research was supported in part by the Mouse Phenome Project of The Jackson Laboratories as well as NIH grants AA12714 (D. W.), AA10760 (J. C. C.), a grant from the Department of Veterans Affairs (J. C. C.), and grant 45825 from the Natural Sciences and Engineering Research Council of Canada (D. W.). The authors are grateful to Molly Bogue, Sean F. Cooper, Brandie Moisan, Pam Metten, Chia-Hua Yu, and Jason Sibert for assistance in conducting several of the experiments.

Appendix

Mice

All mice were shipped from The Jackson Laboratory at about 6 weeks of age and tested from 9 to 11 weeks of age. In both laboratories they were housed in 29 × 18 × 13 cm Ancare cages with 6 mm Bed-o-cob bedding, fed Purina or PMC 5001 chow *ad libitum*, and given free access to local tap water. Both colonies were kept at about 21 °C and had the lights on from 0600 to 1800. All behavioral tests were conducted during the light phase of the cycle. Every experiment involved eight or nine genetic groups of mice, seven of which were always the inbred strains from priority List A of the Mouse Phenome Database (129S1/SvImJ, A/J, BALB/cByJ, C3H/HeJ, C57BL/6J, DBA/2J, FVB/NJ). In addition, some experiments included strain BTBR T+ tf/J that was originally in List A. When that strain was not available owing to short supply, the hybrid B6D2F1/J was included. In a few instances, both BTBR and B6D2F1 were available and were tested. Sample sizes were four males and four females per strain in Edmonton and six of each sex in Portland.

For the reliability trials, five kinds of tests were done in the Edmonton laboratory, while the accelerating rotarod and grid tests were done in Portland. Data are also presented here for the accelerating rotarod test done in both labs with 21 strains

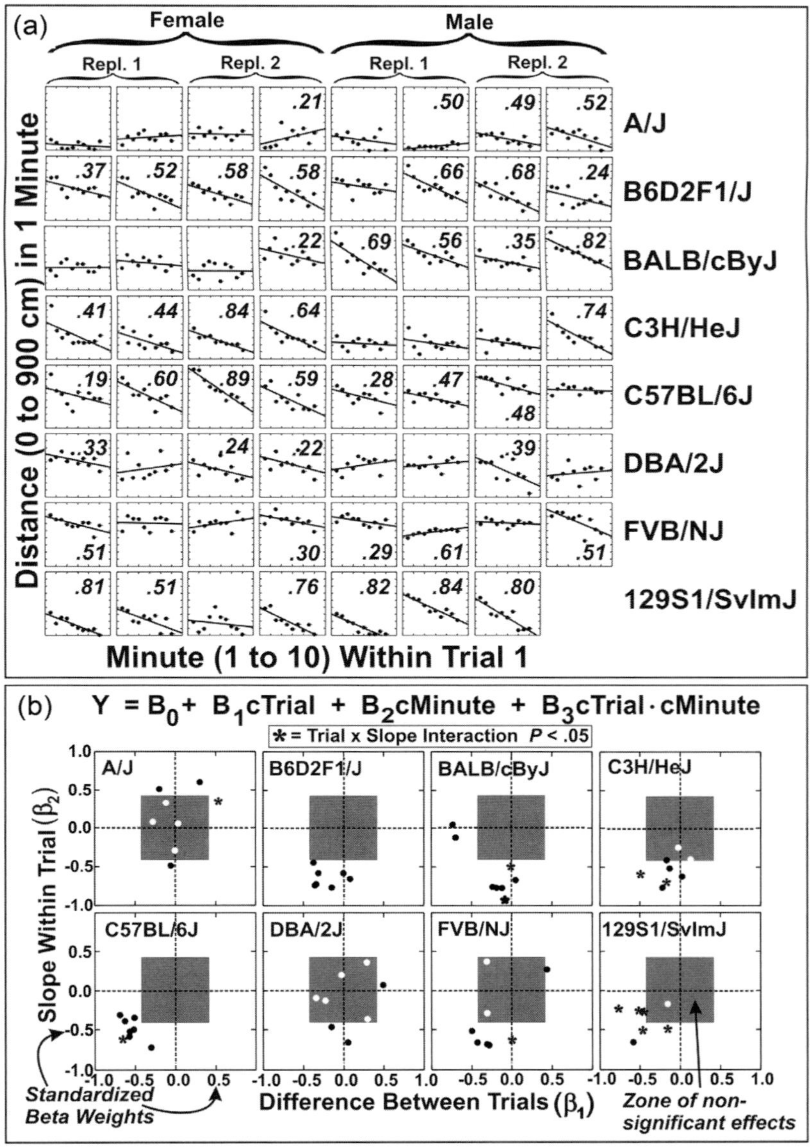

Figure 4.10 (a) Changes in distance in the open field across 10 minutes on the first trial for individual mice in the eight groups shown in Figure 4.8. Numbers are multiple R^2 that indicate the proportion of variance attributable to the linear change across time. Panels with no numbers showed $P > 0.05$. (b) Multiple regression analysis for each mouse shown in Figure 4.10a. The model uses effect coding for trial and minute within trial, so that the interaction term for difference in slope between trials is independent of trial and average slope effects. Values in the gray zone have $P > 0.05$ for both minute and trial for that individual. Values shown as an asterisk also involve a significant ($P < 0.05$) difference in slope between days.

as part of the Mouse Phenome Project; these results have been described previously in Rustay et al. (2003).

Open field

Mice were tested in a 40 × 40 × 30 cm high clear Lexan box with no floor. The floor was a fresh sheet of waxed pink butcher's paper for each animal. The box was housed in a 50 × 50 × 65 cm cubicle that provided about 25 Lux illumination, and behavior was monitored with a VideoScan video-tracking system from AccuScan, Inc. Each mouse was given one 10-minute trial on each of 2 successive days for the purpose of assessing reliability. Data from the VideoScan system consisted of a separate text file of type ATK for each trial of each mouse in a standard format. The TransATK program from MusWare Technology, Inc. was used to combine all ATK files from an entire experiment and format the data in four different arrangements that facilitated different kinds of analysis between and within subjects.

Y-maze

A maze with three identical arms measuring 15 × 6 × 6 cm high was constructed with no floor so that a clean floor paper could be used on each trial. The maze was covered by a lid made of 6 mm mesh. The maze was housed in the same cubicle as was used for the open field and behavior was tracked with VideoScan. One 10-minute trial was given on two successive days. The SEQ file produced by VideoScan was modified by the TransSEQ program from MusWare Technology, Inc. to obtain a more accurate assessment of the sequence of arm entries. An entry was defined as the center of the image of a mouse extending at least 2 cm into an arm, and no new entry was scored until the animal had either entered another arm or remained in the center zone for at least 1 second. This procedure eliminated rapidly repeated entries into a single arm when the mouse hovered near the edge of a zone.

Water escape

A four-arm water escape apparatus with arms 10.5 cm wide was inserted into a 70 cm diameter tank. A clear mesh platform 10 cm in diameter was submerged 5 mm below the surface of 26 ^{N}C water rendered opaque by white Crayola paint. The apparatus was the same as version 2 of the water escape maze described in detail by Wahlsten et al. (2005). After 1 day of pretraining, mice were given 4 days of four trials per day with a trial limit of 60 seconds and an intertrial interval of 30 seconds. Each trial began from one of three randomly chosen compass positions, never pointing towards the platform.

Eating test

Mice were weighed daily for 5 days. A weighed amount of food was given in a glass dish and amount consumed was recorded daily. On the 2nd and 4th days, food was removed for 24 hours and then the mouse was given a 2-minute eating test using a single pre-weighed test pellet of PMC 5001 chow. A separate sample of mice was given a 5-minute trial in the open field with food present, and consumption was recorded by weighing the pellet.

Elevated plus maze

The maze had a black plastic floor with four arms 6 × 25 cm and was 50 cm above the table top. Open arms had a 0.5 cm clear plastic rim, while closed arms had 15 cm clear plastic walls. One trial of 5 minute duration was given, and the mouse was tracked with the VideoScan system under 100 Lux illumination. Further details of the test are provided elsewhere (Crabbe et al., 1999; Wahlsten et al., 2003).

Accelerating rotarod

The four-lane AccuRod apparatus from AccuScan, Inc. was modified to have a 6.5 cm diameter rod that we covered with 320 grit sandpaper. Ten daily trials were given with a 120 seconds trial limit on 2 successive days for reliability testing. Acceleration rate was 20 RPM/min. The fall latency was recorded by photocells in the bedding trough, backed up by a stopwatch for instances when the mouse bypassed the photocells. Further details are provided in Rustay et al. (2003).

Grid test

A 15 × 15 × 20 cm high box with a 1.25 cm mesh floor had a metal touch plate 1 cm below the grid to detect foot slips electronically. The box had clear plastic walls and was inserted into a DigiScan photocell activity monitor from AccuScan Inc. Promptly after receiving an injection of 1.5 g/kg ethanol, each mouse was given one 20-minute trial in the apparatus. Distance traveled and number of foot slips was recorded in each minute. Further details are provided elsewhere (Crabbe et al., 2003).

References

Bausell, R.B. and Li, Y.-F. (2002) *Power Analysis for Experimental Research: A Practical Guide for the Biological, Medical, and Social Sciences*. Cambridge University Press, Cambridge.

Beck, J.A., Lloyd, S., Hafexparast, M., Lennon-Pierce, M., Eppig, J.T., Festing, M.F.W., et al. (2000) Genealogies of mouse inbred strains. *Nat Genet* 24: 23–25.

Chesler, E.J., Wilson, S.G., Lariviere, W.R., Rodriguez-Zas, S.L., and Mogil, J.S. (2002) Influences of laboratory environment on behavior. *Nat Neurosci* 5: 1101–1102.

Cohen, J. (1988) *Statistical Power Analysis for the Behavioral Sciences*. Erlbaum, Hillsdale, NJ, USA.

Crabbe, J.C., Wahlsten, D., and Dudek, B.C. (1999) Genetics of mouse behavior: interactions with laboratory environment. *Science* 284: 1670–1672.

Crabbe, J.C., Metten, P., Yu, C.H., Schlumbohm, J.P., Cameron, A. J., and Wahlsten, D. (2003). Genotypic differences in ethanol sensitivity in two tests of motor incoordination. *J Appl Physiol* 95: 1338–1351.

Dotson, C.D. and Spector, A.C. (2005) Drinking spout orifice size affects licking behavior in inbred mice. *Physiol Behav* 85: 655–661.

Grubb, S.C., Churchill, G.A., and Bogue, M.A. (2004) A collaborative database of inbred mouse strain characteristics. *Bioinformatics* 20: 2857–2859.

Izidio, G.S., Lopes, D.M., Spricigo, L., Jr., and Ramos, A. (2005) Common variations in the pretest environment influence genotypic comparisons in models of anxiety. *Genes Brain Behav* 4: 412–419.

Kafkafi, N., Benjamini, Y., Sakov, A., Elmer, G.I., and Golani, I. (2005) Genotype–environment interactions in mouse behavior: a way out of the problem. *Proc Natl Acad Sci USA* 102: 4619–4624.

Murphy, K.R. and Myors, B. (2004) *Statistical Power Analysis: A Simple and General Model for Traditional and Modern Hypothesis Tests*. Lawrence Erlbaum Associates, Mahwah, NJ, USA.

Paigen, K. and Eppig, J.T. (2000) A mouse phenome project. *Mamm Genome* 11: 715–717.

Petkov, P.M., Ding, Y.M., Cassell, M.A., Zhang, W.D., Wagner, G., Sargent, E.E., et al. (2004) An efficient SNP system for mouse genome scanning and elucidating strain relationships. *Genome Res* 14: 1806–1811.

Rustay, N.R., Wahlsten, D., and Crabbe, J.C. (2003) Assessment of genetic susceptibility to ethanol intoxication in mice. *Proc Natl Acad Sci USA* 100: 2917–2922.

Shavelson, R.J. and Webb, N.M. (1991) *Generalizability Theory. A Primer.* Sage Publications, Newbury Park, CA, USA.

Tordoff, M.G., Alarcon, L.K., Byerly, E.A., and Doman, S.A. (2005) Mice acquire flavor preferences during shipping. *Physiol Behav* 86: 480–486.

Traub, R.E. (1994) *Reliability for the Social Sciences. Theory and Applications.* Sage Publications, Newbury Park, CA, USA.

Wahlsten, D. (1990) Insensitivity of the analysis of variance to heredity-environment interaction. *Behav Brain Sci* 13: 109–120.

Wahlsten, D. (2006) Sample size requirements for experiments on laboratory animals. In Jones, B.C. and Mormède, P. (eds.), *Neurobehavioral Genetics: Methods and Applications*, 2nd edn. Taylor and Francis, Boca Raton, FL, USA.

Wahlsten, D., Cooper, S.F., and Crabbe, J.C. (2005) Different rankings of inbred mouse strains on the Morris maze and a refined four-arm water escape task. *Behav Brain Res* 165: 36–51.

Wahlsten, D., Metten, P., Phillips, T.J., Boehm, S.L., II, Burkhart-Kasch, S., Dorow, J., et al. (2003) Different data from different labs: lessons from studies of gene–environment interaction. *J. Neurobiol* 54: 283–311.

Wang, H., Tranguch, S., Xie, H., Hanley, G., Das, S.K., and Dey, S.K. (2005) Variation in commercial rodent diets induces disparate molecular and physiological changes in the mouse uterus. *Proc Natl Acad Sci USA* 102: 9960–9965.

Winer, B.J. (1991) *Statistical Principles in Experimental Design*, 10th edn. McGraw-Hill Book Co., New York.

Chapter 5

Audition

James F. Willott

Introduction

The term, "auditory perception," whether applied to humans or other mammals, encompasses a wide array of functions including detection and identification of auditory properties (pitch, loudness, location, timbre), interpretation of the auditory signal, its integration with memory systems, integration with other modalities, self-monitoring (e.g., control of vocalizations), and extraction of information for reproductive, parenting, and other types of behavior. These processes, or aspects of them, are presumably experienced at a high cognitive level and they are also modulated by attention, habituation, sleep, arousal, emotion, and context. Processing of acoustic stimuli underlying homeostatic, autonomic, middle ear reflex systems, etc., need not be "experienced" prior to their activation and are typically viewed as not necessarily "perceived."

Human individuals with normal hearing sensitivity can differ greatly from one another in the way they respond to and process the same auditory stimuli. This is manifested on virtually all tests of auditory perceptual ability, which typically exhibit a substantial range of scores above and below the mean, even when the population is restricted to individuals with normal hearing sensitivity (e.g., Bellis and Ferre, 1999; Schönweiler et al., 1998). Presumably, some of this variance is genetic, and indeed, evidence has been obtained on humans for heritability of some phenotypes that have a strong auditory component. Examples are reading disabilities, specifically "phonological awareness" and single-word reading (DeFries et al., 1987) and perfect (absolute) pitch perception (Baharloo et al., 2000; Keenan et al., 2001). These abilities are not purely auditory but cognitive as well, and this obscures the role of variance within the auditory system.

If cognitive and other "non-auditory" influences on perception are experimentally controlled for, the factors that determine whether an individual's auditory performance is abnormally low, average, or unusually high are likely to include the efficacy of neural processing by the *central auditory system* (CAS). This should be true of mice as well as humans, suggesting the high potential of mouse models. Thus, research on the CAS may at least provide a manageable focus for genetic research on auditory perception that can utilize mice.

An understanding of the genetics of auditory perception is important in its own right, but research on the genetics of normal auditory processes actually has relevance regarding *disorders* of auditory perception that lend the topic extra *gravitas*. First, understanding the biology of normal perception provides goals and standards for clinical approaches to emulate. Second, performance at the extremes of a normal distribution of perceptual abilities is often problematic and requires some sort of clinical treatment or amelioration.

Mouse models

It is extremely difficult to perform genetic research on humans when the phenotype is an aspect of normal behavior that is subject to numerous environmental and age-related influences, and for which an appropriate database is not available. As is the case for most biologically oriented studies that involve the nervous system, a good animal model is needed. To investigate genes affecting auditory behavioral/perceptual phenomena, one needs to have access to animal subjects that vary reliably across a range of performance, so that the factors contributing to these differences can be studied. The use of carefully selected inbred mouse strains and mutants uniquely fills this bill, as is evident throughout this volume. Because a huge body of data is available on the auditory system, genetics, and biology of mice (e.g., Mouse Phenome Database, http://phenome.jax.org/; Willott, 2001), powerful tools of genetics, neuroscience, and experimental psychology can be brought to bear on issues related to auditory processing.

Because perception implies an experience, its measurement, verification or validation can be difficult in mice. Nonetheless, a number of tests allow inferences to be made about aspects of mouse auditory perception or experience that are in some ways analogous to those shared by humans. Studying the genetics of auditory perception in mice may use several basic approaches. First, the *acoustic startle response* (ASR), and its modulation by *prepulse inhibition* (PPI) can be evaluated. The strength or magnitude of these behaviors reflects the salience of the eliciting stimuli. Such auditory behaviors have the advantage of not requiring training, and thus they can be performed quickly, easily, and economically. Consequently, some genetic research has

Behavioral Genetics of the Mouse: Volume I. *Genetics of Behavioral Phenotypes*, eds. Wim E. Crusio, Frans Sluyter, Robert T. Gerlai, and Susanna Pietropaolo. Published by Cambridge University Press.

been done using these methods. The downside of these methods is that they provide limited insight about perception. Second, *fear-potentiated startle* (FPS) is a form of fear conditioning that can utilize sounds as conditioned stimuli. Fear-potentiated startle has the disadvantage of requiring time (usually days) for training and testing, but does permit the emotional dimension of perceptions to be tapped. The FPS method has also provided some preliminary genetic research using mice. Third, other methods have the potential for auditory behavior genetic research but have yet to be exploited by researchers. These include associative learning for determination of discrimination or other auditory abilities, observation of natural behaviors in response to sounds, and evaluation of anatomical or physiological properties of the CAS.

Hearing loss

Conditions associated with pathological cochleas obviously affect auditory perception by limiting or distorting the input to the CAS. Moreover, sensorineural damage to the cochlea can result in substantial reorganization and plasticity of the CAS that presumably alters perception (Carlson and Willott, 1996, 1998; Willott, 1986; Willott and Turner, 2000; Willott et al., 1994; Zettel et al., 2001, 2003). This chapter, however, focuses on "normal" hearing mice – young adult mice without apparent cochlear damage or dysfunction. The limited focus excludes the vast majority of research on the genetics of mouse hearing and auditory function, which has addressed cochlear pathology and related issues (e.g., Johnson et al., 1997, 2000) covered in other chapters. Indeed some of the most commonly used inbred strains such as C57BL/6, BALB/c, 129, and DBA/2 exhibit relatively early age-related hearing loss. Because most data on mice have been derived from these strains, research on "normal hearing" issues is problematic.

Age considerations

The *auditory system* of *Mus musculus* is functionally mature by age 1 to 1.5 months (Saunders et al., 1980; Shnerson and Pujol, 1982; Willott and Shnerson, 1978). In strains without hearing loss, auditory physiology and behavior do not appear to differ significantly from that of 5 or 6 month-olds (Henry, 1983; Parham and Willott, 1988; Willott et al., 1994). Therefore, testable cohorts of mice can be generated relatively rapidly, adding to their attractiveness as subjects for behavior genetics research.

ASR and PPI

Acoustic startle response

The ASR (see also Chapter 18) is a fundamental auditory behavior, occurring in mice and humans (Davis, 1984). In the present context the ASR may be viewed as a behavioral indicator of responsiveness to intense sound. The ASR, a jerk-like

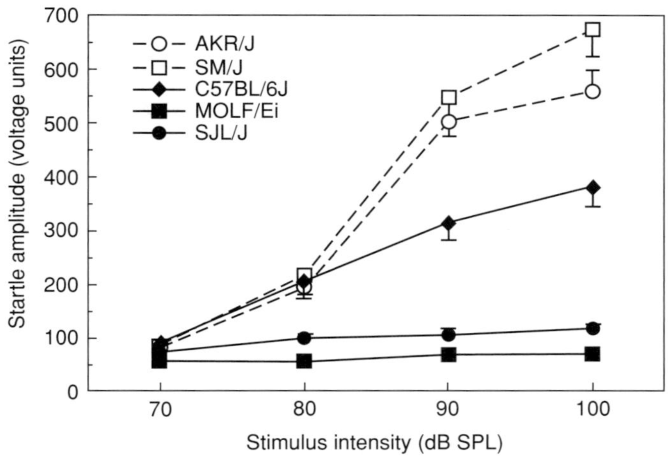

Figure 5.1 Acoustic startle amplitude for five inbred strains. Acoustic startle responses (ASRs) were evoked by brief noise bursts of 70–100 dB sound pressure level (SPL) (abscissa). Despite having excellent hearing sensitivity, MOLF/Ei and SJL/J mice exhibit minimal ASRs. By comparison, SM/J and AKR/J mice exhibit large amplitude ASRs.

motor reflex, is reliably elicited by bursts of noise or tones having sound pressure levels (SPLs; re 20 μPa) of 80–90 dB and greater. The primary neural pathway for the ASR resides in the lower brainstem, with auditory neurons in the cochlear nucleus and/or eighth nerve root projecting to startle-triggering neurons of the reticular formation (e.g., Davis et al., 1982; Huffman and Henson, 1990; Kandler and Herbert, 1991; Koch, 1999; Lingenhöhl and Friauf, 1994; Yeomans and Frankland, 1996). The ASR is an interesting and important auditory behavior in its own right, but also serves as a behavioral "probe," as discussed in later sections. Acoustic startle response amplitude has been shown to vary among strains of mice (Falls et al., 1997b; Willott et al., 2003b) and rats (Glowa and Hansen, 1994).

We have evaluated the ASR in 40 inbred strains of mice using standard procedures (Willott et al., 2003b). Whereas some of the strains are known to exhibit early hearing loss, many have been shown to have low thresholds for the auditory brainstem response (ABR; see below). Among these normal-hearing strains, ASR amplitudes vary greatly, suggesting that factors other than cochlear sensitivity are playing a role. Figure 5.1 presents examples of ASR amplitude-intensity functions from five inbred strains. Four of these, SJL/J, MOLF/Ei, AKR/J, and SM/J have ABR thresholds that are as low as those of any inbred strain (Zheng et al., 1999); C57BL/6J (B6) mice exhibit age-related hearing loss, but at the age tested (1–2 months) have near-normal hearing. Mice of the AKR/J and SM/J strains have large ASRs; mice of the MOLF/Ei and SJL/J strains have minimal ASRs, even with 100 dB SPL stimuli; and B6 mice are intermediate.

Prepulse inhibition

Prepulse inhibition (see also Chapter 18) involves a type of central processing whereby the perceptual impact or salience of sensory stimuli is regulated by the nervous system. Prepulse

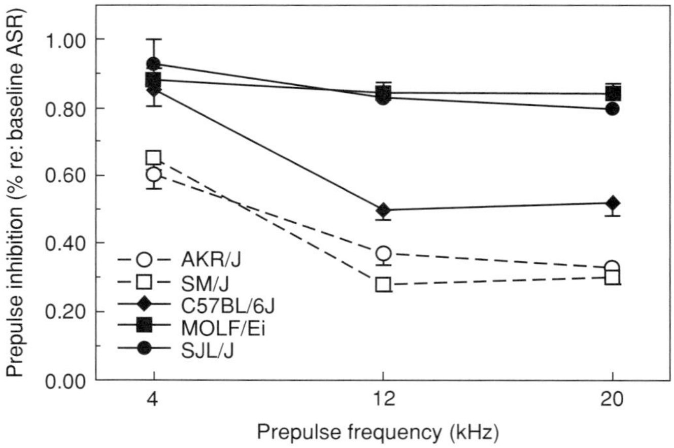

Figure 5.2 Prepulse inhibition (PPI) for five inbred strains obtained using brief 70 dB sound pressure level (SPL) tone pulses of 4, 12, or 20 kHz. Higher values = weak PPI. As was the case of acoustic startle responses (ASRs), MOLF/Ei and SJL/J mice exhibited weak auditory behavior, SM/J and AKR/J mice had weak behavior, and C57BL/6J mice were intermediate.

inhibition is manifested as follows: An audible "prepulse" stimulus (S1) is presented about 100 ms before an intense startle-eliciting stimulus (S2). Whether the S1 is a tone burst, a gap in the background noise, or any other acoustic change, the S1 is processed by the central auditory system (at least up to the midbrain level). Neural output from the auditory system then activates other components of the PPI neural circuitry, including pathways that ultimately descend to the reticular formation and inhibit the neurons that trigger the startle reflex (e.g., Carlson and Willott, 1998; Davis, 1984; Hoffman and Ison, 1980; Ison, 2001; Koch, 1999; Li et al., 1998; Willott et al., 1994). Activation of the PPI circuit inhibits the startle pathway for a period lasting several hundred ms, resulting in an "inhibited" ASR with a reduced amplitude. Prepulse inhibition can be expressed by the simple ratio of ASR amplitude when a prepulse has been presented (S1–S2) to ASR amplitude without the prepulse (S2 only).

Inbred strains of mice differ greatly with respect to the strength of PPI, as seen in the study of 40 inbred strains mentioned earlier (Willott et al., 2003b). Figure 5.2 shows PPI for the same five strains shown in Figure 5.1. Prepulses were brief 70 dB SPL tones of 4 kHz, 12 kHz, and 20 kHz. The relative strength of PPI across strains was similar to that observed for the ASR, with AKR/J and SM/J mice exhibiting strong PPI, MOLF/Ei and SJL/J exhibiting weak (minimal) PPI, and B6 in the middle.

The strength of PPI can be affected in two (not mutually exclusive) ways. First is the "quality" of central auditory processing. Because the initial portion of the PPI neural circuit is the auditory system, the quality of central auditory processing can determine whether PPI is weak or strong. For example, in a particular individual, sounds that are psychophysically salient (i.e., well-processed by the CAS) make the most effective prepulses (Reiter and Ison, 1979; Young and Fechner, 1983). Thus, PPI would directly reflect central auditory processing of the S1s,as demonstrated by the direct relationship between PPI magnitude and intensity of the prepulse stimulus. We can speculate that central auditory processing differences subject to genetic regulation might involve excitatory and/or inhibitory synaptic mechanisms, specific auditory pathways (e.g., number of neurons), or metabolic support of auditory neurons. The second way PPI could be affected is by how effectively the descending portion of the PPI circuit "drives" the behavior, irrespective of the ascending, auditory part of the circuit. Thus, auditory processing could be unremarkable, but PPI could be weak or strong depending on the efficacy of the descending circuit's ability to inhibit the ASR pathway. There is a body of literature showing that sensory gating/PPI is deficient in human subjects with neurobehavioral disorders such as schizophrenia and in rodent models of these disorders (Braff and Geyer, 1990; Cadenhead et al., 1993; Swerdow and Geyer, 1998; Swerdow et al., 1995), and this probably is not caused by deficits in auditory processing. The first possibility for regulation of PPI (quality of auditory processing) is most relevant for auditory research, whereas the latter (post-auditory PPI circuitry) is relevant for neurobehavioral disorders that involve "gating." Of course, the two need not be mutually exclusive.

Several studies have demonstrated genetic influences on PPI in mice (Bell et al., 1998; Bullock et al., 1997; Hitzemann et al., 2001); Kline et al., 1998; Paylor and Crawley, 1997). As mentioned, PPI might be affected by genes related to factors other than central auditory processing. For example, rodent models which have induced alterations of dopaminergic processes exhibit deficient PPI (Ellenbroek et al., 1995; Rigdon, 1990; Zhang et al., 2000), and PPI is affected by several other neurotransmitter systems including serotonergic systems (Geyer, 1998). Work on knockout mice show that mice lacking *Dv11*(a mouse homolog of a *Drosophila* gene) (Lijam et al., 1997), NCAM–180 (neural cell adhesion molecule)(Wood et al., 1998) or the serotonin-1B receptor (Dulawa et al., 1997) show deficient ASR habituation and/or PPI. *Thus, care must be taken in interpreting the results of genetic experiments on ASR or PPI vis à vis central auditory processing.* It would be valuable to have behavioral profiles for a variety of auditory behaviors other than PPI and ASR, so that the involvement of auditory processing per se (irrespective of the behavioral task) can be evaluated. For example, if a mutant mouse were observed with weak PPI but normal performance on the other auditory tasks, this would suggest genetic action on the PPI pathway (or something that modulates PPI), rather than auditory processing that feeds the PPI neural circuit. Whereas this would be a very exciting finding, it would suggest that the mice did not have a general auditory processing deficit.

Experiments with F1 hybrids and backcrosses

The five inbred strains shown in Figures 5.1 and 5.2 were chosen for genetic experiments. Acoustic startle responses and PPI were obtained from mice of each inbred strain and F1 hybrids generated for all possible combinations of the five target strains (Willott et al., 2003a). F1 groups contained between eight and

Table 5.1 Summary of acoustic startle response (ASR) and prepulse inhibition (PPI) for all F1 combinations of five inbred strains.

ASR (mean and sem) in voltage units					
	AKR	**B6**	**MOLF**	**SJL**	**SM**
AKR	560(40)	306(34)	147(14)	397(74)	260(43)
B6		382(38)	110(9)	140(12)	425(44)
MOLF			70(4)	93(4)	112(16)
SJL				117(7)	349(68)
SM					675(49)
PPI (mean and SEM): sum of scores for three tests (prepulse frequencies of 4, 12, and 20 kH); 3.0 = very poor PPI					
	AKR	**B6**	**MOLF**	**SJL**	**SM**
AKR	1.3 (0.08)	1.93 (0.22)	2.21 (0.15)	1.68 (0.22)	1.4 (0.19)
B6		1.88 (0.10)	2.62 (0.17)	2.53 (0.14)	1.7 (0.17)
MOLF			2.56 (0.08)	2.74 (0.22)	2.61 (0.22)
SJL				2.57 (0.11)	1.8 (0.19)
SM					1.23 (0.05)

ASR: acoustic startle response; PPI: prepulse inhibition; SPL, sound pressure level.

16 mice. These data are presented in Table 5.1; note that the PPI scores are the sum of PPI from all three S1 frequencies (4, 12, 20 kHz) so that a score near 3.0 indicates very poor PPI, whereas lower scores indicate strong PPI. In general, the directions of most F1 hybrid results were similar for both ASR and PPI. This suggests that common components of neural pathways and/or the general "strength" of auditory processing are being affected (it is possible for a mouse to have robust ABRs with minimal PPI or small ABRs and strong PPI because the phenomena utilize different, albeit partially overlapping, neural pathways). Several interesting observations can be made from the F1 hybrid data: (1) When MOLF/Ei, which has very small ASRs and minimal PPI, is crossed with any of the other strains, the F1 hybrids are like MOLF/Ei: they have small ASRs and weak PPI. This suggests that weak auditory behaviors are "dominant" in this case. (2) The situation is more complex for the other poorly responding strain, SJL/J. When crossed with strong-responding AKR/J or SM/J mice, the F1 behavior is intermediate. Thus, the "dominance" of weak behavior seen in MOLF/Ei is not present for SJL/J. (3) When B6 is crossed with strong-responding AKR/J or SM/J, the F1s are close to B6. (4) Crossing the two strong-responding strains, AKR/J and SM/J resulted in smaller ASRs in the F1 offspring, whereas PPI was strong, like the parental strains (so here is a disconnect between ASR and PPI). (5) Crossing the weak responding strains, MOLF/Ei and SJL/J resulted in weak responding F1 offspring.

For backcross experiments (Table 5.2), ASRs and PPI were divided into three categories based on the large data base obtained for the 40 strains (Willott et al., 2003b). Several highlights should be noted. No SJL/J inbreds or B6SJL F1 hybrids exhibited ASRs in the medium or strong categories. However, the backcross of B6SJL and B6 produced 22 of 37 mice with

Table 5.2 Summary of acoustic startle response (ASR) amplitude and prepulse inhibition (PPI) data for three inbred strains, F1 hybrids, and backcrosses.

ASR with 100 dB SPL stimuli (incidence)				
		Weak (<220)	**Medium (220–470)**	**Strong (>470)**
Inbred strains	B6	2	16	2
	SJL	20	0	0
	AKR	1	7	12
F1 hybrid strains	B6SJL	14	0	0
	B6AKR	2	8	1
	AKRSJL	2	7	1
Backcrosses	(B6SJL)B6	15	14	8
	(AKRSJL)SJL	3	11	1
	(AKRSJL)AKR	4	11	14
	(B6AKR)AKR	–	–	–
PPI with 12 kHz prepulses (incidence)				
		Weak (>0.75)	**Medium (0.75–0.38)**	**Strong (<0.38)**
Inbred strains	B6	0	17	4
	SJL	12	8	0
	AKR	1	6	13
F1 hybrid strains	B6SJL	6	8	0
	B6AKR	2	8	1
	AKRSJL	1	5	4
Backcrosses	(B6SJL)B6	6	25	4
	(AKRSJL)SJL	3	8	4
	(AKRSJL)AKR	3	11	16
	(B6AKR)AKR	–	–	–

ASR: acoustic startle response; PPI: prepulse inhibition; SPL, sound pressure level.

ASRs in these categories, suggesting the influence of B6. By the same token, 15 "weak" responders are likely to largely reflect the influence of SJL/J. Prepulse inhibition data (using the 12 kHz S1) for these backcross mice are differentiated almost as well. Importantly, ASR and PPI categories are not concordant for about half of these backcross mice, suggesting partial segregation of ASR and PPI.

The AKR/J and SJL/J strains also show good separation of ASR and PPI scores, albeit with some overlap that is inevitable in behavioral measures. Only 1 of 20 AKR/J mice has weak ASRs compared to 20 of 20 SJL/J mice; similarly, 12 SJL/J (vs. 1 AKR/J) have weak PPI, whereas 13 AKR/J (vs. 0 SJL/J) have strong PPI. Thus, for example, the 16 mice of the (AKRSJL) AKR backcross with strong PPI are likely expressing the AKR/J trait.

Fear-potentiated startle

Fear-potentiated startle is widely used to investigate the neurobiological basis of anxiety but can be used as a tool in the context of auditory perception. In FPS a mouse learns to fear an auditory cue that was paired with footshock; later, when startled in the presence of the cue, startle amplitude is increased as a function of the learned fear. Because FPS utilizes Pavlovian conditioning procedures, it is also used to investigate the

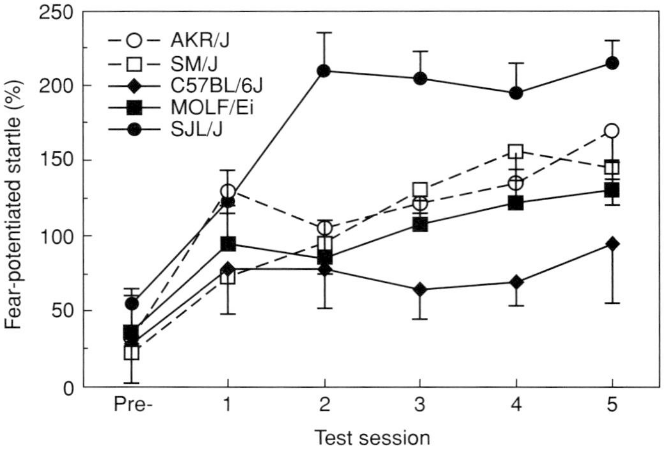

Figure 5.3 Fear-potentiated startle (FPS) for five inbred strains is shown as the startle amplitude % re baseline. Five successive training session are shown after an initial pre-training session. Relative strength of FPS among strains differed from the relative strengths of acoustic startle responses (ASRs) and prepulse inhibition (PPI). For example, despite showing weak ASR and PPI behavior (see Figures 5.1 and 5.2), SJL/J mice exhibited the strongest FPS.

neurobiological basis of learning and memory (Campeau and Davis, 1992, 1995a; Davis and Astrachan, 1978; Falls et al., 1992, 1997b; Gewirtz and Davis, 1997; Miserendino et al., 1990). The amygdala plays a key role in FPS, receiving auditory information from the auditory system, via the auditory thalamus and cortex (e.g., Campeau and Davis, 1995a, 1995b; Falls et al., 1997a; Romanski and LeDoux, 1992; Rosen et al., 1991).

In FPS, conditioned fear is operationally defined as elevated startle amplitude in the presence versus the absence of a stimulus that was previously paired with foot shock. FPS is particularly well suited for examining the acquisition, expression, and retention of aversive learning in mice: it is a highly reliable and sensitive measure of conditioned fear; it is easily and objectively quantified; and a great deal is known about the behavioral parameters and neural circuits for FPS in rodents.

Figure 5.3 summarizes data on FPS from the five strains we have been focusing on with respect to ASR and PPI (Falls et al., 2001). Acquisition of FPS differs among strains, with SJL/J mice exhibiting a pronounced effect and C57BL/6J relatively weak FPS. These data suggest that well-hearing strains differ in their ability to associate tone and shock or to retain this association once acquired. Moreover, this is not related to the relative strengths of PPI or the ASR in a simple way (compare Figures 5.1,5.2, and 5.3), suggesting different genetic mechanisms underlying the behaviors. Because the differences cannot be ascribed to auditory sensitivity, they could be a function of their effectiveness as conditioned stimuli (as in the quality of auditory processing in PPI mentioned earlier) or how readily fear is conditioned irrespective of the conditioned stimuli. In either case, however, the same sounds come to have different emotional tone, which presumably affects how they are perceived by the mice.

Other approaches

Psychophysical measurements using associative conditioning

Directional hearing, gap detection, frequency discrimination, and intensity discrimination tap the fundamental dimensions of hearing: where the sound is coming from, what changes are occurring over brief periods of time, what is the spectral and intensive content of the signal? Whereas a certain degree of peripheral processing is, of course, critical for these tasks, they are ultimately beholden to central processing and are essential aspects of perception. It is generally accepted that directional hearing makes use of binaural comparisons at central auditory nuclei; gap detection requires that central neurons exhibit a response change to the gap; frequency is coded in part by central tonotopic maps; and intensity coding is presumably related to neural input–output functions that may differ substantially in shape among central and peripheral sites. The neural code arriving at the CAS via auditory nerve impulses may be flawless, but if central processes are operating at a subpar level, perceptual problems will arise. By testing strains with normal cochlear sensitivity, we can assume that differences in performance on the auditory tests are a function of central processing.

A good example of the method for assessing auditory behavior is the work of Heffner and colleagues (Heffner and Heffner, 2001; Heffner et al., 2001; Koay et al., 2002) who have developed a *conditioned suppression/avoidance* procedure to test the hearing of mice. They have even used the method on mice homozygous for the med^J mutation that affects sodium channels at the nodes of Ranvier and elsewhere. The *med^J* mutation results in the near absence of Scn8a voltage-gated sodium channels, which severely compromises conduction of action potentials, but when maintained on a genetic background containing a resistant allele at the master locus *Scnm* (from the C3H parent, Sprunger et al., 1999), survival past 1 year of age is common (Meisler et al., 2001). The homozygous med^J mice are weak, drag their hind legs, and display tremors and dystonic postures (Sprunger et al., 1999). Such mice present a severe challenge to any behavioral procedure, but the investigators were successful. Both wildtype and med^J mice performed consistently and produced very reliable psychophysical curves showing that the mutants had normal hearing.

This example shows that the conditioned suppression/avoidance task provides a useful procedure for testing mice even if they have motor disorders. The reliability of the test thresholds coupled with the limited within-strain variation suggest that any strain differences that occur should be demonstrable with relatively small numbers of mice. Thus, the method has potential for genetic studies of mouse hearing, provided that possible deficits in associative learning are controlled for. Examples might include the genetics of discrimination abilities for frequency, intensity, or other acoustic parameters.

Observation of natural behaviors in response to sounds

Observing and scoring the qualitative or quantitative aspects of behaviors in response to acoustic stimuli can be used to assess the biological relevance of the sounds. Responses to vocalizations used in parenting would be an example (reviewed by Nyby, 2001). To date, most work has dealt with production of sounds, and not on genetic influences on the qualitative or quantitative aspects of responses to those sounds. However, pup calls induce maternal behavior that can be quantified (D'Amato et al., 2005; Farrell and Alberts, 2002; Liu et al., 2003), with the potential for such responses to be used as a phenotype in genetic experiments. One potential limitation is that biologically essential auditory behaviors are likely to be strongly expressed by mice in general, so there might not be a great deal of phenotypic variation among inbred strains or mutants. Moreover, one would have to distinguish between differences in the properties of the pup vocalizations versus the responses to them. Hormonal influences might also play a role and be regulated by genes, as well as attachment and social tendencies. Thus, this approach may not be especially promising for research on the genetics of auditory perception.

Physiological methods

Two physiological measures have been widely used to assess hearing in mice and can be performed rapidly enough to allow testing of many mice: the ABR and distortion product otoacoustic emissions (DPOAE). The ABR is an electroencephalographic response usually measured with scalp electrodes. The electrodes pick up voltages produced by neural activity reaching it from throughout the brain, including the auditory brainstem and eighth cranial nerve. The voltages can be revealed by repeatedly presenting brief sounds and "averaging" the stimulus-evoked voltages that are time-locked to each acoustic stimulus (for thorough discussions of ABRs in mice see Parham et al., 2001; Willott, 2006). Several ABR parameters can be measured, including the amplitude of waves, the latency (time elapsed from time zero to wave peak), interpeak latency (an indication of transmission time), and threshold. Threshold provides an excellent tool for assessing auditory sensitivity: stimulus intensity for tones of various frequencies is decreased from some maximum and threshold is defined as the minimum intensity at which a reliable ABR can be obtained. Zheng et al. (1999) tested 80 inbred strains of mice using this method, and additional data are available from the Jackson Laboratory (Mouse Phenome Database, http://phenome.jax.org or http://hearingimpairment.jax.org/screening.html).

Otoacoustic emissions are faint sounds that are generated by outer hair cells (OHCs) of the inner ear and can be measured in the ear canal following the presentation of acoustic stimuli. When two tones are presented to the ear canal, a DPOAE is produced. Tone pairs can be selected to test frequencies within the mouse's hearing range to assess OHC function (for thorough discussions of DPOAEs in mice see Martin et al., 2006; Parham et al. 2001).

Whereas these methods are extremely valuable to assess hearing sensitivity, they tell us little about auditory perception in normal-hearing mice. That being said, it is important to use these tools on mouse models to make sure they are able to adequately hear stimuli being studied.

Anatomical methods

Whereas anatomical measures are somewhat removed from experience, inferences about perception can nonetheless be gleaned. For example, genetic variation in hippocampus structure has been demonstrated among inbred strains and correlated with relevant behaviors (Crusio et al., 1986, 1989). An extreme example with respect to the auditory system is the huge cochlear nucleus of bats, which comports with the special importance of sounds for these mammals. Perhaps more subtle anatomical differences between mouse strains or mutants might provide similar insights. The anterior cochlear nucleus of B6 and CBA/J mice differs in size and dimensions in a manner that may be associated with differences in the audiogram (Willott et al., 1987): B6 mice are most sensitive to a lower range of frequencies than CBA/J mice (Willott, 1986). Thus, it may be feasible to use this type of approach in a comparison of different normal-hearing inbred strains. The limitations of anatomical approaches in genetic research loom large, however. They typically require euthanasia (for anatomy) and technically challenging, time-consuming procedures. Also, it is usually difficult (at best) to establish a correlation between perception and some morphological characteristic of a brain structure within the same species. Thus, even if genetic contributions to variations in a CAS structure were established, the implications with respect to perception would be hard to interpret.

Research approaches

We still know relatively little about the genetic, neural, and other biological variables that underlie the individual differences that are ubiquitous within the auditory domain in the absence of cochlear or central damage. Understanding genetic and other biological factors that regulate auditory processing may ultimately lead to the development of new diagnostic and ameliorative strategies. Whereas the challenges for research in this area are substantial, one can be optimistic that the limitations can be overcome by adept researchers, leading to new breakthroughs in our understanding.

The mouse models allow two approaches. First, by establishing backcrosses using strains expressing certain phenotypes and testing for these behaviors, segregation and determination of *quantitative trait loci* (QTLs) can be evaluated. Quantitative trait loci provide a means for genetic mapping for traits regulated by multiple genes. Many mice need to be tested, and the screening protocols for startle and PPI fill the bill for high throughput (e.g., Willott et al., 2003a, 2003b). The QTL approach has proven somewhat useful for PPI and ASR phenotypes using BXD recombinant inbred strains (Hitzemann et al., 2001). However, the BXD strains all exhibit sensorineural hearing loss (Willott and Erway, 1998), so those findings are not applicable with respect to normal auditory processing. The major problem with respect to the establishment of QTLs with almost any behavior is the phenotypic variance among individuals. Even though PPI and the ASR are among the most reliable of behaviors, it is still difficult to score individual mice with respect to the strength or clarity of the phenotype. This is especially problematic when it comes to weak phenotypes. For example, some individual mice from strains with "average" or "strong" ASR or PPI can exhibit low levels of the behavior on a given test session. This makes it impossible to distinguish them from genetic "weak" responders. Consequently, very large numbers of mice may be required, or multiple testing sessions, with considerable expenditure of time and money and no guarantee of success. Nonetheless, the continuous development of these techniques may hold great promise for future research.

Another approach has made use of mutagenized mice. Chemical mutagens are given to a parent to produce a high rate of random mutations in the offspring. F1 progeny may exhibit dominant traits and alternatively, through backcrossing, offspring can be produced that exhibit recessive traits. ASR, PPI, and other tests are used on all mice in order to identify "phenodeviants." Families that possess the gene for the deviant trait are used to produce a line possessing the mutation and, if heritable, this is made available to the scientific community. Because a single gene is likely to be involved, mutated mice with especially weak or strong phenotypes can be used for mapping the gene. Importantly, once the mutated lines become available for research, they can be tested for the other phenotypes by various laboratories. This can determine whether the gene affects other auditory behaviors. This approach suffers from the same phenotypic-variability problem as the QTL approach because individuals must be identified as having the phenotype which is then tested for heritability with additional breeding. False identification of a phenodeviant would lead to considerable wasted time and effort. A second problem with *mutagenesis* approach is the low likelihood that a single gene will significantly affect auditory perception in a way that would alter behaviors like ASR, PPI, FPS, or auditory learning. Despite these problems, the approach still provides a potentially important approach in providing insights about the genetics of ASR, PPI, or FPS in the absence of cochlear hearing loss.

In summary, performing research on the genetics of normal auditory behavior is quite challenging. Technically, tasks that both reflect perception and are amenable to genetic research are limited, and the possibility of hearing loss must always be considered. A strain or mutant exhibiting poor PPI or FPS, for example, must be shown to have good hearing in order to make inferences about central auditory processing or perception. It is the author's opinion that these issues should not discourage investigations of this type. The potential gains are too important.

Acknowledgments

Preparation of this chapter was supported by NIH grant R01 AG07554. The work on ASRs, PPI, and FPS was supported by NIH grant R01 MH61090, with participation of Dr. Kenneth Johnson, Dr. William Falls, and Leona Gagnon. This chapter was originally written in 2006. Dr. Willott retired in 2009.

References

Baharloo, S., Service, S.K., Risch, N., Gitschier, J., and Freimer, N.B. (2000) Familial aggregation of absolute pitch. *Amer J Hum Genet* 67: 755–758.

Bell, J. III., Hitzemann, R.J., and McCaughran, J.A. (1998) Quantitative trait loci (QTL) analysis of the fear-potentiated startle response (FPS) in the BXD series of recombinant inbred mice. *Soc Neuroci Abstr* 24: 1200.

Bellis, T.J. and Ferre, J.M. (1999) Multidimensional approach to the differential diagnosis of central auditory processing disorders in children. *J Amer Acad Audiol* 10: 319–328.

Braff, D.L. and Geyer, M.A. (1990) Sensorimotor gating and schizophrenia. Human and animal studies. *Arch Gen Psychiat* 47: 181–188.

Bullock, A.E., Slobe, B.S., Vazquez, V., and Collins, A.C. (1997) Inbred mouse strains differ in the regulation of startle and prepulse inhibition of the startle response. *Behav Neurosci* 111: 1353–1360.

Cadenhead, K.S., Geyer, M.A., and Braff, D.L. (1993) Impaired startle prepulse inhibition and habituation in patients with schizotypal personality disorder. *Amer J Psychiat* 150: 1862–1867.

Campeau, S. and Davis, M. (1992) Fear-potentiation of the acoustic startle reflex in rats using noises of various spectral frequencies. *Anim Learn Behav* 20: 177–186.

Campeau, S. and Davis, M. (1995a) Involvement of subcortical and cortical afferents to the lateral nucleus of the amygdala in fear conditioning measured with fear-potentiated startle in rats trained concurrently with auditory and visual conditioned stimuli. *J Neurosci* 15(3): 2312–2327.

Campeau, S. and Davis, M. (1995b) Involvement of the central nucleus and basolateral complex of the amygdala in fear conditioning measured with fear-potentiated startle in rats trained concurrently with auditory and visual conditioned stimuli. *J Neurosci* 15(3): 2301–2312.

Carlson, S. and Willott, J.F. (1996) The behavioral salience of tones as indicated by prepulse inhibition of the startle response: relationship to hearing loss and central neural plasticity in C57BL/6J mice. *Hear Res* 99: 168–175.

Carlson, S. and Willott, J.F. (1998) Caudal pontine reticular formation of C57BL/6J mice: responses to startle stimuli, inhibition by tones, plasticity. *J Neurophysiol* 79: 2603–2614.

Crusio, W.E., Genthner-Grimm, G., and Schwegler, H. (1986) A quantitative-genetic analysis of hippocampal variation in the mouse. *J Neurogenet* 3: 203–214.

Crusio, W.E., Schwegler, H., and van Abeelen, J.H. (1989) Behavioral responses to novelty and structural variation of the hippocampus in mice. II. Multivariate genetic analysis. *Behav Brain Res* 32: 81–88.

D'Amato F.R., Scalera, E., Sarli, C., and Moles, A. (2005) Pups call, mothers rush: does maternal responsiveness affect the amount of ultrasonic vocalizations in mouse pups? *Behav Genet* 35: 103–112.

Davis, M. (1984) The mammalian startle response. In Eaton, R.C. (ed.), *Neural Mechanisms of Startle Behavior*. Plenum, New York, pp. 287–351.

Davis, M. and Astrachan, D.I. (1978) Conditioned fear and startle magnitude: effects of different footshock or backshock intensities used in training. *J Exper Psychol Anim Behav Proc* 4: 95–103.

Davis, M., Gendelman, D.S., Tischler, M.D., and Gendelman, P.M. (1982) A primary acoustic circuit: lesion and stimulation studies. *J Neurosci* 2: 791–805.

DeFries, J.C., Fulker, D.W., and LaBuda, M.C. (1987) Evidence for a genetic aetiology in reading disability in twins. *Nature* 328: 537–539.

Dulawa, S.C., Hen, R., Scearce-Levie, K., and Geyer, M.A. (1997) Serotonin-sub1B receptor modulation of startle reactivity, habituation, and prepulse inhibition in wildtype and serotonin1B knockout mice. *Psychopharmacol* 132(2): 125–134.

Ellenbroeck, B.A., Budde, S., and Cools, A.R. (1995) The role of dopamine D1 and D2 receptors in the medial prefrontal cortex in prepulse inhibition. *Soc Neurosci Abstr* 21: 747.

Falls, W.A., Bakken, K., and Heldt, S. (1997a) Lesions of the perirhinal cortex block conditioned excitation but not conditioned inhibition of fear. *Behav Neurosci* 111: 476–486.

Falls, W.A., Carlson, S., Turner, J., and Willott, J.F. (1997b) Fear-potentiated startle in two strains of inbred mice. *Behav Neurosci* 111: 855–861.

Falls, W.A., Miserendino, M.J.D., and Davis, M. (1992) Extinction of fear-potentiated startle: blockade by infusion of an excitatory amino acid antagonist into the amygdala. *J Neurosci* 12(3): 854–863.

Falls, W.A., Pistell, P.J., Willott, J.F. (2001) *Parametric Analysis of Fear-potentiated Startle in Common Inbred Strains of Mice.* Society for Neuroscience, San Diego, CA, USA.

Farrell, W.J. and Alberts, J.R. (2002) Stimulus control of maternal responsiveness to Norway rat (*Rattus norvegicus*) pup ultrasonic vocalizations. *J Comp Psychol* 116: 297–307.

Gewirtz, J.C. and Davis, M. (1997) Second-order fear conditioning prevented by blocking NMDA receptors in amygdala. *Nature* 388: 471–474.

Geyer, M.A. (1998) Behavioral studies of hallucinogenic drugs in animals: Implications for schizophrenai research. *Pharmacopsychiat* 31: 73–79.

Glowa, J.R. and Hansen, C.T. (1994) Differences in response to an acoustic startle stimulus. *Behav Genet* 24: 79–84.

Heffner, H.E. and Heffner, R.S. (2001) Behavioral assessment of hearing in mice. In Willott, J.F. (ed.), *Handbook of Mouse Auditory Research: From Behavior to Molecular Biology*. CRC Press, New York, pp. 19–29.

Heffner, R.S., Koay, G., and Heffner, H.E. (2001) Sound-localization acuity changes with age in C5BL/6J mice. In Willott, J.F. (ed.), *Handbook of Mouse Auditory Research: From Behavior to Molecular Biology*. CRC Press, New York, pp. 31–35.

Henry, K.R. (1983) Ageing and audition. In Willott, J.F. (ed.), *Auditory Psychobiology of the Mouse*. Charles C. Thomas, Springfield, IL, USA, pp. 470–493.

Hitzemann, R., Bell, J., Rasmussen, E., and McCaughran, J. (2001) Mapping the genes for the acoustic startle response (ASR) and prepulse inhibition of the ASR in the BXD recombinant inbred series: Effect of high-frequency hearing loss and cochlear pathology. In Willott, J.F. (ed.), *Handbook of Mouse Auditory Research: From Behavior to Molecular Biology*. CRC Press, New York, pp. 441–455.

Hoffman, H.S. and Ison, J.R. (1980) Reflex modification in the domain of startle. I. Some empirical findings and their implications for how the nervous system processes sensory input. *Psychol Bull* 87: 175–189.

Huffman, R.F. and Henson, O.W. (1990) The descending auditory pathway and acousticomotor systems: connections with the inferior colliculus. *Brain Res Rev* 15: 295–323.

Ison, J.R. (2001) The acoustic startle response: reflex elicitation and reflex modification by preliminary stimuli. In Willott, J.F. (ed.), *Handbook of Mouse Auditory Research: From Behavior to Molecular Biology*. CRC Press, New York, pp. 59–82.

Johnson, K.R., Erway, L.C., Cook, S.A., Willott, J.F., and Zheng, Q.Y. (1997) A major gene affecting age-related hearing loss in C57BL/6J mice. *Hear Res* 114: 83–92.

Johnson, K.R., Zheng, Q.Y., and Erway, L.C. (2000) A major gene affecting age-related hearing loss is common to at least ten inbred strains of mice. *Genomics* 70: 171–180.

Kandler, K. and Herbert, H. (1991) Auditory projections from the cochlear nucleus to pontine and mesencephalic reticular nuclei in the rat. *Brain Res* 562: 230–242.

Keenan, J.P., Thangaraj, V., Halpern, A.R., and Schlaug, G. (2001) Absolute pitch and planum temporale. *Neuroimage* 14: 1402–1408.

Kline, L., Decena, E., Hintzemann, R., and McCaughran, J. (1998) Acoustic startle, prepulse inhibition, locomotion, and latent inhibition in the neuroleptic-responsive (NR) and neuroleptic-nonresponsive (NNR) lines of mice. *Psychopharmacol* 139: 322–331.

Koay, G., Heffner, R.S., and Heffner, H.E. (2002) Behavioral audiograms of homozygous medj mutant mice with sodium channel deficiency and their unaffected littermates. *Hear Res* 171: 111–118.

Koch, M. (1999) The neurobiology of startle. *Prog Neurobiol* 59: 107–128.

Li, L., Priebe, R.P., and Yeomans, J.S. (1998) Prepulse inhibition of acoustic or trigeminal startle of rats by unilateral electrical stimulation of the inferior colliculus. *Behav Neurosci* 112: 1187–1198.

Lijam, N., Paylor, R., McDonald, M.P., Crawley, J.N., Deng, C.X., Herrup, K., et al. (1997) Social interaction and sensorimotor gating abnormalities in mice lacking DV11. *Cell* 90(5): 895–905.

Lingenhöhl, K. and Friauf, E. (1994) Giant neurons in the rat reticular formation: a sensorimotor interface in the elementary acoustic startle circuit? *J Neurosci* 14: 1176–1194.

Liu, R.C., Miller, K.D., Merzenich, M.M., and Schreiner, C.E. (2003) Acoustic variability and distinguishability among mouse ultrasound vocalizations. *J Acoust Soc Am* 114: 3412–3422.

Martin, G.K., Stagner, B.B., and Lonsbury-Martin, B.L. (2006) Assessment of cochlear function in mice: distortion-product otoacoustic emissions sensitivity. *Curr Protoc Neurosci* Chapter 8, Unit 8.21C.

Meisler, M.H., Kearney, J., Escayg, A., MacDonald, B.T., and Sprunger, L.K. (2001) Sodium channels and neurological disease: insights from Scn8a mutations in the mouse. *Neuroscientist* 7, 136–145.

Miserendino, M.J.D., Sananes, C.B., Melia, K.R., and Davis, M. (1990) Blocking of acquisition but not expression of conditioned fear-potentiated startle by NMDA antagonists in the amygdala. *Nature* 345(6277): 716–718.

Nyby, J. (2001) Auditory communication among adults. In Willott, J.F. (ed.), *Handbook of Mouse Auditory Research: From Behavior to Molecular Biology*. CRC Press, New York, pp. 3–18.

Parham, K., Sun, X.-M., and Kim, D.O. (2001) Noninvasive assessment of auditory function: distortion product otoacoustic emissions and auditory brainstem responses. In Willott, J.F. (ed.), *Handbook of Mouse Auditory Research: From Behavior to Molecular Biology*. CRC Press, New York, pp. 37–58.

Parham, K. and Willott, J.F. (1988) The acoustic startle response in young and aging C57BL/6J and CBA/J mice. *Behav Neurosci* 102: 881–886.

Paylor, R. and Crawley, J.N. (1997) Inbred strain differences in prepulse inhibition of the mouse startle response. *Psychopharmacol* 132: 169–180.

Reiter, L.A. and Ison, J.R. (1979) Reflex modulation and loudness recruitment. *J Aud Res* 19: 201–207.

Rigdon, G. (1990) Differential effects of apomorphine on prepulse inhibition of acoustic startle reflex in two rat strains. *Psychopharmacol* 102: 419–421.

Romanski, L.M. and LeDoux, J.E. (1992) Equipotentiality of thalamo–amygdala and thalamo–cortico–amygdala circuits in auditory fear conditioning. *J Neurosci* 12: 4501–4509.

Rosen, J.B., Hitchcock, J.M., Miserendino, M.J.D., Falls, W.A., Campeau, S., and Davis, M. (1991) Lesions of the perirhinal cortex, but not of the frontal, medial pre-frontal, visual or insular cortex block fear-potentiated startle using a visual conditioned stimulus. *J Neurosci* 12: 4624–4633.

Saunders, J.C., Dolgin, K.G., and Lowery, L.D. (1980) The maturation of frequency selectivity in C57BL/6J mice studied with auditory evoked response tuning curves. *Brain Res* 187: 69–79.

Schönweiler, R., Ptok, M., and Radü, H.-J. (1998) A cross-sectional study of speech and language-abilities of children with normal hearing, mild fluctuating conductive hearing loss, or moderate to profound sensorineural hearing loss. *J Ped Otorhinolaryngol* 44: 251–258.

Shnerson, A. and Pujol, R. (1982) Age-related changes in the C57BL/6J mouse cochlea. I. Physiological findings. *Devel Brain Res* 2: 65–75.

Sprunger, L.K., Escayg, A., Tallaksen-Greene, S., Albin, R.L., and Meisler, M.H. (1999) Dystonia associated with mutation of the neural sodium channel Scn8a and identification of the modifier locus Scnm1 on mouse chromosome 3. *Hum Mol Genet* 8: 471–479.

Swerdow, N.R. and Geyer, M.A. (1998) Using an animal model of deficient sensorimotor gating to study the pathophysiology and new treatment of schizophrenia. *Schizophr Bull* 24(2): 285–301.

Swerdow, N.R., Paulsen, J., Braff, D.L., and Butters, N. (1995) Impaired prepulse inhibition of acoustic and tactile startle response in patients with Huntington's disease. *J Neurol Neurosurg Psychiat* 58(2): 192–200.

Willott, J.F. (1986) Effects of aging, hearing loss, and anatomical location on thresholds of inferior colliculus neurons in C57BL/6 and CBA mice. *J Neurophysiol* 56: 391–408.

Willott, J.F. (2001) (ed.), *Handbook of Mouse Auditory Research: From Behavior to Molecular Biology*. CRC Press, Boca Raton, FL, USA.

Willott, J.F. (2006) Measurement of the auditory brainstem response (ABR) to study auditory. *Curr Protoc Neurosci* Chapter 8, Unit 8.21C.

Willott, J.F., Carlson, S., and Chen, H. (1994) Prepulse inhibition of the startle response in mice: relationship to hearing loss and auditory system plasticity. *Behav Neurosci* 108: 703–713.

Willott, J.F. and Erway, L.C. (1998) Genetics of age-related hearing loss in mice. IV. Cochlear pathology and hearing loss in 25 BXD recombinant inbred mouse strains. *Hear Res* 119: 27–36.

Willott, J.F., Jackson, L.M., and Hunter, K.P. (1987) Morphometric study of the anteroventral cochlear nucleus of two mouse models of presbycusis. *J Comp Neurol* 260: 472–480.

Willott, J.F., Johnson, K.R., and Gagnon, L. (2003a) *Genetic Mapping Study of Startle and Prepulse Inhibition*. Association for Research in Otolaryngology, Daytona Beach, FL, USA.

Willott, J.F. and Shnerson, A. (1978) Rapid development of tuning characteristics of inferior colliculus neurons in mouse pups. *Brain Res* 148: 230–233.

Willott, J.F., Tanner, L., O'Steen, J., Johnson, K.R., Bogue, M.A., and Gagnon, L. (2003b) Startle and prepulse inhibition in 40 inbred strains of mice. *Behav Neurosci* 117: 716–727.

Willott, J.F. and Turner, J.G. (2000) Neural plasticity in the mouse inferior colliculus: relationship to hearing loss, augmented acoustic stimulation, and prepulse inhibition. *Hear Res* 147: 275–281.

Wood, G.K., Tomasiewicz, H., Rutishauser, U., Magnuson, T., Quirion, R., Rochford, J., et al. (1998) NCAM-180 knockout mice display increased lateral ventricle size and reduced prepulse inhibition of startle. *Neuroreport* 9(3): 461–466.

Yeomans, J.S. and Frankland, P.W. (1996) The acoustic startle reflex: neurons and connections. *Brain Res Rev* 21: 301–314.

Young, J.S. and Fechter, L.D. (1983) Reflex inhibition procedures for animal audiometry: a technique for assessing ototoxicity. *J Acoust Soc Am* 73: 1686–1693.

Zettel, M.L., O'Neill, W.E., Trang, T.T., and Frisina, R.D. (2001) Early bilateral deafening prevents calretinin up-regulation in the dorsal cortex of the inferior colliculus of aged CBA/CaJ mice. *Hear Res* 158: 131–138.

Zettel, M.L., O'Neill, W.E., Trang, T.T., and Frisina, R.D. (2003) The effects of early bilateral deafening on calretinin expression in the dorsal cochlear nucleus of aged CBA/CaJ mice. *Hear Res* 183: 57–66.

Zhang, J., Forkstam, C., Engel, J.A., and Svensson, L. (2000) Role of dopamine in prepulse inhibition of acoustic startle. *Psychopharmacol* 149: 181–188.

Zheng, Q.Y., Johnson, K.R., and Erway, L.C. (1999) Assessment of hearing in 80 inbred strains of mice by ABR threshold analysis. *Hear Res* 130: 94–107.

Chapter 6

Behavioral measurement of mouse visual function

Glen T. Prusky and Nazia M. Alam

Introduction

As a nocturnal rodent with small eyes, the laboratory mouse has not traditionally been a popular animal model for vision-based studies. As the mouse emerged as the most powerful mammalian model for molecular genetic manipulations in the 1990s, however, cellular and physiological studies of the mouse visual system gained acceptance, and even prominence. Behavioral studies of mouse vision, on the other hand, lagged, and for at least two reasons: (1) there were inadequate methodologies to quantify mouse vision, and (2) many considered measures of vision as an adjunct to corroborate cellular and physiological studies, rather than as a powerful and necessary tool to understand visual system function. The lack of quantitative behavioral testing methodologies meant that the vast majority of experimental studies of mouse vision employed no behavioral measures of vision at all. In addition, little basic information on normal mouse visual behavior was available, and mouse strains, or animals with naturally-occurring or directed genetic modifications affecting the visual system were utilized, but not screened for vision. Moreover, many mutations that affect the visual system would have consequences for visual development and plasticity, and interpretations of the experiments would benefit from knowledge about these processes. Fortunately, there has been progress on the behavioral methodological front, and the quantification of mouse vision is now practicable. As such, demonstrations of the experimental value of measuring mouse vision have followed and are now present in the literature.

In this chapter, the rationale for measuring mouse vision is presented, the attributes of the best available methodologies for measuring mouse vision are introduced, and data pertinent to behavioral genetic studies of the mouse visual system is presented, as are strategies for incorporating mouse visual behavior into future studies. The goal is to provide a primer on mouse visual behavior to enable researchers to effectively utilize behavioral measurements of mouse vision.

Why measure visual behavior?

A core principle of neurobiology is that the evolutionary selection and survival of neural circuits occur only as their effects are manifested in adaptive behavior. Thus, the mission of all neurophysiological research is the understanding of behavior, whether or not individual researchers directly study behavior. Thus, since visual function is the "essence" of the natural selection of the visual system, visual behavior is the metric of function that is most relevant to the evolutionary history of the system.

Behavioral measures of vision also have important experimental advantages that enable the measurement of function in ways not feasible with other methodologies. For example, experiments that would benefit from repeated-measures design are often precluded when using methodologies that are effectively limited to a single session (i.e., electrophysiological recordings), or are dependent on euthanasia (i.e., histology). Many neurophysiological measurements also require the use of anesthetics, which can alter cellular responses in the visual system (Pham et al., 2004) and can produce detrimental cumulative effects when used repeatedly.

Another advantage is that behavioral measures of vision are often more sensitive than physiological methods in detecting visual function or dysfunction. For example, behavioral methods are able to measure visual thresholds in rodents long after the electroretinogram (ERG) disappears (McGill et al., 2004). The likely reasons for this disparity include that vision is a system-wide product, not solely a function of the retina (i.e., "vision" is the collective function of the visual system), and that the circuitry underlying most visual functions is widely distributed, with most physiological methodologies sampling function in only a subset of these circuits. In addition and as mentioned above, anesthetic, analgesic, and paralytic agents that are often necessary for electrophysiological studies can dampen or otherwise alter cellular responses.

Among the most important experimental advantages of measuring visual behavior is that such measures are the most relevant to the identification and treatment of visual diseases. Many diseases of the visual system are manifested as, and defined by, abnormalities in vision (i.e., age-related macular degeneration, amblyopia, retinitis pigmentosa, glaucoma, myopia, etc.), and it is almost always the lack of normal vision that brings individuals to the clinic. The practical implication of this is that an animal model of a visual disease must include

Behavioral Genetics of the Mouse: Volume I. *Genetics of Behavioral Phenotypes*, eds. Wim E. Crusio, Frans Sluyter, Robert T. Gerlai, and Susanna Pietropaolo. Published by Cambridge University Press.

measures of vision to ensure that it is relevant to the human condition. Many visual diseases are characterized by great variability in the presentation of symptoms (i.e., expressivity) and an animal model that does not embody this feature may not be a useful model, or alternatively, the definition of the human disease may need to be revised. The conceptual implication is that a successful treatment for a visual disorder will be defined not solely as a gain or replacement of circuits or physiological function, but as a benefit to, or replacement of, vision; i.e., it will be no help to a patient if a "successful" therapy at a cellular or physiological level does not translate into improved vision.

Considerations when testing mouse vision

Although there are distinct experimental advantages in measuring vision, there are also idiosyncrasies to whole animal behavior that should not be ignored when doing so. It is not possible to measure visual sensation directly; rather sensation can only be indirectly inferred from actions. Thus, when behavioral signals are the dependent variables in an experiment, the environmental circumstances under which the experiment is conducted and the physiology and genotype of the animal, are important concerns. For example, mice are nocturnal rodents that almost certainly use their visual systems to solve different problems under "light" and "dark" conditions. As light levels decrease from daylight conditions, receptive fields in the retina are effectively transformed from band-pass filters, to low-pass filters. As a consequence, scotopic vision (roughly approximated with rod-dominated, dark-cycle vision) is characterized by enhanced light detection at the expense of pattern detection, whereas, photopic vision (roughly approximated with cone-dominated vision during the light-cycle) is characterized by the ability to detect patterned stimuli at the expense of light sensitivity. Thus, asking pattern vision questions under scotopic conditions, or light sensitivity questions in photopic conditions, should be carefully considered. It should also be noted that most behavioral experiments that utilize visual cueing (i.e., spatial learning and memory tasks) in mice are conducted under photopic conditions and say little about the capacity of animals to use the visual system to solve the problems under scotopic conditions.

Mouse behavior is often modulated by light-driven circadian cycles, and mice are most active around the transitions between light and dark phases of the cycle. In most rodent vivaria, the transition between light and dark is abrupt and not gradual as it typically is in the real world, and the ability to measure behavior effectively under these "hyperactive" conditions is often problematic. In addition, mice are social animals that function at a high metabolic level and need social contact and regular sustenance in order to maintain normal behavioral function. Novel odors and sounds, an animal's sex, age, housing conditions, position in the estrous cycle, etc., should also be brought under experimental control to ensure stable baseline behavior. Genotype is also an important factor that can affect behavioral measures of vision. Gene products almost always support functions beyond the visual system, and so-called "performance effects" on behavior can compromise the ability to measure vision. If a mutation affects a component of the motor system that is required for a behavioral readout in a visual task, the measurement of vision may be impossible, or the data generated may be misleading.

As with any complex adaptive function, there is also no single measure that completely "captures" visual behavior. This lesson can be drawn from the measurement of visual acuity, which is probably the most widely used visual metric. The most widely accepted operational definition of acuity is that it is the perceptual threshold to distinguish two, high-contrast items as distinct (e.g., the ability to differentiate Mizar and Alcor, adjacent stars on the "handle" of the Big Dipper, is one of the first documented examples; a modern surrogate is the measure of Snellen acuity using size-decrementing lines of high contrast letters). Many think of visual acuity as the best measure of vision that distills to one number. However, the visual system is not one system, but a collection of many systems organized in a serial and parallel fashion, and acuity measures only a subset of the functions of those systems under optimal lighting and contrast conditions. As impressive as this function can be (i.e., acuity in raptors), an equally impressive function is the ability of the visual system to maintain relatively constant spatio-temporal perception under radically (on the order of 100 million-fold) different lighting conditions, which measures of contrast and temporal sensitivity best capture. Consequently, measures of acuity should be considered as one of many possible measures of function to capture a visual phenotype, and the appropriate visual measure should be tailored to the specific experimental question. Additionally, the spatial frequency (SF) thresholds of different visual behaviors (see below) do not all measure acuity.

Experimental questions are also often explicitly directed at the function of a specific structure and are not compatible with a system-wide measure of function. If a manipulation of the lateral geniculate nucleus (LGN) is the critical variable in an experiment, for example, a measurement of vision that depends on the collective integrity of the retina, LGN and visual cortex (V1; as perceptual measures of acuity do), may not provide useful information, since manipulations of any of the structures may result in the same behavioral result. Likewise, an unanesthetized and freely moving animal is not compatible with many techniques that require stability of the visual system (i.e., electrophysiological recordings, optical imaging, etc.), or when temporal and spatial restriction of a visual stimulus is required (i.e., mapping cellular receptive-field properties). Technical advances may enable such techniques to be applied in awake, behaving animals in the future, but at present, a moving animal is not compatible with such methods.

As with any useful technique, measuring visual behavior requires an investment in technology and technical expertise. We have often encountered the preconception that measuring behavior is easy, that generating data should be fast and simple, and that the measurement of behavior should be easily automated. Although behavioral measurements of vision can be relatively fast and simple, and technological advances have been applied effectively (see below), such advances are usually the result of the hard work of talented and trained people and the development of sophisticated technology; fast and simple measurements of vision carried-out without attention to detail can be misleading at best, and possibly worse than not measuring vision at all. As a consequence, anyone deciding to include measures of visual behavior in their experimental repertoire should be prepared to invest significant resources to the effort. There are solid grounds for including measures of vision in many experiments today, including in the phenotyping of mouse visual behavior, and this rationale will not subside in the future for the reasons outlined above. Consequently, a laboratory that makes a long-term investment in behavioral methodologies to measure vision will undoubtedly reap significant rewards.

Testing methodologies: measures of visual perception in reinforcement-based tasks

The most practical way to approach the behavioral measurement of mouse vision is to differentiate between tests that measure *whether* an animal can see, and *what* it can see. Often, the test of a hypothesis only requires an endpoint measure of whether a mouse's visual system can detect light, and simple tests such as the pupillary reflex, light avoidance response, light-induced circadian phase shifts, etc., may suffice. In most cases, however, a measure of the quality of vision would be most informative.

The most common approach to quantify behavioral function is to use a reward (reinforcer; i.e., food, water, escape, etc.) as a means to shape a desired behavior. The first reinforcement-based task for measuring vision in rodents was the Lashley jumping stand (Lashley, 1930), which had many pitfalls as a practical method for the measurement of rat vision and to our knowledge was never successfully adapted to the mouse. As mice have supplanted rats as experimental rodent models, numerous attempts have been made to convert "rat" tasks to mouse versions, apparently on the assumption that mice are simply small versions of rats. However, mice and rats are very different animals with different adaptive traits, and mice seem to be less prepared to learn many behavioral tasks (Whishaw, 1995). Consequently, far fewer reinforcement-based behavioral methodologies have been successfully applied in mice than in rats.

Despite this, a number of clever instrumental learning tasks have been developed for mice. These include an adaptation of a three-alternative, forced-choice discrimination task (Jacobs et al., 1999) with which mouse spectral sensitivities have been measured (Jacobs et al., 2004). This and other similar tasks (i.e., Gianfranceschi et al., 1999) have not gained wide popularity, however, likely because the shaping and testing procedures, even in the most cooperative mice, require a considerable amount of time (weeks to months), and in our hands, many strains are effectively untrainable in such appetitive-based tasks.

The Morris water maze (Morris et al., 1982), which was originally developed as a rat visuo-spatial learning task, has also been modified to measure mouse vision. Laboratory rodents are instinctive swimmers and are motivated to escape from water to a solid, submerged platform in a large circular pool of water, the position of which is relatively stable with reference to the constellation of surrounding visual cues. Animals can readily learn to use the configuration of these cues to locate the platform and swim to it. Since no appetitive motivation is required, adaptations of the task provided for the possibility of accelerating the efficiency of visual measures.

The pioneering experiments that measured place learning and memory in the water maze anticipated that interpretations of the data could be confounded by abnormal visual function (i.e., "performance" effects). Thus, a configuration of the task using a visually cued platform was developed to control for reduced visual function. The rationale was that if animals were impaired in swimming directly to the visual cue, they probably had visual impairments; conversely, if animals could swim directly to the cued platform, they were thought to be relatively free of visual deficits, at least to the extent that a deficit would affect the interpretation of learning and memory results. We explicitly tested this hypothesis in rats by measuring place learning in the water maze in animals with an independently measured 30% reduction of vision (Prusky et al., 2000b), and found that they had a significant impairment in learning a spatial version of the task, but were not impaired in their ability to locate a cued platform. Thus, animals with even a modest reduction of acuity can show "performance" effects, and the cued platform version of the Morris water maze task is not sufficient to identify them.

Problems with the cued platform task include that there is little stimulus control and viewing distances are hard to monitor in a moving animal, which are important variables for quantifying spatial vision and thus, the task is essentially a signal detection task not a visual discrimination task. By example, any image can be deconstructed as a set of sine waves of different SFs, and an organizational principle of the visual system is that it contains receptive fields organized to respond to the magnitude and location of the different SF components. Thus, the ability to identify cues and locate a place in visual space is limited by sensitivity to the highest spatial frequencies. In a typical cued platform task, however, the single platform cue typically appears dark against a white pool wall, and that object can be decomposed into many frequencies, any of which can be used to identify the platform. In the absence of distractors or multiple visual cues that must be discriminated then, detection of a single large cue is possible with vision utilizing only the lowest SF detectors. In addition, performance in the cued platform task is not a good measure of visual acuity because the cue is usually too large and there is little control of the viewing distance. For example, even from the most distant viewing position in a typical 1.5 m pool, a 10 cm cue will subtend about 5° of visual angle, roughly corresponding to a grating acuity of 0.1 cycles/degree (c/d), i.e., well below the threshold that normal mice can discriminate. In an effort to control these variables, a configuration of the Morris water task and an accompanying training procedure was developed to measure visual acuity in mice (Robinson et al., 2001). With it, mice were trained to discriminate between two cards displaying square-wave gratings of very different SFs, and the cards were presented in adjacent quadrants of the pool separated by a barrier, with an escape platform located in front of the card displaying the higher SF. When cards were systematically exchanged to make the discrimination more difficult and the swimming pattern of the animals was analyzed, the threshold for mice was estimated to be ~1.3 c/d (Robinson et al., 2004).

Although swimming to an exit platform and using visual cues in the Morris water maze is natural for the mouse and enables the relatively rapid training and testing of visual thresholds, the configuration described above still has drawbacks for measuring visual acuity. These include that even with the analysis of swimming patterns, it is not known at what distance from the cues animals make their choices. That the reported acuity of mice in this task is about twice that reported with other tasks (Gianfranceschi et al., 1999; Prusky et al. 2000c), indicates that mice are making decisions much closer to the stimuli than the video analysis indicated. In addition, printed cards restrict the range of stimulus options, and the manual exchanging of the cards practically limits the throughput of the task.

Our realization that the motivation and reward structure of a water-based escape task could enable more efficient training and testing of animals, and that computer-controlled stimulus presentation could better manage experimental logistics, prompted us to develop the visual water task (VWT; Prusky et al., 2000c). As shown in Figure 6.1, a trapezoidal-shaped tank is configured into a Y-maze with the addition of a central divider, and computer monitors are situated behind a glass wall and project into the end of each arm. A platform is submerged below the monitor that displays a stimulus that will be reinforced (i.e., sine wave grating), and nothing is placed below the other monitor, which will display a non-reinforced stimulus (i.e., gray). Over trials, mice learn that the stimuli on the screens predict the location of the platform, and they either stop at the end of the divider and inspect both screens, or pause near the end of the divider while alternately looking at the screens, before making the choice to swim into one of the arms. Thus, the end of the divider sets a choice plane enabling the length of the divider to set the SF of the stimuli. In addition to providing a motivator, the water aids in dispersing odor trails and focuses the mouse's attention on the visual stimuli without generating a great deal of stress. On land, mice have multiple sensory inputs to consider, but in water, vision is the only potential cue that can be used to distinguish the cues that are at a distance. The use of computer-generated stimuli is also advantageous, since it allows the presentation of stimuli that are difficult if not impossible to print. In addition, software control facilitates the automatic interleaving of stimuli and animals over trials, which greatly increases experimental throughput. The collective benefit of these features is that the measurement of visual thresholds can be completed relatively efficiently, and the VWT has

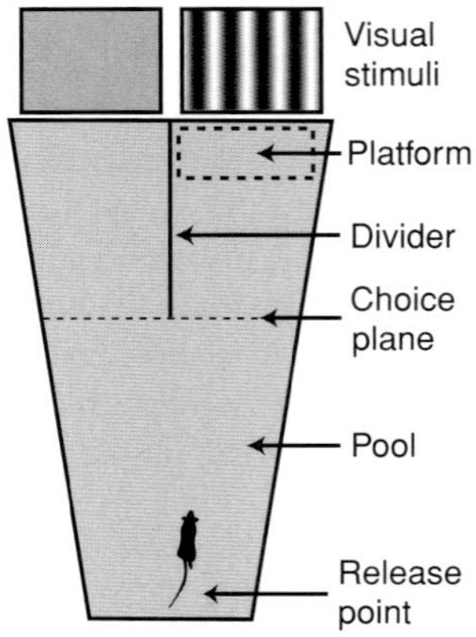

Figure 6.1 Visual water task schematic. A trapezoidal-shaped pool with a midline divider bisecting the wide end creates a Y-shaped maze. Water is added to the pool and monitors displaying visual stimuli face into each arm through a glass wall. A submerged platform is positioned below the stimulus that will be reinforced (i.e., grating). The left/right location of the reinforced and non-reinforced (i.e., equiluminant gray) stimuli is varied pseudo-randomly over trials. Mice are released into the pool at the narrow end and then swim toward the stimuli. They inspect each stimulus from a vantage near the end of the divider and then choose to swim into one of the arms. If they select the arm that displays the reinforced stimulus, they are rewarded quickly with escape from water and the trial is scored as correct. If they choose the non-reinforced stimulus (by crossing the choice plane), they are compelled to swim until they find the escape platform on the opposite side, and the trial is scored as incorrect. (Adapted from Prusky et al., 2000.)

been widely embraced as a standard measure of mouse visual perception.

We have measured spatial contrast sensitivity (see Figure 6.7d, later in this chapter), as well as grating acuity with two different sets of discriminanda, and have found that the thresholds are limited by the function of visual cortex (Figure 6.2a; Prusky and Douglas, 2004). Others have used analogous configurations of the task to screen mouse strains for spatial visual function/dysfunction (Wong and Brown 2006, 2007, 2008), and have found significant strain variation. We have also used the discrimination of moving dot kinematograms, revealing that mice possess local and global visual motion processing systems (Douglas et al., 2006) that are analogous to those in "higher" mammals. In addition, we have measured the threshold for mice to discriminate the size (visual angle) of sine wave gratings to gain insight into areal processing in the retina (Figure 6.2b).

That C57BL/6 mouse acuity at ~0.56 c/d is over 50 times lower than human 20/20 visual acuity of ~30 c/d, at first glance appears to confirm the widespread belief that mice are poor visual models. However, the relatively low acuity does not mean that mouse vision is not useful. For example, much of the visual information that is relevant to a mouse is within 30 cm, much closer than what humans routinely utilize to guide their behaviors. Small objects at a distance may not be as salient to a mouse, but large objects may be helpful for navigation in photopic conditions, as is evident from mouse performance in the Morris water maze. In Figure 6.3, we model stimulus visibility at an intermediate distance, 1 m, and illustrate what a cat and a mouse

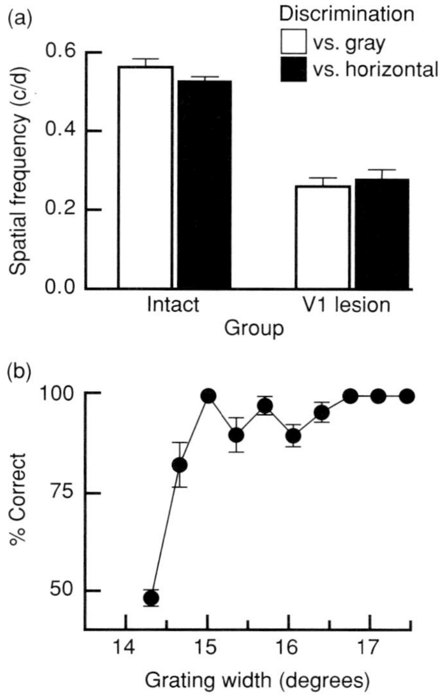

Figure 6.2 Examples of visual water task data. (a) Vertically oriented sine wave grating (reinforced stimulus) versus equiluminant gray (non-reinforced stimulus; open bars), and vertical grating (reinforced) versus horizontal grating (non-reinforced; black bars) discrimination spatial frequency (SF) thresholds (acuity) in intact adult C57BL/6 mice, and the effect of bilateral V1 lesions. Thresholds between the tasks were similar and were reduced by ~50% following lesions, indicating that acuity is limited by the function of visual cortex. Vertical lines at the top of each bar in this and other histograms represent +SEM. (Adapted from Prusky and Douglas, 2004.) (b) Effect of varying the two-dimensional size of the discriminanda (vertical grating (0.315 cycles/degrees) vs. gray) on task performance. Performance fell when the dimensions of the grating fell below 15°, revealing the likely minimum retinal surface area required for normal computations related to SF processing. Vertical lines on the data points in this and other scatterplots represent ±SEM.

Figure 6.3 Model of the vision of the mouse (adult C57BL/6) vision, and of a common mouse predator, the cat. Mouse acuity is about 10-fold below that of the cat. Panels (a) and (d) are original photographs and maintain as much resolution as printing makes possible. Panels (b) and (c) have had the spatial frequencies (SFs) above the two species' thresholds removed, assuming a viewing distance of 1 m; the approximate distance a cat might pounce on a mouse. This was achieved in Photoshop by reducing the size in pixels of each image to 2 × (width in degrees)/SF, and then expanding the image to the original size using bicubic interpolation. This ignores changes in contrast sensitivity as a function of SF, but nonetheless enables a direct comparison of what the two animals could see. The image of the cat (b) would be discernible to the mouse. A second set of comparisons can be made horizontally. Panel (c) shows what a mouse would look like to another mouse at 10 cm, whereas (b) is how a cat would appear to another cat 10 m away. These simple simulations suggest that vision can be very useful to the mouse in dealing with its local environment. Courtesy of Dr. Robert M. Douglas.

might see of each other, by removing all SFs in an image above each species' threshold. Whereas the cat's view of the mouse (Figure 6.3c) is better than the mouse's view of the cat (Figure 6.3b), in fact the mouse can see the cat surprisingly well. Furthermore, Figure 6.3c also presents what a mouse might see of another mouse 10 cm away: such vision may well be used adaptively by mice in their social interactions or to interact with objects in their niche.

Testing methodologies: measures of automated visual responses

The visual system is most often characterized as a sensory system. However, from a functional point of view, the visual system is more accurately defined as an integrated sensory-motor system. Normal function is not possible without a variety of visuomotor reflexes that enable relative image stability on the retina. Because these reflexes are automatic, they provide the opportunity to measure visual function under circumstances in which reinforcement training is either impractical or precluded. One of these reflexes in rodents, optokinetic tracking (OKT), compensates for the motion of the visual field, using the movement of a visual scene on the retina to induce head movements in the same direction and at about the same velocity. Rodent OKT has been studied for some time using a mechanical apparatus (Cowey and Franzini, 1979) consisting of a drum with printed stimuli on the inside wall and which rotates around the animal, generating compensatory head movements that track the rotation. Post-hoc video analysis of mouse tracking behavior in this task has been used to screen for visual abnormalities under scotopic and photopic conditions (Abdeljalil et al., 2005). A variant of the task has also been used to automatically track the angular speed of mice in relation to the center of the drum, showing that mice are most responsive to bright stimuli, as well as the surprising finding that rods are not saturated in the mesopic or low photopic range (Schmucker et al., 2005). In addition, other research has shown that the mouse OKT system has sufficient spatial resolution to respond to refractive errors induced with spectacle lenses or diffusers (Schmucker and Schaeffel, 2006).

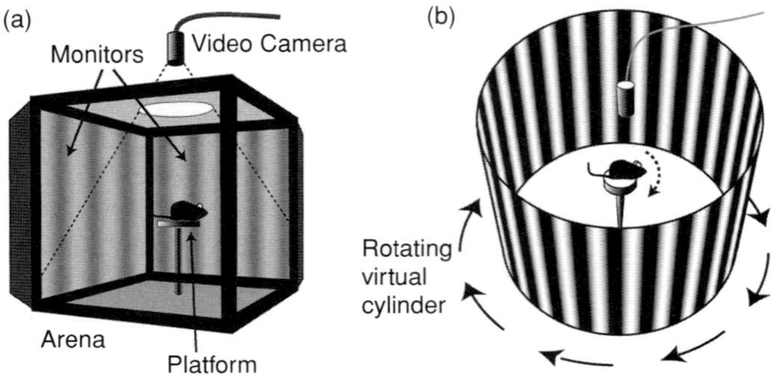

Figure 6.4 Virtual optokinetic system. (a) Cutaway schematic diagram. A testing arena is created by mounting four, inward-facing LCD computer monitors, on a frame. An unrestrained mouse stands on a circular platform in the epicenter of the chamber. A camera mounted above is used to image the animal. (b) Optokinetic stimulus. A virtual cylinder comprising a sine wave grating is calculated by computer software (OptoMotry; CerebralMechanics.com) in three-dimensional coordinate space and projected by the monitors. At the animal's viewing position, an illusory cylinder rotates outside the arena. The spatial frequency (SF), contrast, and direction of rotation of the grating can be changed in real-time. An experimenter uses the video image to maintain the hub of cylinder between the eyes as the animal moves about the platform (maintaining SF), and to judge when the animal tracks (dotted arrow) the cylinder movement. (Adapted from Douglas et al., 2005.)

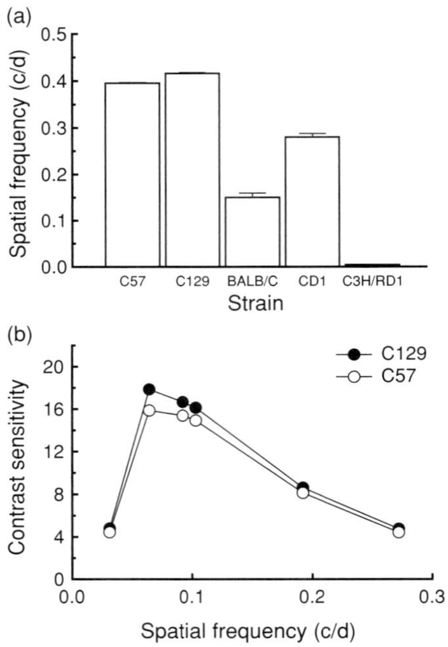

Figure 6.5 Strain comparisons of visual thresholds generated in the virtual optokinetic system. (a) Spatial frequency (SF) thresholds for adult C57BL/6 (most common experimental mouse), C129 (common background strain for knockout and transgenic manipulations), BALB/c (albino), CD1 (albino), and C3H/rd1 (retinal degeneration) mouse strains. Albinism in rodents is known to independently cause reduced SF thresholds (Prusky et al., 2002), as is retinal degeneration (McGill et al., 2004). No optokinetic tracking threshold can be recorded in adult C3H/rd1 mice. (b) Contrast sensitivity comparison of C57BL/6 and C129 strains reveals that vision can vary as a function of SF (i.e., peak contrast sensitivity and spatial frequency threshold (a) is better in C129, whereas contrast sensitivity at other spatial frequencies is similar between the strains).

Notwithstanding the utility of these tasks, the difficulty of controlling the speed of the drum and its position in relation to the animal, combined with the problem of reproducing precise visual stimuli and exchanging them rapidly, has limited the practicality and thus, the popularity of mechanical OKT devices for testing mouse vision. We have addressed these problems by developing a virtual optokinetic system (VOS; Prusky et al., 2004; Figure 6.4). A virtual cylinder displaying a vertical sine wave grating is projected in three-dimensional coordinate space on a quadrangle of computer monitors around a testing arena. An unrestrained mouse stands on a platform in the epicenter of the arena, the mouse is imaged from above and the head is tracked continuously with a computer mouse and a crosshair superimposed on a video image of the arena. The positional coordinates of the crosshair in the video frame are used to center the hub of the virtual cylinder, enabling the cylinder wall to be maintained at a constant virtual "distance" from the animal's eyes and thereby "clamping" the SF of the grating at the animal's viewing position. When the cylinder is rotated and the mouse follows with head and neck movements tied to the stimulus movement (OKT), it is judged by an observer that the animal's visual system can distinguish the grating. The tracking behavior is relatively stereotyped and we find the variability in thresholds to be lower than that generated with the VWT.

In order to measure the SF threshold of OKT, we normally project a homogeneous gray stimulus on the cylinder at the beginning of each testing session, and the experimenter waits until the animal stops moving, and then gray is switched to a low SF (~0.04 c/d), high-contrast (~100%) sine wave grating moving in one direction. The animal is then assessed for tracking behavior for a few seconds, after which the gray stimulus is restored. After multiple examples of tracking are observed, the SF of the grating is increased incrementally, with multiple tracking episodes at each SF, until a SF is reached when tracking can no longer be elicited. We have reported that the OKT SF threshold of normal C57BL/6 mice is ~0.4 c/d (Figure 6.5a; Prusky et al., 2004, 2006). This value is lower than that obtained with the VWT (left bars Figure 6.2a vs. left bar Figure 6.5a), which is not a surprise since the threshold reflects the function of subcortical visual circuitry (Douglas et al., 2005), rather than cortical function as the VWT does (Figure 6.2). Reduced thresholds do not reflect a lower sensitivity, however; contrast sensitivities measured in the VOS are significantly better than those generated in the VWT (see Figure 6.7d, later in this chapter). Another major difference between the two measures is in the speed to obtain thresholds; the contrast sensitivity curve from the VWT required testing on the order of months, whereas the VOS curve was generated on the order of minutes (see Figure 6.7d, later in this chapter).

Thresholds through each eye can be measured independently in the VOS, simply by reversing the rotation of the cylinder. By example, when one eye is temporarily occluded, only motion in the temporal-to-nasal direction for the other eye evokes tracking; when the maximal SF capable of driving a response is measured separately under monocular and binocular viewing conditions, the monocular threshold is identical to the binocular threshold measured with the same direction of rotation (Douglas et al., 2005).

Although there are currently several thousand strains of mice (including knockouts), the most commonly employed experimental pigmented strain is C57BL/6 (Charles River; B6), and we have used it for most of our behavioral experiments. We have measured the VOS SF threshold for adult B6 mice numerous times and have found thresholds consistently near 0.4 c/d (Figure 6.5). The B6129F1/Tac (C129) mouse is the most commonly used strain for the production of genetic knockouts (which are often bred back into B6 stock), and we have found their threshold to be comparable, if not slightly better than that of B6 animals. We have also measured contrast sensitivity in the strains (Figure 6.5b), and have found differences that vary as a function of SF; at SFs near the peak (0.06 c/d), contrast sensitivity is better in C129, whereas at other SFs, the values are similar. This variation likely reflects discrepancies between the two strains in the organization and function of SF channels, indicating that identifying differences in vision between strains can be complicated.

We have previously reported that albinism leads to reduced SF thresholds in rats (Prusky et al., 2002), and report here the VOS SF threshold in two albino mouse strains, BALB/cAnNCrl (BALB/c) and Crl:CD1(ICR) (CD1). Whereas the SF threshold in both strains is lower than that recorded in pigmented C57 and C129 strains (Figure 6.5a), CD1 mice have a substantially better threshold than BALB/C mice. Mutations that affect retinal degeneration in mammals are common and we have also measured VOS SF thresholds in C3H/rd1 strain (with massive retinal degeneration by mice that were derived from juvenile life natural mutation) and found no detectable OKT in either strain in adulthood, confirming their visual dysfunction. Although the list of strains in Figure 6.5 is not exhaustive, the selection serves to illuminate that substantial strain variation in visual function is to be expected in naturally-occurring and mutant mouse strains, and that albinism and retinal degeneration can dramatically reduce thresholds. Thus, behavioral measures, such as the measurement of OKT in the VOS, can be used as a relatively rapid method to characterize visual phenotypes.

Another potential source of experimental variance is that which can result from the use of different experimental procedures within the same behavioral task. An example of this is manifested in the apparent discrepancy between VOS SF thresholds generated using "direct" observation of tracking behavior (tracking judgements are made while gratings are visible; i.e., Prusky et al. 2004a, b, c; Douglas et al., 2005), and experimentally "blind" testing procedures (experimenter's view of gratings is occluded when judging OKT; i.e., Umino et al., 2006, 2008). Although "blind" testing procedures have the potential to reduce observer bias, they require much more time to execute in the VOS and have generally resulted in higher thresholds and greater variation than direct observations. We set out to understand the source of these differences, hypothesizing that movement of the animal during the typical 5-second observation period in the "blind" testing procedure, could induce a measurement error. To do this, we varied the observation period during "blind" testing from 5 to 1 seconds (the approximate duration of a tracking event in animals that are not otherwise moving), and compared the values to those generated in the same animals using the "direct" observations (Figure 6.6). We found that a five-second testing epoch generated higher thresholds and greater variation with blind testing than with the "direct" mode of observation. In addition, as the duration of the viewing epoch was reduced incrementally to 1 second, the thresholds and variation approached those of the "direct" measurement procedure. This provides strong evidence that movement of the animal during the observation period can place them at varied distances from the viewing screens (most movements are lateral or forward, placing them closer to the screens), or that behaviors not directly tied to a retinal image stabilizing reflex can introduce measurement error. Whereas there is probably no "best" way to measure visual thresholds in VOS because "direct" and "blind" methods have complementary strengths and weaknesses, the above example serves to illustrate that the evaluation method should be carefully considered when choosing a procedure and when comparing data in the literature.

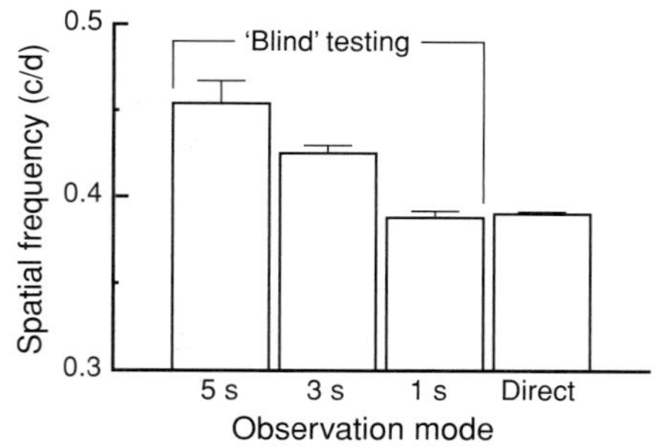

Figure 6.6 Visual thresholds obtained in the virtual optokinetic system (VOS) can vary with experimental procedures. Variation in spatial frequency (SF) thresholds of mice recorded in the VOS have been reported between "blind" testing procedures (i.e., experimenter view of gratings occluded when judging optokinetic tracking) and procedures in which tracking judgements are made while gratings are visible. Although "blind" testing procedures are desirable for reducing the potential for observer bias, they take more time to execute and have generally resulted in higher thresholds and greater variation than direct observations when using the VOS. We investigated the nature of these differences by varying the interval of time in which behavior is observed before making a judgment (typically a 5 second epoch) in a "blind" configuration, and with a method in which the stimuli were not occluced (direct). We confirmed that a 5-second epoch in the "blind" configuration generated higher thresholds and greater variation (SEMs) than using the "direct" mode of observation. However, as the duration of the viewing epoch was reduced to 1 second (the approximate duration of a tracking event when the animal is immobile; normal mice in the VOS rarely remain still for more than a few seconds), the thresholds and variation were also reduced. These data indicate that movement of the animals during the observation epoch (placing them at varied distances from the viewing screens and possibly inducing behaviors not directly tied to a retinal image stabilizing reflex) when using "blind" testing procedures measures can induce a measurement bias.

The potential use of the VOS is not limited to the measurement of SF thresholds, and other visual thresholds can be quantified. We noticed, for example, that mice often look directly forward at information of interest (i.e., "foveate"), and thus selectively use their binocular field. In order to quantify whether binocular and monocular visual functions differ, we have used a modification of the VOS testing configuration to measure SF thresholds within restricted segments of the visual field. The general procedures described above were used, with the addition that the grating was limited to a selected region of the visual field. We monitored the gaze of the animals by tracking the horizontal orientation of the head with the aid of a line superimposed on the video frame and oriented along the long axis of the head. The vector generated by this procedure was then used by the software to maintain in real time, the grating patch at a constant angle relative to the head. The central 30–40° of each visual hemifield (~110° from midline) in normal mice receives input from both eyes (Wagor et al., 1980), and since cells in V1 with receptive fields outside the central 25° in each hemifield are less likely to be driven by the ipsilateral eye (Gordon and Stryker, 1996), we limited our stimulation to 50° straddling the midline. For measuring thresholds within the monocular zone we centered the grating 65° from the midline (Prusky et al., 2006). Figure 6.7a illustrates the measurement of SF thresholds in various visual fields in normal adult C57BL/6 mice, revealing that the monocular field threshold is lower than that in the binocular field, confirming a functional specialization in the frontal visual field. This is interesting in light of the general lack of evidence for a fovea in the mouse.

Since the virtual cylinder in the VOS can be rotated in any orientation around the animal, we have used a modification of the VOS to measure the orientation sensitivity of OKT. The data in Figure 6.7b provides evidence of an "oblique effect" in the SF threshold, with decreased responsivity around 45° from horizontal or vertical, indicating that there is a specialization in the OKT system for receptive fields tuned to vertical orientations.

Stimulus velocity can also be varied in the VOS and Figure 6.7c illustrates that SF thresholds do not vary appreciably between 2 and 20°/s, and then decrease with faster speeds. We have typically used 12°/s as the cylinder velocity to measure VOS thresholds in our experiments, and these data indicate that experimental variations in the SF threshold (i.e., Figure 6.7b) are not due to changes in temporal frequency tuning. As referenced above, contrast sensitivity curves can also be produced by measuring minimum stimulus contrast over a range of SFs that generated OKT. Figure 6.7d plots a contrast sensitivity curve generated from C57BL/6 mice in the VWT as a reference (adapted from Prusky and Douglas, 2004), as well as a VOS-generated curve. Although the general characteristics of the curves are comparable (i.e., typical mammalian inverted "U" shape), the overall and peak sensitivities differ between the two, suggesting that each methodology measures the function of different visual circuits. Since a contrast sensitivity curve can take several months to generate using the VWT, the VOS methodology provides the only practical method to measure contrast sensitivity in mice behaviorally.

Combining measures of vision with experimental manipulations of cellular or experiential function can also provide a

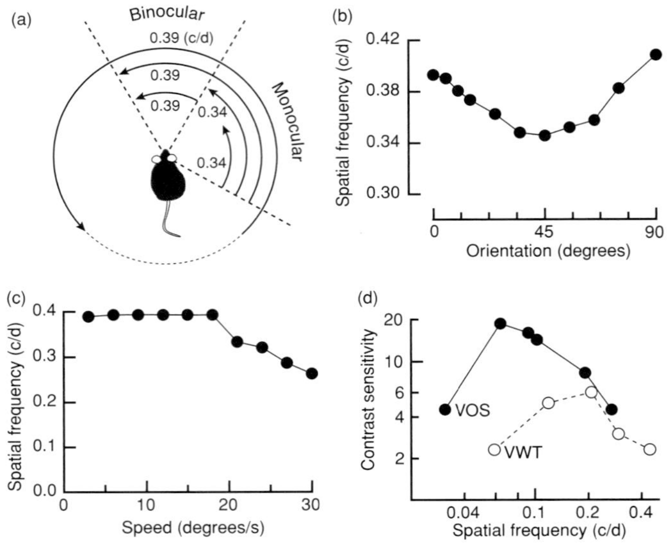

Figure 6.7 Alternative visual thresholds generated in the virtual optokinetic system (VOS) and comparison with the visual water task (VWT). (a) Selective visual-field responses of normal C57BL/6 mice recorded by monitoring gaze, in schematic illustration. (Adapted from Prusky et al., 2006.) The length of arching arrows represent the portion of the visual field in which the grating was present during testing (binocular = binocular field; monocular = monocular field), and the adjacent numbers represent spatial frequency (SF) thresholds in cycles/degree (c/d). Full-field (outside arrow; dotted segment represents the region of the field not visible to the eyes) stimulation generated a threshold of 0.39 c/d. Limiting the stimulus to the binocular-field resulted in the same threshold as full-field stimulation (0.39 c/d); monocular-field responses (0.34 c/d) were lower regardless of the size of the stimulus within the field. Combining monocular and binocular stimulation resulted in the same threshold (0.39 c/d) as binocular alone or full-field stimulation. These data indicate that the central-temporal retina (which views the binocular field) may have a higher acuity than the nasal retina. (b) Effect of varying the orientation of the grating in the VOS from vertical (0°) to horizontal (90°), through oblique (45°), on SF threshold. Threshold is reduced at oblique orientations, and is slightly higher at horizontal, indicating orientation tuning of the optokinetic tracking (OKT) system. (c) Effect of stimulus velocity on SF threshold. Threshold does not vary between 2 and 20°/s, but decreases thereafter, indicating that experimental variations in the SF threshold at 12°/s (the most convenient velocity to use for OKT experiments) do not represent changes in temporal frequency tuning. (d) Contrast sensitivity curve generated in intact adult C57BL/6 mice. (Adapted from Prusky et al., 2004.) Dotted line represents contrast sensitivity curve generated with the VWT. (Adapted from Prusky and Douglas, 2004.) General and peak sensitivity, and SF thresholds (predicted points at which curves intersect the *x* axis) differ between tasks (i.e., the VWT and the VOS measure different visual system functions).

simple means to expand the characterization of visual function in mouse strains with naturally-occurring or directed genetic modifications. For example, systemic (scopolamine) or ocular (atropine) treatments with drugs can be implemented in a repeated measures design in the VOS, providing a rapid and quantifiable way to distinguish effects on central or retinal function. In addition, we have reported (Prusky et al., 2006) that simple experiential manipulations of visual function, such as eyelid suturing, can lead to long-term changes in vision. For example, monocular and binocular deprivation during a "critical" period in juvenile life can lead to loss of acuity (amblyopia; Prusky and Douglas, 2003) as measured in the VWT. In addition, we have reported that monocular deprivation in adult mice can lead to enduring enhancement of OKT if the deprivation is combined with selective OKT testing experience (Prusky et al., 2006).

Distinct vision-testing methodologies

Perceptual and reflexive measures of mouse vision are practicable, but each evaluates different visual functions and has inherent strengths and weaknesses. Perceptual methodologies, such as the VWT, currently have the advantage that they measure thresholds most comparable with clinical measures of vision, and the SF threshold (i.e., acuity) is limited by visual cortex function. Reinforcement-based tasks in general, however, require a significant investment of time (on the order of weeks for each threshold) and training young mice, or mice with low vision, can take even longer. In contrast, measuring automated responses, such as the OKT SF threshold, can be accomplished in only a few minutes (depending on the experimental procedure; i.e., Figure 6.6) and be repeated daily. However, the SF threshold is mediated through subcortical, not cortical circuits, it is not a measure of visual perception (possibly less clinical relevance) and the thresholds are lower than that generated in the VWT.

Practical application of mouse visual behavior

The relatively recent introduction of behavioral methodologies to measure mouse vision has facilitated a surge of research, and has brought the realization that the mouse has a sophisticated visual system with great potential as an experimental model; it is currently the mammal and brain system in which the widest range of experimental manipulations and quantitative measures of function can be undertaken.

One of the most important applications of behavioral tests of mouse vision in the future will be for the characterization of visual phenotypes. Too often this process has involved large numbers of animals, and there has been a trade-off between the time it takes to test animals, and the rigor or quality of visual testing (i.e., it is often not feasible to generate a psychometric function, since it requires a number of weeks to complete). In an effort to balance these tradeoffs, we suggest a "staircase" approach to testing mouse vision as a practical alternative, in which it is first determined *whether* a mouse can see, and then the stringency of the tests are incrementally increased to ultimately determine *what* it can see. Optokinetic tests are probably best suited to the question of *whether* a mouse can see, because if an unanesthetized, freely moving animal is placed in an OKT device and it tracks a salient stimulus (i.e., low SF grating), it is clear that the animal can see. The process would be fast (seconds–minutes), and would identify animals with any gross visual problems. More rigorous and specific testing of OKT could then follow, such as measuring SF thresholds, contrast sensitivity, contrast function, visual fields, temporal, or orientation functions. If visual abnormalities were not identified by these measures, then more time-consuming tests of visual perception, such as acuity, might be appropriate. Such a strategy would avoid devoting time to animals with obvious abnormalities (unless that was the goal of the experiment), which would significantly improve throughput. Other applications of the staircase strategy could be to evaluate the effect on visual function of drug treatments with known actions or the response of the visual system to experiential manipulations.

The methodologies and approaches outlined in this chapter provide a means for behavioral measures of vision not only to corroborate the results of cellular or physiological studies, but also to enable the generation of specific hypotheses to drive experiments at another level of analysis, enabling the discovery of novel visual system functions. Such an approach can be applied to the mouse, which is currently the only mammalian model that provides an opportunity to address research questions from molecules to behavior.

References

Abdeljalil, J., Hamid, M., Abdel-Mouttalib, O., Stephane, R., Raymond, R., Johan, A., et al. (2005) The optomotor response: a robust first-line visual screening method for mice. *Vision Res* 45: 1439–1446.

Cowey, A. and Franzini, C. (1979) The retinal origin of uncrossed optic nerve fibres in rats and their role in visual discrimination. *Exp Brain Res* 35: 443–455.

Douglas, R.M., Alam, N.M., Silver, B.D., McGill, T.J., Tschetter, W.W., and Prusky, G.T. (2005) Independent visual threshold measurements in the two eyes of freely moving rats and mice using a virtual-reality optokinetic system. *Vis Neurosci* 22: 677–684.

Douglas, R.M., Neve, A., Quittenbaum, J.Q., Alam, N.M., and Prusky, G.T. (2006) Perception of motion coherence by rats and mice. *Vision Res* 46: 2842–2847.

Gianfranceschi, L., Fiorentini, A., and Maffei, L. (1999) Behavioural visual acuity of wildtype and bcl2 transgenic mouse. *Vision Res* 39: 569–574.

Gordon, J.A. and Stryker, M.P. (1996) Experience-dependent plasticity of binocular responses in the primary visual cortex of the mouse. *J Neurosci* 16: 3274–3286.

Jacobs, G.H., Fenwick, J.C., Calderone, J.B., and Deeb, S.S. (1999) Human cone pigment expressed in transgenic mice yields altered vision. *J Neurosci* 19: 3258–3265.

Jacobs, G.H., Williams, G.A., and Fenwick, J.A. (2004) Influence of cone pigment coexpression on spectral sensitivity and color vision in the mouse. *Vision Res* 44: 1615–1622.

Lashley, K.S. (1930) The mechanism of vision: I. A method for rapid analysis of pattern vision in the rat. *J Gen Psychol* 37: 453–460.

McGill, T.J., Lund, R.D., Douglas, R.M., Wang, S., Lu, B., and Prusky, G.T. (2004) Preservation of vision following cell-based therapies in a model of retinal degenerative disease. *Vision Res* 44: 2559–2566.

Morris, R.G., Garrud, P., Rawlins, J.N., and O'Keefe, J. (1982) Place navigation impaired in rats with hippocampal lesions. *Nature* 297: 681–683.

Pham, T.A., Graham, S.J., Suzuki, S., Barco, A., Kandel, E.R., Gordon, B., et al. (2004) A semi-persistent adult ocular dominance plasticity in visual cortex is stabilized by activated CREB. *Learn Mem* 11: 738–747.

Prusky, G.T., Alam, N.M., Beekman, S., and Douglas, R.M. (2004) Rapid quantification of adult and developing mouse spatial vision using a virtual optomotor system. *Invest Ophthalmol Vis Sci* 45: 4611–4616.

Prusky, G.T., Alam, N.M., and Douglas, R.M. (2006) Enhancement of vision by monocular deprivation in adult mice. *J Neurosci* 26: 11554–11561.

Prusky, G.T. and Douglas, R.M. (2003) Developmental plasticity of mouse visual acuity. *Eur J Neurosci* 17: 167–173.

Prusky, G.T. and Douglas, R.M. (2004) Characterization of mouse cortical spatial vision. *Vision Res* 44: 3411–3418.

Prusky, G.T., Harker, K.T., Douglas, R.M., and Whishaw, I.Q. (2002) Variation in visual acuity within pigmented, and between pigmented and albino strains of rats. *Behav Brain Res* 136: 339–348.

Prusky, G.T., Reidel, C., and Douglas, R.M. (2000a) Environmental enrichment from birth enhances visual acuity but not place learning in mice. *Behav Brain Res* 114: 11–15.

Prusky, G.T., West, P.W.R., and Douglas, R.M. (2000c) Behavioral assessment of visual acuity in mice and rats. *Vision Res* 40: 2201–2209.

Prusky, G.T., West, P.W.R., and Douglas, R.M. (2000b) Reduced visual acuity impairs place but not cued learning in the Morris water task. *Behav Brain Res* 116: 135–140.

Robinson, L., Bridge, H., and Riedel, G. (2001) Visual discrimination learning in the water maze: a novel test for visual acuity. *Behav Brain Res* 119: 77–84.

Robinson, L., Harbaran, D., and Riedel, G. (2004) Visual acuity in the water maze: sensitivity to muscarinic receptor blockade in rats and mice. *Behav Brain Res* 151: 277–286.

Schmucker, C. and Schaeffel, F. (2006) Contrast sensitivity of wildtype mice wearing diffusers or spectacle lenses, and the effect of atropine. *Vision Res* 46: 678–687.

Schmucker, C., Seeliger, M., Humphries, P., Biel, M., and Schaeffel, F. (2005) Grating acuity at different luminances in wild-type mice and in mice lacking rod or cone function. *Invest Ophthalmol Vis Sci* 46: 398–407.

Umino, Y., Everhart, D., Solessio, E., Cusato, K., Pan, J.C., Nguyen, T.H., et al. (2006) Hypoglycemia leads to age-related loss of vision. *Proc Natl Acad Sci USA* 103: 19541–19545.

Umino, Y., Solessio, E., and Barlow, R.B. (2008) Speed, spatial, and temporal tuning of rod and cone vision in mouse. *J Neurosci* 2: 189–198.

Wagor, E., Mangini, N.J., and Pearlman, A.L. (1980) Retinotopic organization of striate and extrastriate visual cortex in the mouse. *J Comp Neurol* 193: 187–202.

Whishaw, I.Q. (1995) A comparison of rats and mice in a swimming pool place task and matching to place task: some surprising differences. *Physiol Behav* 58: 687–693.

Wong, A.A. and Brown, R.E. (2006) Visual detection, pattern discrimination and visual acuity in 14 strains of mice. *Genes Brain Behav* 5: 389–403.

Wong, A.A. and Brown, R.E. (2007) Age-related changes in visual acuity, learning and memory in C57BL/6J and DBA/2J mice. *Neurobiol Aging* 28: 1577–1593.

Wong, A.A. and Brown, R.E. (2008) Visual and other sensory abilities of mice and their influence on behavioral measures of cognitive function. In Chalupa, L.M. and Williams R.L. (eds.), *Eye, Retina, and Visual System of the Mouse*. MIT Press, MA, USA, pp. 13–34.

Chapter 7

Tactile system and nociception

Shad B. Smith and Jeffrey S. Mogil

Introduction

Individual differences

The systematic study of subjective pain experience has provided evidence of robust individual variability, in both humans and animals (see Mogil, 1999). The existence of individual differences in human pain perception has been appreciated since the work of Libman (1934), who reported the differential sensitivity in his clinical patients to pressure stimulation of the styloid process of the temporal bone towards the mastoid. Although rat strain differences have been repeatedly observed (see Xu and Wiesenfeld-Hallin, 2004), the genetics of pain has been studied much more extensively and systematically in the mouse. The use of the mouse has enabled researchers to reduce or eliminate many confounding environmental variables, and the wide range of genetic mouse models available (artificially selected lines; inbred and recombinant inbred strains; congenic and recombinant congenic lines; spontaneous, induced, and transgenic mutants) offers powerful tools to dissect complex phenotypes like pain.

Pain genes

One popular form of pain genetics attempts to identify all genes of relevance to nociceptive transmission and modulation. In this regard transgenic "knockout" mice have been useful. Over 195 different genes are now known to affect nociception or antinociception (analgesia) as determined by a knockout "phenotype." We have reviewed these experiments (and their limitations) on a number of occasions (Lariviere et al., 2001; Mogil and Grisel, 1998; Mogil et al., 2000a). The number of known "pain genes" is also increasing rapidly due to the publication of relevant microarray gene expression profiling studies (Barr et al., 2005; Costigan et al., 2002; Kim et al., 2001; Rabert et al., 2004; Sun et al., 2002; Valder et al., 2003; Wang et al., 2002; Xiao et al., 2002).

However, the genetics of pain as discussed below is concerned with the search for pain *variability* genes – those genes which have common allelic variants and small but additive and interacting effects on pain perception and/or pathology. Knockout and microarray studies are of limited utility to identify pain variability genes; only linkage (quantitative trait loci; QTLs) mapping and genetic association methodologies are directly relevant. Although at present only a small number of pain-relevant QTLs are known, and only in a subset of them have strong candidate genes been identified, our and others' efforts continue, guided by a number of principles of pain "phenomics" that we have discovered using Mendelian genetic techniques (see below).

Measuring nociception in the mouse

Pain can be described in terms of its duration (acute to chronic), intensity (mild to severe), location (somatic or visceral), qualitative nature (e.g., stabbing, burning), and emotional reactivity (e.g., annoying, frightening) (Melzack, 1975). Such an extensive pain vocabulary illustrates the heterogeneous nature of the human pain experience, and evidence suggests that qualitatively different neuroanatomical, neurochemical, and genetic mechanisms mediate different types of pain. Clinically, the main problems are chronic inflammatory pain (e.g., osteoarthritis) and chronic neuropathic pain (i.e., pain due to nerve damage), since acute pain states (e.g., postoperative pain) can generally be managed well with existing treatment strategies.

Nociceptive assays in laboratory rodents have been developed to model the different types of human pain; in all cases behaviors are observed and scored for their activation threshold, latency, frequency, and/or duration (Mogil, 2009; Wilson and Mogil, 2001). Pain modality may be classified as evoked – whether by mechanical, chemical, electrical or thermal stimuli – or spontaneous (i.e., nocifensive or recuperative) behavior. The type and intensity of a stimulus (e.g., temperature, pressure, acid concentration) will determine which primary afferent neurons (Aδ or C fibers) are engaged to transmit the sensory information to the central nervous system. The location of the stimulus may also determine how the nociceptive information is processed, as cutaneous, subcutaneous, and visceral tissues are innervated by different types of afferent fibers expressing and releasing different neuromodulators. The nature and duration of the stimulus, whether acute, tonic or chronic, will determine whether sensitization (i.e., plasticity) occurs in peripheral and central pathways. When

Behavioral Genetics of the Mouse: Volume I. *Genetics of Behavioral Phenotypes*, eds. Wim E. Crusio, Frans Sluyter, Robert T. Gerlai, and Susanna Pietropaolo. Published by Cambridge University Press.

sensitization occurs, it may result in behavioral hypersensitivity: hyperalgesia (i.e., increased response to a noxious stimulus) and/or allodynia (i.e., response to a non–noxious stimulus) to thermal or mechanical stimuli. The behavioral reaction to the noxious stimulus may consist of reflex withdrawal behaviors (e.g., flexion/extension, writhe, flinch) or organized responses (e.g., vocalization, licking/biting, escape) to suprathreshold stimulation.

Major nociceptive assays that have been studied genetically are shown in Table 7.1. The reader is directed to recent papers from our group (Lariviere et al., 2002; Mogil et al., 1999a, 1999b; Wilson and Mogil, 2001) for further discussion of these techniques and their relevance and limitations.

Heritability of nociception: mouse studies

The mid 1960s saw the first use of inbred mouse strains in the exploration of the role of genetics in producing variability in nociception and analgesia (Takagi et al. 1966). Our group was the first to study this issue systematically, using a large set of strains. Current evidence suggests that nociceptive phenotypes are moderately to highly heritable in mice. Over the past decade, we have compared a panel of 12 inbred strains, on more than 20 nociceptive assays (Bon et al., 2003; LaCroix-Fralish et al., 2005; Lariviere et al., 2002; Mogil and Adhikari, 1999; Mogil et al., 1998, 1999a; Smith et al., 2004), representing a true "phenome project" (Paigen and Eppig, 2000) of pain. The 12 strains were chosen well in advance of the Mouse Phenome Project (http://phenome.jax.org), but include five members of the Mouse Phenome Database (MPD) "A" priority group (A/J, BALB/cJ, C3H/HeJ, C57BL/6J, DBA/2J), three members of the "B" group (AKR/J, C58/J, SM/J), and three members of the "C" group (C57BL/10J, CBA/J, RIIIS/J). We have also used 129P3/J as a representative of the 129 strains. Heritability estimates for strain surveys performed on assays shown in Table 7.1 range from $h^2 = 0.30$ for carrageenan-induced thermal hyperalgesia, to $h^2 = 0.76$ in the abdominal constriction (acetic acid) test, with a median of $h^2 = 0.52$. These values agree rather well with heritability estimates observed in twin studies of human pathological pain states, such as migraine ($h^2 = 0.39$–0.58), menstrual pain ($h^2 = 0.55$), back pain ($h^2 = 0.50$), sciatica ($h^2 = 0.21$) and functional bowel disorder ($h^2 = 0.57$) (see MacGregor, 2004). To date, there are three published human twin studies of acute experimental pain. MacGregor and colleagues (1997) used a pressure pain stimulus applied to the forehead of identical and fraternal twins; they observed similar concordance rates in both types of twins, suggestive of a genetic contribution to the variance of only 10%. However, in a more recent twin study by this same group investigating thermal pain sensitivity, chemical (acid) pain sensitivity, and hypersensitivity after burn injury, heritability estimates from 20 to 90% were obtained for the various traits examined (Norbury et al., 2007). Heritability estimates in the same range as in mice (25–60%) were also found by a different group (Nielsen et al., 2008), studying thermal pain sensitivity and cold pressor (ischemic) pain.

The strain surveys of experimental pain also allowed us to estimate the number of effective genetic loci mediating the phenotypic responses (Falconer and Mackay, 1996). These values range from 1.9 (von Frey test) to 3.6 (mechanical hypersensitivity following peripheral nerve injury), although these are probably underestimates.

Heritability and number of effective genetic loci can also be estimated using selective breeding. To date, nociception has not been directly studied using selective breeding in mice (but see also Devor and Raber, 1990), but two bidirectional artificial selection projects have been undertaken to explore stress-induced analgesia and opioid analgesia. The rapidity of selection and segregation analyses also yielded evidence for surprisingly few effective genetic loci (Mogil et al., 1995a, 1995b). Very interesting findings regarding genetic correlations were also obtained from these breeding projects (see Mogil, 1999; Mogil et al., 1996a for a review), but the selected genes have never been identified.

Pain phenomics in the mouse

In advance of identifying any responsible genes, a number of strain surveys have been performed over the past decade. These have led to a number of general principles that seem to characterize the genetics of this particular biological domain.

Qualitative strain differences

Mouse strains differ not just in the magnitude of behavioral response to a given noxious stimulus, but sometimes in qualitative aspects of the response as well (Belknap et al., 1990). Striking strain differences in pain processing mechanisms have also been noted. For example, it appears that the antinociceptive effects of heroin are mediated by the μ-opioid receptor in some strains, and by the δ- or κ-opioid receptor in others (Rady et al., 1999). Release of endogenous opioid peptides is responsible for stress-induced analgesia in some strains, while in other strains, comparable analgesic effects are produced by non-opioid mechanisms (Mogil and Belknap, 1997). Naloxone produces analgesia in the formalin test in BALB/c, but not C57BL/6 or CD-1 mice (Vaccarino et al., 1988).

Genetic mediation of nociception

The same strains are not sensitive or resistant, respectively, to all types of pain. As described above, we have subjected a panel of 12 inbred strains to 22 assays of nociceptive sensitivity in order to systematically study genetic correlations among these phenotypes, hoping to observe "clusters" or "types" of correlated assays which share genetic and thus mechanistic features (Lariviere et al., 2002; Mogil et al., 1999a). From the observed patterns of strain sensitivities, we were able to conclude that the genetic mediation of nociception is specific to the *modality* of the noxious stimulus. Using multivariate analyses, we found five distinct clusters, in which the same strains were generally

Table 7.1 Characteristics, heritability, and range of variation of 22 nociceptive assays studied in 12 inbred mouse strains.

Assay	Etiology	Stimulus modality	Location	Duration[a]	Nocifensive response	Dependent measure	h^{2} [b]	Fold-change[c]
Abdominal constriction – acetic acid (AC_{AA})	Chemical	Spontaneous	Subcutaneous/visceral	Subacute	Reflexive	Number of writhes	0.76	2.8
Abdominal constriction – magnesium sulfate (AC_{MS})	Chemical	Spontaneous	Subcutaneous/visceral	Acute	Reflexive	Number of writhes	0.59	2.3
Automy (AUT)	Neuropathic	Spontaneous	PNS/CNS[d]	Chronic	Organized	Autotomy rating	0.63	2.2
Bee venom (BV)	Chemical	Spontaneous	Subcutaneous	Subacute	Organized	Time spent licking	0.39	11.5
Bee venom-induced thermal hypersensitivity (BV_{HT})	Chemical	Thermal	Subcutaneous	Tonic	Reflexive/organized	Withdrawal latency	0.47	1.7
Bee venom-induced thermal hypersensitivity (contralateral) (BV_{CON})	Chemical	Thermal	PNS/CNS	Tonic	Reflexive/organized	Withdrawal latency	0.48	9.2
Capsaicin (CAP)	Chemical	Spontaneous	Subcutaneous	Subacute	Organized	Time spent licking	0.39	5.0
Capsaicin-induced thermal hypersensitivity (CAP_{TW})	Chemical	Thermal	Subcutaneous	Subacute	Reflexive	Withdrawal latency	0.60	2.4
Carrageenan-induced thermal hypersensitivity (CAR_{HT})	Chemical	Thermal	Subcutaneous	Tonic	Reflexive/organized	Withdrawal latency	0.30	3.4
Dynorphin-induced mechanical hypersensitivity (DYN_{VF})	Neuropathic	Mechanical	PNS/CNS	Chronic	Reflexive	Pressure threshold	0.46	2.4
Formalin, early phase (F_{early})	Chemical	Spontaneous	Subcutaneous	Acute	Organized	Time spent licking	0.39	3.4
Formalin, late phase (F_{late})	Chemical	Spontaneous	Subcutaneous	Subacute	Organized	Time spent licking	0.46	2.8
Hot-plate (HP)	Nociceptive	Thermal	Cutaneous (paw)	Acute	Organized	Latency to lick	0.59	2.5
Hargreaves' paw-withdrawal (HT)	Nociceptive	Thermal	Cutaneous (paw)	Acute	Reflexive/organized	Withdrawal latency	0.72	2.9
Peripheral nerve injury-induced thermal hypersensitivity (PNI_{HT})	Neuropathic	Thermal	PNS/CNS	Chronic	Reflexive/organized	Withdrawal latency	0.45	2.3
Peripheral nerve-injury induced mechanical hypersensitivity (PNI_{VF})	Neuropathic	Mechanical	PNS/CNS	Chronic	Reflexive (withdrawal)	Pressure threshold	0.45	3.6
Tail-clip (TC)	Nociceptive	Thermal	Cutaneous (tail)	Acute	Organized	Latency to bite clip	0.69	8.3
Tail-flick from radiant heat source (TF)	Nociceptive	Thermal	Cutaneous (tail)	Acute	Reflexive (withdrawal)	Latency to withdrawal	0.31	0.4
Tail-withdrawal, −15°C ethanol (TW_{-15})	Nociceptive	Thermal	Cutaneous (tail)	Acute	Reflexive (withdrawal)	Withdrawal latency	0.43	3.4
Tail-withdrawal, 47.5°C water ($TW_{47.5}$)	Nociceptive	Thermal	Cutaneous (tail)	Acute	Reflexive (withdrawal)	Withdrawal latency	0.50	2.2
Tail-withdrawal, 49°C water (TW_{49})	Nociceptive	Thermal	Cutaneous (tail)	Acute	Reflexive (withdrawal)	Withdrawal latency	0.41	2.9
Von Frey monofilament (VF)	Nociceptive	Thermal	Cutaneous (paw)	Acute	Reflexive (withdrawal)	Pressure threshold	0.69	1.9

[a] Acute: seconds to minutes; Subacute: <1 hour; Tonic: hours to days; Chronic: days to months.

[b] Narrow-sense heritability.

[c] Ratio of most sensitive strain compared to least sensitive strain out of 12 strains tested.

[d] PNS/CNS: thought to be due to plasticity in the peripheral nervous system (PNS) or central nervous system (CNS).

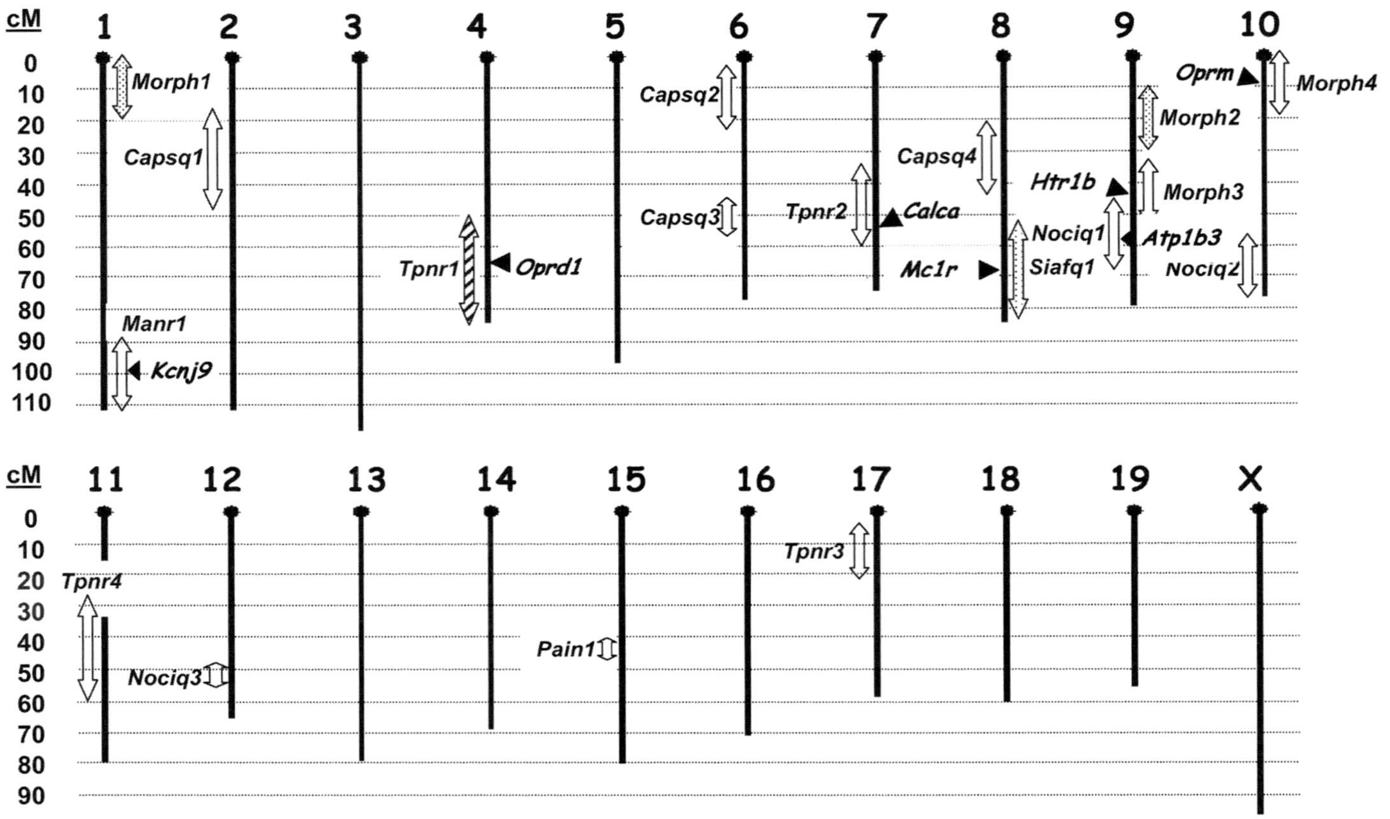

Figure 7.1 Known quantitative trait loci (QTLs) for nociception and analgesia. The confidence intervals describing the location (in cM from the centromere) of the gene(s) underlying variability in each trait are illustrated by vertical arrows; "pain genes" to the left of the chromosome and "analgesia genes" to the right. Filled arrowheads opposite to confidence intervals indicate the location of candidate genes for which supporting functional data has been collected (see text). Stippled arrows represent female-specific QTLs; cross-hatched arrows represent male-specific QTLs. The QTL phenotypes and mapping populations are described in Table 7.2.

sensitive or resistant on each test (Lariviere et al., 2002; Mogil et al., 1999b; see Figure 7.1). All tests employing a thermal stimulus (whether hot or cold) to elicit a nociceptive response were highly genetically correlated; likewise, all tests that measured a spontaneous response to a chemical agent formed a tight group. Two of the other clusters involved hypersensitivity caused by an inflammatory or neuropathic injury, but they were distinguished by the modality of the acute evoking stimulus. Tests which used von Frey filaments to measure mechanical hypersensitivity (i.e., "punctate" mechanical allodynia) were highly genetically correlated, but completely uncorrelated with tests that measured thermal hyperalgesia. It appears that the genes underlying hypersensitivity are not those which regulate inflammatory or neuropathic pain per se, but rather are specific to the particular *symptom* of pain being assessed, regardless of its etiology. A fifth cluster was composed of tests with a component of spinal sensitization mediated by peripheral afferent activity (Lariviere et al., 2002; Mogil et al., 1999b).

These findings, of course, make specific predictions about the QTLs and genes that will be identified in due course. That is, we would expect to find QTLs that are common to nociceptive assays within a cluster (i.e., genetically correlated assays), but also to find completely distinct QTLs for assays in different clusters. These predictions are starting to be borne out, as will be described below.

Genetic mediation of analgesia

An association between nociception and its inhibition by endogenous analgesic mechanisms was first uncovered in artificially selected lines (Panocka and Marek, 1986). An inverse genetic relationship between nociception and opioid drug analgesia has been observed in inbred strains tested for morphine sensitivity as well (Elmer et al., 1998; Mogil et al., 1996b): strains sensitive to nociception are resistant to opioid analgesia, whereas strains resistant to nociception exhibit high analgesic responses. We recently demonstrated that this genetic correlation may hold for inhalation anesthesia as well (Mogil et al., 2005a, 2005b). It is tempting to attribute individual differences in morphine analgesia to inheritance of variants of the μ-opioid receptor gene, *Oprm*, but analgesic strain sensitivities are entirely dependent on the nociceptive assay being used to measure analgesia. That is, strains sensitive to morphine inhibition of thermal nociception are not necessarily the same strains sensitive to morphine inhibition of chemical nociception (Elmer et al., 1998; Mogil et al., 1996b), which rules out the simplest explanation that some strains simply have more (or more efficient) μ-opioid receptors in pain-relevant areas. This dissociation of analgesic sensitivity according to the type of pain has also been shown to be true of the neuropathic pain drugs, gabapentin and pregabalin (Chesler et al., 2003).

Analgesia may be produced by activating receptors other than the μ-opioid receptor, of course, and we have demonstrated moderate-to-high genetic correlations between drugs binding to the κ-opioid, α_2-adrenergic, cannabinoid, and nicotinic receptors (Wilson et al., 2003). This finding suggests that the genetic mediation of analgesia is not mainly due to variants of genes specific to individual drugs (e.g., genes coding for the drugs' molecular binding sites), but rather to genetic variation downstream of the activated receptors, where mechanisms may converge (Basbaum and Fields, 1984). A "master analgesia gene" (or genes) may therefore control sensitivity to analgesia across drug classes (Wilson et al., 2003). The negative correlation of nociceptive sensitivity with analgesic drug sensitivity suggests this "master" gene may in fact mediate nociception directly, and analgesia only indirectly.

Genetic mediation of pruritis

Animal models of experimental pruritis have only recently been developed (Kuraishi et al., 1995; Nojima and Carstens, 2003), and involve the injection of an itch-inducing substance into the mid-scapular region and the subsequent quantification of scratching behavior. It was noted that outbred ICR mice displayed a far more robust behavioral and electrophysiological response to histamine injection than outbred ddY mice (Kuraishi et al., 2000; Maekawa et al., 2000). Inagaki et al. (2001) observed that ICR mice displayed the highest response of nine inbred and outbred strains to 50 nmol histamine or 50 nmol serotonin. Since it is increasingly appreciated that pruritogens feature complex, inverted-U-shaped dose response relationships, we tested 11 inbred mouse strains for their scratching behavior to histamine and the antimalarial drug, chloroquine (which produces intolerable itching as a side effect in 8–20% of human patients; Sowunmi et al., 1989), injected at a range of doses. Robust genetic effects were demonstrated ($h^2 = 0.45$–0.51), and an intriguing negative genetic correlation of itch sensitivity and thermal nociception was also observed (Green et al., 2006).

Sex × genotype interactions

Many of the genotypic differences described above were found to interact with sex (see Mogil, 2003). For example, males of the AKR/J, C57BL/6J, and outbred Swiss Webster strains are more sensitive to morphine inhibition of thermal pain than female mice of these strains, whereas CBA/J females are more sensitive than males, and other strains show no sex difference at all (Kest et al., 1999; Mogil et al., 2000b). These findings have led us to the expectation that sex-specific QTLs might be uncovered, as will be described below.

Pain genes: results of linkage mapping studies

We and others have used standard QTL mapping strategies – using F2 intercrosses, recombinant inbred (RI) strains, congenic lines, and haplotype mapping – to identify a number of pain and analgesia-relevant QTLs on the mouse genome. These are described in Table 7.2. In many cases, candidate gene approaches have provided compelling evidence for the genes underlying these QTLs (i.e., the quantitative trait genes). We have been less successful in identifying the responsible DNA variants, but feel strongly that for this particular biological domain this task is largely moot, since direct human translational studies are readily performed.

Thermal nociception QTLs

A two-part strategy was used to identify QTLs responsible for the difference in sensitivity to thermal nociception between DBA/2 (D2; resistant) and C57BL/6 (B6; sensitive) mice on the 54°C hot-plate test (Mogil et al., 1997a). Initially, the BXD/Ty panel of RI strains was phenotyped, and the resulting strain distribution patterns were correlated with published microsatellite genotype data. Six chromosomal regions were observed to meet a significance criterion of $\alpha < 0.01$, and the most promising of these was subsequently confirmed using an B6D2F2 intercross. This QTL (dubbed *Tpnr1*, for "*t*hermal *p*ai*n* *r*esponse"), located on chromosome 4 between 50–80 cM, was more highly associated with hot-plate latencies in male compared to female mice. Among genes in the QTL region, the *Oprd1* gene (65 cM), encoding the δ-opioid receptor, seemed a likely candidate given the important role of opioid receptors in pain processing. Pharmacological experiments using selective antagonists administered systemically demonstrated that blocking δ-opioid receptors lowered hot-plate latencies in D2 but not B6 mice, and in males but not females. Intriguingly, sex differences in the effects of δ-opioid receptor (*OPRD1*) alleles on threshold sensitivity and unpleasantness of thermal pain stimuli in humans have been observed (Kim et al., 2004), mirroring our hot-plate QTL results.

A larger-effect thermal sensitivity QTL was found using a different assay, the radiant heat paw-withdrawal test of Hargreaves et al. (1988). B6 mice are much more sensitive on this test than are AKR mice, and a region of chromosome 7 (34–67 cM) accounted for almost 40% of the genetic variance in a small B6AKRF2 intercross (peak LOD >6.0) (Mogil et al., 2005b). The location of this QTL was confirmed and refined with a separate intercross (B6CF2) and the A.B6-Tyr^+/J congenic strain to a 12-cM region. We evaluated *Calca* (54 cM) as a likely candidate gene. *Calca* codes for the calcitonin gene-related peptide, alpha (CGRPα), a neuropeptide well known for its involvement in neurogenic inflammation and central sensitization, and known to be released from nociceptors following noxious stimulation (see van Rossum et al., 1997). A series of behavior pharmacological, electrophysiological, gene expression, and mRNA knockdown studies all provided compelling evidence

Table 7.2 Known quantitative trait loci (QTLs) for nociception and analgesia.

QTL name	Chrom.	Location (cM)[a]	Phenotype	Population[b]	Candidate gene	Reference
Nociception						
Capsq1	2	17–47.5	Capsaicin intake	KJRB6F2		Furuse et al., 2003
Tpnr1	4	50–80	Hot-plate	BXD/Ty RI, B6D2F2	*Oprd1*	Mogil et al., 1997b
Capsq2	6	3.5–24.5	Capsaicin intake	KJRB6F2		Furuse et al., 2003
Capsq3	6	44–57.5	Capsaicin intake	KJRB6F2		Furuse et al., 2003
Tpnr2	7	34–62	Hargreaves' test, tail-withdrawal	B6AKRF2, B6CF2, A.B6-*Tyr*$^{+}$/J congenic	*Calca*	Mogil et al., 2005b
Capsq4	8	21–43	Capsaicin intake	KJRB6F2		Furuse et al., 2003
Nociq1	9	44–68	Formalin (early)	B6AF2, haplotype	*Atp1b3*	Wilson et al., 2002
Nociq2	10	69–70	Formalin (late)	B6AF2, AcB/BcA RC, ISSC		Wilson et al., 2002
Tpnr4	11	28–59	Tail-withdrawal	B6129F2		Unpublished data
Nociq3	12	50–54	Formalin (late)	B6.MOLF congenic, ISSC		Unpublished data
Pain1	15	41–46	Autotomy	AXB/BXA RI		Seltzer et al., 2001
Tpnr3	17	6.5–19	Hot-plate, tail-withdrawal, Hargreaves' test	B6.MOLF congenic		Unpublished data
Analgesia						
Morph1	1	0–20	Morphine	BXD/Ty RI, B6D2F2		Bergeson et al., 2001
Manr1	1	90–110	Clonidine, morphine, cannabinoid analgesia	B6129F2, haplotype, *Fcgr* KO congenic	*Kcnj9*	Smith et al., 2008
Siafq1	8	51–81	Stress-induced analgesia, U50,488	BXD/Ty RI, B6D2F2	*Mc1r*	Mogil et al., 2003
Morph2	9	10–30	Morphine	BXD/Ty RI, B6D2F2		Bergeson et al., 2001
Morph3	9	32–52	Morphine	BXD/Ty RI, B6D2F2	*Htr1b*	Bergeson et al., 2001
Morph4	10	0–18	Morphine	BXD/Ty RI, B6D2F2	*Oprm*	Bergeson et al., 2001

All of the QTLs shown above are significant based on permutation analysis, with LOD scores >3 in every case.
[a] Confidence interval (95%) based on a 1-LOD drop off. In some cases, confidence intervals are based on known congenic donor regions.
[b] Mapping population abbreviations are by standard nomenclature or as stated in the text.
Other abbreviations: *Fcgr* KO congenic: the knockout of *Fcgr* used as a congenic line (B6.129X1); haplotype: *in silico* haplotype mapping using inbred strain means; ISSC: interval-specific subcongenic; RC: recombinant congenic strains.

for *Calca*'s candidacy (Mogil et al., 2005b), consistent with the hypothesis that mouse strains resistant to thermal nociception show deficient CGRP release from nociceptors secondary to lower *Calca* gene expression. These studies featured a complete "rescue" of the AKR versus B6 strain difference, either by peripheral injection of CGRP itself to the resistant AKR strain or antisense knockdown of *Calca* expression in the sensitive B6 strain. Moreover, injection of 5 μg CGRP into the plantar hind-paw was sufficient to completely (albeit temporarily) abolish *all* strain differences among the 13 strains tested (Mogil et al., 2005b). Corroborating our prediction that similar genes underlie the variable responses elicited in multiple assays within the same "pain cluster," we have also recently identified a QTL (peak LOD >10) on mid chromosome 7 that mediates thermal sensitivity on the 49flC tail-withdrawal test (Smith et al., unpublished data); whether *Calca* also underlies this QTL is not yet known.

One may have predicted that the proteins responsible for heat transduction would contribute to individual variability in thermal nociception. The multi-functional ion channels (especially the TRPV1 channel) that are activated by heat as well as protons and the vanilloid capsaicin, are critical for thermal nociception (Caterina et al., 2000; Davis et al., 2000), but they have not as yet been implicated in mediating strain differences in thermal nociceptive sensitivity. We note, however, that a significant QTL for thermal tail-withdrawal latency on chromosome 11 (Smith et al., 2008) contains the mouse *Trpv1* gene within its confidence interval. Surprisingly, QTL mapping of voluntary capsaicin drinking failed to link the "capsaicin receptor" TRPV1 channel to observed strain differences. Furuse and colleagues (2003) crossed B6 mice, which (sensibly) avoid capsaicin, with KJR mice apparently insensitive to oral capsaicin burn. They found four significant QTLs (*Capsq1-Capsq4*), none of which contain any of the *Trpv* genes. These results serve as a reminder that pain variability genes are not limited to those at the initiation of the pain signal, but may act at any level of nociceptive processing.

Chemical/inflammatory nociception QTLs

The formalin test of chemical/inflammatory nociception (Dubuisson and Dennis, 1977) comprises two distinct phases, an early/acute phase (0–10 min post-injection) in which nociceptors are activated directly, and a late/tonic phase (20–90 min post-injection) in which pain behaviors – mostly licking and biting, in the mouse – are triggered by ongoing inflammation and possibly central sensitization. We bred B6AF2 intercross to map QTLs for both phases of the formalin test. Chromosome 9 appeared to harbor an early phase-specific locus (*Nociq1*, 44–68 cM), whereas a QTL with significant effects on both phases was identified on distal chromosome 10 (*Nociq2*, >58 cM) (Wilson et al., 2002).

Both QTLs have been positionally cloned in our laboratory. Using a haplotype mapping strategy (Grupe et al., 2001), we have identified a gene – *Atp1b3* – near the peak of linkage in *Nociq1* showing a high association with strain variation in a set of 16 strains. *Atp1b3* codes for the β_3 subunit of the sodium-potassium ATPase (pump). Using a combination of electrophysiological, immunohistochemical and RNA interference strategies, we showed that differential expression of the β_3 subunit is inversely correlated with licking behavior in the early phase of the formalin test, due to an effect of the β_3 subunit incorporation into pumps on resting membrane potentials (and thus action potential thresholds) in dorsal root ganglion neurons (LaCroix-Fralish et al., 2009). The protein hereby implicated in pain processing is extremely novel, with only a handful of existing published papers describing its existence and no prior role in any biological phenomenon ever attributed to it.

To identify the gene underlying *Nociq2*, we have employed AcB/BcA recombinant congenic strains (Fortin et al., 2001), marker-assisted congenic and interval-specific subcongenic lines derived from them in order to reduce the QTL region to <10 Mb, and microarray profiling and knockout mouse testing, to help decide among the ≈100 annotated genes in this region. Of note is that *Nociq2* overlaps with an unpublished QTL (Hain and Belknap, personal communication) for sensitivity on the acetic acid writhing test, another assay of chemical nociception, supporting our prediction that genes contributing to the formalin response may also play a role in other tests of chemical/inflammatory pain.

Finally, using a congenic line with MOLF/Ei genome introgressed on a B6 background, we have identified another late phase formalin test QTL (*Nociq3*). Construction of interval-specific subcongenic lines have reduced the QTL region to <5 Mb on chromosome 12, a region containing only 14 genes. We are currently running quantitative reverse-transcription polymerase chain reaction (RT-PCR) experiments on all 14 genes in order to identify the one responsible for the strain difference.

Neuropathic pain QTLs

A locus contributing to neuropathic pain was mapped using the autotomy model, in which denervation of a hind limb leads to varying levels of self-injurious behaviors in different mouse strains (Mogil et al., 1999a). Seltzer and colleagues (2001), using 23 RI strains from the AXB/BXA panel, localized a QTL to a surprisingly small length of chromosome 15, but one with a surprisingly large number of potential candidate genes. This QTL, referred to as *pain1*, has recently been replicated in a backcross between C58 and C3H/He (Devor et al., 2005). We have recently completed a survey of the variable mechanical allodynia exhibited by 22 inbred mouse strains after a spared nerve injury, and using haplotype mapping (Grupe et al., 2001) have identified a number of potential candidates (unpublished data).

Analgesia QTLs

Morphine analgesia against thermal nociception on the hot-plate test has been mapped to four different locations (*Morph1-Morph4*) using a combination of genetic models, including the BXD/Ty RI strain set, B6D2F2 mice, artificially selected lines, and congenic strains (Bergeson et al., 2001). A QTL on proximal chromosome 10 is likely the *Oprm* gene (Belknap et al., 1995), which encodes the μ-opioid receptor, the primary binding site of morphine (Matthes et al., 1996). Pharmacogenetic investigations of genes within the mid chromosome 9 QTL interval have provided evidence for the candidacy of *Htr1b*, coding for the serotonin 1B-receptor gene (Hain et al., 1999). Two other QTLs were strongly female-specific, indicating the existence of genes relevant to analgesic variability in female but not male mice (Bergeson et al., 2001).

In fact, experiments in our laboratory dating back to the early 1990s have provided evidence for a sex-specific neural circuitry mediating endogenous non-μ-opioid analgesia (Mogil et al., 1993). In males, this analgesia can be blocked by *N*-methyl-D-aspartate (NMDA) receptor antagonists, whereas females (unless ovariectomized) are wholly insensitive to NMDA blockade. This qualitative sex difference was investigated using the BXD/Ty RI panel and B6D2F2 mice, revealing a female-specific QTL (called *Siafq1*) on chromosome 8 (52–84 cM) (Mogil et al., 1997b). Analgesia produced by the selective κ-opioid receptor agonist, U50,488, also mapped to this region in a new B6D2F2 population, but again this linkage was observed only in female mice (Mogil et al., 2003). A candidate gene, *Mc1r* (encoding the melanocortin-1 receptor, MC1R) was investigated using pharmacological, gene expression and mutant approaches, and convergent evidence was obtained supporting its candidacy (Mogil et al., 2003). The human *MC1R* gene is the "redhead gene" (Valverde et al., 1995), and much is known about the several common human variants that lead to MC1R loss of function. We designed a small translational genetic association study, and demonstrated that, exactly as seen in mice, women (but not men) with non-functional MC1Rs (i.e., redheads) displayed increased sensitivity to the analgesic effects of a κ-opioid-acting drug (pentazocine) against experimental pain (Mogil et al., 2003). Follow-up studies using recessive yellow (C57BL/6-*Mc1r*$^{e/e}$) spontaneous mutants revealed a strong influence of MC1R functionality on both basal nociception and μ-opioid analgesia (although these effects were curiously *not* sex-specific), and both findings were simultaneously confirmed in humans (see, however, Liem et al., 2005; Mogil et al., 2005c).

With respect to other analgesic modalities, we recently provided convergent evidence – using simultaneously performed F2 hybrid screening of multiple analgesics followed by gene expression and transgenic data – for the role of *Kcnj9* (and/or *Kcnj10*) in the mediation of variable analgesia from clonidine (an α_2-adrenergic agonist) and WIN55,212-2 (a cannabinoid CB-1 agonist) (Smith et al., 2008). *Kcnj9* encodes the GIRK3 (Kir3.3) inwardly rectifying potassium channel subunit. This is a somewhat surprising finding since there exists considerable evidence (mostly transgenic) for the involvement of another gene, GIRK2 (*Kcnj6*), in the mediation of analgesic response (e.g., Ikeda et al. 2000). We propose that although GIRK2 is clearly involved, it is GIRK3, which complexes with GIRK2, that provides the variability across strain.

References

Barr, G.A., Gao, P., Wang, S., Cheng, J., Qin, J., Sibille, E.L., et al. (2005) Microarray analysis of gene expression following the formalin test in the infant rat. *Pain* 117: 6–18.

Basbaum, A.I. and Fields, H.L. (1984) Endogenous pain control systems: brainstem spinal pathways and endorphin circuitry. *Annu Rev Neurosci* 7: 309–338.

Belknap, J.K., Lamé, M., and Danielson, P.W. (1990) Inbred strain differences in morphine-induced analgesia with the hot plate assay: a reassessment. *Behav Genet* 20: 333–338.

Belknap, J.K., Mogil, J.S., Helms, M.L., Richards, S.P., O'Toole, L.A., Bergeson, S.E., et al. (1995) Localization to chromosome 10 of a locus influencing morphine analgesia in crosses derived from C57BL/6 and DBA/2 mouse strains. *Life Sci* 57: PL117–124.

Bergeson, S.E., Helms, M.L., O'Toole, L.A., Jarvis, M.W., Hain, H.S., Mogil, J.S., et al. (2001) Quantitative trait loci influencing morphine antinociception in four mapping populations. *Mamm Genome* 12: 546–553.

Bon, K., Lichtensteiger, C.A., Wilson, S.G., and Mogil, J.S. (2003) Characterization of cyclophosphamide cystitis, a model of visceral and referred pain in the mouse: species and strain differences. *J Urol* 170: 1008–1012.

Caterina, M.J., Leffler, A., Malmberg, A.B., Martin, W.J., Trafton, J., Petersen-Zeitz, K.R., et al. (2000) Impaired nociception and pain sensation in mice lacking the capsaicin receptor. *Science* 288: 306–313.

Chesler, E.J., Ritchie, J., Kokayeff, A., Lariviere, W.R., Wilson, S.G., and Mogil, J.S. (2003) Genotype-dependence of gabapentin and pregabalin sensitivity: the pharmacogenetic mediation of analgesia is specific to the type of pain being inhibited. *Pain* 106: 325–335.

Costigan, M., Befort, K., Karchewski, L., Griffin, R.S., D'Urso, D., Allchorne, A., et al. (2002) Replicate high-density rat genome oligonucleotide microarrays reveal hundreds of regulated genes in the dorsal root ganglion after peripheral nerve injury. *BMC Neurosci* 3: 16.

Davis, J.B., Gray, J., Gunthorpe, M.J., Hatcher, J.P., Davey, P.T., Overend, P., et al. (2000) Vanilloid receptor-1 is essential for inflammatory thermal hyperalgesia. *Nature* 405: 183–187.

Devor, M., Gilad, A., Arbilly, M., Yakir, B., Raber, P., Pisante, A., et al. (2005) *pain1*: A neuropathic pain QTL on mouse chromosome 15 in a C3HxC58 backcross. *Pain* 116: 289–293.

Devor, M. and Raber, P. (1990) Heritability of symptoms in an experimental model of neuropathic pain. *Pain* 42: 51–67.

Dubuisson, D. and Dennis, S.G. (1977) The formalin test: a quantitative study of the analgesic effects of morphine, meperidine, and brain stem stimulation in rats and cats. *Pain* 4: 161–174.

Elmer, G.I., Pieper, J.O., Negus, S.S., and Woods, J.H. (1998) Genetic variance in nociception and its relationship to the potency of morphine-induced analgesia in thermal and chemical tests. *Pain* 75(1): 129–140.

Falconer, D.S. and Mackay, T.F.C. (1996) *Introduction to Quantitative Genetics*, 4th edn. Longman, Essex, UK.

Fortin, A., Diez, E., Rochefort, D., Laroche, L., Malo, D., Rouleau, G.A., et al. (2001) Recombinant congenic strains derived from A/J and C57BL/6J: a tool for genetic dissection of complex traits. *Genomics* 74: 21–35.

Furuse, T., Miura, Y., Yagasaki, K., Shiroishi, T., and Koide, T. (2003) Identification of QTLs for differential capsaicin sensitivity between mouse strains KJR and C57BL/6. *Pain* 105: 169–175.

Green, A.D., Young, K.K., Lehto, S.G., Smith, S.B., and Mogil, J.S. (2006) Influence of genotype, dose and sex on pruritogen-induced scratching behavior in the mouse. *Pain* 124(1–2): 50–58.

Grupe, A., Germer, S., Usuka, J., Aud, D., Belknap, J.K., Klein, R.F., et al. (2001) *In silico* mapping of complex disease-related traits in mice. *Science* 292: 1915–1918.

Hain, H.S., Belknap, J.K., and Mogil, J.S. (1999) Pharmacogenetic evidence for the involvement of 5-hydroxytryptamine (serotonin)-1B receptors in the mediation of morphine antinociceptive sensitivity. *J Pharmacol Exp Ther* 291(2): 444–449.

Hargreaves, K., Dubner, R., Brown, F., Flores, C., and Joris, J. (1988) A new and sensitive method for measuring thermal nociception in cutaneous hyperalgesia. *Pain* 32: 77–88.

Ikeda, K., Kobayashi, T., Kumanishi, T., Niki, H., and Yano, R. (2000) Involvement of G-protein-activated inwardly rectifying K+ (GIRK) channels in opioid-induced analgesia. *Neurosci Res* 38: 113–116.

Inagaki, N., Nagao, M., Igeta, K., Kawasaki, H., Kim, J.F., and Nagai, H. (2001) Scratching behavior in various strains of mice. *Skin Pharmacol Appl Skin Physiol* 14: 87–96.

Kest, B., Wilson, S.G., and Mogil, J.S. (1999) Sex differences in supraspinal morphine analgesia are dependent on genotype. *J Pharmacol Exp Ther* 289(3): 1370–1375.

Kim, D.-S., Lee, S.-J., Park, S.-Y., Yoo, H.-J., Kim, S.-H., Kim, K.-J., et al. (2001) Differentially expressed genes in rat dorsal root ganglia following peripheral nerve injury. *Neuroreport* 12(15): 3401–3405.

Kim, H., Neubert, J.K., Rowan, J.S., Brahim, J.S., Iadarola, M.J., and Dionne, R.A. (2004) Comparison of experimental and acute clinical pain responses in humans as pain phenotypes *J Pain* 5: 377–384.

Kuraishi, Y., Nagasawa, T., Hayashi, K., and Satoh, M. (1995) Scratching behavior induced by pruritogenic but not algesiogenic agents in mice. *Eur J Pharmacol* 275: 229–233.

Kuraishi, Y., Yamaguchi, T., and Miyamoto, T. (2000) Itch-scratch responses induced by opioids through central mu opioid receptors in mice. *J Biomed Sci* 7: 248–252.

LaCroix-Fralish, M.L., Mo, G., Smith, S.B., Sotocinal, S.B., Ritchie, J., Austin, J.-S., et al. (2009) The $\beta3$ subunit of the Na+,K+-ATPase affects pain sensitivity. *Pain* 144: 294–302.

LaCroix-Fralish, M.L., Mogil, J.S., Weinstein, J.N., Rutkowski, M.D., and DeLeo, J.A. (2005) The magnitude of mechanical allodynia in a rodent model of lumbar radiculopathy is dependent on strain and sex. *Spine* 30: 1821–1827.

Lariviere, W.R., Chesler, E.J., and Mogil, J.S. (2001) Transgenic studies of pain and analgesia: mutation or background phenotype? *J Pharmacol Exp Ther* 297: 467–473.

Lariviere, W.R., Wilson, S.G., Laughlin, T.M., Kokayeff, A., West, E.E., Adhikari, S.M., et al. (2002) Heritability of nociception. III. Genetic relationships among commonly used assays of nociception and hypersensitivity. *Pain* 97: 75–86.

Libman, E. (1934) Observations on individual sensitiveness to pain. *J Am Med Assoc* 102: 335–341.

Liem, E.B., Joiner, T.V., Tsueda, K., and Sessler, D.I. (2005) Increased sensitivity to thermal pain and reduced subcutaneous lidocaine efficacy in redheads. *Anesthesiology* 102: 509–514.

MacGregor, A.J. (2004) The heritability of pain in humans. In Mogil, J.S. (ed.), *The Genetics of Pain, Progress in Pain Research and Management*, Vol. 28. IASP Press, Seattle, WA, USA, pp. 151–170.

MacGregor, A.J., Griffiths, G.O., Baker, J., and Spector, T.D. (1997) Determinants of pressure pain threshold in adult twins: evidence that shared environmental influences predominate. *Pain* 73: 253–257.

Maekawa, T., Nojima, H., and Kuraishi, Y. (2000) Itch-associated responses of afferent nerve innervating the murine skin: different effects of histamine and serotonin in ICR and ddY mice. *Jpn J Pharmacol* 84: 462–466.

Matthes, H.W.D., Maldonado, R., Simonin, F., Valverde, O., Slowe, S., Kitchen, I., et al. (1996) Loss of morphine-induced analgesia, reward effect and withdrawal symptoms in mice lacking the μ-opioid-receptor gene. *Nature* 383: 819–823.

Melzack, R. (1975) The McGill Pain Questionnaire: major properties and scoring methods. *Pain* 1: 277–299.

Mogil, J.S. (1999) The genetic mediation of individual differences in sensitivity to pain and its inhibition. *Proc Natl Acad Sci USA* 96: 7744–7751.

Mogil, J.S. (2003) Interaction between sex and genotype in the mediation of pain and pain inhibition. *Sem Pain Med* 1: 197–205.

Mogil, J.S. (2009) Animal models of pain: progress and challenges. *Nat Rev Neurosci* 10: 283–294.

Mogil, J.S., and Adhikari, S.M. (1999) Hot and cold nociception are genetically correlated. *J Neurosci* 19: 21–25.

Mogil, J.S., and Belknap, J.K. (1997) Sex and genotype determine the selective activation of neurochemically distinct mechanisms of swim stress-induced analgesia. *Pharmacol Biochem Behav* 56: 61–66.

Mogil, J.S., Chesler, E.J., Wilson, S.G., Juraska, J.M., and Sternberg, W.F. (2000b) Sex differences in thermal nociception and morphine antinociception in rodents depend on genotype. *Neurosci Biobehav Rev* 24: 375–389.

Mogil, J.S., Flodman, P., Spence, M.A., Sternberg, W.F., Kest, B., Sadowski, B., et al. (1995a) Oligogenic determination of morphine analgesic magnitude: a genetic analysis of selectively bred mouse lines. *Behav Genet* 25: 397–406.

Mogil, J.S., and Grisel, J.E. (1998) Transgenic studies of pain. *Pain* 77: 107–128.

Mogil, J.S., Kest, B., Sadowski, B., and Belknap, J.K. (1996b) Differential genetic mediation of sensitivity to morphine in genetic models of opiate antinociception: influence of nociceptive assay. *J Pharmacol Exp Ther* 276: 532–544.

Mogil, J.S., Lichtensteiger, C.A., and Wilson, S.G. (1998) The effect of genotype on sensitivity to inflammatory nociception: characterization of resistant (A/J) and sensitive (C57BL/6) inbred mouse strains. *Pain* 76: 115–125.

Mogil, J.S., Marek, P., Flodman, P., Spence, M.A., Sternberg, W.F., Kest, B., et al. (1995b) One or two genetic loci mediate high opiate analgesia in selectively bred mice. *Pain* 60: 125–135.

Mogil, J.S., Meirmeister, F., Seifert, F., Strasburg, K., Zimmermann, K., Reinold, H., et al. (2005b) Variable sensitivity to noxious heat is mediated by differential expression of the CGRP gene. *Proc Natl Acad Sci USA* 102: 12938–12943.

Mogil, J.S., Richards, S.P., O'Toole, L.A., Helms, M.L., Mitchell, S.R., and Belknap, J.K. (1997a) Genetic sensitivity to hot-plate nociception in DBA/2J and C57BL/6J inbred mouse strains: possible sex-specific mediation by $\delta2$-opioid receptors. *Pain* 70: 267–277.

Mogil, J.S., Richards, S.P., O'Toole, L.A., Helms, M.L., Mitchell, S.R., Kest, B., et al. (1997b) Identification of a sex-specific quantitative trait locus mediating nonopioid stress-induced analgesia in female mice. *J Neurosci* 17: 7995–8002.

Mogil, J.S., Ritchie, J., Smith, S.B., Strasburg, K., Kaplan, L., Wallace, M.R., et al. (2005c) Melanocortin-1 receptor gene variants affect pain and μ-opioid analgesia in mice and humans. *J Med Genet* 42: 583–587.

Mogil, J.S., Smith, S.B., O'Reilly, M.K., and Plourde, G. (2005a) Influence of nociception and stress-induced antinociception on genetic variation in isoflurane anesthetic potency among mouse strains. *Anesthesiology* 103(4): 751–758.

Mogil, J.S., Sternberg, W.F., Kest, B., Marek, P., and Liebeskind, J.C. (1993) Sex differences in the antagonism of swim stress-induced analgesia: effects of gonadectomy and estrogen replacement. *Pain* 53: 17–25.

Mogil, J.S., Sternberg, W.F., Marek, P., Sadowski, B., Belknap, J.K., and Liebeskind, J.C. (1996a) The genetics of pain and pain inhibition. *Proc Natl Acad Sci USA* 93: 3048–3055.

Mogil, J.S., Wilson, S.G., Bon, K., Lee, S.E., Chung, K., Raber, P., et al. (1999a) Heritability of nociception. I. Responses of eleven inbred mouse strains on twelve measures of nociception. *Pain* 80: 67–82.

Mogil, J.S., Wilson, S.G., Bon, K., Lee, S.E., Chung, K., Raber, P., et al. (1999b) Heritability of nociception. II. "Types" of nociception revealed by genetic correlation analysis. *Pain* 80: 83–93.

Mogil, J.S., Wilson, S.G., Chesler, E.J., Rankin, A.L., Nemmani, K.V.S., Lariviere,

W.R., et al. (2003) The melanocortin-1 receptor gene mediates female-specific mechanisms of analgesia in mice and humans. *Proc Natl Acad Sci USA* 100: 4867–4872.

Mogil, J.S., Yu, L., and Basbaum, A.I. (2000a) Pain genes? Natural variation and transgenic mutants. *Annu Rev Neurosci* 23: 777–811.

Nielsen, C.S., Stubhaug, A., Price, D.D., Vassend, O., Czajkowski N, and Harris, J.R. (2008) Individual differences in pain sensitivity: genetic and environmental contributions. *Pain* 2008 136: 21–29.

Nojima, H. and Carstens, E. (2003) Quantitative assessment of directed hind limb scratching behavior as a rodent itch model. *J Neurosci Meth* 126: 137–143.

Norbury, T.A., MacGregor, A.J., Urwin, J., Spector, T.D., and McMahon, S.B. (2007) Heritability of responses to painful stimuli in women: a classical twin study. *Brain* 130: 3041–3049.

Paigen, K. and Eppig, J.T. (2000) A Mouse Phenome Project. *Mamm Genome* 11: 715–717.

Panocka, I. and Marek, P. (1986) Inheritance of stress-induced analgesia in mice. *Ann N Y Acad Sci* 467: 441–443.

Rabert, D., Xiao, Y., Yiangou, Y., Kreder, D., Sangameswaran, L., Segal, M.R., et al. (2004) Plasticity of gene expression in injured human dorsal root ganglia revealed by GeneChip oligonucleotide microarrays. *J Clin Neurosci* 11: 289–299.

Rady, J.J., Elmer, G.I., and Fujimoto, J.M. (1999) Opioid receptor selectivity of heroin given intracerebroventricularly differs in six strains of inbred mice. *J Pharmacol Exp Ther* 288: 438–445.

Seltzer, Z., Wu, T., Max, M.B., and Diehl, S.R. (2001) Mapping a gene for neuropathic pain-related behavior following peripheral neurectomy in the mouse. *Pain* 93: 101–106.

Smith, S.B., Crager, S.E., and Mogil, J.S. (2004) Paclitaxel-induced neuropathic hypersensitivity in mice: responses in 10 inbred mouse strains. *Life Sci* 74: 2593–2604.

Smith, S.B., Marker, C.L., Perry, C., Liao, G., Sotocinal, S.G., Austin, J.-S., et al. (2008) Quantitative trait locus and computational mapping identifies *Kcnj9* (GIRK3) as a candidate gene affecting analgesia from multiple drug classes. *Pharmacogenet Genom* 18: 231–241.

Sowunmi, A., Walker, O., and Salako, L.A. (1989) Pruritus and antimalarial drugs in Africans. *Lancet* 2: 213.

Sun, H., Xu, J., Della Penna, K.B., Benz, R.J., Kinose, F., Holder, D.J., et al. (2002) Dorsal horn-enriched genes identified by DNA microarray, *in situ* hybridization and immunohistochemistry. *BMC Neurosci* 3: 11.

Takagi, H., Inukai, T., and Nakama, M. (1966) A modification of Haffner's method for testing analgesics. *Jpn J Pharmacol* 16: 287–294.

Vaccarino, A.L., Tasker, R.A.R., and Melzack, R. (1988) Systemic administration of naloxone produces analgesia in BALB/c mice in the formalin pain test. *Neurosci Lett* 84: 103–107.

Valder, C.R., Liu, J.-J., Song, Y.-H., and Luo, Z.D. (2003) Coupling gene chip analyses and rat genetic variances in identifying potential target genes that may contribute to neuropathic allodynia development. *J Neurochem* 87: 560–573.

Valverde, P., Healy, E., Jackson, I., Rees, J.L., and Thody, A.J. (1995) Variants of the melanocyte-stimulating hormone receptor gene are associated with red hair and fair skin in humans. *Nat Genet* 11: 328–330.

van Rossum, D., Hanisch, U.-K., and Quirion, R. (1997) Neuroanatomical localization, pharmacological characterization and functions of CGRP, related peptides and their receptors. *Neurosci Biobehav Rev* 21: 649–678.

Wang, H., Sun, H., Della Penna, K.B., Benz, R.J., Xu, J., Gerhold, D.L., et al. (2002) Chronic neuropathic pain is accompanied by global changes in gene expression and shares pathobiology with neurodegenerative diseases. *Neuroscience* 114: 529–546.

Wilson, S.G., Chesler, E.J., Hain, H.S., Rankin, A.L., Schwarz, J.Z., Call, S.B., et al. (2002) Identification of quantitative trait loci for chemical/inflammatory nociception in mice. *Pain* 96: 385–391.

Wilson, S.G. and Mogil, J.S. (2001) Measuring pain in the (knockout) mouse: big challenges in a small mammal. *Behav Brain Res* 125: 65–73.

Wilson, S.G., Smith, S.B., Chesler, E.J., Melton, K.A., Haas, J.J., Mitton, B.A., et al. (2003) The heritability of antinociception: common pharmacogenetic mediation of five neurochemically distinct analgesics. *J Pharmacol Exp Ther* 304: 547–559.

Xiao, H.-S., Huang, Q.-H., Zhang, F.-X., Bao, L., Lu, Y.-J., Guo, C., et al. (2002) Identification of gene expression profile of dorsal root ganglion in the rat peripheral axotomy model of neuropathic pain. *Proc Natl Acad Sci USA* 99: 8360–8365.

Xu, X.-J. and Wiesenfeld-Hallin, Z. (2004) Individual differences in pain: rat models. In Mogil, J.S. (ed.), *The Genetics of Pain, Progress in Pain Research and Management*, Vol. 28, IASP Press, Seattle, WA, USA, pp. 107–121.

Chapter 8

Olfactory acuity in *Mus musculus*

Heather Schellinck and Burton Slotnick

Mice, like many other mammals, live in a world dominated by olfactory cues. Information regarding the availability of food and the presence of predators is odor mediated (Apfelbach et al., 2005; Maruniak et al., 1975). Moreover, odor signals released from specialized scent glands, urine, and feces may be used to determine the individual identity, age, sex, territory, and dominance status of a conspecific (Brown and Mcdonald, 1985). Olfaction is the first sensory system to become functional in rodents and the first bouts of suckling depend upon olfaction (Abate et al., 2002; Pederson and Blass, 1982). Studies in rats have indicated that pups exposed to odors prenatally and early postnatally can learn to associate odors with food and other reinforcement (Hepper, 1990; Smotherman, 1982; Sullivan and Leon, 1987), and that this perinatal odor conditioning may influence the processing of odors at both the neural and behavioral levels (Wilson et al., 2004). Because of this extreme reliance on odor detection for feeding and social interaction, it seems likely that mice, like other rodents, have developed an extraordinary sense of smell.

Nonetheless, although the mouse has been used to study the molecular aspects of olfactory coding, e.g., they have been employed to examine the representation of odors in the brain (Korsching, 2001; Mori et al., 2006; Zou et al., 2005) and to investigate the specificity of odorant receptor expression in the olfactory bulb (Twsuboi et al., 2006; Zufall and Munger, 2001), little is known about their olfactory acuity. The purpose of this chapter is to review the experimental literature on olfactory acuity in inbred mice and to provide an overview of the factors that should be considered when interpreting the outcome of such studies. Acuity is defined as the threshold at which an odor may be detected.

We first discuss the methods used to assess acuity and the role of methodological differences in influencing the outcome of experimental results. We then describe inbred strain differences in odor detection thresholds, specific anosmias and the role of variables such as age, gender, odor exposure, and receptor stimulation in modulating olfactory acuity. Odor quality perception, i.e., the determination of the perceived similarity between odors has been primarily examined in rats (see Cleland and Narla, 2003) and will not be addressed here. The assessment of olfactory acuity in transgenic mice will be reviewed in Volume II of this Handbook.

Methodology

A variety of paradigms are being used to assess olfactory acuity in mice and rats. They range from simple digging tasks used to assess anosmia to sophisticated odor delivery systems that provide very precise information regarding olfactory acuity.

Buried food test

As exploratory digging is a natural behavior in mice, minimal training is required in this task. Sometimes called the "buried cookie test," the paradigm requires that a food-restricted animal find a food particle by digging in a substrate of sand, gravel or bedding material. The test has not been shown to be a good predictor of anosmia as animals that have not dug for buried food have subsequently been found to detect odors in other tasks (B. Slotnick, unpublished data) and we do not recommend it for use other than in olfactory memory paradigms or as a broad measure of olfactory capability.

Habituation–dishabituation

The habituation–dishabituation test is based upon the observation that successive presentations of the same stimulus odor, e.g., 0.1% amyl acetate, will result in a decrease in investigatory behavior such as sniffing. This is referred to as habituation (Sundberg et al., 1982). Then, when a different odor stimulus is presented to the same subject, e.g., 0.2% amyl acetate, the previously habituated response, e.g., sniffing will reoccur; this is referred to as dishabituation. In some instances, the test begins with a no-odor adaptation period during which the subject is put in the test chamber and presented with the odor vehicle, e.g., water, several times (Brown and Mcdonald, 1985; Schellinck et al., 1995). Some investigators prefer to call this procedure a habituation–discrimination task because the second stimulus is different from the first odor and thus the animal's response is considered to be a discrimination rather than a dishabituation. Others only refer to habituation–discrimination if the second odor is presented simultaneously with the first

Behavioral Genetics of the Mouse: Volume I. *Genetics of Behavioral Phenotypes*, eds. Wim E. Crusio, Frans Sluyter, Robert T. Gerlai, and Susanna Pietropaolo. Published by Cambridge University Press.

odor, following the habituation trials (Halpin, 1974). We have considered that the "dishabituation" refers to the response rather than the stimulus, and so have referred to the test as habituation–dishabituation after the description by Sundberg et al. (1982).

Although originally designed to assess odor discrimination in rats, this test requires little or no modification for use with mice. For example, Lee et al. (2003) used the task to demonstrate that male BALB/c mice had a greater olfactory sensitivity than male 129/S1s and that male C57BL/6 mice were the least sensitive of the three strains when tested with urinary odors of males from the other strains. The principal benefits of this test arise from the simple design, minimal need for equipment, and speed and ease of testing. One investigator may complete a nine-trial session for one mouse in less than 30 minutes.

When dishabituation occurs it is clear that the two odors presented can be discriminated. It is not so clear what a failure to dishabituate means. This could reflect a failure to discriminate or a disinterest in the test odor. It is also possible that the test odor is discriminable but not sufficiently different to produce a clear dishabituation effect. Also, the sniffing response, the usual measure of investigative behavior, is not often defined and this in turn makes it hard to compare the results from different studies. As a consequence, although the test may be used to compare the responses of different strains to specific odors, it is not an appropriate method of assessing olfactory acuity.

Conditioned digging task

Investigators have developed a number of "digging" tasks to assess both olfactory capacity and memory (Berger-Sweeney et al., 1998; Mihalick et al., 2000; Schellinck et al., 2001a; Zagreda et al., 1999). In the latter task, Schellinck et al. (2001b) trained male and female CD1 mice and male C57BL6/J and DBA mice to discriminate odors associated with food (CS+) from odors without any food (CS–). They remembered these discriminations for up to 120 days. This paradigm is more appropriate to assess anosmia than the "buried cookie" and habituation–dishabituation tests. For example, when NCAM null mice were tested in a habituation–dishabituation test, it was found that they were unable to discriminate between two odors (Gheusi et al., 2000). In contrast, Schellinck et al. (2004) tested the same strain of mice in an operant-digging task and the ability of mice of the same strain to discriminate between odors was not found to be impaired. The different results likely reflect the increased motivation of the mice in the operant task.

Conditioned avoidance learning

In a landmark study (Garcia et al., 1974), rats were shown to exhibit one-trial learning when presented with flavoured drinking water, i.e., the conditioned stimulus (CS), followed by exposure to a toxic substance such as lithium chloride, i.e., the unconditioned stimulus (UCS). Mice also may be conditioned in this manner (Horowitz and Whitney, 1975) and this learning paradigm may be adapted to provide a measurement of relative sensitivity to odors. Wysocki et al. (1977) used several variations of this method to assess specific anosmia for isovaleric acid in mice. In one experiment they used the following procedure. After a daily habituation period of 1 hour to a vehicle saccharin solution for several days, mice were presented with one of four different concentrations of isovaleric acid in saccharin solution (CS) followed by a toxic injection of lithium chloride. Twenty-four hours later the mice were given a two bottle preference test in which they had the opportunity to choose between a saccharin solution containing the odorant and a saccharin solution alone. If the animal avoided the solution containing the odorant, the authors concluded that the odorant was detectable. C57BL/6J and C57BL/10J mice were less sensitive to the isovaleric acid at a 10^{-3} concentration compared to mice from A/J, AKR/J, BALB/cJ, C3HeB/FeJ, DBA/2J, SJL/J, and SWR/J strains. The investigators also conditioned subjects from C57BL/6J and C57BL/10J strains of mice to a second control odor to ensure that the anosmia was indeed a specific rather than a general deficit. More recently, the conditioned odor avoidance paradigm was used to verify that genomic regions on chromosomes 4 and 6 code for the receptor proteins responsible for the recognition of isovaleric acid (Griff and Reed, 1995). While this testing method should not be used to determine absolute olfactory thresholds, it should be adequate to screen mice for relative sensitivity to one or many odors.

Electroolfactogram

A unique method of assessing olfactory activity, the electro-olfactogram, has been used to demonstrate that strains of mice that have a specific anosmia may show an improvement in sensitivity following repeated exposure to the odor. Wang et al. (1993) measured changes in voltage in the olfactory epithelium of NZB/B1NJ mice following a lengthy exposure to androstenone or isoamyl acetate (i.e., 2–6 weeks). Compared with controls that did not receive odor exposure, the experimental animals demonstrated significantly greater peak amplitudes to androstenone but not to isoamyl acetate. A replication of the experiment with C57BL/6J and AKR/J strains revealed a similar result, in that exposure to isovaleric acid increased the responses of the C57BL/6J mice but not the AKR/Js. The authors concluded that this induction of olfactory receptor sensitivity is genetically mediated.

Operant-olfactometry

Both Laska et al. (2006) and Bodyak and Slotnick (1999) used a Knosys olfactometer to deliver the odors assessed in their acuity experiments. This equipment provides control of the concentration, flow, and temporal parameters of monomolecular odors and is relatively easy to construct. It is combined with operant conditioning methods such that a mouse is systematically trained to insert its nose in an odor sampling tube until an odor is delivered. The mouse signals the presence of the odor by licking at a tube to receive water reinforcement. One problem with the use of olfactometers in the past has been the concern that odor contamination within the system could nullify experimental results. The multiple channel liquid dilution system can be easily cleaned with the use of pinch valves and disposable tubing and odor saturator containers. Specifying the precise molecular concentration of the stimulus is very difficult and may require assumptions about vapor saturation levels and other factors that are not easy to confirm (Dravnieks, 1975). For a multi-stage air dilution olfactometer, Bodyak and Slotnick (1999) suggest that concentration should be reported as the percent dilution of the odorant material, taking into account the original concentration of the odorant, the amount of air passed over the surface of the odorant material and the amount of clean air with which it is delivered. Olfactometers and training methods designed for research with rodents have been described in detail by Slotnick and Schellinck (2002).

Odor detection thresholds

Absolute olfactory threshold is defined as the lowest concentration of an odorant that can be detected (Doty and Kobal, 1995). The procedure for assessing odor detection threshold varies. A subject may be presented with odors sequentially from high to low concentrations (descending limits) or from low to high concentrations (ascending limits). A so-called staircase method in which the concentration is increased following an error in detection or decreased following detection of the odorant is also used. Clevenger and Restrepo (2006) have described the advantages and disadvantages of this methodology in detail. Organic compounds such as alcohols, aldehydes, esters, and acids are generally used to assess olfactory threshold (Passe and Walker, 1985; Walker and Jennings, 1991).

Most investigations of odor thresholds in mice have provided data for single odorants. For example, Walker and O'Connell (1986) determined that outbred laboratory mice could detect n-amyl acetate at a concentration between 1×10^{-12} and 1×10^{-13} molar when trained with both positive and negative reinforcement in an automated olfactometer. CF-1 mice were trained in an operant olfactometer to detect ethyl acetate that had been diluted to 0.00005% (Bodyak and Slotnick, 1999). Interestingly, rats have been reported to have a similar threshold for ethyl acetate (Apfelbach et al.,1991). The olfactory acuity for butanol was tested in two strains of mutant mice. The non-mutant littermates of staggerer mutants (C57BL/6 background) discriminated between 6.7×10^{-3} M amyl acetate and 5.5×10^{-3} M butanol (Deiss and Baudoin, 1997). Both wild-type and heterozygous reeler mice were able to detect 0.00003% butanol (Larson et al., 2003).

In a comprehensive examination of olfactory sensitivity, Laska et al. (2006) assessed the ability of male CD-1 mice to detect aliphatic aldehydes. Compared with an odorless solvent, the mice could discriminate extraordinarily low concentrations of n-butanal (1:300 million), n-pentanal (1:540 million), n-hexanal (1:4.2 million), n-heptanal (1:5.7 million, n-octanal (1:1.1 million) and n-nonanal (1:1.4 million). The authors noted that in contrast with findings in humans (Cometto-Muniz et al., 1998), the mice did not show a steep decline in olfactory sensitivity with increased carbon chain length of the aldehydes. One could speculate that this disparity may be a result of the differences in olfactory receptor numbers and organization between mice and humans. Further evidence of the superior ability of mice to detect odors has been found by Hu et al. (2007). Using an operant-olfactometer task, they demonstrated that C57Bl/6 mice could detect carbon dioxide (CO_2) at near atmospheric concentrations (e.g., 0.05%), whereas humans are only able to detect the compound trigeminally and then only at very high concentrations (i.e., 30%). This is also an instance where mice seem to have a lower threshold than rats. Several strains of rats had an average threshold of 0.5% CO_2 concentration when tested for acuity in an operant olfactometer (Ferris et al., 2007; Youngentob et al., 1991).

Specific anosmias

Some strains of mice show a lack of sensitivity to a particular odorant and are considered to have a specific anosmia. For example, C57BL/6J and C57BL/10J male mice are relatively insensitive to isovaleric acid, compared with A/J, AKR/J, BaLBC/cJ, C3HeB/feJ, DBA/2J, SJL.J, and SWR/J mice (Wysocki et al., 1977; see also Pourtier and Sicard, 1990). When compared with CBA/J mice, male NZB/B1NJ mice have a specific anosmia for androstenone (Wysocki et al., 1977). Hyposmias have also been described in mutant strains of mice. For example, reeler mice have a high threshold for several odors (Caviness and Sidman, 1972) and compared with their non-mutant littermates, staggerer male mice had higher behavioral thresholds for both an aversive odor, butanol, and an attractive odor, vanillin, in that they did not avoid as high concentrations of butanol and they did not find as low concentrations of vanillin attractive (Deiss and Baudoin, 1997). As other odorants were not tested it is not clear how specific these anosmias are. Nonetheless, it should not be surprising that genetically altered mice such as the former have specific hyposmias. The genes coding for olfactory receptors occur in clusters on 18 of the 20 mouse chromosomes with over half on chromosome 7, 9, and 11 but none on chromosome 12 or the Y chromosome (Zhang and Firestein, 2002). As a result of this broad distribution, as well as the large numbers of olfactory genes, it seems possible that during the construction of mutant strains of mice specific olfactory receptor genes could be deleted resulting in anosmia for particular odorants.

Mediating factors

In humans, the ability to detect odors is influenced by age, gender, and hormonal status. Females have a lower detection threshold for odors than males (Brand and Millet, 2001; Doty, 1986; Koegla and Koster, 1974) and the ability to detect odors also declines with normal aging (Kaye et al., 1994; Kovacs, 2004; Stevens and Cain, 1987; Stevens and Dardarwala, 1993). Moreover, olfactory acuity varies with the phase of the menstrual cycle such that women are more sensitive during ovulation (Doty et al., 1981; Navarrette-Palacios et al., 2003). In mice, the influence of these factors is not as clear-cut.

Age

Walker and O'Connell (1986) reported that little change in acuity was found when adult male mice were retested in an operant-olfactometer on their amyl acetate threshold a year following their original testing, i.e., at approximately 14 months of age. Rats tested at 24 months of age in an air-dilution olfactometer had a lower sensitivity than 13-month-old animals (Kramer and Apfelbach, 2004), so it is possible that a reduction in acuity would also occur in mice after sufficient time had passed. Electrophysiological responses to odors declined with age in the senescence accelerated mouse (SAM-PI) as the number of receptor cells for the specific odor used (vanillin) decreased with age (Nakayasu et al., 2000). In a digging task, C57Bl6/J and DBA mice learned to discriminate between phenyethyl acetate and linalool equally well at 6, 12, and 24 months of age but in this instance moderately high concentrations of odorants were used (Wong and Brown, 2006). Given the differences in methodology and odor concentrations used in the latter two experiments, no firm conclusions can be drawn at this time regarding the possibility of deleterious effects of aging on acuity in mice.

Sex

In most studies, only the olfactory acuity of male mice has been examined as sensitivity to odors varies with state of estrus in females. When the sensitivity of both males and females to odors has been assessed, the primary interest has been the role of gender in determining the response to urinary odors of conspecifics. In a habituation–dishabituation paradigm, female CBA × C57BL/6 F1 hybrids were shown to respond to lower concentrations of urinary odors from conspecifics than males (Baum and Keverne, 2002). The sensitivity of behaving male and female mice to urinary "pheromones" has also been assessed through electrophysiological recordings in the main olfactory system (Lin et al., 2005). Responses in males and females differed depending upon the predominant component in the urine. Because of the complex composition of a urinary stimulus, it is impossible to generalize these findings to differences in acuity for non-social odors.

Hormonal status

Both estrogen and testosterone have been implicated in the modulation of olfactory perception. In female albino mice, olfactory sensitivity was found to correlate with stage of estrus cycle (Schmidt and Schmidt, 1980). Evoked potential readings from permanently implanted electrodes in the olfactory bulb demonstrated that maximum odor sensitivity was during proestrus; in metestrus, the olfactory thresholds for geraniol, butyric acid, and butyric methyl ester were one million times higher than during proestrous (Schmidt and Schmidt, 1980). It is not clear how these results are related to behavior. Female mice have been reported to be attracted to presentations of isovaleric acid during proestrous but found it aversive during estrus and diestrus. They ignored this odor during metestrus (Gladysheva et al., 1992). In intact male BALB mice, the presentation of 1 mM isovaleric acid produced aversive behavior whereas in gonadectomized mice no such effect was found (Troitskaya and Gladysheva, 1997).

Thyroxine has also been found to influence olfactory acuity. Mice made hypothyroid with propylthiouracil were found to be anosmic as assessed by time spent sniffing food versus water odors following fasting (Beard and Mckay-Sim, 1987).

Previous experience

In mice, as well as in other mammals, extensive pre-exposure may increase sensitivity to an odorant (Dalton et al., 2002; Voznessenskaya et al., 1999; Yee and Wysocki, 2001). In NMRI mice, preweanlings reared with geraniol demonstrated an increased olfactory bulb EEG response to this odor during adulthood (Schmidt and Eckert, 1988). Appropriate attention to the odors selected for testing should prevent confounding of experimental results with odor experience. If social odors such as urine are used as test stimuli, however, such factors as dominance status and previous sexual experience lead to an enhanced olfactory preference that may be a result of increased acuity (Brown and Mcdonald, 1985). Thus, the subject's social characteristics must be taken into consideration when interpreting the results.

The possibility of adaptation to odor stimuli should also be taken into consideration when assessing acuity. If the interval between odor presentations is too short, an animal may adapt to an odor and become less sensitive to its presence. Adaptation and cross-adaptation for qualitatively similar odorants has been studied primarily in humans (e.g., Pierce et al., 1993), but adaptation has also been demonstrated in mice (Kadohisa and Wilson, 2006).

Trigeminal detection

Assessment of olfactory perception may be confounded by cues that are mediated by trigeminal receptors in the nasal vault or cornea and sensory receptors in the trachea that can respond to certain vapors. Trigeminal receptors respond to irritants and, while they generally have a much higher threshold than olfactory sensory receptors, can respond to high concentrations of many commonly used odorants such as menthol, cineole, acetic acid, acetone, ethanol, and propanol (Laska et al., 1997). We suggest that to minimize or eliminate the possible contribution of these systems, experimenters should use odorant concentrations that are non-irritating to human judges or odorants for which the concentrations are known to be below the threshold for these receptors (Slotnick and Schellinck, 2002). Trigeminal and respiratory thresholds have been established for a number of odorants (Cometto-Muniz et al., 2002; Nielsen et al., 1984; Silver, 1992; Silver et al., 1986).

Conclusions

Determination of olfactory acuity, defined as absolute detection threshold, requires accurate measurement and reporting of the concentration of the odorant, precise control of odor delivery, and standardized training methods. This review of the literature would suggest that, currently, the wide variation in procedures and methodology does not provide us with the opportunity to know if acuity is always being appropriately measured. In addition, differences in motivation as well as genetic influences on taste and olfaction may confound the results. To advance our understanding of the capacity of mice to detect odors, it would be appropriate to adopt more standard methods of experimentation so that interlaboratory comparisons of experimental findings are possible. A number of recent studies (Clevenger and Restrepo, 2006; Laska et al., 2006; Pho et al., 2005; Wesson et al., 2006) that used olfactometers to measure acuity also carefully calculated odorant concentrations and are illustrative of the appropriate methodology for assessing olfactory acuity in mice.

Acknowledgments

This work was supported by an NSERC Discovery Grant to H. S. Michael J. Wong provided editorial assistance.

References

Abate, P., Varlinskaya, E.I., Cheslock, S.J., Spear, N.E., and Molina, J.C. (2002) Neonatal activation of alcohol-related prenatal memories: impact on the first suckling response. *Alcsm Clin Exp Res* 26: 1512–1522.

Apfelbach, R., Blanchard, C.D., Blanchard, R.J., Hayes, R.A., and McGregor, I.S. (2005) The effects of predator odors in mammalian prey species: a review of field and laboratory studies. *Neurosci Biobehav Rev* 29: 1123–1144.

Apfelbach, R., Russ, D., and Slotnick, B. (1991) Olfactory development and sensitivity to odors in rats. *Chem Senses* 16: 209–218.

Baum, M.J. and Keverne, E.B. (2002) Sex difference in attraction thresholds for volatile odors from male and estrous female mouse urine. *Horm Behav* 41: 213–219.

Beard, M.D. and Mckay-Sim, A. (1987) Loss of sense of smell in adult, hypothyroid mice. *Brain Res* 433: 181–189.

Berger-Sweeney, J., Libbey, M., Arters, J., Junagadhwalla, M., and Hohmann, C.F. (1998) Neonatal monoaminergic depletion in mice (*Mus musculus*) improves performance of a novel odor discrimination task. *Behav Neurosci* 112: 1318–1326.

Bodyak, N. and Slotnick, B. (1999) Performance of mice in an automated olfactometer: odor detection, discrimination and odor memory. *Chem Senses* 24: 637–645.

Brand, G. and Millet, J.L. (2001) Sex differences in human olfaction: between evidence and enigma. *Q J Exp Psychol B* 54: 259–270.

Brown, R.E. and Mcdonald, D.W. (1985) *Social Odours in Mammals*. Oxford University Press, Oxford.

Caviness, V.S. and Sidman, R.L. (1972) Olfactory structures of the forebrain in the reeler mutant mouse. *J Comp Neurol* 145: 54–104.

Cleland, T.A. and Narla, V.A. (2003) Intensity modulation of olfactory acuity. *Behav Neurosci* 117: 1434–1440.

Clevenger, A.C and Restrepo, D. (2006) Evaluation of the validity of a maximum likelihood adaptive staircase procedure for measurement of olfactory detection threshold in mice. *Chem Senses* 31: 9–26.

Cometto-Muniz, J.E., Cain, W.S., and Abraham, M.H. (1998) Nasal pungency and odor of homologous aldehydes and carboxylic acids. *Exp Brain Res* 118: 180–188.

Cometto-Muniz, J.E., Cain, W.S., Abraham, M.H., and Gola, J.M. (2002) Psychometric functions for the olfactory and trigeminal detectbility of butyl acetate and toluene. *J Appl Toxicol* 22: 25–30.

Dalton, P., Doolittle, N., and Breslin, P.A. (2002) Gender-specific induction of enhanced sensitivity to odors. *Nat Neurosci* 5: 199–200.

Deiss, V. and Baudoin, C. (1997) Hyposmia for butanol and vanillin in mutant staggerer male mice. *Physiol Behav* 61: 209–213.

Doty, R.L. (1986) Odor-guided behavior in mammals. *Experientia* 42: 257–271.

Doty, R.L. and Kobal, G. (1995) Current trends in the measurement of olfactory function. In Doty, R.L. (ed.), *Handbook of Olfaction and Gustation*, Marcel Dekker Inc., New York, p. 196.

Doty, R.L., Snyder, P., Huggins, G., and Lowry, L.D. (1981) Endocrine, cardiovascular, and psychological correlates of olfactory sensitivity changes during the human menstrual cycle. *J Comp Physiol Psychol* 95: 45–60.

Dravnieks, A. (1975) Instrumental aspects of olfactometry. In Moulton, D.G., Turk, A., and Johnston, J.W. (eds.), *Methods in Olfactory Research*, Academic Press, New York, pp. 1–61.

Ferris, K.E., Clark, R.D. and Coates, E.L. (2007) Topical inhibition of nasal carbonic anhydrase affects the CO_2 detection threshold in rats. *Chem Senses* 32: 263–271.

Garcia, J., Hankins, W.G., and Rusiniak, K.W. (1974) Behavioral regulation of the milieu interne in man and rat. *Science* 185: 824–831.

Gheusi, G., Cremer, H., McLean, H., Chazal, G., Vincent, J.D., and Lledo, P.M. (2000) Importance of newly generated neurons in the adult olfactory bulb for odor discrimination. *Proc Natl Acad Sci USA* 97: 1823–1828.

Gladysheva, O.S., Troitskaya, V.T., and Georgieve, T.V. (1992) Behavioral reactions to isovaleric acid in female mice during the estrus cycle. *Sens Syst* 6: 254–257.

Griff, I.C. and Reed, R.R. (1995) The genetic basis for specific anosmia to isovaleric acid in the mouse. *Cell* 83: 407–414.

Halpin, Z.T. (1974) Individual differences in the biological odors of the Mongolian gerbil (*Meriones unguiculatus*). *Behav Biol* 11: 253–259.

Hepper, P.G. (1990) Feotal olfaction. In Macdonald, D.W., Muller-Schwarze, D., and Natynczuk, S.E. (eds.), *Advances in Chemical Signals in Vertebrates V*. Oxford University Press, Oxford, pp. 282–286.

Horowitz, G.P. and Whitney, G. (1975) Alcohol-induced conditioned aversion: genotypic specificity in mice (*Mus musculus*). *J Comp Physiol Psychol* 89: 340–346.

Hu, J., Zhong, C., Ding, C., Chi, Q., Walz, A., Mombaerts, P., et al. (2007) Detection of near-atmosphere concentrations of CO_2 by an olfactory subsystem of the mouse. *Science* 317: 953–957.

Kadohisa, M. and Wilson, D.A. (2006) Olfactory cortical adaptation facilitates detection of odors against background. *J Neurophysiol* 95: 1888–1896.

Kaye, J.A., Oken, B.S., Howieson, D.B., Howieson, J., Holm, L.A., and Dennison, K. (1994) Neurologic evaluation of the optimally healthy oldest old. *Arch Neurol* 51: 1205–1211.

Koegla, H.S. and Koster, E.P. (1974) Some experiments on sex differences in odor perception. *Ann N Y Acad Sci* 237: 234–246.

Korsching, S.I. (2001) Odor maps in the brain: spatial aspects of odor representation in sensory surface and olfactory bulb. *Cell Mol Life Sci* 58: 520–530.

Kovacs, T. (2004) Mechanisms of olfactory dysfunction in aging and neurodegenerative disorders. *Ageing Res Rev* 3: 215–232.

Kramer, S. and Apfelbach, R. (2004) Olfactory sensitivity, learning and cognition in young adult and aged Wistar rats. *Physiol Behav* 81: 435–442.

Larson, J., Hoffman, J.S., Guidotti, A., and Costa, E. (2003) Olfactory discrimination learning in heterozygous reeler mice. *Brain Res* 971: 40–46.

Laska, M., Distel, H., and Hudson, R. (1997) Trigeminal perception of odorant quality in congenitally anosmic subjects. *Chem Senses* 22: 447–456.

Laska, M., Joshi, D., and Shepherd, G.M. (2006) Olfactory sensitivity for aliphatic aldehydes in CD-1 mice. *Behav Brain Res* 167: 349–354.

Lee, A.W., Emsley, J.G., Brown, R.E., and Hagg, T. (2003) Marked differences in olfactory sensitivity and apparent speed of forebrain neuroblast migration in three inbred strains of mice. *Neurosci* 118: 263–270.

Lin, D.Y., Zhang, S.G., Block, E., and Katz, L.C. (2005) Encoding social signals in the mouse main olfactory bulb. *Nature* 434: 470–477.

Maruniak, J.A., Darney, K.J., and Bronson, F.H. (1975) Olfactory perception of the nonsocial environment by male house mice. *Behav Biol* 14: 237–240.

Mihalick, S.M., Langlois, J.C., Krienke, J.D., and Dube, W.V. (2000) An olfactory discrimination procedure for mice. *J Exp Anal Behav* 73: 305–318.

Mori, K., Takahashi, Y.K., Igarashi, K.M., and Yamaguchi, M. (2006) Maps of odorant molecular features in the mammalian olfactory bulb. *Physiol Rev* 86: 409–433.

Nakayasu, C., Kanemura, F., Hirano, Y., Shimizu, Y., and Tonosaki, K. (2000) Sensitivity of the olfactory sense declines with the aging in senescence-accelerated mouse (SAM-P1). *Physiol Behav* 70: 135–139.

Navarrette-Palacios, E., Hudson, R., Reyes-Guerrero, G., and Guevara-Guzmán, R. (2003) Correlation between cytological characteristics of the nasal epithelium and the menstrual cycle. *Arch Otolaryngol Head Neck Surg* 129: 460–463.

Nielsen, G.D., Bakbo, J.C., and Holst, E. (1984) Sensory irritation and pulmonary irritation by airborne allyl acetate, allyl alcohol, and allyl ether compared to acrolein. *Acta Pharmacol Toxicol* 54: 292–298.

Passe, D.H. and Walker, J.C. (1985) Odor psychophysics in vertebrates. *Neurosci Biobehav Rev* 9: 431–467.

Pedersen, P.E. and Blass, E.M. (1982) Prenatal and postnatal determinants of the first suckling episode in albino rats. *Dev Psychobiol* 15: 349–355.

Pho, V., Butman, M.L., and Cherry, J.A. (2005) Type 4 phosphodiesterase inhibition impairs detection of low odor concentration in mice. *Behav Brain Res* 161: 245–253.

Pierce, J.D., Wysocki, C.J., and Aronov, E.V. (1993) Mutual cross-adaptation of the volatile steroid androstenone and a non-steroidal functional analog. *Chem Senses* 18: 245–256.

Pourtier, L. and Sicard, G. (1990) Comparison of the sensitivity of C57BL/6J and AKR/J mice to airborne molecules of isovaleric acid and amyl acetate. *Behav Genet* 20: 499–509.

Schellinck, H.M, Arnold, A., and Rafuse, V.F. (2004) Neural cell adhesion molecule (NCAM) null mice do not show a deficit in odour discrimination learning. *Behav Brain Res* 152: 327–334.

Schellinck, H.M., Forestell, C.A., Dill, P., Lolordo, V.M., and Brown, R.E. (2001a) The development of an associative olfactory test of learning and memory. In Apfelbach, R., and Muller-Schwarze, D. (eds.), *Chemical Signals in Vertebrates VII*. Pergamon Press, New York, pp. 459–466.

Schellinck, H.M., Forestell, C., and Lolordo, V. (2001b) A simple and reliable test of olfactory learning and memory in mice. *Chem Senses* 26: 663–672.

Schellinck, H.M., Rooney, E., and Brown, R.E. (1995) Odors of individuality of germfree mice are not discriminated by rats in a habituation–dishabituation procedure. *Physiol Behav* 57: 1005–1008.

Schmidt, C. and Schmidt, U. (1980) Changes of olfactory sensitivity during the estrus cycle in female laboratory mice. *Chem Senses* 5: 359–365.

Schmidt, U. and Eckert, M. (1988) The influence of early odour experience on the neural response of the olfactory bulb in laboratory mice. *J Comp Physiol* 16: 771–776.

Silver, W.L. (1992) Neural and pharmacological basis for nasal irritation. *Ann N Y Acad Sci* 641: 152–163.

Silver, W.L., Mason, J.R., Adams, M.A., and Smeraski, C.A. (1986) Nasal trigeminal chemoreception: responses to n-aliphatic alcohols. *Brain Res* 376: 221–229.

Slotnick, B.M. and Schellinck, H. (2002) Methods in olfactory research with rodents. In Simon, S.A. and Nicolelis, M. (eds.), *Frontiers and Methods in Chemosenses*. CRC Press, New York, pp. 21–61.

Smotherman, W.P. (1982) Odor aversion learning by the rat fetus. *Physiol Behav* 29: 769–771.

Stevens, J.C. and Cain, W.S. (1987) Old-age deficits in the sense of smell as gauged by thresholds, magnitude matching, and odor identification. *Psychol Aging* 2: 36–42

Stevens, J.C. and Dardarwala, A.D. (1993) Variability of olfactory threshold and its role in assessment of aging. *Percept Psychophys* 54: 296–302.

Sullivan, R.M. and Leon, M. (1987) One trial learning enhances olfactory bulb responses to an appetitive conditioned odor in 7-day-old rats. *Brain Res* 432: 301–311.

Sundberg, H., Doving, K., Novikov, S., and Ursin, H. (1982) A method for studying responses and habituation to odors in rats. *Behav Neural Biol* 34: 113–119.

Troitskaya, V.T. and Gladysheva, O.S. (1997) Effect of gonadectomy and subsequent injection of testosterone on the olfactory behavior of male mice. *Sens Syst* 11: 156–162.

Twsuboi, A., Miyazaki, T., Imai, T., and Sakano, H. (2006) Olfactory sensory neurons expressing class I odorant receptors converge their axons on an anterodorsal domain of the olfactory bulb in the mouse. *Eur J Neurosci* 23: 1436–1444.

Voznessenskaya, V.V., Wysocki, C.J., Chukhrai, E.S., Poltorack, O.M., and Atyaksheva, F. (1999) Long lasting effects of chemical exposures in mice. In Johnston, R.E. (ed.), *Advances in Chemical Signals in Vertebrates.* Plenum Press, New York, pp. 563–571.

Walker, J.C. and Jennings, R.A. (1991) Comparison of odor perception in humans and animals. In Laing, D.G., Doty, R.L., and Breipohl, W. (eds.), *The Human Sense of Smell.* Springer, Berlin, pp. 261–280.

Walker, J.C. and O'Connell, R.J. (1986) Computerized odor psychophysical testing in mice. *Chem Senses* 11: 439–453.

Wang, H.W., Wysocki, C.J., and Gold, G.H. (1993) Induction of olfactory receptor sensitivity in mice. *Science* 260: 998–1000.

Wesson, D.W., Keller, M., Douhard, Q., Baum, M.J., and Bakker, J. (2006) Enhanced urinary odor discrimination in female aromatase knockout (ArKO) mice. *Horm Behav* 49: 580–586.

Wilson, D.A., Best, A.R., and Sullivan, R.M. (2004) Plasticity in the olfactory system: lessons for the neurobiology of memory. *Neuroscientist* 10: 513–524.

Wong, A. and Brown, R.E. (2007) Age-related changes in visual acuity, learning and memory in C57BL/6J and DBA/2J mice. *Neurobiol Aging* 28: 1577–1593.

Wysocki, C.J., Whitney, G., and Tucker, D. (1977) Specific anosmia in the laboratory mouse. *Behav Gen* 7: 171–187.

Yee, K.K. and Wysocki, C.J. (2001) Odorant exposure increases olfactory sensitivity: olfactory epithelium is implicated. *Physiol Behav* 72: 705–711.

Youngentob, S., Hornung, D., and Mozell, M. (1991) Determination of carbon dioxide in detection thresholds in trained rats. *Physiol Behav* 49: 21–16.

Zagreda, L., Goodman, J., Druin, D.P., McDonald, D., and Diamond, A. (1999) Cognitive deficits in a genetic mouse model of the most common biochemical cause of human mental retardation. *J Neurosci* 19: 6175–6182.

Zhang, X. and Firestein, S. (2002) The olfactory receptor gene superfamily of the mouse. *Nat Neurosci* 5: 24–133.

Zou, Z., Li, F., and Buck, L.B. (2005) Odor maps in the olfactory cortex *Proc Natl Acad Sci USA* 102: 7724–7729.

Zufall, F. and Munger, S.D. (2001) From odor and pheromone transduction to the organization of the sense of smell. *Trends Neurosci* 24: 191–193.

Chapter 9

Taste genetics

John D. Boughter, Jr.

Introduction: taste is a complex trait

The search for a genetic basis to taste ability began over 75 years ago with the serendipitous discovery of human differences in sensitivity to the bitter-tasting compound phenylthiocarbamide (PTC) (Fox, 1931, 1932). This discovery prompted considerable research into the genetic basis of taste in both man and mouse, which may be considered to culminate in the recent discovery of gene families for receptors for bitter and sweet taste (for recent reviews, see Drayna, 2005; Roper, 2007; Scott, 2005). Linkage analyses of bitter and sweet solution intake in mice identified gene intervals that were indicated in later studies to harbor G-protein-coupled receptors (GPCRs). The function and specificity of taste receptors continues to be elucidated based on either *in vivo* expression assays or transgenic approaches. Current studies build on the discovery of taste receptors by further dissection of common elements in taste cell signaling pathways (Finger et al., 2005; Roper, 2009), examining how information is transmitted to specific populations of taste neurons in the brain (Lemon and Katz, 2007; Sugita and Shiba, 2005), and the role these specific receptors and second messengers play in taste-specific behaviors such as stimulus recognition and discrimination (Delay et al., 2006; Dotson et al., 2005).

Beyond its inherent interest as a model of sensory function, it is also important to place the gustatory system, and indeed the genetic basis of this evolutionarily fundamental sense, in the context of the larger set of behaviors to which it is inextricably linked. The taste system functions in the initial monitoring of foods and liquids and as such is anatomically situated at the start of the alimentary canal. It is a key player in larger homeostatic systems dealing with food and fluid intake (Figure 9.1). Consider an extreme example of perturbing taste, and its effect on human health: clinical studies demonstrate that loss of taste sensitivity or *ageusia*, resulting from surgery or disease, decreases appetite and may lead to anorexia (Woschnagg et al., 2002). Taste, feeding, and reward are linked in terms of reciprocal connectivity between brain centers (Yamamoto, 2006), and in recent studies that provide important insights into the basis of disorders such as obesity and diabetes (Donaldson et al., 2009; Hajnal et al., 2005; Tepper and Seldner, 1999). Intriguingly, evidence also suggests a link between taste sensitivity and addiction to orally ingested drugs such as alcohol or nicotine (Blednov et al., 2008; Oliveira-Maia et al., 2009), a connection that was first noted in the 1960s by G. E. McClearn (Rodgers and McClearn, 1964).

The focus of this chapter is on the genetic basis of gustatory behavior, in the most common models, humans and rodents. In reviewing these studies of taste in man and mouse it is of great importance to try to understand the genetic basis of behaviors in the context of an animal's natural habitat and surroundings. It should also be pointed out that *Drosophila* is a promising model for studying taste genetics, as recent advances have been made both in elucidating fly taste receptor gene families, and in understanding the organization of the fly central nervous system (CNS) taste pathways (Colomb and Stocker, 2007; Gordesky-Gold et al., 2008; Scott et al., 2001).

Summary of the taste system

The gustatory system recognizes sapid stimuli classified into several qualities, including sweet, salty, bitter, and sour. A fifth quality, called "umami," refers to the savory taste evoked by certain cyclic nucleotides and L-type amino acids such as glutamic acid (monosodium glutamate; MSG). The initial events in taste processing occur in taste buds, structures found in the epithelia of the tongue, palate, larynx, and epiglottis of mammals, which contain approximately 50–100 cells of neuroepithelial origin (Kinnamon, 1987). Taste buds contain morphologically distinct cell types including elongate, spindle-shaped cells that function as taste receptor cells (TRCs). The TRCs possess apical processes with microvilli that protrude into the oral cavity and interact with taste stimuli. These microvilli contain taste receptors and taste-responsive ion channels. Tight junctions between the cells in a taste bud restrict the access of most stimuli to the apical membranes of the cells. Multiple signaling cascades in TRCs have been described, and there are fundamental differences in the transduction pathways associated with each stimulus quality (Scott, 2005; Roper, 2007). For example, salt and acid stimuli directly permeate or gate apical ion channels, causing TRC depolarization, while umami-, sweet-, and bitter-tasting stimuli are thought to predominantly activate taste receptors. Current molecular and

Behavioral Genetics of the Mouse: Volume I. *Genetics of Behavioral Phenotypes*, eds. Wim E. Crusio, Frans Sluyter, Robert T. Gerlai, and Susanna Pietropaolo. Published by Cambridge University Press.

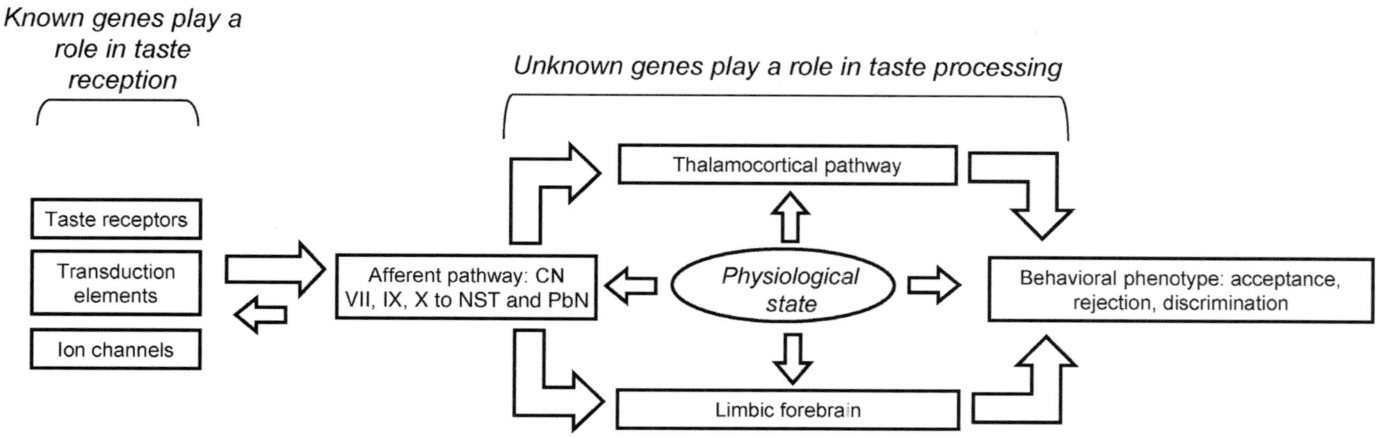

Figure 9.1 Schematic diagram of the gustatory pathway, and the search for genes underlying variation in gustatory behaviors. Taste receptor cells are innervated by one of three cranial nerves (VII, IX, or X), which project into the nucleus of the solitary tract (NST); ascending fibers connect to the parabrachial nuclei (PbN) of the pons, from which two parallel pathways emerge. One pathway carries taste information to the insular cortex (IC) via the ventral posterior medial nucleus, parvicellularis (VPMpc) of the thalamus. The other pathway projects into areas of the limbic forebrain involved in food and water regulation, reinforcement, reward, and stress. Thus, taste processing is subject to modulation by changes in physiological state. The search for the genetic basis of taste has primarily focused on taste receptors and other components of the taste cell signaling pathway. Other genes must play a role in the complex interactions between central nervous system taste centers, physiological systems dealing with motivation, environmental factors, and behavioral outcomes.

genetic/genomic approaches have focused heavily on the discovery and functional significance of taste transduction mechanisms, especially GPCRs.

Taste receptors can be divided into two families, T1R (sweet and umami taste) and T2R (bitter taste). Three *Tas1r* genes in both humans and mice encode three receptors, T1R1, T1R2, and T1R3. The T1Rs are class C-like GPCRs and share a large N-terminal domain, which comprises about half of the protein length. Expression studies suggest T1Rs function primarily as heteromultimeric complexes, with T1R2 + T1R3 comprising a receptor for sweet-tasting stimuli, while T1R1 + T1R3 functions as a receptor for umami stimuli. In contrast, there are about 30 functional *Tas2r* genes in mammals that encode a family of T2R bitter taste receptors. The T2Rs are similar to class A GPCRs and are characterized by a short N-terminus. At present, only a subset of T2Rs have been linked to the detection of particular bitter-tasting compounds (Behrens and Meyerhoff, 2009).

Taste information from taste buds is carried to the brain via cranial nerves, which synapse in the nucleus of the solitary tract (NST) in the brainstem (Lundy and Norgren, 2004). Gustatory neurons in the NST of rodents send second-order projections to the parabrachial nuclei (PBN) of the pons (Norgren and Leonard, 1973), which is an obligatory relay in the taste system of most mammals. In these species, the gustatory projection parallels that of the visceral afferent system. The close association between taste and motivational systems is most clearly seen in the anatomy of the forebrain projections in rodents. From the PBN, the ascending gustatory projection follows two pathways: a thalamocortical pathway and a limbic forebrain pathway (Lundy and Norgren, 2004). Like other sensory systems, taste projects into a thalamic nucleus, specifically the ventral posterior medial nucleus, parvicellularis (VPMpc). From the VPMpc, a cortical projection goes to the insular cortex. In addition to this thalamocortical projection, there exist reciprocal connections from the NST and PBN with several areas of the limbic forebrain, including the lateral hypothalamus, the central nucleus of the amygdala, and the bed nucleus of the stria terminalis (Yamamoto, 2006). These areas are important in food and water regulation, responses to stress, conditioned taste aversion learning, and mechanisms of reinforcement, reward, and drug addiction. Electrophysiological studies have shown the existence of taste-responsive neurons in several of these areas (Fontanini et al., 2009; Nishijo et al., 1998; Norgren, 1970; Yamamoto et al., 1989).

Bitter taste genetics

Variation in taste sensitivity for the bitter anti-thyroid compound PTC is a classic example of an inherited trait in humans, and has been the source of much genetic and sensory research since the 1930s (Bartoshuk et al., 1994; Guo and Reed, 2001; Kim and Drayna, 2005). The discovery that some humans could taste PTC while others could not is usually attributed to Fox (1931). Early studies established a dimorphism among human subjects for this substance, i.e., "tasters" or "non-tasters," and that the inheritance of this phenotype was controlled by a single autosomal locus with one dominant (taster) and one recessive (non-taster) allele (Blakeslee, 1932; Fox, 1932; Snyder, 1931). Non-tasters were said to have "taste blindness" – the phenomenon thought to be analogous perhaps to color blindness. Differences between tasters and non-tasters were not necessarily discrete, however (Blakeslee, 1932; Harris and Kalmus, 1949), and more recent studies postulated that PTC inheritance was better described by a polygenic model (Olson et al., 1989; Reddy and Rao, 1989).

A development of obvious importance to the study of the genetic basis of PTC taste was the discovery in 2000, through data mining of mouse and human genome databases, of the putative bitter taste receptor gene family (Adler et al., 2000; Matsunami et al., 2000). A Quantitative trail loci (QTL) study conducted in Utah CEPH (Centre d'Etude du Polymorphisme Humaine) families demonstrated linkage of PTC taste sensitivity to chromosome 7q, with a secondary locus on Chr 16p (Drayna et al., 2003). Further genotyping and a candidate search narrowed down the search to a solitary bitter taste receptor gene, dubbed *TAS2R38* (Kim et al., 2003). Like other *Tas2rs*, this gene contained a single coding exon, encoding a 333 amino acid, 7 transmembrane-domain GPCR. Within the single coding region of the gene, three common single neuclotide polymorphisms (SNPs) were described that result in amino acid changes in the protein, differentiating taster and non-taster haplotypes, and accounting for 55–85% of the variance in human PTC taste sensitivity. The most common two, PAV and AVI, are the major taster and non-taster haplotypes, respectively. Seven different haplotypes have now been described among worldwide human populations, determined by SNPs at five different positions (Wooding et al., 2004).

Because the majority of human taste genetics studies centered around the PTC polymorphism, investigators searched for a parallel finding using animal models. There was some evidence for a PTC dimorphism in chimpanzees, and the existence of threshold differences within other primate species suggested that taste polymorphisms were not restricted to humans (Chiarelli, 1963; Fischer et al., 1939). Richter and Clisby (1941) reported individual variation in PTC avoidance thresholds among rats, with a threshold comparable to humans; a few rats had thresholds in the non-taster range.

In mice, the search for taste polymorphisms has traditionally involved measuring intake of a taste solution using 24-hour two-bottle tests (solution vs. water). The taste quality and/or intensity of the stimulus are then inferred based on level of avoidance or preference for the solution bottle. Such a procedure is particularly amenable for testing the large numbers of mice required for quantitative analysis, although measuring behavior over this relatively long time frame does not dissociate an animal's ability to recognize or discriminate a stimulus from post-ingestive consequences (St. John and Spector, 2008). Taste-salient, "brief-access" tests minimize post-ingestive cues by restricting sampling times to short durations (<30 s). Such tests have been recently modified for use in high-throughput screening of taste function in mouse populations (Boughter et al., 2002; Nelson et al., 2005).

The measurement of PTC solution intake in mice provided a compelling example of the difficulty of accurately relating taste phenotypes in humans with those in rodents. Klein and DeFries (1970) examined PTC two-bottle intake and found "threshold" (i.e., the concentration where avoidance begins) differences among inbred strains extending into the human taster range. A Mendelian cross (F1, F2, F3, B1, and B2 generations) between BALB/cIbg (taster) and C57BL/Ibg (non-taster) strains yielded segregation ratios consistent with phenotypic control by a single autosomal locus with a dominant taster allele. As later attempts were made to replicate this finding, it was realized that PTC avoidance by BALB mice only developed gradually across 10–12 consecutive days of testing (Whitney and Harder, 1986). Phenylthiocarbamide is extremely toxic to mice (oral LD_{50} 10 mg/kg), comparable to strychnine (oral LD_{50} 2 mg/kg). A direct comparison of two-bottle intake tests and brief-access procedures demonstrated that differential aversion across inbred strains to PTC depended on the quantity of the stimulus that was consumed, and not on immediate taste cues (Nelson et al., 2003). Injected intraperitoneally, PTC was nearly as effective as LiCl in conditioning a taste aversion (CTA) to sucrose (St. John et al., 2005). The strain difference, and its relatively simple genetic basis, likely reflects other variables besides sensory ability, such as sensitivity to toxicity or to the goitrogenic effects of the stimulus, or differences in the ability or speed of acquisition of the CTA (Risinger and Cunningham, 2000).

An interesting comparison can be made for two other less toxic bitter taste compounds, quinine and sucrose octaacetate (SOA; a synthetic acetylated sugar). Substantial inbred strain variation in two-bottle intake/aversion exists for sensitivity to either of these compounds; for SOA, the strain difference is particularly robust as it occurs over a 2–3 log-step M concentration range (Lush, 1981; Warren and Lewis, 1970). Sucrose octaacetate aversion appears to be a fairly unambiguous model of monogenic control, whereas quinine intake/aversion has been demonstrated to be under polygenic control. Both phenotypes are linked to a region of distal mouse chromosome 6 that contains a cluster of bitter taste receptor *Tas2r* genes and two genes encoding salivary proline-rich proteins (PRPs), although the specific genes exerting a major effect on these phenotypes have yet to be elucidated (Bachmanov et al., 2001a; Nelson et al., 2005).

For SOA aversion, ratios among segregating generations of mice were consistent with variation at a single autosomal locus (*Soa*), with the taster allele dominant over either a "demitaster" (partial sensitivity) or non-taster allele (Harder et al., 1992; Lush, 1981), although at least one study indicated polygenic control of SOA taste (Le Roy et al., 1999). Unlike PTC, results of both taste-salient tests and peripheral electrophysiology yielded essentially concentration-by-concentration similarity with two-bottle data, confirming the gustatory-based nature of the strain difference (Boughter et al., 2002; Inoue et al., 2001). This is perhaps not surprising in light of the low toxicity and lack of apparent odor of this compound to mice. *Soa* was subsequently mapped to the same location as *Prp* on distal mouse chromosome 6 (Capeless et al., 1992; Lush et al., 1995). However, the salivary proteins encoded by the *Prp* genes (*Prp2* and *Prh1*) do not appear to play any role in modulating bitter taste (Harder et al., 2000).

If SOA is a synthetic compound not present in nature, what is the adaptive advantage among mice of a specific gene conferring sensitivity to it? Important clues could be discerned from studies using congenic mice. A congenic strain is one that is identical to a progenitor inbred strain except for the substitution of a particular locus of interest (in this case *Soa*), created through serial backcrossing (Flaherty, 1981). The newly created congenic mice can then be used to evaluate an allelic variant against a particular inbred strain background. To investigate possible pleotropic effects of *Soa*, Whitney and colleagues used as progenitor strains SW (taster), B6 (non-taster), or C3HeB/FeJ (demi-taster). Results of testing SOA congenics indicated that the chromosome 6 locus strongly affected aversion to other acetylated sugars, as well as a few other bitter compounds (Boughter and Whitney, 1998; Harder et al., 1996; Whitney et al., 1990). Intriguingly, the only compounds other than some acetylated sugars for which identical effects were found were strychnine and brucine, extremely toxic plant alkaloids. There are approximately 24 bitter taste receptor genes at the *Soa* locus in mice, but only one of these has been de-orphanized with respect to a functional ligand (*Tas2r105*; cycloheximide) (Chandrashekar et al., 2000). While it is certainly possible that unique taste receptor genes in this cluster underlie sensitivity to SOA and strychnine or brucine, the behavioral effects of these compounds (brucine is dimethoxystrychnine) in strain distribution patterns (SDPs), including congenics, at least raises the possibility that the gene in this case is linked to the detection of these toxic plant compounds.

Although a single locus affecting quinine intake/avoidance was also proposed on the basis of segregating generations (Lush, 1984), the balance of evidence suggests a polygenic mode of inheritance. Cross-breeding studies were not consistent with the ratios expected for a single gene, and SDPs for CXB/By, BXD/Ty, and BXH RI strains contained intermediate values and appeared to support polygenicity (Bachmanov et al., 1996; Boughter et al., 1992; Harder and Whitney, 1998; Lush, 1984). However, these and other studies demonstrated significant linkage of quinine two-bottle intake to the *Prp*⁻ (and subsequently *Tas2r*⁻) -containing region on chromosome 6 (Blizard et al., 1999; Lush and Holland, 1988). Moreover, this linkage was verified in a study of BXD RI mice using a highly taste-salient assay, suggesting that a great deal of the strain variation in quinine taste sensitivity was due to a polymorphism in a *Tas2r* bitter taste receptor gene (Nelson et al., 2005). The quinine taste QTL was restricted to a ~5 Mb interval that included 24 of the 25 discovered intact *Tas2r* genes. Notably, nearly all of these genes were polymorphic between the parent strains C57BL and DBA/2, and within each BXD RI line, every *Tas2r* originated from the same parent and was highly correlated with taste phenotype, leading the authors to conclude that the entire gene cluster was a haplotype for quinine taste. However, the polygenic nature of quinine intake also suggests the existence of other, possibly *Tas2r* independent, mechanisms that modulate acceptance and rejection behaviors to bitter compounds.

Sweet and umami taste genetics

Genetic variation affecting rodent preference for compounds classified as tasting sweet was reported decades ago (Hoshishima et al., 1962; Nachman, 1959). Rodgers and McClearn described a strong relationship between sucrose and ethanol intake in several strains of mice. C57BL mice displayed a significantly greater avidity for sucrose than did DBA/2 mice; this increased avidity went in-hand with a greater propensity to ingest ethanol (Rodgers and McClearn, 1962, 1964). Similarly, Fuller (1974) found that DBA/2 mice were indifferent to certain concentrations of sodium saccharin in two-bottle tests, whereas C57BL/6 mice strongly preferred it. Segregation ratios in a Mendelian cross of these strains were reported to be consistent with a two-locus model, and an autosomal gene (*Sac*), with a dominant allele for preference, was proposed. This genetic variation was thought to affect hedonic response rather than threshold sensitivity. Studies by Lush and others suggested that the *Sac* locus influenced sensitivity to sugars, some sweet-tasting amino acids, and several other non-caloric sweeteners (Capeless and Whitney, 1995; Lush, 1989; Peltz et al., 1973; Ramirez and Fuller, 1976). Thus, studies were supportive of a common genetic basis underlying sweet taste in mice.

Mapping studies postulated at least 1–2 loci on mouse chromosome 4 (Blizard et al., 1999; Lush et al., 1995; Ninomiya et al., 1992; Phillips et al., 1994) influencing sweet preference. Bachmanov et al. (1997) reported two distinct loci on chromosome 4 (including *Sac*) in a C57BL (preferrer) by 129 (non-preferrrer) cross that together accounted for more than 50% of the genetic variability in sucrose intake; these loci were shown to directly affect the magnitude and threshold of the peripheral electrophysiological response. Using a positional approach, the *Tas1r3* taste receptor gene was localized within the *Sac* interval (Bachmanov et al., 2001b). In the mouse, three genes, *Tas1r1*, *Tas1r2*, and *Tas1r3*, are located (but not tightly clustered) on chromosome 4; these genes encode three taste receptor subunits that collectively account for much of sweet taste sensitivity (Kitagawa et al., 2001; Max et al., 2001; Montmayeur et al., 2001; Nelson et al., 2001; Reed et al., 2004; Sainz et al., 2001). Heterologous expression studies confirmed that a T1R2 + T1R3 heterodimer responds to a broad variety of sweet-tasting compounds in a species-specific manner (Li et al., 2002; Nelson et al., 2001). In contrast, the T1R1 + T1R3 heterodimer responds to a battery of L-type amino acids, which are umami stimuli (Li et al., 2002; Nelson et al., 2002). The gene in common, *Tas1r3*, has two major haplotypes differentiated by one crucial SNP; these haplotypes segregate among inbred mouse strains according to behavioral sensitivity (Nie et al., 2005; Reed et al., 2004). Both subunits in the T1R2 + T1R3 heterodimer are crucial for binding sweet-tasting stimuli (Nie et al., 2005). With the exception of frogs, all vertebrates possess T1R1 and T1R3 genes. However, members of the family Felidae (cats) lack functional T1R2 genes (Li et al., 2005a). It has long been known that members of the cat family do not show behavioral sensitivity towards sugars.

Despite the broad ligand sensitivity of the T1R receptors, evidence abounds that other gene products play a role in sweet and savory tastes, acting either at the level of TRCs, or elsewhere. Knockout studies with *Tas1r* genes have produced equivocal results: heteromerization is necessary to create fully functional receptors, but T1Rs also likely function *in vivo* as less effective homomeric receptors (Damak et al., 2003; Nie et al., 2005; Zhao et al., 2003). Still, mice lacking T1R3 have similar detection thresholds to sucrose as wildtypes, and can discriminate among sweet and savory stimuli (Delay et al., 2006). Neither behavioral nor neural responses to glycine are strongly influenced by the *Tas1r3* genotype (Eylam and Spector, 2004; Inoue et al., 2004), suggesting the possibility of a non-T1R mechanism. D-phenylalanine is another sweet-tasting amino acid for which sensitivity in mice has been linked to a locus on chromosome 4, termed *Dpa*; evidence suggests this locus may also act on sweet taste independently of T1Rs (Ninomiya et al., 1992; Shigemura et al., 2005). Other possibilities for gene effects on sweet taste may be found upstream; in particular, sweet taste detection in mammals has been shown to be heavily modulated by post-ingestive feedback, including neural or hormonal factors (e.g., Atchley et al., 2005; Curtis et al., 2005; Hajnal et al., 1999, 2005). This feedback extends to the level of the TRCs and afferent nerves (Shigemura et al., 2004; Simon et al., 2003).

Salt and sour taste

Comparatively few studies have reported genetic variation affecting preference/aversion or taste sensitivity for salty or sour substances. Several ion channel species may be thought of as receptors for ionic taste stimuli: these channels interact directly with the stimulus and participate in the initial steps of taste transduction (Gilbertson and Boughter, 2003). A large portion of NaCl or salt taste is dependent on the influx of sodium through the selective epithelial channel ENaC. This type of channel is a member of the degenerin/ENaC superfamily of ion channels. Taste receptor cells express amiloride-blockable, Na^+-selective channels that share a number of physiological properties with the ENaCs. Sour (acid) taste stimuli are thought to depolarize TRCs either by activating cation channels or by directly permeating ion channels, including ENaC (Ishimaru et al., 2006).

Inbred strain differences have been reported for NaCl in both mice (Beauchamp, 1990; Hoshishima et al., 1962; Ninomiya et al., 1989) and rats (Midkiff et al., 1985). One strain difference of potential interest was related to both intake behavior and peripheral responsiveness (taste nerve recordings) to NaCl. Ninomiya et al. (1989) reported that C57BL and BALB mice possessed a greater level of aversion to high concentrations of NaCl than DBA/2 or C3H mice, and this variation appeared to be related to the amiloride (ENaC antagonist) sensitivity of the neural response. Moreover, a recent study found that phenotypic differences in salt taste among mice are linked to variation in the ENaC alpha subunit gene (Shigemura et al., 2008).

Taste and other complex phenotypes

As mentioned above, it has long been known that one of the most consistent predictors of ethanol intake in rodents is proclivity for consumption of sweet-tasting compounds. Positive correlations between ethanol and sweet intake have been noted in multiple strains of rats and mice, and a common genetic basis is strongly supported by a robust association of these phenotypes in selected lines and segregating populations (Bachmanov et al., 1996; Belknap et al., 1993; Dess et al., 1998; McClearn and Rodgers, 1959; Overstreet et al., 1993; Rodgers and McClearn, 1962; Stewart et al., 1994, 2003). *Tas1r3* null mice do not prefer ethanol relative to wildtype controls, further arguing for a direct role of the sweet taste receptor in mediating ethanol intake (Blednov et al., 2008). In rats, ethanol applied to the tongue selectively activates a subpopulation of sweet-responsive taste neurons, and this activity is blocked by application of gurmarin, a peripheral antagonist of sweet taste (Lemon et al., 2004). Evidence also exists for an association between alcohol drinking and sweet taste in humans (Yamamoto et al., 1991). Studies have shown that alcoholics prefer higher concentrations of sucrose compared with non-alcoholic control subjects, and enhanced sweet preference may be a factor in genetic risk for alcoholism (Kampov-Polevoy et al., 2003, 2004).

Other effects on orally ingested drugs have been found for variation at *PTC* (*TAS2R38*). Subjects possessing non-taster haplotypes rate alcoholic beverages as less bitter, and more preferred (Duffy et al., 2004; Lanier et al., 2005). Similarly, there is a greater percentage of bitter-insensitive individuals among smokers than non-smokers, suggesting these non-tasters are more at risk for nicotine addiction (Enoch et al., 2001). A more recent study found an association of intermediate, but not taster (PAV) sensitivity with reduced smoker prevalence (Cannon et al., 2005). Women may quit smoking during pregnancy because during this period cigarettes taste unpleasant (Pletsch and Kratz, 2004). On the other hand, long-term smoking or oral tobacco use may contribute to a decrement in overall taste acuity, although evidence for this is equivocal (Khan et al., 2003; Sato et al., 2002). Finally, there is a strong relationship between smoking and alcohol use in humans (Perkins et al., 2000; Pomerleau et al., 2004); taste may be a common factor in both addictions. In mice, oral nicotine consumption is correlated with ethanol consumption (Li et al., 2005b).

It has been long appreciated that taste preferences are a predictor of propensity to consume sweet or high-fat foods (e.g., Drewnowski et al., 1985, 1998; Salbe et al., 2004). Weight loss is facilitated in dieting subjects instructed to concentrate on taste and sensory sensations throughout a meal (Poothullil, 1999). Unexpectedly, haplotypes at the *TAS2R38* gene were significantly related to children's preference for sucrose and sweet-tasting foods and beverages: those possessing taster alleles tended to be sweet-preferrers; the authors speculate that the PTC gene may be in disequilibrium with sweet taste genes (Mennella et al., 2005). Furthering this possible connection between bitter taste and sweet consumption and its consequences, a haplotype within the bitter taste receptor gene *TAS2R9* was found to be associated with altered glucose and insulin homeostasis within the Amish Family Diabetes Study (Dotson et al., 2008).

Conclusions

Historically, behavioral genetic studies of taste and intake behaviors have been instrumental in the eventual discovery of G-protein-coupled taste receptors. In particular, the discovery of the PTC polymorphism in humans stimulated research into the genetics basis of taste behavior. In rodents, strain differences to bitter and sweet-tasting stimuli, primarily in two-bottle tests, provided linkage to chromosome regions later found to harbor taste receptor genes. Still, the balance of evidence suggests that intake behaviors are under polygenic control, and other genes have yet to be elucidated. The path to further gene discovery potentially leads upstream, where taste processing interacts with homeostatic systems involved in the regulation of feeding, appetite, fluid balance, and reward to produce taste-related behaviors.

References

Adler, E., Hoon, M.A., Mueller, K.L., Chandrashekar, J., Ryba, N.J.P., and Zuker, C.S. (2000) A novel family of mammalian taste receptors. *Cell* 100: 693–702.

Atchley, D.P., Weaver, K.L., and Eckel, L.A. (2005) Taste responses to dilute sucrose solutions are modulated by stage of the estrous cycle and fenfluramine treatment in female rats. *Physiol Behav* 86: 265–271.

Bachmanov, A.A., Li, X., Li, S., Neira, M., Beauchamp, G.K., and Azen, E.A. (2001a) High-resolution genetic mapping of the sucrose octaacetate taste aversion (Soa) locus on mouse Chromosome 6. *Mamm Genome* 12: 695–699.

Bachmanov, A.A., Li, X., Reed, D.R., Ohmen, J.D., Li, S., Chen, Z., et al. (2001b) Positional cloning of the mouse saccharin preference (Sac) locus. *Chem Senses* 26: 925–933.

Bachmanov, A.A., Reed, D.R., Ninomiya, Y., Inoue, M., Tordoff, M.G., Price, R.A., et al. (1997) Sucrose consumption in mice: major influence of two genetic loci affecting peripheral sensory responses. *Mamm Genome* 8: 545–548.

Bachmanov, A.A., Reed, D.R., Tordoff, M.G., Price, R.A., and Beauchamp, G.K. (1996) Intake of ethanol, sodium chloride, sucrose, citric acid, and quinine hydrochloride solutions by mice: a genetic analysis. *Behav Genet* 26: 563–573.

Bartoshuk, L.M., Duffy, V.B., and Miller, I.J. (1994) PTC/PROP tasting: anatomy, psychophysics, and sex effects. *Physiol Behav* 56: 1165–1171.

Beauchamp, G.K. (1990) Genetic control over salt preference in inbred strains of mice. *Chem Senses* 15: 551.

Behrens, M. and Meyerhof, W. (2009) Mammalian bitter taste perception. *Results Probl Cell Differ* 47: 1–18.

Belknap, J.K., Crabbe, J.C., and Young, E.R. (1993) Voluntary consumption of ethanol in 15 inbred mouse strains. *Psychopharmacology (Berl)* 112: 503–510.

Blakeslee, A.F. (1932) Genetics of sensory thresholds: taste for phenyl thio carbamide. *Proc Natl Acad Sci USA* 18: 120–130.

Blednov, Y.A., Walker, D., Martinez, M., Levine, M., Damak, S., and Margolskee, R.F. (2008) Perception of sweet taste is important for voluntary alcohol consumption in mice. *Genes Brain Behav* 7: 1–13.

Blizard, D.A., Kotlus, B., and Frank, M.E. (1999) Quantitative trait loci associated with short-term intake of sucrose, saccharin and quinine solutions in laboratory mice. *Chem Senses* 24: 373–385.

Boughter, J.D., Jr., Harder, D.B., Capeless, C.G., and Whitney, G. (1992) Polygenic determination of quinine sensitivity among mice. *Chem Senses* 17: 427–434.

Boughter, J.D., Jr., St. John, S.J., Noel, D.T., Ndubuizu, O., and Smith, D.V. (2002) A brief-access test for bitter taste in mice. *Chem Senses* 27: 133–142.

Boughter, J.D., Jr. and Whitney, G. (1998) Behavioral specificity of the bitter taste gene Soa. *Physiol Behav* 63: 101–108.

Cannon, D.S., Baker, T.B., Piper, M.E., Scholand, M.B., Lawrence, D.L., Drayna, D.T., et al. (2005) Associations between phenylthiocarbamide gene polymorphisms and cigarette smoking. *Nicotine Tob Res* 7: 853–858.

Capeless, C.G. and Whitney, G. (1995) The genetic basis of preference for sweet substances among inbred strains of mice: preference ratio phenotypes and the alleles of the Sac and dpa loci. *Chem Senses* 20: 291–293.

Capeless, C.G., Whitney, G., and Azen, E.A. (1992) Chromosome mapping of Soa, a gene influencing gustatory sensitivity to sucrose octaacetate in mice. *Behav Genet* 22: 655–663.

Chandrashekar, J., Mueller, K.L., Hoon, M.A., Adler, E., Feng, L., Guo, W., et al. (2000) T2Rs function as bitter taste receptors. *Cell* 100: 703–711.

Chiarelli, B. (1963) Sensitivity to PTC (phenyl-thio-carbamide) in primates. *Fol Primat* 1: 88–94.

Colomb, J. and Stocker, R.F. (2007) Combined rather than separate pathways for hedonic and sensory aspects of taste in fly larvae? *Fly (Austin)* 1: 232–234.

Curtis, K.S., Stratford, J.M., and Contreras, R.J. (2005) Estrogen increases the taste threshold for sucrose in rats. *Physiol Behav* 86: 281–286.

Damak, S., Rong, M., Yasumatsu, K., Kokrashvili, Z., Varadarajan, V., Zou, S., et al. (2003) Detection of sweet and umami taste in the absence of taste receptor T1r3. *Science* 301: 850–853.

Delay, E.R., Hernandez, N.P., Bromley, K., and Margolskee, R.F. (2006) Sucrose and monosodium glutamate taste thresholds and discrimination ability of T1R3 knockout mice. *Chem Senses* 31: 351–357.

Dess, N.K., Badia-Elder, N.E., Thiele, T.E., Kiefer, S.W., and Blizard, D.A. (1998) Ethanol consumption in rats selectively bred for differential saccharin intake. *Alcohol* 16: 275–278.

Donaldson, L.F., Bennett, L., Baic, S., and Melichar, J.K. (2009) Taste and weight: is there a link? *Am J Clin Nutr* 90: 800S–803S.

Dotson, C.D., Roper, S.D., and Spector, A.C. (2005) PLCbeta2-independent behavioral avoidance of prototypical bitter-tasting ligands. *Chem Senses* 30: 593–600.

Dotson, C.D., Zhang, L., Xu, H., Shin, Y.K., Vigues, S., Ott, S.H., et al. (2008) Bitter taste receptors influence glucose homeostasis. *PLoS One* 3: e3974.

Drayna, D. (2005) Human taste genetics. *Annu Rev Genomics Hum Genet* 6: 217–235.

Drayna, D., Coon, H., Kim, U.K., Elsner, T., Cromer, K., Otterud, B., et al. (2003) Genetic analysis of a complex trait in the Utah Genetic Reference Project: a major locus for PTC taste ability on chromosome 7q and a secondary locus on chromosome 16p. *Hum Genet* 112: 567–572.

Drewnowski, A., Henderson, S.A., and Barratt-Fornell, A. (1998) Genetic sensitivity to 6-n-propylthiouracil and sensory responses to sugar and fat mixtures. *Physiol Behav* 63: 771–777.

Drewnowski, A.J., Brunzell, J.D., Sande, K., Iverius, P.H., and Greenwood, M.R. (1985) Sweet tooth reconsidered: taste responsiveness in human obesity. *Physiol Behav* 35: 617–622.

Duffy, V.B., Davidson, A.C., Kidd, J.R., Kidd, K.K., Speed, W.C., Pakstis, A.J., et al. (2004) Bitter receptor gene (TAS2R38), 6-n-propylthiouracil (PROP) bitterness and alcohol intake. *Alcohol Clin Exp Res* 28: 1629–1637.

Enoch, M.A., Harris, C.R., and Goldman, D. (2001) Does a reduced sensitivity to bitter taste increase the risk of becoming nicotine addicted? *Addict Behav* 26: 399–404.

Eylam, S. and Spector, A.C. (2004) Stimulus processing of glycine is dissociable from that of sucrose and glucose based on behaviorally measured taste signal detection in Sac "taster" and "non-taster" mice. *Chem Senses* 29: 639–649.

Finger, T.E., Danilova, V., Barrows, J., Bartel, D.L., Vigers, A.J., Stone, L., et al. (2005) ATP signaling is crucial for communication from taste buds to gustatory nerves. *Science* 310: 1495–1499.

Fischer, R.A., Ford, E.B., and Huxley, J. (1939) Taste testing the anthropoid apes. *Nature* 144: 750.

Flaherty, L. (1981) Congenic strains. In Foster, H.L., Small, J.D., and Fox, J.G. (eds.), *The Mouse in Biomedical Research. History, Genetics, and Wild Mice*, Vol. 1. Academic Press, New York, pp. 215–222.

Fontanini, A., Grossman, S.E., Figueroa, J.A., and Katz, D.B. (2009) Distinct subtypes of basolateral amygdala taste neurons reflect palatability and reward. *J Neurosci* 29: 2486–2495.

Fox, A.L. (1931) Six in ten "tasteblind" to bitter chemical. *Sci News Lett* 9: 249.

Fox, A.L. (1932) The relationship between chemical constitution and taste. *Proc Natl Acad Sci USA* 18: 115–120.

Fuller, J.L. (1974) Single-locus control of saccharin preference in mice. *J Hered* 65: 33–36.

Gilbertson, T.A. and Boughter, J.D., Jr. (2003) Taste transduction: appetizing times in gustation. *Neuroreport* 14: 905–911.

Gordesky-Gold, B., Rivers, N., Ahmed, O.M., and Breslin, P.A. (2008) Drosophila melanogaster prefers compounds perceived sweet by humans. *Chem Senses* 33: 301–309.

Guo, S.W. and Reed, D.R. (2001) The genetics of phenylthiocarbamide perception. *Ann Hum Biol* 28: 111–142.

Hajnal, A., Covasa, M., and Bello, N.T. (2005) Altered taste sensitivity in obese, prediabetic OLETF rats lacking CCK-1 receptors. *Am J Physiol Regul Integr Comp Physiol* 289: R1675–1686.

Hajnal, A., Takenouchi, K., and Norgren, R. (1999) Effect of intraduodenal lipid on parabrachial gustatory coding in awake rats. *J Neurosci* 19: 7182–7190.

Harder, D.B., Azen, E.A., and Whitney, G. (2000) Sucrose octaacetate avoidance in nontaster mice is not enhanced by two type-A Prp transgenes from taster mice. *Chem Senses* 25: 39–45.

Harder, D.B., Capeless, C.G., Maggio, J.C., Boughter, J.D., Jr., Gannon, K.S., Whitney, G., et al. (1992) Intermediate sucrose octaacetate sensitivity suggests a third allele at mouse bitter taste locus Soa and Soa-Rua identity. *Chem Senses* 17: 391–401.

Harder, D.B., Gannon, K.S., and Whitney, G. (1996) SW.B6-Soab nontaster congenic strains completed and a sucrose octaacetate congenic quartet tested with other bitters. *Chem Senses* 21: 507–517.

Harder, D.B. and Whitney, G. (1998) A common polygenic basis for quinine and PROP avoidance in mice. *Chem Senses* 23: 327–332.

Harris, H. and Kalmus, H. (1949) The measurement of taste sensitivity to phenylthiourea (PTC) *Annals of Eugenics* 15: 24–31.

Hoshishima, K., Yokoyama, S., and Seto, K. (1962) Taste sensitivity in various strains of mice. *Am J Physiol* 202: 1200–1204.

Inoue, M., Li, X., McCaughey, S.A., Beauchamp, G.K., and Bachmanov, A.A. (2001) Soa genotype selectively affects mouse gustatory neural responses to sucrose octaacetate. *Physiol Genomics* 5: 181–186.

Inoue, M., Reed, D.R., Li, X., Tordoff, M.G., Beauchamp, G.K., and Bachmanov, A.A. (2004) Allelic variation of the Tas1r3 taste receptor gene selectively affects behavioral and neural taste responses to sweeteners in the F2 hybrids between C57BL/6ByJ and 129P3/J mice. *J Neurosci* 24: 2296–2303.

Ishimaru, Y., Inada, H., Kubota, M., Zhuang, H., Tominaga, M., and Matsunami, H. (2006) Transient receptor potential family members PKD1L3 and PKD2L1 form a candidate sour taste receptor. *Proc Natl Acad Sci USA* 103: 12569–12574.

Kampov-Polevoy, A.B., Eick, C., Boland, G., Khalitov, E., and Crews, F.T. (2004) Sweet liking, novelty seeking, and gender predict alcoholic status. *Alcohol Clin Exp Res* 28: 1291–1298.

Kampov-Polevoy, A.B., Garbutt, J.C., and Khalitov, E. (2003) Family history of alcoholism and response to sweets. *Alcohol Clin Exp Res* 27: 1743–1749.

Khan, G.J., Mehmood, R., Salah-ud-Din, and Ihtesham-ul-Haq. (2003) Effects of long-term use of tobacco on taste receptors and salivary secretion. *J Ayub Med Coll Abbottabad* 15: 37–39.

Kim, U.K. and Drayna, D. (2005) Genetics of individual differences in bitter taste perception: lessons from the PTC gene. *Clin Genet* 67: 275–280.

Kim, U.K., Jorgenson, E., Coon, H., Leppert, M., Risch, N., and Drayna, D. (2003) Positional cloning of the human quantitative trait locus underlying taste sensitivity to phenylthiocarbamide. *Science* 299: 1221–1225.

Kinnamon, J.C. (1987) Organization and innervation of taste buds. In Finger, T.E. and Silver, W.L. (eds.), *Neurobiology of Taste and Smell.* Wiley, New York, pp. 277–297.

Kitagawa, M., Kusakabe, Y., Miura, H., Ninomiya, Y., and Hino, A. (2001) Molecular genetic indentification of a candidate receptor gene for sweet taste. *Biochem Biophys Res Commun* 283: 236–242.

Klein, T.W. and DeFries, J.C. (1970) Similar polymorphism of taste sensitivity to PTC in mice and men. *Nature* 225: 555–557.

Lanier, S.A., Hayes, J.E., and Duffy, V.B. (2005) Sweet and bitter tastes of alcoholic beverages mediate alcohol intake in of-age undergraduates. *Physiol Behav* 83: 821–831.

Lemon, C.H., Brasser, S.M., and Smith, D.V. (2004) Alcohol activates a sucrose-responsive gustatory neural pathway. *J Neurophysiol* 92: 536–544.

Lemon, C.H. and Katz, D.B. (2007) The neural processing of taste. *BMC Neurosci* 18 (Suppl 3): S5.

Le Roy, I., Pager, J., and Roubertoux, P.L. (1999) Genetic dissection of gustatory sensitivity to bitterness (sucrose octaacetate) in mice. *C R Acad Sci III* 322: 821–831.

Li, X.C., Karadsheh, M.S., Jenkins, P.M., and Stitzel, J.A. (2005b) Genetic correlation between the free-choice oral consumption of nicotine and alcohol in C57BL/6JxC3H/HeJ F2 intercross mice. *Behav Brain Res* 157: 79–90.

Li, X., Li, W., Wang, H., Cao, J., Maehashi, K., Huang, L., et al. (2005a) Pseudogenization of a sweet-receptor gene accounts for cats' indifference toward sugar. *PLoS Genet* 1: 27–35.

Li, X., Staszewski, L., Xu, H., Durick, K., Zoller, M., and Adler, E. (2002) Human receptors for sweet and umami taste. *Proc Natl Acad Sci USA* 99: 4692–4696.

Lundy, R.F. and Norgren, R. (2004) The gustatory system. In Paxinos, G. (ed.), *The Rat Nervous System*, 3rd edn. Elsevier Academic Press, San Diego, CA, USA, pp. 891–921.

Lush, I.E. (1981) The genetics of tasting in mice. I. Sucrose octaacetate. *Genet Res* 38: 93–95.

Lush, I.E. (1984) The genetics of tasting in mice. III. Quinine. *Genet Res* 44: 151–160.

Lush, I.E. (1989) The genetics of tasting in mice. VI. Saccharin, acesulfame, dulcin and sucrose. *Genet Res* 53: 95–99.

Lush, I.E. and Holland, G. (1988) The genetics of tasting in mice. V. Glycine and cycloheximide. *Genet Res* 52: 207–212.

Lush, I.E., Hornigold, N., King, P., and Stoye, J.P. (1995) The genetics of tasting in mice. VII. Glycine revisited, and the chromosomal location of Sac and Soa. *Genet Res* 66: 167–174.

Matsunami, H., Montmayeur, J.-P., and Buck, L.B. (2000) A family of candidate taste receptors in human and mouse. *Nature* 404: 601–604.

Max, M., Shanker, Y.G., Huang, L., Rong, M., Liu, Z., Campagne, F., et al. (2001) Tas1r3, encoding a new candidate taste receptor, is allelic to the sweet responsiveness locus Sac. *Nat Genet* 28: 58–63.

McClearn, G.E. and Rodgers, D.A. (1959) Differences in alcohol preference among inbred strains of mice. *Quart J Stud Alcohol* 20: 691–695.

Mennella, J.A., Pepino, M.Y., and Reed, D.R. (2005) Genetic and environmental determinants of bitter perception and sweet preferences. *Pediatrics* 115: e216–222.

Midkiff, E.E., Bernstein, I.L., Fitts, D.A., and Simpson, J.B. (1985) Salt preference and salt appetite of Fischer-344 rats. *Chem Senses* 10: 446.

Montmayeur, J.P., Liberles, S.D., Matsunami, H., and Buck, L.B. (2001) A candidate taste receptor gene near a sweet taste locus. *Nat Neurosci* 4: 492–498.

Nachman, M. (1959) The inheritance of saccharin preference. *J Comp Physiol Psychol* 52: 451–457.

Nelson, G., Chandrashekar, J., Hoon, M.A., Feng, L., Zhao, G., Ryba, N.J.P., et al. (2002) An amino-acid taste receptor. *Nature* 416: 199–202.

Nelson, G., Hoon, M.A., Chandrashekar, J., Zhang, Y., Ryba, N.J.P., and Zuker, C.S. (2001) Mammalian sweet taste receptors. *Cell* 106: 381–390.

Nelson, T.M., Munger, S.D., and Boughter, J.D., Jr. (2003) Taste sensitivities to PROP and PTC vary independently in mice. *Chem Senses* 28: 695–704.

Nelson, T.M., Munger, S.D., and Boughter, J.D., Jr. (2005) Haplotypes at the Tas2r locus on distal chromosome 6 vary with quinine taste sensitivity in inbred mice. *BMC Genet* 6: 32.

Nie, Y., Vigues, S., Hobbs, J.R., Conn, G.L., and Munger, S.D. (2005) Distinct contributions of T1R2 and T1R3 taste receptor subunits to the detection of sweet stimuli. *Curr Biol* 15: 1948–1952.

Ninomiya, Y., Nomura, T., and Katsukawa, H. (1992) Genetically variable taste sensitivity to D-amino acids in mice. *Brain Res* 596: 349–352.

Ninomiya, Y., Sako, N., and Funakoshi, M. (1989) Strain differences in amiloride inhibition of NaCl responses in mice, *Mus musculus*. *J Comp Physiol* 166: 1–5.

Nishijo, H., Uwano, T., Tamura, R., and Ono, T. (1998) Gustatory and multimodal neuronal responses in the amygdala during licking and discrimination of sensory stimuli in awake rats. *J Neurophysiol* 79: 21–36.

Norgren, R. (1970) Gustatory responses in the hypothalamus. *Brain Res* 21: 63–71.

Norgren, R. and Leonard, C.M. (1973) Ascending central gustatory pathways. *J Comp Neurol* 150: 217–238.

Oliveira-Maia, A.J., Stapleton-Kotloski, J.R., Lyall, V., Phan, T.H., Mummalaneni, S., Melone, P., et al. (2009) Nicotine activates TRPM5-dependent and independent taste pathways. *Proc Natl Acad Sci USA*. 106: 1596–1601.

Olson, J.M., Boehnke, M., Neiswanger, K., Roche, A.F., and Siervogel, R.M. (1989) Alternative genetic models for the inheritance of the phenylthiocarbamide taste deficiency. *Genet Epidemiol* 6: 423–434.

Overstreet, D.H., Kampov-Polevoy, A.B., Rezvani, A.H., Murrelle, L., Halikas, J.A., and Janowsky, D.S. (1993) Saccharin intake predicts ethanol intake in genetically heterogeneous rats as well as different rat strains. *Alcohol Clin Exp Res* 17: 366–369.

Peltz, W., Whitney, G., and Smith, J.C. (1973) Genetic influences on saccharin preference of mice. *Physiol Behav* 10: 263–265.

Perkins, K.A., Fonte, C., and Grobe, J.E. (2000) Sex differences in the acute effects of cigarette smoking on the reinforcing value of alcohol. *Behav Pharmacol* 11: 63–70.

Phillips, T.J., Crabbe, J.C., Metten, P., and Belknap, J.K. (1994) Localization of genes affecting alcohol drinking in mice. *Alcohol Clin Exp Res* 18: 931–941.

Pletsch, P.K. and Kratz, A.T. (2004) Why do women stop smoking during pregnancy? Cigarettes taste and smell bad. *Health Care Women Int* 25: 671–679.

Pomerleau, C.S., Marks, J.L., Pomerleau, O.F., and Snedecor, S.M. (2004) Relationship between early experiences with tobacco and early experiences with alcohol. *Addict Behav* 29: 1245–1251.

Poothullil, J.M. (1999) Maintenance of weight loss using taste and smell sensations. *J Womens Health* 8: 109–113.

Ramirez, I. and Fuller, J.L. (1976) Genetic influence on water and sweetened water consumption in mice. *Physiol Behav* 16: 163–168.

Reddy, B.M. and Rao, D.C. (1989) Phenylthiocarbamide taste sensitivity revisited: complete sorting test supports residual family resemblance. *Genet Epidemiol* 6: 413–421.

Reed, D.R., Li, S., Li, X., Huang, L., Tordoff, M.G., Starling-Roney, R., et al. (2004) Polymorphisms in the taste receptor gene (Tas1r3) region are associated with saccharin preference in 30 mouse strains. *J Neurosci* 24: 938–946.

Richter, T.A. and Clisby, K.H. (1941) Phenylthiocarbamide taste thresholds of rats and human beings. *Am J Physiol* 134: 157–164.

Risinger, F.O. and Cunningham, C.L. (2000) DBA/2J mice develop stronger lithium chloride-induced conditioned taste and place aversions than C57BL/6J mice. *Pharmacol Biochem Behav* 67: 17–24.

Rodgers, D.A. and McClearn, G.E. (1962) Mouse strain differences in preference for various concentrations of alcohol. *Q J Stud Alcohol* 23: 26–33.

Rodgers, D.A. and McClearn, G.E. (1964) Sucrose versus ethanol appetite in inbred strains of mice. *Q J Stud Alcohol* 25: 26–35.

Roper, S.D. (2007) Signal transduction and information processing in mammalian taste buds. *Pflugers Arch* 454: 759–776.

Roper, S.D. (2009) Parallel processing in mammalian taste buds? *Physiol Behav* 97: 604–608.

Sainz, E., Korley, J.N., Battey, J.F., and Sullivan, S.L. (2001) Identification of a novel member of the T1R family of putative taste receptors. *J Neurochem* 77: 896–903.

Salbe, A.D., DelParigi, A., Pratley, R.E., Drewnowski, A., and Tataranni, P.A. (2004) Taste preferences and body weight changes in an obesity-prone population. *Am J Clin Nutr* 79: 372–378.

Sato, K., Endo, S., and Tomita, H. (2002) Sensitivity of three loci on the tongue and soft palate to four basic tastes in smokers and non-smokers. *Acta Otolar* 2002: 74–82.

Scott, K. (2005) Taste recognition: food for thought. *Neuron* 48: 455–464.

Scott, K., Brady, R., Jr., Cravchik, A., Morozov, P., Rzhetsky, A., Zuker, C., et al. (2001) A chemosensory gene family encoding candidate gustatory and olfactory receptors in Drosophila. *Cell* 104: 661–673.

Shigemura, N., Ohkuri, T., Sadamitsu, C., Yasumatsu, K., Yoshida, R., Beauchamp, G.K., et al. (2008) Amiloride-sensitive NaCl taste responses are associated with genetic variation of ENaC alpha-subunit in mice. *Am J Physiol Regul Integr Comp Physiol* 294: R66–75.

Shigemura, N., Ohta, R., Kusakabe, Y., Miura, H., Hino, A., Koyano, K., et al. (2004) Leptin modulates behavioral responses to sweet substances by influencing peripheral taste structures. *Endocrinology* 145: 839–847.

Shigemura, N., Yasumatsu, K., Yoshida, R., Sako, N., Katsukawa, H., Nakashima, K., et al. (2005) The role of the dpa Locus in mice. *Chem Senses* 30 Suppl 1: i84-i85.

Simon, S.A., Liu, L., and Erickson, R.P. (2003) Neuropeptides modulate rat chorda tympani responses. *Am J Physiol Regul Integr Comp Physiol* 284: R1494–1505.

Snyder, L.H. (1931) Inherited taste deficiency. *Science* 74: 151–152.

St. John, S.J., Pour, L., and Boughter, J.D., Jr. (2005) Phenylthiocarbamide produces conditioned taste aversions in mice. *Chem Senses* 30: 377–382.

St. John, S.J. and Spector, A.C. (2008) Behavioral analysis of taste function in rodent models. In Firestein, S. (ed.), *The Senses: A Comprehensive Review*. Elsevier, London, pp. 409–427.

Stewart, R.B., Bice, P., Foroud, T., Lumeng, L., Li, T.K., and Carr, L.G. (2003) Correlation of saccharin and ethanol intake in the F2 progeny of HAD2 and LAD2 crosses. *Alcohol Clin Exp Res* 27: 49A.

Stewart, R.B., Russell, R.N., Lumeng, L., Li, T.K., and Murphy, J.M. (1994) Consumption of sweet, salty, sour, and bitter solutions by selectively bred alcohol-preferring and alcohol-nonpreferring lines of rats. *Alcohol Clin Exp Res* 18: 375–381.

Sugita, M. and Shiba, Y. (2005) Genetic tracing shows segregation of taste neuronal circuitries for bitter and sweet. *Science* 309: 781–785.

Tepper, B.J. and Seldner, A.C. (1999) Sweet taste and intake of sweet foods in normal pregnancy and pregnancy complicated by gestational diabetes mellitus. *Am J Clin Nutr* 70: 277–284.

Warren, R.P. and Lewis, R.C. (1970) Taste polymorphism in mice involving a bitter sugar derivative. *Nature* 227: 77–78.

Whitney, G. and Harder, D.B. (1986) PTC preference among laboratory mice: understanding of a previously “unreplicated” report. *Behav Genet* 16: 409–416.

Whitney, G., Maggio, J.C., and Harder, D.B. (1990) Manifestations of the major gene influencing sucrose octaacetate (SOA) tasting among mice: classic taste qualities. *Chem Senses* 15: 243–252.

Wooding, S., Kim, U.K., Bamshad, M.J., Larsen, J., Jorde, L.B., and Drayna, D. (2004) Natural selection and molecular evolution in PTC, a bitter-taste receptor gene. *Am J Hum Genet* 74: 637–646.

Woschnagg, H., Stollberger, C., and Finsterer, J. (2002) Loss of taste is loss of weight. *Lancet* 359: 891.

Yamamoto, T. (2006) Neural substrates for the processing of cognitive and affective aspects of taste in the brain. *Arch Histol Cytol* 69: 243–255.

Yamamoto, M.E., Block, G.D., and Ishii, E. (1991) Food pattens among adolescents: influence of alcohol consumption. *Alcohol Clin Exp Res* 15: 359.

Yamamoto, T., Matsuo, R., Kiyomitsu, Y., and Kitamura, R. (1989) Response properties of lateral hypothalamic neurons during ingestive behavior with special reference to licking of various taste solutions. *Brain Res* 481: 286–297.

Zhao, G.Q., Shang, Y., Hoon, M.A., Chandrashekar, J., Erlenbach, I., Ryba, N.J.P., et al. (2003) The receptors for mammalian sweet and umami taste. *Cell* 115: 255–266.

Chapter 10

Motor coordination in inbred mouse strains and the crucial role of the cerebellum

Robert Lalonde and Catherine Strazielle

The main purpose of this chapter is to summarize differences in motor coodination in different mouse strains. The principal tests currently used will be introduced, followed by analyses of those mouse strains that have so far been compared. We close the chapter by emphasizing the important role of the cerebellum in motor coordination, as exemplified in mutant mice.

Testing methods

The principal methods used for evaluating motor capabilities in mice are stationary beam, vertical grid, suspended wire, and rotorod tests (Lalonde and Strazielle, 1999). Additional tests include the parallel rod floor, whereby foot slips and ambulation are monitored (Kamens et al., 2005), as well as foot-print analyses of stride length and width (Brooks et al., 2004). The coat-hanger is a variation of the suspended wire test with diagonal side-bars added for climbing. The main measure used in all tests is the latency before falling, the validity of which being assured when mice are motivated to remain on the apparatus. Because of excessive fear or inappropriate task parameters (e.g., insufficient height or a beam rotating at velocities far beyond their capacity), mice may deliberately jump away. A voluntary fall is observed whenever mice jump up instead of dropping straight down. When this response occurs, the trial should not count. Instead, the trial should be repeated until the mouse cooperates. If not, that mouse should be removed from data analyses. It is necessary to pad the landing surface to prevent any risk of injury. To gauge motor speed, distance traveled on the stationary beam and movement times before climbing on grid or coat-hanger may be used.

Inbred mouse strains

Motor performance of inbred mouse strains

The comparison between inbred mouse strains is the initial step in identifying quantitative loci (QTLs) and also facilitates data interpretation from mutagenesis, transgenesis, and targeted null mutations (Rogers et al., 1999; Tarantino et al., 2000). Because null mutants are often generated by injecting 129 mouse strain cells on C57BL/6 blastocysts, these two strains have most often been compared. We present the current nomenclature of 129 mouse strains (www.informatics.jax.org) with the designation the authors used in parenthesis.

Stationary beam

Mice placed on a narrow non-rotating beam must maintain their balance without falling. Because some mice remain immobile for most of the test period, distance traveled provides an auxiliary measure of sensorimotor abilities. The time spent before reaching the end of the beam to a dark refuge or to a platform has also been used. Because these alternative measures likely depend on anxiety levels, in particular photophobia, they do not exclusively reflect motor control or speed.

The C57BL/6 strain maintained balance on the stationary beam for a longer period than the 129P3 (129/J) strain (Montkowski et al., 1997). The same result was found when C57BL/6J mice were compared to 129S2/Sv (Voikar et al., 2004). On the contrary, 129P2 (129/Ola) and 129T2 (129/Sv-ter) strains did not differ from C57BL/6 in this test. These results underline the importance of analyzing specific substrains. When motor speed was examined by latencies before reaching the end of the stationary beam, the C57BL/6JOlaHsd strain was faster than 129S2/SvHsd (Contet et al., 2001).

Few data are available on other mouse strains performing the stationary beam test. One exception is a study comprising FVB and C3H strains, in which no group difference in falling latencies was found (Dubois et al., 2002).

Vertical grid

Mice placed on a vertically positioned grid are evaluated for latencies before falling or before reaching the top of it. From postnatal days 9 and 13, the NZB/BlNJ strain developed more quickly than C57BL/6By in the vertical grid test (Le Roy et al., 1999). Interbreeding between these strains revealed two provisional QTLs on chromosomes 1 and 7 (Le Roy et al., 1999, 2001). These results provide a start in understanding the genetic basis of motor abilities in mice, which may eventually be extended to the human population.

Behavioral Genetics of the Mouse: Volume I. *Genetics of Behavioral Phenotypes*, eds. Wim E. Crusio, Frans Sluyter, Robert T. Gerlai, and Susanna Pietropaolo. Published by Cambridge University Press.

Suspended wire or coat-hanger

In the suspended wire test, mice are placed upside-down on a horizontal bar. Because the mice must support their own body weight, this test depends on muscle strength to a greater degree than the previous ones. As mice are motivated to reach an upright position, movement times on the coat-hanger are measured by the time taken before reaching or climbing the side-bar.

The C57BL/6By strain was compared to NZB/BlNJ in the suspended wire test on postnatal days 10 and 13 (Le Roy et al., 1999). In an opposite manner to vertical grid data, C57BL/6By developed more quickly than NZB/BlNJ mice. This study illustrates the test-specific nature of interstrain differences. The interbreeding between the two strains revealed two provisional QTLs on chromosomes 14 and 15 (Le Roy et al., 1999, 2001).

At 2 months of age, C57BL/10 mice stayed on the horizontal bar for a longer time than C57BL/6J (Deacon et al., 2007). At 10 months of age, the A/J strain was faster than the C57BL/6J strain before climbing the diagonal bar of the coat-hanger, although the groups did not differ with regard to latencies before falling (Thifault et al., 2002). Likewise, despite equivalent latencies before falling, the C3H line was faster than the FVB line before reaching the midportion and the end of the horizontal bar (Dubois et al., 2002). In contrast, latencies before climbing on the diagonal bar did not differ between C57BL/6J and 129S2/Sv strains (Voikar et al., 2004). Overall, these results justify the use of response speed in addition to latencies before falling in evaluating motor capabilities of mouse strains.

Rotorod

In the rotorod test, mice are trained to locomote forward in synchrony with a rotating beam. The apparatus may be turned at a constant speed or else accelerating in smooth, graduated steps to progressively higher velocities. When a mouse clings to the rod without moving, it is judicious to retrieve it at once and to consider the onset of passive rotation as equivalent to a fall. Otherwise, mice with well-coordinated movements will obtain the same scores as those unable to keep up.

Several experimenters have reported a superiority of C57BL/6 over 129 mouse strains on the rotorod test in a substrain-specific fashion (Table 10.1). In particular, C57BL/6J mice remained on the accelerating rotorod for a longer time than 129/SvJ (Homanics et al., 1999) or 129S1/SvImJ (Brown and Wong, 2009) mice. The same result was obtained when comparing C57BL/6JOlaHsd to 129S2/SvHsd (Contet et al., 2001) or C57BL/6J@ico to 129/SvPas@ico (Serradj and Jamon, 2007) on the same type of apparatus. These results were reproduced when comparing the C57BL/6J strain to 129S2/Sv (Voikar et al., 2004) and the C57BL/6 strain to 129/Sv (Kelly et al., 1998) on the fixed-speed rotorod. Moreover, C57BL/6NTac but not C57BL/6J mice were better than several 129 substrains on a mixed version, i.e., accelerated to a fixed speed (Bothe et al., 2005). On the contrary, no difference was noted between the C57BL/6JOlaHsd strain and 129S2/SvHsd on the fixed-speed rotorod (Voikar et al., 2001). Moroever, the C57BL/6 strain did not differ from 129/Sv on the accelerating rotorod (Rogers et al., 1999). Likewise, neither C57BL/6J nor C57BL/6NTac strains differed from 129S6/SvEvTac on the mixed version (Bothe et al., 2004). In addition, there was no difference between the C57BL/6J strain and 129S2/Sv on both accelerating and fixed-speed rotorods (Brooks et al., 2004).

Table 10.1 Comparisons between C57BL/6 and 129 mouse strains on fixed-speed, accelerating, or mixed (accelerated to a constant speed) rotorod tests.

Apparatus	Results	References
Accelerating	C57BL/6J > 129X1 (129/SvJ)	Homanics et al., 1999
Accelerating	C57BL/6J > 129S1/SvImJ mice	Brown and Wong, 2009
Accelerating	C57BL/6JOlaHsd > 129S2/SvHsd	Contet et al., 2001
Accelerating	C57BL/6J@ico > 129S2 (129/SvPas@ico)	Serradj and Jamon, 2007
Fixed-speed	C57BL/6J > 129S2/Sv	Voikar et al., 2004
Fixed-speed	C57BL/6 > 129S1 (129/Sv)	Kelly et al., 1998
Mixed	C57BL/6NTac >129S1, 129S6, 129P3,129T2,129X1	Bothe et al., 2005
Fixed-speed	C57BL/6JOlaHsd = 129S2/SvHsd	Voikar et al., 2001
Accelerating	C57BL/6 = 129S1 (129/Sv)	Rogers et al., 1999
Mixed	C57BL/6J = C57BL/6NTac = 129S6/SvEvTac	Bothe et al., 2004
Both fixed-speed and accelerating	C57BL/6J = 129S2/Sv	Brooks et al., 2004

Substrain differences have been found on the B6 background. Indeed, C57BL/6J mice outperformed C57BL/10 (Deacon et al., 2007). Moreover, the former outperformed C57BL/6NCrl or C57BL/6NTac mice, while C57BL/6NCrl outperformed C57BL/6NTac (Bryant et al., 2008).

An analysis of six inbred strains revealed lower scores for the C3H line relative to C57BL/6 and CBA/Ca (Rogers et al., 1999). Likewise, C3H/HeHNsd performance was poorer than that of CBA/Ca (Brooks et al., 2004) and C3H poorer than FVB (Dubois et al., 2002) or DBA/2 (Bothe et al., 2005). In turn, the FVB/NTac strain was outperformed by C57BL/6NTac and 129S6/SvEvTac (Bothe et al., 2004) and the BALB/c strain by C57BL/6JIco (Lepicard et al., 2003). C57BL/6J was the topmost line in a panel of eight inbred strains (McFadyen et al., 2003). Gene expression patterns of 10 inbred mouse strains revealed significant correlations between rotorod performance and over 2000 transcripts in cerebellum (Nadler et al., 2006).

Natural mutants with atrophy of the cerebelum

Motor performance depends on a functional cerebellum, as indicated by the severe deficits observed in ataxic mutants with atrophy of this brain region, including *Grid2*^{Lc} (*Lurcher*),

$Grid2^{ho}$ (*hot-foot*), *$Rora^{sg}$* (*staggerer*), *$Agtpbp1^{pcd}$* (*Purkinje cell degeneration*), and *nervous*, as well as the *Dst^{dt}* (*dystonia musculorum*) spinocerebellar mutant.

Neuropathology

Selective degeneration of cerebellar cortex

The autosomal semi-dominant *Lurcher* mutation causes a gain-in-malfunction of the *Grid2* gene, encoding the GluRδ2 ionotropic glutamate receptor (Zuo et al., 1997). GluRδ2 mRNA is predominantly expressed in cerebellar Purkinje cells (Takayama et al., 1996). The depolarized membrane potential of *$Grid2^{Lc}$* Purkinje cells (Zuo et al., 1997) is probably responsible for their nearly complete disappearance from postnatal week 2 to 4 (Caddy and Biscoe, 1979). The massive degeneration of *$Grid2^{Lc}$* granule cells is secondary to the missing trophic influence normally exerted by Purkinje cells (Vogel et al., 1991; Wetts and Herrup, 1982). The 60–75% loss of inferior olive neurons (Caddy and Biscoe, 1979; Heckroth and Eisenman, 1991) and the 30% loss of deep cerebellar nuclei (Heckroth, 1994) also appear to be secondary consequences of Purkinje cell atrophy.

Two autosomal recessive *hot-foot* alleles (4J and Nancy) cause different deletions in the coding sequences of *Grid2* (Lalouette et al. 1998, 2001). For the *ho-4J* allele, the truncated GluRδ2 protein was expressed in Purkinje cell soma without being transported to the surface (Matsuda and Yuzaki, 2002). The main neuropathological hallmark of *$Grid2^{ho\text{-}Nancy}$* mutants is defective parallel fiber-Purkinje cell innervation and mild depletion of cerebellar granule cells, resulting in cerebellar ataxia and a hopping gait reminiscent of mice walking on a hot plate (Guastavino et al., 1990).

The autosomal recessive *staggerer* mutation deletes the *Rora* gene highly expressed in Purkinje cells and encoding a retinoid-like nuclear receptor involved in neuronal differentiation and maturation (Hamilton et al., 1996; Sashihara et al., 1996). In *$Rora^{sg}$* homozygotes, Purkinje cells declined in number before postnatal day 5 and by the end of the 1st postnatal month only 25% of them remained (Herrup and Mullen, 1979; Vogel et al., 2000). The granule cell loss is secondary to Purkinje cell degeneration, begins soon after their migration (Herrup, 1983), and by the end of the 1st postnatal month is nearly total (Landis and Sidman, 1978). Despite Purkinje cell loss, deep cerebellar nuclei were present in normal numbers (Roffler-Tarlov and Herrup, 1981). But presumably because of Purkinje cell loss, inferior olive neurons decreased by 60% as early as postnatal day 24 (Shojaeian et al., 1985) and remained lower than normal in the adult stage (Blatt and Eisenman, 1985).

The autosomal recessive *Purkinje cell degeneration* (*pcd*) mutation affects the *Agtpbp* gene, encoding ATP/GTP binding protein 1 (Fernandez-Gonzalez et al., 2002). In normal mouse brain, Agtpbp1 mRNA was prominent in Purkinje cells. Several alleles of the affected gene have been discovered, including *1J* on C57BR/cdJ, *2J* on SM/J, and *3J* on BALB/cByJ backgrounds, respectively. The allele was subsequently transferred to B6C3-a/a and C57BL/6J backgrounds and the *2J* allele to C57BL/6J. The predominant pathology of *$Agtpbp1^{pcd\text{-}1J}$* mutants concerns the almost complete disappearance of Purkinje cells from the 3rd to the 4th postnatal week (Mullen et al., 1976). The late onset degeneration of granule cells (Ghetti et al., 1978) and of deep nuclei (Triarhou et al., 1987) seen in *$Agtpbp1^{pcd\text{-}1J}$* mutants appear to be secondary consequences of Purkinje cell loss. In addition, there is a retrograde degeneration of the inferior olive in this mutant, reaching 20% by the 3rd postnatal week and 50% at 10 months of age (Ghetti et al., 1987; Triarhou and Ghetti, 1991).

The main pathology of the autosomal recessive *nervous* (*nr*) mutation of an undiscovered gene is the loss of Purkinje cells, the granule cell degeneration being less severe and of later onset (Sidman and Green, 1970). From 38 weeks of age, 90% of Purkinje cells disappeared in hemispheres but only 50% in vermis, leading to cerebellar ataxia and a nervous-looking agitation (Sidman and Green, 1970). Purkinje cell loss is presumably responsible for inferior olive degeneration (Zanjani et al., 2004).

Spinocerebellar degeneration

The autosomal recessive *dystonia musculorum* mutation deletes the *Dst* gene, encoding dystonin, a cytoskeletal organizing protein associated with neurofilaments and microtubules (Brown et al., 1995). *Dst* expression is especially marked in dorsal root and cranial nerve ganglia, main sites of neurodegeneration in the mutant (Bernier et al., 1995). Alleles of this mutation include Jackson (*J*) on B6C3, Albany (*Alb*) on BALB/c, and Orleans (*Orl*) on DBA/2 backgrounds, respectively. Duchen (1976) first described the degeneration of spinocerebellar and spinothalamic tracts in the mutant, as well as peripheral sensory nerves, dorsal root ganglia, and cranial nerve ganglia 5, 7, 9, and 10. The degeneration of spinocerebellar and somatosensory tracts observed in *$Dst^{dt\text{-}J}$* and *$Dst^{dt\text{-}Orl}$* alleles resembles the pathology of patients with Friedreich's ataxia (Sotelo and Guénet, 1988).

Motor performance

Stationary beam

Four types of cerebellar mutant were compared to non-ataxic controls on the same rectangular-shaped stationary beam (Table 10.2). Latencies before falling decreased in *$Rora^{sg}$* (Lalonde, 1987) and *$Dst^{dt\text{-}J}$* (Lalonde et al., 1994) mutants with cerebellar ataxia. But despite ataxia, latencies remained equivalent to controls in *$Grid2^{Lc}$* (Lalonde et al., 1992) and *$Agtpbp1^{pcd\text{-}1J}$* (Le Marec and Lalonde, 1997) mutants.

The inferior performance of *$Dst^{dt\text{-}J}$* may be ascribed to spinocerebellar and lemniscal damage, leading to dystonic movements and crawling, not seen in cerebellar mutants. The inferior performance of *$Rora^{sg}$* mutants is a paradox in view of the 75% Purkinje cell loss seen in this mutant (Herrup and Mullen, 1979), while in *$Grid2^{Lc}$* (Caddy and Biscoe, 1979) and

Table 10.2 Latencies before falling from the same stationary beam, vertical grid, and coat-hanger by young adult *Grid2*^{Lc} (*Lc*), *Rora*^{sg} (*sg*), *Agtpbp1*^{pcd} (*pcd*), *nervous* (*nr*), and *Dst*^{dt} (*dt*) mutants versus controls.

Mutant	Rectangular beam	Vertical grid	Coat-hanger	References
Lc	NS	↓	↓	Lalonde et al., 1992
sg	↓	↓	↓	Lalonde, 1987; Lalonde et al., 1996
pcd	NS	NS	↓	Le Marec and Lalonde, 1997
nr	?	?	NS	Lalonde and Strazielle, 2003
dt	↓	↓	↓	Lalonde et al., 1994

↓ Decreased performance; NS: not significant; ? not yet tested.

Table 10.3 Latencies before falling from different rotorods by young adult *Grid2*^{Lc} (*Lc*), *Grid2*$^{ho\text{-}Nancy}$ (*ho*), *Rora*^{sg} (*sg*), *Agtpbp1*^{pcd} (*pcd*), *nervous* (*nr*), NFH-LacZ, *Grm1*, and *Prkcc* mutants versus controls.

Mutant	Rotorod	References
Lc	↓	Hilber and Caston, 2001; Lalonde et al., 1995; Le Marec et al., 1997; Strazielle et al., 1998
ho	↓	Krémarik et al., 1998; Lalonde et al., 1995, 2003; Lalouette et al., 2001
sg	↓	Deiss et al., 2000; Lalonde et al., 1995
pcd	↓	Le Marec and Lalonde, 1997
nr	↓	Lalonde and Strazielle, 2003
NFH/LacZ transgenic	↓	Dubois et al., 2002; Lalonde et al., 1999
Grm1 knockout	↓	Aiba et al., 1994
Prkcc knockout	↓	Chen et al., 1995

↓ Decreased performance versus controls.

Agtpbp1$^{pcd\text{-}1J}$ (Mullen et al., 1976) it is nearly total. Dysfunctional Purkinje cells may lead to poorer motor control than their disappearance.

Vertical grid

The same cerebellar mutants were compared to non-ataxic controls on a vertical grid (Table 10.2). As seen in the previous test, *Rora*^{sg} (Lalonde, 1987) and *Dst*$^{dt\text{-}J}$ (Lalonde et al., 1994) mutants were impaired, as were *Grid2*^{Lc} mutants (Lalonde et al., 1992). On the contrary, *Agtpbp1*$^{pcd\text{-}1J}$ mutants performed normally (Le Marec and Lalonde, 1997). Once more, the apparently unhealthy Purkinje cells in *Rora*^{sg} mutants lead to a poorer outcome than their disappearance in *Agtpbp1*$^{pcd\text{-}1J}$ mutants. The superiority of *Agtpbp1*$^{pcd\text{-}1J}$ over *Grid2*^{Lc} mutants may be ascribed to the relative preservation of cerebellar granule cells in the former. The performance of *Dst*$^{dt\text{-}J}$ mutants is expected to be poor in view of their dystonic movements and inability to stand.

Suspended wire or coat-hanger

Six types of cerebellar mutant were compared to non-ataxic controls on the same coat-hanger (Table 10.2). *Grid2*^{Lc} (Lalonde and Thifault, 1994; Lalonde et al., 1992, 1996,), *Grid2*$^{ho\text{-}Nancy}$ (Lalonde et al., 1996, 2003), *Rora*^{sg} (Lalonde et al., 1996), *Agtpbp1*$^{pcd\text{-}1J}$ (Le Marec and Lalonde, 1997), and *Dst*$^{dt\text{-}J}$ (Lalonde et al., 1994) fell sooner than their respective controls from the suspended wire. The only mutant with normal values was *nervous* (Lalonde and Strazielle, 2003). This result is attributable to the relatively mild (50%) loss of Purkinje cells in midline cerebellum (Sidman and Green, 1970), crucially involved in posture and equilibrium, together with relative preservation of granule cells. Nevertheless, the *nervous* mutants had higher movement times before climbing than controls (Lalonde and Strazielle, 2003). Once more, this result illustrates the interest in estimating speed of responding in phenotyping mouse motor performances.

On a similar apparatus, *Grid2*$^{ho\text{-}Nancy}$ mutants fell sooner from a suspended wire than wildtype controls, a result that was not seen in mice with the *Grid2*$^{ho\text{-}4J}$ allele (Lalouette et al., 2001).

Rotorod

The rotorod test is sensitive to a wide range of mouse models with cerebellar atrophy. *Grid2*^{Lc} mutants fell sooner than wildtype from rotorods of different size and speed (Hilber and Caston, 2001; Lalonde et al., 1995; Le Marec et al., 1997; Strazielle et al., 1998) (Table 10.3). Likewise, *Grid2*$^{ho\text{-}Nancy}$ and *Grid2*$^{ho\text{-}4J}$ mutants had lower latencies before falling from the rotorod than controls (Krémarik et al., 1998; Lalonde et al., 1995, 2003; Lalouette et al., 2001). Latencies before falling also diminished in *Rora*^{sg} (Deiss et al., 2000; Lalonde et al., 1995), *Agtpbp1*$^{pcd\text{-}1J}$ (Le Marec and Lalonde, 1997), and *nervous* (Lalonde and Strazielle, 2003) mutants. Thus, the rotorod is the only sensorimotor test so far considered that invariably detects impairments in cerebellar mutants with ataxia.

This test is sensitive as well to neurofilament maldistribution in Purkinje cells of NFH/LacZ transgenic mice with action tremor but without ataxia (Dubois et al., 2002; Lalonde et al., 1999). The same result was found with a gain-of-malfunction mutation of *Gabra1*, encoding the αA subunit of the γ-amino butyric acid subunit A ($GABA_A$) receptor (Homanics et al., 2005). This receptor is susceptible to diminished binding in dysgranular mutant mice (Rotter et al., 1988).

Null mutants for genes highly expressed in Purkinje cells are sensitive to the rotorod test. For example, mice with a *Grm1* null mutation were deficient on the rotorod test (Aiba et al., 1994). *Grm1* encodes MgluR1, a subtype of metabotropic glutamate receptors. Mice with a *Prkcc* null mutation were also impaired on the rotorod test (Chen et al., 1995). *Prkcc* encodes the γ isoform of protein kinase C.

These results underline the particular sensitivity of the rotorod task for detecting motor abnormalities in mice with a dysfunctional cerebellum irrespective of the presence of ataxia.

Acknowledgments

This research was funded by Natural Sciences and Engineering Research Council of Canada (NSERC).

References

Aiba, A., Kano, M., Chen, C., Stanton, M.E., Fox, G.D., Herrup, K., et al. (1994) Deficient cerebellar long-term depression and impaired motor learning in mGluR1 mutant mice. *Cell* 79: 377–388.

Bernier, G., Brown, A., Dalpe, G., De Repentigny, Y., Mathieu, M., and Kothary, R. (1995) Dystonin expression in the developing nervous system predominates in the neurons that degenerate in *dystonia musculorum* mutant mice. *Mol Cell Neurosci* 6: 509–520.

Blatt G.J. and Eisenman, L.M. (1985) A qualitative and quantitative light microscopic study of the inferior olivary complex in the adult *staggerer* mutant mouse. *J Neurogenet* 2: 51–66.

Bothe, G.W.M., Bolivar, V.J., Vedder, M.J., and Geistfeld, J.G. (2004) Genetic and behavioral differences among five inbred mouse strains commonly used in the production of transgenic and knockout mice. *Genes Brain Behav* 3: 149–157.

Bothe, G.W.M., Bolivar, V.J., Vedder, M.J., and Geistfeld, J.G. (2005) Behavioral differences among fourteen inbred mouse strains commonly used as disease models. *Comp Med* 55: 325–333.

Brooks, S.P., Pask, T., Jones, L., and Dunnett, S.B. (2004) Behavioural profiles of inbred mouse strains used as transgenic backgrounds. I: Motor tests. *Genes Brain Behav* 3: 206–215.

Brown, A., Bernier, G., Mathieu, M., Rossant, J., and Kothary, R. (1995) The mouse *dystonia musculorum* gene is a neural isoform of bullous pemphigoid antigen 1. *Nature Genet* 10: 301–306.

Brown, R.E. and Wong, A.A. (2009) The influence of visual ability on learning and memory performance in 13 strains of mice. *Learn Mem* 14: 134–144.

Bryant, C.D., Zhang, N.N., Sokoloff, G., Fanselow, M.S., Ennes, H.S., Palmer, A.A., et al. (2008) Behavioral differences among C57BL/6 substrains: implications for transgenic and knockout studies. *J Neurogenet* 22: 315–332.

Caddy, K.W.T. and Biscoe T.J. (1979) Structural and quantitative studies on the normal C3H and *Lurcher* mutant mouse. *Philos Trans Roy Soc Lond (Biol)* 287: 167–201.

Chen, C., Kano, M., Abeliovich, A., Chen, L., Bao, S., Kim, J.J., et al. (1995) Impaired motor coordination correlates with persistent multiple climbing fiber innervation in PKCγ mutant mice. *Cell* 83: 1233–1242.

Contet, C., Rawlins, J.N., and Deacon, R.M. (2001) A comparison of 129S2/SvHsd and C57BL/6JOlaHsd mice on a test battery assessing sensorimotor, affective and cognitive behaviours: implications for the study of genetically modified mice. *Behav Brain Res* 124: 33–46.

Deacon, R.M.J., Thomas, C.L., Rawlins, J.N.P., and Morley, B.J. (2007) A comparison of the behavior of C57BL/6 and C57BL/10 mice. *Behav Brain Res* 179: 239–247.

Deiss, V., Strazielle, C., and Lalonde, R. (2000) Regional brain variations of cytochrome oxidase activity and motor co-ordination in *staggerer* mutant mice. *Neuroscience* 95: 903–911.

Dubois, M., Strazielle, C., Eyer, J., and Lalonde, R. (2002) Sensorimotor functions in transgenic mice expressing the neurofilament/heavy-LacZ fusion protein on two genetic backgrounds. *Neuroscience* 112: 447–454.

Duchen, L.W. (1976) *Dystonia musculorum*, an inherited disease of the nervous system of the mouse. *Adv Neurol* 14: 353–365.

Fernandez-Gonzalez, A., La Spada, A.R., Treadaway, J., Higdon, J.C., Harris, B.S., Sidman, R.L., et al. (2002) Purkinje cell degeneration (*pcd*) phenotypes caused by mutations in the axotomy-induced gene, *Nna1*. *Science* 295: 1904–1906.

Ghetti, B., Alyea, C.J., and Muller, J. (1978) Studies on the *Purkinje cell degeneration* (*pcd*) mutant mice: primary pathology and transneuronal changes. *J Neuropathol Exp Neurol* 37: 617.

Ghetti, B., Norton, J., and Triarhou, L.C. (1987) Nerve cell atrophy and loss in the inferior olivary complex of "*Purkinje cell degeneration*" mutant mice. *J Comp Neurol* 260: 409–422.

Guastavino, J.-M., Sotelo, C., and Damez-Kinselle, I. (1990) *Hot-foot* murine mutation: behavioral effects and neuroanatomical alterations. *Brain Res* 523: 199–210.

Hamilton, B.A., Frankel, W.N., Kerrebrock, A.W., Hawkins, T.L., Fitzhugh, W., Kusumi, K., et al. (1996) Disruption of the nuclear hormone receptor ROR in *staggerer* mice. *Nature* 379: 736–739.

Heckroth, J.A. (1994) Quantitative morphological analysis of the cerebellar nuclei in normal and *Lurcher* mutant mice. I. Morphology and cell number. *J Comp Neurol* 343: 173–182.

Heckroth, J.A. and Eisenman, L.M. (1991) Olivary morphology and olivocerebellar atrophy in adult *Lurcher* mutant mice. *J Comp Neurol* 312: 641–651.

Herrup, K. (1983) Role of *staggerer* gene in determining cell number in cerebellar cortex. I. Granule cell death is an indirect consequence of *staggerer* gene action. *Dev Brain Res* 11: 267–274.

Herrup, K. and Mullen, R.J. (1979) Regional variation and absence of large neurons in the cerebellum of the *staggerer* mouse. *Brain Res* 172: 1–12.

Hilber, P. and Caston, J. (2001) Motor skills and motor learning in *Lurcher* mutant mice during aging. *Neuroscience* 102: 615–623.

Homanics, G.E., Elsen, F.P., Ying, S.W., Jenkins, A., Ferguson, C., Sloat, B., et al. (2005) A gain-of-function mutation in the GABA receptor produces synaptic and behavioral abnormalities in the mouse. *Genes Brain Behav* 4: 10–19.

Homanics, G.E., Quinlan, J.J., and Firestone, L.L. (1999) Pharmacologic and behavioral responses of inbred C57BL/6J and strain 129/SvJ mouse lines. *Pharmacol Biochem Behav* 63: 21–26.

Kamens, H.M., Phillips, T.J., Holstein, S.E., and Crabbe, J.C. (2005) Characterization of the parallel rod floor apparatus to test motor coordination in mice. *Genes Brain Behav* 4: 253–266.

Kelly, M.A., Rubinstein, M., Phillips, T.J., Lessov, C.N., Burkhart-Kasch, S., Zhang, G., et al. (1998) Locomotor activity in D2 dopamine receptor-deficient mice is determined by gene dosage, genetic background, and developmental adaptations. *J Neurosci* 18: 3470–3479.

Krémarik, P., Strazielle, C., and Lalonde, R. (1998) Regional brain variations of cytochrome oxidase activity and motor coordination in *hot-foot* mutant mice. *Eur J Neurosci* 10: 2802–2809.

Lalonde, R. (1987) Motor abnormalities in *staggerer* mutant mice. *Exp Brain Res* 68: 417–420.

Lalonde, R., Bensoula, A.N., and Filali, M. (1995) Rotorod sensorimotor learning in cerebellar mutant mice. *Neurosci Res* 22: 423–426.

Lalonde, R., Botez, M.I., Joyal, C.C., and Caumartin, M. (1992) Motor deficits in *Lurcher* mutant mice. *Physiol Behav* 51: 523–525.

Lalonde, R., Dubois, M., Strazielle, C., and Eyer, J. (1999) Motor coordination and spatial orientation are affected by neurofilament maldistribution: correlations with regional brain activity of cytochrome oxidase. *Exp Brain Res* 126: 223–234.

Lalonde, R., Filali, M., Bensoula, A.N., and Lestienne, F. (1996) Sensorimotor learning in three cerebellar mutant mice. *Neurobiol Learn Mem* 65: 113–120.

Lalonde, R., Hayzoun, K., Selimi, F., Mariani, J., and Strazielle, C. (2003) Motor coordination in mice with *hot-foot*, *Lurcher*, and double mutations of the *Grid2* gene encoding the delta-2 excitatory amino acid receptor. *Physiol Behav* 80: 333–339.

Lalonde, R., Joyal, C.C., and Botez, M.I. (1994) Exploration and motor coordination in *dystonia musculorum* mutant mice. *Physiol Behav* 56: 277–280.

Lalonde, R. and Strazielle, C. (1999) Motor performance of spontaneous murine mutations with cerebellar atrophy. In Crusio, W.E. and Gerlai, R.T. (eds.), *Handbook of Molecular-Genetic Techniques for Brain and Behavior Research*, Vol. 13. Elsevier, Amsterdam, pp. 627–637.

Lalonde, R. and Strazielle, C. (2003) Motor coordination, exploration, and spatial learning in a natural mouse mutation (*nervous*) with Purkinje cell degeneration. *Behav Genet* 33: 59–66.

Lalonde, R. and Thifault, S. (1994) Absence of an association between motor coordination and spatial orientation in *Lurcher* mutant mice. *Behav Genet* 24: 497–501.

Lalouette, A., Guénet, J.-L., and Vriz, S. (1998) *Hot-foot* mutations affect the $\delta 2$ glutamate receptor gene and are allelic to *Lurcher*. *Genomics* 50: 9–13.

Lalouette, A., Lohof, A., Sotelo, C., Guénet, J., and Mariani, J. (2001) Neurobiological effects of a null mutation depend on genetic context: comparison between two *hotfoot* alleles of the delta-2 ionotropic glutamate receptor. *Neuroscience* 105: 443–455.

Landis, D.M. and Sidman, R.L. (1978) Electron microscopic analysis of postnatal histogenesis in the cerebellar cortex of *staggerer* mutant mice. *J Comp Neurol* 179: 831–863.

Le Marec, N., Caston, J., and Lalonde, R. (1997) Impaired motor skills on static and mobile beams in *Lurcher* mutant mice. *Exp Brain Res* 116: 131–138.

Le Marec, N. and Lalonde, R. (1997) Sensorimotor learning and retention during equilibrium tests in *Purkinje cell degeneration* mutant mice. *Brain Res* 768: 310–316.

Lepicard, E.M., Venault, P., Negroni, J., Perez-Diaz, F., Joubert, C., Nosten-Bertrand, M., et al. (2003) Posture and balance responses to a sensory challenge are related to anxiety in mice. *Psychiatry Res* 118: 273–284.

Le Roy, I., Carlier, M., and Roubertoux, P.L. (2001) Sensory and motor development in mice: genes, environment and their interactions. *Behav Brain Res* 125: 57–64.

Le Roy, I., Perez-Diaz, F., Cherfouh, A., and Roubertoux, P.L. (1999) Preweanling sensorial and motor development in laboratory mice: quantitative trait loci mapping. *Dev Psychobiol* 34: 139–158.

Matsuda, S. and Yuzaki, M. (2002) Mutation in *hotfoot-4J* mice results in retention of $\delta 2$ glutamate receptors in ER. *Eur J Neurosci* 16: 1507–1516.

McFadyen, M.P., Kusek, G., Bolivar, V.J., and Flaherty, L. (2003) Differences among eight inbred strains of mice in motor ability and motor learning on a rotorod. *Genes Brain Behav* 2: 214–219.

Montkowski, A., Poettig, M., Mederer, A., and Holsboer, F. (1997) Behavioural performance in three substrains of mouse strain 129. *Brain Res* 762: 12–18.

Mullen, R.J., Eicher, E.M., and Sidman, R.L. (1976) *Purkinje cell degeneration*: a new neurological mutation in the mouse. *Proc Natl Acad Sci USA* 73: 208–212.

Nadler, J.J., Zhou, F., Huang, H. Moy, S.S., Lauder, J., Crawley, J.N., et al. (2006) Large-scale gene expression differences across brain regions and inbred mouse strains correlate with a behavioral phenotype. *Genetics* 174: 1229–1236.

Roffler-Tarlov, S. and Herrup, K. (1981) Quantitative examination of the deep cerebellar nuclei in the *staggerer* mutant mouse. *Brain Res* 215: 49–59.

Rogers, D.C., Jones, D.N., Nelson, P.R., Jones, C.M., Quilter, C.A., Robinson, T.L., et al. (1999) Use of SHIRPA and discriminant analysis to characterise marked differences in the behavioural phenotype of six inbred mouse strains. *Behav Brain Res* 105: 207–217.

Rotter, A., Gorenstein, C., and Frostholm, A. (1988) The localization of GABAA receptors in mice with mutations affecting the structure and connectivity of the cerebellum. *Brain Res* 439: 236–248.

Sashihara, S., Felts, P.A., Waxman, S.G., and Matsui, T. (1996) Orphan nuclear receptor *ROR* gene: isoform-specific spatiotemporal expression during postnatal development of brain. *Mol Brain Res* 42: 109–117.

Serradj, N. and Jamon, M. (2007) Age-related changes in the motricity of the inbred mouse strains 129/sv and C57BL/6j. *Behav Brain Res* 177: 80–89.

Shojaeian, H., Delhaye-Bouchaud, N., and Mariani, J. (1985) Decreased number of cells in the inferior olivary nucleus of the developing *staggerer* mouse. *Dev Brain Res* 21: 141–146.

Sidman, R.L. and Green, M.C. (1970) "*Nervous*," a new mutant mouse with cerebellar disease. In Sabourdy, M. (ed.), *Les Mutants Pathologiques chez l'Animal*. CNRS, Paris, pp. 69–79.

Sotelo, C. and Guénet, J.L. (1988) Pathologic changes in the CNS of *dystonia musculorum* mutant mouse: an animal model for human spinocerebellar ataxia. *Neuroscience* 27: 403–424.

Strazielle, C., Krémarik, P., Ghersi-Egea, J.-F., and Lalonde, R. (1998) Regional brain variations of cytochrome oxidase activity and motor coordination in *Lurcher* mutant mice. *Exp Brain Res* 121: 35–45.

Takayama, C., Nakagawa, S., Watanabe, M., Mishina, M., and Inoue, Y. (1996) Developmental changes in expression and distribution of the glutamate receptor channel delta 2 subunit according to the Purkinje cell maturation. *Dev Brain Res* 92: 147–155.

Tarantino, L.M., Gould, T.J., Druhan, J.P., and Bucan, M. (2000) Behavior and mutagenesis screens: the importance of baseline analysis of inbred strains. *Mamm Genome* 11: 555–564.

Thifault, S., Lalonde, R., Sanon, N., and Hamet, P. (2002) Comparisons between C57BL/6J and A/J mice in motor activity and coordination, hole-poking, and spatial learning. *Brain Res Bull* 58: 213–218.

Triarhou, L.C. and Ghetti, B. (1991) Stabilisation of neurone number in the inferior olivary complex of aged "*Purkinje cell degeneration*" mutant mice. *Acta Neuropathol* 81: 597–602.

Triarhou, L.C., Norton, J., and Ghetti, B. (1987) Anterograde transsynaptic degeneration in the deep cerebellar nuclei of *Purkinje cell degeneration* (*pcd*) mutant mice. *Exp Brain Res* 66: 577–588.

Vogel, M.W., McInnes, M., Zanjani, H.S., and Herrup, K. (1991) Cerebellar Purkinje cells provide target support over a limited spatial range: evidence from *Lurcher* chimeric mice. *Dev Brain Res* 64: 87–94.

Vogel, M.W., Sinclair, M., Qiu, D., and Fan, H. (2000) Purkinje cell fate in *staggerer* mutants: agenesis versus cell death. *J Neurobiol* 42: 323–337.

Voikar, V., Koks, S., Vasar, E., and Rauvala, H. (2001) Strain and gender differences in the behavior of mouse lines commonly used in transgenic studies. *Physiol Behav* 72: 271–281.

Voikar, V., Vasar, E., and Rauvala, H. (2004) Behavioral alterations induced by repeated testing in C57BL/6J and 129S2/Sv mice: implications for phenotyping screens. *Genes Brain Behav* 3: 27–38.

Wetts, R. and Herrup, K. (1982) Interaction of granule, Purkinje and inferior olivary neurons in *Lurcher* chimeric mice. II. Granule cell death. *Brain Res* 250: 358–362.

Zanjani, H., Herrup, K., and Mariani, J. (2004) Cell number in the inferior olive of *nervous* and *leaner* mutant mice. *J Neurogenet* 18: 327–339.

Zuo, J., De Jager, P.I., Takahashi, K.A., Jiang, W., Linden, D.J., and Heintz, N. (1997) Neurodegeneration in *Lurcher* mutant mice caused by mutation in the delta 2 glutamate receptor gene. *Nature* 388: 769–773.

Chapter 11

Reflex development

Francesca Cirulli and Enrico Alleva

Introduction: about reflexology and innateness

Before addressing the issue of suggesting specific strategies to evaluate reflex-ontogeny in rodents, with particular reference to mice, it seems important to define what reflex-ontogeny may mean in the emerging field of behavioral neurosciences, particularly for laboratory-based animal studies.

When addressing the issue of the study of reflexes, a few points need to be established. The history of reflexology is a long one (Prochàzka et al., 2000), dating since Descartes (1664) and Willis (1664). In their treatises, these authors were interested in and described machine-like or automatic control of behavior. Descartes, in particular, proposed a model according to which sensory stimulation was transmitted to the pineal gland, which selected and opened neural tubes conveying *spiritus animus* to muscles, causing them to contract. In humans this process was governed by the *soul*, an entity distinct from the brain. The term reflex was first defined by Georgiy Prochàzka (1784) as a behavior in response to an excitation, mediated by separate motor and sensory nerves. The reflexes had the important function to maintain "individual conservation."

The present contribution contains a first part aimed at scoring a series of neonatal and infantile responses, providing information on patterns of neurobehavioral development. A battery of tests is therefore described.

The second part of this chapter will include behavioral patterns controlling rodent maturational processes. Those patterns have been considered "innate," mainly because they allow neonatal survival on a not-learned basis.

Several authors (Lorenz, 1943, 1981; Tinbergen, 1989) defined "instinctual" and/or "innate" behavioral patterns, very often referring to stable and consistent modules present during early stages of postnatal life. However, this terminology evolved in the last decades, and similarities between human and animal neonatal patterns have been recognized (Bateson, 1979; Caro and Bateson, 1986; Eibl-Eibesfeldt, 1989; Hinde, 1975, 1982; Lorenz, 1943, 1981; Prechtl, 1974; Prochàzka et al., 2000; Tinbergen, 1989).

Basic considerations when studying early behavioral development of mice

Changes in pup development due to the behavior of the dam

Parental care is one of the most important variables influencing the course of physical and psychological development of the offspring. Indeed, a generalized delay in pup development may simply reflect a change in maternal behavior (Bignami, 1996; Chiarotti et al., 1987; D'Udine and Alleva, 1988).

Parity has been shown to affect maternal behavior. In mice maternal care differs between primiparous and biparous dams (Cohen-Salmon, 1987). It has been shown that the development of a number of behavioral responses is delayed in third, compared with second, litters (Crusio and Schmitt, 1996). In addition, it has been shown that in mice-specific items of maternal behavior or the latency to perform a single behavior may markedly differ between strains (Carlier et al., 1982).

Pilot studies on maternal behavior of transgenic dams are warmly suggested, particularly if a laboratory is going to invest on a multi-annual project using a particular transgenic strain (Alleva et al., 1989; Capone et al., 2005).

Cross-fostering, in-fostering and un-fostering

Cross-fostering procedures, aimed at controlling for postnatal maternal effects, have been shown to be highly useful in discerning the effects of a prenatal treatment (Bignami, 1996). Most studies use unidirectional fostering of both control and treated litters to untreated dams, eliminating postnatal maternal effects such as those produced by changes in maternal functions (e.g., milk production) or behavior (Chiarotti et al., 1987).

Mother and littermates are the most important components of the individual pup's environment and can provide important sources of thermal, somatosensory kinaesthetic, olfactory, and auditory stimulation. Each individual pup interacts with the mother and the littermates and is, in turn, stimulated by them; an example of the complexity of this three-way dyadic

Behavioral Genetics of the Mouse: Volume I. *Genetics of Behavioral Phenotypes*, eds. Wim E. Crusio, Frans Sluyter, Robert T. Gerlai, and Susanna Pietropaolo. Published by Cambridge University Press.

postnatal relationship is provided by the observation that arched-back nursing and milk ejection by the rat dam depend upon the combined suckling stimulation provided by several pups (Alleva et al., 1989; Rosenblatt and Lehrman, 1963; Stern, 1996).

Thus, even subtle deficits in maternal care may represent a not irrelevant bias when assessing neonatal or infantile behavioral patterns. A complete experimental design should include cross-fostering, in-fostering, and un-fostering procedures (Alleva and Sorace, 2000). In some cases, adoption of pups derived from feeble strains by parents belonging to outbred strains characterized by robust parental behavior should be considered, as well as paternal behavior or communal (e.g., a three dams' set) nesting (D'Udine and Alleva, 1988).

Litter culling

Pup culling is a delicate issue. Indeed, the necessity to obtain a minimal amount of tissue and/or biological samples (as required in a molecular biology study that also includes basic behavioral endpoints) is often in conflict with the requirements of a high-standard and exhaustive behavioral analysis. For many scientists performing breeding studies in a regulatory context, the culling of litters to a standard size shortly after birth (generally 3–5/gender/litter) is an accepted practice, although it is often seen as a non-naturalistic strategy. While the origins of this habit seem to be lost in the mists of time, the perceived benefits (reduction in workload and costs, reduction in litter size variability and in the growth and development of pups during the postnatal period) appear to justify the random discarding of some animals. Dams nursing large litters were observed to be away from the nest more often than those nursing smaller litters (Grota and Ader, 1969; Priestnall, 1972), and the decrease of the total time that the rodent dams spend with their litters decreased more rapidly in litters of 12 pups (Grota and Ader, 1969).

The point that should not be missed by behavioral neuroscientists is that, when performing developmental work, data quality, reliability, and avoiding important experimental bias, all reside in keeping pup development within a normal range. Maintaining litter size constant is a good prerequisite when studying pup development. Operationally, outbred mouse lines having been selected for being highly prolific easily overcome the natural number of "really wild" (feral) mice (5–6 pups, up to 12 for females at their 2nd, 3rd or 4th delivery). In outbred Swiss-albino mice, which have been selected for several decades to be efficient breeders under laboratory conditions, the litter size may easily reach 10–12 pups without major reflections on the normal range of development of the neonatal pups. For inbred mice, this number is often halved. The same applies to transgenic mouse lines in the absence of any relevant indication of these being particularly feeble.

In several cases, due to physical impairment or to unexpected deficits in maternal care, or both, the culling of the pups within a litter deserves particular attention. In fact, the number of pups should not exceed the capability of the dam to provide a sufficient amount of maternal care, both qualitatively and quantitatively (Alleva et al., 1989; Capone et al., 2005; Cirulli and Laviola, 2000; Cirulli et al., 2003b; D'Udine and Alleva, 1988; Hall and Rosenblatt, 1978; Laviola and Alleva, 1995).

A final very relevant consideration is that, in the case of significant retardations in somatic and/or behavioral development, the first explanation (and the most parsimonious, *sensu* Occam's razor) should be that maternal care by the dam is impaired. Any hypothesis attributing pup's deficits to a developmental change should be at least accompanied in the discussion by this alternative explanation.

Sex ratio in the litter

Mouse maternal behavior depends consistently on the sex ratio of the litter (Alleva et al., 1989; Cirulli and Laviola, 2000; Laviola and Alleva, 1995; Moore and Morelli, 1979) and this characteristic is present in both rats and mice. In particular, rat and mouse dams have been shown to interact differently with male and female pups, tending to take care of male pups more frequently and for longer periods in comparison with females. When litter culling leads to a particular unbalance in the gender ratio, one of the two sexes being mainly or exclusively represented, it has to be taken into account that some items of maternal behavior may be over-represented. Indeed, some behaviors, such as ano-genital licking, are particularly sensitive to the presence of males in the litter as the mother performs these more often towards the male than towards the female offspring (Laviola and Alleva, 1995; Moore and Morelli, 1979). In the case of feeble transgenic strains, elicitation of parental behavior, and thus pup survival, may then also depend upon the sex ratio, particularly in the case of very small litters. A high mortality level, or major bias in sex survival, can be avoided by using a strategic combination of sex-dependent culling (Wainwright, 1999).

Behavioral development of the mouse as assessed through test batteries

The rat and the mouse are altricial species, that is pups are born in a highly immature condition after a short pregnancy (18–22 days, depending on species and strain). At birth, the eyes and the ears are closed, the pup is able to crawl and to get attached to the nipple to suckle. Contact with the mother is required in order to thermoregulate and feed. Several reflexes and responses appear at successive postnatal stages, in parallel with somatic changes increasing pup's sensory and motor abilities.

The time of occurrence of specific somatic changes and the time of first appearance and subsequent complete maturation of various reflexes and responses show a remarkable regularity, providing the experimenter with an important and effective tool to assess whether somatic and neurobehavioral development is modified by prenatal, postnatal, or genetic manipulations. An isolated newborn rat has the limbs extended slightly away from the main axis of the body, the head being on the supporting surface. However, lateral movements can be recorded and when the animal is placed on its back, it is able to right itself. There is a rostro-caudal gradient of maturation. Right after birth, at rest, the pup is able to make some pivoting movements of the head and to initiate some crawling sequences (Clarac et al., 1998). Later on, pups raise their head, sniffing in the horizontal plane. During the second week, walking occurs and eyes are open.

Fox (1965) performed the first major investigation of sensorimotor and/or reflex development in mice. This classic study has served as the foundation of most of the work that has been subsequently done, e.g., assessing the behavioral effects of developmental neurotoxicants in this species (Alleva et al., 1985, 1987; Calamandrei and Alleva, 1989; Petruzzi et al., 1995).

This scale has been designed to include a number of endpoints that are representative of various components of neural and behavioral development in the first postnatal weeks, without imposing an excessive test burden on either the pup or the experimenter. The Fox battery has several limitations, however, such as an emphasis on early (preweaning) assessment and an absence of measures of either locomotor activity or learning. A number of mouse test batteries have therefore being developed to overcome these limitations (Bignami, 1996). As an example, the mouse test battery by Kallman and Condie (1985), while still focusing on early development, adds a test of muscular coordination and a passive avoidance task to the basic measures of sensorimotor function (i.e., righting reflex, forelimb placing, rooting reflex, auditory startle, cliff drop, and bar holding). The description given below represents a slightly modified version of the original scale since it also includes some additional measures (such as strong and weak tactile stimulation test with von Frey hairs, inducing a "wind-up-like pain") (Bignami et al., 1992).

The general procedure involves testing the animals daily to avoid time lags in the detection of a maturational event. It is essential that each experimenter establishes and verifies control baselines specific for the strain or conditions used in the study. In addition to scoring the behavioral endpoints described below, it is essential to perform in parallel an assessment of somatic developmental endpoints, including at least body weight gain and time of eye opening, ear opening, and incisors eruption.

The behavioral test battery that we have consistently adopted since the late 1970s in our developmental work with animals is as follows (Figure 11.1):

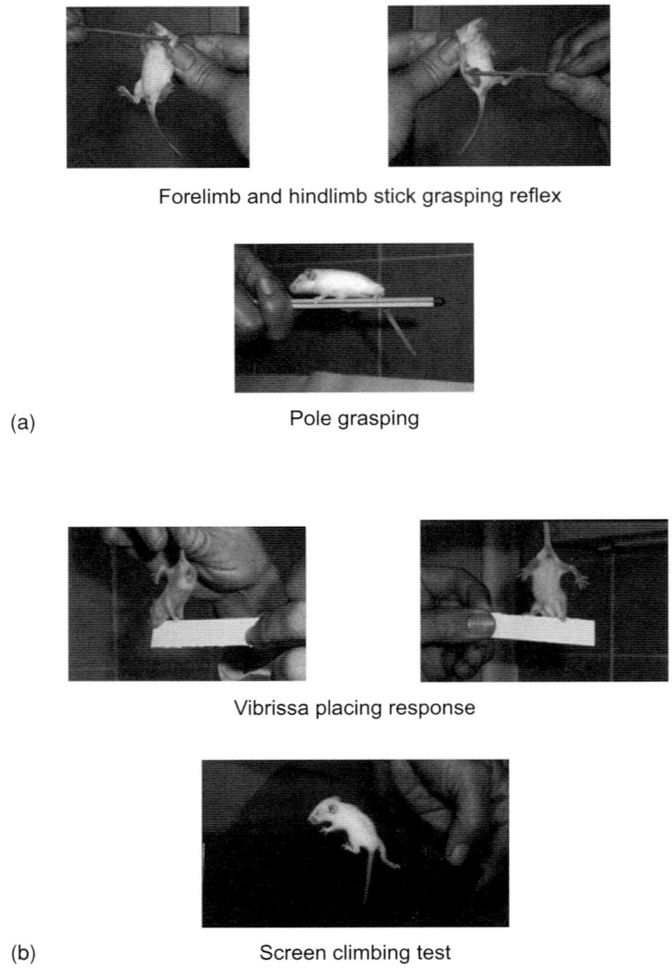

Figure 11.1 Scoring "reflex" development of neonatal mice (a and b).

1. *Righting reflex* – pup returns rapidly to its feet when placed on its back.
2. *Cliff aversion* – pup withdraws from the edge of a flat surface when its snout and forepaws are placed over the cliff.
3. *Forelimb and hind-limb stick grasping reflex* – pup grasps the shaft of a toothpick when it is stimulated on the palm of each paw.
4. *Forelimb and hind-limb placing reflex* – pup raises and places its forepaw or hind-paw on the surface of the edge of an object when stroked on the dorsum of the paw.
5. *Weak and strong tactile stimulation* – a head-turning response is triggered by the application of tactile stimuli (von Frey hairs of 0.35 or 0.05 g) in the perioral area on both sides of the head.
6. *Vibrissa placing response* – pup places its forepaw on a cotton swab stroked across its vibrissae.
7. *Level and vertical screen tests* – pup holds onto a wire-mesh (55 mm) when it is dragged across it horizontally or vertically by the tail.
8. *Screen climbing test* – pup climbs up the vertical screen using both forepaws and hind-paws.
9. *Pole grasping* – pup grips a wooden pencil with its forepaws.
10. *Auditory startle response* – pup shows a whole-body startle response when a loud clap of the hands occurs less than 10 cm away.

The reflexological and behavioral tests employed in the battery can give considerable information on the ontogeny of the nervous system of the mouse. Indeed, five well-defined periods of neurological development can be defined (Fox, 1965). These periods have been designated as *perinatal* (birth to 3 days), *neonatal* (3–9 days), *postnatal* transition (9–15 days), *postnatal* infantile or *pre-juvenile* (15–26 days), and *juvenile* (from 26 days until sexual maturity). It must be noted that the duration of these neurological periods of development is somewhat arbitrary, since some responses persist or are modified within a period so that in certain respects there is considerable overlap between one period and another. Minor strain differences in development occur very often. As an example, it has been shown that C3H/He mice develop faster than BALB/c, 120/S mice being the slowest (Kodama and Sekiguchi, 1984). In addition there is an inverse proportionality of litter size to the rate of postnatal growth.

Genetic and nutritional factors may affect neurobehavioral ontogeny. Nutritional factors can affect development severely. For instance, nutritionally deprived pups appear smaller than their littermates, hair growth is retarded, and the eyes open much later. Weak body righting and locomotion, together with persistent primitive responses, such as rooting and pivoting and weak bar-holding and grasp responses at 18 days of age, indicate that muscular weakness is coupled with delayed ontogenesis of adult behavior patterns and persistence of neonatal reflexes and behavior (Fox, 1965).

Suckling test

It is useful to reflect on the fact that young animals, particularly neonates, are considered animals deprived of postnatal life experience and thus surviving thanks to "autonomic patterns of spontaneous behaviors," which serve primary survival needs. The expression of a suckling pattern (not rarely the suckling phase is used to define a specific phase of postnatal development) is a very good example of a reflex-type behavioral pattern displayed by neonatal rodents.

Among mammals, suckling is the only behavior that is universal and characteristic (Blass and Teicher, 1980). Suckling is a peculiar form of feeding which is specifically expressed during early postnatal life and differs substantially from adult feeding behavior (Hall, 1975; Hall and Rosenblatt, 1978). In addition to allowing the newborn to obtain nutrients and fluids it serves other functions as well, such as reaching maternal contact.

The natural suckling situation is an interactive one. In order to distinguish maternal contribution from pup's own engagement, a procedure has been developed that involves anesthetizing the mother and then studying the aspects of pup suckling behavior in the absence of any maternal contribution to the situation.

Even poorly developed and motorically immature newborn pups are capable of locating and attaching, unaided, to the nipples of an anesthetized mother (Hall, 1975). Using this nipple-attachment capability, suckling behavior has been described in detail. In particular, the pup accomplishes nipple attachment by a process of searching the mother's ventral surface with sweeping head movements defined as "rooting." In the "rooting" response the infant, using swimming-like motions, moves quickly across the mother's body, scanning the surface by moving its head from side to side until the area where a nipple is located. Rooting thus stops and the pup probes into the ventrum with its snout to establish oral contact with a nipple (Pedersen and Blass, 1982). As contact with the nipple is made by the lips and tongue, the nipple is vigorously brought into the mouth by a process of licking and grasping. Olfaction provides a primary cue in eliciting nipple attachment (Teicher and Blass, 1977). Changes in olfactory sensitivity might affect survival in feeble transgenic strains.

Paddling and treading are important components of the suckling behavior pattern, helping pups to maintain contact with the dam's ventrum. Coordinated limb movements during the first postnatal week of life reflect the degree of motor competence; they are also used as an index of pup activation (Calamandrei et al., 1991).

A test to evaluate suckling in neonatal mice has been previously described (Blass and Teicher, 1980; Calamandrei et al., 1991; Ristine and Spear, 1984). In particular, the test can be performed at different ages after birth and uses an anesthetized pup's own dam in order to assess the responses to the suckling stimulus; these include attachment to the nipple and several other behaviors such as paddling with forelimbs, treading with hind-limbs, nipple shifting, and displacing a sibling from a nipple (Calamandrei et al., 1991).

In multiparous mammals, such as rats and mice, another neonatal necessity is to select and somehow to defend one particular nipple. In fact, neonatal rodents actively compete for milk. It has to be remembered that under natural conditions (as well as in the case of experiments which involve a reduced feeding schedule) this behavioral pattern is strictly modulated by the amount of milk produced by a single dam (Teicher and Blass, 1977; Teicher et al., 1978).

Although strain differences have not been described in this test, it might be used in the future to discriminate neurobehavioral development in different mouse strains and relate these differences to transgenics.

Patterns of ultrasonic emissions

Among innate behavioral patterns (so far no indication of learning has been reported) vocal signals – ultrasonic vocalizations (USVs) – emitted by neonates of altricial rodent species can be enlisted. Produced by mice and rats, these are whistle-like sounds characterized by frequencies ranging between 30 and 90 kHz, with a duration of 10–200 ms, and sound pressures of 60–100 dB (Bell et al., 1972; Branchi et al., 1998, 2001; Nyby and Whitney, 1978). Infant rodents emit also clicks characterised by a very short duration and a broadband frequency, and audible sounds with or without ultrasonic components normally produced in stress situations. Ultrasounds are characterized by a

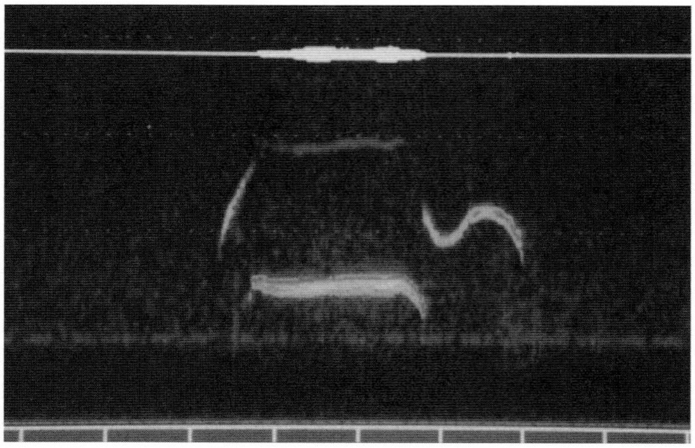

Figure 11.2 Spectrogram (frequency, kHz; time, ms; amplitude, dB) of ultrasonic calls of infant CD-1 mice illustrating the frequency steps sound category. An oscillogram (amplitude, dB; time, ms) is superimposed on the spectrogram. (Reproduced with permission from Branchi et al., 1998. © John Wiley and Sons, 1998.)

high rate of attenuation with distance and are easily deflected by very small objects such as blades of grass or twigs; they are also attenuated by humidity and dust particles in the air (Griffin, 1971; Sales and Pye, 1974).

For most mouse strains (e.g., CS-1, CD-1, C3H) the rate of USV calling follows an ontogenetic profile, increasing during the first 5–6 days of life, reaching a peak around postnatal day (PND) 6–7, then starting to decrease, and completely disappearing around the end of the second postnatal week (Elwood and Keeling, 1982). This pattern is highly strain-specific, and a wide variation has been shown. For example, C57BL6 and BALB/c show a peak of ultrasound rate on PND 3 (Bell et al., 1972).

Ultrasonic vocalization analysis is a reliable tool to investigate behavioral neurodevelopment. In mice USV production is an ethologically–ecologically relevant behavior, playing an important role in mother–offspring dyadic communication. Indeed, it strongly contributes to survival of the pup and to its physiological and behavioral development.

The test, which can easily be included in routine developmental behavioral testing, seems to be more appropriate than traditional methods used for the assessment of emotional reactivity during early postnatal life, such as changes in arousal/locomotory activity or latency time to emerge from the nest box (Adams, 1982; Adams et al., 1983; Cuomo et al., 1996). Indeed, the accurate analysis of USV emission profile presents the following advantages: (1) USVs are one of the few responses produced by very young mice that can be quantitatively analysed and be elicited by quantifiable stimuli; (2) USV production follows a strain-dependent ontogenetic profile from birth to PND 14–15, thus allowing longitudinal neurobehavioral analysis during very early postnatal life; (3) this technique has the advantage of minimal-animal-handling requirements (Adams, 1982). A more detailed evaluation is provided by spectrographic analysis (Figure 11.2).

A detailed description of the ultrasonic vocalization pattern considering the spectrographic structure of sounds has been performed both for mice (Branchi et al., 1998; Brudzynski et al., 1999) and rats (Brudzynski et al., 1999). Upon consideration of the spectrogram typology, it appears that the conditions under which vocalizations are emitted (i.e., exposure to low temperature, to the nest odor, to a conspecific adult-male odor, or to tactile stimulation) strongly influence the qualitative characteristics of ultrasounds production. In particular, it has been shown that the greatest change in acoustic signal emission due to exposure to different conditions occurs in frequency modulated sounds and that this modulation concerns mainly low-frequency ultrasounds (Branchi et al., 1998). For more details on the spectrographic analysis of mice see Branchi et al. (2001).

Genetic factors affecting reflex development

With the purpose of investigating genes-driven changes in developmental pathways, developmental batteries, such as Fox scales (Fox, 1965), have been usefully adapted for inbred or mutant mice in order to assess differences in neurodevelopment.

The sensory motor development of mouse pups from birth to weaning has been documented in several inbred strains of laboratory mice (Carlier et al., 1983; Kodama, 1993; Roubertoux et al., 1985; van Abeelen, 1980; van Abeelen and Schoones, 1977; Ward, 1980). It has been demonstrated that several reflexes are present already in the perinatal period in mice thus indicating that there is continuity from the fetal to the neonatal period (Kodama and Sekiguchi, 1984). By using cesarean-delivered fetuses, crossed extensor reflex, rooting reflex, and righting reflex were tested in different mouse strains. During the perinatal period, strain differences were detected in the duration of spontaneous body movements but not in the occurrence of reflexes. In more detail, crossed extensor reflex and rooting reflex were already observed in 30% of the subjects belonging to the ICR, C3H/He, and BALB/c strains on day 18 of gestation. Almost all subjects showed crossed extensor reflex on day 19 of gestation and rooting reflex on day 19.5 (Kodama 1993).

Strain differences have been documented in the postnatal period. As an example, it has been shown that C3H/He mice develop faster than BALB/c, 120/S mice being the slowest (Kodama and Sekiguchi, 1984). In the search for the genetic determinants of early innate motor responses, van Abeelen and Schoones (1977) studied the development of 14 reflex responses in two strains of mice genetically selected for high (SRH) and low (SRL) rearing response frequency. Data indicate that the strain selected for higher rearing frequency (SHR) showed high levels of activity only at later developmental stages, being less active initially.

Differences between strains may be related to differences in the maternal care received by the mother. Direct genetic and maternal influences on behavior and growth have indeed been found. The work performed by Ward (1980) on inbred mouse strains has indicated an important effect of maternal behavior on early postnatal development, pups receiving less maternal attention developing more rapidly. Carlier et al. (1983), in an important study on gene × environment interactions, have analyzed the effects of pup and mother strain and their interaction on reflex development in mice. Two strains NZB (N) and CBA/H (H) were selected as their reflex responses had been systematically described and found to differ, the N being one of the strains developing the slowest, the H being the one which develops the fastest. A strain effect was observed in 50% of the cases examined. The N development was slower than that of H (rooting and cross-extensor persisted longer in N than in H and the other reflexes appeared later in the N compared to the H strain). The effect of the mother strain was observed in about 30% of cases. In the case of the H mother strain, this contributed either towards slowing development or towards accelerating development of reflexes destined to persist (hind-limb grasping). It is important to keep in mind, as underlined by these authors, that pup strain and pup genotype are not the same thing. More precisely, the effect of strain is the result of several components including the genotype, the prenatal as well as the postnatal maternal environment, and the interaction of all these factors. One hypothesis put forward to explain these results is that because H mothers exhibit more frequent pup care, the pups they rear develop faster. Indeed it has been shown that early manipulation, in the form of handling, has a positive effect on neurobehavioral development in rodents (Cirulli et al., 2003a).

Again, Roubertoux et al. (1985) investigated the genetic and environmental bases for differences in rate of development in two other inbred strains of mice: C57BL/6By (B) and BALB/cBy (C). Twelve motor responses were selected from those described by Fox (1965) and used to measure these differences, in addition to individual body weight. A genetic analysis was then performed by using the recombinant inbred strains method. Over dominance was present in two variables alone (eye opening and weight at 10 and 20 days). In most cases, each of the response reflexes (locomotion, hind-limb, crossed extensor, righting, vibrissae placing, bar holding) was found to be associated with several genes. Differences across strains were associated with one segregating unit for rate of disappearance of the rooting. Since the strain distribution pattern differed for each sensory motor response, not a single general genetic factor of development was found. Maternal effects were found for four variables (grasping, forelimb placing, eye opening, and weight). For two responses, the F1 pups developing the fastest were reared by mothers from the slowest developing parental strain, leading to the hypothesis that not only the mothers differ as to the quality of the environment they furnish to their young but, in turn, pups differ in their ability to benefit from these environments (Roubertoux et al., 1985).

It cannot be excluded that mothers from different inbred strains might provide milk differing in quantity and biochemical characteristics, and that these differences may affect development of the central nervous system of the pups (Carlier et al., 1983).

The rate of behavioral development depends upon complex interactions between the genotype and the environment (Le Roy et al., 1998, 2001; Roubertoux et al., 1985; Wahlsten, 1974). Molecular genetic studies have helped unraveling such interplay. Properties such as homologies (identity of structure or function of a gene between species) and synthenies (highly conserved order of genes across mammalian genomic maps) have been described to occur in mice and could be usefully exploited to identify critical regions corresponding to the position of the target gene on the human chromosome, which should allow the development of appropriate animal models for human developmental disorders.

In a number of studies testing neurobehavioral development in different mouse strains, consistency coefficients and test–retest correlations indicate, with few exceptions (cliff aversion, crossed extensor, vertical clinging), that reliability is sufficiently high to expect stable behavioral measures. Such stability is of course a prerequisite for testing the effects of genes, environments, or treatments on the mean score of the strains or to map genes linked with these phenotypes. Using this approach, measures of sensory and motor development in mice, reliability scores and chromosomal position of putative loci have been described (Le Roy et al., 2001; Roubertoux et al., 1985). The use of sophisticated statistical techniques can be of great help in unravelling complex gene × environment interactions. Factor analysis has been successfully used to test the independence of behavioral measures employed to assess neurobehavioral development: in the F2 population deriving from C57BL/6By and NZB/B1NJ mice the sensory and motor variables modulated by growth factors were characterized by high loading on the same factor (Le Roy et al., 2001). Other factors were specifically linked to bar-holding measurement. By contrast, measures of symmetrical coordination, neonate motor behavior, and acquisition of coordination from cranial to caudal parts of the body were found to contribute differently to the variance in the two analyses. Overall these data have indicated that there is no general factor for sensory and motor development in mice, confirming results obtained with lines derived from divergent selections of mutants (Le Roy et al., 2001). Spontaneous mutants have been exploited to screen the relationship between a specific allele and the behavioral phenotype. In one of such studies, four mutant co-isogenic strains of mice with the same genetic background (C57BL/6JOrl) were used and important effects of the different genes tested on mice development were found, but the direction of these effects (delay or advance) differed across tests (Noel, 1989). These results underscore the importance of assessing multiple developmental measures rather than a composite index, since subtle differences might be missed. Another useful approach to establish causal relationships between genes and phenotypes is to use a wide genome scan with segregating populations (Le Roy et al., 2001). This strategy calls for identifying strains of mice presenting a large difference for the trait of interest and by deriving segregating populations. Alternatively, recombinant inbred strains (RIS) of mice can be used. Using RIS and backcrosses derived from C57BL/6 and BALB/cBy a transmission for age of disappearance of the rooting response fitted with one locus model was found. Furthermore, the age of disappearance of such response cosegregates with two different loci previously mapped on chromosome 4 (*Mup1* and *H-21*) (Roubertoux et al., 1987). This approach has ultimately proved that there is no general genetic factor for sensory and motor development (Le Roy et al., 2001).

Interestingly, it has been described that, as a general rule, the contribution of genotype to the phenotypic variance of sensory and motor measures of development appears low, thus calling for an investigation on other sources of variation. Indeed, results obtained from ovary transplantation, embryo transfer, and fostering protocols indicate that different components of the maternal environment can contribute to the behavioral phenotype (Le Roy et al., 2001; Roubertoux et al., 2003). As outlined previously, the contributions of the mother to the phenotype of the offspring represents a complex source of variation that needs to be taken into account in order to explain the mechanisms underlying individual differences (Roubertoux et al., 2003). In addition to nuclear genes, which are contributed by both parents, the mother, but not the father, provides the mitochondrial DNA to the offspring. The chromosomes that are inherited by the offspring contain several regions that are able to inhibit the expression of the homologous paternal genes (genomic imprinting). In addition, the mother contributes both prenatal and postnatal sources of variation which can affect the offspring's phenotype, both in addition and/or in interaction with the genotype. It has been shown, as an example, that mitochondrial DNA can affect cognitive abilities through an interaction with the nuclear genome and age in mice (Roubertoux et al., 2003).

Overall, the well-organized development of the nervous system relies on the fundamental interplay between the genes of each individual and the environment. Recent advances in the field of behavioral development have clearly shown that the nervous system of mammals is characterized by an intrinsic plasticity that enables complex information from the environment to become incorporated into the genome by means of molecular mechanisms, such as DNA methylation, which are able to produce stable differences in gene expression throughout the lifespan (environmental programming) (Meaney and Szyf, 2005). Such epigenetic variation in gene expression and function, while bringing exciting implications for the study of behavioral genetics, adds a further level of complexity to the system which will need to be taken into account in the future.

Final considerations

It is important to take into account that measuring behavior and reflex development is in itself a manipulation that can affect somatic and neurological development of the infant (Cirulli et al., 2003a; Kuhn et al., 1978; Levine et al., 1957). Indeed, simply removing pups from the mother for a few minutes until weaning results in animals which show increased exploration, less defecation, and urination in an open field and important changes in neuroendocrine reactivity at adulthood (Levine, 1967; Levine et al., 1957). This does not mean that one should not test young animals, but that appropriate control groups need to be added to take into account for such effects. In some cases this manipulation can be advantageous, especially if one is dealing with weak strains, since it might stimulate maternal behavior upon returning the pups to the mother, leading to a bout of nursing and maternal behavior.

Acknowledgments

We acknowledge the help of Nadia Francia for helping in collecting references and warmly thank Igor Branchi for the section on ultrasonic vocalizations. Work supported by the Italian Ministry of Health (grant on Neurodegenerative Diseases ex art. 56 to F. C. and E. A.) and by ISS (collaborative project ISS-NIH Rif. 530B, grant to E. A. and F. C.).

References

Adams, J. (1982) Ultrasonic vocalizations as diagnostic tools in studies of developmental toxicity: an investigation of the effects of hypervitaminosis A. *Neurobehav Toxicol Teratol* 4: 299–304.

Adams, J., Miller, D.R., and Nelson, C.J. (1983) Ultrasonic vocalizations as diagnostic tools in studies of developmental toxicity: an investigation of the effects of prenatal treatment with methylmercuric chloride. *Neurobehav Toxicol Teratol* 5: 29–34.

Alleva, E., Aloe, L., and Calamandrei, G. (1987) Nerve growth factor influences neurobehavioral development of newborn mice. *Neurotoxicol Teratol* 9: 271–275.

Alleva, E., Caprioli, A., and Laviola, G. (1989) Litter gender composition affects maternal behavior of the primiparous mouse dam (*Mus musculus*). *J Comp Psychol* 103: 83–87.

Alleva, E., Laviola, G., Tirelli, E., and Bignami, G. (1985) Short-, medium-, and long-term effects of prenatal oxazepam on neurobehavioural development of mice. *Psychopharmacology (Berl)* 87: 434–441.

Alleva, E. and Sorace, A. (2000) Important hints in behavioural teratology of rodents. *Curr Pharm Des* 6: 99–126.

Bateson, P. (1979) How do sensitive periods arise and what are they for. *Anim Behav* 27: 470–486.

Bell, R.W., Nitschke, W., and Zachman, T.A. (1972) Ultra-sounds in three inbred strains of young mice. *Behav Biol* 7: 805–814.

Bignami, G. (1996) Economical test methods for developmental neurobehavioral toxicity. *Environ Health Perspect* 104 (Suppl 2): 285–298.

Bignami, G., Alleva, E., Chiarotti, F., and Laviola, G. (1992) Selective changes in mouse behavioral development after prenatal benzodiazepine exposure: a progress report. *Prog Neuropsychopharmacol Biol Psychiatry* 16: 587–604.

Blass, E.M. and Teicher, M.H. (1980) Suckling. *Science* 210: 15–22.

Branchi, I., Santucci, D., and Alleva, E. (2001) Ultrasonic vocalisation emitted by infant rodents: a tool for assessment of neurobehavioural development. *Behav Brain Res* 125: 49–56.

Branchi, I., Santucci, D., Vitale, A., and Alleva, E. (1998) Ultrasonic vocalizations by infant laboratory mice: a preliminary spectrographic characterization under different conditions. *Dev Psychobiol* 33: 249–256.

Brudzynski, S.M., Kehoe, P., and Callahan, M. (1999) Sonographic structure of isolation-induced ultrasonic calls of rat pups. *Dev Psychobiol* 34: 195–204.

Calamandrei, G. and Alleva, E. (1989) Epidermal growth factor has both growth-promoting and growth-inhibiting effects on physical and neurobehavioral development of neonatal mice. *Brain Res* 477: 1–6.

Calamandrei, G., Valanzano, A., and Alleva, E. (1991) NGF and cholinergic control of behavior: anticipation and enhancement of scopolamine effects in neonatal mice. *Brain Res Dev Brain Res* 61: 237–241.

Capone, F., Bonsignore, L.T., and Cirulli, F. (2005) Methods in the analysis of maternal behavior in the rodent. In Maines, M., Costa, L., Reed, D., and Sassa, S. (eds.), *Current Protocols in Toxicology*. John Wiley and Sons, Hoboken, NJ, USA, pp. 13.9–13.9.16.

Carlier, M., Roubertoux, P., and Cohen-Salmon, C. (1982) Differences in patterns of pup care in *Mus musculus domesticus* I – comparisons between 11 inbred strains. *Behav Neural Biol* 35: 205–210.

Carlier, M., Roubertoux, P., and Cohen-Salmon, C. (1983) Early development in mice: I. Genotype and post-natal maternal effects. *Physiol Behav* 30: 837–844.

Caro, T. and Bateson, P. (1986) Organization and ontogeny of alternative tactics. *Anim Behav* 34: 1483–1499.

Chiarotti, F., Alleva, E., and Bignami, G. (1987) Problems of test choice and data analysis in behavioral teratology: the case of prenatal benzodiazepines. *Neurotoxicol Teratol* 9: 179–186.

Cirulli, F., Berry, A. and Alleva, E. (2003a) Early disruption of the mother-infant relationship: effects on brain plasticity and implications for psychopathology. *Neurosci Biobehav Rev* 27: 73–82.

Cirulli, F., Bonsignore, L.T., Venerosi, A., Valanzano, A., Chiarotti, F., and Alleva, E. (2003b) Long-term effects of acute perinatal asphyxia on rat maternal behavior. *Neurotoxicol Teratol* 25: 571–578.

Cirulli, F. and Laviola, G. (2000) Paradoxical effects of d-amphetamine in infant and adolescent mice: role of gender and environmental risk factors. *Neurosci Biobehav Rev* 24: 73–84.

Clarac, F., Vinay, L., Cazalets, J.R., Fady, J.C., and Jamon, M. (1998) Role of gravity in the development of posture and locomotion in the neonatal rat. *Brain Res Brain Res Rev* 28: 35–43.

Cohen-Salmon, C. (1987) Differences in patterns of pup care in *Mus musculus domesticus*. VIII. Effects of previous experience and parity in XLII inbred mice. *Physiol Behav* 40: 177–180.

Crusio, W.E. and Schmitt, A. (1996) Prenatal effects of parity on behavioral ontogeny in mice. *Physiol Behav* 59: 1171–1174.

Cuomo, V., De Salvia, M.A., Petruzzi, S., and Alleva, E. (1996) Appropriate end points for the characterization of behavioral changes in developmental toxicology. *Environ Health Perspect* 104 (Suppl 2): 307–315.

Descartes, R. (1664) *Traitè de l'homme*. Le Gras, Paris. English translation by Hall, T.S. (1972) *Treatise of Man*. [French text with translation and commentary], 1972 edn. Harvard University Press, Cambridge.

D'Udine, B. and Alleva, E. (1988) The *Acomys cahirinus* (spiny mouse) as a new model for biological and neurobehavioural studies. *Pol J Pharmacol Pharm* 40: 525–534.

Eibl-Eibesfeldt, I. (1989) *Human Ethology*. Aldine de Gruyter, New York.

Elwood, R.W. and Keeling, F. (1982) Temporal organization of ultrasonic vocalizations in infant mice. *Dev Psychobiol* 15: 221–227.

Fox, W.M. (1965) Reflex-ontogeny and behavioural development of the mouse. *Anim Behav* 13: 234–241.

Griffin, D. (1971) The importance of atmospheric attenuation for the echolocation of bats (*Chiroptera*). *Anim Behav* 19: 55–61.

Grota, L.J. and Ader, R. (1969) Effects of litter size on emotionality, adrenocortical reactivity, and susceptibility to gastric erosions in the rat. *Psychol Rep* 24: 547–549.

Hall, W.G. (1975) Weaning and growth of artificially reared rats. *Science* 190: 1313–1315.

Hall, W.G. and Rosenblatt, J.S. (1978) Development of nutritional control of food intake in suckling rat pups. *Behav Biol* 24: 413–427.

Hinde, R. (1975) Motivation of human and animal behavior (review of). *Q Rev Biol* 50: 352.

Hinde, R. (1982) *Ethology: Its Nature and Relations with Other Sciences*. Oxford University Press, Oxford.

Kallman, M.J. and Condie, L.W., Jr. (1985) A test battery for screening behavioral teratogens in mice. *Neurobehav Toxicol Teratol* 7: 727–731.

Kodama, N. (1993) Behavioral development and strain differences in perinatal mice. *J Comp Psychol* 107: 91–98.

Kodama, N. and Sekiguchi, S. (1984) The development of spontaneous body movements in prenatal and perinatal mice. *Developmental Psychobiology* 17: 139–150.

Kuhn, C.M., Butler, S.R., and Schanberg, S.M. (1978) Selective depression of serum growth hormone during maternal deprivation in rat pups. *Science* 201: 1034–1036.

Laviola, G. and Alleva, E. (1995) Sibling effects on the behavior of infant mouse litters (*Mus domesticus*). *J Comp Psychol* 109: 68–75.

Le Roy, I., Carlier, M., and Roubertoux, P.L. (2001) Sensory and motor development in mice: genes, environment and their interactions. *Behav Brain Res* 125: 57–64.

Le Roy, I., Roubertoux, P.L., Jamot, L., Maarouf, F., Tordjman, S., Mortaud, S., et al. (1998) Neuronal and behavioral differences between *Mus musculus domesticus* (C57BL/6JBy) and *Mus musculus castaneus* (CAST/Ei). *Behav Brain Res* 95: 135–142.

Levine, S. (1967) Maternal and environmental influences on the adrenocortical response to stress in weanling rats. *Science* 156: 258–260.

Levine, S., Alpert, M., and Lewis, G.W. (1957) Infantile experience and the maturation of the pituitary adrenal axis. *Science* 126: 1347.

Lorenz, K. (1943) The innate forms of potential experience. *Zeitschrift für Tierpsychologie* 5: 235–409.

Lorenz, K. (1981) *The Foundations of Ethology*, Springer-Verlag, New York.

Meaney, M.J. and Szyf, M. (2005) Environmental programming of stress responses through DNA methylation: life at the interface between a dynamic environment and a fixed genome. *Dialogues Clin Neurosci* 7: 103–123.

Moore, C.L. and Morelli, G.A. (1979) Mother rats interact differently with male and female offspring. *J Comp Physiol Psychol* 93: 677–684.

Noel, M. (1989) Early development in mice: V. Sensorimotor development of four coisogenic mutant strains. *Physiol Behav* 45: 21–26.

Nyby, J. and Whitney, G. (1978) Ultrasonic communication of adult myomorf rodents. *Neurosci Biobehav Rev* 2: 1–14.

Pedersen, P. and Blass, E. (1982) Olfactory control over suckling in albino rats. In Aslin, R., Alberts, J., and Peterson, M. (eds.), *Development of Perception: Psychobiological Perspectives*, Vol. 1. Academic Press, New York, pp. 359–381.

Petruzzi, S., Fiore, M., Dell'Omo, G., Bignami, G., and Alleva, E. (1995) Medium and long-term behavioral effects in mice of extended gestational exposure to ozone. *Neurotoxicol Teratol* 17: 463–470.

Prechtl, H.F. (1974) The behavioural states of the newborn infant (a review). *Brain Res* 76: 185–212.

Priestnall, R. (1972) Effects of litter size on the behavior of lactating female mice (*Mus musculus*). *Anim Behav* 20: 286–294.

Prochàzka, A., Clarac, F., Loeb, G.E., Rothwell, J.C., and Wolpaw, J.R. (2000) What do reflex and voluntary mean? Modern views on an ancient debate. *Exp Brain Res* 130: 417–432.

Prochàzka, G. (1784) *De functionibus systematis nervosi. Commentatio.* Wolfgang Gerle, Prague. English translation: Laycock, T. (1851) *A Dissertation on the Functions of the Nervous System. The Principles of Physiology, Prochaska on the Nervous System.* The Sydenham Society, London.

Ristine, L.A. and Spear, L.P. (1984) Effects of serotonergic and cholinergic antagonists on suckling behavior of neonatal, infant, and weanling rat pups. *Behav Neural Biol* 41: 99–126.

Rosenblatt, J.S. and Lehrman, D.S. (1963) Maternal behavior of the laboratory rat. In Rheingold, H. (ed.), *Maternal Behavior in Mammals.* Wiley, New York, pp. 8–57.

Roubertoux, P., Semal, C., and Ragueneau, S. (1985) Early development in mice: II. Sensory motor behavior and genetic analysis. *Physiol Behav* 35: 659–666.

Roubertoux, P.L., Baumann, L., Ragueneau, S., and Semal, C. (1987) Early development in mice. IV. Age at disappearance of the rooting response: genetic analysis in newborn mice. *Behav Genet* 17: 453–464.

Roubertoux, P.L., Sluyter, F., Carlier, M., Marcet, B., Maarouf-Veray, F., Cherif, C., et al. (2003) Mitochondrial DNA modifies cognition in interaction with the nuclear genome and age in mice. *Nat Genet* 35: 65–69.

Sales, G.D. and Pye, J.D. (1974) *Ultrasonic Communication by Animals.* Chapman and Hall, London.

Stern, J. (1996) Somatosensation and maternal care in Norway rat. In Rosenblatt, J.S. and Snowdon, C. (ed.), *Parental Care: Evolution, Mechanisms and Adaptative Significance.* Academic Press, San Diego, CA, USA, pp. 243–294.

Teicher, M.H. and Blass, E.M. (1977) First suckling response of the newborn albino rat: the roles of olfaction and amniotic fluid. *Science* 198: 635–636.

Teicher, M.H., Flaum, L.E., Williams, M., Eckhert, S.J., and Lumia, A.R. (1978) Survival, growth and suckling behavior of neonatally bulbectomized rats. *Physiol Behav* 21: 553–561.

Tinbergen, N. (1989) *The Study of Instinct.* Clarendon Press, New York.

van Abeelen, J.H. (1980) Direct genetic and maternal influences on behavior and growth in two inbred mouse strains. *Behav Genet* 10: 545–551.

van Abeelen, J.H. and Schoones, A.H. (1977) Ontogeny of behavior in two inbred lines of selected mice. *Dev Psychobiol* 10: 17–23.

Wahlsten, D. (1974) A developmental time scale for postnatal changes in brain and behavior of B6D2F2 mice. *Brain Res* 72: 251–264.

Wainwright, P. (1999) Methodological issues in the assessment of behavioral development in laboratory mice. In Crusio, W. and Gerlai, R. (eds.), *Handbook of Molecular-Genetic Techniques for Brain and Behavior Research*, Vol. 13. Elsevier, Amsterdam.

Ward, R. (1980) Some effects of strain differences in the maternal behavior of inbred mice. *Dev Psychobiol* 13: 181–190.

Willis, T. (1664) *Cerebri Anatome cui accessit nervorum descriptio et usus.* English translation: Feindel, W. (1965) *The Anatomy of the Brain and Nerves.* McGill University Press, Montreal.

Chapter 12

Feeding and drinking

Richard J. Bodnar, Sarah R. Lewis-Levy, and Benjamin Kest

Systematic analyses of rodent, and particularly mouse, strain differences are important sources of information regarding the genetic control of all aspects of ingestive behavior (see reviews: Reed et al., 1997; West and York, 1998). Such studies not only indicate the presence of genetic variance in ingestive responses, but may also identify strains with divergent sensitivities for quantitative trait loci (QTLs) analyses to localize chromosomal regions, and ultimately genes, critically involved in such differences. In addition to studies examining food and water intake per se and/or macronutrient choice, another behavioral approach has been employed to assess this genetic variance using preference tests between a given ingestive stimulus and a control (e.g., water). A number of studies, particularly the older ones, used either two or a small number of inbred strains to make specific preference comparisons between "sensitive" and "insensitive" strains for a particular response. This approach is also used in many more recent genetic QTL approaches to define gene loci. A second and more recent approach uses large numbers of inbred strains to insure reliability, assess heritability estimates, and facilitate the identification of strains with highly divergent responses in order to increase the success of subsequent QTL mapping. Collectively, recent studies, particularly over the past decade and reviewed in detail below, have demonstrated marked strain differences in food and water intake as well as for the following ingestive stimuli: (a) salts, (b) bitter tastants, (c) saccharin, (d) sugars, (e) ethanol, (f) glutamate/umami, and (g) fats. This review also focuses on mechanisms of the above forms of intake as they relate to the development of obesity. The final section of this chapter will examine a series of studies recently conducted in our laboratory to evaluate and compare strain differences in sweet (sucrose) and fat (Intralipid) intake relative to the feeding responses elicited by glucoprivic (2-deoxy-D-glucose: 2DG) and lipoprivic (mercaptoacetate: MA) stimuli.

Salts

Genetic variance between mouse strains was initially observed for intake of moderate concentrations (0.075–0.150 M) of sodium chloride (saline) solutions, with 129/J mice preferring this solution to water in 48 hour preference tests and C57BL/6J mice rejecting this range of solutions (Beauchamp and Fisher, 1993; see also, Bachmanov et al., 1996a, 1998b; Gannon and Contreras, 1993; Kotlus and Blizard, 1998). Genetic variance is quite marked in large analyses of strains as well (Bachmanov et al., 2002a). Increasing choice by using more than two bottles enhanced the salt preferences in 129X1/SvJ mice relative to C57BL/6J mice (Tordoff and Bachmanov, 2003a), and this effect was further enhanced by repeated testing (Tordoff and Bachmanov, 2002). However, in analyzing saline preferences in mice maintained on different maintenance diets, C57BL/6J mice showed a greater preference for a 75 nmol/L sodium chloride solution, particularly on purified diets, than 129X1/SvJ mice, indicating that there might be differences due to both the substrain and the underlying diet (Tordoff et al., 2002). Another factor in assessing multiple concentrations of solutions to evaluate strain sensitivity is the order of presentation. Thus, five mouse strains exposed to progressively increasing or decreasing sodium chloride solutions displayed greater preferences for low concentrations of the saline solutions in the ascending as compared to the descending series of stimulus testing. The NZB/B1NJ strain displayed greater sodium chloride acceptance than 129/J, SM/J, and C57BL/6ByJ strains that in turn displayed stronger responses than the CBA/J strain. Whereas NZB/B1NJ and 129/J mice displayed strong preferences at low concentrations, only the former groups showed persistent preferences at the highest concentrations (Bachmanov et al., 1998b). Further, a subsequent evaluation of 28 mouse strains confirmed that NZB/B1NJ mice avidly consumed high amounts of concentrated sodium, but not potassium or calcium, chloride solutions. At lower sodium chloride concentrations, CAST/Ei mice showed the strongest preferences whereas CBA/J, C3H/HeJ, and AKR/J mice showed the strongest avoidance in preference tests (Bachmanov et al., 2002a). A 40 mouse strain survey of water and sodium intake revealed strains with high preference (129S1/SvIm, MA/MyJ, NZW/LacJ, and SWR/J) or indifference (A/J, C57BL/6J, FVB/NJ, and SEA/GnJ) for sodium at all concentrations (Tordoff et al., 2007a). Analysis of normotensive (BPN/3), hypertensive (BPH/2), and hypotensive (BPL/1) mouse strains revealed that hypertensive mice consumed greater fluid and water intakes that normotensive controls, but consumed lower amounts of sodium chloride and

Behavioral Genetics of the Mouse: Volume I. *Genetics of Behavioral Phenotypes*, eds. Wim E. Crusio, Frans Sluyter, Robert T. Gerlai, and Susanna Pietropaolo. Published by Cambridge University Press.

potassium chloride solutions. In contrast, hypotensive mice consumed higher intakes of potassium chloride at moderate concentrations and lower amounts of calcium chloride (Bachmanov et al., 1998a). A survey of 40 mouse strains revealed strains with consistently high avidity (PWK/PhJ, BTBR T^{+}tf/J, JF1/Ms) and low avidity–high avoidance (KK/H1J, C57BL/10J, CE/J, C58/J) for a wide range of calcium solutions (Tordoff et al., 2007b). A genome screen involving the F2 generation of C57BL/6J mice with low avidity for calcium and PWK/PhJ mice with high avidity for calcium revealed 30 QTLs of which six involved consumption of calcium chloride (Tordoff et al., 2008a). In addition to its importance in sweet intake, the T1R3 receptor also functions as a gustatory calcium–magnesium receptor (Tordoff et al., 2008b). Thus, whereas genetic variance plays a major role in the consumption of salts, it is apparent that a number of methodological factors need to be controlled in assessing such differences; this latter point is reiterated in analyses of other forms of ingestion.

Bitter tastants

Following the initial observation that mice homozygous for the allele Soa^{a} display greater aversions to sucrose octa-acetate (SOA) and strychnine than heterozygous counterparts (Warren and Lewis, 1970), Lush (1981, 1982) characterized the behavioral genetics of tasting of SOA and strychnine in multiple strains of mice. He indicated that SWR/J mice possess taster ability of SOA, and that two allelic forms (Soa^{a}: avoidance and Soa^{b}: indifference) were proposed to explain these genetic differences. SWR/J "taster" mice displayed both strong SOA avoidance in behavioral tests, and potent neural responses to SOA in the chorda tympani and glossopharyngeal nerves in neurophysiological assessments. In contrast, "non-taster" SWR.B6 mice displayed behavioral indifference and weak neural responses to SOA (Inoue et al., 2001). Subsequent assessment of 29 strains consuming either a single concentration (0.8 mM) or a range (0.4–1.6 mM) of quinine solutions in a study by Lush (1984) indicated that A2G, DBA/2, and BALB/cBy strains displayed a moderate quinine aversion (30–43%) in a two-bottle taste test, whereas SWR, 129/Sv, and C57BL/6By mice displayed powerful aversions (96–98%) over this range. Subsequent description of a third intermediate demitaster category of mice was provided by Harder et al. (1992), indicating differentiation in avoidance responses between the 0.1 and 1.0 mM range of SOA concentrations, suggesting a third allele (Soa^{c}). A common polygenic basis for quinine and propylthiouracil (PROP) avoidance was subsequently described as well (Blizard et al., 1999; Harder and Whitney, 1998). Interestingly, insertion of two type-A Prp transgenes from taster mice failed to alter SOA avoidance in non-taster mice (Harder et al., 2000). C57BL/6J and C57BL/6ByJ mice displayed greater intake of and preference for citric acid and quinine solutions relative to 129/J and 129Xi/SvJ mice (Bachmanov et al., 1996a), an effect enhanced by the use of purified diets (Tordoff et al., 2002). Both hypertensive (BPH/2) and hypotensive (BPL/1) strains of mice consumed significantly less quinine than their normotensive (BPN/3) counterparts (Bachmanov et al., 1998a). The underlying genetics and recently described polymorphisms in the Soa gene are discussed in Chapter 9.

Saccharin

Strain differences have been observed for saccharin intake in a wide series of studies (e.g., Blizard et al., 1999; Capeless and Whitney, 1995; Inoue et al., 2004b; Reed et al., 2004; Tordoff et al., 2002). An early study using rats by Nachman (1959) found that F1 and F2 generation progeny of saccharin-preferring animals displayed saccharin preferences comparable to those of their parents, whereas water-preferring parents and their progeny failed to display saccharin preferences. In subsequent murine studies by Pelz, a strong preference for a 0.1% saccharin solution relative to water was observed in BALB/cJ, C57BL/6J, I^{S}/Bi mice, but not in 101Bag/R1 mice (Pelz et al., 1973). Correspondingly, Fuller (1974) found that C57BL/6J mice displayed greater intake of the same (0.1%) saccharin solution than DBA/2J mice. Genetic factors accounted for 78% of the genetic variation associated with consumption of this concentration of saccharin in one outbred and seven inbred strains. This effect was extended to four saccharin concentrations with a rank-order of strain preference scores of 129P3/J, C57BL/6J, BALB/cJ, C3H/HeJ, 129P3/J, and DBA/2J mice (Blizard et al., 1999; Capeless and Whitney, 1995). Intake for a single 1.6 mM saccharin solution in 26 inbred mouse strains by Lush (1989) revealed a pattern of stronger saccharin preferences in A/2G, C57BL/6Ty, C57BL/10, and SWR strains (73–93%) than in AKR, CBA/Ca, C3H/He, DBA/2Ty, and 129/Sv strains (53–59%). Extreme responding strains identified for saccharin intake served as progenitors for quantitative trait loci (QTLs) and subsequently, the identification of trait relevant genes. Previous QTLs for saccharin intake using C57BL/6J crosses with DBA/2J or 129P3/J mice revealed both a saccharin (Sac) preference locus and a sweet taste receptor gene,Tas1r3, that are described in detail in Chapter 9.

Sucrose

Similarly to saccharin, genetic variance among mouse strains has been observed for sucrose intake across a wide range of studies (e.g., Bachmanov et al., 1997; Blizard et al., 1999; Inoue et al., 2004b; Lewis et al., 2005). C57BL6/J mice displayed greater intake of five (0.005–1.000 M) glucose and sucrose concentrations than 101Bag/R1 mice in an early study by Stockton and Whitney (1974), and of a 4% sucrose solution than 129P3/J mice (Bachmanov et al., 1997). The greater sensitivity of C57BL/6J to sweetened solutions like saccharin and sucrose relative to 129P3/J mice was extended to maltose, acesulfame-K, sucralose, and SC-45647 as well as to the amino acids, D-phenylalanine, D-tryptophan, L-proline, and glycine (Bachmanov et al., 2001b). Genetic factors accounted for 83% of the genetic variation associated with consumption of 3% sucrose in outbred and seven inbred strains in an early study by Ramirez and Fuller (1976). In examining up to 30 strains of mice for sucrose (50 mM) intake, Lush (1989) found that the patterns of strong preferences for sucrose, like saccharin, were greater in A/J, C57BL/6J, C57BL/10J, and SWR/J strains (73–97%) than in AKR/J, CBA/J, C3H/HeJ, DBA/2J, and 129P3/J strains (51–61%). SWR/J mice displayed increased flavor preferences and lick activity for both sucrose and corn oil compared to AKR/J mice, suggesting greater sensitivity to orosensory flavor factors (Smith et al., 2001). A recent evaluation (Pothion et al., 2004) of sucrose intake at seven supra-threshold (1–50%) concentrations in 11 mouse strains revealed a difficulty in determining clear-cut strain differences because virtually every strain consumed far more sucrose than water for at least one of these higher concentrations. Conditioned flavor preferences induced by intragastric infusions of either a 16% sucrose solution or a 5.6% soybean oil solution appeared to be stronger in C57BL/6J mice than 129 mice, whereas both strains consumed similar amounts of an isosweet solution in preference tests, suggesting that C57BL/6J mice may possess a stronger orosensory response to sugar and fat (Sclafani and Glendinning, 2005). C57BL/6J mice also display conditioned flavor preferences to the intragastric effects of a 8% maltodextrin solution (Sclafani and Glendinning, 2003). Similarly, C57BL/6ByJ mice display higher consumption and lower preference thresholds for the sweet amino acids, l-glutamine, l-alanine and l-threonine, the monosaccharides glucose and fructose, and malto-oigosaccharide (Bachmanov and Beauchamp, 2008). Age fails to contribute significantly to a wide range of taste preferences observed in C57BL/6J and 129X1/SvJ mice (Tordoff, 2007). Finally, the use of trpm5–/– mice, which lack the cellular machinery for sweet taste transduction, can develop a robust preference for sucrose solutions based solely on caloric content (de Araujo et al., 2008).

Ethanol

In addition to sucrose, ethanol-related phenotypes have been identified in animal models (see reviews: Crabbe et al., 1994, 1999), and the two-bottle choice test in mice appears to produce data relevant to human alcoholism. Interestingly, in preference studies using more than two (e.g., up to six) bottles, alcohol intake is positively and persistently related to the number of alcohol bottles available, and inversely related to the number of water bottles available (Tordoff and Bachmanov, 2003b). Further, restricted, relative to continuous, access to ethanol resulted in greater consumption of ethanol in C57BL/6J and WSC strains (Finn et al., 2005). Moreover, the former strain is also more effective than DBA/2J mice in displaying an animal model of intoxication using blood ethanol concentration as a measure (Rhodes et al., 2005). Indeed, the C57BL/6J and DBA/2J strains have been respectively identified in high and low consumption of ethanol in such preference tests (Belknap et al., 1997; Melo et al., 1996; Phillips et al., 1994, 1998; Tarantino et al., 1998; Whatley et al., 1999). Indeed, the former strain displayed higher ethanol preferences than 129P3/J mice in a manner similar to that observed for sucrose and citric acid (Bachmanov et al., 1996b). Analyses of 15 mouse strains over a range (3–10%) of ethanol concentrations revealed that C57BL/6J, C57BR/cdJ, and C57L/J mice consumed the greatest amounts of ethanol, whereas DBA/1J and DBA/2J strains consumed the least (Belknap et al., 1993). The C57BL/6J strain avidly consumes ethanol in a drinking-in-the-dark paradigm that fits in well as an animal model for human alcoholism based on its sensitivity to acamprosate (Gupta et al., 2008). Further, supplementing the liquid alcohol diet with chow enhanced alcohol intake in C57BL/6 mice (Anji and Kumari, 2008).

Quantitative trait loci studies of ethanol preference (3% and 10% concentrations) in C57BL/6By × 129P3/J F2 hybrids, identified two loci on distal chromosome 4 (Ap3q) and proximal chromosome 7 (Ap7q); their presence strongly affected ethanol intake at the high, but not low concentration. Further, an identified male-specific locus on chromosome 8 (Ap8q) affected ethanol preference at the low, but not the high concentration. Additional linkages on chromosomes 2, 9, 12, 13, 17, and 18 were found as well (Bachmanov et al., 2000a). A meta-analysis (Belknap and Atkins, 2001) of eight studies suggested consistent QTLs for ethanol for chromosome 2 (proximal to mid), 3 (mid to distal), 4 (distal), and 9 (proximal to mid). The Ap3q locus on chromosome 4 contains both the saccharin (Sac) preference locus and corresponds to the sweet taste receptor gene, Tas1R3 (Blizard et al., 1999; Phillips et al., 1994). Quantitative trait loci mapping has also revealed a role for the syntaxin binding protein 1 gene for an ethanol preference locus on mouse chromosome 2 (Fehr et al., 2005). In contrast, gene mapping of chronic withdrawal from ethanol revealed loci on chromosomes 1 (proximal), 4 (mid), 8 (mid), 11 (proximal), and 14 (mid) (Bergeson et al., 2003). A recent meta-analysis for alcohol preference in mice based on QTL analysis revealed eight candidate genes, the expression of which was localized to the olfactory zone, limbic areas, and the orbitofrontal cortex (Tabakoff et al., 2008).

Glutamate/umami

Although relatively few studies have explored taste preferences for glutamate and umami-type stimuli, the data appear very consistent. Initial acceptance studies using one-bottle tests demonstrated that glutamate could be distinguished from the four other taste substances in mice in early studies by Ninomiya and Funakoshi (1989a, 1989b). C57BL/6J mice: (a) display lower preference thresholds for monosodium glutamate (MSG); (b) prefer MSG over a greater range of concentrations; and (c) consume greater amounts of MSG at high concentrations than 129/J mice. Prior experience with MSG, but not with saccharin, enhanced the subsequent expression of MSG acceptance, an effect also observed with inosine-5'-monophosphate (Bachmanov et al., 2000b). Use of F2 generations of C57BL/6By and 129P3/J mice bred for sucrose preference failed to reveal corresponding changes in MSG relative to sucrose intake preferences, suggesting a unique genetic mechanism for this taste (Bachmanov et al., 2000b; Beauchamp et al., 1998). Interestingly, the increased ingestive responses to umami taste in C57BL/6J as compared to 129/J mice are accompanied by either unchanged or decreased neural responses in the chorda tympani or glossopharyngeal nerves, an effect sharply divergent from that described previously for other taste stimuli (Inoue et al., 2004a).

Fat intake and obesity-prone and obesity-resistant animals

The intake of dietary fat also systematically varies as a function of genetic predisposition among a host of other variables (see review: West and York, 1998). Indeed, the analysis of genetic variance has led to the identification of dietary resistance and susceptibility phenotypes in inbred and outbred strains of mice (e.g., West et al., 1992, 1995). These studies led to the identification of particular mouse strains in which only moderate intake of a high-fat diet promoted weight gain and obesity (e.g., AKR/J mice), and other strains in which large intake of the high-fat diet was not accompanied by weight gain (e.g., SWR/J). Although such effects were largely due to variation in the dietary fat content, this latter variable weakly correlated with total energy intake. The AKR/J and SWR/J strains displayed similar effects on intake and compensatory weight changes whether the fat source was shortening, lard or powder, and whether the high- and low-fat diets were isocaloric (Smith-Richards et al., 1999). Indeed, whereas AKR/J and C57BL/6J mice self-selected the highest proportion of fat in macronutrient diet selection with ependymal fat correlating with fat consumption, SWR/J and CAST/Ei strains consumed a great deal of fat that was inversely correlated with ependymal fat (Smith et al., 2000). Moreover, whereas the diet-sensitive AKR/J and DBA/2J strains (a) consumed greater amounts of fat; (b) displayed more adiposity; and (c) displayed elevated levels of leptin and insulin; the C57BL/6J strain showed an equal preference between protein and fat, and displayed normal insulin and leptin levels (Alexander et al., 2006). In contrast, the obesity-resistant SWR/J and A/J mice that consumed more fat than carbohydrate, yet failed to gain weight, did so potentially because of (a) lower insulin levels; (b) increased capacity of skeletal muscle to metabolize fat; (c) enhanced paraventricular galanin; and/or (d) reduced arcuate Neuropeptide Y (Leibowitz et al., 2005). Maintenance on a very high-fat diet (60%) resulted in type 2 diabetes for C57BL/6J mice and normoglycemic responses in A/J mice (Gallou-Kabani et al., 2007). Mapping of a series of multiple genetic loci (mob 1–4), located on chromosomes 9 and 15, appeared to explain some of these genetic variations for fat and obesity (e.g., Bachmanov et al., 2001a; Fisler et al., 1993; Smith-Richards et al., 2002; Warden et al., 1995; West et al., 1994a, 1994b). A 40 mouse strain survey of body composition revealed profound genetic variance in percentage of body fat that ranged from 16 (C58/J) to 39% (NON/LtJ) (Reed et al., 2007). Increased body fat has also been associated with a novel p-locus fat-associated ATPase on mouse chromosome 7 (Dhar et al., 2000). Moreover, loci on chromosomes 2, 4, 9, and 16 have been identified for body weight, body length, and adiposity in a genome scan of an F2 intercross between the 129P3/J and C57BL/6ByJ mouse strains (Reed et al., 2003). Using a C57BL/6J × PWK/PhJ mouse intercross, a genetic loci analysis identified 28 suggestive or significant linkages for four traits (body weight, adjusted lean and fat weight, and percent fat) (Shao et al., 2007). Moreover, using C57BL/6ByJ × 129P3/J F2 hybrids, absolute deposit weight was linked to chromosomes 5, 11, and 14, relative deposit weight was linked to chromosomes 9, 15, and 16, and both types were linked to chromosomes 2 and 7 (Reed et al., 2006). Finally, using mouse lines divergently selected for food intake, chromosomes 4 and 19 were associated with white and brown adipose issue, and chromosome 9 was associated with white adipose tissue deposits (Rance et al., 2007). Therefore, fat intake, and its attendant changes (or lack thereof), appears to be under the influence of genetic variability.

Recent strain survey studies from our laboratory

Our laboratory examined strain differences among 11 inbred (A/J, AKR/J, BALB/cJ, CBA/J, C3H/HeJ, C57BL/6J, C57BL10/J, DBA/2J, SJL/J, SWR/J, 129P3/J) and one outbred (CD-1) mouse strains in four different paradigms that analyzed sweet (sucrose) intake (Lewis et al., 2005) and fat (Intralipid) intake (Lewis et al., 2007), as well as feeding responses elicited by glucoprivic (2DG: Lewis et al., 2006a) and lipoprivic (MA: Lewis et al., 2006b) stimuli. In the first pair of studies (Lewis et al., 2005, 2007), we presented animals in each strain a choice of nine different subthreshold, threshold, and supra-threshold sucrose (0.0001–20.0%) and Intralipid (0.00001–5.0%) concentrations using two-bottle 24-hour preference tests. We controlled for relevant methodological variables such as ascending and descending presentations of different sucrose/Intralipid concentrations (Harder et al., 1989), the relative positions of the two bottles containing sucrose/Intralipid and water (Bachmanov et al., 2002b), measured kilocalorie intake as sucrose/Intralipid or chow, and examined such effects in absolute terms or relative to body weight. In the second pair of studies (Lewis et al., 2006a, 2006b), we used either the anti-metabolic glucose analog, 2DG or the free fatty acid oxidation inhibitor, MA to elicit respective glucoprivic or lipoprivic states to examine whether genetic variance was present in the ingestive responses to these regulatory challenges. We employed systemic dose ranges of 2DG (200–800 mg/kg) and MA (50–100 mg/kg) that in previous studies both elicited feeding, while controlling for the order of ascending and descending 2DG and MA doses. Finally, we examined the presence of potential relationships in common or differential genetic variance between sweet and fat intake and between glucoprivic and lipoprivic responses.

Table 12.1 Baseline water (ml, ±SEM) and chow (g, ±SEM) intake and body weight (g, ±SEM) in 12 mouse strains in analyses of sucrose and Intralipid intakes.

Strain	Sucrose water (ml)	Intralipid water (ml)	Sucrose chow (g)	Intralipid chow (g)	Sucrose weight (g)	Intralipid weight (g)
A/J	5.6 (0.7)	4.6 (0.2)	4.4 (0.3)	3.8 (0.2)	26.2 (1.4)	19.6 (0.5)
AKR/J	5.7 (0.6)	8.6 (0.2)	4.6 (0.1)	4.1 (0.1)	33.6 (1.4)	26.6 (1.0)
BALB/cJ	5.6 (0.1)	7.4 (0.2)	5.4 (0.2)	6.7 (0.2)	27.9 (0.8)	23.5 (0.2)
C57BL/6J	4.9 (0.3)	4.8 (0.4)	4.1 (0.1)	3.6 (0.2)	27.9 (0.6)	27.8 (0.8)
C57BL/10J	5.5 (0.3)	6.9 (0.5)	3.7 (0.1)	4.8 (0.2)	26.8 (0.5)	22.7 (0.3)
CBA/J	5.3 (0.3)	6.7 (0.2)	3.8 (0.1)	4.0 (0.2)	32.3 (1.5)	22.6 (0.5)
CD-1	7.8 (0.4)	9.3 (0.8)	5.8 (0.3)	4.1 (0.4)	37.7 (0.8)	37.0 (1.0)
C3H/HeJ	5.2 (0.4)	5.5 (0.2)	4.2 (0.1)	4.4 (0.2)	28.4 (0.5)	18.5 (1.1)
DBA/2J	5.1 (0.1)	6.5 (0.2)	4.4 (0.2)	5.3 (0.2)	27.0 (0.8)	23.7 (0.2)
SJL/J	5.7 (0.3)	6.6 (0.1)	3.5 (0.2)	3.6 (0.2)	25.6 (0.2)	21.1 (0.3)
SWR/J	7.3 (0.3)	9.2 (0.3)	4.6 (0.2)	3.9 (0.1)	27.2 (0.3)	18.3 (0.6)
129P3/J	7.1 (0.4)	5.9 (0.4)	NA	5.0 (0.4)	30.5 (0.4)	22.6 (0.4)
Correlation	$r = 0.65$	$P < 0.05$	$r = 0.42$	NS	$r = 0.77$	$P < 0.05$

NA: not applicable; NS: not significant, SEM: standard error of mean.

Baseline food, water, and weight responses

A previous analysis of food intake and body weight in male mice from 28 inbred strains indicated high narrow-sense heritability estimates, particularly for body weight ($h^2 = 0.87$: Bachmanov et al., 2002b). Baseline 24 hour intake of chow and water (across two bottles) as well as body weight were measured for the 12 tested strains (Lewis et al., 2005, 2007). Table 12.1 summarizes the means of these three variables for both studies; significant correlations were observed for water intake ($r = 0.65$) and body weight ($r = 0.77$) as a function of the 12 strains in these two studies. Indeed, water intake in strains tested by both Bachmanov et al. (2002b) and our laboratory (Lewis et al., 2007) yielded positive and significant correlations as well ($r = 0.67$). Such data indicate that baseline responses elicited by the 12 different strains appear consistent between studies both within and across laboratories.

Sucrose

Our evaluation (Lewis et al., 2005) of genetic variance in sucrose intake revealed strong and marked differences in terms of sensitivity to sucrose concentrations (Figure 12.1c), the absolute magnitude of sucrose intake (Figure 12.1a), and the evaluation of the amount of kilocalories consumed as sucrose (Figure 12.1b). In this regard, A/J, C57BL/6J, CD-1, and SWR/J strains consumed the greatest (11.6–22.0 ml) amounts of sucrose (Figure 12.1a), whereas the A/J, C57BL/10J, SJL/J, and SWR/J strains consumed the greatest (44–56%) percentages of kilocalories as sucrose (Figure 12.1b). Among these strains only the CD-1 and SWR/J consumed significantly more sucrose than water at each of the nine concentrations tested (Figure 12.1c). The BALB/cJ and 129P3/J strains displayed intermediate responsiveness in terms of sensitivity to sucrose concentrations (Figure 12.1c). In contrast, the AKR/J, CBA/J, C3H/HeJ, and DBA/2J strains consumed the least (6.9–7.9 ml) amount of sucrose (Figure 12.1a), and displayed low (20–30%) percentages of kilocalories consumed as sucrose (Figure 12.1b). Correspondingly, the DBA/2J and C3H/HeJ strains significantly increased sucrose intake over water intake only at the two highest concentrations, indicating less sensitivity (Figure 12.1c). The consistently higher sucrose responses observed in C57BL6/J mice relative to 129P3/J mice is consistent with other recent findings (Sclafani, 2006a, 2006b). A number of previous studies have employed as a measure of preference the percentage of sweetener consumed as a function of total fluid intake over a very restricted range of sucrose concentrations (e.g., Capeless and Whitney, 1995; Fuller, 1974; Lush, 1989; Pothion et al., 2004). This appeared not to be a reliable measure in our study because strains that showed both large (e.g., C57BL/6J, C57BL/10J, SWR/J, SJL/J) and small (e.g., C3H/HeJ) magnitudes of sucrose intake invariably showed very high ($\geq$95%) preferences for sucrose. This underlines the importance of studying multiple strains across a greater range of sucrose concentrations.

A further noteworthy finding of our first study (Lewis et al., 2005) was that 24 hour sucrose over-consumption produced strain-dependent effects on overall kilocaloric intake. Whereas A/J, C57BL/6J, C57BL/10J, CD-1, SWR/J, and SJL/J strains all displayed the most pronounced compensatory decreases in chow intake as the percentage of kilocalories consumed as sucrose increased, the AKR/J, C3H/HeJ, and DBA/2J strains failed to significantly alter chow intake even at high sucrose concentrations. This very rapid compensation to the energy provided by sucrose suggests that some strains have either a greater sensitivity to changes in energy and/or a quicker ability to both adapt and respond to these changes in energy.

Divergent responders may be a model for studying and identifying genetic substrates associated with this ability to regulate kilocalorie intake across a variety of energy sources. The fact that AKR/J, C3H/HeJ, and DBA/2J strains persist in consuming normal chow intake in addition to their increased kilocalories consumed at high sucrose concentrations make these strains potential models in chronic studies that might show increased weight gain, obesity, and diabetic symptoms. By testing a sufficient number of randomly chosen inbred strains, this study design also allows for the valid estimation of genetic correlations (Hegmann and Possidente, 1981). Thus, in determining whether sucrose consumption in the present study correlated with Tas1R3 polymorphisms in mouse strains (Reed et al., 2004), significant correlations were observed between these polymorphisms and moderate (0.01%: $r = 0.83$; 0.1%: $r = 0.91$; 2.5%: $r = 0.86$), but not higher (5–20%) sucrose concentrations. Thus, marked genetic variance was observed for the sensitivity to and consumption of sucrose.

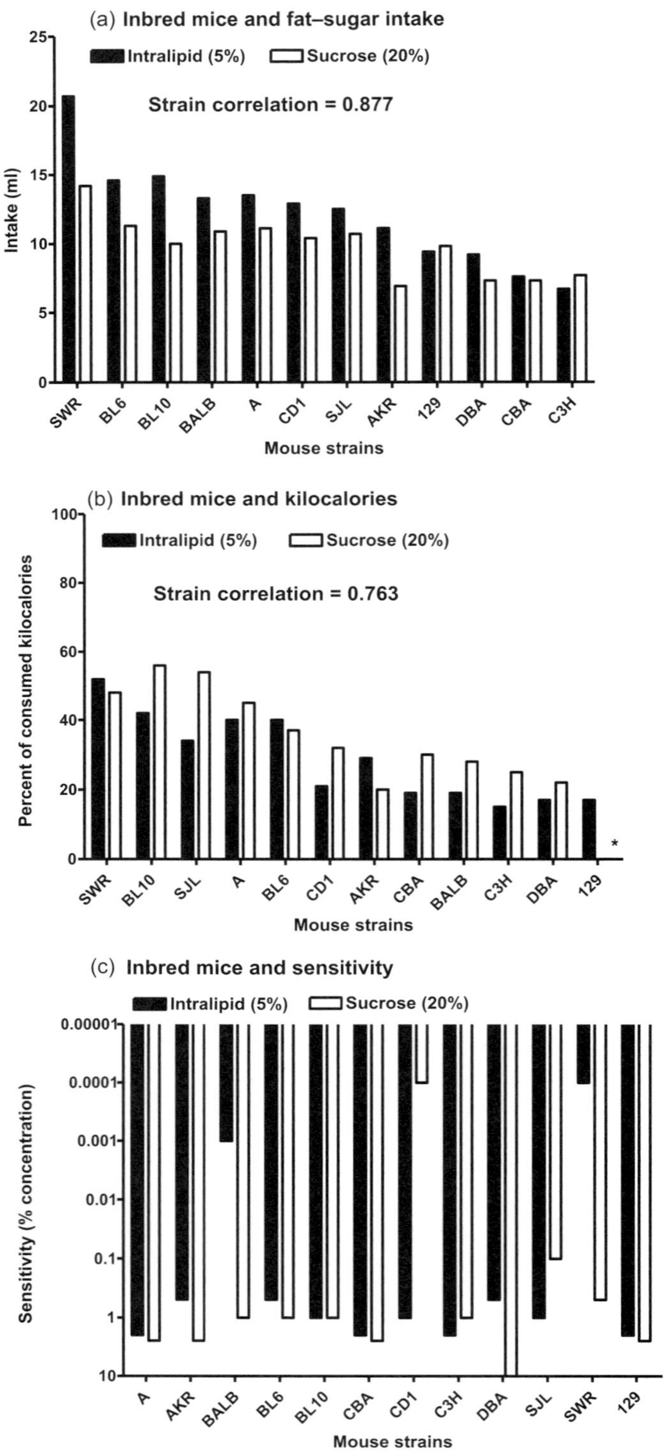

Figure 12.1 Comparison of inbred mouse strains in their responsiveness to intake of Intralipid (5%) fat and sucrose (20%) solutions across three dimensions. Panel (a) displays systematic strain-specific differences in Intralipid and sucrose intake over 24 hours, and indicates a highly significant ($r = 0.87$) correlation among strains in their intakes of the fat and sugar solutions. Panel (b) displays systematic strain-specific differences in the percentage of kilocalories consumed as Intralipid and sucrose over 24 hours, and indicates a corresponding and highly significant ($r = 0.76$) correlation among strains in their ability to consume fat and sugar solutions as part of their total caloric daily intake. Panel (c) displays the sensitivity to nine different concentrations of either Intralipid (0.001–5%) or sucrose (0.0001–20%) wherein these forms of intake over 24 hours were significantly higher than water in two-bottle preference tests. In contrast to intake per se and percentage of kilocalories consumed, significant strain-specific differences in sensitivity failed to correlate between sucrose and Intralipid ($r = -0.06$). (*Spillage in the sucrose study in 129P3/J mice precluded careful measurement of chow intake, and therefore the percentage of kilocalorie intake consumed as sucrose could not be ascertained.)

Fat (Intralipid)

A goal of a parallel study (Lewis et al., 2007) was to assess similarities or differences in genetic variance observed for sucrose intake relative to genetic variance in fat intake. However, unlike sucrose, typical difficulties in using different liquefied fat sources presented at different concentrations include their inability to stay in solution over a time course (e.g., 24 h) that is necessary to study murine intake. Intralipid is an emulsified fat solution (20%) made almost exclusively from soybean oil (20 g in 100 ml), and thereby insures that the fat is equally distributed in solution across a wide range of concentrations. Indeed, Intralipid solutions are readily consumed in a manner similar to sucrose and other palatable solutions (e.g., Higgs and Cooper, 1998a, 1998b). Further, a number of strains previously evaluated for sucrose intake (Lewis et al., 2005) also display three divergent patterns of fat intake: high fat intake with weight gain (e.g., AKR/J, C57BL/6J, DBA/2J: Alexander et al., 2006; Smith-Richards et al., 1999; West et al., 1992, 1995), high fat intake without weight gain (e.g., A/J, SWR/J: Leibowitz et al., 2005; Smith et al., 2000), and low fat intake (BALB/cJ, C3H/HeJ SJL/J, 129/J: Alexander et al., 2006; Smith et al., 2000). Our second study (Lewis et al., 2007) employed Intralipid as the source of fat.

First, it was clear that all strains displayed significant increases in Intralipid intake relative to water intake in

24-hour, two-bottle preference tests (Figure 12.1a). As expected, strong and systematic strain differences were observed for Intralipid preference and intake. Thus, sensitivity analyses (Figure 12.1c) revealed significant increases in Intralipid relative to concomitantly offered water intake to the greatest degree (0.001–5.0% Intralipid concentrations) in BALB/cJ mice, and to progressively lesser degrees in AKR/J, C57BL/6J, DBA/2J, and SWR/J inbred strains (0.5–5.0%), in outbred CD-1 and inbred C57BL/10J and SJL/J strains (1–5%), and to the least degree in A/J, CBA/J, C3H/HeJ, and 129P3/J inbred strains (2–5%). Congruent sensitivity data were observed for the percentage of fluid intake consumed as Intralipid with significant increases noted in BALB/cJ mice at the seven highest concentrations, in SWR/J mice at the five highest concentrations, in C57BL/6J, C57BL/10J, and DBA/2J mice at the four highest concentrations, and in A/J, AKR/J, and SJL/J mice at the two highest concentrations (Figure 12.1a). However, the percentage of fluid intake consumed as Intralipid was only significantly greater at the highest concentration in outbred CD-1, C3H/HeJ, and 129P3/J mice, but failed to differ at any concentration in CBA/J mice (Figure 12.1b). Moreover, Intralipid intake per se (Figure 12.1a) or Intralipid intake adjusted for body weight (ml/30 g body weight) indicated that SWR/J mice (25.9 ml adjusted) consumed by far the most among inbred strains, followed by A/J, BALB/cJ, C57BL/10J, and C57BL/6J mice (12.9–15.6 ml adjusted), followed then in turn by SJL/J, AKR/J, and 129P3/J mice (10.9–14.7 ml adjusted), and finally by DBA/2J, C3H/HeJ, and CBA/J mice (8.0–9.5 ml adjusted). Correspondingly, strains differed in the percentage of kilocalories consumed as Intralipid across concentrations (Figure 12.1b) with SWR/J mice displaying significantly greater consumption of Intralipid as a function of total intake at the three highest concentrations, A/J, AKR/J, C57BL/6J, CBA/J, and SJL/J mice displaying significantly greater consumption of Intralipid as a function of total intake at the two highest concentrations, and the other six strains displaying this effect at only the highest concentration. Further, compensatory decreases in chow intake were noted at the highest Intralipid concentration only in A/J, AKR/J, BALB/cJ, C57BL/10J, and SWR/J strains. Finally, we observed significant positive correlations for both the magnitude of intake and the percentage of kilocalories consumed as Intralipid among the four highest (0.5, 1.0, 2.0, and 5.0%) concentrations, indicating consistency of the effects. Interestingly, a more recent study (Glendenning et al., 2008) using Intralipid demonstrated that the Tas1r3 genotype does not modulate orosensory stimulation from oil, and that orosensory and post-ingestive mechanisms respectively modulate dilute and concentrated Intralipid solutions.

We then systematically analyzed whether genetic variance in sucrose intake was related to genetic variance in Intralipid intake. Although the threshold sensitivity for the 12 strains for sucrose intake and for Intralipid intake failed to display significant relationships ($r = -0.06$, ns; Figure 12.1c), a highly significant positive correlation ($r = 0.87$, $P < 0.01$; Figure 12.1a) for the peak magnitude of sucrose intake and Intralipid intake was noted among the 12 strains. Moreover, significant positive correlations were also observed when comparing Intralipid (5%) intake with sucrose intake at concentrations of 5% ($r = 0.82$), 10% ($r = 0.85$), and 20% ($r = 0.88$). An identical pattern of positive correlational effects was observed when one analyzed the percentage of kilocalories consumed as Intralipid (5%) relative to the percentage of kilocalories consumed as sucrose at concentrations of 5% ($r = 0.81$), 10% ($r = 0.89$), and 20% ($r = 0.76$, Figure 12.1b). These data support the notion that genetic variance in the consumption of sweets and fats are highly related to each other.

Glucoprivic and lipoprivic responses

Most of the above studies examining genetic variance in ingestive responses employed hedonic and/or orosensory stimuli in distinguishing responsiveness across murine strains. To extend the analysis of genetic variance in ingestive responses to homeostatic mechanisms, our laboratory (Lewis et al., 2006a, 2006b) examined whether different mouse strains varied in their feeding responses induced by glucoprivation and lipoprivation. Glucoprivic feeding can be induced by the anti-metabolic glucose analog, 2DG. Our first study (Lewis et al., 2006a) surveyed 11 inbred and one outbred strain for variations in feeding responses following a wide range of systemic 2DG doses (200–800 mg/kg) across a 4 hour time course (Table 12.2). Similar to outbred CD-1 mice that displayed orderly time and dose-dependent increases in 2DG-induced feeding, genetic variability was observed in the inbred strains with dose-dependent increases in 2DG-induced feeding observed across all four doses (DBA/2J), across the three highest doses (BALB/cJ, SJL/J, and 129P3/J), and across the two highest doses (CBA/J and AKR/J). In contrast, some mouse strains (A/J and C3H/HeJ: 800 mg/kg; C57BL/6J: 400 mg/kg) displayed very limited instances of 2DG-induced feeding, failed to show any increase (C57BL/10J), or actually significantly reduced intake (SWR/J). Such effects could not be predicted by any difference in baseline intakes. Moreover, although there was significant cross-correlation between 2DG doses of 200, 400, and 600 mg/kg, they in turn failed to correlate with the highest 800 mg/kg 2DG dose. Interestingly, significant correlations between sucrose intake (Lewis et al., 2005) and 2DG food intake failed to occur across the 11 inbred strains. Thus, although both experimental paradigms are thought to provide insight into glucosensing processes, the present differential pattern of strain sensitivity in each suggests differential genetic organization.

Using the free fatty acid oxidation inhibitor, MA, that significantly increases food intake following systemic administration, our second study in this series (Lewis et al., 2006b) surveyed the 11 inbred and one outbred strain for variations in feeding responses following a wide range of systemic MA doses (5–100 mg/kg) across a 4 hour time course (Table 12.2). Strain-specific effects for MA-induced feeding were observed following the three highest (35–100 mg/kg) MA doses in inbred DBA/2J mice

Table 12.2 Comparison of sensitivity (minimum dose) to feeding responses to 2-deoxy-D-glucose (2DG) and mercaptoacetate (MA) and the greatest magnitude (g) of food intake following the glucoprivic and lipoprivic stimuli. Sensitivity is defined as that dose that significantly increases intake over vehicle values after 4 hours. Magnitude is defined as the increased intake after 4 hours following 2DG or MA over vehicle values.

Strain	2DG sensitivity (mg/kg)	MA sensitivity (mg/kg)	2DG magnitude (g)	MA magnitude (g)
A/J	800	>100	0.5	0.04
AKR/J	600	35	0.4	0.4
BALB/cJ	400	100	0.7	0.7
C57BL/6J	800	100	0.2	0.1
C57BL/10J	>800	>100	0.1	0.3
CBA/J	600	100	0.5	0.1
CD-1	200	70	0.5	0.7
C3H/HeJ	800	5	0.6	0.4
DBA/2J	200	35	0.5	0.6
SJL/J	400	100	0.4	0.2
SWR/J	>800	>100	0.01	0.2
129P3/J	400	>100	0.3	0.3
Correlation	$r = 0.26$	NS	$r = 0.48$	NS

2DG: 2-deoxy-D-glucose; MA: mercaptoacetate; NS, not significant.

and the two highest (70–100 mg/kg) doses in outbred CD-1 mice. Dose-specific increases in intake were observed following the two middle (35–70 mg/kg) MA doses in AKR/J mice, only the 5 mg/kg MA dose in C3H/HeJ mice, only the 35 mg/kg MA dose in BALB/Cj and CBA/J mice, only the 70 mg/kg dose in SJL/J and SWR/J mice, and only the 100 mg/kg dose in C57BL/6J mice. In contrast, MA failed to significantly increase food intake at any dose in this wide range in A/J, C57BL/10J, and 129P3/J mice.

Functional comparisons between lipoprivic (MA) and glucoprivic (2DG) feeding in rats have revealed similar c-fos responses in the nucleus of the solitary tract, lateral parabrachial nucleus, central nucleus of the amygdala, and the dorsal motor nucleus of the vagus (Ritter and Dinh, 1994) as well as elevated sympathoadrenal plasma levels of epinephrine and norepinephrine (Scheurink and Ritter, 1993). However, there are marked differences in sensitivity to different physiological manipulations between MA-induced and 2DG-induced feeding responses following vagotomy (Ritter and Taylor, 1990) and lesions placed in the lateral parabrachial nucleus (Calingasan and Ritter, 1993) or the central nucleus of the amygdala (Ritter and Hutton, 1995). Importantly, our correlational analyses of genetic differences in feeding responses elicited by 2DG and MA failed to find significant strain-specific relationships in terms of either sensitivity (the dose range to elicit feeding) or magnitude (the actual amount of intake elicited by each dose) of effects (Table 12.2). Similarly, significant genetic correlations between Intralipid intake and MA-induced intake failed to occur despite the recent finding (Matsumura et al., 2008) that MA attenuated the oral acceptance of fat in BALB/c mice. Thus, the differences among diverse mouse strains in their ingestive responses to lipoprivation and glucoprivation suggest that they employ different neural circuitry, and indeed provide evidence that these two homeostatic responses operate via different genetic mechanisms of action.

Conclusions

This chapter has evaluated a great deal of recent empirical evidence examining genetic variance in inbred, outbred, and cross-bred murine strains across a wide array of ingestive behaviors. However, for other ingestive states, the complexity of the response, the differences in procedures, and the wide variety of tested strains precluded a more thorough investigation of the genetic substrates of these responses, particularly involving fat intake and obesity. Yet, it is becoming increasingly clear that the two major approaches in examining these responses, either the use of a small number (2–4) of mouse strains for "sensitive" and "insensitive" responders on behavioral and QTL analyses, or the use of large numbers of strains across a wide array of ingestive stimuli and concentrations, have together provided insights into the complex, multi-genomic mechanisms mediating feeding behavior. Together with studies using knockout and knockdown genetic approaches, feasible strategies for the analysis of genetic contributions of normal, intact genetically varied strains in evaluating feeding responses have clearly emerged, allowing us to analyze complex genetic × environmental interactions in the etiology of both normal and disordered feeding behaviors.

Acknowledgments

Preparation of this paper was supported in part by a CUNY Collaborative Grant (80209-03-09) to R. J. B. and B. K. We thank Dr. Anthony Sclafani for his helpful comments.

References

Alexander, J., Chang, G.Q., Dourmashkin, J.T., and Leibowitz, S.F. (2006) Distinct phenotypes of obesity-prone AKR/J, DBA/2J and C57BL/6J mice compared to control strains. *Int J Obesity* 30: 50–59.

Anji, A. and Kumari, M. (2008) Supplementing the liquid alcohol diet with chow enhances alcohol intake in C57/BL/6 mice. *Drug Alcohol Depend* 97: 86–93.

Bachmanov, A.A. and Beauchamp, G.K. (2008) Amino acid and carbohydrate preferences in C57BL/6ByJ and 129P3/J mice. *Physiol Behav* 93: 37–43.

Bachmanov, A.A., Beauchamp, G.K., and Tordoff, M.G. (2002a) Voluntary consumption of NaCl, KCl, CaCl2, and NH4Cl solutions by 28 mouse strains. *Behav Genet* 32: 445–457.

Bachmanov, A.A., Reed, D.R., Beauchamp, G.K., and Tordoff, M.G. (2002b) Food intake, water intake, and drinking spout side preference of 28 mouse strains. *Behav Genet* 32: 435–443.

Bachmanov, A.A., Reed, D.R., Li, X., Li, S., Beauchamp, G.K., and Tordoff, M.G. (2000a) Voluntary ethanol consumption by mice: genome-wide analysis of quantitative trait loci and their interactions in a C57BL/6ByJ and 129P3/J F2 intercross. *Genome Res* 12: 1257–1268.

Bachmanov, A.A., Reed, D.R., Ninomiya, Y., Inoue, M., Tordoff, M.G., Price, R.A., et al. (1997) Sucrose consumption in mice: major influence of two genetic loci affecting peripheral sensory responses. *Mamm Genome* 8: 545–548.

Bachmanov, A.A., Reed, D.R., Tordoff, M.G., Price, R.A., and Beauchamp, G.K. (1996a) Intake of ethanol, sodium chloride, sucrose, citric acid and quinine hydrochloride solutions by mice: a genetic analysis. *Behav Genet* 26: 563–573.

Bachmanov, A.A., Reed, D.R., Tordoff, M.G., Price, R.A., and Beauchamp, G.K. (2001a) Nutrient preference and diet-induced adiposity in C57BL/6JByJ and 129P3/J/J mice. *Physiol Behav* 72: 603–613.

Bachmanov, A.A., Schlager, G., Tordoff, M.G., and Beauchamp, G.K. (1998a) Consumption of electrolytes and quinine by mouse strains with different blood pressures. *Physiol Behav* 64: 323–330.

Bachmanov, A.A., Tordoff, M.G., and Beauchamp, G.K. (1996b) Ethanol consumption and taste preferences in C57BL/6ByJ and 129/J mice. *Alcohol Clin Exp Res* 20: 201–206.

Bachmanov, A.A., Tordoff, M.G., and Beauchamp, G.K. (1998b) Voluntary sodium chloride consumption by mice: differences among five inbred strains. *Behav Genet* 28: 117–124.

Bachmanov, A.A., Tordoff, M.G., and Beauchamp, G.K. (2000b) Intake of umami-tasting solutions by mice: a genetic analysis. *J Nutr* 130: 935S–941S.

Bachmanov, A.A., Tordoff, M.G., and Beauchamp, G.K. (2001b) Sweetener preference of C57/6ByJ and 129P3/J mice. *Chem Senses* 26: 905–913.

Beauchamp, G.K., Bachmanov, A.A., and Stein, L.J. (1998) Development and genetics of glutamate taste preference. *Ann N Y Acad Sci* 855: 412–416.

Beauchamp, G.K. and Fisher, A.S. (1993) Strain differences in consumption of saline solutions by mice. *Physiol Behav* 54: 179–184.

Belknap, J.K. and Atkins, A.L. (2001) The replicability of QTLs for murine alcohol preference drinking behavior across eight independent studies. *Mamm Genome* 12: 893–899.

Belknap, J.K., Crabbe, J.C., and Young, E.R. (1993) Voluntary consumption of alcohol in 15 inbred mouse strains. *Psychopharmacol* 112: 503–510.

Belknap, J.K., Richards, S.P., O'Toole, L.A., Helms, M.L., and Phillips, T.J. (1997) Short-term selective breeding as a tool for QTL mapping: ethanol preference drinking in mice. *Behav Genet* 27: 55–66.

Bergeson, S.E., Kyle Warren, R., Crabbe, J.C., Metten, P., Gene Erwin, V., and Belknap, J.K. (2003) Chromosomal loci influencing chronic alcohol withdrawal severity. *Mamm Genome* 14: 454–463.

Blizard, D.A., Kotlus, B., and Frank, M.E. (1999) Quantitative trait loci associated with short-term intake of sucrose, saccharin and quinine solutions in laboratory mice. *Chem Senses* 24: 373–385.

Calingasan, N.Y. and Ritter, S. (1993) Lateral parabrachial subnucelus lesions abolish feeding induced by mercaptoacetate but not by 2-deoxy-D-glucose. *Am J Physiol* 265: R1168–R1178.

Capeless, C.G. and Whitney, G. (1995) The genetic basis of preference for sweet substances among inbred strains of mice: preference ratio phenotypes and the alleles of the Sac and dpa loci. *Chem Senses* 20: 291–298.

Crabbe, J.C., Belknap, J.K., and Buck, K.J. (1994) Genetic animal models of alcohol and drug abuse. *Science* 264: 1715–1723.

Crabbe, J.C., Phillips, T.J., Buck, K.J., Cunningham, C.L., and Belknap, J.K. (1999) Identifying genes for alcohol and drug sensitivity: recent progress and future directions. *Trends Neurosci* 22: 173–179.

De Araujo, I.E., Oliveira-Maia, A.J., Sotnikova, T.D., Gainetdinov, R.R., Caron, M.G., Nicolelis, M.A., et al. (2008) Food reward in the absence of taste receptor signaling. *Neuron* 57: 930–941.

Dhar, M., Webb, L.S., Smith, L., Hauser, L., Johnson, D., and West, D.B. (2000) A novel ATPase on mouse chromosome 7 is a candidate gene for increased body fat. *Physiol Genomics* 4: 93–100.

Fehr, C., Shirley, R.L., Crabbe, J.C., Belknap, J.K., Buck, K.J., and Phillips, T.J. (2005) The syntaxin binding protein 1 gene (Stxbp1) is a candidate for an ethanol preference drinking locus on mouse chromosome 2. *Alcohol Clin Exp Res* 29: 708–720.

Finn, D.A., Belknap, J.K., Cronise, K., Yoneyama, N., Murillo, A., and Crabbe, J.C. (2005) A procedure to produce high alcohol intake in mice. *Psychopharmacol* 178: 471–480.

Fisler, J.S., Warden, C.H., Pace, M.J., and Lusis, A.J. (1993) BSB: a new mouse model of multigenic obesity. *Obesity Res* 1: 271–280.

Fuller, J.L. (1974) Single-locus control of saccharin preference in mice. *J Hered* 65: 33–36.

Gallou-Kabani, C., Vige, A., Gross, M.S., Rabes, J.P., Boileau, C., Larue-Achagiotis, C., et al. (2007) C57BL/6J and A/J mice fed a high-fat diet delineate components of metabolic syndrome. *Obesity (Silver Spring)* 15: 1996–2005.

Gannon, K. and Contreras, R.J. (1993) Sodium intake linked to amiloride-sensitive gustatory transduction in C57BL/6J and 129/J mice. *Physiol Behav* 57: 231–239.

Glendenning, J.I., Feld, N., Goodman, L., and Bayor, R. (2008) Contribution of orosensory stuimulation to strain differences in oil intake by mice. *Physiol Behav* 95: 476–483.

Gupta, T., Syed, Y.M., Revis, A.A., Miller, S.A., Martinez, M., Cohn, K.A., et al. (2008) Acute effects of acamprosate and MPEP on ethanol drinking-in-the-dark in male C57BL/6J mice. *Alcohol Clin Exp Res* 32: 1992–1998.

Harder, D.B., Azen, E.A., and Whitney, G. (2000) Sucrose octaacetate avoidance in nontaster mice is not enhanced by two type-A Prp transgenes from taster mice. *Chem Senses* 25: 39–45.

Harder, D.B., Capeless, C.G., Maggio, J.C., Boughter, J.D., Gannon, K.S., et al. (1992) Intermediate sucrose octa-acetate sensitivity suggests a third allele at mouse bitter taste loci Soa and Soa-Rua identity. *Chem Senses* 17: 391–401.

Harder, D.B., Maggio, J.C., and Whitney, G. (1989) Assessing gustatory detection capabilities using preference procedures. *Chem Senses* 14: 547–564.

Harder, D.B. and Whitney, G. (1998) A common polygenic basis for quinine and PROP avoidance in mice. *Chem Senses* 23: 327–332.

Hegmann, J.P. and Possidente, B. (1981) Estimating genetic correlations from inbred strains. *Behav Genet* 11: 103–114.

Higgs, S. and Cooper, S.J. (1998a) Evidence for early opioid modulation of licking responses to sucrose and Intralipid: a microstructural analysis in the rat. *Psychopharmacol* 139: 342–355.

Higgs, S. and Cooper, S.J. (1998b) Effects of benzodiazepine receptor ligands on the ingestion of sucrose, Intralipid and maltodextrin: an investigation using a microstructural analysis of licking behavior in a brief contact test. *Behav Neurosci* 112: 447–457.

Inoue, M., Beauchamp, G.K., and Bachmanov, A.A. (2004a) Gustatory neural responses to umami taste stimuli in C57BL/6ByJ and 129P3/J mice. *Chem Senses* 29: 789–795.

Inoue, M., Li, X., McCaughey, S.A., Beauchamp, G.K., and Bachmanov, A.A. (2001) Soa genotype selectively affects mouse gustatory neural responses to sucrose octaacetate. *Physiol Genomics* 5: 181–186.

Inoue, M., Reed, D.R., Li, X., Tordoff, M.G., Beauchamp, G.K., and Bachmanov, A.A. (2004b) Allelic variation of the Tas1r3 taste receptor gene selectively affects behavioral and neural taste responses to sweeteners in the F2 hybrids between C57BL/6JByJ and 129P3/J/J mice. *J Neurosci* 24: 2296–2303.

Kotlus, B.S. and Blizard, D.A. (1998) Measuring gustatory variation in mice. *Physiol Behav* 64: 37–47.

Leibowitz, S.F., Alexander, J., Dourmashkin, J.T., Hill, J.O., Gayles, E.C., and Chang, G.-Q. (2005) Phenotypic profile of SWR/J and A/J mice compared to control strains: possible mechanisms underlying resistance to obesity on a high-fat diet. *Brain Res* 1047: 137–147.

Lewis, S.R., Ahmed, S., Dym, C., Khaimova, E., Kest, B., and Bodnar, R.J. (2005) Inbred mouse strain survey of sucrose intake. *Physiol Behav* 85: 546–556.

Lewis, S.R., Ahmed, S., Khaimova, E., Israel, Y., Singh, A., Kandov, Y., et al. (2006a) Genetic variance contributes to ingestive processes: a survey of 2-deoxy-D-glucose-induced feeding in eleven inbred mouse strains. *Physiol Behav* 87: 595–601.

Lewis, S.R., Dym, C., Chai, C., Singh, A., Kest, B., and Bodnar, R.J. (2007) Genetic variance contributes to ingestive processes: a survey of Intralipid (fat) feeding in 11 inbred mouse strains. *Physiol Behav* 90: 82–94.

Lewis, S.R., Dym, C., Ginzberg, M., Kest, B., and Bodnar, R.J. (2006b) Genetic variance contributes to ingestive processes: a survey of mercaptoacetate-induced feeding in eleven inbred and one outbred mouse strains. *Physiol Behav* 88: 516–522.

Lush, I.E. (1981) The genetics of tasting in mice. I. Sucrose octaacetate. *Genet Res* 38: 93–95.

Lush, I.E. (1982) The genetics of tasting in mice. II. Strychnine. *Chem Senses* 7: 93–98.

Lush, I.E. (1984) The genetics of tasting in mice. III Quinine. *Genet Res* 44: 151–160.

Lush, I.E. (1989) The genetics of tasting in mice. VI. Saccharin, acesulfame, dulcin and sucrose. *Genet Res* 53: 95–99.

Matsumura, S., Saitou, K., Miyaki, T., Yoneda, T., Mizushige, T., Eguchi, A., et al. (2008) Mercaptoacetate inhibition of fatty acid beta-oxidation attenuates the oral acceptance of fat in BALB/c mice. *Am J Physiol Regul Integr Comp Physiol* 295: R82–R91.

Melo, J.A., Shendure, J., Pociask, K., and Silver, L.M. (1996) Identification of sex-specific quantitative trait loci controlling alcohol preference in C57BL/6J mice. *Nature Genet* 13: 147–153.

Nachman, M. (1959) The inheritance of saccharin preference. *J Comp Physiol Psychol* 52: 451–457.

Ninomaya, Y. and Funakoshi, M. (1989a) Behavioural discrimination between glutamate and the four basic taste substances in mice. *Comp Biochem Physiol* 92: 365–370.

Ninomaya, Y. and Funakoshi, M. (1989b) Peripheral neural basis for behavioural discrimination between glutamate and the four basic taste substances in mice. *Comp Biochem Physiol* 92: 371–376.

Pelz, W.E., Whitney, G., and Smith, J.C. (1973) Genetic influences on saccharin preference of mice. *Physiol Behav* 10: 263–265.

Phillips, T.J., Belknap, J.K., Buck, K.J., and Cunningham, C.L. (1998) Genes on mouse chromosomes 2 and 9 determine variation in ethanol consumption. *Mamm Genome* 9: 936–941.

Phillips, T.J., Crabbe, J.C., Metten, P., and Belknap, J.K. (1994) Localization of genes affecting alcohol drinking in mice. *Alcohol Clin Exp Res* 18: 931–941.

Pothion, S., Bizot, J.-C., Trovero, F., and Belzung, C. (2004) Strain differences in sucrose preference and in the consequences of unpredictable chronic mild stress. *Behav Brain Res* 155: 135–146.

Ramirez, I. and Fuller, J.L. (1976) Genetic influence on water and sweetened water consumption in mice. *Physiol Behav* 16: 163–168.

Rance, K.A., Hambly, C., Dalgleish, G., Fustin, J.M., Bunger, L., and Speakman, J.R. (2007) Quantitative trait loci for regional adiposity in mouse lines divergently selected for food intake. *Obesity (Silver Spring)* 15: 2994–3004.

Reed, D.R., Bachmanov, A.A., Beauchamp, G.K., Tordoff, M.G., and Price, R.A. (1997) Heritable variation in food preferences and their contribution to obesity. *Behav Genet* 27: 373–387.

Reed, D.R., Bachmanov, A.A., and Tordoff, M.G. (2007) Forty mouse strain survey of body composition. *Physiol Behav* 91: 593–600.

Reed, D.R., Li, S., Li, X., Huang, L., Tordoff, M.G., Starling-Roney, R., et al. (2004) Polymorphisms in the taste receptor gene (Tas1r3) region are associated with saccharin preference in 30 mouse strains. *J Neurosci* 24: 938–946.

Reed, D.R., Li, X., McDaniel, A.H., Lu, K., Li, S., Tordoff, M.G., et al. (2003) Loci on chromosomes 2, 4, 9 and 16 for body weight, body length, and adiposity identified in a genome scan of an F2 intercross between the 129P3/J and C57BL/6ByJ mouse strains. *Mamm Genome* 14: 302–313.

Reed, D.R., McDaniel, A.H., Li, X., Tordoff, M.G., and Bachmanov, A.A. (2006) Quantitative trait loci for individual adipose depot weights in C57BL/6ByJ × 129P3/J F2 mice. *Mamm Genome* 17: 1065–1077.

Rhodes, J.S., Best, K., Belknap, J.K., Finn, D.A., and Crabbe, J.C. (2005) Evaluation of a simple model of ethanol drinking to intoxication in C57BL/6J mice. *Physiol Behav* 84: 53–63.

Ritter, S. and Dinh, T.T. (1994) 2-Mercaptoacetate and 2-deoxy-D-glucose induce fos-like immunoreactivity in rat brain. *Brain Res* 641: 111–120.

Ritter, S. and Hutton, R. (1995) Mercaptoacetate-induced feeding is impaired by central nucleus of the amygdala lesions. *Physiol Behav* 58: 1215–1220.

Ritter, S. and Taylor, J.S. (1990) Vagal sensory neurons are required for lipoprivic but not glucoprivic feeding in rats. *Am J Physiol* 258: R1395–R1401.

Scheurink, A. and Ritter, S. (1993) Sympathoadrenal responses to glucoprivation and lipoprivation in rats. *Physiol Behav* 53: 995–1000.

Sclafani, A. (2006a) Sucrose motivation in sweet "sensitive" (C57BL/6J) and "subsensitive" (129P3/J) mice measured by progressive ratio licking. *Physiol Behav* 87: 734–744.

Sclafani, A. (2006b) Enhanced sucrose and polycose preference in sweet "sensitive" (C57BL/6J) and "subsensitive" (129P3/J) mice after experience with these saccharides. *Physiol Behav* 87: 745–756.

Sclafani, A. and Glendinning, J.I. (2003) Flavor preferences conditioned in C57BL/6J mice by intragastric carbohydrate self-infusion. *Physiol Behav* 79: 783–788.

Sclafani, A. and Glendinning, J.I. (2005) Sugar and fat conditioned flavor preferences in C57BL/6J and 129 mice: oral and postoral interactions. *Am J Physiol* 289: R712–R720.

Shao, H., Reed, D.R., and Tordoff, M.G. (2007) Genetic loci affecting body weight and fatness in a C57BL/6J × PWK/PhJ mouse intercross. *Mamm Genome* 18: 839–851.

Smith, B.K., Andrews, P.K., and West, D.B. (2000) Macronutrient diet selection in 13 mouse strains. *Am J Physiol* 278: R797–R805.

Smith, B.K., Volaufova, J., and West, D.B. (2001) Increased flavor preference and lick activity for sucrose and corn oil in SWR/J vs. AKR/J mice. *Am J Physiol* 281: R596–R606.

Smith-Richards, B.K., Andrews, P.K., York, D.A., and West, D.B. (1999) Divergence in proportional fat intake in AKR/J and SWR/J mice endures across diet paradigms. *Am J Physiol* 277: R776–R785.

Smith-Richards, B.K., Belton, B.N., Poole, A.C., Mancuso, J.J., Churchill, G.A., Li, R., et al. (2002) QTL analysis of self-selected macronutrient diet intake: fat, carbohydrate and total kilocalories. *Physiol Genomics* 11: 205–217.

Stockton, M.D. and Whitney, G. (1974) Effects of genotype, sugar and concentration on sugar preference of laboratory mice (*Mus musculus*). *J Comp Physiol Psychol* 86: 62–68.

Tabakoff, B., Saba, L., Kechris, K., Hu, W., Bhave, S.V., Finn, D.A., et al. (2008) The genomic determinants of alcohol preference in mice. *Mamm Genome* 19: 352–365.

Tarantino, L.M., McClearn, G.E., Rodriguez, L.A., and Plomin, R. (1998) Confirmation of quantitative trait loci for alcohol preference in mice. *Alcohol Clin Exp Res* 22: 1099–1105.

Tordoff, M.G. (2007) Taste solution preferences of C57BL/6J and 129X1/SvJ mice: influence of age, sex, and diet. *Chem Senses* 32: 655–671.

Tordoff, M.G. and Bachmanov, A.A. (2002) Influence of test duration on the sensitivity of the two-bottle choice test. *Chem Senses* 27: 759–768.

Tordoff, M.G. and Bachmanov, A.A. (2003a) Mouse taste preference tests: why only two bottles? *Chem Senses* 28: 315–324.

Tordoff, M.G. and Bachmanov, A.A. (2003b) Influence of the number of alcohol and water bottles on murine alcohol intake. *Alcohol Clin Exp Res* 27: 600–606.

Tordoff, M.G., Bachmanov, A.A., and Reed, D.R. (2007a) Forty mouse strain survey of water and sodium intake. *Physiol Behav* 91: 620–631.

Tordoff, M.G., Bachmanov, A.A., and Reed, D.R. (2007b) Forty mouse strain survey of voluntary calcium intake, blood calcium and bone mineral content. *Physiol Behav* 91: 632–643.

Tordoff, M.G., Pilchak, D.M., Williams, J.A., McDaniel, A.H., and Bachmanov, A.A. (2002) The maintenance diets of C57BL/6JJ and 129/JXi/SvJ mice influence their taste solution preferences: implications for large-scale phenotyping projects. *J Nutr* 132: 2288–2297.

Tordoff, M.G., Reed, D.R., and Shao, H. (2008a) Calcium taste preferences: genetic analysis and genome screen of C57BL/6J × PWK/PhJ hybrid mice. *Genes Brain Behav* 7: 618–628.

Tordoff, M.G., Shao, H., Alarcon, L.K., Margolskee, R.F., Mosinger, B., Bachmanov, A.A., et al. (2008b) Involvement of the T1R3 in calcium-magnesium taste. *Physiol Genomics* 34: 338–348.

Warden, C.H., Fisler, J.S., Shoemaker, S.M., Wen, P.Z., Svenson, K.L., Pace, M.J., et al. (1995) Identification of four chromosomal loci determining obesity in a multifactorial mouse model. *J Clin Invest* 95: 1545–1552.

Warren, R.P. and Lewis, R.C. (1970) Taste polymorphism in mice involving a bitter sugar derivative. *Nature* 227: 77–78.

West, D.B., Boozer, C.N., Moody, D.L., and Atkinson, R.L. (1992) Obesity induced by a high-fat diet in nine strains of inbred mice. *Am J Physiol* 262: R1025–R1032.

West, D.B., Goudey-Lefevre, J., York, B., and Truett, G.E. (1994a) Dietary obesity linked to genetic loci on chromosomes 9 and 15 in a polygenic mouse model. *J Clin Invest* 94: 1410–1416.

West, D.B., Waguespack, J., and McCollister, S. (1995) Dietary obesity in the mouse:

interaction of strain with diet composition. *Am J Physiol* 268: R658–R665.

West, D.B., Waguespack, J., York, B., Goudey-Lefevre, J., and Price, G.A. (1994b) Genetics of dietary obesity in AKR/J × SWR/J mice: segregation of the trait and identification of a linked locus on chromosome 4. *Mamm Genome* 5: 546–552.

West, D.B. and York, B. (1998) Dietary fat, genetic predisposition, and obesity: lessons from animal models. *Am J Clin Nutr* 67: 505S–512S.

Whatley, V.J., Johnson, T.E., and Erwin, V.G. (1999) Identification and confirmation of quantitative trait loci regulating alcohol consumption in congenic strains of mice. *Alcohol Clin Exp Res* 23: 1262–1271.

Chapter 13

Getting it right

Learning and memory determines hand-preference behavior in the mouse

Fred G. Biddle and Brenda A. Eales

Introduction

In 1968 Rob Collins described paw-preference behavior in laboratory mice (Collins, 1968, 1969) and, 30 years later, we were surprised to discover a dynamic process of learning and memory underlying the behavior (Biddle and Eales, 1999). We can now describe paw preference as an adaptive behavior because it is a directed response to changing environment and an individual's future performance depends on its past experience. However, to see the adaptive behavior, we had to return to fundamental principles of complex trait analysis and document both the dynamics and the kinetics of the behavior. Laboratory mice reveal a research model of right-hand and left-hand usage that is novel and arresting, and we believe that it will have implications for the analysis of other behaviors that exhibit asymmetry and lateralization. In this chapter, we set out a personal perspective of the behavior that may lead to a systems analysis at the individual level, not just the population level. Unpacking the complexity in the research model and understanding the true nature of the behavior is taking time (Biddle and Eales, 2006; Ribeiro et al., 2010).

We provide an overview to illustrate the patterns of paw preference among genetically different mouse strains and what was believed to be the heritable variation in the behavior. We use our own data to show that informative aspects of the behavior are buried in numerical summaries of behavioral scores. Attention to fundamental genetic principles provided insight into paw-preference behavior and those principles guide our current efforts to identify relevant genes and function by attaching the heritable alternatives in the behavior to the genome. We use Smolin's three criteria to assess the value of our current attempt to reorganize the analysis of paw-preference behavior: surprise (or did we already know the answer); new insights that generate new hypotheses, which drive progress in understanding; new predictions, confirmed by experiments that only make sense in the light of the new theory (Smolin, 2006).

For the curious biologist, "variation is an endless source of challenging questions" (Mayr, 2005) and variation is significant to an organism when it can be altered by mutation. Therefore, understanding the naturally-occurring, heritable variation in any behavior at the individual level and the population level has the practical purpose of providing a framework to recognize and assess behaviors that we might consider as different or abnormal. This framework is also the context to assess and interpret the effects of environmental manipulation as well as both random and gene-directed mutagenesis. Unfortunately, contemporary analysis often leaves an impression that specific behaviors are already understood and that it is a trivial matter to assign genes and function to the cause of behavior.

We support our statements by showing that paw-preference behavior of mice is contextual and variation in paw preference is contingent on both genes ***and*** environment rather than a dichotomy of genes ***or*** environment. Genes act in the context of other genes and in different environments, and the genetics of hand preference tells us as much about nurture (the functional role of the environment) as we believe it is telling us about nature (the functional role of genetic change). We bring this review to the point where heritable differences in hand-preference behavior among individuals are, not surprisingly, a reflection of heritable differences in a process of learning and memory.

We show that genetics enters the regulation of paw preference at two, functionally different levels. First, different genotypes (i.e., different mouse strains) learn at different rates in response to paw reaching. Second, some genotypes may differ in their constitutive ability to learn more from a left- or right-hand reach than they learn from a reach with their opposite hand (or they may have a constitutive asymmetry in their paw usage). To support this, we briefly describe current analysis with stochastic agent-based models that reproduce paw-preference behavior at both the individual and population levels (Ribeiro et al., 2010). Preliminary results show how the characteristic patterns of the noisy distributions of left- and right-paw usage can emerge from the deterministic learning ability of different genotypes and the probabilistic nature of paw preference and, more importantly, they show how individuals and populations of individuals respond to change in environment. With a clearer definition of phenotype and differences in phenotype, forward genetics may be a practical method to identify biological

Behavioral Genetics of the Mouse: Volume I. *Genetics of Behavioral Phenotypes*, eds. Wim E. Crusio, Frans Sluyter, Robert T. Gerlai, and Susanna Pietropaolo. Published by Cambridge University Press.

Figure 13.1 Unbiased or U-world (food tube centered) and a mouse reaching for food with its right paw. In biased test chambers, the food tube is located to the right side or left side (R-world or L-world, respectively) from the perspective of the mouse. (Reproduced with permission from Biddle et al., 1993. © 2008 NRC Canada or its licensors.)

mechanisms that drive and regulate paw-preference behavior. Our goal is a more fundamental question: how does an individual use the act of paw reaching to inform the next reach and what are the functions of "genes" that regulate the process?

Historical overview of mouse hand preference

What does the behavior look like in the unbiased world?

After a light fasting, all mice will reach with their right and left hands or forepaws to retrieve food from a tube in the face of a small testing chamber (Figure 13.1) and, when they do reach, their paw preference provides a quantifiable behavioral trait. We use the words "hand" and "paw" interchangeably. Originally, the food tube was placed at an appropriate height, equidistant from the right and left sides (Collins, 1968, 1969) and, later, this was described as an unbiased or U-world (Collins, 1975). The U-world does not appear to hamper the mouse's ability to reach with either paw and direction of paw preference is simply the relative amount of right-paw usage in a set number of paw reaches and expressed as a right-paw entry (RPE) score (Collins 1968). If a mouse is allowed 50 reaches, which is the number used by Collins (1968), 25 is the midpoint of the measurement scale. A RPE score of 0–24 indicates more left-paw usage and a RPE score of 26–50 indicates more right-paw usage. Except where indicated, our assessments are based on a 50-reach test.

Wahlsten (1999) remarked "[f]ew measures in neurobehavioral genetics involve scales of measurement that are readily understood by nonexperts." Hand preference of mice is no exception. We will try to reconnect the measurement scale (the RPE score) to the behavior of individuals because a clear understanding of the source of the numbers will show how the numbers are used to describe phenotypic heterogeneity. If genes are involved in phenotypic heterogeneity, a clear path should be visible from the numbers of the behavioral score to the biological function and to the genes that regulate it. In other words, if genes are involved in hand preference, what are their function, especially since all mice reach with their right and left paws?

Different strains of mice express complex, noisy distributions or patterns of right- and left-paw usage in the U-world. Figure 13.2 compares the patterns of RPE scores from U-world tests with 50 reaches of the C57BL/6J, SWV, CDS/Lay, and DBA/2J inbred strains (Biddle et al., 2001). These distributions are characteristic of different strains and RPE scores of individual mice within a strain are highly reliable when the individuals are retested in the same test-world (Biddle and Eales, 1999; Collins, 1968; Takeda and Endo, 1993).

Figure 13.2 illustrates the problem of linking paw preference of an individual mouse to its population (i.e., to the genetically identical individuals in a specific strain). For example, a mouse with a RPE score of 35 right-paw reaches in a test with 50 reaches is more right-pawed than left-pawed, but that individual looks like any other individual with a RPE score of 35, regardless of genotype (and, potentially, regardless of how it got to that number of 35). Knowing the RPE score of an individual mouse does not uniquely predict its genotype and samples of genetically identical mice from the same strain exhibit phenotypic heterogeneity as illustrated by noisy patterns of RPE scores (Figure 13.2). If the RPE score does not predict an individual's genotype, what property of paw-usage behavior is predictive? Predictability or uncertainty in the correspondence between genotype and phenotype is the core problem in complex trait analysis.

Figure 13.2 also illustrates the first stage of our contention that biologically relevant information is buried in numerical summaries of paw-preference scores. Left or right direction of paw usage in the U-world does not appear to be the genetically determined trait because each strain has approximately equal numbers of left-pawed and right-pawed mice. Location of the mean or average RPE score, on the scale of 0–50 right-paw reaches, is numerically similar in the four strains; that is, regardless of the shapes of the original population distributions in Figure 13.2, average RPE scores from small and replicated random samples of mice from each of the four strains will have normally distributed mean scores of approximately 25 on the 0–50 RPE scale. (Means of random sample means will always be normally or Gaussian distributed, if sampling is truly random, regardless of the shape of the original population distribution; this fact is what we have often called "the trap of the central limit theorem.") In the sense of exploratory data analysis (Magnusson and Mourao, 2004; Tukey, 1977), it is the

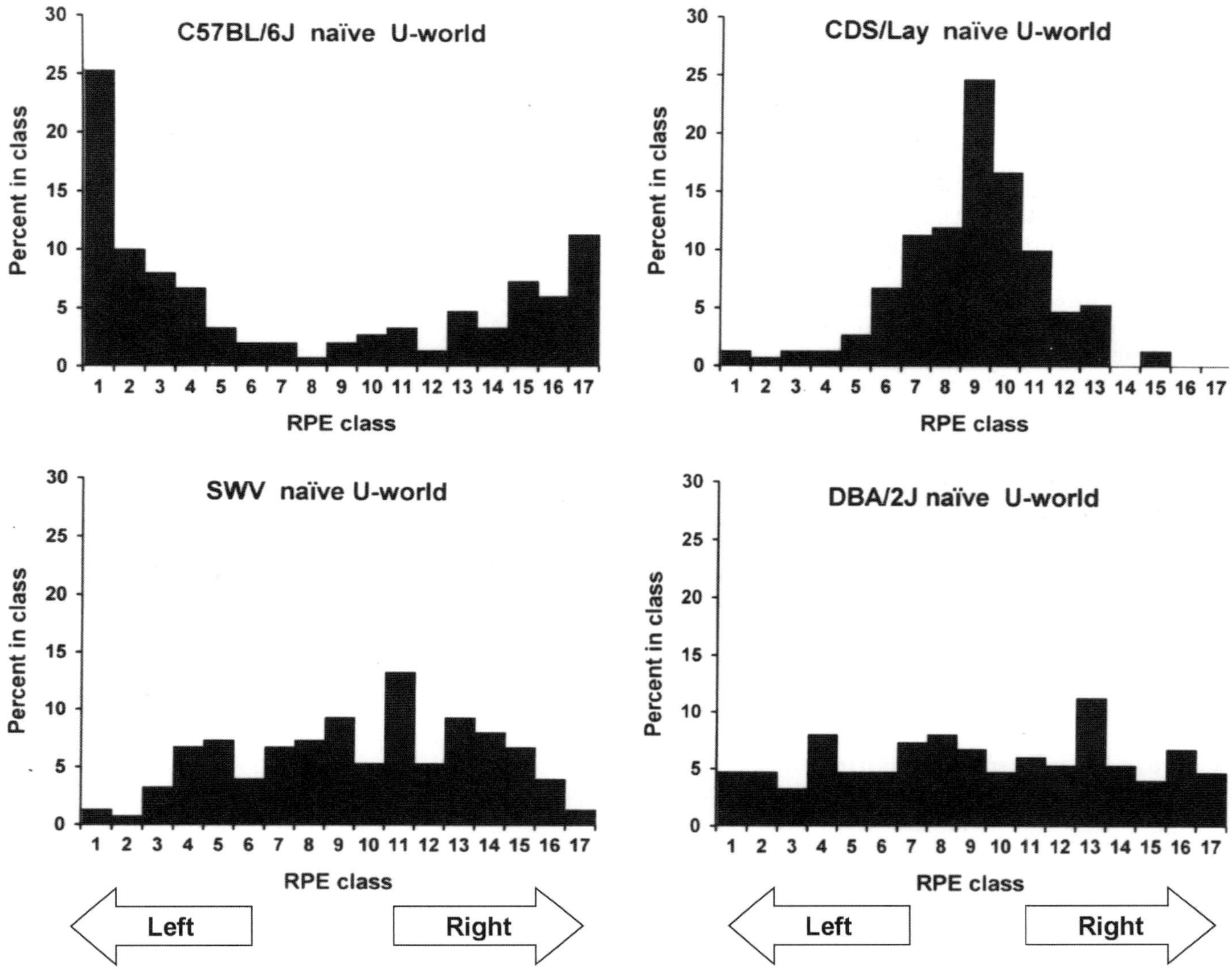

Figure 13.2 Distributions of right-paw entry (RPE) scores from an unbiased world (U-world) test of previously untested C57BL/6J, SWV, CDS/Lay, and DBA/2J mice (150 mice and equal numbers of each sex from each strain). The mice were allowed 50 reaches and their 51 RPE scores (0–50) were grouped in 17 equal-sized classes of 3 RPE scores. Relative direction of paw usage is indicated. (Reproduced with permission from Biddle et al., 2001. © 2008 NRC Canada or its licensors.)

shape of the distribution of the population of RPE scores that is qualitatively different and informative among the four strains in Figure 13.2. Obviously, C57BL/6J is bimodal and CDS/Lay is unimodal and Gaussian-like; however, this information in the different patterns is lost when the means of small samples of RPE scores and the variances of the sample means are tabulated.

Degree or strength of lateralization of the preferred paw, regardless of its right or left direction, was initially used to capture this qualitative variation. One measure is the preferred-paw entry (PPE) score (Collins, 1985), which is the larger of the number of right- or left-paw reaches. For a 50-paw-reach test, it has numerical values of 25 to 50. An alternative, but equivalent, measure of degree of lateralization is the absolute difference (i.e., without regard for sign) between the numbers of right- and left-paw reaches made by an individual mouse (e.g., Roubertoux et al., 2003). For 50 paw reaches, this measure of degree of lateralization has a minimum numerical value of zero in a mouse that is ambilateral or equal handed and a maximum value of 50 in an individual that uses exclusively its right or left paw.

Transformation of the RPE score to a lateralization score, by either method, is equivalent to folding the RPE scale in half at the median value of 25 on the 0–50 RPE scale (see Figure 13.2). In effect, one numerical score, the RPE, is converted into a second numerical score, the PPE, and the underlying assumption is that variation in degree of lateralization has one developmental cause. However, folding the distribution at the midpoint of the RPE scale abolishes our ability to see genetically different patterns of hand preference (e.g., Figure 13.2). Later, we will show that it also blinds us to the potential for heritable differences in directional asymmetry of hand preference because some genotypes of mice have an inherent or constitutive left-handedness (and others have a right-handedness) and, therefore, they cannot be "folded" symmetrically because their median value is not at the midpoint of 25 on the 0–50 RPE scale.

Surveys of different strains of the laboratory mouse as well as strains of different species and subspecies of the genus *Mus* revealed significant heterogeneity in degree of lateralization of hand preference with a U-world testing chamber (e.g., Betancur et al., 1991; Biddle et al., 1993; Biddle and Eales, 1996; Gruber et al., 1991; Signore et al., 1991; Waters and Denenberg, 1994). These differences in lateralization of the different strains must be heritable because true breeding, highly lateralized (HI) and weakly lateralized (LO) strains were quickly obtained after only a few generations of selection from a constructed, genetically heterogeneous mouse stock (Collins, 1985, 1991). Unfortunately, most of the eight strains, used to construct the genetically heterogeneous mouse stock for this selection study, were never individually assessed for paw-preference behavior. The later strain surveys suggest the selected highly lateralized and weakly lateralized phenotypes of the respective HI and LO strains were most likely existing phenotypes among the foundation strains. Also, the speed by which the HI and LO phenotypes were selected suggests the heritable factors (alternate allelic forms of genes) controlling the phenotype may be few in number. Quantitative trait loci (QTL) analysis has detected a positive association between the degree of lateralization of paw usage and marker genes on mouse chromosome 4 (Roubertoux et al., 2003), but it remains to be validated by a test of transmission of alternate "genes" and recovery of predicted alternate phenotypes.

We should also note the historical context of a selection experiment for left-pawed and right-pawed mice within the highly lateralized C57BL/6J strain (Collins, 1969). Selected left- or right-pawed mice, over three generations, produced only bimodal distributions of both left- and right-pawed offspring (see C57BL/6J distribution in Figure 13.2). The experiment demonstrated that the apparent phenotypic heterogeneity of the bimodal distribution of RPE scores is the genetically determined phenotype of the C57BL/6J genotype in the U-world. The experiment was done to counter a popular belief that phenotypic heterogeneity in a highly inbred strain must reflect some residual genetic heterogeneity (Dr. R. L. Collins, personal communication). Nevertheless, the experiment continues to be interpreted as a demonstration that direction of hand preference cannot be selected in mice (e.g., Vallortigara and Rogers, 2005). No one has yet reported a successful genetic selection for left- or right-direction per se.

Associations of other developmental traits with paw preference

There have been many suggestions for functional associations between other developmental traits and laterality of paw-preference behavior in mice. Except to indicate their intriguing variety, we will not review them because they need to be assessed in greater depth for causal relationship. For example, morphological variation in size of the intercerebral hemispheric fiber tract of the corpus callosum, including its complete absence, has been associated with differences in laterality of paw preference in some, but not all, strains of mice (Biddle and Eales, 1996; Bulman-Fleming et al., 1992; Gruber et al., 1991; Ward et al., 1987). A positive association has been suggested between relative size of the intra/infrapyramidal mossy fiber (IIP-MF) projections in the hippocampus with weakly lateralized versus highly lateralized paw preference among different inbred strains as well as between the Collin's LO and HI lines (Lipp et al., 1996), but the association is greatly diminished when C57BL/6J is removed from the comparisons among nine different strains. Variation in different measures of immune reactivity is associated with a difference in paw preference in some but not all strains that were assessed and, in some comparisons, it is observable in only one, but not both sexes (Denenberg et al., 1991a, 1991b; Neveu et al., 1988, 1991). And, still other strain comparisons and reciprocal cross differences suggest potential maternal uterine effects on paw preference (Denenberg et al., 1992).

Genetic analysis reveals complexity in paw preference

Our experiments uncovered greater complexity in hand-preference behavior (Biddle and Eales, 1999). From a survey of mouse resources, the large difference in degree of lateralization in the U-world between C57BL/6J and CDS/Lay mice appeared to be suitable for genetic assessment by QTL analysis (Biddle and Eales, 1996) (see difference in distribution of RPE scores in Figure 13.2). Two forces conspired to make the temptation irresistible. First, QTL analysis is simple in concept (Lynch and Walsh, 1998) and genome-wide association studies became practical with accessible molecular technology and dense gene-marker maps for the mouse. Second, mouse behaviors are often regarded as complex polygenic systems in which genetic variation in some unknown biological process results in heritable variation in a continuously distributed, numerical behavioral score (e.g., Wehner and Balogh, 2003; Wehner et al., 2001), and we considered hand preference to be no different.

Whether RPE scores are evaluated as in Figure 13.2 or they are transformed to a lateralization score for the preferred paw, C57BL/6J is clearly highly lateralized and CDS/Lay is ambilateral or very weakly lateralized. We produced reciprocal F1 generations and a segregating F2 generation from each

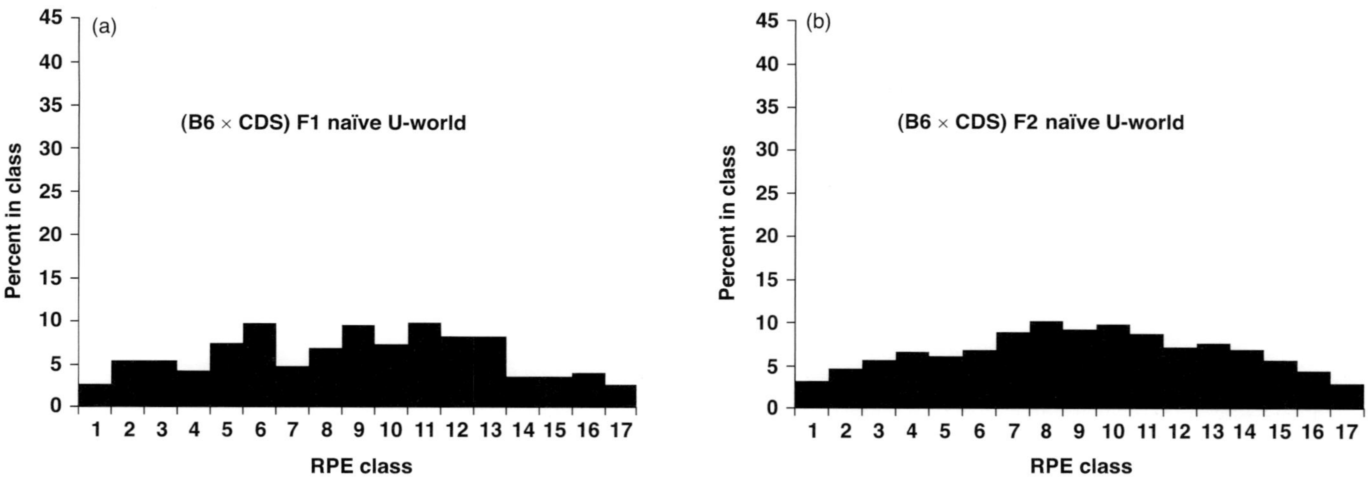

Figure 13.3 Distributions of right-paw entry (RPE) scores from a unbiased world (U-world) test of the (a) B6 × CDS F1 ($n = 300$) and (b) B6 × CDS F2 ($n = 1000$) generations, derived from reciprocal crosses between C57BL/6J and CDS/Lay. The mice were allowed 50 reaches. (Reproduced with permission from Biddle and Eales, 1999. © 2008 NRC Canada or its licensors.)

F1 generation and assessed their paw preference in a U-world (Figure 13.3) (Biddle and Eales, 1999). Parametric means (averages) of the RPE scores of the F2 and F1 generations did not differ, but that was expected for any phenotypic variation in a quantitative character that is determined by the segregation of simple Mendelian factors. However, we did not anticipate that the variances of the distributions of the F2 and F1 generations would also be identical (Figure 13.3). In plain language, the 1000-mouse F2 generation showed no increase in the variance of the numerical behavioral score, over what was observed in the F1 generation; that is, there was no clear evidence that more mice with the highly lateralized C57BL/6J phenotype and the weakly lateralized CDS/Lay phenotype were present in the F2 distribution than what was observed in the F1 hybrid generation. This means there was no phenotypic evidence for segregation of simple "additive genetic factors" in the F2 generation that might be responsible for the obvious quantitative difference in degree of lateralization between the C57BL/6J and CDS/Lay parental strains. And, if there is no evidence for an increase in phenotypic variance in the F2 generation, it was not clear what we expected to detect in the genome with a QTL analysis, regardless of the numerical index of the behavior.

We returned to explore the obvious qualitative differences in the distributions of RPE scores among the different mouse strains (see Figure 13.2). We had lost sight of the dramatically obvious differences in shape of the distributions of RPE scores among mouse strains (Figure 13.2) when we transformed the RPE score to a degree of lateralization and further averaged the lateralization scores. Even the summary statement of mean (average) RPE score and its sampling variance obscures the qualitative information in the histograms of Figure 13.2. This is what we mean by stepping into the trap of the central limit theorem (Gonick and Smith, 1993). In order to unpack the complexity, we looked at paw-preference behavior in other ways.

Biased testing chambers reveal a process of learning and memory

Phenotypic reaction norms and the genetic objective

We assessed hand preference of C57BL/6J and CDS/Lay in asymmetrical or biased testing chambers (Biddle and Eales, 1999), in which the food tube is placed flush to the left or right side and described as a L-world or R-world, respectively (Figure 13.4). Collins (1975) had used biased testing chambers only to assess the variation in direction of hand preference within the C57BL/6J inbred strain, but features of the learning and memory process were missed. We found that direction of paw preference in previously untested C57BL/6J mice moves in response to reaching in a left- or right-biased world, but it did not appear to change in CDS/Lay (Figure 13.4). Mean (average) RPE scores, plotted against direction of the testing chamber, show the phenotypic reaction norms of the two strains (Figure 13.5a) and there is obvious genotype × environment interaction because the norms of reaction intersect. When viewed as histograms in Figure 13.5b, C57BL/6J is relatively more left-handed than CDS/Lay in a L-world, but the strains switch order of paw preference in the R-world, where CDS/Lay is relatively more left-handed than C57BL/6J.

In our enthusiasm to find genes, we had forgotten the importance of norms of reaction to the analysis of phenotypic variation in quantitative characters (e.g., Falconer and Mackay, 1996; Schlichting and Pigliucci, 1998; and see introductory texts by Futuyma, 2005; Griffiths et al., 2000). Reaction norms of

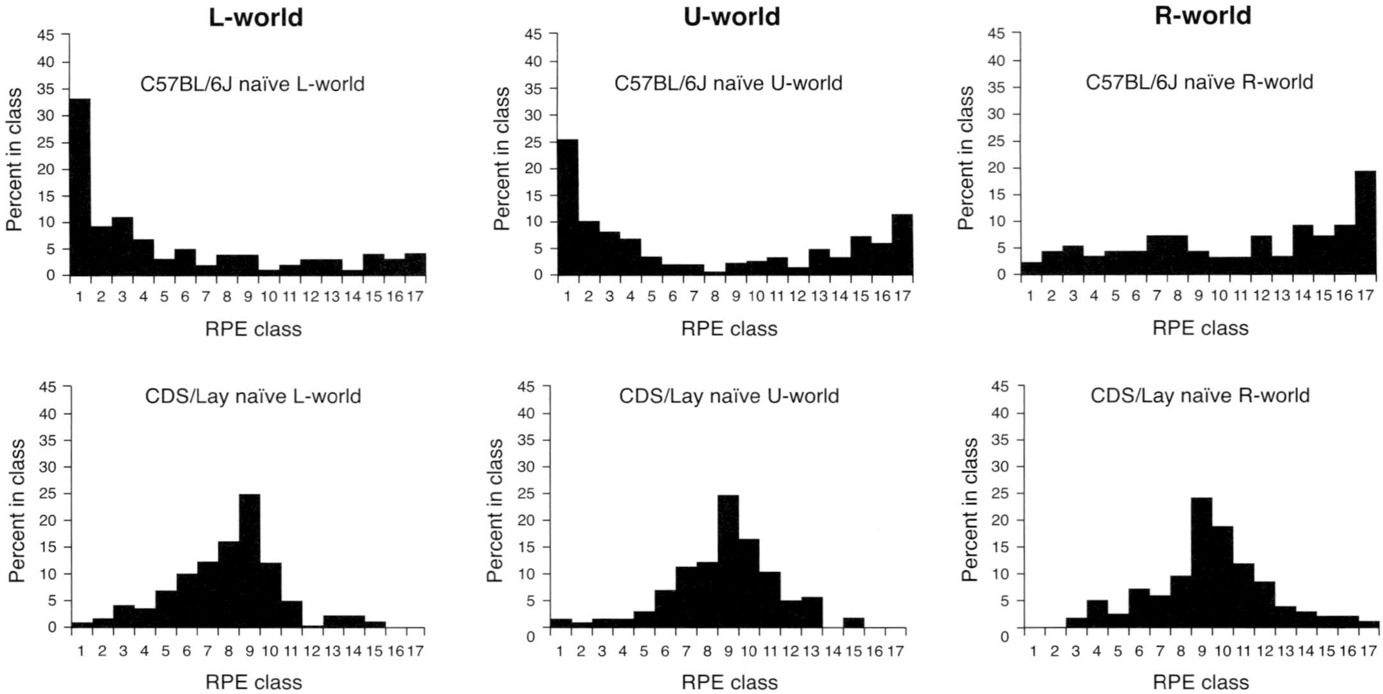

Figure 13.4 Distributions of right-paw entry (RPE) scores from biased left-sided world (L-world) and right-sided world (R-world) tests of previously untested C57BL/6J and CDS/Lay mice compared with RPE scores from unbiased world (U-world) tests. The mice were allowed 50 reaches and there were 150 mice in each group. (Reproduced with permission from Biddle and Eales, 1999. © 2008 NRC Canada or its licensors.)

C57BL/6J and CDS/Lay (Figure 13.5a) have immediate value in nature × nurture discussions. They reinforce the genetic principle that was elaborated, for example, in the landmark analysis of the amount of developmental variation in natural populations of *Drosophila* (Gupta and Lewontin, 1982), namely that norms of reaction provide an appreciation of the array of phenotypes that are possible for different genotypes in different environments. In contrast, phenotypic variation that is expressed by individual genotypes in only one environment, such as the patterns of RPE scores of the four different mouse strains in the U-world (see Figure 13.2) or the catalog of average preference scores from 26 different strains and stocks (Biddle and Eales, 1996), is of limited genetic value.

Lines joining the mean RPE values in Figure 13.5a visually relate the direction of the testing chamber to phenotype. Borrowing vocabulary used by Gupta and Lewontin (1982) and not speculating about genes and function, the C57BL/6J and CDS/Lay genotypes are simply "rules that transform environment into phenotypes." Average RPE scores in Figure 13.5 reveal some of the genetic variation in direction of hand preference in the L-world and R-world, but they show the more important gene × environment interaction because the reaction norms intersect when going from the L-world to the R-world (Figure 13.5a). The two genotypes differ in their phenotypic expression in different environments and the environments differ in their phenotypic effect on the two genotypes. Most obviously, the difference in average RPE score between the two strains essentially disappears in the U-world or where the phenotypic reaction norms intersect. Therefore, changing the environment from biased worlds to a U-world "destroys" evidence for genetic variation in phenotype when it is measured simply as a mean or average RPE score; the mean score would detect no difference in the U-world.

The objective of the genetic study became clear with the norms of reaction. It is to determine how and why the rules are qualitatively different between C57BL/6J and CDS/Lay when each strain (the genotype) transforms environment (direction of the testing chamber) into phenotype (direction of hand preference). Whether genotype (nature) or environment (nurture) plays a greater or lesser role in phenotype is not relevant to the research model.

Retesting in oppositely biased worlds reveals learning and memory

Two qualitatively different behaviors appeared to be uncovered by retesting mice in the oppositely biased worlds, one week after their test in left- or right-biased worlds (Biddle and Eales, 1999). Here, we use the mean RPE score as a measure of phenotype and the test–retest paradigm is illustrated in Figure 13.6 to convey what we were comparing on the RPE scale. Initially, the one-week interval between test and retest was logistically convenient; later, we show that it was fortuitous because it allowed for consolidation of memory of the first test. The mean RPE score of previously untested C57BL/6J mice moved in response to the direction of left- or right-biased testing chamber and

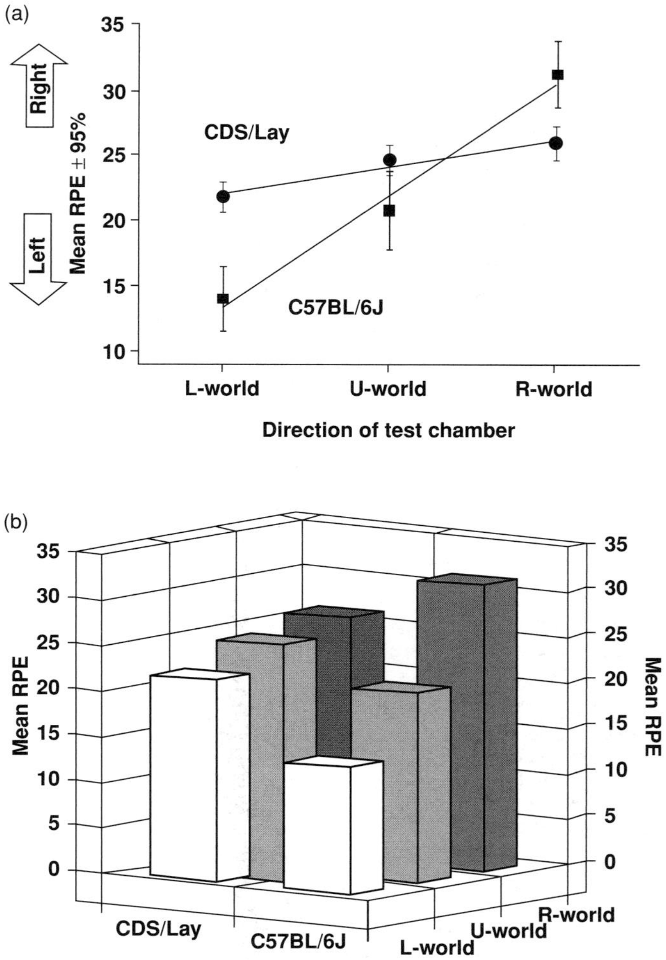

Figure 13.5 Phenotypic reaction norms of hand preference in response to reaching in a left-sided world (L-world), unbiased world (U-world), or right-sided world (R-world). (a) Mean right-paw entry (RPE) scores (±95% confidence limits) are from previously untested C57BL/6J and CDS/Lay mice that were allowed 50 reaches. Relative direction of paw usage is indicated. (Modified and reproduced with permission from Biddle and Eales, 1999. © 2008 NRC Canada or its licensors.) (b) When viewed as histograms, the strain difference in hand preference and its response to different test worlds is clear but the interaction is not as conspicuous.

did not return to the expected baseline mean RPE score when they were retested in the oppositely biased world. We could not detect a change in mean RPE scores in CDS/Lay by this testing paradigm. This suggested that paw preference in C57BL/6J mice is conditioned by the experience of reaching, and the direction or context of the testing chamber determines the left or right direction of hand preference. In contrast, CDS/Lay mice appeared to have a constitutive or innate behavior because the direction of their paw preference did not appear to respond to biased worlds, and it was not statistically different from direction of paw preference in the U-world (see Figure 13.2).

The two test–retest sequences, conducted in opposite directions and diagramed in Figure 13.6, suggested a simpler method to assess and distinguish between experience-conditioned and constitutive paw preference (Biddle and Eales, 2001). The

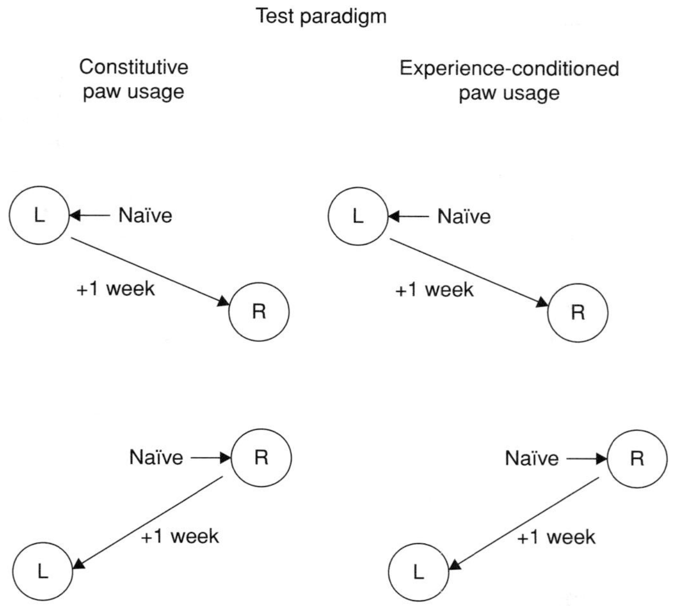

Figure 13.6 Testing protocol to identify constitutive and conditional paw preference. Previously untested mice are tested in a biased left-sided or right-sided world (L- or R-world) chamber (circled L or R), and one week later they are retested in the oppositely biased world. The figure attempts to express the observed difference between constitutive and conditional paw-preference behavior. In mice with a constitutive paw preference, the right-paw entry (RPE) scores may move in response to the direction of the testing chamber and, on retesting in the oppositely biased world, they return to the expected value in the oppositely biased world. In mice with a conditional behavior, the RPE scores move in response to the direction of the testing chamber but do not return to their expected value in the oppositely biased world. (Reproduced with permission from Biddle and Eales, 2001. © 2008 NRC Canada or its licensors.)

average of the two RPE scores, measured in individual mice from their biased test and oppositely biased retest, can be compared between independent samples of mice of the same genotype that are assessed in opposite test–retest sequences. Mice with a constitutive paw usage are expected to have the same mean averaged RPE score, regardless of whether they were assessed in a left-to-right or a right-to-left test–retest sequence (i.e., the difference between the means of the two averages, ± its confidence interval, should include zero). In contrast, mice with experience-conditioned behavior are expected to have significantly different averaged RPE scores from the two test–retest sequences (i.e., the difference between the means of the two averages, ± its confidence interval, are expected to exclude zero). See Wahlsten (1999) for a succinct discussion of effect size when comparing a group difference or the strength of an experimental treatment.

We documented the phenotypic complexity in hand-preference behavior among inbred strains and selected F1 hybrid generations, using the testing paradigm of Figure 13.6 (Biddle and Eales, 2001; Biddle et al., 2001). In particular, we included the four strains of C57BL/6J, SWV, CDS/Lay and DBA/2J strains from Figure 13.2 and their respective F1 hybrids. Each test of paw preference was based on 50 paw reaches.

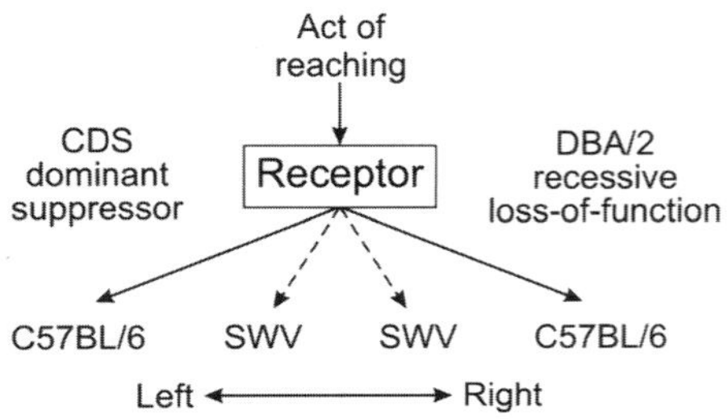

Figure 13.7 Regulated model of learning and memory using a metaphor of receptor–ligand interaction. C57BL/6J and SWV have a "receptor-like" process that allows them to learn a paw preference in response to reaching. C57BL/6J and SWV may be, respectively, a strong and a poor learner and, when given a choice in the unbiased world (U-world), individual C57BL/6J mice are highly lateralized, left or right, and SWV mice are weakly lateralized, left or right. DBA/2J may have a recessive loss of function in the putative receptor; since it cannot learn a direction of preference, DBA/2J has a random paw preference. CDS/Lay has a functional receptor, but it also has a dominantly acting suppressor, so that it expresses an equal or ambilateral paw preference. (Reproduced with permission from Biddle et al., 2001. © 2008 NRC Canada or its licensors.)

Genetically dominant and recessive forms of constitutive hand preference

Using biased testing chambers, only two phenotypic classes of paw preference were apparent, namely experience-conditioned and constitutive, but F1 hybrids revealed two different genetic causes of constitutive behavior. Both C57BL/6J and SWV strains have experience-conditioned behaviors because they learn a direction of hand preference and both CDS/Lay and DBA/2J strains have a constitutive behavior because they do not appear to learn a direction of hand preference. In F1 hybrids, the constitutive behavior of CDS/Lay is phenotypically dominant to the conditioned behavior of both C57BL/6J and SWV; however, the constitutive behavior of DBA/2J is phenotypically recessive to the conditioned behavior of C57BL/6J, but it is phenotypically dominant to the conditioned behavior of SWV.

We used the metaphor of receptor–ligand interaction (Figure 13.7) to interpret the observed phenotypic response to reaching in the C57BL/6J, SWV, CDS/Lay, and DBA/2J strains and the information that is contained in their respective F1 hybrids. The metaphor may be an exaggeration, but it provides a useful heuristic to consider gene action and interaction in paw-preference behavior and the metaphor led us to a kinetic analysis of the learning and memory process. Here, we superimpose this interpretation on the phenotypic variation in the patterns of U-world RPE scores of the parental strains that are pictured in Figure 13.2. Among experience-conditioned strains, C57BL/6J is highly lateralized (left and right) as seen in the U-world test, and SWV is weakly lateralized (left and right) and their B6 × SWV F1 hybrid appeared to be intermediate in degree of lateralization. Therefore, C57BL/6J and SWV might have respectively a "strong" and a "weak" allele in their ability to learn a direction of hand preference.

The experience-conditioned strains of C57BL/6J and SWV might be able to identify a functional difference between the constitutive behaviors of CDS/Lay and DBA/2J. Building on the metaphor, DBA/2J may have a loss of function at the same locus as the "strong" and "weak" allele and this loss of function allele of DBA/2J is recessive to the strong C57BL/6J allele in the B6 × D2 F1 compound heterozygote. Since SWV has a weak allele, the SWV × D2 F1 compound heterozygote may be below the threshold to detect conditional behavior and the constitutive behavior of DBA/2J appears to be phenotypically dominant to the conditional behavior of SWV. We suggested that CDS/Lay has a dominant allele at a second locus that suppresses or regulates the learning process in all F1 hybrids. Consequently, phenotypic evidence for the putative dominant suppressor in CDS/Lay and a recessive loss of function in DBA/2J would only be detected in the F1 hybrid generations with a strong learner, such as C57BL/6J.

With this receptor–ligand model, the complex distributions of hand preference in the U-world (Figure 13.2) began to make biological sense. C57BL/6J, SWV, and CDS/Lay may be ambilateral (i.e., they may have approximately equal-paw usage). With a difference in the putative receptor, C57BL/6J learns a lot and SWV learns a little and, when allowed 50 paw reaches in the U-world, C57BL/6J is highly lateralized left and right, whereas SWV is only weakly lateralized left and right. If CDS/Lay also has a functional receptor as well as a dominantly acting suppressor, CDS/Lay may reach with its left and right paws, but it is suppressed or regulated in its ability to learn a direction from this reaching and, therefore, it is expected to be *ambilateral* and weakly lateralized in the U-world. Moreover, the distribution of RPE scores in CDS/Lay is also expected to show a good fit to a normal or Gaussian distribution (Biddle and Eales, 1999; Biddle et al., 2001). (This is an *equal*, not random, hand usage; the distribution appears to be Gaussian because of stochastic variation in development and sampling.)

In contrast, if DBA/2J has a true loss of function in the putative receptor, its hand preference might be *randomly* determined in development (not equal-handed). Therefore, the uniform (flat or "platykurtic") distribution of RPE scores of DBA/2J mice in the U-world would be expected if there was an equal probability of having any RPE score on the 0–50 RPE scale (Biddle et al., 2001). This interpretation suggested that, when DBA/2J mice are allowed 50 paw reaches in the U-world, each mouse behaves as if it is a "51-sided fair-coin toss" with an equal probability of expressing any one of the 51 RPE scores from 0 to 50.

Key elements of the learning and memory process in hand preference

The metaphor of a receptor–ligand model suggested that a systematic and detailed kinetic analysis of the learning process in paw-preference behavior might provide an objective framework for a genetic and functional analysis. Moreover, it might distinguish mice that are true non-learners from those that are weak learners and below a threshold of detection. The kinetic analysis is discussed in some detail because it provided the foundation

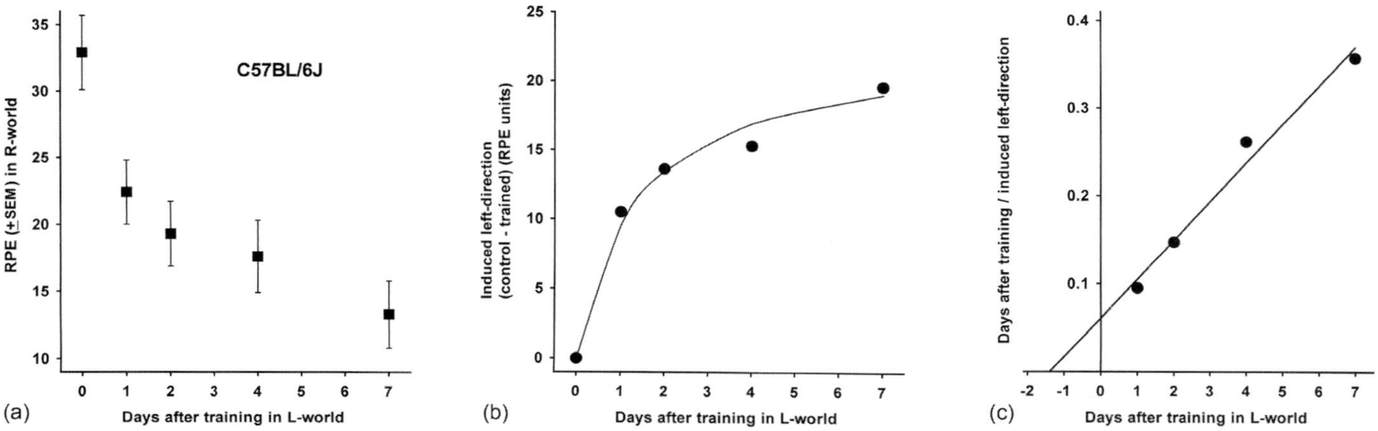

Figure 13.8 Consolidation of memory of 50 left-sided world (L-world) training reaches in C57BL/6J. (a) Direction of hand preference in the right-sided world (R-world) (mean right-paw entry (RPE) ±SE in 50 paw reaches) changes from being right-handed to being left-handed in response to elapsed time after L-world training. (b) Induced left-preference is the difference in mean RPE score in the R-world between untrained and L-world trained mice, and it increases to a maximum value over time after training. (Curve is from the regression analysis with the linear transformation in panel (c).) (c) Least-squares regression of the ratio of number of days after training divided by induced left-preference and plotted against number of days after training. Inverse of the slope estimates that 50 L-world training reaches induce a maximum of approximately 22.8 RPE units of preference and the *x*-intercept estimates that half of this maximum is consolidated in 1.4 days. (Reproduced with permission from Biddle and Eales, 2006. © 2008 NRC Canada or its licensors.)

for our stochastic agent-based model of the behavior and current efforts to uncover its gene regulatory network.

C57BL/6J appeared to be a good learner of hand preference and we assessed the learning of a left-paw preference (i.e., a left bias) by measuring how number of training reaches in the L-world alters the reaching behavior in the R-world (Biddle and Eales, 2006). All mice were previously untested. We adopted a definition of learning as simply "the acquisition of an altered behavioral response due to an environmental stimulus" (e.g., Sweatt, 2003) and, for the present purpose, memory is the contextual memory of L-world training (or procedural learning) that we can detect when L-world-trained mice are tested for hand preference in the opposite R-world. We chose the left direction, but preliminary results had predicted that training in either the L-world or R-world, followed by testing in the oppositely biased world, would lead to the same inferences about the learning and memory process in the behavior (Biddle and Eales, 1999, 2001; Biddle et al., 2001).

We documented the usual characteristics of a learning and memory process, namely consolidation of memory with time after training, blocking of this consolidation by anisomycin (an inhibitor of protein synthesis), retention of memory (rate of decay of memory), and the rate of learning in response to number of training-reaches. Then, we compared the rate of learning among different mouse strains and their F1 hybrids, derived from matings with C57BL/6J. Greater detail is found in the original publication (Biddle and Eales, 2006).

Consolidation of memory in response to elapsed time after L-world training

In response to elapsed time after training in the L-world with 50 training reaches, direction of paw preference in the R-world changed gradually from being right-handed to being left-handed (Figure 13.8a) (Biddle and Eales, 2006). The difference in mean RPE score in the R-world between untrained mice and the L-world-trained mice is an estimate of the amount of hand preference in the left-direction that is caused by the L-world training (Figure 13.8b). Change in magnitude of this difference follows a rectangular hyperbola in response to elapsed time and asymptotically achieves a maximum value. Since the maximum value of such a response cannot be measured directly, a linear transformation of the change in magnitude of the difference (Figure 13.8c) was done, similar to the familiar Hanes–Woolf transformation of the Henri–Michaelis–Menten-type kinetic equations (Dixon and Webb, 1964; Segel, 1975). The inverse of the slope estimates that a maximum of 22.8 RPE units is induced in the left direction in C57BL/6J mice by the 50 L-world training-reaches and the *x*-intercept of the regression estimates that 1.4 ± 0.5 (SE) days are required to consolidate half of this estimated maximum.

Anisomycin blocks consolidation of memory

Anisomycin prevents consolidation of long-term memory in mice (e.g., Abel et al., 1997) and is known to inhibit cerebral protein synthesis (Davis and Squire, 1984). Anisomycin blocked consolidation of training in a dose-dependent manner when C57BL/6J mice were treated with anisomycin, immediately after their L-world training (Biddle and Eales, 2006). This supported the hypothesis that a process of learning and long-term memory determines conditional paw preference.

Retention of memory and dynamics of the learning process

If contextual memory of L-world training is consolidated with time after training, how quickly does it decay? Independent

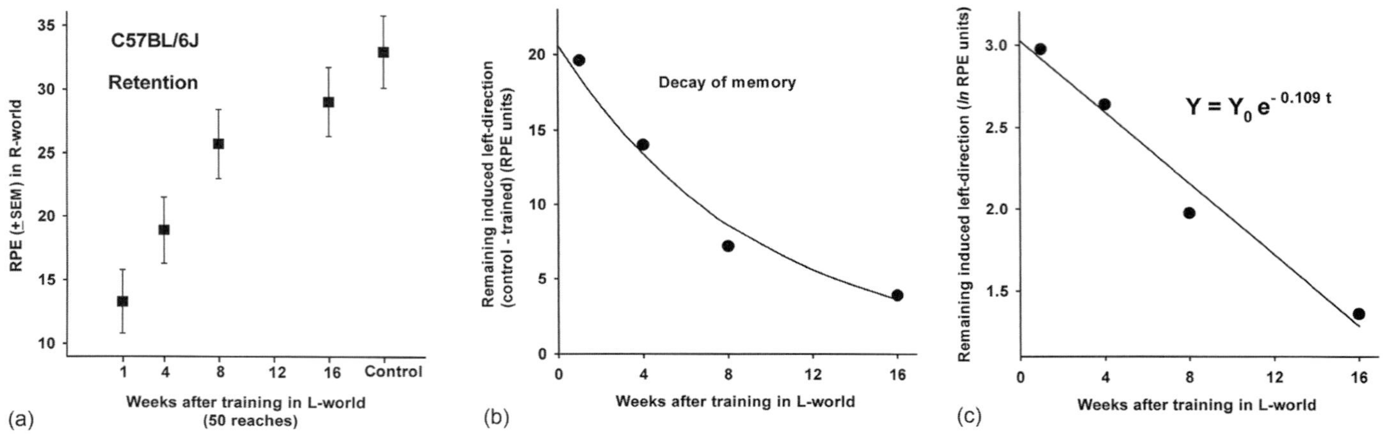

Figure 13.9 Retention of memory of left-sided world (L-world) training reaches in C57BL/6J. (a) With the passage of time, direction of hand preference of the L-world-trained mice returned to the preference of untrained mice in the right-sided world (R-world) (mean right-paw entry (RPE) ± SE in 50 paw reaches). (b) Induced left-preference is the difference in mean RPE score in the R-world between untrained and L-world trained mice and the remaining amount of left-preference decayed at a constant (exponential) rate. (Curve is from the regression analysis in panel c.) (c) Least-squares regression of the natural logarithm (*ln*) of remaining left-preference at different times after L-world training. Estimated half-life is 6.4 weeks. (Reproduced with permission from Biddle and Eales, 2006. © 2008 NRC Canada or its licensors.)

groups of C57BL/6J mice were trained in the L-world with 50 reaches and we tested them in the R-world with 50 reaches at different times beyond the one-week period (Figure 13.9) (Biddle and Eales 2006). Paw preference in L-world-trained mice gradually returned to being right-handed, like the preference of untrained C57BL/6J mice in the R-world (Figure 13.9a). The difference in mean RPE score in the R-world between untrained and L-world trained mice provides a measure of the remaining memory of the L-world training-reaches (and expressed in RPE units) (Figure 13.9b) and a least-squares regression revealed an exponential decay (Figure 13.9c). Therefore, at any time after L-world training, memory of that training is being lost at a constant rate with an estimated half-life of 6.4 weeks. In order for a L-world-trained C57BL/6J mouse to be considered similar to an untrained mouse in the R-world, we might want it to lose greater than 95% of its induced left-hand preference and that would require five half-lives or approximately 32 weeks in real time. This has consequences for the assessment of this behavior and implications for other behaviors if individuals are being retested or repeatedly tested.

L-world-trained mice, which were tested for memory retention, provided further evidence for the dynamic nature of the learning and memory process in paw-preference behavior. One week after the R-world test, which assessed the retention of the memory of L-world training-reaches (Figure 13.9), the mice were given another test with 50 reaches in the original L-world (Biddle and Eales, 2006). This second test in the L-world demonstrated that the L-world trained mice can be "re-conditioned" in the right-direction by their R-world test; the mice simply require time to lose sufficient memory of their original L-world training before we are able to detect the effect of their R-world test.

Rate of learning and limit of learning

We assessed the rate of learning a bias in paw preference of C57BL/6J by allowing previously untested mice different numbers of L-world-training reaches and, one week later, testing them in the R-world with 50 reaches (Figure 13.10) (Biddle and Eales, 2006). Direction of paw preference in the R-world changed from being right-handed to being left-handed in response to number of L-world training reaches (Figure 13.10a) and it reached a limit where more L-world training had no measurable effect on the retest in the R-world. The limit is expected because rate of memory loss during the one-week interval plays a role and there will be an equilibrium state between continued learning due to training and forgetting. The difference in mean preference in the R-world between untrained and L-world-trained mice estimates the amount of left-preference induced by the L-world training (Figure 13.10b), and the rate of change in magnitude of this difference follows a rectangular hyperbola in response to number of training reaches. The linear transformation (Figure 13.10c) was similar to the Hanes–Woolf transformation of the Henri–Michaelis–Menten-type kinetic equations (Dixon and Webb, 1964; Segel, 1975) that we had used to assess the rate of memory consolidation. The inverse of the slope estimates that a maximum of 21.5 RPE units is induced in C57BL/6J mice and the x-intercept estimates that 10.4 ± 4.0 (SE) training-reaches are required to induce half of this maximum.

Since speed of learning a preference is a hyperbolic response to an increasing number of training reaches (Figure 13.10b), it is not possible to determine the limit of learning by the point of contact to the asymptote. The linear transformation used to estimate this maximum, as well as number of training reaches

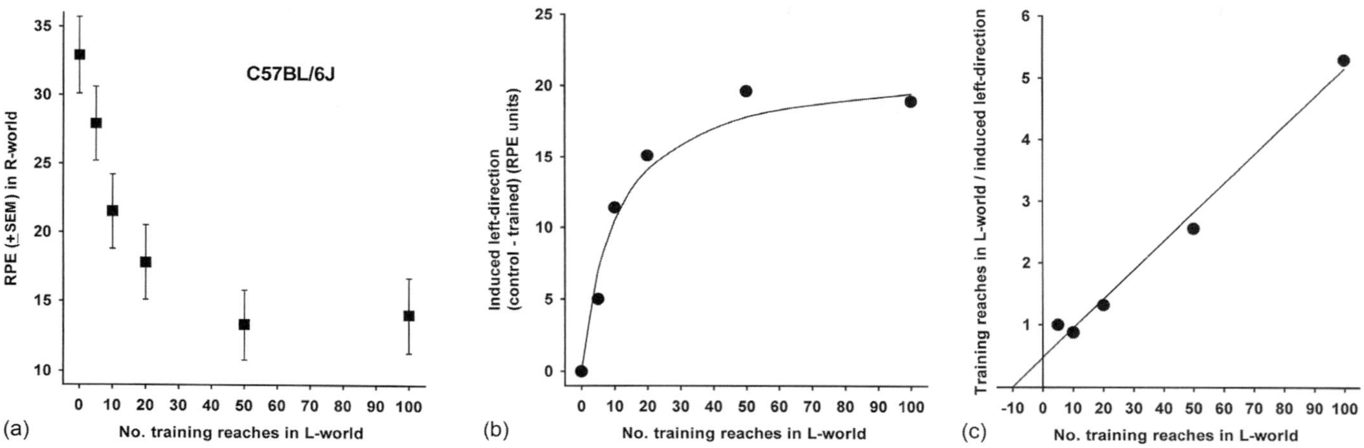

Figure 13.10 Two parameters of "capacity" and "ability" describe the rate of learning a direction of hand preference in C57BL/6J in response to left-sided world (L-world) training reaches. (a) Direction of hand preference in the right-sided (R-world) (mean right-paw entry (RPE) ± SE in 50 paw reaches) decreased in response to number of L-world training reaches. (b) Induced left-preference is the difference in hand preference (RPE) in the R-world between untrained and L-world trained mice and it increases to a maximum value with increasing number of training reaches. (Curve is from the regression analysis with the linear transformation in panel c.) (c) Least-squares regressions of the ratio of number of L-world training reaches divided by the induced left-preference and plotted against number of training reaches. The inverse of the slope estimates that 21.5 RPE units is the "capacity" or maximum induced preference and the *x*-intercept estimates that 10.4 training reaches is the "ability" to achieve half the capacity. (Reproduced with permission from Biddle and Eales, 2006. © 2008 NRC Canada or its licensors.)

to achieve it, is worth considering here to "keep the numbers connected to the behavior" and, perhaps, to see if other mouse behaviors might be amenable to kinetic analysis. The basic equation is

$$(a - y)(b + x) = constant$$

and the limits are a and $-b$. We wrote the equation for speed of learning as

$$(D_{max} - d)(K_r + r) = D_{max}K_r$$

and, rearranged to a linear form, it is

$$r/d = r/D_{max} + K_r/D_{max}.$$

The value (d) is the observed difference between untrained and trained mice in their mean RPE scores and it is the amount of learned preference in response to number of training-reaches (r). D_{max} is the maximum amount of learned preference and K_r is the number of training-reaches that is required for half of this maximum. When the ratio r/d is plotted against r, the asymptote D_{max} is estimated efficiently from the reciprocal of the slope ($1/D_{max}$) and K_r is estimated efficiently by extending the regression to the x-intercept. (Note: Time to consolidate memory of L-world training was assessed by a similar analysis (Figure 13.8).)

Kinetic analysis simply demonstrates that the biological process, underlying hand-preference behavior of C57BL/6J mice, responds to training reaches and is saturable by them. D_{max} is the maximum preference and K_r is the measure of how much training is required to achieve half the maximum preference. The parameter D_{max} is the "capacity" to learn in the sense of an amount of preference and the parameter K_r is the "ability" to learn in the sense of fast versus slow.

If environmental and genetic factors influence the system of paw-preference behavior, we can characterize them by their measurable effects on the numerical parameters of capacity (D_{max}) and ability (K_r). Environmental factors can influence the behavior because anisomycin blocks consolidation of memory of training reaches in a dose-dependent manner (Biddle and Eales, 2006). We turned to characterize the effect of genetic factors.

Genetic factors influence capacity and ability to learn hand preference

We applied a kinetic analysis to the conditional and constitutive paw preference of C57BL/6J, SWV, CDS/Lay, and DBA/2J (see Figure 13.7 and Biddle et al., 2001). Responses to L-world training are illustrated in Figure 13.11 and numerical estimates of capacity and ability are listed in Table 13.1.

Differences among strains with conditional behavior

SWV has conditional hand preference like C57BL/6J, but appears to learn less than C57BL/6J (Biddle and Eales, 2001; Biddle et al., 2001). Response to training demonstrated that SWV has significantly less capacity than C57BL/6J to learn a preference (smaller D_{max}) as well as significantly less ability to learn it (larger K_r) (Figure 13.11a and b; Table 13.1). Furthermore, the B6 × SWV F1 hybrid has both a capacity and an ability to learn that are approximately intermediate between the values of the C57BL/6J and SWV parental strains. From the kinetics perspective, determinants of hand preference from the two parental strains could be described as active and competing for training reaches in the B6 × SWV F1 hybrid, in the same manner that might be expected for additive mutations in the metaphor of a receptor–ligand binding process.

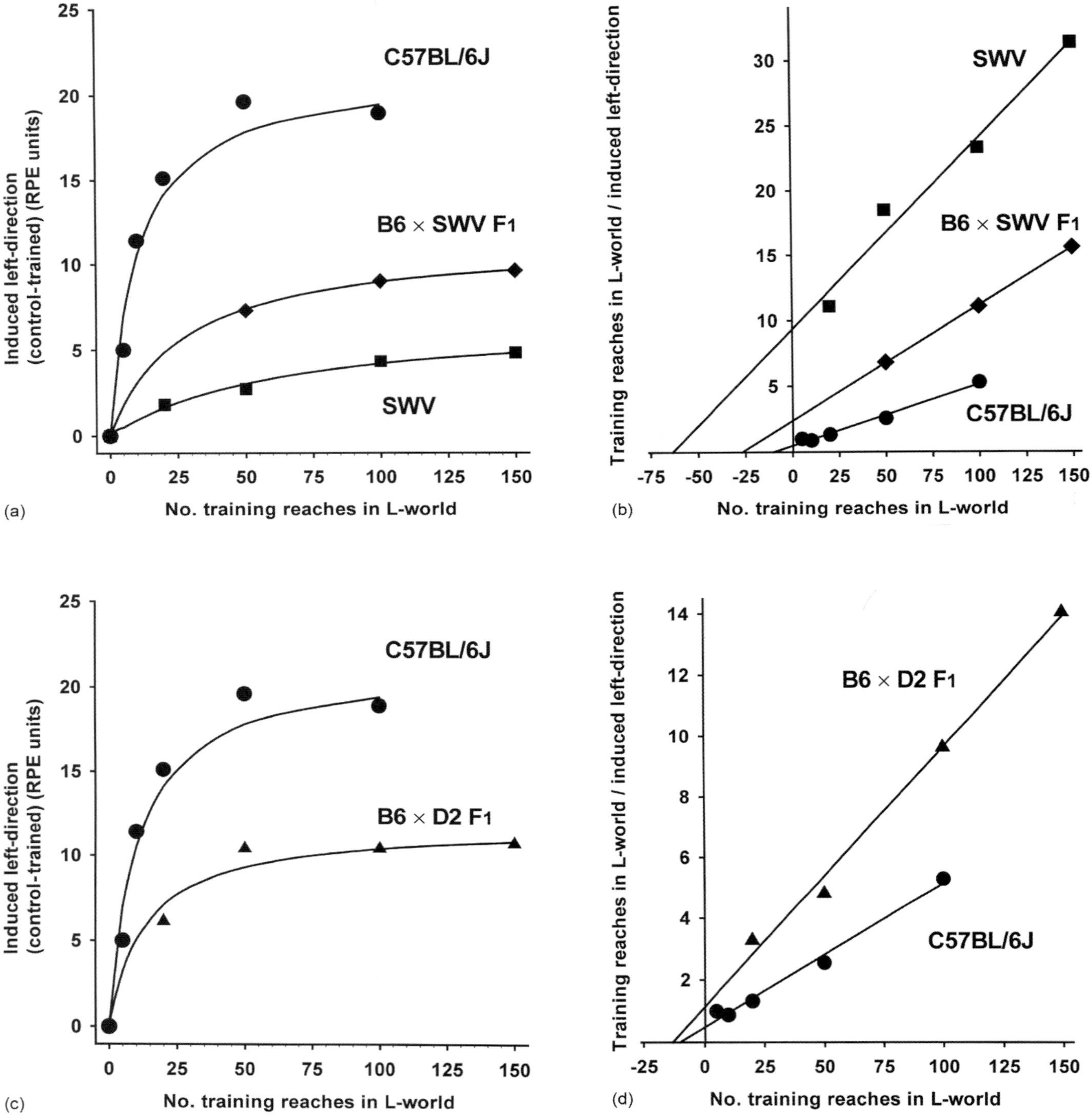

Figure 13.11 Capacity and ability to learn a direction of hand preference in SWV and DBA/2J compared with C57BL/6J. (a) Rates of learning a left-preference in response to left-sided world (L-world) training reaches in SWV and B6 × SWV F1 compared with C57BL/6J. (b) Least-squares regressions of the linear transformed learning curves in panel (a) estimate the capacity and ability parameters (summary in Table 13.1). (c) Rate of learning a left-preference in response to L-world training reaches in B6 × D2 F1 compared with C57BL/6J. (d) Least-squares regressions of the linear transformed learning curves in panel (c) estimate that B6 × D2 F1 has the same ability as C57BL/6J to learn a preference but only half the capacity to learn it (see summary in Table 13.1). (Reproduced with permission from Biddle and Eales, 2006. © 2008 NRC Canada or its licensors.)

Table 13.1 Capacity and ability to learn a direction of paw preference in C57BL/6J, SWV, B6 × SWV F1, and B6 × D2 F1.

Genotype	Capacity (RPE units) D_{max}	Ability (no. reaches) K_r
C57BL/6J	21.5	10.4
SWV	6.8	63.7
B6 × SWV F1	11.4	26.9
B6 × D2 F1	11.8	13.3

Modified from Biddle and Eales (2006).
RPE: right-paw entry.
Note: Capacity (D_{max}) is derived from the inverse of the slopes ($1/D_{max}$) and ability (K_r) is derived from the *x*-intercepts of the respective least-squares regression analyses of the linear-transformed learning curves in Figures 13.10c, 13.11b, and 13.11d.

Resolution of dominant and recessive constitutive hand preference

Genetically dominant constitutive hand preference of CDS/Lay was interpreted to be the result of a suppressor or regulator of the learning and memory process (Biddle et al., 2001). A dominantly acting constitutive behavior simply means that we could not detect a conditional behavior in either the strain or its F1 hybrid from mating with a strong learner. Dominantly acting constraints on a behavior may reflect a sensitive and, perhaps selectively precise, method to regulate the learning and memory process underlying specific behaviors (Abel et al., 1998). They may provide a framework to understand the regulation of specific behaviors in the same way that dominantly acting tumor suppressors brought order to tumor biology and cell-cycle regulation (Knudson, 1971). Therefore, analysis of genetically dominant constitutive paw preference should provide insight to the behavior.

On the other hand, genetically recessive constitutive paw preference, as found with the DBA/2J strain, presents an analytical problem. Phenotypically recessive effects may be caused by a true null or loss of function in the learning process; alternatively, they may be a very weak conditional behavior that we cannot detect by the usual testing procedure. Instead of trying to see whether we could detect a learning response in the DBA/2J strain following a large and, perhaps, unrealistic number of training reaches, we used the predicted kinetic properties of the learning process of the B6 × D2 F1 hybrid to distinguish between the two hypotheses.

If DBA/2J has a loss of function in the learning process, only C57BL/6J alleles are expected to determine paw-preference learning in the B6 × D2 F1 hybrid. Since C57BL/6J alleles are present in only half the amount, the B6 × D2 F1 heterozygote should have the same ability parameter (K_r) to learn a preference as the C57BL/6J parent, but only half the capacity (D_{max}) to learn it. Alternatively, DBA/2J may have a very weak learning response, and the difference in the learning response between SWV and C57BL/6J (Figure 13.11a and b) provides the framework for the alternative hypothesis. In that case, the B6 × D2 F1 would be predicted to have a significantly larger K_r than C57BL/6J because ability to learn would be the average of an undetectable weak ability of DBA/2J and the measurable ability of C57BL/6J.

DBA/2J appears to have a recessive loss of function in the learning process (Figure 13.11c and d and Table 13.1). Ability of the B6 × D2 F1 to learn is similar to C57BL/6J, but capacity is only half.

Independence of capacity and ability parameters

Genetic factors influence hand preference by their independent effects on the kinetic parameters of capacity (D_{max}) and ability (K_r) to learn a paw preference (Table 13.1). B6 × SWV F1 and B6 × D2 F1 have a similar capacity to learn but they have very different abilities to learn because they require significantly different numbers of training reaches to achieve the same half-maximum capacity. In contrast, C57BL/6J and B6 × D2 F1 have a similar ability to learn a preference, but they learn significantly different amounts (different capacities) with that same apparent ability.

Stochastic agent-based simulation of hand-preference behavior

Kinetic analyses in biased worlds revealed the genetically determined, dynamic process of learning and memory underlying paw-reaching behavior of mice. The genetic variation comes to attention as different patterns of paw-preference behavior, expressed by different genotypes of mice in the U-world (Figure 13.2) and in left- and right-biased worlds (Figure 13.4). Phenotypic heterogeneity is also obvious in the individual-to-individual variation in RPE score among genetically identical individuals within an inbred strain. In other words, paw-preference phenotypes are extremely noisy. Also, distributions of RPE scores of a population are not continuous and rarely are they Gaussian-like. In collaboration with Dr. Andre S. Ribeiro, we proposed that critical information about regulatory mechanisms and gene × environment interactions is contained in the patterns of phenotypic diversity of paw-preference behavior among genetically identical individuals, and the noisy distributions may provide an insight that is not captured by mean (or average) values or by various descriptors of the shapes of the RPE distributions. Moreover, a simulator that reproduces the diversity in paw-preference behavior and matches and/or predicts changes in behavior in different experiments might provide a deeper understanding of the gene × environment interactions.

A surprisingly simple, stochastic agent-based model of paw preference was produced and it matched paw reaching at the individual and the population levels from different genotypes (Ribeiro et al., 2010). Briefly, each model mouse has a probabilistic paw preference; learning causes the biasing of paw preference and decay of learning causes its unbiasing. Memory of a previously successful reach might determine the probability of

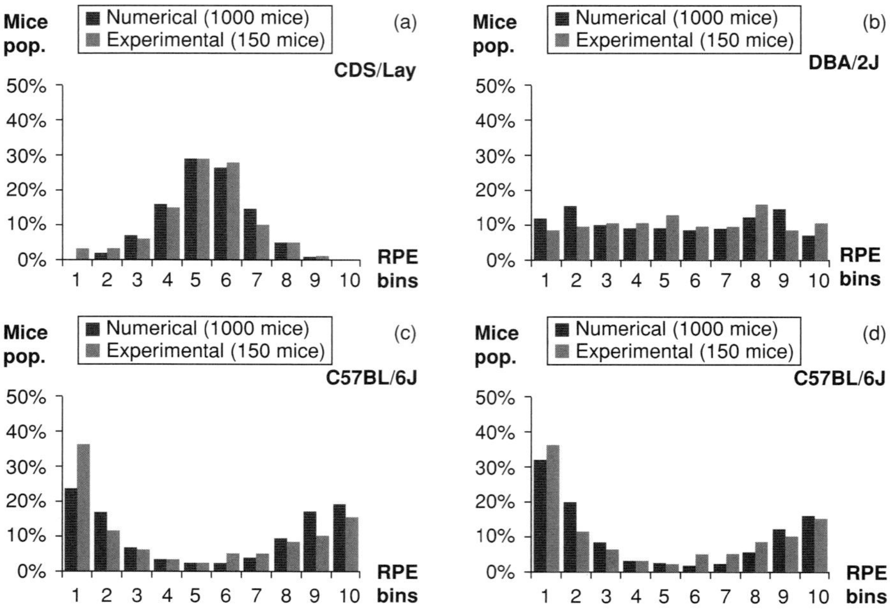

Figure 13.12 Comparisons of the right-paw entry (RPE) distributions from 50 paw reaches in an unbiased world (U-world) of previously untested, numerically simulated model mice, and experimental mice. Right-paw entry scores are binned in 10 equal-sized classes. (a) Model of CDS/Lay with a training rate of (25,25) in the respective left- and right-paw. (b) Model of DBA/2J with a training rate of (60,60). (c) Model of C57BL/6J with a training rate of (95,95). (d) Model of C57BL/6J with a biased training rate of (105,85). (Reproduced with permission from Ribeiro et al., 2010.)

using the right or left paw in a subsequent reach. If all model mice have the same number of memory units that are divided into right and left, the fraction of right- and left-memory units might define the probability of using the right or left paw at the next reach. Then, at each reach, a successful right or left reach causes a change in number of right or left memory units that, in turn, changes the probability of using the right or left paw in the next reach.

By tuning the two parameters of rate of learning and rate of decay of learning, simulations of the model matched the unbiased U-world distributions of populations of previously untested CDS/Lay, DBA/2J, and C57BL/6J mice (Figure 13.12a and b). The models identified a constitutive left bias in the paw preference of C57BL/6J, which we were able to match by altering the rate of learning with the left paw relative to the right paw (Figure 13.12c and d). Without any need to further tune the models, numerical simulations matched the experimental measures of paw preference when mice are tested in left and right-biased worlds as well as when they are tested in biased or unbiased worlds after previous training.

Unexpectedly, the models and their simulation identified a significant learning ability in the CDS/Lay and DBA/2J strains, which we could not detect with previous experiments. The models are now showing how to detect and describe learning ability with new measures of the behavior (A. S. Ribeiro and F. G. Biddle, unpublished data). Simulations with agent-based models are also showing what might be expected for paw reaching by a true null or non-learner mouse and how to detect it. This is important for functional genetic analysis because such an individual would reach for food and express a numerical score, but it would not learn.

Kinetic response to number of prior training reaches is a hallmark of the learning and memory process in paw

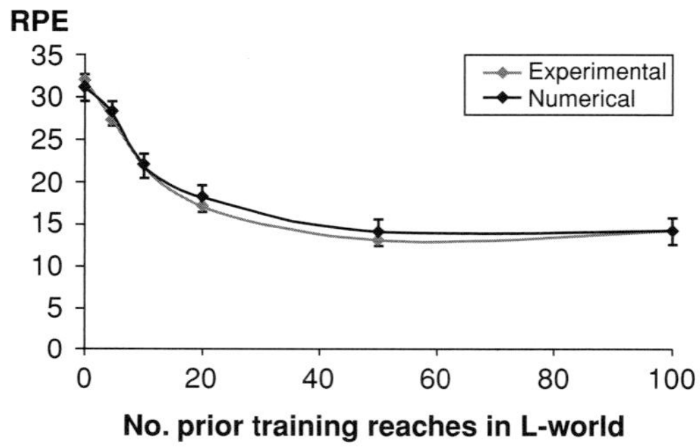

Figure 13.13 Learning curves for experimental and model C57BL/6J mice. Average right-paw entry (RPE) scores are shown for 50 reaches in a right-sided world (R-world), one week after a specified number of training reaches in a left-sided world (L-world) (experimental mice are from Figure 13.10a). The *y*-axis error bars in model mice are calculated from 100 independent simulations of a population of model mice. (Reproduced with permission from Ribeiro et al., 2010.)

preference (Figure 13.10). The saturation point or limit is where more training reaches, for example in a L-world, have no measurable effect on direction of paw preference in the opposite R-world after a one-week interval. Therefore, the ideal and perhaps a critical test of our paw-preference model is whether simulations with it can predict and match such a limit of learning where decay of memory during the one-week interval plays a major role. Without any need to further tune the model, numerical simulations with 100 independent populations of C57BL/6J model mice (Figure 13.13) exactly matched the previously observed limit of learning in C57BL/6J experimental mice (Figure 13.10). Experimental measurements fall within the error bars of the numerical simulations.

The model appears to capture biologically realistic features of paw-reaching behavior. Model mice retain information from their training and use it to influence subsequent paw-reaching events that are similar to real mice. The time course of the reaching events also appears to be well tuned in the simulation. Therefore, stochastic agent-based models and their simulations will help to explore how the different noisy phenotypes of the behavior are regulated by deterministic changes in the genome. In that regard, heritable changes in noisy phenotypes are not always obvious and a major problem for analysis, such as kinetic analyses (Figure 13.11), has been the need for large sample size, making experiments costly or daunting to execute and often producing observations that are not always well understood. We anticipate the agent-based models will be extremely useful in designing meaningful experiments and in estimating sample sizes that are required to detect biological, not just statistical, effects.

Challenging questions and persistent loose ends

We began with the statement that "variation is an endless source of challenging questions" (Mayr, 2005). Embracing the complexity in the phenotypic variation in hand preference reveals a biologically more accurate model of the mouse behavior. Phenotypic heterogeneity in the behavior is becoming understandable in a framework of learning and memory and we are mapping heritable differences onto the genome by transmission genetics (Biddle et al., unpublished data) and assessing recombinant strains (e.g., see review in Hitzemann, 2005). A challenging question is why paw-preference behavior exhibits classic features of learning and memory. Systems biology may provide an answer as we learn how the biology works to produce phenotypic heterogeneity and how to assess the functions of relevant genes in other biological contexts (Kirschner, 2005). A beginning may be the dominantly acting memory suppressor that regulates learning of hand preference in CDS/Lay mice and comes to attention when it is assessed in F1 hybrids with other strains. Nevertheless, some loose ends impinge on the question and they are addressed here.

Hand-preference patterns are predicted emergent properties

Kinetic analysis of mouse paw preference in biased test-worlds revealed the underlying process of learning and memory and the ability to replicate experiments with samples of genetically defined strains and their hybrids gave credibility to the model. In concept, reliable paw preference in biased environments has three basic causes: some mice clearly learn a direction of their paw preference in response to reaching and they can learn different genetically specified amounts. Other mice, in some contexts, appear unable to learn a preference in response to reaching and their constitutive preference is either a genetically recessive or dominant trait, leading to the respective metaphors of a "loss of function" and a "suppressor or regulator." Stochastic variation in the form of probabilistic hand choice affects both conditional and constitutive paw preference. From our present understanding of the behavior, supported now by agent-based simulations, the different shapes of the distributions of paw-preference scores (Figures 13.2 and 13.4) are the expected emergent properties of gene × gene and gene × environment interactions that are expressed by the replicate samples of the same single genotypes.

Research problem, the genetic task, and the biological question

The mindset that dominates human hand preference created a fundamental problem for the analysis of the mouse behavior and it is illustrated with the following "mind-experiment." The four distributions of RPE scores in our well-worn Figure 13.2 are respectively the U-world phenotypes expressed by four different genotypes of mice; in reality, they are simply the replicate samples of four mice. If an unknown individual mouse has a RPE score of 10 in 50 reaches, it reached 10 times with its right hand and 40 times with its left and it should be placed in RPE class 4 in Figure 13.2. That mouse might be described as left-handed; however, based solely on its RPE score of 10, we could only assign it probabilistically to a specific genotype. Moreover, without knowing its prior experience, the RPE score 10 does not tell us whether the individual learned its hand preference in response to reaching or is simply expressing an innate behavior. If we cannot yet answer these simple questions, why were we talking about genes and genetics in this behavior?

The mouse shows us that phenotype of a single genotype (i.e., an individual strain or its F1 hybrid) is the pattern or distribution of RPE scores, such as the U-world scores in Figure 13.2, not the RPE score of one individual mouse. Our genetic task is to map the alternative patterns of hand preference onto the genome. Since the U-world "destroys" the genetic variation in hand preference (e.g., Figure 13.5), the method to accomplish the genetic task is clear: don't give the mouse a choice by testing it in a U-world; test it in some other way, such as in a biased-world. Nevertheless, our biological question is really the reverse: how does biology generate the variation in hand usage of the different single genotypes that we see as the patterns of hand preference and the differences in patterns among the different genotypes in Figure 13.2?

Instructive value of the mouse – the replicated individual

We have presented mouse paw-preference behavior as a developmental system in order to contrast it with other models of hand-preference behavior. Human handedness has been embroiled in a long-standing nature versus nurture debate (e.g., Provins, 1997; Corballis, 1997) and we suggest that it is a troublesome stereotype, in the sense discussed by Sherman et al. (1997), which is in serious need of rethinking. Some believe that "people are right- or left-handed because of the genes they carry" (McManus, 2002), and compelling biallelic, single-gene models of human hand preference are almost too easy to construct from population and family studies. Definitions of handedness vary from simply the left or right hand that is preferred for writing to complex inventories derived from different kinds of single-hand tasks, which has provoked the remark that "[t]here are almost as many theories concerning hand preference as there are investigators" (Perelle and Ehrman, 2005).

Human hand preference has two basic models or three, if structural "pathologies" are included. First, some models are based on chance or random events in development, with elements of genetically determined directional bias (Klar, 1996; McManus and Bryden, 1992), sometimes referred to as a right-shift (Annett, 1998, 2000). Lateralization of cerebral dominance for language and motor skills lies at their foundation (Geschwind and Galaburda, 1985), leading some to believe that hand preference is a proxy for cerebral lateralization. Nevertheless, all genetic models have been refractory to candidate and marker gene analysis (Van Agtmael et al., 2002, 2003). A restricted definition of handedness may have an association with direction of rotation of scalp hair whorls, suggesting a common developmental genetic mechanism (Klar, 2005), but marker analysis is conspicuously absent. Second, other models include some degree of learned preference (Perelle and Ehrman, 2005).

In concept, people may be no different from mice in their hand preference: some may learn a bias and some may express a constitutive bias. After an individual has revealed their hand preference in both people and the mouse, it is impossible to tell whether preference is a learned or an innate behavior and, when "retested," the individuals reliably express the same preference. The difference between mouse and human models is that the mouse provides an ability to replicate the previously untested individual and, therefore, to distinguish unambiguously between learned and innate behaviors. For us, the instructive value of the mouse model is that replicated individuals reveal the phenotypic plasticity within single genotypes and the phenotypic reaction norms among different genotypes reveal the hidden process of learning and memory in hand preference. The mouse has allowed us to see that different individuals can have the same hand-preference score for genetically different reasons and, if they do have the same preference score, we have methods to assign (or map) them to different genotypes without error.

Directional asymmetry – a "left-shift" or "right-shift"

Paw preference scores from U-world tests of most mice are distributed symmetrically, although not necessarily in a Gaussian manner, at the midpoint of the RPE measurement scale (i.e., 25 on the RPE scale of 50 reaches). Lest the mouse be dismissed because it has no evidence for a "right shift" model, popular for human hand preference (e.g., Annett, 1998, 2000), a few strains do have a directional asymmetry (see Palmer, 2005, for asymmetry terminology). For example, U-world RPE scores of C57BL/6J are "shifted" to more left-hand usage in Figure 13.2 and similar impressions of directional asymmetry have been noted in strain surveys (e.g., Biddle and Eales, 1996, 2001; Bulman-Fleming et al., 1997; Signore et al., 1991; Waters and Denenberg, 1994). Therefore, some mouse strains express a directional asymmetry in the form of a "left shift" or a "right shift" in the bias of their paw preference, but the relationship between developmental asymmetry and the learning and memory process remains to be explored. We avoid universal conclusions because the few strain surveys represent the equivalent of only a few genotypes from the hundreds of possible mouse strains (e.g., Beck et al., 2000).

Physical variation in brain structure – cause or coincidence?

Since mouse paw preference is a genetically regulated adaptive behavior, deficiencies and asymmetries in brain structure, found in strain surveys, may or may not be relevant to paw-preference behavior. Functional connections need to be made between variation in brain structure and variation in the learning and memory phenotype and a geneticist might ask for a test of the association by a test of transmission. Since weakly lateralized paw-preference phenotype is found in different strains with apparently normal corpus callosum (e.g., Biddle and Eales, 1996, 2001), the occurrence of weakly lateralized paw-preference behavior and absence of the corpus callosum in I/LnJ mice (e.g., Gruber et al., 1991) may be coincidental. Perhaps both phenotypes cosegregated and are together by chance in the I/LnJ strain, along with many other mouse traits. Similarly, the asserted relation between cerebral asymmetries in gene-expression patterns and asymmetry of hand preference (Sun et al., 2005) may be inappropriate without direct experimental evidence, especially if gene-expression patterns are assessed in sagittal sections of hemispheric cortices from one-day-old random-bred "Swiss Webster" mice that are unknown for hand-preference behavior. We make these remarks cautiously in the hope that the paradigm of learning and memory engenders the functional analysis of paw preference.

Conclusions

Mice reach for food with their left and right forepaws, and simple kinetic analysis revealed the quantitative features of a dynamic process of learning and memory that changes the bias in paw preference and shapes the behavior. Direction of the bias in paw preference is contingent on both genes and environment and, in this mouse behavior, "context and interaction are [certainly] of the essence" (Lewontin, 1974). As geneticists, we believe the role for genetics is to identify and characterize allelic forms of genes that influence the parameters of the learning and memory process in paw preference.

We anticipate this knowledge will provide a better understanding of how the biology works and gives rise to phenotypic heterogeneity in the behavior. Recently, Wahlsten drew attention to Sewall Wright's regret that some biologists discard complex traits because they are considered "inconvenient for genetic analysis" (Wright, 1934), and we support his reinforcement of the fact that "extreme phenotypic variability among animals with the same genotype, while complicating genetic analysis, can help us perceive and understand developmental interactions more clearly" (Wahlsten et al., 2005). Knowledge from a deeper understanding of the mouse behavior should help to resolve complexity in other models of hand preference that we cannot possibly assess with any experimental rigor; at the very least, it will rattle some of our beliefs.

Acknowledgments

The Alberta Children's Hospital Research Foundation supported this work. We thank the Life and Environmental Sciences Animal Resource Centre and the Department of Biological Sciences of the University of Calgary for accommodating and facilitating this research. We especially acknowledge the friendship and vigorous research interaction with our collaborator, Dr. Andre S. Ribeiro, and his team in Computational Systems Biology at Tampere University of Technology. We greatly appreciate the opportunity to contribute to this Cambridge Encyclopedia of Behavior Genetics and thank the editors for their patience. To our colleagues who said, "Complexity of evaluating the behavioral endpoint raised the issue of feasibility of the project," we reply, "That is why we call it research."

References

Abel, T., Martin, K.C., Bartsch, D., and Kandel, E.R. (1998) Memory suppressor genes: inhibitory constraints on the storage of long-term memory. *Science* 279: 338–341.

Abel, T., Nguyen, P.V., Barad, M., Deuel, T.A.S., Kandel, E.R., and Bourtchouladze, R. (1997) Genetic demonstration of a role for PKA in the late phase of LTP and in hippocampus-based long-term memory. *Cell* 88: 615–626.

Annett, M. (1998) Handedness and cerebral dominance: the right shift theory. *J Neuropsychiat* 10: 459–469.

Annett, M. (2000) Predicting combinations of left and right asymmetries. *Cortex* 36: 485–505.

Beck, J.A., Lloyd, S., Hafezparast, M., Lennon-Pierce, M., Eppig, J.T., Festing, M.F.W., et al. (2000) Genealogies of mouse inbred strains. *Nature Genet* 24: 23–25.

Betancur, C., Neveu, P.J., and Le Moal, M. (1991) Strain and sex differences in the degree of paw preference in mice. *Behav Brain Res* 45: 97–101.

Biddle, F.G., Coffaro, C.M., Zeihr, J.E., and Eales, B.A. (1993) Genetic variation in paw preference (handedness) in the mouse. *Genome* 36: 935–943.

Biddle, F.G. and Eales, B.A. (1996) The degree of lateralization of paw usage (handedness) in the mouse is defined by three major phenotypes. *Behav Genet* 26: 391–406.

Biddle, F.G. and Eales, B.A. (1999) Mouse genetic model for left-right hand usage: context, direction, norms of reaction, and memory. *Genome* 42: 1150–1166.

Biddle, F.G. and Eales, B.A. (2001) Lateral asymmetry of paw usage: phenotypic survey of constitutive and experience-conditioned paw-usage behaviours among common strains of the mouse. *Genome* 44: 539–548.

Biddle, F.G. and Eales, B.A. (2006) Hand preference training in the mouse reveals key elements of its learning and memory process and resolves the phenotypic complexity in the behaviour. *Genome* 49: 666–677.

Biddle, F.G., Jones, D.A., and Eales, B.A. (2001) A two-locus model for experience-conditioned direction of paw usage in the mouse is suggested by dominant and recessive constitutive paw usage behaviours. *Genome* 44: 872–882.

Bulman-Fleming, B., Wainwright, P.E., and Collins, R.L. (1992) The effects of early experience on callosal development and functional lateralization in pigmented BALB/c mice. *Behav Brain Res* 50: 31–42.

Bulman-Fleming, M.B., Bryden, M.P., and Rogers, T.T. (1997) Mouse paw preference: effects of variations in testing protocol. *Behav Brain Res* 86: 79–87.

Collins, R.L. (1968) On the inheritance of handedness. I. Laterality in inbred mice. *J Hered* 59: 9–12.

Collins, R.L. (1969) On the inheritance of handedness. II. Selection for sinistrality in mice. *J Hered* 60: 117–119.

Collins, R.L. (1975) When left-handed mice live in right-handed worlds. *Science* 187: 181–184.

Collins, R.L. (1985) On the inheritance of direction and degree of asymmetry. In Glick, S.D. (ed.), *Cerebral Lateralization in Nonhuman Species*. Academic Press, Orlando, FL, USA, pp. 41–71.

Collins, R.L. (1991) Reimpressed selective breeding for lateralization of handedness in mice. *Brain Res* 564: 194–202.

Corballis, M.C. (1997) The genetics and evolution of handedness. *Psychol Rev* 104: 714–727.

Davis, H.P. and Squire, L.R. (1984) Protein synthesis and memory: a review. *Psychol Bull* 96: 518–559.

Denenberg, V.H., Mobraaten, L.E., Sherman, G.F., Morrison, L., Schrott, L.M., Waters, N.S., et al. (1991b) Effects of the autoimmune uterine/maternal environment upon cortical ectopias, behavior and autoimmunity. *Brain Res* 563: 114–122.

Denenberg, V.H., Sherman, G.F., Morrison, L., Schrott, L., Waters, N.S., Rosen, G.D., *et al.* (1992) Behavior, ectopias and immunity in BD/DB reciprocal crosses. *Brain Res* 571: 323–329.

Denenberg, V.H., Sherman, G.F., Schrott, L.M., Rosen, G.D., and Galaburda, A.M. (1991a) Spatial learning, discrimination learning, paw preference and neocortical ectopias in two autoimmune strains of mice. *Brain Res* 562: 98–114.

Dixon, M. and Webb, E.C. (1964) *Enzymes*, 2nd edn. Longmans Green, London.

Falconer, D.S. and Mackay, T.F.C. (1996) *Introduction to Quantitative Genetics*, 4th edn. Longmans Green, Harlow, Essex, UK.

Futuyma, D.J. (2005) *Evolution*. Sinauer Associates, Sunderland, MA, USA.

Geschwind, N. and Galaburda, A.M. (1985) Cerebral lateralization. *Arch Neurol* 42: 521–552.

Gonick, L. and Smith, W. (1993) *The Cartoon Guide to Statistics*. Harper Perennial, New York.

Griffiths, A.J.F., Miller, J.H., Suzuki, D.T., Lewontin, R.C., and Gelbart, W.M. (2000) *An Introduction to Genetic Analysis*, 7th edn. W. H. Freeman and Co., New York.

Gruber, D., Waanders, R., Collins, R.L., Wolfer, D.P., and Lipp, H.-P. (1991) Weak or missing paw lateralization in a mouse strain (I/LnJ) with congenital absence of the corpus callosum. *Behav Brain Res* 46: 9–16.

Gupta, A.P. and Lewontin, R.C. (1982) A study of reaction norms in natural populations of *Drosophila pseudoobscura*. *Evolution* 36: 934–948.

Hitzemann, R.J. (2005) Genetics and behavior. In Eisen, E.J. (ed.), *The Mouse in Animal Genetics and Breeding Research*. Imperial College Press, London, pp. 177–203.

Kirschner, M.W. (2005) The meaning of systems biology. *Cell* 121: 503–504.

Klar, A.J.S. (1996) A single locus, RGHT, specifies preference for hand untilization in humans. *Cold Spring Harb Symp Quant Biol* 61: 59–65.

Klar, A.J.S. (2005) A 1927 study supports a current genetic model for inheritance of human hair-whorl orientation and hand-use preference traits. *Genetics* 170: 2027–2030.

Knudson, A.G. (1971) Mutation and cancer: statistical study of retinoblastoma. *Proc Natl Acad Sci USA* 68: 820–823.

Lewontin, R.C. (1974) *The Genetic Basis of Evolutionary Change*. Columbia University Press, New York, p. 318.

Lipp, H-P., Collins, R.L., Hausheer-Zarmakupi, Z., Leisinger-Trigona, M-C., Crusio, W.E., Nosten-Bertrand, M., *et al.* (1996) Paw preference and intra-/infrapyramidal mossy fibers in the hippocampus of the mouse. *Behav Genet* 26: 379–390.

Lynch, M. and Walsh, B. (1998) *Genetics and Analysis of Quantitative Traits*. Sinauer Associates, Sunderland, MA, USA.

Magnusson, W.E. and Mourao, G. (2004) *Statistics without Math*. Sinauer Associates, Sunderland, MA, USA.

Mayr, E. (2005) Foreword. In Hallgrimsson, B. and Hall, B.K. (eds.), *Variation. A Central Concept in Biology*. Elsevier, Amsterdam, p. vii.

McManus, C. (2002) *Right Hand, Left Hand. The Origins of Asymmetry in Brains, Bodies, Atoms and Cultures*. Harvard University Press, Cambridge, MA, USA.

McManus, I.C. and Bryden, M.P. (1992) The genetics of handedness, cerebral dominance and lateralization. In Rapin, I. and Segalowitz, S.J. (eds.), *Handbook of Neuropsychology*, Vol. 6. Elsevier Science, Amsterdam, pp. 115–144.

Neveu, P.J., Barneoud, P., Vitiello, S., Betancur, C., and Le Moal, M. (1988) Brain modulation of the immune system: association between lymphocyte responsiveness and paw preference in mice. *Brain Res* 457: 392–394.

Neveu, P.J., Betancur, C., Vitiello, S., and Le Moal, M. (1991) Sex-dependent association between immune function and paw preference in two substrains of C3H mice. *Brain Res* 559: 347–351.

Palmer, A.R. (2005) Antisymmetry. In Hallgrimsson, B. and Hall, B.K. (eds.), *Variation. A Central Concept in Biology*. Elsevier, Amsterdam, pp. 359–397.

Perelle, I.B. and Ehrman, L. (2005) On the other hand. *Behav Genet* 35: 343–350.

Provins, K.A. (1997) Handedness and speech: a critical reappraisal of the role of genetic and environmental factors in the cerebral lateralization of function. *Psychol Rev* 104: 554–571.

Ribeiro, A.S., Lloyd-Price, J., Eales, B.A., and Biddle, F.G. (2010) Dynamic agent-based model of hand-preference behavior patterns in the mouse. *Adapt Behav* 18: 116–131.

Roubertoux, P.L., Le Roy, I., Tordjman, S., Cherfou, A., and Migliore-Samour, D. (2003) Analysis of quantitative trait loci for behavioral laterality in mice. *Genetics* 163: 1023–1030.

Schlichting, C.D. and Pigliucci, M. (1998) *Phenotypic Evolution: A Reaction Norm Perspective*. Sinauer Associates, Sunderland, MA, USA.

Segel, I.H. (1975) *Enzyme Kinetics: Behavior and Analysis of Rapid Equilibrium and Steady-state Enzyme Systems*. Wiley, New York.

Sherman, S.L., DeFries, J.C., Gottesman, I.I., Loehlin, J.C., Meyer, J.M., Pelias, M.Z., et al. (1997) Recent developments in human behavioral genetics: past accomplishments and future directions. *Am J Hum Genet* 60: 1265–1275.

Signore, P., Chaoui, M., Nosten-Bertrand, M., Perez-Diaz, F., and Marchaland, C. (1991) Handedness in mice: comparison across eleven inbred strains. *Behav Genet* 21: 421–429.

Smolin, L. (2006) *The Trouble with Physics*. Houghton Mifflin Harcourt, Boston.

Sun, T., Patoine, C., Abu-Khalil, A., Visvader, J., Sum, E., Cherry, T.J., *et al.* (2005) Early asymmetry of gene transcription in embryonic human left and right cerebral cortex. *Science* 308: 1794–1798.

Sweatt, J.D. (2003) *Mechanisms of Memory*. Elsevier Academic Press, Amsterdam.

Takeda, S. and Endo, A. (1993) Paw preference in mice: a reappraisal. *Physiol Behav* 53: 727–730.

Tukey, J.W. (1977) *Exploratory Data Analysis*. Addison-Wesley, Reading, MA, USA.

Vallortigara, G. and Rogers, L.J. (2005) Survival with an asymmetrical brain: advantages and disadvantages of cerebral lateralization (and open peer commentary). *Behav Brain Sci* 28: 575–633.

Van Agtmael, T., Forrest, S.M., and Williamson, R. (2002) Parametric and nonparametric linkage analysis of several candidate regions for genes for human handedness. *Eur J Hum Genet* 10: 623–630.

Van Agtmael, T., Forrest, S.M., Del-Favero, J., Van Broeckhoven, C., and Williamson, R. (2003) Parametric and nonparametric genome scan analysis for human handedness. *Eur J Hum Genet* 11: 779–783.

Wahlsten, D. (1999) Experimental design and statistical inference. In Crusio, W.E. and Gerlai, R.T. (eds.), *Handbook of Molecular-Genetic Techniques for Brain and Behavioral Research*. Elsevier, Amsterdam, pp. 40–57.

Wahlsten, D., Bishop, K.M., and Ozaki, H.S. (2005) Recombinant inbreeding in mice reveals thresholds in embryonic corpus callosum development. *Genes Brain Behav* 5: 170–188.

Ward, R., Tremblay, L., and Lassonde, M. (1987) The relationship between callosal variation and lateralization in mice is genotype-dependent. *Brain Res* 424: 84–88.

Waters, N.S. and Denenberg, V.H. (1994) Analysis of two measures of paw preference in a large population of inbred mice. *Behav Brain Res* 63: 295–204.

Wehner, J.M. and Balogh, S.A. (2003) Genetic studies of learning and memory in mouse models. In Plomin, R., Defries, J.C., Craig, I.W., and McGuffin, P. (eds.), *Behavioral Genetics in the Postgenomic Era*. American Psychological Association, Washington, DC, pp. 103–121.

Wehner, J.M., Radcliffe, R.A., and Bowers, B.J. (2001) Quantitative genetics and mouse behavior. *Annu Rev Neurosci* 24: 845–867.

Wright, S. (1934) On the genetics of subnormal development of the head (otocephaly) in the guinea pig. *Genetics* 19: 494.

Chapter 14

Rhythms and sleep

Circadian and seasonal activity patterns

Bernard Possidente

Preface

Temporal organization on a daily basis is a general characteristic of behavior, most conspicuously visible in the sleep–wake cycle. The persistence of daily biological rhythms in the absence of external time cues, and the heritability of individual variations in daily cycles reflects the existence of endogenous circadian clock mechanisms. Inbred strain comparisons, artificial selection, and quantitative trait loci (QTL) analyses have identified heritable variation in many different behavioral circadian rhythms and in the circadian clock mechanism itself. Seasonal rhythms have received less direct attention from behavioral geneticists, but since many seasonal rhythms are induced by circadian mechanisms that monitor changes in day length our understanding of circadian mechanisms should translate to analysis of seasonal rhythms in appropriate model organisms. Temporal variation and covariation adds an important functional dimension to behavioral analysis, and behavioral geneticists should consider circadian phase as a variable in experimental design that may interact with genetic components of variation.

Introduction

Historical perspective

Colin Pittendrigh, more than any other figure, has been the "Mendel" of biological clocks with his insightful construction of functional models for *biological clocks* in the absence of any knowledge of their mechanism at the cellular and molecular levels (see also Pittendrigh, 1981a). Pittendrigh's analyses of biological clock function were derived from observations and manipulations of functional behavioral phenotypes and naturally-occurring genetic variation, from which he developed formal rules of biological clock behavior that have proven to be of broad general relevance. Artificial selection on peak time of eclosion in *Drosophila*, and comparative analysis of three mouse species and hamsters (Daan and Pittendrigh, 1976a, 1976b; Pittendrigh and Daan, 1976a, 1976b, 1976c) contributed to Pittendrigh's functional analyses of circadian and seasonal biological clock mechanisms.

During the 1990s, direct molecular genetic analysis of biological clock mechanisms identified a core set of clock genes and an equally elegant model that provides a cellular and molecular understanding of circadian clock function (Sehgal, 2004). Predictably, our understanding of biological clocks at a molecular level was based on genetic dissection of induced mutations that block or disrupt biological clock function, combined with gene mapping, sequencing, and expression studies (Dunlap, 1996). This review is devoted to genetic analysis of biological clock function based on naturally-occurring genetic variation associated with phenotypic variation in wild populations of mice and laboratory populations derived from them, and therefore it has much more in common with Pittendrigh's formal analysis of biological clock function and organization than with analysis of biological clock mechanisms at the molecular level. One difference, however, between the earliest behavioral genetic studies of biological clocks and current research is that genome-wide methods of molecular analysis are increasingly capable of identifying QTLs underlying behavioral traits (Clayton et al., 2001; Salathia et al., 2002; Takahashi et al. 2008). Another difference is that molecular and genomic analyses are identifying new genetic and epigenetic mechanisms regulating gene expression. Copy number variation, alternative splicing pathways, non-coding RNA, chromatin remodelling, and genetic imprinting, for example (Dardente and Cermakian, 2007), may enrich our more traditional understanding of heritable variation in complex behavioral phenotypes evident from strain comparisons and response to selection beyond the level of allelic differences in coding sequences of DNA.

These formal and mechanistic approaches to the analysis of biological clock behavior have converged to create a body of general theory that describes "rules" of biological clock function that are causally tied to tractable molecular, cellular, and physiological mechanisms. Together, the combined genetic and phenotypic analysis of biological clocks has established an exciting, young, interdisciplinary field that cuts across every major taxonomic group, every level of biological organization, and most biological functions including many behavioral rhythms with direct relevance to human health (see also Bhattacharjee, 2007; McClung, 2007; Roybal et al., 2007; Turek et al., 2005).

Behavioral Genetics of the Mouse: Volume I. *Genetics of Behavioral Phenotypes*, eds. Wim E. Crusio, Frans Sluyter, Robert T. Gerlai, and Susanna Pietropaolo. Published by Cambridge University Press.

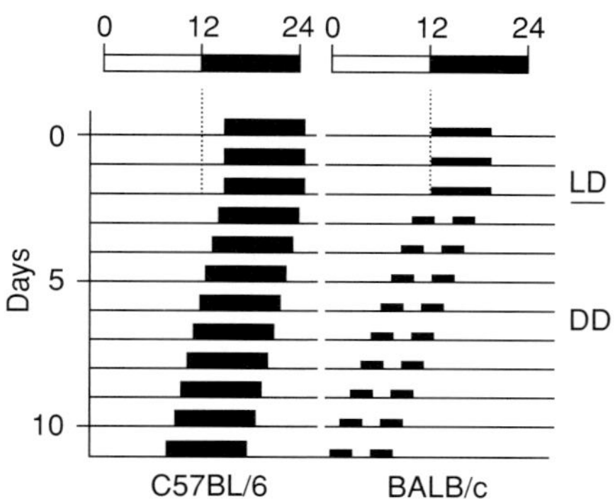

Figure 14.1 An "actogram" displaying entrained activity patterns in a 12:12 light–dark (LD) cycle, followed by free-running circadian activity patterns in constant dark (DD) for typical C57BL/6 versus BALB/c inbred mice. The C57 pattern has activity throughout the night and a relatively long circadian period that is slightly less than 24 hours. The BALB/c pattern is more fragmented, with a shorter free-running period closer to 23 hours. (Reproduced with permission from Salathia et al., 2002.)

Biological clocks concepts, mechanisms, and relevance

Biological clocks are endogenous mechanisms that generate self-sustaining rhythms in organisms (Dunlap et al., 2004; Sehgal, 2004; e.g., Figure 14.1). They synchronize to exogenous rhythms with a similar period, and to each other. The most common biological clocks match ecologically relevant geophysical cycles such as solar, lunar, tidal, and seasonal periods. An important distinction is made between an overt, observable rhythm in a particular function (e.g., activity, mating, hormone level) and the covert biological clock mechanism driving such a rhythm that is not directly observable except through observation of its mechanism (e.g., gene expression) or its functional properties. Since an observed rhythm may be driven, or modified directly, by an environmental cycle such as photoperiod or temperature changes, the demonstration and measurement of biological clock behavior requires that a rhythm be expressed in the absence of any exogenous periodicity (e.g., Figure 14.2). Only a "*free-running*" rhythm that persists in such aperiodic conditions (e.g., constant darkness, constant temperature) can be assumed to derive its period from an underlying biological clock, and the period itself is a direct measure of the state of the biological clock. The relationship of a biological clock to a rhythm is, therefore, analogous to the relationship between a gene and a trait.

Under constant conditions, biological clock cycles typically approximate, but do not exactly match the relevant external synchronizer – thus the name "circa"-dian (about a day), "circa"-annual, etc., but the endogenous periods typically only differ from their relevant exogenous synchronizers by a small fraction of a cycle. A typical *circadian* period, for instance, of an inbred laboratory mouse (*Mus musculus*) is approximately 23.5 hours (Pittendrigh and Daan, 1976a; Schwartz and Zimmerman, 1990). There is, however, significant variation in mean circadian period among different inbred strains, and individual mice, which typically vary symmetrically by about 30 minutes around the 23.5 hour modal period (see also Schwartz and Zimmerman, 1990).

A second functional property of biological clocks that can also be directly observed is "phase-response," which is their ability to shift the phase of the intrinsic oscillation to synchronize with the phase of an external cycle ("*Zeitgeber*"). The process of synchronization is typically called "*entrainment*," and it occurs incrementally until a steady state equilibrium is reached between the synchronizing phase shifts and the difference in period between the biological clock and the synchronizing oscillation. The sudden onset of jet-lag and its gradual resolution is a familiar example of this process. A clock that is entrainable by an exogenous cycle and then entrains, in turn, other internal oscillators is defined as a "*pacemaker*." The most well-understood circadian pacemaker in mammals is the light-sensitive *suprachiasmatic nucleus* (SCN) in the hypothalamus that drives the locomotor activity rhythm and numerous others as well (e.g., body temperature, sleep, metabolic functions: Dunlap et al., 2004). The SCN receives information about the photoperiod exclusively through retinal photoreceptors that include additional pigments beyond those used for visual image formation (Ruan et al., 2006).

The present review of behavioral genetics and mouse biological clock function will address genetic variation underlying circadian and seasonal behavioral rhythms. Analysis of circadian rhythms will be emphasized since they are the most prominent and widely studied rhythms in laboratory mice (e.g., rest–activity, temperature, mating, ingestive behavior). A glance at the literature on mouse behavioral genetics in the biological clocks field might give the impression that wheel running activity is the only meaningful behavior mice display. Wheel running activity is, in fact, an ideal "reporter" trait for assaying the state of the internal clock mechanism associated with the main circadian pacemaker in mammals, especially under aperiodic conditions where the phase and period of circadian rhythms must be set by the underlying clock mechanism. Laboratory mice typically run on wheels spontaneously during their waking hours, and it is not unusual for a mouse to generate 5000 wheel revolutions per day in the laboratory (personal observation). Since running wheel activity is spontaneous in laboratory mice and relatively easy to automate it is a sensitive and informative trait for analysis of circadian clock behavior under most experimental conditions (Siepka and Takahashi, 2005).

Although the ultimate adaptive function of endogenous biological clocks must be to time overt rhythms under entrained conditions, variation in any particular rhythm is of limited interest in proportion to the importance of timing that specific function, whereas variation in clock mechanism (e.g., period and phase-response) is of much greater intrinsic interest since it will translate into variation in all the rhythms controlled by that clock. The concept of a single "master" clock is an oversimplification as there are several anatomically and functionally distinct circadian clock mechanisms in mammals (e.g., see "Food entrainable rhythms" below), and numerous

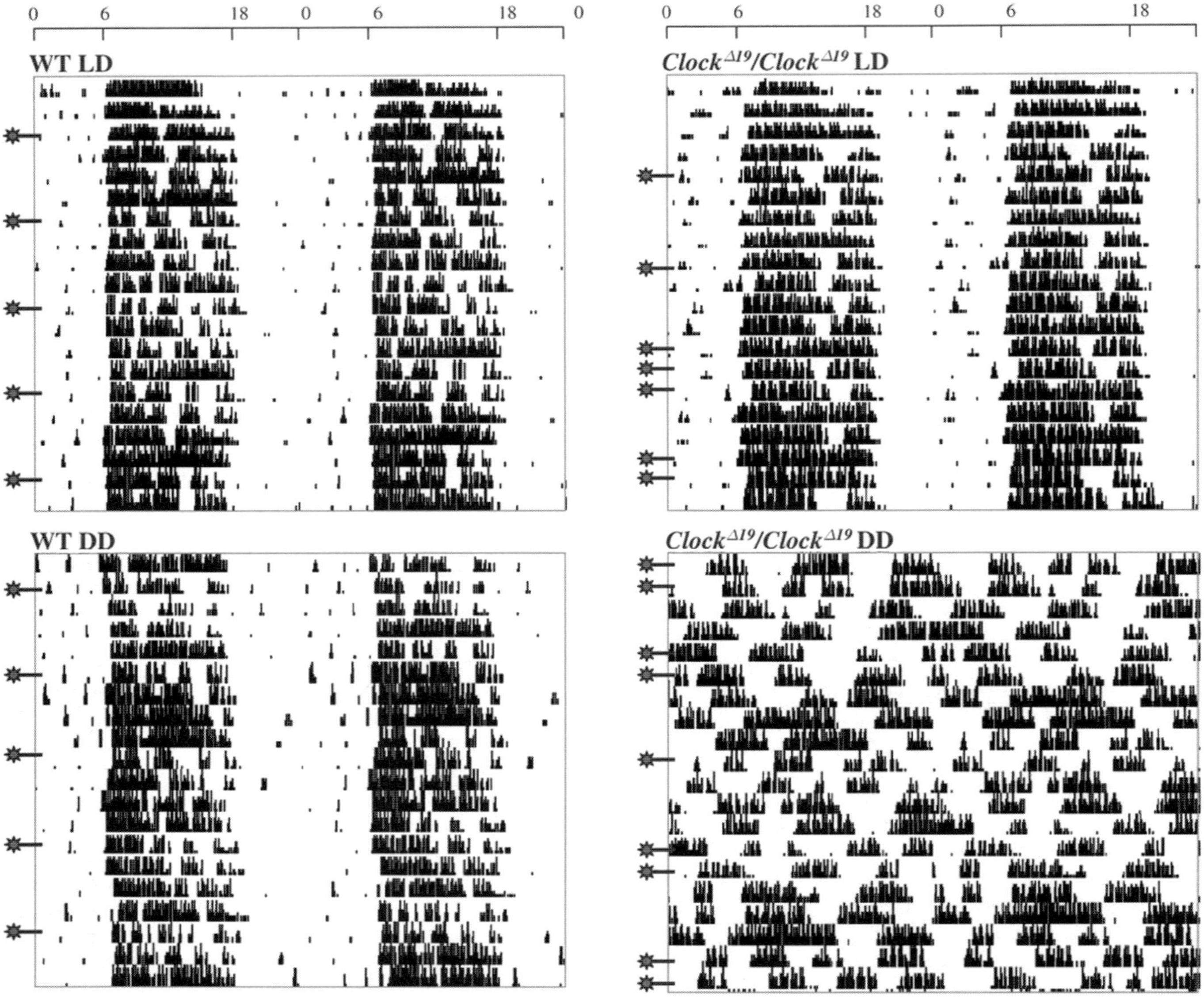

Figure 14.2 The top left panel (WT LD) shows nocturnal activity in a 12:12 light–dark (LD) cycle for a wildtype mouse that continues into constant dark (DD) (bottom left). The top right panel shows a mutant (Clock/Clock) mouse. The activity pattern displays a daily rhythm in the 12:12 LD cycle, driven by masking (suppression) of activity in the light phase. DD (bottom right) reveals an arrhythmic phenotype in the mutant mouse in the absence of the masking effect of the photoperiod. The wildtype mouse has an intact circadian clock, revealed in DD, that organizes activity into a circadian pattern regardless of the presence or absence of the photoperiod, although in the LD cycle it is also subject to the masking effect of bright light. (Reproduced with permission from Dolatshad et al., 2006.)

peripheral oscillators (e.g., liver, heart, gut, brain) that represent a network of coupled oscillators driving rhythms in response to several exogenous temporal cues to constitute a circadian "system" (see also Lowrey and Takahashi, 2004; Stratmann and Schibler, 2006). There is, currently, a strong model of light-entrained circadian clock function at the molecular level, but relatively little mechanistic understanding of other central clocks and the coupling of peripheral oscillators to central pacemakers.

Molecular clock mechanisms were dissected, primarily, by mutational analysis in *Drosophila* and mice and they display extraordinary homology at the molecular level among fruit fly, mouse, and human circadian oscillators (Sehgal, 2004). Quantitative genetic analysis of intact clock mechanisms can, however, complement mutational analysis by partitioning phenotypic variance and correlation for complex functional relationships among myriad circadian system components that are not well understood. This approach may also, through QTL analysis, identify new genetic loci underlying these complex clock mechanisms (e.g., food entrainable pacemakers, coordination among central pacemakers and peripheral oscillators and rhythms, development and plasticity of temporal organization, changes during aging and interaction with relevant exogenous cycles and conditions).

Photoperiodism and seasonal rhythms

Seasonal rhythms have received relatively little attention from behavioral geneticists, primarily because laboratory mice are not seasonal breeders and typically do not display prominent *seasonal rhythms* in the lab. Alternative mouse models, for instance *Peromyscus* species, are promising alternatives to *Mus* in this regard (see also Heidemann, 2004). Although there is evidence for endogenous "circannual" clock mechanisms that may play an important role in seasonal cycles of behavior (Dunlap et al., 2004), seasonal rhythms are currently best understood as a direct function of circadian mechanisms that monitor time of year by measuring the daily interval from sunrise to sunset and mediate the induction of seasonal events (Elliott, 1976; Pittendrigh, 1981b). General models for the induction of seasonal rhythms by specific photoperiods ("*photoperiod induction*") include "*external coincidence*" in which a seasonal response is induced when a particular phase of a circadian cycle is subjected to light, depending on the day length, and "*internal coincidence*" in which different circadian rhythms whose phases shift differentially with changes in day length achieve a critical phase relationship that induce a seasonal response. Physiological mechanisms mediating photoperiodic induction in mammals typically depend on the timing of a nocturnal peak for melatonin, which serves as an internal endocrine marker of the dark phase of the photoperiod (Goldman, 2001; Hazlerigg and Wagner, 2006).

Circadian regulation of sleep

Sleep is one of the most prominent and complex daily rhythms controlling behavior, and it is of particular relevance to basic biomedical research in psychiatry as clinical depression is typically associated with changes in sleep patterns (Goodwin and Jamison, 1990). Behavior genetic analysis of sleep may be useful in unraveling some of the mechanisms underlying its various phases, and elucidating some of its functions. Some common genetic loci have been identified between circadian clock function and sleep regulation in humans (Gottleib et al., 2007) and mice (Shimomura et al., 2001). The emergence of *Drosophila* as a new model organism for sleep research (Shaw and Franken, 2003) combined with mice as a mammalian model may provide some synergy leading to advances in sleep research from genetic analysis in both organisms, building on homologies between their circadian systems.

Behavioral genetic studies of biological rhythms in mice

Behavioral genetic approaches to mouse biological clock research began to identify naturally-occurring heritable variation in mammalian circadian rhythms among inbred strains in the 1970s (Ebihara and Tsuji, 1976; Ebihara et al., 1978; Pittendrigh and Daan, 1976a; Wax, 1977), and have continued to play a useful role in demonstrating heritable variation for most aspects of biological clock function. Recombinant inbred strain comparisons, and linkage to segregating markers in F2 offspring have served as the basis for QTL analysis of candidate loci altering circadian clock function, and as models for unusual phenotypes fixed in particular strains, such as splitting of rhythms into two peaks (Abe et al., 1999). Behavioral genetic studies of circadian rhythmicity have, to date, focused strongly on biological clock properties such as free-running period, and among traits, on sleep cycle characteristics. They have, however, been under-utilized for analysis of other circadian system components and for analyzing genetic and functional correlations among multiple circadian system variables. Quantitative trait loci studies of circadian function have become as common as first order inbred strain comparisons and suggest that there are many more interesting genetic loci relevant to circadian regulation to be discovered (Salathia et al., 2002). No circadian rhythm or circadian clock function has been subject to analysis by direct artificial selection in mice, although selection experiments have been carried out for reproductive response to short-day photoperiod in deer mice (Desjardins et al., 1986; Heideman and Bronson, 1991).

Circadian period for locomotor activity

Running wheel activity has been especially useful for characterizing genetic variation altering circadian clock period with a reasonable level of consistency among the same genetic strains in different studies and different laboratories (Hofstetter et al., 1999; Possidente and Stephan, 1988; Schwartz and Zimmerman, 1990). Some investigators have also used infrared beam disruption to monitor circadian activity rhythms (see also Hofstetter et al., 1995) with similar results to those of running wheels. Hofstetter et al. (1999) observed a significant correlation of 0.65 between strain means for the same set of 26 BXD recombinant inbred strains for circadian period measured in one experiment (Hofstetter et al., 1995) with infrared beam crossings, and another (Hofstetter at al., 1999) using running wheels. Some difference is expected, as mice and rats both have been observed to shorten their circadian periods for activity in running wheels compared to activity measured using alternative methods (see also Edgar et al., 1991; Yamada et al., 1988). Koteja et al. (2003) have observed a shorter circadian period in mice subjected to selection for high mean levels of locomotor activity in running wheels, consistent with the observation of a negative correlation between levels of running wheel activity and circadian period among individuals within strains (see also Shiori et al., 1991).

The combined assays of inbred mouse strain comparisons for circadian period cited above, plus those using recombinant inbred sets of strains (see also Hofstetter et al., 1995, 1999; Mayeda et al., 1996; Shimomura et al., 2001), show a range of mean circadian periods of approximately an hour, with the shortest just below 23 hours (typically including the Balb/cByJ strain) and the longest just below 24 hours (typically including the C57BL/6J strain). Schwartz and Zimmerman (1990) examined an approximate 50 minute period difference between Balb/cByJ and C57BL/6J mice further, and showed that neither differences in previous light exposure, the albino mutation, nor circulating levels of testosterone accounted for the strain effect on circadian period. A few studies (Possidente and Stephan, 1988; Schwartz and Zimmerman, 1990; Shimomura et al., 2001) assessed dominance effects in derived F1 strains with different results, depending on the parent strains. Similarly, Schwartz and Zimmerman (1990) reported a small but significant maternal effect (a difference of approximately 15 min) between BALB/cByJ and C57BL/6J strains, but Possidente and Stephan (1988) found no maternal effect in two different pairs of parental strains.

Overall, these studies indicate relatively small, but reasonably reliable, significant differences among inbred strains for circadian period, with variable dominance outcomes depending on the strains crossed, relatively little maternal influence, and polygenic inheritance. *Sex differences* are under-assessed, since many studies only include males (since breeding females are less available and male data are not affected by estrus cycles). No sex differences were reported in the studies cited here that included both sexes (Koteja et al., 2003; Possidente and Stephan, 1988; Shimomura et al., 2001). These results are consistent with a mix of polygenic factors altering circadian period from different loci in different inbred strains acting through different components of genetic variance and phenotypic pathways: in short, just what one would expect from a complex quantitative genetic trait.

Inbred strain comparisons, without derived crosses, allow estimates of narrow-sense heritability, and in the case of multiple traits and strains, genetic correlation (Blizard and Baily, 1979; Hegmann and Possidente, 1980). Heritability estimates for circadian period of wheel running are comparable across studies: 0.21 from Hofstetter et al. (1999); 0.51 from Suzuki et al. (2000); 0.55 from Hofstetter et al. (1995, using infrared monitors), and 0.43 from Possidente and Stephan (1988) indicating significant additive effects and potential for response to selection.

Quantitative trait loci analysis using either established sets of recombinant inbred lines and their progenitor parent strains, or panels of F2 individuals derived from any two inbred strains, have suggested many candidate loci, most of which map apart from known clock genes (Hofstetter et al., 1995, 1999, 2003b, 2007; Mayeda and Hofstetter, 1999; Mayeda et al., 1996; Shimomura et al., 2001; Suzuki et al., 2000, 2001). Many candidate loci are unique, but some overlap among different studies and are of greater interest for further analysis.

Shimomura et al. (2001) examined linkage between F2 markers and measures of period, phase, amplitude, mean, and coherence of the activity pattern in mice derived from C57BL/6J by BALB/c parent lines (Figure 14.1). They identified 14 mostly novel candidate QTLs affecting the various traits, including one with significant main effects on both phase and period. They also employed a method for detecting epistatic interactions among pairs of marker loci, and identified relatively small but significant interactions for at least two loci affecting each trait. Five of the QTLs identified by Shimomura et al. (2001) mapped near chromosomal regions previously identified in earlier studies, and five corresponded to regions identified as QTL sites in analysis of sleep cycles, including one QTL that was in both of these sets.

Hofstetter et al. (2003a) expanded on this method by examining linkage between circadian period measured by infrared beam crossing, and also by running wheels in the same mice, for an F2 derived from P1 parents representing both the CXB (C57BL/6 × BALB/c) and BXD (BALB/c × DBA/2) recombinant inbred lines. They identified five significant QTLs for the circadian period measured by beam crossing, and two based on wheel running activity. Four of these were novel, and three overlapped with QTLs identified in previous experiments, each explaining approximately 5–10% of the total phenotypic variance. Only one QTL (D13mit20) was found in common between circadian periods measured by beam crossing versus running wheels, suggesting that there is either significant genetic differentiation between the circadian pacemaker mechanism driving each type of activity rhythm, or that most of these QTLs are measuring relatively small but significant factors peripheral to the suprachiasmatic nuclei that may regulate the observed period; for example, food-entrained pacemaker input, age differences, non-photic feedback specific to wheel-running, or differences in sleep cycle dynamics. None of the QTLs identified in the studies by either Shimomura et al. (2001) or Hofstetter et al. (2003a) mapped close to any of the known clock genes.

Only three candidate QTLs from recombinant inbred line and F2 segregation linkage mapping have been independently confirmed and mapped more precisely using alternative methods. Mayeda and Hofstetter (1999) used a congenic strain to confirm that the distal region of chromosome 1 carries an allelic difference between the BALB/c and DBA/2 inbred strains with a significant effect on circadian period. Kerneck et al. (2004, 2006) used congenic strains to show that allelic differences at the *carbonic anhydrase II* locus on chromosome three alter the circadian period by approximately 0.2 hours against the C57BL/6J background. Hofstetter et al. (2007) also used congenic strains to map a QTL-altering circadian period, and distribution of activity during the dark phase of the photoperiod, on chromosome 12 to a region spanning only 15 known genes.

Although QTL analysis is relatively new and it is still difficult to translate its outcomes into more precise information, results to date are intriguing in demonstrating quantitative genetic variation for an interesting variety of circadian system variables that map apart from known clock loci. These initial studies are promising since the results of QTL analyses are cumulative across studies, and methods for more precise measurement and follow-up analysis are constantly improving.

Phase response

Circadian clocks entrain to the photoperiod by adjusting their phase in response to the transitions from dark to light at dawn, and light to dark at dusk (Pittendrigh, 1981b). These signals are mimicked, in the laboratory, by discrete light pulses, typically from 10 to 30 minutes in duration, against a background of constant darkness. Following the light pulse, phase shifts in the rhythm from the single pulse are observed in subsequent cycles in constant darkness. A "*phase response curve*" represents the differential response to the light pulse in the magnitude of phase shifts across the circadian cycle that allows the clock to find an equilibrium condition to accomplish entrainment. This is a key property of the clock itself, and variation in phase response alters the dynamics of all the rhythms under the clock's control. Schwartz and Zimmerman (1990) published the only inbred strain comparison, to date, of a complete phase response curve. The curves were qualitatively similar, but the Balb/cByJ strain was less responsive in the area of the curve generating phase advances. Freedman et al. (1999) demonstrated that mice lacking both rod and cone photoreceptors showed phase shifts in response to light that were indistinguishable from those of wildtype mice, demonstrating that these photoreceptors are not required for circadian phase response. Yoshimura et al. (2002) compared two inbred strains, both fixed for the rd/rd retinal degeneration mutation, for a single circadian time point (ct16). They observed that the C57BL/6J rd/rd strain generated a significantly larger phase response than CBA/J rd/rd, consistent with evidence that circadian phase response does not depend on visual system photoreceptors and indicating background strain differences that significantly alter the phase response magnitude. These effects may be mediated through alleles that alter clock function, differences in sensitivity of the clock to light, or the strength of signal transduction of light perception to the clock through the retina (Ruan et al., 2006). Yoshimura et al. (2002) identified five QTLs on three chromosomes, with several mapping near previously identified QTLs for circadian variables.

Aschoff's rule

Nocturnal mammalian species typically increase circadian period under exposure to constant light, and diurnal species more often shorten their period. This pattern, known as "*Aschoff's rule*," was given a functional explanation by Daan and Pittendrigh (1976b) based on corresponding differences in phase response curve capacity, and therefore genetic variation altering the magnitude of period change may represent a simple assay of important underlying clock properties for entrainment. Possidente and Hegmann (1982) demonstrated significant differences among seven inbred strains in the effect of constant light on the circadian period. Hofstetter et al. (1995) demonstrated comparable differences among 26 recombinant inbred BxD strains and their progenitor strains with heritability estimated at 0.38, and also among the CxB recombinant inbred strain set (Hofstetter and Mayeda, 1998). They identified several QTLs, including significant markers in both studies on chromosomes 8 and 11. These results suggest significant variation in the underlying phase response ability of different inbred strains.

Phase of circadian locomotor activity rhythms

Mouse activity rhythms typically have a sharp onset, and are also amenable to best-fit curve analysis to estimate time of peak activity. Both of these measures provide estimates of the *phase* (position in time) of circadian activity rhythms. Shimomura et al. (2001) describe a 3 hour phase difference between the C57BL/6J and BALB/cByJ inbred strains with dominance of the earlier Balb phenotype in F1 offspring. Suzuki at al. (2000) define a similar 3 hour difference in activity onset relative to lights-on between the inbred strains SM/J and A/J, and an approximate 4 hour range among a set of derived recombinant inbred strains with a heritability of 0.42. Both groups identified significant QTLs associated with the strain difference in phase. Wisor et al. (2007) identified a QTL from an F2 cross derived from the C57BL/6 and Cast/Ei inbred strains associated with an extreme phase advance in the time of activity onset. The same QTL was also associated with diminished phase delay responses to light pulses.

Amplitude of circadian locomotor rhythms

Several studies have compared the pattern of entrained locomotor activity in a light–dark cycle among inbred strains. *Amplitude* is commonly defined as the difference in activity between the peak level and the mean level, but here it refers to any variation in the waveform of the circadian activity pattern. Beau (1991) and Toth and Williams (1999a) describe differences between the C57BL/6J and BALB/cByJ inbred strains and derived recombinant inbred strains, including the ratio of activity in the light to the dark phases, and Shimomura et al. (2001) compare these same inbred strains and their F1 and F2 generations for amplitude, including a measure of fragmentation of activity. All three studies find significant strain differences and evidence of multifactorial inheritance for the differences, and identify candidate QTLs. Shimomura et al. (2001) note that for this trait, their largest QTL effect was an epistatic interaction. Sans-Fuentes et al. (2005) describe an association between Robertsonian translocations among mice in a polymorphic zone on the Iberian Peninsula and the waveform of the activity pattern, but with no underlying difference in circadian period, suggesting that entrained temporal patterns of activity may exert selection pressure on these chromosomal polymorphisms.

Sleep

Sleep is one of the least understood behaviors with respect to both function and mechanism, but one of the most important in regulating daily patterns of behavior and maintaining health. Investigating circadian regulation of sleep, particularly through genetic methods, may be helpful in providing new insights into basic questions about its biological foundations and significance. Recent reviews provide excellent overviews of genetic analyses of sleep in mice and other model systems (O'Hara et al., 2007; Pace-Schott and Hobson, 2001; Shaw and Franken,

2003; Tafti and Franken, 2002; Toth, 2001). There are a relatively large number of behavioral genetic studies of sleep behavior, despite the more complex technical demands compared to assaying locomotor activity. One recent development that may prove valuable for mouse sleep research is the development of *Drosophila* models (Shaw and Franken, 2003). Translating from *Drosophila* to mice could be useful in identifying circadian and other mechanisms regulating sleep if there is homology comparable to that for circadian clock function at the molecular level.

Friedmann (1974) provided one of the earliest indications of significant genetic variance for sleep behavior among inbred mouse strains using diallel analysis. A series of more recent recombinant inbred strain analyses have extended these findings to identify a number of candidate QTLs for different components of the sleep cycle, including *short wave sleep*, *paradoxical sleep*, wakefulness, and total sleep (Franken et al., 1999; Tafti et al., 1999; Toth and Williams, 1999b) and for short wave sleep after infection with influenza virus (Toth and Williams, 1999c). Toth and Verhulst (2003), and Toth et al. (1995) demonstrated significant differences in sleep among a panel of influenza-infected inbred strains of mice, which is of particular interest with respect to understanding immune system modulation of sleep. Additional genetic variation has been reported for myriad aspects of sleep behavior, including influence of contextual fear (Sanford et al., 2003), early development of strain differences spanning adolescence (Daszuta et al., 1983), sleep response to physical restraint (Meerlo et al., 2001), effects of environmental novelty (Tang et al., 2005), and effects of ambient temperature (Roussel et al., 1984). New methods have been developed and validated, using inbred strain comparisons, to facilitate more extensive genetic studies of sleep behavior (Veasey et al., 2000, 2004). Apart from analysis of generic circadian clock mechanisms, sleep behavior is the most active area of mouse behavioral genetic research on temporal regulation of behavior.

Shimomura et al. (2001) identified QTL factors for circadian rhythm characteristics in mice that mapped in some of the same regions as the QTLs reported in these studies of sleep behavior, indicating that the circadian mechanism regulating locomotor behavior also regulates sleep cycles. A similar overlap between QTLs for sleep and circadian function was reported for human subjects as well (Gottlieb et al., 2007), which should make mouse models of sleep regulation more valuable for understanding human sleep function. Koehl et al. (2003) reported inbred strain differences for sleep behavior in female mice, and concluded that genetic factors played a much larger role in contributing to strain differences than influences of the estrous cycle.

Food entrainable rhythms

Mice and other mammals can entrain wheel-running activity to a 24 hour restricted feeding schedule, even in the absence of a functional light-entrainable clock (Stephan, 2002), revealing a circadian system that entrains to periodic feeding rather than photoperiod based on a clock mechanism that is anatomically distinct from the suprachiasmatic nuclei. *Feeding-entrained circadian activity* rhythms in CLOCK mice with genetically lesioned light-entrained circadian pacemakers also display entrainment to periodic feeding cycles (Pitts et al., 2003). Behavior genetic analysis of this circadian behavior is almost non-existent, other than the description of feeding entrained circadian rhythmicity in individual inbred strains (Holmes and Mistlberger, 2000; Marchant and Mistlberger, 1997) and one comparison between CS and C57BL/6J mice (Abe et al., 1989). Intact mice are typically capable of displaying simultaneous activity rhythms with different periods, one entrained by *feeding cycles* and the other either entrained by a photoperiod or free-running in constant dark. Abe et al. (1989) showed that intact B6 mice sustain both rhythms independently under *food restriction* in constant dark, but CS mice only display a single activity rhythm entrained to periodic feeding. This observation suggests that the coupling between activity and the two underlying circadian systems is altered by differences between these strains, and suggests that further investigation of feeding entrained circadian behavior in mice would be productive for identifying additional genetic variation. There is a growing body of research on food-entrained rhythms employing lesioning methods and genetic dissection with known clock mutants, but little general agreement on the mechanisms mediating feeding-entrained circadian behavior (Mendoza, 2007). Feeding-entrained circadian behavior is well established in rodent models, so behavior genetic analysis using inbred strain variation and QTL linkage analysis might be particularly useful for generating a clearer understanding of its functional and mechanistic properties. One caveat to note is a dependency of food-entrainable wheel-running activity in C57BL/6J mice on housing conditions where mice in cages isolated from other subjects did not display the behavior (de Groot and Rusak, 2004).

Methamphetamine-sensitive circadian oscillator

Methamphetamine administration can significantly alter the expression of circadian rhythms in intact rodents, and, like feeding cycles, can induce circadian organization even in suprachiasmatic nucleus-lesioned animals without a functional light-entrainable circadian pacemaker (Honma et al., 1987). Tataroğlu et al. (2006) have shown that methamphetamine induces circadian behavior in arrhythmic SCN-lesioned mice, and that methamphetamine-sensitive circadian behavior differs significantly between C57BL/6J and C3H inbred strains. Non-photic circadian systems, such as this one and the feeding-entrained system, are ripe for further behavior genetic analysis as little is known about their underlying mechanisms and the dynamics of how apparently independent circadian pacemakers sensitive to different cues are organized in intact animals.

Development and aging

Circadian rhythms present an example of traits that change with age but may or may not play a causal role in the *aging* process. Inbred strains of mice developed as a model for accelerated aging resynchronize their circadian activity rhythm faster than strains that age more slowly (Sanchez-Barceló et al., 1997), and circadian period was longer in older male mice in a strain-dependent manner (Mayeda et al., 1997), but not in females measured longitudinally (Kopp et al., 2005). Recent proposals link circadian mechanisms to caloric restriction attenuation of aging (Froy and Miskin, 2007) and to non-circadian roles in the aging process for circadian clock proteins (Kondratov, 2007).

Development of circadian clock function appears to be relatively impervious to environmental manipulations of photoperiod (Davis and Menaker, 1981), but there are few studies on the development of genetically associated differences in circadian rhythms or clock function. One such study compares the development of sleep cycle differences between BALB/c and C57BL mice (Daszuta et al., 1983) from three to 14 weeks of age. Behavioral genetic studies of development and aging may be useful regardless of whether or not circadian changes over time are causes or consequences of aging as long as temporal organization has functional significance.

CS inbred mice: an anomalous strain with rhythm splitting and long period

The CS inbred strain of mice is an outlier, primarily because it displays spontaneous splitting of the circadian activity rhythm in constant darkness into two components (Abe et al., 1999), but it is also unusual in many other respect such as high mean activity level and a circadian period longer than 24 hours (Suzuki et al., 2001), and unusual sleep cycle dynamics compared to more typical inbred strains (Ebihara et al., 2003). Analysis of clock gene expression in the suprachiasmatic nuclei compared to other brain regions (Abe et al., 2001) and peripheral organs (Watanabe et al., 2006) indicate that the unusual circadian profile of CS mice is associated with mechanisms outside the suprachiasmatic nuclei. Suzuki et al. (2001) mapped several novel period-lengthening candidate QTLs in F2 mice derived from crosses of CS mice with C57BL/6J and MSM inbred mice.

Body temperature and thermoregulation

Circadian regulation of *body temperature* is of interest as a potential driver of circadian rhythms in temperature-dependent behavior, including thermoregulatory behaviors, and as a measure of general health. There are few studies of genetic differences altering circadian body temperature cycles in mice (see also Connolly and Lynch, 1981), but there are some interesting analyses of responses of temperature rhythms to artificial selection on related traits (Bult et al., 1993; Castillo et al., 2005; Mousel et al., 2001). Connolly and Lynch (1981) observed strain differences in the daily rhythm of body temperature between C57BL/6J and C3H/21bg inbred mice in a light–dark cycle, no significant dominance in their F1 offspring, and no strain difference for mean body temperature, indicating genetic effects on timing independent of mean level. Bult et al. (1993) observed that mice selected for extremes in thermoregulatory nest size produced a small nesting line with a higher amplitude of wheel-running activity, a later peak of activity and a significantly altered phase response curve to light pulses, indicating that correlated responses to selection for nest size included significant changes in circadian clock function and circadian timing of behaviors. Castillo et al. (2005) also reported higher amplitude of the body temperature rhythm in the small-nest size lines.

Mousel et al. (2001) demonstrated differences in waveform of daily rhythms for mean activity and body temperature in mouse lines artificially selected for high and low heat loss, with the high-loss lines exhibiting higher amplitudes and mean levels in a light–dark cycle. Roussel et al. (1984) assayed C57BL/6J and BALB/c mice for sleep cycle differences at high and low temperatures, and observed similar patterns of change in slow-wave and paradoxical sleep, but with different amplitudes, suggesting that temperature changes alter sleep dynamics differentially in these strains.

Ingestive behaviors: food and water consumption

Daily rhythms for *feeding* (Kowal et al., 2002) and *drinking* (Saito et al., 1980) have been compared among inbred strains in light–dark cycles and in constant light (Possidente and Hegmann, 1980). Kowal et al. (2002) demonstrated a dependency of strain differences on age from 30 to more than 120 days, and Possidente and Hegmann (1980) showed that daily timing of food and water consumption is highly correlated, both in the light–dark cycle and in constant light.

Ethanol metabolism

Circadian rhythms interact with metabolic and behavioral effects of *alcohol* (Rosenwasser, 2001), but relatively little research has been focused on genetic variation meditating these effects. Hofstetter et al. (2003a) identified a shorter circadian period in mice selected for high alcohol consumption, compared to the low-consumption line, suggesting a correlated response to selection, or, alternatively, genetic drift based on genetic variation inherent in the base population for the selection experiment. Gilliam and Collins (1983) examined daily rhythms for alcohol metabolism and sleep time in selected lines of mice for short versus long sleep times in response to an injection of ethanol. The long sleep line displayed daily rhythms for sleep time and blood and brain levels of alcohol, while short-sleep line mice, given two different doses of alcohol by injection, showed no daily rhythm for the higher dose but displayed daily rhythms for blood and brain alcohol levels at a lower dose. These studies are consistent with a significant level of genetic correlation between behaviors related to alcohol consumption and metabolism and circadian organization. Further comparisons using additional approaches, such as inbred strain comparisons could test these associations further.

Photoperiodism and seasonal rhythms

Seasonal changes in behavior and physiology are thought to be induced by changes in day length across seasons that interact with circadian mechanisms tracking day length to trigger seasonal events when critical phase relationships occur (Elliott, 1976; Pittendrigh, 1981a,b). Genetic variation in circadian mechanism, therefore, is potentially correlated with variation in *seasonal rhythms*. Since *Mus musculus* breed year-round in the laboratory regardless of photoperiod, alternative mouse species have been used to investigate genetic factors influencing seasonal breeding as a model for photoperiodic induction. Desjardins et al. (1986) observed individual differences among deer mice (*Peromyscus maniculatus*), and were able to significantly alter the frequency of responders who stopped breeding in short day photoperiods in just two generations of artificial selection. Heideman and Bronson (1991) achieved a similar rapid response with *Peromyscus leucopus*, and observed a significantly longer circadian period in constant dark in the non-responsive mice. Non-responders to short days, however, also did not respond to constant darkness by delaying puberty, suggesting that their lack of response to short days may not be a function of circadian measurement of critical day length, but rather an alteration downstream in the neuroendocrine system that overrides circadian clock mediated regulation of breeding (Majoy and Heideman, 2000). Avigdor et al. (2005) report approximately 50% more neurons labeled by antibodies to mature gonadotropin-releasing hormone in the hypothalamus of non-responsive strain mice compared to the responsive strain. These results suggest that non-circadian mechanisms may mediate response to selection for short-day inhibition of breeding, possibly because altering central circadian mechanisms would have widespread correlated effects on many other biological clock functions, whereas neuroendocrine mechanisms essential for regulating seasonal breeding might be more specific (Majoy and Heideman, 2000). A similar conclusion was suggested by Carlson et al. (1989) who compared circadian organization between short-day responsive *Peromyscus leucopus* from Connecticut to a non-responsive population further south from Georgia in the eastern USA. They found no differences in circadian period or phase of activity, but pinealectomy abolished the short-day response in the Connecticut mice. These results suggest that genetic variation altering reproductive mechanisms downstream from a central clock mechanism may mediate variation in seasonal breeding in these natural populations. Vitaterna and Turek (1993) drew a similar conclusion from comparisons among inbred strains of photoperiodic hamsters. One possibility is that such downstream mechanisms represent peripheral circadian oscillators that depend on internal coincidence of multiple circadian cycles achieved under specific day lengths, rather than direct external coincidence of circadian cycles with light stimuli (Pittendrigh, 1981a). Daily rhythms in melatonin may play a key role in mediating seasonal responses to changes in photoperiod since nocturnal melatonin rhythms typically track the dark phase of the light–dark cycle (Goldman, 2001). Puchalski and Lynch (1986), for example, showed that responders and non-responders for ceasing reproduction in chronic short-day photoperiods in a population of hamsters displayed differences in both activity and melatonin rhythms. The activity and melatonin rhythms for responders tracked differences in day length between long and short days, while the rhythms of non-responders failed to conform to the short day length. This observation suggests that seasonal changes in circadian organization are coupled with seasonal breeding rhythms.

Notes on methods

Illumination levels

Comparative studies of biological clock function are rarely standardized for equivalency of light sensitivity, making it difficult to know how relevant the results of an experiment in one species might be to another. Bullough et al. (2006) have provided a context for calculating equivalent absolute and spectral sensitivities of the circadian systems of humans and nocturnal rodents. Calibrating, standardizing, and noting such information will improve the design of circadian experiments in animal models, and also facilitate comparative analysis across species, as well as the accurate assessment of circadian and genetic variation in sensitivity of circadian systems to photic stimuli. Many researchers in the biological clocks field maintain a constant, monochromatic dim red light background during circadian rhythm experiments, in order to facilitate observation and care of the animals in the dark, under the assumption that the circadian system of rodents is not responsive to dim, pure red light. Hofstetter et al. (2005), however, have shown that the circadian period of mice is altered by dim red light illumination levels in the range of 1 lux or greater. Dim red light less than 1 lux is still adequate to facilitate animal care without altering experimental outcomes relative to complete darkness.

Reverse photoperiods

Some researchers maintain their nocturnal rodent colonies on a reverse photoperiod so that the active phase of the mice coincides with that of the research staff. Beeler et al. (2006) have shown that "low-amplitude entrainment", consisting of a photoperiod of alternating bright light and dim red light provides an adequate stimulus for stable entrainment, facilitating behavioral testing during the subject's dark phase, whether it is reversed or not relative to the ambient photoperiod. While Hofstetter et al. (2005) showed that dim red light can alter free-running period, it is less problematic for analysis of daily rhythms in overt traits under stable entrainment. Hossain et al. (2004) examined mice from three inbred strains for a variety of behavioral tests at different times of day and night and concluded that dark phase testing can improve strain discrimination on some measures, light phase is more discriminating for others, and some are neutral. Low-amplitude entrainment combined with photoperiod reversal may facilitate behavioral genetic analysis of circadian behaviors, other than free-running circadian period.

Temporal responses to artificial selection

The design and interpretation of *artificial selection* experiments on behaviors should take circadian variation in activity into account in two ways. The most obvious is that any selected trait that is measured at one time of day may respond to selection for altered mean level simply by shifting the phase of the rhythm to a higher or lower amplitude depending on the direction of selection. A second factor is the possibility that mean levels of a trait depend on a particular state of underlying circadian organization such that altering circadian organization may facilitate a response to selection even if selection is not applied at a specific time in the circadian cycle. These factors can be taken into account by measuring circadian rhythmicity of the selected trait in the base population before selection is implemented, and by measuring circadian rhythmicity of the same trait after selection, including free-running period to assess any changes in the underlying circadian clock. More generally, temporal organization of behavior may facilitate or constrain behavioral adaptation in this manner through genetic correlation. Assays of circadian regulation of the mouse transcriptome indicates that a substantial percentage of transcripts are clock-controlled (see also McCarthy et al., 2007) and that many are clock-controlled in a tissue-specific manner (Miller et al., 2007), suggesting that response to selection for complex traits, in general, is likely to be mediated, at least in part, through allelic variation in clock-controlled genes.

Cage enrichment

Evidence that mice raised in standard laboratory cages may be developmentally and behaviorally impaired, and increasing attention to laboratory animal welfare, has increased the use of *cage enrichment* methods (see also Wolfer et al., 2004). There is the possibility that results of circadian behavioral assays on mice from standard cage environments may differ from those obtained from mice reared under enriched conditions.

Reebs and Maillet (2003) observed that phase and amplitude of hamster locomotor activity was slightly modified by cage enrichment consisting of wooden objects and tunnels linking multiple cages, but free-running circadian period was not. No such studies have been published, to date, assessing effects of cage enrichment on mouse circadian behavior assayed by running wheel activity. Providing a running wheel can, itself, be a form of cage enrichment (Pang et al., 2006; Pietropaolo et al., 2006) suggesting that it is important to control for access to running wheels as a potential treatment effect in experiments using them to assay circadian variables.

Quantitative genetic background in transgenic and congenic lines

The evaluation of single-locus allelic variation, whether it represents a spontaneous mutation, a knockout, knock-in, or some other genetically engineered mutation, will always be conditionally dependent on the *genetic background* of the host organism. Ultimately, a complete model of genetic mediation of biological clock function will need to take both genomic and quantitative genetic measures into account. Recent reviews have begun to emphasize these considerations and recommend more systematic, standardized experimental designs and attention to genetic background effects (see also Van Gelder and Hogenesch, 2004). Pugh et al. (2004) compared, for example, the FVB/N mouse strain that is widely used for generating transgenic mice, to the C57BL/6J inbred strain for a variety of behaviors, including circadian activity and observed that the FVB/N strain displays a significant amount of activity in the light phase of the light–dark cycle, and fragmented activity patterns compared to C57BL/6J mice. Just the FVB/N strain background, therefore, can have a significant confounding effect on the evaluation of transgenic manipulations involving circadian clock function. Schalkwyk et al. (2007) emphasizes the confounding potential of the "*congenic footprint*" when backcrossing transgenic mutations or other loci into a standard genetic background. Bae et al. (2001) report that mice homozygous for a *mPer2* clock gene mutation on an SV129 inbred strain background display shorter circadian periods followed by loss of rhythmicity in DD, compared to Xu et al. (2007a) who report that the *mPer2* genotype against a B6 background is similar to wildtype controls in period length and persists in DD. These complications are manageable, but need to be taken into account in the interpretation of data from experiments where they are potential factors.

Summary of QTL data

Seventeen studies cited above include over 200 QTLs for more than a dozen circadian and sleep phenotypes. They are based on screens of the BxD, CxB, and SMXA recombinant inbred line panels, segregating F2 generations derived from seven different inbred parental strains, and selected congenic strains. These studies represent only a small fraction of potential allelic variation for biological rhythm function, and have been conducted over a time when QTLs and their confirmational methods have only begun to be developed and standardized. Although the potential reservoir of quantitative genetic variation altering biological rhythm phenotypes and the array of relevant phenotypes of interest is almost unlimited, eventually the cumulative data at both genetic and phenotypic levels will saturate the degrees of freedom for meaningful variation in the genome and patterns of behavior. Quantitative trait loci analyses will, at that point, most likely include genome sequencing, nanotechnology, and three-dimensional imaging for behavioral monitoring systems and chronoinformatic computational tools. At the

Table 14.1 Chromosome 1 listing of QTL for biological rhythms.

Map position (cM)	Marker name	Trait	Method	Reference
7–14	D1Nds4	tauDDb[a]	BXD	Hofstetter et al., 1995
		tauDDw[b]	BXD	Hofstetter et al., 1999
20	D1Mit122	tauDDw	BXD	Hofstetter et al., 1999
37	Ugt1al	tauDDb	BXD	Hofstetter et al., 1995
43	D1Mit132	SWS[c]	CXB	Toth and Williams, 1999b
48–77[d]	–	PS[e]	BXD	Tafti et al., 1999
52	Ugt1a1	tauDDw	BXD	Hofstetter et al., 1999
71	Fcgr2	tauDDb	BXD	Hofstetter et al., 1995
81	D1mit33	Amplitude	F2:CXB	Shimomura et al., 2001
87	D1mit150	tauDDb	F2:CXBxBXD	Hofstetter et al., 2003a,b
92	cgr2b/Apoa2	tauDDw	BXD	Hofstetter et al., 1999
93	Apoa-2	tauDDb	CXB	Mayeda et al., 1996
94	Sap	tauDDw	BXD	Hofstetter et al., 1999
		tauDDb	CXB	Mayeda et al., 1996
89–106	Cplaq3	tauDDb	Congenic	Mayeda and Hofstetter, 1999

[a] tauDDb describes circadian period in constant dark for activity measured by infrared-beam crossings.
[b] tauDDw describes circadian period in constant dark for activity measured by wheel running.
[c] SWS describes time spent in short-wave sleep.
[d] No marker name listed.
[e] PS describes time spent in paradoxical sleep.

present stage, variation in methods, inherent imprecision in QTL mapping, provisional status of almost all QTLs identified to date, and new data in the pipeline make it difficult to generalize beyond the descriptive level with respect to QTLs for biological rhythms. The fact that there are so many apparently non-overlapping QTLs for the traits analyzed so far should not be surprising given the complexity of biological clock mechanisms that regulate, directly or indirectly, most biological functions. Quantitative trait loci analysis is, however, a useful and promising tool for behavioral genetic research because it permits functional variation in complex traits to be readily mapped to genetic loci across the genome. Given that we know so little about the pathways mediating causal connections between DNA sequence variation and behavioral variation, the potential information that can be gained from QTL analysis of complex traits is enormous. Chronobiologists already think of temporal regulation in terms of "systems" of interacting components, and QTL methods of analysis are ideal for genetic analysis at the systems level. The QTLs cited above for various inbred strain lineages, biological rhythm traits, and experiments are summarized (see Tables 14.1–14.19) by published chromosomal map location.

Table 14.2 Chromosome 2 listing of QTL for biological rhythms.

Map position (cM)	Marker name	Trait	Method	Reference
6	Il2ra	tauDDw[a]	BXD	Hofstetter et al., 1999
45	D2MDS1	tauDDb[b]	BXD	Hofstetter et al., 1995
45	D2Mit329	SWS h 0–6[c]	CXB	Toth and Williams, 1999c
46	Ea 6	tauDDw	CXB	Hofstetter et al., 1995
69	D2Mit423	REMS:SWS ratio[d,e]	CXB	Toth and Williams, 1999b
65–74[f]	–	PS[g]	CXB	Tafti et al., 1999
71	Src	tauDDb	BXD	Hofstetter et al., 1995
83	D2Mc1	tauDDb	BXD	Hofstetter et al., 1995
87	Odc-rs2	tauDDw	BXD	Hofstetter et al., 1999
105	D2Mit265	SWS[e]	CXB	Toth and Williams, 1999b

[a] tauDDw describes circadian period in constant dark for activity measured by wheel running.
[b] tauDDb describes circadian period in constant dark for activity measured by infrared-beam crossings.
[c] SWS h 0–6 induced by influenza infection during hours 0–6 of photoperiod.
[d] REMS describes time spent in rapid-eye-movement sleep.
[e] SWS describes time spent in short-wave sleep.
[f] No marker name listed.
[g] PS describes time spent in paradoxical sleep.

Table 14.3 Chromosome 3 listing of QTL for biological rhythms.

Map position (cM)	Marker name	Trait	Method	Reference
0–3	car2	tauDDb[a]	Congenic	Kernek et al., 2006
0–5	car2	tadDDb	Congenic	Kernek et al., 2006
4	Car-2	tauDDb	BXD	Hofstetter et al., 1995
14	vil	tauDDw[b]	BXD	Hoffstetter et al., 1999
28	D3Mit120	SWS h 12–24[c]	CXB	Toth and Williams, 1999c
44	D3Mit28	REMS:SWS dark ratio[d]	CXB	Toth and Williams, 1999b
64	H23	tauDDb	CXB	Mayeda et al., 1996
65	Amy1	tauDDw	CXB	Hofstetter et al., 1995
67	Gbp1	tauDDb	CXB	Mayeda et al., 1996
78	lapls2–14	tauLLb-tauDDb[e]	BXD	Hofstetter and Mayeda, 1998

[a] tauDDb describes circadian period in constant dark for activity measured by infrared-beam crossings.
[b] tauDDw describes circadian period in constant dark for activity measured by wheel running.
[c] SWS h 12–24 describes time spent in short-wave sleep induced by influenza infection during hours 12–24 of the photoperiod.
[d] REMS:SWS dark ratio describes the ratio of time spent in rapid-eye-movement sleep to short-wave sleep in the dark phase of the light–dark cycle.
[e] Circadian period in constant light, measured by infrared-beam crossings, minus tauDDb.

Table 14.4 Chromosome 4 listing of QTL for biological rhythms.

Map position (cM)	Marker name	Trait	Method	Reference
6	Mos	tauDDw[a]	CXB	Hofstetter et al., 1995
		tauLLb[b]	BXD	Hofstetter et al., 1995
24	Pmv-30	tauDDb[c]	BXD	Hofstetter et al., 1995
29	D4Mit288	REMS[d]	CXB	Toth and Williams, 1999b
36	4Mit178	tauDDw	F2:CXB	Shimomura et al., 2001
43	4Mit27	Amplitude	F2:CXB	Shimomura et al., 2001
48–55[e]	–	TST[f], SWS[g]	CXB	Tafti et al., 1999
55	Iapls3-10	tauDDw	BXD	Hofstetter et al., 1999
		tauLLb-tauDDb	BXD	Hofstetter and Mayeda, 1998
71	D4Mit13	REMS	CXB	Toth and Williams, 1999b

[a] tauDDw describes circadian period in constant dark for activity measured by wheel running.
[b] tauLLb describes period in constant light (infrared-beam crossings).
[c] tauDDb describes circadian period in constant dark for activity measured by infrared-beam crossings.
[d] REMS describes time spent in rapid-eye-movement sleep.
[e] No marker name listed.
[f] TST describes time spent in total sleep time.
[g] SWS describes time spent in short-wave sleep.

Table 14.5 Chromosome 5 listing of QTL for biological rhythms.

Map position (cM)	Marker name	Trait	Method	Reference
10	D5Ncvs52	tauLLb[a]	BXD	Hofstetter et al., 1995
26	Mpmv-7	tauDDb[b]	CXB	Hofstetter et al., 1995
40	D5Mit7	tauLLb	BXD	Hofstetter et al., 1995
53	Pmv-12	tauLLb	BXD	Hofstetter et al., 1995
28–53[c]	–	PS, TST, SWS[d]	CXB	Tafti et al., 1999
46–60	D5Mit24	SWS	CXB	Toth and Williams, 1999b
64	D5Mit188	tauDDw[e]	F2:CXB	Shimomura et al., 2001
85	D5Rik77	tauLLb-tauDDb	BXD	Hofstetter and Mayeda, 1998

[a] tauLLb describes circadian period in constant light (infrared-beam crossings).
[b] tauDDb describes circadian period in constant dark for activity measured by infrared-beam crossings.
[c] No marker name listed.
[d] PS, TST, and SWS describe time spent in paradoxical sleep, total sleep and short-wave sleep, respectively.
[e] tauDDw describes circadian period in constant dark for activity measured by wheel running.

Table 14.6 Chromosome 6 listing of QTL for biological rhythms.

Map position (cM)	Marker name	Trait	Method	Reference
0	D6Mit86	tauDDw[a]	BXD	Hofstetter et al., 1999
18	D6Mit275	tauDDb[b]	F2:(BXD)X(BXC)	Hofstetter et al., 2003
26	D6Mit184	SWS h 3–6[c]	CXB	Toth and Williams, 1999c
28	D6Mit316	SWS[d]	CXB	Toth and Williams, 1999c
		SWS h 0–12[e]		
48–53	D6Bir2	tauLLb[f]-tauDDb	BXD	Hofstetter and Mayeda, 1998
		tauLLb		Hofstetter et al., 1995
65	Ea-10	tauDDb	BXD	Hofstetter et al., 1995

[a] tauDDw describes circadian period in constant dark for activity measured by wheel running.
[b] tauDDb describes circadian period in constant dark for activity measured by infrared-beam crossings.
[c] SWS h 3–6 describes SWS, caused by influenza inoculation, during hours 3–6 of photoperiod.
[d] SWS describes time spent in short wave sleep.
[e] SWS h 0–12 describes SWS, caused by influenza inoculation, during hours 0–12 of photoperiod.
[f] tauLLb describes circadian period in constant light (infrared-beam crossings).

Table 14.7 Chromosome 7 listing of QTL for biological rhythms.

Map position (cM)	Marker name	Trait	Method	Reference
9	Pmv18	tauLLb[a]-tauDDb	CXB	Hofstetter and Mayeda, 1998
11	Gpi1	tauDDb[b]	CXB	Mayeda et al., 1996
5–26[c]	–	PS[d]	CXB	Tafti et al., 1999
27	Gas2	tauDDb	CXB	Mayeda et al., 1996
35	D7Rik104	Phase[e]	SMXA	Suzuki et al., 2001
36	D7Ncvs57	tauDDw[f]	BXD	Hofstetter et al., 1999
37	D7Mit30	Phase	F2:CXB	Shimomura et al., 2001
37	D7Mit90	SWS[g] SWS 24 h[h]	CXB	Toth and Williams, 1999c
43	Mod-2	tauLLb	BXD	Hofstetter et al., 1995
49	Hbb	tauLLb	BXD	Hofstetter et al., 1995
57	Ras14	tauDDw	CXB	Hofstetter et al., 1995
66	D7Mit10	tauDDw	SMXA	Suzuki et al., 2000
74	Fis-1	tauDDw	CXB	Hofstetter et al., 1995

[a] tauLLb describes circadian period in constant light (infrared-beam crossings).
[b] tauDDb describes circadian period for activity in constant dark measured by infrared-beam crossings.
[c] No marker name listed.
[d] PS describes time spent in paradoxical sleep.
[e] Phase describes the time of activity onset during entrainment.
[f] tauDDw describes circadian period for activity in constant dark measured by wheel running.
[g] SWS describes time spent in short-wave sleep.
[h] SWS 24 h describes time spent in short-wave sleep, caused by inoculation with influenza.

Table 14.8 Chromosome 8 listing of QTL for biological rhythms.

Map position (cM)	Marker name	Trait	Method	Reference
22	Xmv12	tauLLb[a]-tauDDb[b]	CXB	Hofstetter and Mayeda, 1998
29	D8Ncsv34	tauLLb-tauDDb	CXB	Hofstetter and Mayeda, 1998
29	D8Mit8	tauLLb	BXD	Hofstetter et al., 1995
39	D8Mit75	Phase response[c]	F2:C57BL/6rd/rdXCBArd/rd	Yoshimura et al., 2002
44	8Mit242	Phase response	F2:C57BL/6rd/rdXCBArd/rd	Yoshimura et al., 2002
54	D8Mit213	SWS h 12–24[d]	CXB	Toth and Williams, 1999c
57	D8Mit318	SWS h 12–24	CXB	Toth and Williams, 1999c
67	D8Mit13	Phase[e]	F2:CXB	Shimomura et al., 2001
67	D8Mit49	SWS light phase[f]	CXB	Toth and Williams, 1999b
		SWS[g]	CXB	Toth and Williams, 1999c
		SWS 24 h[h]	CXB	Toth and Williams, 1999c

[a] tauLLb describes circadian period in constant light measured by infrared-beam crossings.
[b] tauDDb describes circadian period in constant dark measured by infrared-beam crossings.
[c] Phase response describes phase shift in a free-running rhythm induced by a light pulse.
[d] SWS h 12–24 describes time spent in short-wave sleep, induced by influenza inoculation, during hours 12–24 of the photoperiod.
[e] Phase describes time of activity onset of the entrained rhythm.
[f] SWS light phase describes time spent in short-wave sleep in the light phase of the light–dark cycle.
[g] SWS describes time spent in short-wave sleep.
[h] SWS 24 h describes time spent in short-wave sleep induced by influenza inoculation.

Table 14.10 Chromosome 10 listing of QTL for biological rhythms.

Map position (cM)	Marker name	Trait	Method	Reference
61	D10Mit70	tauLLb[a]-tauDDb[b]	CXB	Hofstetter and Mayeda, 1998

[a] tauLLb describes circadian period in constant light, measured by infrared-beam crossings.
[b] tauDDb describes circadian period in constant dark, measured by infrared-beam crossings.

Table 14.9 Chromosome 9 listing of QTL for biological rhythms.

Map position (cM)	Marker name	Trait	Method	Reference
15	Est1	tauLLb-tauDDb[a]	CXB	Hofstetter and Mayeda, 1998
26	D9Mit19	SWS h 0–12[b]	CXB	Toth and Williams, 1999c
35	D9Mit104/308	SWS h 0–6[c]	CXB	Toth and Williams, 1999c
31–37[d]	–	TST[e], SWS[f]	BXD	Tafti et al., 1999
44–88	D9Ncvs44	tauDDw[g]	BXD	Hofstetter et al., 1999
71	D9Mit18	tauDDb	CXB	Mayeda et al., 1996

[a] tauDDb describes circadian period in constant dark for activity measured by infrared-beam crossings.
[b] SWS h 0–12 describes time spent in short-wave sleep, induced by influenza inoculation, during 0–12 hours of the photoperiod.
[c] SWS h 0–6 describes time spent in short-wave sleep, induced by influenza inoculation, during hours 0–6 of the photoperiod.
[d] No marker name listed.
[e] TST describes total sleep time.
[f] SWS describes total time spent in short-wave sleep.
[g] tauDDw describes circadian period in constant dark for activity measured by wheel running.

Table 14.11 Chromosome 11 listing of QTL for biological rhythms.

Map position (cM)	Marker name	Trait	Method	Reference
26	D11Ncvs44	tauDDw[a]	BXD	Hofstetter et al., 1999
40	D11Ncvs77	tauLLb[b]-tauDDb[c]	CXB	Hofstetter and Mayeda, 1998
43	D11Mit320	SWS[d]	CXB	Toth and Williams, 1999b
45	D11Mit34	tauDDw	BXD	Hofstetter et al., 1999
52	D11Nut327	REMS:SWS ratio light Phase[e]	CXB	Toth and Williams, 1999b
53	Xmv42	tauLLb-tauDDb	CXB	Hofstetter and Mayeda, 1998

[a] tauDDw describes circadian period in constant dark for activity measured by wheel running.
[b] tauLLb describes circadian period in constant light measured by infrared-beam crossings.
[c] tauDDb describes circadian period in constant dark for activity measured by infrared-beam crossings.
[d] SWS describes time spent in short-wave sleep.
[e] REMS:SWS ratio light phase describes the ratio of time spent in rapid-eye-movement sleep to the time spent in short-wave sleep, during the light phase of the photoperiod.

Table 14.12 Chromosome 12 listing of QTL for biological rhythms.

Map position (cM)	Marker name	Trait	Method	Reference
5	D12Nyu7	tauDDw[a]	BXD	Hofstetter et al., 1999
6	D12Mit170	Phase response[b]	F2:C57BL/6rd/rdXCBArd/rd	Yoshimura et al., 2002
7	Es25	tauDDw	CXB	Hofstetter et al., 1995
11	D12Mit221	tauDDb	F2:(BXD) X(CXB)	Hofstetter et al., 2003a,b
13	D12Mit81	Phase[c]	F2:CXB	Shimomura et al., 2001
8–24[d]	–	PS	BXD	Tafti et al., 1999
23	D12Mit54	Phase response	F2:C57BL/6rd/rdXCBArd/rd	Yoshimura et al., 2002
23	lapls2–10	tauLLb-tauDDb[e]	CXB	Hofstetter and Mayeda, 1998
23	D12Nyu15	SWS LD ratio[f]	CXB	Toth and Williams, 1999b
29	D12Mit251	Dissociation of activity[g]	F2:CXB	Shimomura et al., 2001
34	D12Mit44	tauDDw	F2:CSXC57BL6	Suzuki et al., 2001
35–41	D12mit60	tauDDb	Congenic	Hofstetter et al., 2007
		Time of nocturnal activity minimum[h]		
50	Spi-2	tauLLb[i]	BXD	Hofstetter et al., 1995
58	D12Mit263	Phase	F2:CXB	Shimomura et al., 2001

[a] tauDDw describes circadian period in constant dark for activity measured by wheel running.
[b] Phase response describes phase shift of a free running rhythm in response to a light pulse.
[c] Phase describes time of activity onset during entrainment.
[d] No marker name listed.
[e] tauDDb describes circadian period in constant dark for activity measured by infrared-beam crossings.
[f] SWS LD ratio describes ratio of light-phase short-wave sleep to dark phase in a light–dark cycle.
[g] Dissociation of activity describes fragmentation of the activity pattern.
[h] Time of nocturnal activity minimum describes time of the low point of activity in the dark phase of the entrained rhythm.
[i] tauLLb describes circadian period in constant light measured by infrared-beam crossings.

Table 14.13 Chromosome 13 listing of QTL for biological rhythms.

Map position (cM)	Marker name	Trait	Method	Reference
35	D13Mit20	tauDDw[a]	F2(DXB)X(CXB)	Hofstetter et al., 2003a,b
72	lapls1–8	tauDDb[b]	BXD	Hofstetter et al., 1995
		tauDDw	CXB	Hofstetter et al., 1999

[a] tauDDw describes circadian period in constant dark for activity measured by wheel running.
[b] tauDDb describes circadian period in constant dark for activity measured by infrared-beam crossings.

Table 14.14 Chromosome 14 listing of QTL for biological rhythms.

Map position (cM)	Marker name	Trait	Method	Reference
40	D14Mit160	tauDDb[a]	BXD	Hofstetter et al., 2003a,b
60	D14Mit266	SWS light phase[b]	CXB	Toth and Williams, 1999c
		SWS h 0–12[c]		

[a] tauDDb describes circadian period in constant dark for activity measured by infrared-beam crossings.
[b] SWS light phase describes time spent in short-wave sleep in the light phase of the photoperiod.
[c] SWS h 0–12 describes short-wave sleep, induced by influenza inoculation, during hours 0–12 of the photoperiod.

Table 14.15 Chromosome 15 listing of QTL for biological rhythms.

Map position (cM)	Marker name	Trait	Method	Reference
12	D15Mit226	SWS h 0–6[a]	CXB	Toth and Williams, 1999c
6–30[b]	–	TST[c], SWS[d]	CXB	Tafti et al., 1999
44	D15Mit28	Dissociation[e]	F2:CXB	Shimomura et al., 2001
51	Spt-2	tauDDb[f]	BXD	Hofstetter et al., 1995

[a] SWS h 0–6 describes time spent in short-wave sleep, caused by inoculation with influenza.
[b] No marker name listed.
[c] TST describes total sleep time.
[d] SWS describes time spent in short-wave sleep.
[e] Dissociation describes fragmentation of the activity pattern.
[f] tauDDb describes circadian period in constant dark for activity measured by infrared-beam crossings.

Table 14.16 Chromosome 16 listing of QTL for biological rhythms.

Map position (cM)	Marker name	Trait	Method	Reference
36	D16Mit171	REMS dark phase[a]	CXB	Toth and Williams, 1999b
		REMS:SWS ratio[b]		
43	D16Mit47	REMS[c]	CXB	Toth and Williams, 1999b
68	Pmv16	tauDDb[d]	CXB	Mayeda et al., 1996

[a] REMS dark phase describes time spent in rapid-eye-movement sleep in the dark phase of the photoperiod.
[b] REMS:SWS ratio describes the ratio of time spent in rapid-eye-movement sleep to time spent in short wave sleep.
[c] REMS describes time spent in rapid-eye-movement sleep.
[d] tauDDb describes circadian period in constant dark for activity measured by infrared-beam crossings.

Table 14.18 Chromosome 18 listing of QTL for biological rhythms.

Map position (cM)	Marker name	Trait	Method	Reference
12	D18Mit34	tauDDw[a]	SMXA	Suzuki et al., 2000
16	D18Mit15	tauDDw	SMXA	Suzuki et al., 2000
16	D18Mit59	SWS h 0–24[b]	CXB	Toth and Williams, 1999c
16	D18Mit120	SWS h 0–24	CXB	Toth and Williams, 1999c
28	D18Byu22	tauDDb[c]	BXD	Hofstetter et al., 1995
41	D18Mit184	Phase[d]	F2:CXCast/Ei	Wisor et al., 2007
		Phase response[e]		
56	D18Mit4	tauDDb	BXD	Hofstetter et al., 1995

[a] tauDDw describes circadian period in constant dark for activity measured by wheel running.
[b] SWS h 0–24 describes time spent in short-wave sleep caused by influenza inoculation.
[c] tauDDb describes circadian period in constant dark for activity measured by infrared-beam crossings.
[d] Phase describes time of activity onset during entrainment.
[e] Phase response describes shift in activity pattern, induced by a light pulse, of a free-running rhythm.

Table 14.17 Chromosome 17 listing of QTL for biological rhythms.

Map position (cM)	Marker name	Trait	Method	Reference
4	D17Mit164	SWS h 12–24[a]	CXB	Toth and Williams, 1999c
8	Tcp1	REMS[b]	CXB	Toth and Williams, 1999b
		REMS:SWS ratio[c]		
11	D17Mit113	Phase response	F2:C57B/6rd/rdXCBArd/rd	Yoshimura et al., 2002
11	Glo2	REMS	CXB	Toth and Williams, 1999b
		REMS:SWS ratio		
12	D17Mit46	REMS	CXB	Toth and Williams, 1999b
12	D17Tu36	tauLLb[d]-tauDDb[e]	CXB	Hofstetter and Mayeda, 1998
33	Ckb-rs2	tauLLb-tauDDb	CXB	Hofstetter and Mayeda, 1998
42	D17Nds14	tauDDw[f]	BXD	Hofstetter et al., 1999
8–47[g]	–	PS[h]	BXD	Tafti et al., 1999

[a] SWS h 12–24 describes short-wave sleep, induced by influenza inoculation, during hours 12–24 of the photoperiod.
[b] REMS describes time spent in rapid-eye-movement sleep.
[c] REMS:SWS ratio describes ratio of time spent in rapid-eye-movement sleep to time spent in short-wave sleep.
[d] tauLLb describes circadian period in constant light measured by infrared-beam crossings.
[e] tauDDb describes circadian period in constant dark for activity measured by infrared-beam crossings.
[f] tauDDw describes circadian period in constant dark for activity measured by wheel running.
[g] No marker name listed.
[h] PS describes time spent in paradoxical sleep.

General conclusions

Behavioral genetic analysis of mouse biological clock function is still relatively new, highly interdisciplinary, and just starting to move away from a focus on core circadian clock function to a more complete analysis of circadian system components and functional relationships. Perennial advances in molecular biological methods, increased automation of behavioral monitoring, and homology with other model systems at the genetic level should facilitate acceleration of progress and interest in the field. Genes involved in generating core circadian clock oscillations and mediating entrainment to light–dark cycles are highly conserved among *Drosophila*, mice, and humans (see also Allada et al., 2001; Wager-Smith and Kay, 2000). Mice, therefore, can play a key role in translating from basic genetic research in *Drosophila* to mammalian models more directly relevant to human biological clock function. Mice are an ideal model organism for basic research on biological clock mechanisms and outputs unique to mammals as they possess multiple pacemakers and synchronizers, and peripheral circadian oscillators throughout the organism that impose circadian rhythms on most traits, including traits of interest for biomedical research (Lowrey and Takahashi, 2004).

As our understanding of circadian organization increases, and our models of circadian mechanisms become more complex, with a proliferation of clock genes, oscillators, pacemakers, and rhythms, multivariate and quantitative genetic

Table 14.19 Chromosome 19 listing of QTL for biological rhythms.

Map position (cM)	Marker name	Trait	Method	Reference
0[a]	–	tauDDw[b]	F2:CSXMSM	Suzuki et al., 2001
1	D19Ncvs30	tauDDw	CXB	Hofstetter et al., 1999
0–16[a]		PS[c]	BXD	Tafti et al., 1999
26	D19Mit19	SWS h 12–24[d]	CXB	Toth and Williams, 1999c
		SWS light phase[e]	CXB	Toth and Williams, 1999b
34[a]	–	tauDDw	F2:CSXMSM	Suzuki et al., 2001
52[a]	–	tauDDw	F2:CSXMSM	Suzuki et al., 2001

[a] No marker name listed.
[b] tauDDw describes circadian period in constant dark for activity measured by wheel running.
[c] PS describes time spent in paradoxical sleep.
[d] SWS h 12–24 describes short-wave sleep, caused by influenza inoculation, from hours 12–24 of the photoperiod.
[e] SWS light phase describes short-wave sleep during the light phase of the photoperiod.

analysis will become more prevalent by necessity. Advantages of quantitative genetic methods, such as the ability to work at the phenotypic level in functional systems, tends to favor complex traits and functional questions, while genetic dissection is better suited to lesioning functional systems to reveal critical mechanisms. The rapid development of genomics and proteomics and information technology, combined with more interdisciplinary training, and significant progress in understanding developmental and neural mechanisms that link genetic and behavioral variation, are promoting the convergence of methods, researchers, and questions across fields and subdisciplines in an exciting and productive manner. Only a few decades ago, for example, genetic variation such as response to artificial selection or differences among inbred strains implied DNA sequence variation. Recent years have brought a deeper understanding of mechanisms that regulate gene expression (e.g., chromatin remodeling, *miRNA*, mRNA splicing, gene imprinting) that, in turn, enriches the significance of behavioral genetic research on inbred strain differences, response to artificial selection, and QTLs. Behavioral genetic variation may be mediated through any of these epigenetic mechanisms, as indicated by current research linking miRNA regulation of gene expression and *chromatin remodeling* to regulation of circadian mechanisms (Cheng et al., 2007; Crosio et al., 2000; Dardente and Cermakian, 2007; Nakahata et al., 2007; O'Neil and Hastings, 2007; Xu et al., 2007b).

Our understanding of circadian system function and mechanism is still elementary, but it won't be for long, and the field is sure to become integrated into every area of biological and behavioral science and medicine in the near future. This makes sense if only because most biological functions are regulated directly or indirectly by circadian clock mechanisms. Molecular genetics has provided the main tools and experimental breakthroughs that led to the development of a strong model for biological clock function at the molecular level. Behavioral and quantitative geneticists have a similar opportunity to contribute methods and models for a useful understanding of complex circadian traits and systems at a functional level as well. Most associations of circadian rhythm variation with pathology, particularly in psychiatry involving sleep cycles, is quantitative in nature (see also Goodwin and Jamison, 1990), and modeling complex behaviors such as sleep will require the full complement of methods and insights that all areas of genetics and related disciplines have to offer. It has been exciting to observe the dramatic growth of both fields, genetics and biological clocks, over the past half century. It is gratifying to know that basic contributions to date will provide a foundation for much more to come. Even relatively simple behavioral genetic analysis of biological clock function, such as the characterization of inbred strains, can provide "off the shelf" models for numerous aspects of mammalian circadian systems and serve as a basis for further research from molecular to behavioral levels.

References

Abe, H., Honma, K., Honma, T., Suzuki, S., and Ebihara, S. (1999) Functional diversities of two activity components of circadian rhythm in genetical splitting mice (CS strain). *J Comp Physiol A* 184: 243–251.

Abe, H., Honma, S., Namihira, M., Masubuchi, S., Ikeda, M., Ebihara, S., et al. (2001) Clock gene expressions in the suprachiasmatic nucleus and other areas of the brain during rhythm splitting in CS mice. *Mol Brain Res* 87: 92–99.

Abe, H., Kida, M., Tsuji, K., and Mano, T. (1989) Feeding cycles entrain circadian rhythms of locomotor activity in CS mice but not in C57BL/6J mice. *Physiol Behav* 45: 397–401.

Allada, R., Emery, P., Takahashi, J.S., and Rosbach, M. (2001) Stopping time: the genetics of fly and mouse circadian clocks. *Ann Rev Neurosci* 24: 1091–1119.

Avigdor, M., Sullivan, S.D., and Heideman, P.D. (2005) Response to selection for photoperiod responsiveness on the density and location of mature GnRH-releasing neurons. *Am J Physiol Regul Integr Comp Physiol* 288: R1226–R1236.

Bae, K., Jin, X., Maywood, E.S., Hastings, M.H., Reppert, S.M., and Weaver, D.R. (2001) Differential functions of mPer1, mPer2, and mPer3 in the SCN circadian clock. *Neuron* 30: 525–536.

Beau, J. (1991) Activity rhythms in inbred mice. I. Genetic analysis with recombinant inbred strains. *Behav Genet* 21: 117–129.

Beeler, J.A., Prendergast, B., and Zhuang, X. (2006) Low amplitude entrainment of mice and the impact of circadian phase on behavior tests. *Physiol Behav* 87: 870–880.

Bhattacharjee, Y. (2007) Psychiatric research: is internal timing key to mental health? *Science* 317: 1488–1490.

Blizard, D.A. and Baily, D.W. (1979) Genetic correlation between open-field activity and defecation: analysis with the C × B recombinant-inbred-strain. *Behav Genet* 9: 349–357.

Bullough, J.D., Rea, M.S., and Figueiro, M.G. (2006) Of mice and women: light as a circadian stimulus in breast cancer research. *Cancer Causes Control* 17: 375–383.

Bult, A., Heistand, L., Van der Zee, E.A., and Lynch, C.B. (1993) Circadian rhythms differ between selected mouse lines: a model to study the role of vasopressin neurons in the suprachiasmatic nuclei. *Brain Res Bull* 32: 623–627.

Carlson, L.L., Zimmerman, A., and Lynch, G.R. (1989) Geographic differences for delay of sexual maturation in *Peromyscus leucopus*: effects of photoperiod, pinealectomy, and melatonin. *Biol Reprod* 41: 1001–1013.

Castillo, M.R., Hochstetler, K.J., Greene, D.M., Firmin, S.I., Tavernier, R.J., Raap, D.K., et al. (2005) Circadian rhythm of core body temperature in two laboratory mouse lines. *Physiol Behav* 86: 538–545.

Cheng, H.M., Papp, J.W., Varlamova, O., Dziema, H., Russell, B., Curfman, J.P., et al. (2007) MicroRNA modulation of circadian-clock period and entrainment. *Neuron* 54: 813–829.

Clayton, J.D., Kyriacou, C.P., and Reppert, S.M. (2001) Keeping time with the human genome. *Nature* 409: 829–831.

Connolly, M.S. and Lynch, C.B. (1981) Circadian variation of strain differences in body temperature and activity in mice. *Physiol Behav* 27: 1045–1049.

Crosio, C., Cermakian, N., Allis, C.D., and Sassone-Corsi, P. (2000) Light induces chromatin modification in cells of the mammalian circadian clock. *Nat Neurosci* 3: 1241–1247.

Daan, S. and Pittendrigh, C.S. (1976a) A functional analysis of circadian pacemakers in nocturnal rodents. II. The variability of phase response curves. *J Comp Physiol* 106: 253–266.

Daan, S. and Pittendrigh, C.S. (1976b) A functional analysis of circadian pacemakers in nocturnal rodents. III. Heavy water and constant light. Homeostasis of frequency. *J Comp Physiol* 106: 267–290.

Dardente, H. and Cermakian, N. (2007) Molecular and circadian rhythms in central and peripheral clocks in mammals. *Chronobiol Int* 24: 195–213.

Daszuta, A., Gambarelli, F., and Ternaux, J.P. (1983) Sleep variations in C57BL and BALBc mice from 3 weeks to 14 weeks of age. *Brain Res* 283: 87–96.

Davis, F.C. and Menaker, M. (1981) Development of the mouse circadian pacemaker: independence from environmental cycles. *J Comp Physiol A Neuroethol Sens Neural Behav Physiol* 143: 527–539.

de Groot, M.H. and Rusak, B. (2004) Housing conditions influence the expression of food-anticipatory activity in mice. *Physiol Behav* 83: 447–457.

Desjardins, C., Bronson, R.H., and Blank, J.L. (1986) Genetic selection for reproductive photoresponsiveness in deer mice. *Nature* 322: 172–173.

Dolatshad, H., Campbell, E.A., O'Hara, L., Maywood, E.S., Hastings, M.H., and Johnson, M.H. (2006) Developmental and reproductive performance in circadian mutant mice. *Hum Reprod* 21: 68–79.

Dunlap, J. (1996) Genetic and molecular analysis of circadian rhythms. *Ann Rev Genet* 30: 579–601.

Dunlap, J.C., Loros, J.J., and DeCoursey, P.J. (2004) *Chronobiology: Biological Timekeeping*. Sinauer Associates, Sunderland, MA, USA.

Ebihara, S., Miyazaki, S., Sakamaki, H., and Yoshimura, T. (2003) Sleep properties of CS mice with spontaneous rhythm splitting in constant darkness. *Brain Res* 980: 212–127.

Ebihara, S. and Tsuji, K. (1976) Strain differences in the mouse's wheel-running behavior. *Jpn J Psychol Res* 18: 20–29.

Ebihara, S., Tsuji, K., and Kondo, K. (1978) Strain differences of the mouse's free-running circadian rhythm in continuous darkness. *Physiol Behav* 20: 795–799.

Edgar, D.M., Martin, C.E., and Dement, W.C. (1991) Activity feedback to the mammalian circadian pacemaker: influence on observed measures of rhythm period length. *J Biol Rhythms* 6: 185–199.

Elliott, J.A. (1976) Circadian rhythms and photoperiodic time measurement in mammals. *Fed Proc* 35: 2339–2346.

Franken, P., Malafosse, A., and Tafti, M. (1999) Genetic determinants of sleep regulation in inbred mice. *Sleep* 22: 155–169.

Freedman, M.S., Lucas, R.J., Soni, B., von Schantz, M., Munoz, M., David-Gray, Z., et al. (1999) Regulation of mammalian circadian behavior by non-rod, non-cone, ocular photoreceptors. *Science* 284: 502–504.

Friedmann, J.K. (1974) A diallel analysis of the genetic underpinnings of mouse sleep. *Behav Genet* 12: 169–175.

Froy, O. and Miskin, R. (2007) The interrelations among feeding, circadian rhythms and ageing. *Prog Neurobiol* 82: 142–150.

Gilliam, D.M. and Collins, A.D. (1983) Circadian and genetic influences on tissue sensitivity and sleep time to ethanol in LS and SS mice. *Pharmacol Biochem Behav* 18: 803–808.

Goldman, B. (2001) Mammalian photoperiodic system: formal properties and neuroendocrine mechanisms of photoperiodic time measurement. *J Biol Rhythms* 16: 283–301.

Goodwin, F.K. and Jamison, K.R. (1990) *Manic-Depressive Illness*. Oxford University Press, New York.

Gottlieb, D.J., O'Connor, G.T., and Wilk, J.B. (2007) Genome-wide association of sleep and circadian phenotypes. *BMC Med Genet* 8 (Suppl. 1): S9.

Hazlerigg, D.G. and Wagner, G.C. (2006) Seasonal photoperiodism in vertebrates: from coincidence to amplitude. *Trends Endocrinol Metab* 17: 83–91.

Hegmann, J.P. and Possidente, B. (1980) Estimating genetic correlations from inbred strains. *Behav Genet* 11: 103–114.

Heideman, P.D. (2004) Top-down approaches to the study of natural variation in complex physiological pathways using the white-footed mouse (*Peromyscus leucopus*) as a model. *ILAR J* 45: 4–13.

Heideman, P.D. and Bronson, F.H. (1991) Characteristics of a genetic polymorphism for reproductive photoresponsiveness in the white-footed mouse (*Peromyscus leucopus*). *Biol Reprod* 44: 1180–96.

Hofstetter, J.R., Graham, N.J., and Mayeda, A.R. (2003a) Circadian activity rhythms in high-alcohol-preferring and low-alcohol-preferring mice. *Alcohol* 30: 81–85.

Hofstetter, J.R., Hofstetter, A.R., Hughes, A.M., and Mayeda, A.R. (2005) Intermittent long-wavelength red light increases the period of daily locomotor activity in mice. *J Circadian Rhythms* 3: 8.

Hofstetter, J.R. and Mayeda, A.R. (1998) Provisional quantitative trait loci (QTL) for the Aschoff effect in RI mice. *Physiol Behav* 64: 97–101.

Hofstetter, J.R., Mayeda, A.R., Possidente, B., and Nurnberger, J.I., Jr. (1995) Quantitative trait loci (QTL) for circadian rhythms of locomotor activity in mice. *Behav Genet* 25: 545–556.

Hofstetter, J.R., Possidente, B., and Mayeda, A.R. (1999) Provisional QTL for circadian period of wheel running in laboratory mice: quantitative genetics of period in RI mice. *Chronobiol Int* 16: 269–279.

Hofstetter, J.R., Svihla-Jones, D.A., and Mayeda, A.R. (2007) A QTL on mouse chromosome 12 for the genetic variance in free-running period between inbred strains of mice. *J Circadian Rhythms* 5: 7.

Hofstetter, J.R., Trofatter, J.A., Kernek, K.L., Nurnberger, J.I., and Mayeda, A.R. (2003b) New quantitative trait loci for the genetic variance in circadian period of locomotor activity between inbred strains of mice. *J Biol Rhythms* 18: 450–462.

Holmes, M.M. and Mistlberger, R.E. (2000) Food anticipatory activity and photic entrainment in food-restricted BALB/c mice. *Physiol Behav* 68: 655–666.

Honma, K., Honma, S., and Hiroshige, T. (1987) Activity rhythms in the circadian domain appears in suprachiasmatic nuclei lesioned rats given methamphetamine. *Physiol Behav* 40: 767–774.

Hossain, S.M., Wong, B.K.Y., and Simpson, E.M. (2004) The dark phase improves genetic discrimination for some high throughput mouse behavioral phenotyping. *Genes Brain Behav* 3: 167–177.

Kernek, K.L., Trofatter, J.A., Mayeda, A.R., and Hofstetter, J.R. (2004) A locus for circadian period of locomotor activity on mouse proximal chromosome 3. *Chronobiol Int* 21: 343–352.

Kernek, K.L., Trofatter, J.A., Mayeda, A.R., Lahiri, D.K., and Hofstetter, J.R. (2006) A single copy of *carbonic anhydrase* 2 restores wild-type circadian period to carbonic anhydrase II-deficient mice. *Behav Genet* 36: 301–308.

Koehl, M., Battle, S.E., and Turek, F.W. (2003) Sleep in female mice: a strain comparison across the estrous cycle. *Sleep* 26: 267–272.

Kondratov, R.V. (2007) A role of the circadian system and circadian proteins in aging. *Ageing Res Rev* 6: 12–27.

Kopp, C., Ressel, V., Wigger, E., and Tobler, I. (2005) Influence of estrus cycle and ageing on activity patterns in two inbred mouse strains. *Behav Brain Res* 167: 165–174.

Koteja, P., Swallow, J.G., Carter, P.A., and Garland, T., Jr. (2003) Different effects of intensity and duration of locomotor activity on circadian period. *J Biol Rhythms* 18: 491–501.

Kowal, M., Buda-Lewandowska, D., Plytyca, B., and Styrna, J. (2002) Day/night food consumption in mice is strain and age-dependent. *Folia Biol (Krakow)* 50: 1–3.

Lowrey, P.L. and Takahashi, J.S. (2004) Mammalian circadian biology: elucidating genome-wide levels of temporal organization. *Ann Rev Genomics Hum Genet* 5: 407–441.

Majoy, S.B. and Heideman, P.D. (2000) Tau differences between short-day responsive and short-day nonresponsive white-footed mice (*Peromyscus leucopus*) do not affect reproductive photoresponsiveness. *J Biol Rhythms* 15: 501–513.

Marchant, E.G. and Mistlberger, R.E. (1997) Anticipation and entrainment to feeding time in intact and SCN-ablated C57BL/6J mice. *Brain Res* 765: 273–282.

Mayeda, A.R. and Hofstetter, J.R. (1999) A QTL for the genetic variance in free-running period and level of locomotor activity between inbred strains of mice. *Behav Genet* 29: 171–176.

Mayeda, A.R., Hofstetter, J.R., Belknap, J.K., and Nurnberger, J.I., Jr. (1996) Hypothetical quantitative trait loci (QTL) for circadian period of locomotor activity in CXB recombinant inbred mice. *Behav Genet* 26: 505–511.

Mayeda, A.R., Hofstetter, J.R., and Possidente, B. (1997) Aging lengthens taudd in C57BL/6J, DBA/2J, and outbred SWR male mice (*Mus musculus*). *Chronobiol Int* 14: 19–23.

McCarthy, J.J., Andrews, J.L., McDearmon, E.L., Campbell, K.S., Barber, B.K., Miller, B.H., et al. (2007) Identification of the circadian trancriptome in adult mouse skeletal muscle. *Physiol Genomics* 31: 86–95.

McClung, C.A. (2007) Circadian genes, rhythms and the biology of mood disorders. *Pharmacol Ther* 114: 222–232.

Meerlo, P., Easton, A., Bergmann, B.M., and Turek, F.W. (2001) Restraint increases prolactin and REM sleep in C57BL/6J mice but not in BALB/cJ mice. *Am J Physiol Regul Integr Comp Physiol* 281: R846–R854.

Mendoza, J. (2007) Circadian clocks: setting time by food. *J Neuroendocrinol* 19: 127–137.

Miller, B.H., McDearmon, E.L., Panda, S., Hayes, K.R., Zhang, J., Andrews, J.L., et al. (2007) Circadian and CLOCK-controlled regulation of the mouse transcriptome and cell proliferation. *PNAS* 104: 3342–3347.

Mousel, M.R., Stroup, W.W., and Nielsen, M.K. (2001) Locomotor activity, core body temperature, and circadian rhythms in mice selected for high or low heat loss. *J Anim Sci* 79: 861–868.

Nakahata, Y., Grimaldi, B., Sahar, S., Hirayama, J., and Sassone-Corsi, P. (2007) Signaling to the circadian clock: plasticity by chromatin remodeling. *Curr Opin Cell Biol* 19: 230–237.

O'Hara, B.F., Ding, J., Bernat, R.L., and Franken, P. (2007) Genomic and proteomic approaches towards an understanding of sleep. *CNS Neurol Disord Drug Targets* 6: 71–81.

O'Neil, J.S. and Hastings, M.H. (2007) Circadian clocks: timely interference by microRNAs. *Curr Biol* 17: R760–R762.

Pace-Schott, E.F. and Hobson, J.A. (2001) The neurobiology of sleep: genetics, cellular physiology and subcortical networks. *Nat Rev Neurosci* 3: 591–605.

Pang, T.Y., Stam, N.C., Nithianantharajah, J., Howard, M.L., and Hannan, A.J. (2006) Differential effects of voluntary exercise on behavioral and brain-derived neurotrophic factor expression deficits in Huntington's disease transgenic mice. *Neuroscience* 141: 569–584.

Pietropaolo, S., Feldon, J., Alleva, E., Cirulli, F., and Yee, B.K. (2006) The role of voluntary exercise in enriched rearing: a behavioral analysis. *Behav Neurosci* 120: 787–803.

Pittendrigh, C.S. (1981a) Circadian systems: entrainment. In Aschoff, J. (ed.), *Handbook of Behavioral Neurobiology, Biological Rhythms*, Vol. 4. Plenum, New York, pp. 95–124.

Pittendrigh, C.S. (1981b) Circadian organization and the photoperiodic phenomena. In: Follett, B.K. and Follett, D.E. (eds.), *Biological Clocks in Seasonal Reproductive Cycles*. John Wright, Bristol, UK, pp. 1–35.

Pittendrigh, C.S. and Daan, S. (1976a) A functional analysis of circadian pacemakers in nocturnal rodents. I. The stability and lability of spontaneous frequency. *J Comp Physiol* 106: 223–252.

Pittendrigh, C.S. and Daan, S. (1976b) A functional analysis of circadian pacemakers in nocturnal rodents. IV. Entrainment: pacemaker as clock. *J Comp Physiol* 106: 291–332.

Pittendrigh, C.S. and Daan, S. (1976c) A functional analysis of circadian pacemakers in nocturnal rodents. V. Pacemaker structure: a clock for all seasons. *J Comp Physiol* 106: 333–355.

Pitts, S., Perone, E., and Silver, R. (2003) Food-entrained circadian rhythms are sustained in arrhythmic Clk/Clk mutant mice. *Am J Physiol Regul Integr Comp Physiol* 285: R57–R67.

Possidente, B. and Hegmann, J.P. (1980) Circadian complexes: circadian rhythms under common gene control. *J Comp Physiol* 139: 212–124.

Possidente, B. and Hegmann, J.P. (1982) Gene differences modify Aschoff's rule in mice. *Physiol Behav* 28: 199–200.

Possidente, B. and Stephan, F.K. (1988) Circadian period in mice: analysis of genetic and maternal contributions to inbred strain differences. *Behav Genet* 18: 109–117.

Puchalski, W. and Lynch, G.R. (1986) Evidence for differences in the circadian organization of hamsters exposed to short day photoperiod. *J Comp Physiol A* 159: 7–11.

Pugh, P., Ahmed, S., Smith, M., Upton, N., and Hunter, J. (2004) A behavioural characterization of the FVB/N mouse strain. *Behav Brain Res* 155: 283–289.

Reebs, S.G. and Maillet, D. (2003) Effect of cage enrichment on the daily use of running wheels by Syrian hamsters. *Chronobiol Int* 20: 9–20.

Rosenwasser, A.M. (2001) Alcohol, antidepressants, and circadian rhythms. Human and animal models. *Alcohol Res Health* 25: 126–135.

Roussel, B., Turrillot, P., and Kitahama, K. (1984) Effect of ambient temperature on the sleep-waking cycle in two strains of mice. *Brain Res* 294: 67–73.

Roybal, K., Theobold, D., Graham, A., DiNieri, J.A., Russo, S.J., Krishnan, V., et al. (2007) Mania-like behavior induced by disruption of CLOCK. *PNAS* 104: 6406–6411.

Ruan, G., Zhang, D., Zhou, T., Yamazaki, S., and McMahon, D.G. (2006) Circadian organization of the mammalian retina. *PNAS* 10: 9703–9708.

Saito, T.R., Katsuyama, M., Ebino, K.Y., and Takahashi, K.W. (1980) Activity patterns of diurnal drinking behavior in male mice. *Jikken Dobutsu* 29: 359–363.

Salathia, N., Edwards, K., and Millar, A.J. (2002) QTL for timing: a natural diversity of clock genes. *Trends Genet* 18: 115–118.

Sanchez-Barceló, E.J., Megias, M., Verduga, R., and Crespo, D. (1997) Differences between the circadian system of two strains of senescence-accelerated mice (SAM). *Physiol Behav* 62: 1225–1229.

Sanford, L.D., Yang, L., and Tang, X. (2003) Influence of contextual fear on sleep in mice: a strain comparison. *Sleep* 26: 527–540.

Sans-Fuentes, M.A., Lopez-Fuster, M.J., Ventura, J., Diez-Noguera, A., and Cambras, T. (2005) Effect of Robertsonian translocations on the motor activity rhythm in the house mouse. *Behav Genet* 35: 603–613.

Schalkwyk, L.C., Fernandes, C., Nash, M.W., Kurrikoff, K., Vasar, E., and Köks, S. (2007) Interpretation of knockout experiments: the congenic footprint. *Genes Brain Behav* 6: 299–303.

Schwartz, W.J. and Zimmerman, P. (1990) Circadian timekeeping in BALB/c and C57BL/6 inbred mouse strains. *J Neurosci* 10: 3685–3694.

Sehgal, A. (2004) *Molecular Biology of Circadian Rhythms*. Wiley-Liss, Hoboken, NJ, USA.

Shaw, P.J. and Franken, P. (2003) Perchance to dream: solving the mystery of sleep through genetic analysis. *J Neurobiol* 54: 179–202.

Shimomura, K., Low-Eddies, S.S., King, D.P., Steeves, T.D.L., Whiteley, A., Kushla, J., et al. (2001) Genome-wide epistatic interaction analysis reveals complex genetic determinants of circadian behavior in mice. *Genome Res* 11: 959–980.

Shiori, T., Takahashi, K., Yamada, N., and Takahashi, S. (1991) Motor activity correlates negatively with free-running period, while positively with serotonin contents in SCN in free-running rats. *Physiol Behav* 49: 779–786.

Siepka, W.E. and Takahashi, J.S. (2005) Methods to record circadian rhythm wheel running activity in mice. *Methods Enzymol* 393: 230–239.

Stephan, F.K. (2002) The "other" circadian system: food as a zeitgeber. *J Biol Rhythms* 17: 284–292.

Stratmann, M. and Schibler, U. (2006) Properties, entrainment, and physiological functions of mammalian peripheral oscillators. *J Biol Rhythms* 21: 494–506.

Suzuki, T., Ishikawa, A., Nishimura, M., Yoshimura, T., Namikawa, T., and Ebihara, S. (2000) Mapping quantitative trait loci for circadian behavioral rhythms in SMXA recombinant inbred strains. *Behav Genet* 30: 447–453.

Suzuki, T., Ishikawa, A., Yoshimura, T., Namikawa, T., Abe, H., Honma, K., et al. (2001) Quantitative trait locus analysis of abnormal circadian period in CS mice. *Mamm Genome* 12: 272–277.

Tafti, M., Cholet, D., Valatx, J., and Franken, P. (1999) Quantitative trait loci approach to the genetics of sleep in recombinant inbred mice. *J Sleep Res* 8 (Suppl 1): 37–43.

Tafti, M. and Franken, P. (2002) Functional genomics of sleep and circadian rhythm invited review: genetic dissection of sleep. *J Appl Physiol* 92: 1339–1347.

Takahashi, J.S., Shimomura, K., and Kumar, V. (2008) Searching for genes underlying behavior: lessons from circadian rhythms. *Science* 322: 909–912.

Tang, X., Xiao, J., Parris, B.S., Fang, J., and Sanford, L.D. (2005) Differential effects of two types of environmental novelty on activity and sleep in BALB/cJ and C57BL/6J mice. *Physiol Behav* 85: 419–429.

Tataroğlu, Ö., Davidson, A.J., Benvenuto, L.J., and Menaker, M. (2006) The methamphetamine-sensitive circadian oscillator (MASCO) in mice. *J Biol Rhythms* 21: 185–194.

Toth, L.A. (2001) Identifying genetic influences on sleep: an approach to discovering the mechanisms of sleep regulation. *Behav Genet* 31: 39–46.

Toth, L.A., Rehg, J.E., and Webster, R.G. (1995) Strain differences in sleep and other pathophysiological sequelae of influenza virus infection in naïve and immunized mice. *J Neuroimmunol* 58: 89–99.

Toth, L.A. and Verhults, S.J. (2003) Strain differences in sleep patterns of healthy and influenza-infected inbred mice. *Behav Genet* 33: 325–336.

Toth, L.A. and Williams, R.W. (1999a) A quantitative genetic analysis of locomotor activity in CXB recombinant inbred mice. *Behav Genet* 29: 319–328.

Toth, L.A. and Williams, R.W. (1999b) A quantitative genetic analysis of slow-wave sleep and rapid-eye movement sleep in CXB recombinant inbred mice. *Behav Genet* 29: 329–338.

Toth, L.A. and Williams, R.W. (1999c) A quantitative genetic analysis of slow-wave sleep in influenza-infected CXB recombinant inbred mice. *Behav Genet* 29: 339–348.

Turek, F.W., Joshu, C., Kohsaka, A., Lin, E., Ivanova, G., McDearnom, E., et al. (2005) Obesity and metabolic syndrome in circadian clock mutant mice. *Science* 308: 1043–1045.

Van Gelder, R.N. and Hogenesch, J.B. (2004) Clean thoughts about dirty genes. *J Biol Rhythms* 19: 3–9.

Veasay, S.C., Valladares, O., Fenik, P., Kapfhamer, D., Sanford, L., Bennington, J., et al. (2000) An automated system for recording and analysis of sleep in mice. *Sleep* 23: 1025–1040.

Veasay, S.C., Yeou-Jey, H., Thayer, P., and Fenik, P. (2004) Murine multiple sleep latency test: phenotyping sleep propensity in mice. *Sleep* 27: 388–393.

Vitaterna, M.H. and Turek, F.W. (1993) Photoperiodic responses differ among inbred strains of golden hamsters (*Mesocricetus auratus*). *Biol Reprod* 49: 496–501.

Wager-Smith, K. and Kay, S.A. (2000) Circadian rhythm genetics: from flies to mice to humans. *Nat Genet* 26: 23–27.

Watanabe, T., Kojima, M., Tomida, S., Nakamura, T.J., Yamamura, T., Nakao, N., Yasuo, S., Yoshimura, T., and Ebihara, S., (2006) Peripherial clock gene expression in CS mice with bimodal locomotor rhythms. *Neurosci Res* 54(4): 295–301.

Wax, T.M. (1977) Effects of age, strain, and illumination intensity on activity and self-selection of light–dark schedules in mice. *J Comp Physiol Psychol* 91: 51–62.

Wisor, J.P., Striz, M., DeVoss, J., Murphy, G.M., Jr., Edgar, D.M., and O'Hara, B.F. (2007) A novel quantitative trait locus on mouse chromosome 18, "era1," modifies the entrainment of circadian rhythms. *Sleep* 30: 1255–1233.

Wolfer, D.P., Litvin, O., Morf, S., Nitsch, R.M., Lipp, H., and Würbel, H. (2004) Cage enrichment and mouse behaviour. *Nature* 423: 821–822.

Xu, S., Witmer, P.D., Lumayag, S., Kovacs, B. and Valle, D. (2007b) MicroRNA (miRNA) transcriptome of the mouse retina and identification of a sensory organ-specific miRNA cluster. *J Biol Chem* 282: 25053–25066.

Xu, Y., Toh, K.L., Jones, C.R., Shin, J.Y., Fu, Y.H., and Ptácek, L.J. (2007a) Modeling of a human circadian mutation yields insights into clock regulation by PER2. *Cell* 128: 22–23.

Yamada, N., Shimoda, K., Ohi, K., Takahashi, S., and Takahashi, K. (1988) Free-access to a running wheel shortens the period of free-running rhythm in blinded rats. *Physiol Behav* 42: 87–91.

Yoshimura, T., Yokota, Y., Ishikawa, A., Yasuo, S., Hayashi, N., Suzuki, T., et al. (2002) Mapping quantitative trait loci affecting circadian photosensitivity in retinally degenerate mice. *J Biol Rhythms* 17: 512–519.

Chapter 15

The genetics of exploratory behavior

Wim E. Crusio

Introduction

Rodents are attracted by novel stimuli and spend long periods exploring when exposed to a novel environment, even when satiated in every aspect. It was therefore perhaps to be expected that this easily perceptible behavior was one of the first to be analyzed in rodents using genetic tools. Only a few years after Tryon performed his classical selection studies on maze learning (Tryon, 1929, 1930), another founder of the field of behavior genetics, Calvin Hall, selected rats for high or low levels of defecation and urination in a brightly lit novel environment (Hall, 1938, 1940, 1951), the open field. Up till today, this has remained the apparatus of choice to investigate exploratory behavior.

Although he initially used a 7 feet square apparatus (Hall and Ballachey, 1932), he soon replaced this with an arena 8 feet in diameter (Hall, 1934). In his experiments, Hall established that rats tended to defecate and urinate when first exposed to an open field, but that this reaction diminished with repeated daily exposures until it disappeared, a phenomenon that we now call habituation (Hall, 1934). He further established that rats that tended to defecate and urinate during more days also refused to eat a desired food (sunflower seeds) while in the open field (novelty-induced hypophagia), reporting correlations of 0.82 with defecation and 0.70 with urination. From this he concluded that defecation and urination are valid measures of interindividual differences in emotionality (Hall, 1934).

Although Hall initially only studied defecation and urination, he soon became interested in the possible correlation of these behaviors with ambulatory activity. In a paper from 1936, he reported a correlation between defecation and ambulation of 0.29 (Hall, 1936a, 1936b). Together, these experiments eventually led to the establishment of the concept of "emotionality," with highly emotional animals showing low levels of locomotion and high levels of defecation and urination (Broadhurst, 1969). However, it should perhaps be noted here that although Hall did obtain a statistically significant correlation, it only explained about 8–9% of the behavioral variation ($r^2 = 0.084$).

Procedures

From his descriptions, it seems evident that what Hall (1934) termed "emotionality" is nowadays, in a more "translational" age, generally called "anxiety" (see also Chapter 16). However, it has long been clear that anxiety or fear is not the only drive for the behavior that rodents display in an open field and that "curiosity" is an important factor, too (Barnett and Cowan, 1976). Discussions about what exactly the open field measures continue up to this day (Blizard et al., 2007; Overstreet, 2007; Rodgers, 2007; Stanford, 2007a, 2007b), more than 70 years after Hall's seminal experiments.

The interpretation of behavioral data obtained in the open field is not simplified by the fact that open fields exist in many varieties (square, circular, or rectangular) and in many different sizes. The procedures employed are also manifold. The exposition to the novel environment may be forced (the animal is placed in the apparatus without possibility to escape) or, more rarely, free (the subject is given the choice when to enter the arena). The duration of the behavioral measurement may vary from a few minutes (Flint et al., 1995) to 20 minutes (van Abeelen, 1963a) or more (Hall, 1934, used 14 daily trials of 2 minutes each). Generally, only a few behavioral measures, such as activity and defecation, are taken (Flint et al., 1995; Talbot et al., 1999). Others, however, have advocated using an ethogram to quantify exploratory activity. With the help of an ethogram, seemingly continuous behavior is described as a sequence of successive, mutually exclusive, and distinct motor-posture patterns that represent species-specific units of behavior which may be quantified subsequently by measuring their frequency and/or duration. Complex behavioral responses are thus regarded as organized appearances of the behavioral units. Ethograms of behavior displayed in open fields have been devised for numerous species including rodents (see Chapter 3 for references) and fish (e.g., Gerlai et al., 1990).

Several good reviews of the different methods for evaluating exploration are available (Archer, 1973; Barnett and Cowan, 1976; Walsh and Cummins, 1976) and this subject will therefore not be discussed further here. This chapter will concentrate on

Behavioral Genetics of the Mouse: Volume I. *Genetics of Behavioral Phenotypes*, eds. Wim E. Crusio, Frans Sluyter, Robert T. Gerlai, and Susanna Pietropaolo. Published by Cambridge University Press.

results obtained with the open field, as this is by far the most-frequently used test situation.

Defining exploration

Although seemingly simple, some confusion exists on the precise definition of exploratory behavior. Most authors merely equate exploratory behavior with "activity," "open field behavior," or even treat it as the opposite of emotionality. This is a moot point. Some authors feel that even the more precisely defined locomotor activity in a runway or an open field contains a non-exploratory component and they distinguish between "general activity" and exploratory activity (e.g., Foshee et al., 1965; Kelley et al., 1989; Weischer, 1976). In addition, the behavioral repertoire exhibited in an open field is substantially richer than just locomotion and defecation (see Chapter 3). Although almost all behaviors are "activities," not all of them can be classified as exploratory.

The concept of exploration is closely associated with that of novelty (Barnett and Cowan, 1976), which may involve some quality never previously experienced or familiar items arranged in an unfamiliar way. O'Keefe and Nadel (1978) defined novelty within the framework of their cognitive map theory as follows: "An item or place is novel if it does not have a representation in the locale system" and exploration as "a direct response of the animal to the detection of a mismatch by the locale system." The locale system is their term for the cognitive mapping system, presumably located in the hippocampus. In other words, the hippocampal system supposedly signals a lack of information about the current environment. Consequently, one of the processes thought to be associated with exploratory activity is what is called latent learning or exploratory learning (O'Keefe and Nadel, 1978; Renner, 1988). Latent learning occurs without overt reinforcement. If a satiated animal is allowed to explore a novel environment (for instance, a maze) and subsequently made hungry or thirsty, then the animal will quickly learn to go to the proper place to find food or water, more quickly so than an animal lacking such previous experience (Blodgett, 1929). Thus, animals acquire information about their surroundings by means of exploratory movement.

In conclusion then, we have formulated previously the following definition of exploration: "Exploration is evoked by novel stimuli and consists of behavioral acts and postures that permit the collection of information about new objects and unfamiliar parts of the environment" (Crusio and van Abeelen, 1986).

The biological significance of exploration emerges clearly: entering and exploring new places promotes dispersion and improves the chances of finding life necessities (food, shelter, escape routes, etc.). Simultaneously, such behavioral activity will render an animal vulnerable to predation (as long as it is not the top predator in its environment itself). Such a past history of stabilizing selection, where an intermediate expression of the phenotype is more favorable than either higher or lower extremes, is expected to lead to a genetic architecture consisting of large additive genetic variation combined with ambidirectional dominance (Broadhurst and Jinks, 1974), which has indeed been reported in mice (Crusio et al., 1989a; Crusio and van Abeelen, 1986) as well as in paradise fish (Gerlai et al., 1990).

Genetics: selection studies

As alluded to above, the first evidence that genetic factors play a role in determining interindividual differences in open field behavior was obtained by Hall himself, several years after he introduced the open field, by selectively breeding rats for high and low levels of "emotionality" (Hall, 1938, 1940). In mice, the best-known selection study was begun in 1963 by John DeFries and colleagues (DeFries et al., 1970, 1974, 1978), who started with an F3 cross between C57BL/6J and BALB/cJ mice to select two lines for high and two lines for low locomotor activity in an open field. Two randomly bred control lines were also maintained.

Van Abeelen (1970, 1974) selected two lines for high and low rearing-up frequencies from an F2 between the high-scoring C56BL/6J and the low-scoring DBA/2J strains. An unusual feature of his study was that he backcrossed the selected animals for several generations to DBA/2J in an effort to transfer opposite-effect alleles of the gene(s) involved in the regulation of rearing onto the same genetic background. This effort was mainly successful, as both the high (SRH) and the low (SRL) lines only differed from DBA/2 mice in a few genomic regions (Robert Williams, Memphis, TN, USA, personal communication). A quantitative-genetic analysis of a classical cross between the two lines indicated that a difference in one major gene could possibly explain the behavioral difference between the two lines (van Abeelen, 1975). A re-analysis of these data using Collin's non-parametric method (Collins, 1967) supported this conclusion (van Abeelen and Crusio, unpublished results).

Genetics: strain comparisons

As is the case for many behavioral phenotypes, the first selection studies were soon followed by comparisons of different inbred strains. A large, early study of 15 strains was conducted by one of the founding fathers of behavior genetics, William Thompson (1953), who described over 20-fold differences in the mean scores between the most and least active strains (C57BR/a and A/J, respectively). Another early study was carried out by McClearn (1959), who investigated six different strains and the reciprocal F1 hybrids between the two extremes (C57BL/10 and A). Another, more recent, example of a strain comparison study is an experiment carried out by Bolivar et al. (2000), who studied habituation to an open field over 3 consecutive days in eight inbred strains. Not surprisingly, large strain differences were found, with some strains decreasing their activity over repeated exposures, some remaining more or less stable, and others increasing. Similar results were obtained by Crusio and Schwegler (1987).

One important advantage of using inbred strains is that they are standardized. However, genetic drift due to mutation does occur (Crusio et al., 1991; Heimrich et al., 1988; Jamot et al., 1994; van Abeelen and Hughes, 1986) and could conceivably lead to changing patterns of strain differences over time. This issue was investigated in more detail by Wahlsten et al. (2006), who found that strain differences in open field activity remained highly stable over a period of 40–50 years, reporting a strain-means correlation of 0.85 between their results and those of Thompson (1953).

Genetics: studies of genetic architecture

Many studies have investigated the evolutionary history of behaviors in an open field. This is done by analyzing the genetic architecture of a trait, which is, after all, the result of past selection pressures. According to Broadhurst and Jinks (1974), a past history of directional selection (where very low or very high scores confer a selective advantage) will lead to a genetic architecture comprising low levels of additive-genetic variation combined with directional dominance (i.e., dominance points in the same direction for all genes) in the direction of the selection. Stabilizing selection (where intermediate levels of a trait confer optimal fitness) will lead to comparatively higher levels of additive genetic variation combined with ambidirectional dominance (that is, for some genes the alleles leading to lower scores are dominant whereas for other genes this is the reverse).

Henderson (1978, 1986) performed a series of elegant diallel-cross studies of the inheritance of locomotor activity in the open field and showed that its genetic architecture changes over time. In very young animals, there is directional dominance for *low* activity levels. Obviously, if very young pups accidently get out of the nest, the best strategy for these deaf and blind animals is to stay put until a parent retrieves them. Slightly older pups (11 days), that are already able to hear and see, can return to the safety of the nest on their own and, indeed, directional dominance for *high* activity levels was found here. Finally, as evoked above, in adult animals intermediate levels of activity appear to be optimal and a corresponding genetic architecture of ambidirectional dominance has indeed been reported for almost all exploratory components of the ethogram (Crusio et al., 1989a; Crusio and van Abeelen, 1986).

Genetics: genetic correlations

Several studies have explored the multivariate structure of the behavior displayed by mice in an open field. Most of these studies applied factor analysis to a matrix of correlations between different variables. However, a weakness inherent in correlational studies is that a phenotypical correlation between characters does not necessarily reflect a functional relationship. On the other hand, if two independent processes, one causing a positive relationship, the other causing a negative relationship, act simultaneously upon two characters, the effects may cancel each other so that no detectable correlation emerges. These problems can to a large extent be avoided by looking at the genetic correlations, that is, at correlations between the genetic effects that influence certain characters. These correlations are the products of either genes with pleiotropic effects or of linkage disequilibrium. By using inbred strains that are only distantly related, the probability that linkage disequilibrium occurs may be minimized so that a possible genetic correlation will most probably be caused by pleiotropy, that is, there exists a (set of) gene(s) influencing both characters simultaneously. Thus, for these characters, at least part of the physiological pathways leading from genotype to phenotype must be shared and a causal, perhaps also functional, relationship must exist. It is this special property that renders the genetic-correlational approach so uniquely valuable. A more technical discussion of phenotypical and genetic correlations has been presented elsewhere (Crusio, 1992, 2007).

A multivariate analysis of genetic correlations between behavioral variables observed in the open field was reported by Crusio et al. (1989b), who carried out a diallel cross in which five different inbred strains were intercrossed in all possible combinations, producing 25 genetically different populations with a total of 150 males being analyzed. Employing a bivariate extension (Crusio, 1993) of the Hayman analysis of variance (ANOVA) for diallel crosses (Hayman, 1954), additive-genetic correlations were estimated between different components of exploratory behavior. To aid in the interpretation of the large correlation matrices thus obtained, they were factor analyzed, which revealed three different factors. One factor, the second one, is dominated by grooming (both frequency and duration) and can be interpreted as self-maintaining behavior, but the first and third factors are more interesting for the discussion here. The first factor showed positive loadings for behavioral variables that can help a mouse to obtain information about its environment (wall-leaning, object-leaning, locomotion, and rearing) and appears to represent exploration. The third factor has a positive loading on defecation, a behavior that is usually seen as indicating stress or fear (Whimbey and Denenberg, 1967). Rearing-up loads negatively on this factor, in contrast to the similar movement, wall-leaning. A possible reason for this is that rearing-up, away from the cover provided by the wall, is a behavior to be avoided as it may make animals more vulnerable to predators.

Unfortunately, a diallel cross requires a large investment in resources and effort in order to breed and test animals from many different groups. The main alternative for the diallel cross as a tool for the genetic dissection of neural and behavioral phenotypes is the estimation of genetic correlations using a battery of inbred strains (Crusio, 2007). In this approach, we "magnify" individual differences by studying animals from different inbred strains and looking for correlations between the means obtained for different variables (see Crusio et al., 1993, and references therein for some illustrative examples). This latter approach was taken by Roullet and Lassalle (1990). They observed 48 animals from 12 isogenic groups (nine inbred strains and three F1 hybrid groups, aged 8–9 weeks; two males and two females per group) during 5 minutes in a circular arena (diameter 40 cm). Several behaviors were measured, that were also evaluated in the previous study: leaning, rearing, defecation, and locomotor activity. The latter variable was subdivided into two separate variables: central sector crossings (CSC) and total sector crossings (SC). A re-analysis of their results showed a very similar factor structure to the one described above, demonstrating the viability of this procedure (Crusio, 2001).

It should be noted here that although in the diallel cross a strong genetic correlation between locomotion and defecation was observed (data not shown), this did not appear in Roullet and Lassalle's strain study. In addition, these variables loaded on different factors in both studies. Combined with the low correlation originally observed by Hall (1936b), these findings render doubtful the often-used concept of "emotionality," defined by high defecation and low locomotor activity (Flint et al., 1995; Talbot et al., 1999). Evidently, studying the behavior in more detail by observing more elements from the ethogram, as was done in the experiments discussed in the preceding sections, makes it possible to dissect the underlying mechanisms into more detail.

Localizing the genes

Single-gene studies

Although it was recognized early on that open field exploration has a polygenic regulation, it was one of the first behaviors for which researchers attempted to identify underlying genes. Van Abeelen studied the effects of three pigmentation mutations, *dilute (Myo5a), pink-eyed dilution (Oca2)*, and *brown (Tyrp1)*, as well as those of the neurological mutations *looptail (Vangl2), jerker (Espn)*, and *waltzer* (*Cdh23*, van Abeelen, 1963b, 1963c, 1965). In this pre-genotyping era, these genes were chosen because the genotype of an animal can readily be derived from its phenotype. Although the effects found were attributed to direct, pleiotropic effects of these genes, van Abeelen did recognize that these effects could also be due to closely linked polygenes (van Abeelen, 1965). Almost 30 years later, Clément et al. (1994a, 1994b, 1995) used a similar approach to study open field behavior, using six marker genes with easily verifiable phenotypic effects, two of which (*pink-eyed dilution* and *brown*) were the same as those used by van Abeelen. Although they reported effects of these genes on the same behaviors as did van Abeelen, it should be noted that some of these effects were in an opposite direction, thereby strongly suggesting that the effects were not due to direct action of these mutations but of nearby, closely linked genes.

One of the most-replicated findings in this area is the effect of the albino mutation on open field activity, with albino animals being much less active than pigmented ones (Creel, 1980; DeFries, 1969; DeFries et al., 1966; Fuller and Thompson, 1978; Henry and Schlesinger, 1967; Lassalle and LePape, 1981, 1983; van Abeelen and Kroes, 1968). This effect is dependent on the levels of ambient lighting because albino animals are not different under dim red light conditions (DeFries et al., 1966). Surprisingly, several quantitative trait loci (QTL) studies (see below), failed to detect this effect (Wahlsten, 1999).

Although the study of single-gene effects on mouse exploratory behavior was largely abandoned after the early 1980s, a resurgence was caused by the advent of transgenic and knockout technologies (Takahashi et al., 1994). However, those studies have more often a gene-physiological focus and much of the induced genetic variants will not occur naturally and therefore fall outside the scope of the present volume.

Mapping studies

Up till now, mapping studies have wrestled with the problem that all too often results do not replicate from one experiment to another, even if experimental conditions and genetic backgrounds are highly similar. For instance, two studies using recombinant inbred strains (Mathis et al., 1995) and an F2 (Gershenfeld et al., 1997) derived from the inbred strains A/J and C57BL/6J carried out in the same laboratory each identified several significant QTLs, but not a single one was replicated in both studies. Neither of these experiments detected any effects of the albino locus.

Similar problems initially beset the up-till-now largest and best-known attempt to identify the genes underlying individual differences in open field behavior, i.e., the ongoing project carried out since about 15 years ago by Flint and colleagues (Flint et al., 1995; Fullerton et al., 2008; Henderson et al., 2004; Mott et al., 2000; Talbot et al., 1999). In an initial report (Flint et al., 1995), these researchers localized three QTLs on chromosomes 1, 12, and 15 for mouse "emotionality" and concluded that a simple genetic basis underlies this complex psychological trait. As I have stated before (Crusio, 2001), this sweeping conclusion was premature at the least. Apart from the caveat that there might be genes with effects under the detection threshold of the methods employed in this latter study, the follow-up experiment (Talbot et al., 1999) failed to replicate the earlier results. The populations used in both studies were not the same: the 1995 experiment investigated an F2 cross between the DeFries selection lines mentioned above, which themselves had been derived from an F3 cross between C57BL/6J and BALB/cJ. The 1999 study employed HS mice. This is a heterogeneous stock derived many generations ago from an eight-way cross between eight different inbred strains, C57BL/6J and BALB/cJ among them. One might therefore expect that a QTL analysis of the latter population would render several new QTLs, not found in the first study, simply because C57BL/6J and BALB/cJ would not carry different alleles for the genes involved. However, genetic differences between these two strains would be expected to crop up in the HS study, too. This was not the case. Both studies reported a QTL on chromosome 1, the only chromosome investigated by Talbot and colleagues (evidently, although not explicitly stated, based on the results of the previous study), but careful comparison of the data (Figure 1 in Flint et al., 1995, and Figures 1 and 2 in Talbot et al., 1999) shows that the confidence intervals of these putative loci did not overlap. Although

Talbot and colleagues did not address this issue, they later showed that QTLs with opposite effects on the phenotype were associated with the same allele in the HS, leading to this detection failure. More refined methods (Mott et al., 2000) showed two loci on chromosome 1, at a distance of about 20 cM, one of them the locus reported by Turri et al. (2001) and the other the locus reported by Flint and colleagues (Flint, 2003).

Conclusions

The studies discussed in the foregoing make it abundantly clear that, albeit deceptively simple-looking, exploratory behavior has a complex genetic basis. Although some of the first QTL studies seemed to contradict this picture, it has since become clear that this was due to a failure to detect genes of smaller effect or to dissolve multiple genes into multiple QTLs. The early findings of Whimbey and Denenberg (1967) that exploratory behavior as displayed in an open field is multifactorial, with exploration and fear/stress being the main motivational systems underlying the behavioral variation observed have been upheld by most later studies addressing this issue. A complex phenotype with a complex underlying genetic architecture implies a complex behavioral test. Obviously, the interpretation of data obtained in an open field is not easy: behavioral differences may be due to a number of factors (exploration, anxiety) or a combination thereof. The relative ease with which this test can be carried out, however, virtually guarantees that it will remain a useful tool in the behavior geneticist's toolbox.

Acknowledgments

I would like to thank Robert T. Gerlai (Toronto, Canada), Susanna Pietropaolo (Talence, France), and Frans Sluyter (Wakefield, MA, USA) for critically reading the manuscript. W. E. C. was supported by the Centre National de la Recherche Scientifique (UMR 5106, 5227, and 5287) and grants from the Conseil Régional d'Aquitaine, CNRS, the University of Bordeaux I, and the National Institute of Mental Health (MH072920).

References

Archer, J. (1973) Tests for emotionality in rats and mice: a review. *Anim Behav* 21: 205–235.

Barnett, S.A. and Cowan, P.E. (1976) Activity, exploration, curiosity and fear: an ethological study. *Interdisc Sci Rev* 1: 43–62.

Blizard, D.A., Takahashi, A., Galsworthy, M.J., Martin, B., and Koide, T. (2007) Test standardization in behavioural neuroscience: a response to Stanford. *J Psychopharmacol* 21: 136–139.

Blodgett, H.C. (1929) The effect of the introduction of reward upon the maze performance of rats. *Univ Calif Publ Psychol* 4: 113–134.

Bolivar, V.J., Caldarone, B.J., Reilly, A.A., and Flaherty, L. (2000) Habituation of activity in an open field: A survey of inbred strains and F1 hybrids. *Behav Genet* 30: 285–293.

Broadhurst, P.L. (1969) Psychogenetics of emotionality in the rat. *Ann N Y Acad Sci* 159: 806–824.

Broadhurst, P.L. and Jinks, J.L. (1974) What genetical architecture can tell us about the natural selection of behavioural traits. In van Abeelen, J.H.F. (ed.), *The Genetics of Behaviour*. North-Holland, Amsterdam, pp. 43–63.

Clément, Y., Adelbrecht, C., Martin, B., and Chapouthier, G. (1994a) Association of autosomal loci with the grooming activity in mice observed in open-field. *Life Sci* 55: 1725–1734.

Clément, Y., Martin, B., and Chapouthier, G. (1994b) Association of the chromosomal fragment containing the *short-ear* locus with grooming activity in the open-field. *Behav Genet* 24: 509–510.

Clément, Y., Martin, B., Venault, P., and Chapouthier, G. (1995) Involvement of regions of the 4th and 7th chromosomes in the open-field activity of mice. *Behav Brain Res* 70: 51–57.

Collins, R.L. (1967) A general nonparametric theory of genetic analysis I. Application to the classical cross. *Genetics* 56: 551.

Creel, D. (1980) Inappropriate use of albino animals as models in research. *Pharmacol Biochem Behav* 12: 969–976.

Crusio, W.E. (1992) Quantitative genetics. In Goldowitz, D., Wahlsten, D., and Wimer, R.E. (eds.), *Techniques for the Genetic Analysis of Brain and Behavior: Focus on the Mouse*, Vol. 8. Elsevier, Amsterdam, pp. 231–250.

Crusio, W.E. (1993) Bi- and multivariate analyses of diallel crosses: a tool for the genetic dissection of neurobehavioral phenotypes. *Behav Genet* 23: 59–67.

Crusio, W.E. (2001) Genetic dissection of mouse exploratory behaviour. *Behav Brain Res* 125: 127–132.

Crusio, W.E. (2007) An introduction to quantitative genetics. In Jones, B.C. and Mormède, P. (eds.), *Neurobehavioral Genetics: Methods and Applications*, 2nd revised edn. CRC Press, Boca Raton, FL, USA, pp. 37–54.

Crusio, W.E. and Schwegler, H. (1987) Hippocampal mossy fiber distribution covaries with open-field habituation in the mouse. *Behav Brain Res* 26: 153–158.

Crusio, W.E., Schwegler, H., and Brust, I. (1993) Covariations between hippocampal mossy fibres and working and reference memory in spatial and non-spatial radial maze tasks in mice. *Eur J Neurosci* 5: 1413–1420.

Crusio, W.E., Schwegler, H., and van Abeelen, J.H.F. (1989a) Behavioral responses to novelty and structural variation of the hippocampus in mice. I. Quantitative-genetic analysis of behavior in the open-field. *Behav Brain Res* 32: 75–80.

Crusio, W.E., Schwegler, H., and van Abeelen, J.H.F. (1989b) Behavioral responses to novelty and structural variation of the hippocampus in mice. II. Multivariate genetic analysis. *Behav Brain Res* 32: 81–88.

Crusio, W.E., Schwegler, H., and van Abeelen, J.H.F. (1991) Behavioural and neuroanatomical divergence between two sublines of C57BL/6J inbred mice. *Behav Brain Res* 42: 93–97.

Crusio, W.E. and van Abeelen, J.H.F. (1986) The genetic architecture of behavioural responses to novelty in mice. *Heredity* 56: 55–63.

DeFries, J.C. (1969) Pleiotropic effects of albinism on open field behaviour in mice. *Nature* 221: 65–66.

DeFries, J.C., Gervais, M.C., and Thomas, E.A. (1978) Response to 30 generations of selection for open-field activity in laboratory mice. *Behav Genet* 8: 3–13.

DeFries, J.C., Hegmann, J.P., and Halcomb, R.A. (1974) Response to 20 generations of selection for open-field activity in mice. *Behav Biol* 11: 481–495.

DeFries, J.C., Hegmann, J.P., and Weir, M.W. (1966) Open-field behavior in mice: evidence for a major gene effect mediated by the visual system. *Science* 154: 1577–1579.

DeFries, J.C., Wilson, J.R., and McClearn, G.E. (1970) Open-field behavior in mice: selection response and situational generality. *Behav Genet* 1: 195–211.

Flint, J. (2003) Analysis of quantitative trait loci that influence animal behavior. *J Neurobiol* 54: 46–77.

Flint, J., Corley, R., DeFries, J.C., Fulker, D.W., Gray, J.A., Miller, S., et al. (1995) A simple genetic basis for a complex psychological trait in laboratory mice. *Science* 269: 1432–1435.

Foshee, D.P., Vierck, C.J., Meier, G.W., and Federspiel, C. (1965) Simultaneous measure of general activity and exploratory behavior. *Percept Mot Skills* 20: 445–451.

Fuller, J.L. and Thompson, W.R. (1978) *Foundations of Behavior Genetics.* C.V. Mosby Co., St Louis, MO, USA.

Fullerton, J.M., Willis-Owen, S.A., Yalcin, B., Shifman, S., Copley, R.R., Miller, S.R., et al. (2008) Human-mouse quantitative trait locus concordance and the dissection of a human neuroticism locus. *Biol Psychiatry* 63: 874–883.

Gerlai, R., Crusio, W.E., and Csanyi, V. (1990) Inheritance of species-specific behaviors in the paradise fish (*Macropodus opercularis*): a diallel study. *Behav Genet* 20: 487–498.

Gershenfeld, H.K., Neumann, P.E., Mathis, C., Crawley, J.N., Li, X., and Paul, S.M. (1997) Mapping quantitative trait loci for open-field behavior in mice. *Behav Genet* 27: 201–210.

Hall, C.S. (1934) Emotional behavior in the rat. I. Defecation and urination as measures of individual differences in emotionality. *J Comp Psychol* 18: 385–403.

Hall, C.S. (1936a) Emotional behavior in the rat. II. The relationship between need and emotionality. *J Comp Psychol* 22: 61–68.

Hall, C.S. (1936b) Emotional behavior in the rat. III. The relationship between emotionality and ambulatory activity. *J Comp Psychol* 22: 345–352.

Hall, C.S. (1938) The inheritance of emotionality. *Sigma Xi Quart* 26: 17–27.

Hall, C.S. (1940) The inheritance of emotionality in the rat. *Psychol Bull* 37: 432.

Hall, C.S. (1951) The genetics of behavior. In Stevens, S.S. (ed.), *Handbook of Experimental Psychology*, 1st edn. John Wiley and Sons, New York, pp. 304–329.

Hall, C.S. and Ballachey, E.I. (1932) A study of the rat's behavior in a field. A contribution to method in comparative psychology. *Univ Calif Publ Psychol* 6: 1–12.

Hayman, B.I. (1954) The theory and analysis of diallel crosses. *Genetics* 39: 789–809.

Heimrich, B., Schwegler, H., Crusio, W.E., and Buselmaier, W. (1988) Substrain divergence in C3H inbred mice. *Behav Genet* 18: 671–674.

Henderson, N.D. (1978) Genetic dominance for low activity in infant mice. *J Comp Physiol Psychol* 92: 118–125.

Henderson, N.D. (1986) Predicting relationships between psychological constructs and genetic characters: an analysis of changing genetic influences on activity in mice. *Behav Genet* 16: 201–220.

Henderson, N.D., Turri, M.G., DeFries, J.C., and Flint, J. (2004) QTL analysis of multiple behavioral measures of anxiety in mice. *Behav Genet* 34: 267–293.

Henry, K.R. and Schlesinger, K. (1967) Effects of the albino and dilute loci on mouse behavior. *J Comp Physiol Psychol* 63: 320–323.

Jamot, L., Bertholet, J.-Y., and Crusio, W.E. (1994) Neuroanatomical divergence between two substrains of C57BL/6J inbred mice entails differential radial-maze learning. *Brain Res* 644: 352–356.

Kelley, A.E., Cador, M., and Stinus, L. (1989) Exploration and its measurement. A psychopharmacological perspective. In Boulton, A.B., Baker, G.B., and Greenshaw, A.J. (eds.), *Neuromethods, Vol. 13: Psychopharmacology*. Humana Press, Clifton, NJ, USA, pp. 95–144.

Lassalle, J.-M. and LePape, G. (1981) Differential effects of the albino gene on behavior according to task, level of inbreeding, and genetic background. *J Comp Physiol Psychol* 95: 655–662.

Lassalle, J.-M. and LePape, G. (1983) Measurements of the behavioral effects of albino mutation in mice (*Mus musculus*): comparisons of coisogenic inbred and hybrid lines. *J Comp Psychol* 97: 353–357.

Mathis, C., Neumann, P.E., Gershenfeld, H., Paul, S.M., and Crawley, J.N. (1995) Genetic analysis of anxiety-related behaviors and responses to benzodiazepine-related drugs in AXB and BXA recombinant inbred mouse strains. *Behav Genet* 25: 557–568.

McClearn, G.E. (1959) The genetics of mouse behavior in novel situations. *J Comp Physiol Psychol* 52: 62–67.

Mott, R., Talbot, C.J., Turri, M.G., Collins, A.C., and Flint, J. (2000) A method for fine mapping quantitative trait loci in outbred animal stocks. *Proc Natl Acad Sci USA* 97: 12649–12654.

O'Keefe, J. and Nadel, L. (1978) *The Hippocampus as a Cognitive Map*. Clarendon Press, Oxford.

Overstreet, D.H. (2007) The open field test for two. *J Psychopharmacol* 21: 140.

Renner, M.J. (1988) Learning during exploration: the role of behavioral topography during exploration in determining subsequent adaptive behavior. *Int J Comp Psychol* 2: 43–56.

Rodgers, R.J. (2007) More haste, considerably less speed. *J Psychopharmacol* 21: 141–143.

Roullet, P. and Lassalle, J.-M. (1990) Genetic variation, hippocampal mossy fibres distribution, novelty reactions and spatial representation in mice. *Behav Brain Res* 41: 61–70.

Stanford, S.C. (2007a) The open field test: reinventing the wheel. *J Psychopharmacol* 21: 134–135.

Stanford, S.C. (2007b) Open fields (unlike wheels) can be any shape but still miss the target. *J Psychopharmacol* 21: 144.

Takahashi, J.S., Pinto, L.H., and Vitaterna, M.H. (1994) Forward and reverse genetic approaches to behavior in the mouse. *Science* 264: 1724–1733.

Talbot, C.J., Nicod, A., Cherny, S.S., Fulker, D.W., Collins, A.C., and Flint, J. (1999) High-resolution mapping of quantitative trait loci in outbred mice. *Nature Genet* 21: 305–308.

Thompson, W.R. (1953) The inheritance of behaviour: behavioural differences in fifteen mouse strains. *Can J Psychol* 7: 145–155.

Tryon, R.C. (1929) The genetics of learning ability in rats. Preliminary report. *Univ Calif Publ Psychol* 4: 71–89.

Tryon, R.C. (1930) Genetic differences in maze-learning ability in rats. 111–119.

Turri, M.G., Henderson, N.D., DeFries, J.C., and Flint, J. (2001) Quantitative trait locus mapping in laboratory mice derived from a replicated selection experiment for open-field activity. *Genetics* 158: 1217–1226.

van Abeelen, J.H.F. (1963a) Mouse mutants studied by means of ethological methods. I. Ethogram. *Genetica* 34: 79–94.

van Abeelen, J.H.F. (1963b) Mouse mutants studied by means of ethological methods. II. Mutants and methods. *Genetica* 34: 95–101.

van Abeelen, J.H.F. (1963c) Mouse mutants studied by means of ethological methods. III. Results with yellow, pink-eyed dilution, brown, and jerker. *Genetica* 34: 270–286.

van Abeelen, J.H.F. (1965) *An Ethological Investigation of Single-Gene Differences in Mice*. Dept of Genetics, University of Nijmegen, Nijmegen, p. 79.

van Abeelen, J.H.F. (1970) Genetics of rearing behavior in mice. *Behav Genet* 1: 71–76.

van Abeelen, J.H.F. (1974) Genotype and the cholinergic control of exploratory behaviour in mice. In van Abeelen, J.H.F. (ed.), *The Genetics of Behaviour*. North-Holland, Amsterdam, pp. 347–374.

van Abeelen, J.H.F. (1975) Genetic analysis of behavioural responses to novelty in mice. *Nature* 254: 239–241.

van Abeelen, J.H.F. and Hughes, R.N. (1986) A note on behavioral divergence between two substrains of DBA/2 inbred mice. *Behav Genet* 16: 281–284.

van Abeelen, J.H.F. and Kroes, H.W. (1968) Albinism and mouse behaviour. *Genetica* 38: 419–429.

Wahlsten, D. (1999) Single-gene influences on brain and behavior. *Annu Rev Psychol* 50: 599–624.

Wahlsten, D., Bachmanov, A., Finn, D.A., and Crabbe, J.C. (2006) Stability of inbred mouse strain differences in behavior and brain size between laboratories and across decades. *Proc Natl Acad Sci USA* 103: 16364–16369.

Walsh, R.N. and Cummins, R.A. (1976) The open-field test: a critical review. *Psychol Bull* 83: 482–504.

Weischer, M.L. (1976) Eine einfache Versuchsanordnung zur quantitativen Beurteilung von Motilität und Neugierverhalten bei Mäusen. *Psychopharmacol* 50: 275–279.

Whimbey, A.E. and Denenberg, V.H. (1967) Two independent behavioral dimensions in open-field performance. *J Comp Physiol Psychol* 63: 500–504.

Chapter 16

Strains, SNPs, and selected lines

Genetic factors influencing variation in murine anxiety-like behavior

Andrew Holmes

Introduction

The study of fear and anxiety is one of the most active areas of research in psychiatry, neuroscience, and genetics. Anxiety research addresses fundamental questions about how the brain orchestrates complex behavioral, physiological, and endocrine responses to environmental provocation, in this case, in the form of danger or threat of harm. Major investment of resources in this field of research is further driven by the proliferation of anxiety disorders in the general population (Andrade et al., 2003; Kessler et al., 2003). Debilitation caused by anxiety disorders represents a growing burden on the health, welfare, and economy of industrialized nations (Greenberg et al., 2003), yet these conditions remain inadequately understood and treated (Holmes et al., 2003; Insel and Charney, 2003).

Rodent models of fear and anxiety provide the foundation for basic research on neurobiology and genetics of anxiety and anxiety disorders. While rats have traditionally been favored by researchers in this field, the mouse has recently become popular because of the relatively practicable application of powerful molecular genetic techniques, such as gene targeting and viral-mediated gene knockdown, that enable precise experimental manipulation of molecules of interest (Holmes et al., 2004; Tarantino and Bucan, 2000; Thakker et al., 2006). As a result, there are now numerous examples of abnormal fear and anxiety-related phenotypes in genetically modified mice in the literature (see Volume II of this Handbook, and Belzung and Griebel, 2001; Finn et al., 2003; Holmes, 2001; Holmes and Cryan, 2006; Thakker et al., 2006). Prior to "transgenics," "gene knockouts," and RNA interference, the mouse had often been employed in studies aimed at uncovering natural, as opposed to engineered, sources of genetic variation underlying individual differences in behavior (Cryan and Holmes, 2005; Phillips et al., 2002).

Within the normal (i.e., non-pathological) range of human temperament and personality, there is marked variation between individuals in the traits of neuroticism, harm avoidance, and negative affect. At the clinical end of the spectrum, there are individual differences in risk and prevalence of anxiety disorders, and twin studies indicate that there is a major genetic contribution to these differences in trait anxiety and risk for anxiety disorders (Kendler, 2001). Mice also exhibit marked differences in fear and anxiety-like behavior across genetically distinct strains and, depending upon genetic heterogeneity, across individuals within a given strain. Uncovering the gene variants underlying these differences is likely to be highly salient to human emotional states given the marked homology in genetic and biological systems across mammalian species.

Research aimed at identifying these variants has employed and continues to develop various approaches to this end (Mackay, 2001), and the present chapter provides an overview of some of the fruits of that work. This chapter also introduces the use of a naturalistic approach to the study of genetic factors influencing murine fear and anxiety-like behaviors.

Naturalistic approach to the study of murine fear and anxiety in the laboratory

In general terms, "anxiety" can be defined as an adaptive response to impending danger that is integral to an organism's preparations to deal with, or better still avoid, potential environmental threat. While psychiatric classifications of anxiety disorders do not distinguish between "anxiety" and "fear" (DSM-IV, 1994), work in animals typically defines fear as a reaction to explicit, imminent threats, and anxiety responses as responses to less explicit, more generalized threats. Anxiety tends to promote preparedness over a more sustained period by increasing arousal and vigilance, while fear responses are usually short-lived, evoking intense escape and avoidance of danger.

The study of behavioral manifestations of fear and anxiety in animals' naturalistic or semi-naturalistic settings has a long history (Darwin, 1867). The ethological approach to modeling anxiety in the mouse is based on knowledge of the natural behavioral patterns of the species. A major underlying assumption of this approach is that these normal patterns of behavior are not only mediated by the same neural and genetic pathways underlying anxiety in human beings, but that the same biological systems are implicated in clinically abnormal forms of

Behavioral Genetics of the Mouse: Volume I. *Genetics of Behavioral Phenotypes*, eds. Wim E. Crusio, Frans Sluyter, Robert T. Gerlai, and Susanna Pietropaolo. Published by Cambridge University Press.

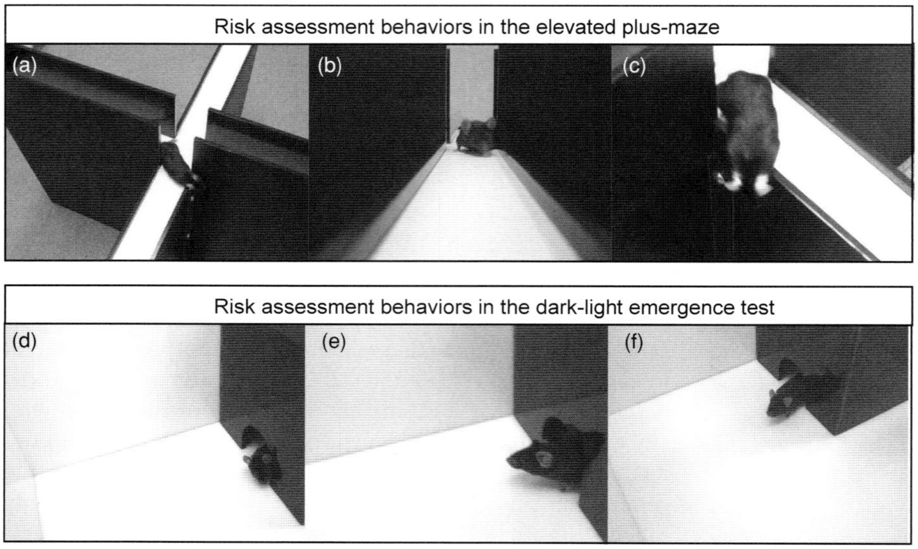

Figure 16.1 Anxiety-related risk assessment behaviors. Mice exhibit readily quantifiable risk assessment behaviors towards potentially dangerous stimuli and environments. Shown here is the "stretched attend posture" directed towards the open arms of the elevated plus-maze (a and b), and, in (c), stretched attend posture with head-dipping), and the light, open area in the dark–light emergence test (d–f).

anxiety, i.e., anxiety disorders are viewed as pathological exaggerations of normal behavioral patterns (for discussion, see Belzung and Griebel, 2001). There are at least two arguments in favor of this concept and in the validity of mouse anxiety models generally. First, in humans extreme anxiety states similar to those seen clinically can be induced in "normal" individuals by various psychological (e.g., public speaking), chemical (e.g., lactate infusion), and pharmacological (e.g., yohimbine) factors (Cryan and Holmes, 2005). Second, "anxiety-like behaviors" in mice can be reduced by treatment with clinically efficacious anxiolytics and are generally good, albeit imperfect, predictors of anxiolytic potential of novel therapeutic drug targets (Holmes et al., 2003).

Numerous ethologically based tests of anxiety-like behavior have been developed in the mouse. Mice have an innate aversion to novel, open, and/or brightly lit spaces (presumably an adaptive response to reduce the risk of predation), but are also a naturally foraging, exploratory species. Hence, exploration-based tests exploit these conflicting tendencies in the form of a test apparatus where the mouse's drive to *approach* is in conflict with the *avoidance* of potential threat (Crawley, 1981; Handley and Mithani, 1984; Takahashi et al., 1989). The so-called exploration-based approach–avoid conflict tasks, such as the elevated plus-maze, novel open field, light–dark exploration, or dark–light emergence test, have proven very popular phenotypic assays for anxiety-like behavior in mice. There are also conflict tests not based on exploration that have been to varying degrees validated for use in mice; these include the Vogel conflict test (Mathiasen and Mirza, 2005), social interaction test (File and Hyde, 1978), shock-probe burying test (De Boer and Koolhaas, 2003), and hyponeophagia test/novelty suppressed feeding test (Bodnoff et al., 1988; Dulawa and Hen, 2005). A discussion of the major methodological issues surrounding these tasks will be found in Volume III of this Handbook.

In addition to informing the development of ethologically based tests for anxiety, observations of wild and laboratory rats in semi-naturalistic environments has identified specific behavioral postures elicited by rodents in response to danger and threat (for reviews, see Blanchard and Blanchard, 2003; Rodgers, 1997; Rodgers et al., 1997; Treit, 1985). Thus, rats exposed to a predator such as a cat exhibited flight or freezing (Blanchard and Blanchard, 1969, 1989), consistent with the aforementioned definition of a fear reaction to a clear and imminent danger. These behaviors are also demonstrated by mice and freezing is commonly used as endpoint measure in tests for emotional learning and memory such as Pavlovian fear conditioning (Davis, 1990; Fanselow and Poulos, 2005; Izquierdo et al., 2006).

Exposure to a less explicit, ambiguous threat stimulus such as cat odor or a place where a cat was previously encountered typically elicits behaviors that serve to acquire information about the level of potential threat in the environment ("risk assessment") and, therefore, to be more anxiety-like. For example, rats will often scan a threatening environment or stimulus from a place of relative safety before venturing out into the open space, and then ambulate toward it with the body low to the ground and hesitantly stretch towards it ("stretched attend posture") (Blanchard and Blanchard, 2003). A number of laboratories have found that risk assessment behaviors can be readily quantified in many of the standard laboratory tests mentioned above (for reviews, see Rodgers, 1997; Treit, 1985). As an illustration of this, Figure 16.1 depicts a C57BL/6J mouse displaying stretched attend postures directed towards the open arms of the elevated plus-maze, and head-out scanning towards the light, open arena in the dark–light emergence test. The incorporation of risk-assessment behaviors as adjunct measures in these tests has met with some success in terms of detecting the actions of anxiolytic drugs and may be of some value in refining the phenotypic analysis in behavior genetics studies (Rodgers et al.,

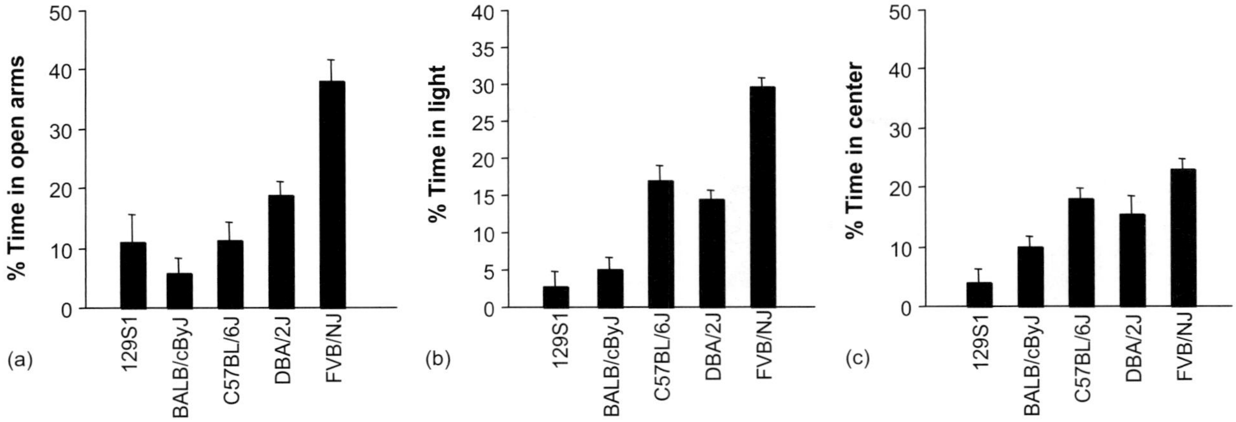

Figure 16.2 Mouse strain differences in anxiety-like behavior. Male 129S1/SvImJ (129S1), BALB/cByJ, C57BL/6J, DBA/2J, and FVB/NJ mice bred in the same vivarium and tested under the same laboratory conditions displayed differences in anxiety-like behavior on the (a) elevated plus-maze, (b) light–dark exploration test, and (c) novel open field test. Note how the pattern of strain differences varied somewhat according to behavioral task. (From Millstein and Holmes, 2007.)

1997). The inclusion of these adjunctive measures aside, the naturalistic ethological approach to measuring fear and anxiety-related behavior currently dominates the study of naturally-occurring genetic variation in these behaviors in mice.

Using between-strain differences to identify genetic factors underlying anxiety

Over the course of a century a plethora of strains of laboratory mice have been generated, such that researchers now have ready access to a range of diverse outbred and inbred strains. The existence of significant strain differences in "emotion-related" phenotypes has long been known (Coburn, 1922), and there have been numerous studies characterizing anxiety-like phenotypes and sensitivity to anxiolytics in some of the strains more commonly used in behavioral research (for reviews see Belzung, 2001a; Crawley et al., 1997; Millstein and Holmes, 2007). Figure 16.2 illustrates differences in anxiety-like behavior between five of the inbred strains that are commonly used in behavioral research (129S1/SvImJ, BALB/cByJ, C57BL/6J, DBA/2J, and FVB/NJ). Because all of these mice were bred in the same vivarium from parental strains obtained from a single supplier (The Jackson Laboratory) and tested under identical laboratory conditions on the elevated plus-maze, light–dark exploration test, and novel open field test (Millstein and Holmes, 2007), the presence of strain differences hints at a strong genetic component of anxiety-related phenotypes in mice.

Of course, these and any ostensibly "genetic differences" between strains or selected lines of mice, must be considered with the caveat that such phenotypic differences cannot be solely attributed to genetic factors because strains also differ in a genotype-related environment (e.g., maternal), which can also significantly affect anxiety-like behaviors (for reviews, see de Kloet et al., 2005; Holmes et al., 2005; Levine, 2000; Meaney, 2001; Newport et al., 2002; Pryce and Feldon, 2003). Notwithstanding, where robust phenotypic differences between strains are identified, these can then be used as a basis for elucidating the underlying genetic contributions. On one level, this can be achieved most by phenotyping a panel of inbred strains and statistically estimating the genetic correlation between phenotypic measures and the proportion of variation contributed by heritability (Owen et al., 1997b). Alternatively, robust differences between strains can lead to a search for polymorphic variation in candidate genes between the strains.

A recent example of how this type of approach can evolve is a comparison of the C57BL/6 and BALB/cBy strains. As shown in Figure 16.2 and other (although certainly not all) studies, the C57BL/6 and BALB/cBy or BALB/c strains have been described as having, respectively, low and high anxiety-like, as well as stress- and depression-related, behaviors (Anisman et al., 2001; Belzung and Griebel, 2001; Bouwknecht and Paylor, 2002; Chapillon et al., 1999; Crawley and Davis, 1982; Ducottet and Belzung, 2005; Francis et al., 2003; Griebel et al., 1993, 2000; Guillot and Chapouthier, 1996; Liu and Gershenfeld, 2001; Lucki et al., 2001; Rogers et al., 1999; Shanks et al., 1990, 1994; Tannenbaum and Anisman, 2003; Trullas and Skolnick, 1993; Trullas et al., 1989).

BALB/cJ have been found to carry a single nucleotide polymorphism (SNP) (C1473G) in the gene encoding for the brain-specific, rate-limiting enzyme synthesizing serotonin, tryptophan hydroxylase 2 (*Tph2*) (Zhang et al., 2004). This variant is associated with considerably lower brain serotonin concentration as compared to other strains carrying SNP C1473C, including C57BL/6J (Zhang et al., 2004). Given the wealth of data implicating serotonergic dysfunction in emotional disorders (Hariri and Holmes, 2006) and evidence specifically linking the a loss-of-function mutation in the human *TPH2* gene to emotional disorders (Zhang et al., 2006), one could speculate that this variant may contribute to differences in anxiety-like behavior between these strains. However, it is less clear whether other strains that are also variant for this SNP, such as 129X1/SvJ

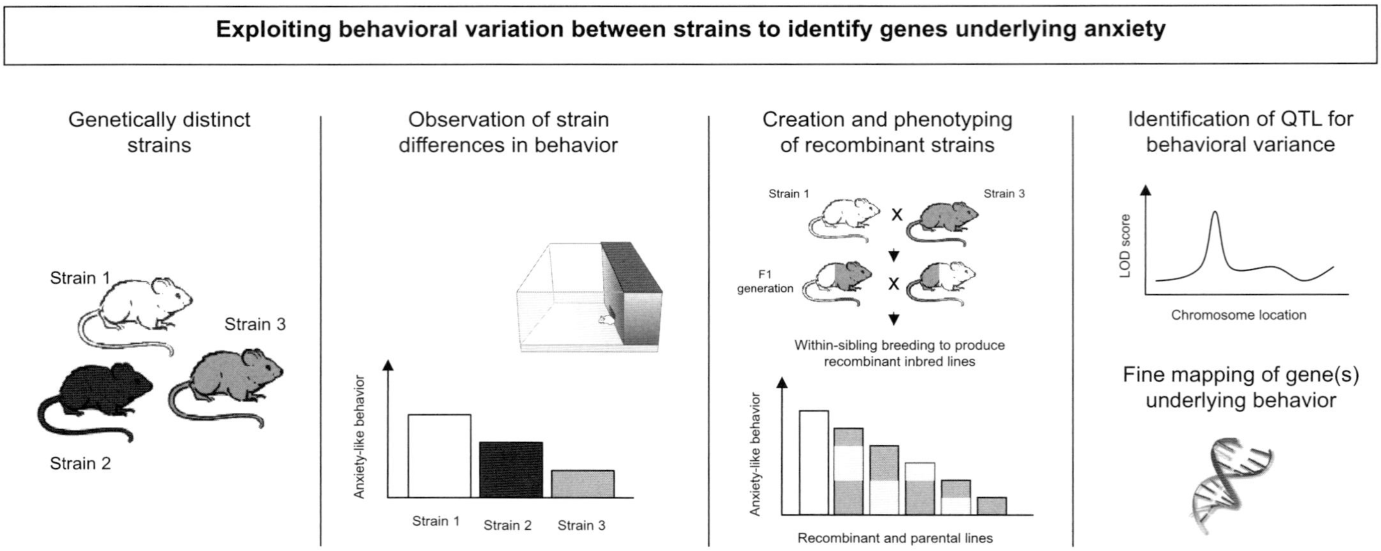

Figure 16.3 Exploiting behavioral variation between strains to identify genes underlying anxiety. Genetically distinct inbred strains can be compared for anxiety-like behaviors in well-validated tests, such as those based on approach–avoid conflict. Two strains exhibiting robust behavioral differences can then be used to create recombinant inbred (RI) strains. Using quantitative trait loci (QTL) and fine-mapping techniques, functional variants underlying differences in anxiety-like behavior between the RI strains can be uncovered.

versus DBA/2J, also exhibit such clear anxiety-related differences, and it will be important to evaluate the influence of variation in this specific gene using additional approaches such as breeding the C1473G SNP into "low-anxiety" background strains and engineering mice with loss of *Tph2* function. Nonetheless, this line of work provides an illustration of the potential power of employing multiple complementary approaches to identifying genetic factors influencing anxiety.

Strain differences provide further opportunities for identifying genes influencing fear and anxiety. Quantitative trait loci (QTL) analysis has had some success in identifying chromosomal regions associated with differences in mouse fear and anxiety-like behavior (for review see Flint, 2003). One method has been to interbreed two inbred strains which may or may not differ on a phenotype of interest, e.g., fear- or anxiety-like behavior. These can then provide the basis for generating a panel of recombinant inbred (RI) lines. Assuming continuous variation in the anxiety-like phenotype, the genetic diversity between the RI lines can be correlated with any observed variation in behavior to the uncover the QTLs and, ultimately, the specific genes involved (Figure 16.3). As an example of this strategy, C57BL/6J × A/J RI lines have been used to detect QTL anxiety-like behavior and sensitivity to the anxiolytic diazepam in the light–dark test (Mathis et al., 1995).

Another approach is to use F2 hybrid populations to identify QTLs. Gershenfeld and colleagues assessed novel open field and light–dark exploration test behaviors in a F2 population of C57BL/6J × A/J hybrid mice and detected two QTLs with moderate to high LOD scores (Gershenfeld and Paul, 1997; Gershenfeld et al., 1997). Using a C57BL/6J × DBA/2J F2 population, Wehner and colleagues have identified multiple QTLs associated with Pavlovian fear conditioning (Owen et al., 1997a; Wehner et al., 1997), and comparable results were independently obtained for QTLs associated with contextual fear conditioning using the C57BL/6J and C3H/HeJ strains (Caldarone et al., 1997).

The fact that multiple QTLs were identified in these various studies strongly suggests that these phenotypic traits are polygenic in nature. In this context, Turri and colleagues evaluated a genetically heterogeneous stock (HS) of mice on multiple tests, including the elevated plus-maze, light–dark exploration test, and novel open field, and found QTLs that were more test-specific, but also QTLs that were common to all tests (Turri et al., 1999). Moreover, at least one of the QTLs identified for Pavlovian fear conditioning (on chromosome 1) is in close proximity to an independently identified QTL for anxiety-related measures in the novel open field and elevated plus-maze in HS of mice. Taken together, these data suggest that, as with any complex trait, fear and anxiety-related phenotypes are almost certainly polygenic in nature, there may be specific genetic factors that exert a broad influence on even diverse measures of "emotionality" (Talbot et al., 1999; see also Flint et al., 1995). In fact, recent work has gone on to pinpoint a potential candidate gene in this region (regulator of G-protein signaling 2, *Rgs2*) by combining QTL analysis in the MF1 outbred strain with the classic behavioral genetics technique of quantitative complementation (Yalcin et al., 2004).

While exploiting strain differences to driving fear and anxiety-related traits is an attractive approach to uncovering naturally-occurring sources of genetic variation there are, as with any approach, caveats associated with QTL analysis (for a fuller discussion, see Flint and Mott, 2001; Mackay, 2001).

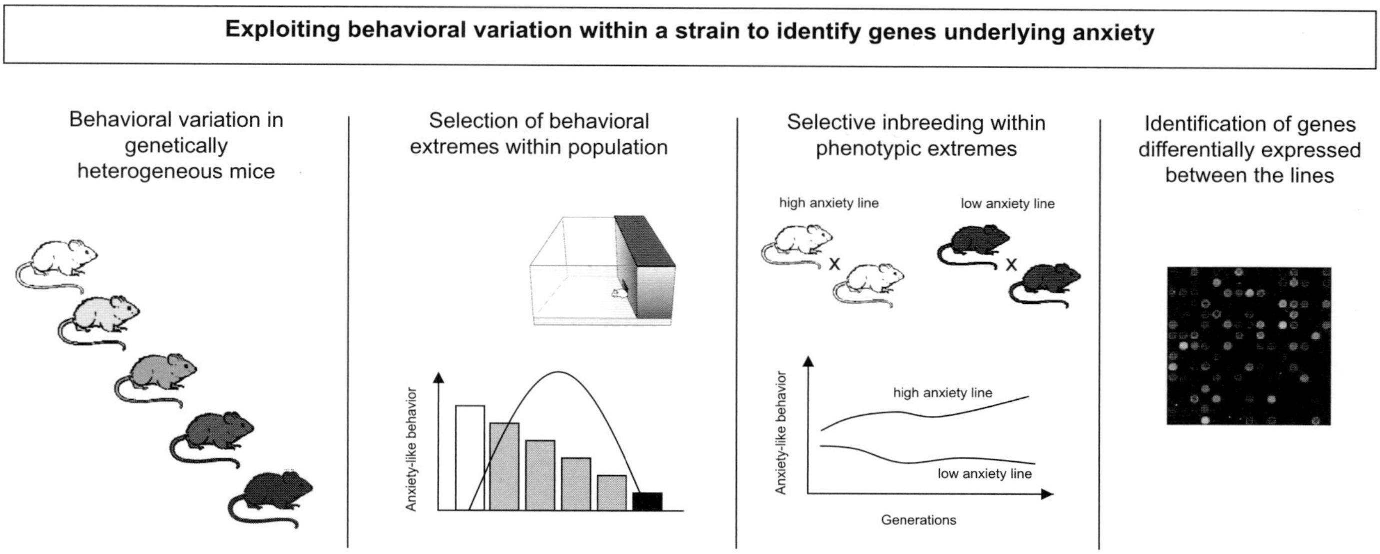

Figure 16.4 Exploiting behavioral variation within a strain to identify genes underlying anxiety. Interindividual variation in behavior within a genetically heterogeneous population such as an outbred strain can be assessed on a well-validated test. Population extremes in behavior can then be selected for over generations to produce "high-anxiety" and "low-anxiety" lines. Using gene and protein expression techniques, functional variants underlying differences in anxiety-like behavior between the lines can be uncovered.

One issue is that, because any given variant influencing a complex quantitative trait, such as anxiety, will exert a relatively small proportion of the overall variation in that trait, rates of false negative findings are likely high. Obtaining positive findings requires a significant investment of breeding and testing to produce what, in the absence of further fine mapping and positional cloning, are statistical associations with large chromosomal regions containing many genetic candidates (Darvasi, 1998). While these regions might contain functional variants affecting anxiety, they do not necessarily harbor the genes themselves (e.g., *trans*- versus *cis*-acting factors). This can lead to guesswork about which genes in the region identified are likely candidates based on other lines of evidence; a circular logic that can dilute the power of this approach for finding novel genes. Finally, it is important to remember that the generalizability of any observed association between a functional locus and anxiety-like behavior to human behavior is wholly contingent upon the validity of the behavioral assay used to measure that phenotype (Belzung, 2001b; Rodgers, 1997).

Using within-strain differences to identify genetic factors underlying anxiety

There is a rich history of selectively breeding rats from a genetically heterogeneous progenitor strain for differences in anxiety-like behavior (Figure 16.4); e.g., the Maudsley reactive line selected for relatively high defecation in a novel open field (Broadhurst, 1975) and the Roman high avoidance line selected for relatively high rates of active shock avoidance (Bignami, 1965). Mice selected for differential sensitivity to the pharmacologically induced convulsions exhibit differences in baseline anxiety-like behavior and responses to anxiolytics (Suaudeau et al., 2000). More recently, a number of groups have bred for differences in depression-related (El Yacoubi et al., 2003) and fear and anxiety-like phenotypes in mice. Thus, Kromer and colleagues exploited variability in elevated plus-maze test profiles in the CD-1 outbred strain to generated lines exhibiting high and low anxiety-like behavior on this test (Kromer et al., 2005). Interestingly, elevated plus-maze differences generalized to the light–dark exploration and pup separation-induced vocalization tests, as well as the forced swim and tail suspension tests for depression-related behavior, suggesting a degree of genetic correlation between these traits.

Short-term selection provides another, and relatively rapid, means to study genetic influences on mouse phenotypes such as anxiety-like behavior (Belknap et al., 1997). Two recent studies illustrate the utility of this approach. Radcliffe and colleagues selected for significant differences in cued and in both cued and contextual fear conditioning over three generations from a C57BL/6J × DBA/2J hybrid foundation population, and showed that candidate QTLs previously observed diverged in parallel with trait selection (Radcliffe et al., 2000). More recently, Ponder and colleagues also selected for differences in contextual fear conditioning four generations from a C57BL/6J × DBA/2J hybrid foundation population (Ponder et al., 2005). This study went on to show that the lines also diverged on another assay of emotional memory (fear potentiated startle), but not reference memory (the Morris water maze) and the novel open field and elevated zero-maze tests for anxiety-like behavior, and mapped a number of QTLs for these phenotypes. Additional studies employing this type of approach have not yet been published. In the future it will be important to generate a corpus of data that will allow the field to detect consistently identified QTLs and then focus on pinpointing the specific genes involved.

Conclusions

The adoption of a naturalistic approach to the study of fear and anxiety-related behaviors in the laboratory, together with the application of various strategies for delineating genetic factors contributing to differences in these behaviors, has produced some interesting findings in this still nascent field of research. Of course, identifying chromosomal loci and even individual genes that associate with a fear- or anxiety-related phenotype ultimately can provide correlation rather than causative evidence of a gene–behavior relationship. The unique power of the mouse as a model system is that it lends itself to the use of multiple, complementary lines of research, and it is the convergence of evidence from these various approaches that can ultimately further our understanding of the genetic basis of anxiety.

Acknowledgments

A. H. is supported by the National Institute on Alcohol Abuse and Alcoholism, Division of Intramural Clinical and Biological Intramural Research Program.

References

Andrade, L., Caraveo-Anduaga, J.J., Berglund, P., Bijl, R.V., De Graaf, R., Vollebergh, W., et al. (2003) The epidemiology of major depressive episodes: results from the International Consortium of Psychiatric Epidemiology (ICPE) Surveys. *Int J Methods Psychiatr Res* 12: 3–21.

Anisman, H., Hayley, S., Kelly, O., Borowski, T., and Merali, Z. (2001) Psychogenic, neurogenic, and systemic stressor effects on plasma corticosterone and behavior: mouse strain-dependent outcomes. *Behav Neurosci* 115: 443–454.

Belknap, J.K., Richards, S.P., O'Toole, L.A., Helms, M.L., and Phillips, T.J. (1997) Short-term selective breeding as a tool for QTL mapping: ethanol preference drinking in mice. *Behav Genet* 27: 55–66.

Belzung, C. (2001a) The genetic basis of the pharmacological effects of anxiolytics: a review based on rodent models. *Behav Pharmacol* 12: 451–460.

Belzung, C. (2001b) Rodent models of anxiety-like behaviors: are they predictive for compounds acting via non-benzodiazepine mechanisms? *Curr Opin Investig Drugs* 2: 1108–1011.

Belzung, C. and Griebel, G. (2001) Measuring normal and pathological anxiety-like behaviour in mice: a review. *Behav Brain Res* 125: 141–149.

Bignami, G. (1965) Selection for high rates and low rates of avoidance conditioning in the rat. *Anim Behav* 13: 221–227.

Blanchard, R.J., and Blanchard, D.C. (1969) Crouching as an index of fear. *J Comp Physiol Psychol* 67: 370–375.

Blanchard, R.J. and Blanchard, D.C. (1989) Antipredator defensive behaviors in a visible burrow system. *J Comp Psychol* 103: 70–82.

Blanchard, R.J. and Blanchard, D.C. (2003) Bringing natural behaviors into the laboratory: a tribute to Paul MacLean. *Physiol Behav* 79: 515–524.

Bodnoff, S.R., Suranyi-Cadotte, B., Aitken, D.H., Quirion, R., and Meaney, M.J. (1988) The effects of chronic antidepressant treatment in an animal model of anxiety. *Psychopharmacology (Berl)* 95: 298–302.

Bouwknecht, J.A. and Paylor, R. (2002) Behavioral and physiological mouse assays for anxiety: a survey in nine mouse strains. *Behav Brain Res* 136: 489–501.

Broadhurst, P.L. (1975) The Maudsley reactive and nonreactive strains of rats: a survey. *Behav Genet* 5: 299–319.

Caldarone, B., Saavedra, C., Tartaglia, K., Wehner, J.M., Dudek, B.C., and Flaherty, L. (1997) Quantitative trait loci analysis affecting contextual conditioning in mice. *Nat Genet* 17: 335–337.

Chapillon, P., Manneche, C., Belzung, C., and Caston, J. (1999) Rearing environmental enrichment in two inbred strains of mice: 1. Effects on emotional reactivity. *Behav Genet* 29: 41–46.

Coburn, C. (1922) Heredity of wildness and savageness in mice. *Behav Monogr* 5: 1–71.

Crawley, J.N. (1981) Neuropharmacologic specificity of a simple animal model for the behavioral actions of benzodiazepines. *Pharmacol Biochem Behav* 15: 695–699.

Crawley, J.N., Belknap, J.K., Collins, A., Crabbe, J.C., Frankel, W., Henderson, N., et al. (1997) Behavioral phenotypes of inbred mouse strains: implications and recommendations for molecular studies. *Psychopharmacology (Berl)* 132: 107–124.

Crawley, J.N. and Davis, L.G. (1982) Baseline exploratory activity predicts anxiolytic responsiveness to diazepam in five mouse strains. *Brain Res Bull* 8: 609–612.

Cryan, J.F. and Holmes, A. (2005) The ascent of mouse: advances in modelling human depression and anxiety. *Nat Rev Drug Discov* 4: 775–790.

Darvasi, A. (1998) Experimental strategies for the genetic dissection of complex traits in animal models. *Nat Genet* 18: 19–24.

Darwin, C.R. (1867) *The Expression of the Emotions in Man and Animals*. Oxford University Press, Oxford.

Davis, M. (1990) Animal models of anxiety based on classical conditioning: the conditioned emotional response (CER) and the fear-potentiated startle effect. *Pharmacol Ther* 47: 147–165.

De Boer, S.F. and Koolhaas, J.M. (2003) Defensive burying in rodents: ethology, neurobiology and psychopharmacology. *Eur J Pharmacol* 463: 145–161.

de Kloet, E.R., Sibug, R.M., Helmerhorst, F.M., and Schmidt, M. (2005) Stress, genes and the mechanism of programming the brain for later life. *Neurosci Biobehav Rev* 29: 271–281.

DSM-IV (1994) *Diagnostic and Statistical Manual of Mental Disorders, Fourth Edition (DSM-IV)*. American Psychiatric Association, Washington, DC.

Ducottet, C. and Belzung, C. (2005) Correlations between behaviours in the elevated plus-maze and sensitivity to unpredictable subchronic mild stress: evidence from inbred strains of mice. *Behav Brain Res* 156: 153–162.

Dulawa, S.C. and Hen, R. (2005) Recent advances in animal models of chronic antidepressant effects: the novelty-induced hypophagia test. *Neurosci Biobehav Rev* 29: 771–783.

El Yacoubi, M., Bouali, S., Popa, D., Naudon, L., Leroux-Nicollet, I., Hamon, M., et al. (2003) Behavioral, neurochemical, and electrophysiological characterization of a genetic mouse model of depression. *Proc Natl Acad Sci USA* 100: 6227–6232.

Fanselow, M.S. and Poulos, A.M. (2005) The neuroscience of mammalian associative learning. *Annu Rev Psychol* 56: 207–234.

File, S.E. and Hyde, J.R. (1978) Can social interaction be used to measure anxiety? *Br J Pharmacol* 62: 19–24.

Finn, D.A., Rutledge-Gorman, M.T., and Crabbe, J.C. (2003) Genetic animal models of anxiety. *Neurogenetics* 4: 109–135.

Flint, J. (2003) Animal models of anxiety. In Plomin, R., DeFries, J., Craig, I.W., and McGuffin, P. (eds.), *Behavioral Genetics in the Postgenomic Era*. American Psychological Association, Washington, D.C., pp. 425–442.

Flint, J., Corley, R., DeFries, J.C., Fulker, D.W., Gray, J.A., Miller, S., et al. (1995) A simple genetic basis for a complex psychological trait in laboratory mice. *Science* 269: 1432–1435.

Flint, J. and Mott, R. (2001) Finding the molecular basis of quantitative traits: successes and pitfalls. *Nat Rev Genet* 2: 437–445.

Francis, D.D., Szegda, K., Campbell, G., Martin, W.D., and Insel, T.R. (2003) Epigenetic sources of behavioral differences in mice. *Nat Neurosci* 6: 445–446.

Gershenfeld, H.K., Neumann, P.E., Mathis, C., Crawley, J.N., Li, X., and Paul, S.M. (1997) Mapping quantitative trait loci for open-field behavior in mice. *Behav Genet* 27: 201–210.

Gershenfeld, H.K. and Paul, S.M. (1997) Mapping quantitative trait loci for fear-like behaviors in mice. *Genomics* 46: 1–8.

Greenberg, P.E., Kessler, R.C., Birnbaum, H.G., Leong, S.A., Lowe, S.W., Berglund, P.A., et al. (2003) The economic burden of depression in the United States: how did it change between 1990 and 2000? *J Clin Psychiatry* 64: 1465–1475.

Griebel, G., Belzung, C., Misslin, R., and Vogel, E. (1993) The free-exploratory paradigm: an effective method for measuring neophobic behaviour in mice and testing potential neophobia-reducing drugs. *Behav Pharmacol* 4: 637–644.

Griebel, G., Belzung, C., Perrault, G., and Sanger, D.J. (2000) Differences in anxiety-related behaviours and in sensitivity to diazepam in inbred and outbred strains of mice. *Psychopharmacology (Berl)* 148: 164–170.

Guillot, P.V. and Chapouthier, G. (1996) Intermale aggression and dark/light preference in 10 inbred mouse strains. *Behav Brain Res* 77: 211–213.

Handley, S.L. and Mithani, S. (1984) Effects of alpha-adrenoceptor agonists and antagonists in a maze-exploration model of "fear"-motivated behaviour. *Naunyn Schmiedebergs Arch Pharmacol* 327: 1–5.

Hariri, A.R. and Holmes, A. (2006) Genetics of emotional regulation: the role of the serotonin transporter in neural function. *Trends Cogn Sci* 10: 182–191.

Holmes, A. (2001) Targeted gene mutation approaches to the study of anxiety-like behavior in mice. *Neurosci Biobehav Rev* 25: 261–273.

Holmes, A. and Cryan, J.F. (2006) Modeling human anxiety and depression in mutant mice. In Fisch, G.S. and Flint, J. (eds.), *Transgenic and Knockout Models of Neuropsychiatric Disorders*. Humana Press, Totowa, NJ, USA, pp. 237–263.

Holmes, A., Heilig, M., Rupniak, N.M., Steckler, T., and Griebel, G. (2003) Neuropeptide systems as novel therapeutic targets for depression and anxiety disorders. *Trends Pharmacol Sci* 24: 580–588.

Holmes, A., Lachowicz, J.E., and Sibley, D.R. (2004) Phenotypic analysis of dopamine receptor knockout mice; recent insights into the functional specificity of dopamine receptor subtypes. *Neuropharmacology* 47: 1117–1134.

Holmes, A., le Guisquet, A.M., Vogel, E., Millstein, R.A., Leman, S., and Belzung, C. (2005) Early life genetic, epigenetic and environmental factors shaping emotionality in rodents. *Neurosci Biobehav Rev* 29: 1335–1346.

Insel, T.R. and Charney, D.S. (2003) Research on major depression: strategies and priorities. *JAMA* 289: 3167–3168.

Izquierdo, A., Wellman, C.L., and Holmes, A. (2006) Brief uncontrollable stress causes dendritic retraction in infralimbic cortex and resistance to fear extinction in mice. *J Neurosci* 26: 5733–5738.

Kendler, K.S. (2001) Twin studies of psychiatric illness: an update. *Arch Gen Psychiatry* 58: 1005–1014.

Kessler, R.C., Berglund, P., Demler, O., Jin, R., Koretz, D., Merikangas, K.R., et al. (2003) The epidemiology of major depressive disorder: results from the National Comorbidity Survey Replication (NCS-R). *JAMA* 289: 3095–3105.

Kromer, S.A., Kessler, M.S., Milfay, D., Birg, I.N., Bunck, M., Czibere, L., et al. (2005) Identification of glyoxalase-I as a protein marker in a mouse model of extremes in trait anxiety. *J Neurosci* 25: 4375–4384.

Levine, S. (2000) Influence of psychological variables on the activity of the hypothalamic-pituitary-adrenal axis. *Eur J Pharmacol* 405: 149–160.

Liu, X. and Gershenfeld, H.K. (2001) Genetic differences in the tail-suspension test and its relationship to imipramine response among 11 inbred strains of mice. *Biol Psychiatry* 49: 575–581.

Lucki, I., Dalvi, A., and Mayorga, A.J. (2001) Sensitivity to the effects of pharmacologically selective antidepressants in different strains of mice. *Psychopharmacology (Berl)* 155: 315–322.

Mackay, T.F. (2001) The genetic architecture of quantitative traits. *Annu Rev Genet* 35: 303–339.

Mathiasen, L. and Mirza, N.R. (2005) A comparison of chlordiazepoxide, bretazenil, L838,417 and zolpidem in a validated mouse Vogel conflict test. *Psychopharmacology (Berl)* 182: 475–484.

Mathis, C., Neumann, P.E., Gershenfeld, H., Paul, S.M., and Crawley, J.N. (1995) Genetic analysis of anxiety-related behaviors and responses to benzodiazepine-related drugs in AXB and BXA recombinant inbred mouse strains. *Behav Genet* 25: 557–568.

Meaney, M.J. (2001) Maternal care, gene expression, and the transmission of individual differences in stress reactivity across generations. *Annu Rev Neurosci* 24: 1161–1192.

Millstein, R.A. and Holmes, A. (2007) Effects of repeated maternal separation on anxiety- and depression-related phenotypes in different mouse strains. *Neurosci Beiobehav Rev* 31: 3–17.

Newport, D.J., Stowe, Z.N., and Nemeroff, C.B. (2002) Parental depression: animal models of an adverse life event. *Am J Psychiatry* 159: 1265–1283.

Owen, E.H., Christensen, S.C., Paylor, R., and Wehner, J.M. (1997a) Identification of quantitative trait loci involved in contextual and auditory-cued fear conditioning in BXD recombinant inbred strains. *Behav Neurosci* 111: 292–300.

Owen, E.H., Logue, S.F., Rasmussen, D.L., and Wehner, J.M. (1997b) Assessment of learning by the Morris water task and fear conditioning in inbred mouse strains and F1 hybrids: implications of genetic background for single gene mutations and quantitative trait loci analyses. *Neuroscience* 80: 1087–1099.

Phillips, T.J., Belknap, J.K., Hitzemann, R.J., Buck, K.J., Cunningham, C.L., and Crabbe, J.C. (2002) Harnessing the mouse to unravel the genetics of human disease. *Genes Brain Behav* 1: 14–26.

Ponder, C.A., Kliethermes C.L., Drew, M.R., Muller, J.M, Das, K., Crabbe, J.C., et al. (2005) *Short-term Selected Lines for Fear Conditioning: QTL, Gene Expression, Learning and Anxiety-like Behaviors.* Society for Neuroscience, 35th Annual Meeting, Washington, D.C.

Pryce, C.R. and Feldon, J. (2003) Long-term neurobehavioural impact of the postnatal environment in rats: manipulations, effects and mediating mechanisms. *Neurosci Biobehav Rev* 27: 57–71.

Radcliffe, R.A., Lowe, M.V., and Wehner, J.M. (2000) Confirmation of contextual fear conditioning QTLs by short-term selection. *Behav Genet* 30: 183–191.

Rodgers, R.J. (1997) Animal models of "anxiety": where next? *Behav Pharmacol* 8: 477–496; discussion 497–504.

Rodgers, R.J., Cao, B.J., Dalvi, A., and Holmes, A. (1997) Animal models of anxiety: an ethological perspective. *Braz J Med Biol Res* 30: 289–304.

Rogers, D.C., Jones, D.N., Nelson, P.R., Jones, C.M., Quilter, C.A., Robinson, T.L., et al. (1999) Use of SHIRPA and discriminant analysis to characterise marked differences in the behavioural phenotype of six inbred mouse strains. *Behav Brain Res* 105: 207–217.

Shanks, N., Griffiths, J., and Anisman, H. (1994) Central catecholamine alterations induced by stressor exposure: analyses in recombinant inbred strains of mice. *Behav Brain Res* 63: 25–33.

Shanks, N., Griffiths, J., Zalcman, S., Zacharko, R.M., and Anisman, H. (1990) Mouse strain differences in plasma corticosterone following uncontrollable footshock. *Pharmacol Biochem Behav* 36: 515–519.

Suaudeau, C., Rinaldi, D., Lepicard, E., Venault, P., Crusio, W.E., Costentin, J., et al. (2000) Divergent levels of anxiety in mice selected for differences in sensitivity to a convulsant agent. *Physiol Behav* 71: 517–523.

Takahashi, L.K., Kalin, N.H., Vanden Burgt, J.A., and Sherman, J.E. (1989) Corticotropin-releasing factor modulates defensive-withdrawal and exploratory behavior in rats. *Behav Neurosci* 103: 648–654.

Talbot, C.J., Nicod, A., Cherny, S.S., Fulker, D.W., Collins, A.C., and Flint, J. (1999) High-resolution mapping of quantitative trait loci in outbred mice. *Nat Genet* 21: 305–308.

Tannenbaum, B. and Anisman, H. (2003) Impact of chronic intermittent challenges in stressor-susceptible and resilient strains of mice. *Biol Psychiatry* 53: 292–303.

Tarantino, L.M. and Bucan, M. (2000) Dissection of behavior and psychiatric disorders using the mouse as a model. *Hum Mol Genet* 9: 953–965.

Thakker, D.R., Hoyer, D., and Cryan, J.F. (2006) Interfering with the brain: use of RNA interference for understanding the pathophysiology of psychiatric and neurological disorders. *Pharmacol Ther* 109: 413–438.

Treit, D. (1985) Animal models for the study of anti-anxiety agents: a review. *Neurosci Biobehav Rev* 9: 203–222.

Trullas, R., Jackson, B., and Skolnick, P. (1989) Genetic differences in a tail suspension test for evaluating antidepressant activity. *Psychopharmacology (Berl)* 99: 287–288.

Trullas, R. and Skolnick, P. (1993) Differences in fear motivated behaviors among inbred mouse strains. *Psychopharmacology (Berl)* 111: 323–331.

Turri, M.G., Talbot, C.J., Radcliffe, R.A., Wehner, J.M., and Flint, J. (1999) High-resolution mapping of quantitative trait loci for emotionality in selected strains of mice. *Mamm Genome* 10: 1098–1101.

Wehner, J.M., Radcliffe, R.A., Rosmann, S.T., et al. (1997) Quantitative trait locus analysis of contextual fear conditioning in mice. *Nat Genet* 17: 331–334.

Yalcin, B., Willis-Owen, S.A., Fullerton, J., Meesaq, A., Deacon, R.M., Rawlins, J.N., et al. (2004) Genetic dissection of a behavioral quantitative trait locus shows that Rgs2 modulates anxiety in mice. *Nat Genet* 36: 1197–1202.

Zhang, X., Beaulieu, J.M., Gainetdinov, R.R., and Caron, M.G. (2006) Functional polymorphisms of the brain serotonin synthesizing enzyme tryptophan hydroxylase-2. *Cell Mol Life Sci* 63: 6–11.

Zhang, X., Beaulieu, J.M., Sotnikova, T.D., Gainetdinov, R.R., and Caron, M.G. (2004) Tryptophan hydroxylase-2 controls brain serotonin synthesis. *Science* 305: 217.

Chapter 17

Genetic influences on infant mouse ultrasonic vocalizations

Robert H. Benno and Martin E. Hahn

Introduction

Ethology of ultrasonic calls

In common with the pups of a number of rodent species, infant mice are altricial and have limited sensory abilities, poorly developed motor skills, and do not thermoregulate (Fox, 1965; Nagy, 1993). Because of these characteristics, pups are dependent on adults to meet their physiological needs in the days following birth. If a pup has a physiological need, it may signal that need to an adult. Thus, in common with the young of a number of rodent species, an infant mouse is able to produce several sounds which may be informative about its physiological state and act as signals in a communication link between the pup and its mother or other adults (Sales and Smith, 1978). These sounds have been cataloged by Haack et al. (1983) and by Ehret and Bernecker (1986). The former (Haack et al., 1983) identified six sounds that are produced by infant mice: wriggling sounds, smacking sounds, cracking sounds, postpartum sounds, rough handling sounds, and isolation sounds. Ehret and Bernecker (1986) argued that only wriggling calls, rough handling sounds, and isolation sounds were signals (or calls) and effective in infant–adult relationships. Wriggling calls have a frequency typically below 10 kHz and elicit licking and nest building by mothers. Rough handling sounds have a broad frequency spectrum with sonic and ultrasonic components that seem to inhibit rough handling by adults. Isolation sounds are typically above 50 kHz and have been shown to elicit maternal searching and retrieval. Isolation sounds are produced by pups that are: isolated from the nest, cold, being handled by adults, falling, or some combination of these situations (Hahn and Lavooy, 2005; Hahn and Schanz, 2002; Okon, 1970a, 1970b). In other words, the infant is signaling a state of distress (Noirot, 1972).

In this chapter, we will focus on the third of the sounds described by Ehret and Bernecker (1986) – ultrasonic vocalizations or calls. While these calls have been referred to in several ways in the literature – they have been called ultrasonic vocalizations, USVs, emissions, distress vocalizations, or isolation calls – the common theme is that these calls are whistles with an average frequency of 40 kHz or higher and have a duration of about 50 ms (Hahn and Lavooy, 2005).

Measures of ultrasonic calls

If a mouse pup is in "distress" and signals that in its vocalizations, what part or parts of the signal carry that information? To quantify distress, what aspects of calls have investigators measured? Ultrasonic calls vary on several dimensions and a pup could encode distress level in one or several of those dimensions. There is wide variation in the number of calls that pups emit in a given time period (rate of calling). Individual calls vary in duration, amplitude, and frequency (pitch) characteristics. Since the pitch of a call typically varies across its duration, those characteristics may combine and individual calls vary in pitch over time, producing the waveform of the call. Calls are almost always emitted in groups. Groups of calls vary in the number per group and the spacing of the calls within and between groups. Thus, the intervals between calls in a group (intercall interval) and the bout structure of an entire set of calls varies. Finally, the call characteristics of individual pups vary across developmental age.

Most investigators have studied the rate of calling. As part of a review article on infant ultrasonic calling, Hahn and Lavooy (2005) surveyed a set of 71 studies – rate data were reported in 70 of them. Duration and or frequency data were reported in 15 studies. The average frequency and the highest and lowest frequencies were often measured and bandwidth was occasionally reported. The amplitude of calls was infrequently measured – only five of 71 studies reported data on amplitude and these five were carried out in the 1970s. Brudzynski and colleagues (Brudzynski, 2005; Brudzynski et al., 1999) characterized the waveforms of the calls of infant rats into a set of categories depending on the number of frequency sweeps or changes. Barron and Gilbertson (2005) used Brudzynski's system to differentiate the calls of infant rats exposed perinatally to ethanol or cocaine or both. Tobon et al. (2005) studied the ultrasonic call waveforms of infant mice using a catalog modeled on Brudzynski et al.'s (1999). They found strain differences in waveform complexity and showed that waveforms differed in complexity as a function of the eliciting stimulus (cold or rotation). Elwood and Keeling (1982) using mice and Myers et al. (2004) using rats looked at the bout structure of calls. In an interesting approach that combined all characteristics of calls, Holy and Guo (2005)

Behavioral Genetics of the Mouse: Volume I. *Genetics of Behavioral Phenotypes*, eds. Wim E. Crusio, Frans Sluyter, Robert T. Gerlai, and Susanna Pietropaolo. Published by Cambridge University Press.

argued that adult male mice produce courtship songs, characterized by different calls, placed in sequences or phrases that are sometimes repeated. The analysis of Holy and Guo has not been applied to infant mice, but observations in our laboratory suggest that infant calling may exhibit similar patterns. Finally, Branchi et al. (2001) and Hahn et al. (1998) have described changes in the rate of calling and other call characteristics across the early days of young mouse pups. Call rate describes a highly reliable, shallow inverted U-shaped function across ages from 2 days through to 10–12 days of age. Hahn et al. (1998) described the developmental trajectories of call duration and frequency.

Previous reviews of ultrasonic vocalization studies

Several reviews are available on topics within the literature on infant mouse ultrasonic vocalizations. Noirot (1972) reviewed the pioneering papers in the field published between the beginning of the field in 1956 and 1972. Geyer and Barfield (1979) organized a special issue of *American Zoologist* which surveyed a number of topics in mouse ultrasonic calling including infant calls. Willott's text (1983) provided a set of chapters which examined the anatomical, physiological, and behavioral aspects of sound use in the mouse. Branchi et al. (2001) surveyed the literature on infant rodent ultrasonic vocalizations, especially in relationship to pup neurobehavioral development. Most recently, Hahn and Thornton (2005) edited a special issue of *Behavior Genetics* focused on various aspects of mouse and rat ultrasonic vocalizations and maternal behavior. Within that issue, Hahn and Lavooy (2005) completed a comprehensive review of the methods used to study infant rodent ultrasonic vocalizations. In the same issue, Ehret (2005) also reviewed the literature on infant rodent ultrasonic vocalizations, and suggested that studies in this area have great potential as a "tool" to study genetic and neural mechanisms of behavioral control. Our objectives in this chapter are to: (1) survey a sample of studies of genetic influences on the ultrasonic calls of infant mice with an eye toward the advantages and disadvantages of the designs employed; and (2) make recommendations for future studies. It would have been illustrative in this paper if we could have included a table with the characteristics of ultrasonic vocalizations of infantile mice and the responses of the mothers to the calls. However, it is not possible to include a meaningful table at the present time due to the paucity of data and the lack of standardization of procedures in reported studies from different laboratories.

Genetic influences on ultrasonic calls of infant mice

Background

Historically, animal behavior geneticists have used four methods to investigate the role of genetic variation on phenotypic variation. The methods are: Mendelian crosses, selective breeding programs, comparisons of inbred strains, and cross-breeding studies (Fuller and Thompson, 1960). The first of the four methods was used to test the notion that single genes had behavioral effects. As Fuller and Thompson stated:

> Theoretically, every gene may contribute to some degree to behavioral variance. This hypothesis must be tested by placing known genes or combinations of known genes upon a uniform genetic background (p. 51).

The remaining three methods were designed to deal with the combined effects of genes at many loci, that is, quantitative characters. Selective breeding experiments were designed to change the mean phenotype of a population by enhancing or reducing the reproduction of some individuals within the population. Inbred strains are the result of brother–sister mating for a minimum of 20 generations and this breeding system produces near genetic homogeneity within strains (Plomin et al., 2001). Crossing inbred strains or selected lines has become almost standard in behavior genetics. As Fuller and Thompson (1960) stated about cross-breeding studies:

> By suitable analytical techniques the inheritance of quantitative characters affected by many genes can be described in terms of additive effects, dominance, and epistasis. Estimates of the number of loci contributing to interstrain variation may also be sought (p. 85).

While behavior genetics has advanced as a field many-fold in the more than four decades since Fuller and Thompson wrote the text which defined the field (Fuller and Thompson, 1960), those four methods designed to estimate the genetic contribution to a trait or determine the means of inheritance remain in use today. In this section of our chapter, we will describe five studies, exemplifying three of the four methods, and note their contributions to the understanding of genetic influences on the characteristics of infant mouse ultrasonic calls and commenting on their benefits and drawbacks.

Strain comparison

Bell et al. (1972) studied the ultrasonic calls of infant mice of three inbred strains (C57BL/6J, BALB/cJ, and C3H/HeJ). Based on previously observed differences among species and subspecies (e.g., Hart and King, 1966), Bell and colleagues set out to complete a strain comparison as a step toward genetic analysis of call characteristics and as a window into established differences in mother–pup interactions. They compared pups of the three strains, at four ages (3, 6, 9, and 12 days of age). The results showed strong strain and age differences in: the rate of calling, highest frequency, and call duration. The C57BL/6J pups had the lowest rates of calling while the other two showed higher rates that were similar to each other. Bell et al. (1972) also observed call rate to roughly describe an inverted U-shaped function over the ages 3, 6, 9, and 12 days. The authors concluded that different rates of pup development among the three strains may have been a function of differential maternal care, which was in turn elicited by differential rates of calling. The Bell et al. (1972) study is the first of which we are aware that demonstrated strain differences and thus genetic influences on infant mouse (*Mus musculus*) ultrasound characteristics.

The laboratory of Pierre Roubertoux completed a series of studies on the interactions between mouse pups and mothers. In study V of the series, Cohen-Salmon et al. (1985) compared the rates of ultrasonic calling by pups and the hearing capacities of adult females of eight inbred strains. They extended the questions of Bell et al. (1972) by asking whether differences in maternal care are directly related to the ultrasound emission rate of their pups and if the mothers have the capacity to hear the calls of their pups. In the first part of their study, they observed an effect of strain on ultrasonic call rate and they were able to divide the set of eight inbred strains into "high" and "low" callers. In a similar manner, they divided adult females of the eight strains into "high" and "low" in terms of their sensitivities to frequencies in the ultrasonic range. In a later phase of the study, they put the pup and adult female data sets to work, by completing a cross-fostering study with two of the original eight strains. High calling pups (C57BL/6JOrl mice) were raised by sensitive and insensitive mothers as were low calling pups (XLII/Orl mice). After the cross-fostering manipulation, they carried out an often-used test of maternal behavior – maternal retrieval. The results showed that pup type (high or low caller) influenced latency to retrieval as high calling C57BL/6J pups were retrieved before the low calling XLII pups, regardless of the strain of the mother who had reared them and was retrieving. Neither the mother type (high or low sensitive) nor its interaction with the pup type was statistically reliable. By using a strain comparison method, the authors were able to further establish genetic influences on infant mouse ultrasonic calls. By cross fostering, the authors were able to test hypotheses about the relationship between ultrasonic calling rate and maternal retrieval. For example, if ultrasounds are in fact cues eliciting retrieval behavior then high emitters should be retrieved more quickly than low emitters (by hearing mothers). In mothers that are insensitive to ultrasounds, there should be no difference in the retrieval latencies of high or low ultrasounding pups. Since the high calling pups were retrieved first regardless of the hearing abilities of the mother, the Cohen-Salmon et al. (1985) study concluded that the reputed positive relationship between ultrasonic calling and maternal retrieval might have been overemphasized in previous studies. They argue that retrieval is likely under the control of multiple sensory modalities.

Taken together, the studies of Bell et al. (1972) and Cohen-Salmon et al. (1985) indicate that genotype (strain) influences the rate of ultrasonic calling by pups. The relationship between ultrasonic call variability and maternal behavior remained elusive.

The results of strain comparison studies allow beginning modeling of the components of total phenotypic variation for any trait, including variability in infant ultrasonic call characteristics.

Based on the well-established model:

$$V_{\text{phenotype}} = V_{\text{genotype}} + V_{\text{environment}}(V_{\text{P}} = V_{\text{G}} + V_{\text{E}})$$

it is possible to estimate broad sense heritability (the coefficient of genetic determination) by dividing the between strain variance by the total variance or $V_{\text{G}}/V_{\text{P}}$.

While neither the Bell et al. (1972) study nor the Cohen-Salmon et al. (1985) study calculated a broad sense heritability, such a calculation would be possible using their data sets. Adding a cross-fostering manipulation to a strain comparison study (as did Cohen-Salmon et al.) and measuring the ultrasonic call characteristics of all the offspring allows an estimation of the amount to which the early rearing environment contributes to variability in call characteristics and a partition of the within (or environmental) component of call variability. Because genetic differences are so commonly observed for behaviors, strain comparison studies are now rarely published without additional analysis.

Finally, a recent study by Wohr et al. (2008) demonstrates how it is possible to take advantage of the latest technology to look at the relative roles of genes and environment in the analysis of isolation-induced ultrasounds. Utilizing an embryo-transfer technique, these authors were able to show the relative roles of genetic background, gender, and early environmental factors on ultrasonic vocalization in isolated mouse pups. This study, carried out in substrains of C57BL/6J mice, indicated that the number of calls produced by the pups are dependent on the dyadic interaction between the mother and pups, while call features, such as frequency modulation, were solely dependent on the genotype and gender of the pup.

Cross-breeding studies

To date, only one laboratory has studied infant mouse ultrasonic calls using crosses of multiple inbred strains. In a series of three studies, Hahn and colleagues (Hahn et al., 1997, 1998; Thornton and Hahn, 2005) examined the patterns of inheritance for several call characteristics using a diallel, cross-breeding design. The design is a complete crossing of a set of inbred strains and allows the estimation of additive and dominance inheritance components, and maternal effects (a component containing both genetic and environmental factors). In the first study (Hahn et al., 1997), call characteristics of 3-day-old pups were analyzed. The second study (Hahn et al., 1998) involved no genetic analysis, but tracked developmental trends of call characteristics of pups aged 2–12 days of age. The third study (Thornton and Hahn, 2005) reported a detailed developmental genetic analysis of the calls of mice as they aged from 3 to 9 days of age. In this section of our chapter, we will discuss the Hahn et al. (1997) study which examined the genetic architecture of call characteristics of 3-day-old inbred mice. The study examined the calls of 246 pups that were the F1 progeny of a complete crossing of four inbred strains (C57BL/10J, DBA/2J, BALB/cJ, and SJL/J), producing 16 genetic groups: four inbred strains, six hybrids, and six reciprocal hybrids. At the core (the diagonal) of an N × N diallel breeding design is a strain comparison of the strains crossed in the diallel. Completing a one-way analysis of variance (ANOVA) on those strains allows the estimation of heritability, where broad sense heritability = $\mathbf{V_G/V_P}$.

The calculated heritabilities were: 0.59 for rate; 0.55 for highest frequency; 0.68 for lowest frequency; and 0.36 for duration. The one-way ANOVAs were significant for strain on each of those call characteristics. In addition to estimating the amount of variability associated with genotype, specific analyses of the diallel allow estimates of the patterns of inheritance of the trait. Looking at call rate, for example, the additive and all dominance components were significant, with the additive component of the model explaining 21% of the variance and the combined dominance components explaining 14%. When the results of the study were interpreted within the theoretical framework of biometrical genetics (see Broadhurst and Jinks, 1974), the results for each of the characteristics, except highest frequency, indicated a character that had been under positive selection pressure. Further, since heritabilities for each of the traits were greater than 0.0, artificial selections on these characteristics would be expected to have positive outcomes using a heterogeneous population that contained the genes of the inbred strains that had been used by Hahn et al. (1997).

Cross-breeding studies provide an overview or survey of the inheritance of a phenotype of interest. If we return to the model we used in the previous section:

$$V_{\text{phenotype}} = V_{\text{genotype}} + V_{\text{environment}}$$

Cross-breeding studies allow the investigator to partition genetic variability into variability associated with additive and dominance components. So the model could be expanded to:

$$V_{\text{phenotype}} = V_{\text{additive}} + V_{\text{dominance}} + V_{\text{environment}}$$

Once estimated, the results could be interpreted within biometric theory as described in the previous paragraph. There are limitations associated with such interpretations and Henderson (1990) provides a thorough analysis of the advantages and disadvantages of cross-breeding studies as they are applied to behavior. Henderson argues that although diallel crosses and related cross-breeding designs are powerful methods to estimate the actions of natural selection on behavioral phenotypes, the assumptions of the methods involving the sampling of genotypes are rarely met. A 4 × 4 diallel design, for example, which produces 16 groups for phenotypic analysis, examines the interactions of only four genotypes. More representative designs, such as 5 × 5 or 6 × 6 diallels, generate 25 and 36 genetic groups respectively. The study of any complex behavior places designs of these sizes outside the feasibility of most investigations.

Selection studies

An artificial selection study is another way of determining whether and to what extent phenotypic variability has a genetic basis. Typically, a bidirectional selection is carried out and the resulting groups with extreme phenotypes are studied, looking for the mechanisms that account for the differences in the phenotype. We know of no artificial selections that have been carried out on characteristics of infant ultrasonic calls in mice. There is, however, an ongoing and effective selection for the rate of ultrasonic calling taking place in rat pups (Brunelli, 2005) and the process and results of that work are instructional. At 20 generations, the selection process had produced a line of rats that produced about 400 calls in 2 minutes and a line that produced fewer than 20 calls in 2 minutes (Brunelli, 2005). Brunelli and her colleagues are now engaged in looking for characteristics in the lines that might underlie the impressive difference in rate.

Two cautions are appropriate here. First, selective breeding studies are not for the "faint of heart." They typically require many years to complete the breeding process and then many more in the search for mechanisms that may underlie the phenotypic differences between selected lines. Second, Henderson (1997) discussed attempts to uncover the mechanisms underlying the phenotypic differences brought about by artificial selection programs, particularly those involving unreplicated selected lines. He cautions that genetic drift may play a role as selected lines are carried across generations and that drift may lead to finding spurious associations between selected lines and mechanisms thought to underlie behavioral differences between lines.

Mendelian crosses and other methods which examine the effects of single genes

The final approach we will discuss attempts to find or isolate individual genes with major influences on a behavior. Two approaches are possible. In the first, a Mendelian cross, the investigator produces F1, F2, B1, B2, and sometimes F3 generations from two inbred strains, as did Mendel with varieties of peas. In early studies on behavior using this approach (e.g., Collins and Fuller, 1968), the intent was to determine if single genes with large effects or polygenic systems having many genes with small effects were influencing phenotypes. In those studies, parental and F1 generations were used to derive expectations for the means of the remaining generations using genetic models that contained various genetic and environmental components. Model fitting tests were employed and the investigator determined which model best fit the generational outcomes. In the case of the Collins and Fuller (1968) study, audiogenic seizure susceptibility was attributed to a specific gene likely located in linkage group VIII of the mouse. In the history of research on infant mouse ultrasonic calls, we know of two studies that have used this approach, Hahn et al. (1987) and Roubertoux et al. (1996), and we will discuss the Roubertoux et al. paper. In that study, the investigators first surveyed the rate of ultrasonic calling in seven inbred strains. Two of the seven strains (CBA/H and NZB/BINJ) produced dramatically different numbers of calls (CBA/H produced high call rates and NZB/BINJ produced low rates) during room temperature/isolation testing and were chosen for crossing and subsequent analysis. After fitting various models, Roubertoux et al. (1996) determined that: (1) the ultrasonic call rate was controlled by a polygenic system; and that (2) those models that contained additive and dominance components provided the best fit. The specifics of the analysis indicated that directional dominance was present and that it favored a high rate of calling.

Two observations are appropriate at this point. First, Roubertoux et al. sampled two genotypes in their study, so the generality of the results is restricted. This is the same criticism that Henderson (1990) applied to diallel cross-studies that used a small sample of inbred strains. Second, in spite of criticisms that could be leveled at each of the three studies, the results of Bell et al. (1972), Roubertoux et al. (1996), and Hahn et al. (1997) are consistent across studies and indicate that there are strong genetic influences on infant mouse ultrasonic call rates and other call characteristics. Further, the genetic analyses of Roubertoux et al. (1996) and of Hahn et al. (1997) using populations of inbred strains that had little overlap indicated the presence of additive and dominance genetic influences consistent with the interpretation that there is strong directional dominance for high rates of calling in those populations. That result indicates that artificial selection would be successful on ultrasonic call rate and that high rates of calling could be a fitness character that has been an important adaptation in infant mice. The successful selection in rats reported by Brunelli (2005) indicates that a similar genetic architecture is likely in place in populations of infant rats.

Konrad Lorenz (1981) discussed the steps required to understand any organic system. He argued that such a system should be understood from the outside–in. That is, the investigator should understand the broad functioning of the system before taking the system apart and attempting to establish causal relationships among its parts. Carrying out strain comparisons, cross-breeding studies, or artificial selections provides information about the overall pattern of inheritance of a specific phenotype. Single gene studies in which genes are substituted or knocked out of a genotype in the absence of information about the components of phenotypic variance or patterns of inheritance of a trait may be uninterpretable.

Summary and recommendations for future studies

For a variety of reasons, the study of infant mouse and rat ultrasonic calls is a growth industry. The calls are interesting from a number of points of view and investigators are beginning to understand their role in infant–mother interactions, in normal pup development, and possibly in preparation for a role in adulthood. The calls are also a convenient model for investigating the roles of pharmaceuticals in human psychopathological conditions, like anxiety. Ultrasonic calls are complex signals in which the basic parameters of individual calls, like rate, length, and frequencies, have been studied and there is a good data base. On the other hand, work is just beginning on the waveforms of calls and how calls are spaced over time in bouts. As we have discussed, inbred strains, artificial selections, crosses among inbred strains, and techniques in which single genes are isolated on standard backgrounds have provided solid information about the roles of genes in variability observed in basic ultrasonic call parameters. Not surprisingly, we have learned that genotype makes a contribution to variation and that a broad sense heritability of over 0.5 has been documented for the rate of calling and some frequency characteristics. Further, the genetic architecture of variation in calls has been shown to change across days from ages day 3 through 9 (Thornton and Hahn, 2005).

In this last section, we make three recommendations about future studies. The first recommendation is that aspects of the ultrasonic calls of mice beyond rate need to be studied in detail. As noted by Branchi et al. (2001) and Brudzynski (2005), focusing on rate of calling alone ignores a potentially rich set of information about the nature of the signals emitted by mouse pups. Holy and Guo (2005) have argued that the ultrasonic calls of adult male mice contain "syllables" and "phrases" – words usually reserved for language. The richness of information present in ultrasounds has also been shown in adolescent mice. For example, Panksepp et al. (2007) was able to demonstrate repetitive bouts of ultrasonic vocalizations that demonstrated significant effect of genotype on several complex categories of the calls. A recent study by Scattoni et al. (2008), utilizing a sophisticated level of analysis, has shown that there is an unusual repertoire of infant vocalizations in the BTBR T + tf/J mouse strain, proposed as a possible model for autism. According to these authors, the BTBR strain exhibits strain-specific patterns of vocalizations that represent different lexicons, or innate variations in complex vocal repertoires, which may relate to the abnormal social communication and interactions previously shown in this strain of mouse (McFarlane et al. 2008). While not passing judgment on the accuracy of those descriptions, it is important to look carefully at ultrasonic calls in search of their meanings. Such investigations have in the past been very time consuming as the analysis depended on the replay of recorded calls and by hand measuring. Current (e.g., Scattoni et al. 2008) and future computer programs should render such detailed analyses feasible.

Second, genetic studies should be combined with pharmacological studies in order to elucidate the underlying neurological substrates responsible for the production of infant mouse ultrasounds. In line with this proposal it has been established that ultrasonic vocalizations during the developmental period are sensitive to pharmacological manipulation and that the changes observed in these calls might be predicted from known behavioral responses to specific drugs (Dirks et al., 2002; Fish et al., 2003; Rowlett et al., 2001; Rupniak et al., 2000). For example, injection of an anxiolytic drug, such as diazepam, into an infant mouse produces a decrease in the number of distress ultrasonic vocalizations relative to controls (Rowlett et al., 2001). In this study, Rowlett et al. (2001) extended earlier investigations showing the role of the GABAergic system in the production of infant mouse ultrasounds (Fish et al. 2000; Oliver et al., 1998), by utilizing three different benzodiazepines (zolpidem, triazolam, and diazepam) and two selective antagonists (flumazenil and β-CCt). This approach allowed the authors to study the role of specific γ-amino butyric acid subunit A ($GABA_A$) receptor subtypes in the distress vocalizations. Their results showed that all three benodiazapine agonists decreased ultrasonic vocalizations, but that the atypical benzodiazepine zolpidem which has a high affinity for the BZ/1 subunit causes suppression of the calls by a non-anxiolytic mechanism, which appears to be hypothermia.

Third, future studies should examine the genome in search of genes which make small contributions to normal variation in infant mouse ultrasonic calls. Behavior genetics is the study of individual differences in behavior and infant ultrasonic calls in mice are rich in intra- and interindividual variation. Some mice make many calls and some just a few. Some calls are very short, on the order of 5 ms, and some are 10 or 15 times as long. Some calls are simple pure tone whistles and some calls are complex with several changes in pitch. The combination of quantitative genetics and single gene approaches provides a means to examine that variation. As nicely described by Plomin et al. (2003), the search for quantitative trait loci (QTLs) is a search for genes with small influences on a behavior of interest. If about 50% of the variability in ultrasonic call rate is associated with genetic variation, individual QTLs might explain 2, or 5, or 7% of the overall 50%. Eventually, all the genetic variation could be tied to individual genes.

In a review on single-gene influences on brain and behavior, Doug Wahlsten (1999) surveyed the merging of fields of investigation of genes, nervous systems, and behaviors. He noted rapid advances in technology but cautioned:

> As the individual research project probes ever more deeply into ever narrower domains of knowledge, there is a growing need to synthesize existing knowledge and make connections among the isolated parts of an expanding discipline. The next major advance must come in the domain of theory (p. 620).

In a way, the study of the influence of single genes upon mouse phenotypes, such as infant ultrasounds, is reminiscent of the early days of brain-ablation studies when researchers tried to lesion a specific region of the brain in order to determine its function. Approximately 60 years ago, Hetherington and Ranson (1942) showed that lesioning of the lateral hypothalamus nucleus produced hypophagia in rodents. These authors interpreted their results to show that there was a feeding center in the lateral hypothalamus that controlled eating. Years later it was discovered that the mechanism behind this lesion induced hypophagia was much more complicated and in part was produced by damage to dopaminergic projection fibers, which pass through the lateral hypothalamus and are involved in control of the animal's motivational state (Bernardis and Bellinger, 1998; Lenard et al., 1988). While there is no doubt that the time has come when genetic manipulation of the mouse will teach us much about single gene influences on brain and behavior, it is imperative that we look to the past to prevent making the same mistakes as those preceding us.

References

Barron, S. and Gilbertson, R. (2005) Neonatal ethanol exposure but not neonatal cocaine selectively reduces specific isolation-induced vocalization waveforms in rats. *Behav Genet* 35: 93–102.

Bell, R.W., Nitschke, W., and Zachman, T. (1972) Ultrasounds in three inbred strains of young mice. *Behav Biol* 7: 805–814.

Bernardis, L.L. and Bellinger, L.L. (1998) The dorsomedial hypothalamic nucleus revisited: 1998 update. *Proc Soc Exp Biol Med* 218: 284–306.

Branchi, I., Santucci, D., and Alleva, E. (2001) Ultrasonic vocalization emitted by infant rodents: a tool for assessment of neurobehavioural development. *Behav Brain Res* 125: 49–56.

Broadhurst, P.L. and Jinks, J.L. (1974) What genetic architecture can tell us about the natural selection of behavioral traits. In van Abeelen, J.H.F. (ed.), *The Genetics of Behavior*. North-Holland, Amsterdam, pp. 43–63.

Brudzynski, S.M. (2005) Principles of rat communication: quantitative parameters of ultrasonic calls in rats. *Behav Genet* 35: 85–92.

Brudzynski, S.M., Kehoe, P., and Callahan, M. (1999) Sonographic structure of isolation-induced ultrasonic calls of rat pups. *Dev Psychobiol* 34: 195–204.

Brunelli, S. (2005) Selective breeding for an infant phenotype: rat pup ultrasonic vocalization (USV). *Behav Genet* 35: 53–66.

Cohen-Salmon, C., Carlier, M., Roubertoux, P., Jouhaneau, J., Semal, C., and Paillette, M. (1985) Differences in patterns of pup care in mice. V. Pup ultrasonic emissions and pup care behavior. *Physiol Behav* 35: 167–174.

Collins, R.L. and Fuller, J.L. (1968) Audiogenic seizure prone (asp): a gene affecting behavior in linkage group 8 of the mouse. *Science* 162: 1137–1139.

Dirks, A., Fish, E.W., Kikusui, T., van der Gugten, J., Groenink, L., Olivier, B., et al. (2002) Effects of corticotropin-releasing hormone on distress vocalizations and locomotion in maternally separated mouse pups. *Pharmacol Biochem Behav* 72: 993–999.

Ehret, G. (2005) A review of the methods of studies on infant ultrasound production and maternal retrieval in small rodents. *Behav Genet* 35: 19–29.

Ehret, G. and Bernecker, C. (1986) Low-frequency sound communication by mouse pups (*Mus musculus*): wriggling calls release maternal behaviour. *Anim Behav* 34: 821–830.

Elwood, R.W. and Keeling, F. (1982) Temporal organization of ultrasonic vocalizations in infant mice. *Dev Psychobiol* 15: 221–227.

Fish, E.W., Faccidomo, S., Gupta, S., and Miczek, K.A. (2003) Anxiolytic-like effects of escitalopram, citalopram, and r-citalopram in maternally separated mouse pups. *J Pharmacol Exp Ther* 308: 474–480.

Fish, E.W., Sekinda, M., Ferrari, P.F., Dirks, A., and Miczek, K. (2000) Distress vocalizations in maternally separated mouse pups: modulation via 5-HT (1A), 5-HT (1B) and GAB (AA) receptors. *Psychopharmacology* 149: 277–285.

Fox, W.M. (1965) Reflex-ontogeny and behavioural development in the mouse. *Anim Behav* 13: 234–241.

Fuller, J.L., and Thompson, W.R. (1960) *Behavior Genetics*. John Wiley and Sons, New York.

Geyer, L.A. and Barfield, R.J. (1979) Introduction to the symposium: ultrasonic communication in rodents. *Amer Zool* 19: 411.

Haack, B., Markl, H., and Ehret, G. (1983) Sound communication between parents and offspring. In Willott, J. F. (ed.), *The Auditory Psychobiology of the Mouse*. C. C. Thomas, Springfield, IL, USA, pp. 57–97.

Hahn, M.E., Hewitt, J.K., Adams, M., and Tully, T. (1987) Genetic influences on ultrasonic vocalizations in young mice. *Behav Genet* 17: 155–166.

Hahn, M.E., Hewitt, J.K., Schanz, N., Weinreb, L., and Henry, A. (1997) Genetic and developmental influences on infant mouse ultrasonic calling. I. A diallel analysis of the calls of 3-day olds. *Behav Genet* 27: 133–143.

Hahn, M.E., Karkowski, L., Weinreb, L., Henry, A., and Schanz, N. (1998) Genetic and developmental influences on infant mouse ultrasonic calling. II. Developmental patterns in the calls of mice 2–12 days of age. *Behav Genet* 82: 315–325.

Hahn, M.E. and Lavooy, M.J. (2005) A review of the methods of studies on infant ultrasound production and maternal retrieval in small rodents. *Behav Genet* 35: 31–52.

Hahn, M.E. and Schanz, N. (2002) The effects of cold, rotation, and genotype on the production of ultrasonic calls in infant mice. *Behav Genet* 32: 267–273.

Hahn, M.E. and Thornton, L.M. (2005) Introduction to the special edition infant mouse and rat ultrasonic vocalizations. *Behav Genet* 35: 1–6.

Hart, F.M. and King, J.A. (1966) Distress vocalizations of young in two subspecies of *Peromyscus maniculatus*. *J. Mammal* 47: 287–293.

Henderson, N.D. (1990) Genetic analysis as a route to understanding the evolution of animal behavior: examples using the diallel cross. In Hahn, M.E., Hewitt, J.K., Henderson, N.D., and Benno, R.H. (eds.), *Developmental Behavior Genetics: Neural, Biometrical, and Evolutionary Approaches*. Oxford University Press, New York, pp. 283–297.

Henderson, N.D. (1997) Spurious associations in unreplicated selected lines. *Behav Genet* 27: 145–154.

Hetherington, A.W. and Ranson, S.W. (1942) The spontaneous activity and food intake of rats with hypothalamic lesions. *Am J Physiol* 136: 609–617.

Holy, T.E., and Guo, Z. (2005) Ultrasonic calls of male mice. *PLoS Biol* 3: 1–10.

Lenard, L., Karadi, Z., Jando, G., Yoshimatsu, H., Hajnal, A., Sandor, P., et al. (1988) Lateral hypothalamic feeding mechanisms: iontophoretic effects of kainic acid, ibotenic acid and 6-hydroxydopamine. *Brain Res Bull* 20: 847–856.

Lorenz, K. (1981) *The Foundations of Ethology*. Simon and Schuster, New York.

McFarlane, H.G., Kusek, G.K., Yang, M., Phoenix, J.L., Bolivar, V.J., and Crawley, J.N. (2008) Autism-like behavioral phenotypes in BTBR T+tf/J mice. *Genes Brain Behav* 7: 152–163.

Myers, M.M., Ali, N., Weller, A., Brunelli, S.A., Tu, A.Y., Hofer, M.A., et al. (2004) Brief maternal interaction increases number, amplitude, and bout size of isolation-induced ultrasonic vocalizations in infant rats (*Rattus norvegicus*). *J Comp Psychol* 118: 95–102.

Nagy, Z.M. (1993) Development of homeothermy in infant C3H mice. *B Psychonomic Soc* 31: 221–224.

Noirot, E. (1972) Ultrasounds and maternal behavior in small rodents. *Dev Psychobiol* 5: 371–387.

Okon, E.E. (1970a) The effect of environmental temperature on the production of ultrasounds by isolated non-handled albino mouse pups. *J Zool Lond* 162: 71–83.

Okon, E.E. (1970b) The ultrasonic responses of albino mouse pups to tactile stimuli. *J Zool Lond* 162: 485–492.

Oliver, B., Molewijk, E., van Oorschot, E., van der Heyden, J., Ronken, E., and Mos, J. (1998) Rat pup vocalization: effects of benzodiazepine receptor ligands. *Eur J Pharmacol* 358: 117–128.

Panksepp, J.B., Jochman, K.A., Kim, J.U., Koy, J.J., and Wilson, E.D. (2007) Affiliative behavior, ultrasonic communication and social reward are influenced by genetic variation in adolescent mice. *PLoS One* 2: e351.

Plomin, R., DeFries, J.C., Craig, I.W., and McGuffin, P. (2003) Behavioral genetics. In Plomin, R., DeFries, J.C., Craig, I.W., and McGuffin, P. (eds.), *Behavioral Genetics in the Postgenomic Era*. American Psychological Association, Washington, D.C., pp. 3–15.

Plomin, R., DeFries, J.C., McClearn, G.E., and McGuffin, P. (2001) *Behavioral Genetics*, 4th edn. Worth, New York.

Roubertoux, P.L., Martin, B., Le Roy, I., Beau, J., Marchaland, C., Perez-Diaz, F., et al. (1996) Vocalizations in newborn mice: genetic analysis. *Behav Genet* 26: 427–437.

Rowlett, J.K., Tornatzky, W., Cook, J.M., Ma, C., and Miczek, K.A. (2001) Zolpidem, triazolam, and diazepam decrease distress vocalizations in mouse pups: differential antagonism by flumazenil and β-carboline-3-carboxylate-t-butyl ester (β-CCt). *J Pharmacol Exp Ther* 297: 247–253.

Rupniak, N.M.J., Carlson, E.C., Harrison, T., Oates, B., Seward, E., Owen, S., et al. (2000) Pharmacological blockade or genetic deletion of substance P (NK1) receptors attenuates neonatal vocalization in guinea-pig and mice. *Neuropharm* 39: 1413–1421.

Sales, G.D. and Smith, J.C. (1978) Comparative studies of the ultrasonic calls of infant murid rodents. *Dev Psychobiol* 11: 595–619.

Scattoni, M.L., Gandhy, S.U., Ricceri, L., and Crawley, J.N. (2008) Unusual repertoire of vocalizations in the BTBR T+tf/J mouse model of autism. *PLoS One* 3: e3067.

Thornton, L.M. and Hahn, M.E. (2005) Genetic and developmental influences on infant mouse ultrasonic calling. III. Patterns of the inheritance in the calls of mice 3–9 days of age. *Behav Genet* 35: 73–84.

Tobon, K., Hahn, M.E., and Schanz, N. (2005) Individual variation in pup vocalization waveforms: effects of genotype, treatment, and prenatal cocaine exposure. *Behav Genet* 35: 823.

Wahlsten, D. (1999) Single-gene influences on brain and behavior. *Ann Rev Psychol* 50: 599–624.

Willott, J.F. (1983) *The Auditory Psychobiology of the Mouse.* C. C. Thomas, Springfield, IL, USA.

Wohr, M., Dahloff, M., Wolf, E., Holsboer, F., Schwarting, R.K.W., and Wotjak, C.T. (2008) Effects of genetic background, gender, and early environmental factors on isolation-induced ultrasonic calling in mouse pups: an embryo-transfer technique. *Behav Genet* 38: 579–595.

Chapter 18

Startle behavior and prepulse inhibition

Claudia F. Plappert and Peter K. D. Pilz

Startle paradigm

Startle behavior

The startle response can be elicited by acoustic, by tactile, or by vestibular stimuli (Pilz et al., 2004; Willott et al., 1979; Yeomans et al., 2002). Startle behavior is characterized by a stereotyped, coordinated motor pattern spreading from the head to the rest of the body: relatively moderate startle stimuli elicit contractions of the head and neck muscles, including eyelid and pinna reflexes. After more intense stimuli, a flexion of the forelimbs and hind-limbs and a crouching of the rump can be observed. The strongest startle response is a jump, during which all four paws leave the floor (Caeser et al., 1989; Horlington, 1968; Tovote et al., 2005).

Different hypotheses exist about the biological significance of startle. Landis and Hunt (1939) show that strong startle responses may lead to a posture where the head and the abdomen are protected. The head protective function is expanded by Yeomans et al. (2002), who found that all sensory systems (i.e., acoustic, tactile, and vestibular) which are activated by a blow to the head strongly add to elicit startle. Pilz and Schnitzler (1996) show that startle very quickly interrupts ongoing behavior, and prepares the motor system for a subsequent reaction with shortened latency; they note that startle induces a good position (e.g., flexed limbs) for an escape reaction.

Neural basis of startle behavior

Acoustic startle is the behavior with the shortest latency. To our knowledge, only the middle ear reflex yields a faster muscle contraction (Pilz et al., 1997). The shortest latency of the startle response can be measured in mice. The minimum time from onset of the stimulus to onset of the response is 7 ms in this species. Figure 18.1 shows an example of such a response, measured behaviorally (this is, to our knowledge, the shortest behavioral latency ever published). Startle measured electromyographically in the muscles of the head and neck region of rats can even have latencies of only 5 ms (Caeser et al., 1989; Cassella et al., 1986; Pilz et al., 1988).

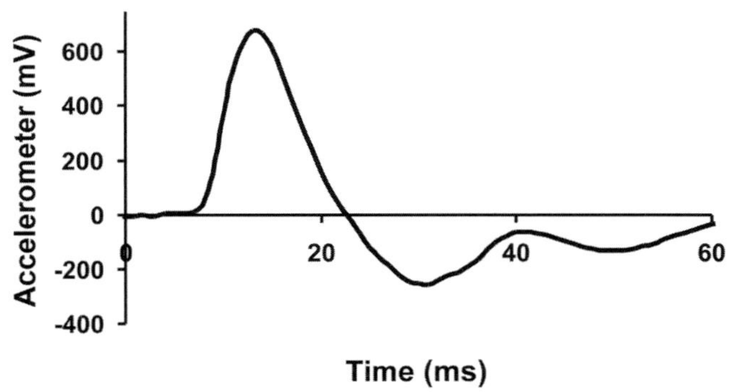

Figure 18.1 Startle 110 dB sound pressure level response to a stimulus measured behaviorally using a standard movement sensitive system (example of a C3H/H mouse; latency: 7 ms).

A response with such a short latency necessitates a very short neural pathway. The primary pathway of the acoustic startle response (ASR) comprises inner hair cells, the neurons of the acoustic nerve (nerve VIII), the nucleus cochlearis complex, the caudal pontine reticular formation (PnC), and the motoneurons (reviewed in Koch, 1999; see Figure 18.5 later). The sensorimotor interface and the most important specific neural substrate of startle are probably giant neurons (cell body diameters >35–40 μm) in the PnC (Koch et al., 1992; Lingenhohl and Friauf, 1992). While most research on the startle pathway has been done with rats, the central role of the PnC for startle has also been shown in mice (Carlson and Willott, 1998; Simons-Weidenmaier et al., 2006). Besides this primary startle pathway, alternative connections have been described, including additional synapses between the nucleus cochlearis complex and the PnC (Wagner et al., 2000; Yeomans and Frankland, 1995), as well as between the PnC and motoneurons (Yeomans and Frankland, 1995).

Startle responses elicited by tactile or vestibular stimuli have been studied less frequently. Tactile stimuli to the face or electrical stimuli to the trigeminal nerve can evoke somatosensory startle. In these cases, tactile input is relayed via trigeminal nuclei to the PnC (reviewed by Yeomans et al., 2002; see also, for the mouse, Simons-Weidenmaier et al., 2006). Vestibular input is also relayed to the PnC. In addition, a direct connection to motoneurons via the vestibulospinal tract also mediates a fast

Behavioral Genetics of the Mouse: Volume I. *Genetics of Behavioral Phenotypes*, eds. Wim E. Crusio, Frans Sluyter, Robert T. Gerlai, and Susanna Pietropaolo. Published by Cambridge University Press.

motor response which probably intensifies the startle response (regarding rat, see Li et al., 2001).

Acoustic startle in mice

Startle stimuli

More than 90% of the published literature on the startle response deals with acoustically elicited startle, on which we will also focus in the following. The ASR is elicited by sudden, loud acoustic stimuli. "Sudden" means that only stimuli with a very short rise time of no more than about 10 ms elicit a strong and reliable ASR. The probability and amplitude of the ASR decrease when the stimulus rise time is longer (Ison, 2001; Willott et al., 1979). "Loud" means that the startle stimulus has to be above the startle threshold, which in most mouse strains lies at a sound pressure level (SPL) between 75 and 85 dB (e.g., Plappert and Pilz, 2002). Standard stimulus parameters are therefore rise times ≤5 ms and SPLs ≥95 dB.

Mouse strain differences

Inbred mouse strains can differ widely in ASRs (Bullock et al., 1997; Crawley et al., 1997; Paylor and Crawley, 1997; Willott et al., 2003). Using inbred strains of mice is one approach for studying the genetic basis of behavior (e.g., Plappert and Pilz, 2002). As an example, here we will compare four of the most often used inbred strains in detail, namely the C57BL/6J ("C57BL/6"), DBA/2N ("DBA/2"), BALB/cAN ("BALB/c"), and C3H/HeN ("C3H/H") mouse strains.

The C57BL/6 strain is the one most often used in startle experiments. It also provides the most commonly used genetic background for knockout mice and can serve as a model for age-dependent sensorineural hearing loss (Willott et al., 1993). One of the interesting features of the DBA/2 strain is its hippocampal deficit (Thinus-Blanc et al., 1996). The BALB/c strain is often used as a model of anxiety (Belzung et al., 2000). The C3H/H strain has none of the above deficits. All four strains have a similar startle threshold of between 74 and 80 dB SPL (estimated with the extrapolation method: Pilz and Schnitzler, 1996; Figure 18.2). In all four strains, startle amplitude increases with startle stimulus intensity (although ASR amplitude decreases in C57BL/6 when the stimulus SPL exceeds about 110 dB; Ison, 2001). However, the strains differ in their maximum startle response in the following order: BALB/c > C3H/H > C57BL/6 > DBA/2 (Bullock et al., 1997). Differences in auditory gating as measured by the auditory evoked potential recorded in the hippocampus (Stevens et al., 1996) and differences in the emotional (anxiety) state of the mouse strains have been discussed as causes for these startle variations (Plappert and Pilz, 2002).

Age-dependent hearing loss occurs not only in the C57BL/6 (>3 months) but also in the DBA/2 (>3 weeks) and BALB/c (>3 months) strains due to a progressive loss of hair cells and spiral ganglion cells in the cochlear base (Willot, 2001, 2009; Willott et al., 1982, 1998). This decreases the ASR (Ison et al. 2007; Ouagazzal et al., 2006). Thus the age of the mice must be considered in comparing the ASR in different strains.

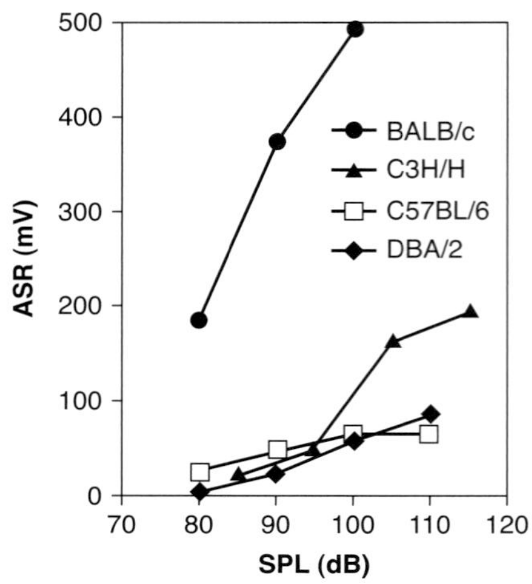

Figure 18.2 Input–output function showing the relationship between sound pressure levels (SPLs) of the startle stimuli and the mean acoustic startle response (ASR) amplitude (output) for four different inbred strains mice ($n = 12–36$).

Sex, estrus cycle

In addition to mouse strain, sex has been found to be a further genetic factor influencing startle behavior. In several different mouse strains it was found that the startle amplitude of males is higher than in females (Logue et al., 1997; Plappert et al., 2005; Ralph et al., 2001; van den Buuse et al., 2003; for rats see e.g., Kinkead et al., 2008). A possible reason for this is the higher body weight of males compared to females. There is a tendency toward increasing startle amplitude when body weight rises (Plappert et al., 2005; Tarantino et al., 2000); this may be due to increased total muscle mass, leading to an increase in motor strength. Another reason may be that male mice are more anxious than female mice. Many behavioral studies have shown greater anxiety in male rats than in female rats in various experimental paradigms (e.g., Johnston and File, 1991), and anxiety has been found to increase the startle response (Belzung et al., 2000; Walker and Davis, 1997). The female ovarian hormones estradiol and progesterone are thought to exert an anti-anxiety effect in rodents (Palanza, 2001); this might explain the fact that startle amplitude is lower in female mice than in male mice. In this case, ASR amplitudes should be increased during metestrus and diestrus, when estradiol and progesterone levels are lower. However, the startle reactivity of female mice is not influenced by the estrus cycle (Meziane et al., 2007; Plappert et al., 2005), indicating that there is no acute anxiolytic effect of these hormones at least in terms of startle amplitude.

Modulations of startle behavior

What makes the startle response especially interesting as a behavioral model are the various modulations of this behavior. Startle can be used as a probe to measure emotional influences or changes by different learning paradigms.

Habituation of the startle response is evoked by repetitive presentation of a startle stimulus, which leads to a decrease in response strength. Habituation occurs both within a test day (short-term habituation; Pilz et al., 2004) and over the course of several test days (long-term habituation; Plappert and Pilz, 2005). Short-term habituation, a simple form of non-associative learning, probably occurs within the primary startle pathway, specifically at the synapses of neurons relaying auditory or tactile input to PnC neurons (Pilz et al., 2004, Simons-Weidenmaier et al., 2006). It has been suggested that habituation is produced at these synapses by synaptic depression in the presynaptic terminals (Simons-Weidenmaier et al., 2006; Weber et al., 2002). In contrast to short-term habituation, long-term habituation involves brain structures extrinsic to the startle pathway (Jordan and Leaton, 1983; Leaton et al., 1985). It has been shown in rats that the cerebellum plays a critical role in the acquisition of long-term habituation (Leaton and Supple, 1986; Lopiano et al., 1990).

Sensitization is elicited by strong aversive stimuli and elevates the general responsiveness of an organism (Davis, 1997). Startle sensitization can be elicited by electric footshocks (Dirks et al., 2001) or by the startle stimuli themselves (Plappert and Pilz, 2002) and leads to an increase in the startle response. For rats it is known that sensitization is mediated by the amygdala (Fendt and Fanselow, 1999; Hitchcock et al., 1989).

In a fear-conditioning paradigm, the animals are trained to associate a neutral (e.g., light or moderate tone) stimulus with an aversive (e.g., electric footshock) stimulus. After a few such pairings, this conditioned stimulus induces a state of fear which is reflected in a potentiated startle response (Falls et al., 1997). Fear conditioning in rats and mice is also mediated by the amygdala (Davis, 1997; Fendt and Fanselow, 1999; Heldt et al., 2000; Herry et al., 2006).

Since prepulse inhibition (PPI) is especially interesting because of its clinical relevance, it will be discussed in greater detail here.

Prepulse inhibition

PPI and prepulse facilitation

Prepulses are stimuli which of themselves do not elicit an ASR. However, they modulate the ASR when administered shortly before a startle stimulus. In the classic paradigm of PPI, which can be applied in different species, a prepulse is given about 100 ms before the startle stimulus. This decreases or "inhibits" the ASR (Figure 18.3). Prepulse inhibition can be observed in many combinations of prepulse modality and startle stimulus modality (e.g., a visual prepulse also inhibits the ASR; Palmer et al., 2000; Taylor et al., 1995). However, most work has been done with acoustic prepulses before acoustic startle stimuli (e.g., Ison, 2001). The PPI paradigm is of special interest because it is affected in several human psychiatric diseases. For example, it has been shown that PPI is an animal model for positive symptoms of schizophrenia (Green et al., 2009; Swerdlow and Geyer, 1998).

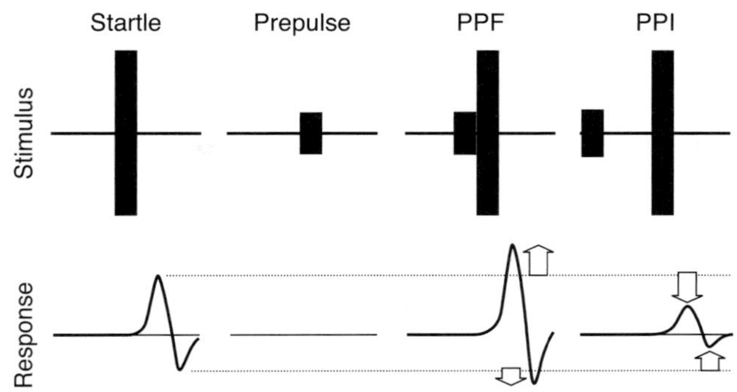

Figure 18.3 Schematic illustration of experimentally induced prepulse inhibition (PPI) and prepulse facilitation (PPF). Either the startle stimuli or prepulses are given alone ("startle," "prepulse"), or the prepulse is given either about 10 ms before the startle stimulus, i.e., directly before it ("PPF") or about 100 ms ("PPI"). In the figure, the respective response appears below each stimulus or stimulus combination. The relative change in PPI or PPF compared to a situation with startle alone is shown by arrows.

Besides PPI, prepulse facilitation (PPF) can be observed if the prepulse is given only briefly, e.g., for only 10 ms, before the startle stimulus (Figure 18.3). Prepulse facilitation in mice can be especially pronounced, depending on the mouse strain (Plappert et al., 2004).

Prepulse inhibition seems to be part of a general mechanism which protects signal analysis from being interrupted if stimuli are given in rapid succession. For example, when one stimulus is presented immediately after another, the analysis of the first stimulus is facilitated by inhibition of the impact of the second stimulus. This is partly due to sensory gating in which the perceived intensity of the second stimulus, i.e., the signal reaching the cortex directly via the auditory pathway, is reduced by a preceding prepulse (e.g., Swerdlow et al., 2005). In addition, PPI is induced by sensorimotor gating (e.g., Geyer et al., 2002) in which the startle-eliciting properties of the second stimulus are inhibited. Since the motor response of startle leads among other things to strong proprioception with short latency, inhibition of the startle response by PPI reduces such additional proprioceptive input which would compete with the analysis of the prepulse (Graham et al., 1975). In schizophrenia, for example, weakened sensory or sensorimotor gating, indicated by an impaired PPI, leads to unfiltered and thus badly analyzed sensory input (Swerdlow and Geyer, 1998). In addition to protection of stimulus processing, Fendt et al. (2001) point out that nuclei involved in PPI (especially SC and PPTg) are also active in approach behavior (orienting towards, foveation, or exploration of novel stimuli), and that prevention of eye closure amongst others improves perceptual processing in a critical time period.

In contrast to PPI, however, no plausible hypothesis has yet been advanced to explain the function of PPF (Ison et al., 1997b). Stimuli of equal strength show summation in about the time window of PPF (e.g., Li and Yeomans, 1999; Marsh et al., 1973). Therefore, PPF could be the result of summation, but this has not yet been compellingly shown with stimuli differing in strength as much as those used in the PPF paradigm.

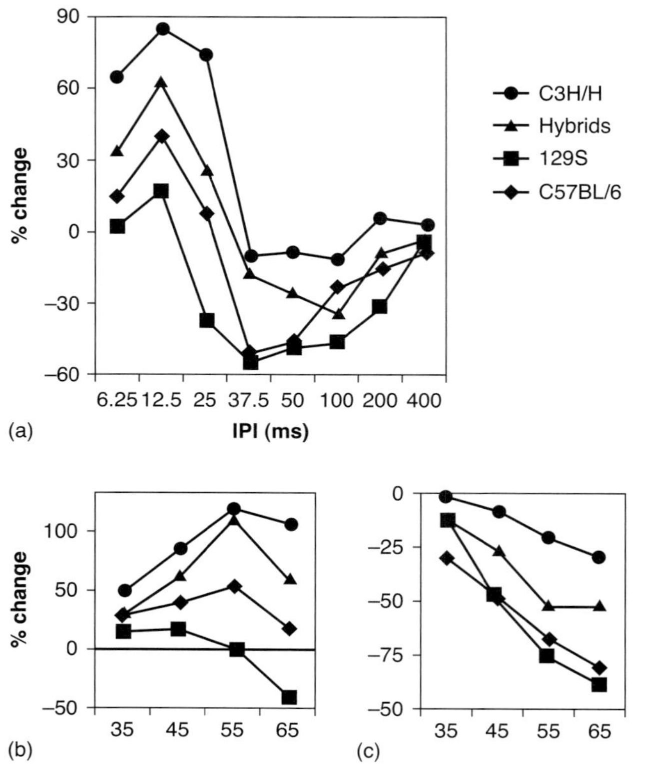

Figure 18.4 Dependence of prepulse inhibition (negative change) and prepulse facilitation (positive change) in three strains of inbred mice and hybrids between NMRI and wild mice (*Mus musculus domesticus*) (a) on the interpulse interval (IPI) between prepulse and startle-eliciting stimulus (prepulses of 45 dB sound pressure level (SPL) were given), (b and c) on the SPL of the prepulse (prepulses were given 12.5 ms (b) and, respectively, 50 ms (c) before the startle stimulus, $n = 12$–16).

Stimulus parameters and mouse strain differences

Whether a prepulse elicits PPI or PPF depends mainly on three factors: (1) the interval between the prepulse and the startle stimulus (Ison, 2001; Plappert et al., 2004); (2) the prepulse intensity (systematically investigated in Bullock et al., 1997; Plappert et al., 2004, Yee et al., 2005); and (3) the inbred mouse strain studied (thoroughly investigated by Willott et al., 2003, with 40 inbred strains, but also by Aubert et al., 2006; Brooks et al., 2004; Bullock et al., 1997; Paylor and Crawley, 1997; Plappert et al., 2004; Popova et al. 2009).

The interpulse interval (IPI) between prepulse and startle stimulus is the factor of the stimulus parameters exerting the strongest impact on PPI. The IPI not only determines the strength of the prepulse effect, but also its direction (Figure 18.4a). Depending on the mouse strain, PPI is elicited at IPIs between about 10 and 1000 ms, with a maximum from 30 to 100 ms. Prepulse facilitation can be observed at IPIs between 1 and about 100 ms, with a maximum between 6 and 12 ms (mice: Ison, 2001; Plappert et al., 2004; note that BALB/cJ mice are an exception from these rules: Aubert et al., 2006; rats: Reijmers and Peeters, 1994).

Raising the prepulse SPL causes a monotonic, nearly linear, increase in PPI (i.e., a decrease in ASR change, Figure 18.4c; here IPI = 50 ms, chosen to evoke PPI). In contrast, prepulse SPL shows an inverted U-shaped influence on PPF, with a maximum at a specific intermediate prepulse SPL (Figure 18.4b; here IPI = 12.5 ms, chosen to elicit PPF). The SPL eliciting maximum PPF depends on the mouse strain. Since prepulse SPL has different effects on PPI and PPF, it can be concluded that these two processes are probably not mediated by the same neural structures (rats: Mansbach and Geyer, 1991; Reijmers and Peeters, 1994; mice: Plappert et al., 2004).

Prepulse inhibition is mostly independent of the startle stimulus SPL (Plappert et al., 2004; however, e.g., Csomor et al., 2006; Yee et al., 2005 critically discussed this issue). This means that PPI does not strongly depend on the startle baseline (however: see Csomor et al., 2008; compare Ison et al., 1997a for PPI in rats). Of course, this is only true if PPI is measured as a relative change (%), since the absolute ASR change induced by a certain prepulse increases with increasing startle stimulus SPL (in the same ratio as the ASR increases). The very weak influence of startle baseline on PPI in percent indicates: (1) that PPI changes the gain of ASR; and (2) that the neural substrates for ASR elicitation and for PPI are largely independent of one another.

Sex and estrus cycle

Sex has no strong influence on PPI and PPF in either mice (Plappert et al., 2005) or rats (Koch, 1998; Kinkead et al., 2008). Furthermore, PPI and PPF do not change during the estrus cycle of female mice (Meziane et al., 2007; Plappert et al., 2005). This is in contrast to findings in humans (Swerdlow et al., 1993), while reports in rats are contradictory (Koch, 1998; Kinkead et al., 2008), suggesting that PPI in rats does not depend on estrous cycle but only on time of day (Adams et al. 2008).

Neuronal circuit of PPI

The brain mechanisms underlying PPI have largely been determined from experiments in rats (Figure 18.5). The decreasing effect of the prepulse on the startle amplitude affects the startle pathway at the level of the PnC (Carlson and Willott, 1998) through activation of the ascending auditory pathway in the midbrain and thereafter the colliculus superior; this in turn activates an inhibitory cholinergic pathway from the pedunculopontine tegmental nucleus (PPTg) to the PnC (Figure 18.5; Koch, 1999). Yeomans et al. (2006) have found evidence for an additional direct input from the inferior colliculus to the PPTg without synapse in the superior colliculus. Takahashi et al. (2007) have confirmed the importance of the PPTg for the mediation of PPI in the mouse. Furthermore, the substantia nigra is also thought to be part of the PPI- mediating pathway, inhibiting the PnC via a direct inhibitory GABAergic projection (Fendt et al., 2001). This inhibition of the PnC causes a reduction of the startle response.

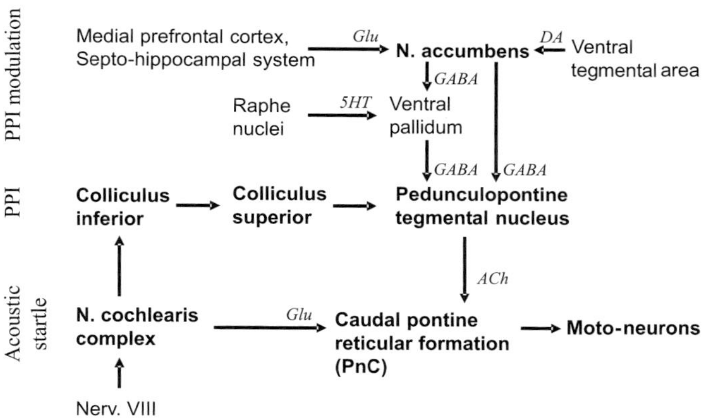

Figure 18.5 Schematic illustration of the primary startle circuit, the circuit of prepulse inhibition (PPI), and part of the circuitry modulating PPI (according to Koch, 1999, modified).

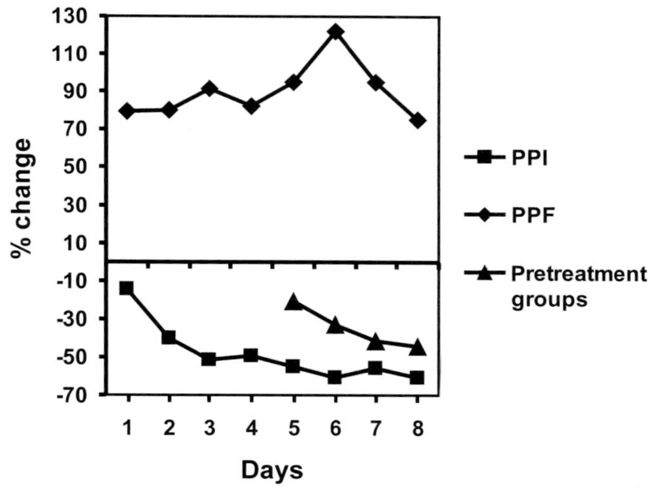

Figure 18.6 Percent change of acoustic startle response (ASR) by prepulses during 8 days of testing in the C3H/HeN strain. Prepulses at a long interpulse interval (IPI) (50 ms) evoke prepulse inhibition (PPI) which increases over days. Prepulse facilitation (PPF) evoked by prepulses presented at a short IPI (12.5 ms) remains almost constant. The PPI increase is only produced by "contingent" presentation of prepulse and startle stimulus. Four days of different pretreatments (pretreatment groups: (1) presentation of test context; (2) startle stimulus alone; (3) prepulse alone; or (4) "non-contingent" presentation of the prepulse and the startle stimulus) do not cause a PPI increase ($n = 12$–23 per group; according to Plappert et al., 2006, modified).

Prepulse inhibition is itself modulated by corticolimbic-striatal circuitry. A reduction of the PPI by abnormalities in this circuitry is regarded as an animal model for the reduced sensorimotor gating found in schizophrenia patients (Braff et al., 2001). The neuronal circuitry modulating PPI has been described in detail in the reviews of Koch (1999) and Swerdlow et al. (2001). In short, the central element of this circuitry is the nucleus accumbens. The nucleus accumbens obtains glutamatergic afferents from the septo-hippocampal system, the medial prefrontal cortex, and dopaminergic afferents from the ventral tegmental area. The nucleus accumbens directly and indirectly projects to the penduncolopotine tegmental nucleus, thereby influencing PPI (Figure 18.5). In addition, in mice, input from the lateral globus pallidus to the PPTg has been shown to be important (Takahashi et al., 2007).

Modulation of PPI by learning

Compared to other species studied, mice show a strong and reliable increase of PPI when it is repetitively elicited over several test days. This PPI increase occurs mainly during the first 4 days and has been found in all mice strains studied thus far (Plappert et al., 2004, 2006). This increase in the inhibitory prepulse effect by experience is not caused by a decrease in PPF but rather by an increase in PPI: the PPI increases in C3H/HeN during the whole testing period while PPF remains constant and in C57BL/6J, despite very low PPF (Plappert et al., 2006). The PPI increase is evoked fully only if prepulses and startle stimuli are repeatedly administered in a temporally paired ("contingent") order. The full PPI increase is not evoked by: (1) presentation of the test context without stimulation; (2) the startle stimulus alone; (3) the prepulse alone; or (4) presentation of the prepulse and startle stimulus together in a variable temporal relationship, i.e., "non-contingently" (Figure 18.6). An added effect is found in the C57BL/6 strain but not in the C3H/H strain: PPI is additionally increased by adaptation to an unchanging context, but not by adaptation to a different context (Plappert et al., 2006). This suggests that the C57BL/6 strain learns over a period of days to recognize the context as familiar, thus permitting the prepulse to "light up" against the contextual stimuli, with increased effectiveness of the prepulse in evoking PPI.

Since paired presentation of the prepulse and the startle stimulus is an essential prerequisite for the observed increase in PPI by experience, an underlying associative learning process is likely. The neuronal site mediating the PPI change needs information about both the prepulse and the startle stimulus, as well as their temporal relationship. All this information is present in the PnC neurons (which mediate startle) and their inhibitory synapses from the PPTg (which mediate PPI). The simplest way to explain our results is a Hebbian mechanism at this site: the number and/or impact of inhibitory (e.g., cholinergic) receptors on the PnC neurons increases when prepulses activate the receptors and the neurons are simultaneously activated by startle stimuli (Plappert et al., 2006).

The finding that PPI can be modulated by learning has implications for using the PPI deficit as an animal model in studying, e.g., schizophrenia. Care must be taken to separate the unlearned PPI component from learning effects.

Summary

The startle response is a valuable tool for assessing the neuronal mechanisms of behavioral plasticity. First, the startle response can be evoked and measured very reliably. Second, the startle response has a non-zero baseline; that is, it can be enhanced as well as attenuated by a variety of modulations. Third, the underlying pathways are relatively simple and well known. This makes the startle response a good starting point for investigating the neural mechanisms of startle modulations (Koch and Fendt, 2003; Plappert and Pilz, 2001). We have elaborated here on PPI of the startle response as an example of such modulations, because a deficit in PPI is a valid animal model of alterations in sensorimotor processing which can also be observed in schizophrenia patients. In mice, the startle response amplitude as well as the amount of PPI varies between different inbred mouse strains. This indicates that the startle response is well suited as a model for studying the influence of genes on behavior.

References

Adams, A.L., Hudson, A., Ryan, C.L., and Doucette, T.A. (2008) Effects of estrous stage and time of day on prepulse inhibition in female rats. *J Neurosci Methods* 173: 295–298.

Aubert, L., Reiss, D., and Ouagazzal, A.M. (2006) Auditory and visual prepulse inhibition in mice: parametric analysis and strain comparisons. *Genes Brain Behav* 5: 423–431.

Belzung, C., Le Grisquet, A.M., and Crestani, F. (2000) Flumazenil induces benzodiazepine partial agonist-like effects in BALB/c but not in C57BL/6 mice. *Psychopharmacology* 148: 24–32.

Braff, D.L., Geyer, M.A., and Swerdlow, N.R. (2001) Human studies of prepulse inhibition of startle: normal subjects, patient groups, and pharmacological studies. *Psychopharmacology* 156: 234–258.

Brooks, S.P., Pask, T., Jones, L., and Dunnett, S.B. (2004) Behavioural profiles of inbred mouse strains used as transgenic backgrounds. I: motor tests. *Genes Brain Behav* 3: 206–215.

Bullock, A.E., Slobe, B.S., Vazquez, V., and Collins, A.C. (1997) Inbred mouse strains differ in the regulation of startle and prepulse inhibition of the startle response. *Behav Neurosci* 111: 1353–1360.

Caeser, M., Ostwald, J., and Pilz, P.K. (1989) Startle responses measured in muscles innervated by facial and trigeminal nerves show common modulation. *Behav Neurosci* 103: 1075–1081.

Carlson, S. and Willott, J.J. (1998) Caudal pontine reticular formation of C57/BL/6J mice: response to startle stimuli, inhibition by tones, and plasticity. *J Neurophysiol* 79: 2603–2614.

Cassella, J.V., Harty, T.P., and Davis, M. (1986) Fear conditioning, pre-pulse inhibition and drug modulation of a short latency startle response measured electromyographically from neck muscles in the rat. *Physiol Behav* 36: 1187–1191.

Crawley, J.N., Belknap, J.K., Collins, A., Crabbe, J.C., Frankel, W., Henderson, N., et al. (1997) Behavioral phenotypes of inbred mouse strains: implications and recommendations for molecular studies. *Psychopharmacology* 132: 107–124.

Csomor, P.A., Yee, B.K., Quednow, B.B., Stadler, R.R., Feldon, J., and Vollenweider, F.X. (2006) The monotonic dependency of prepulse inhibition of the acoustic startle reflex on the intensity of the startle-eliciting stimulus. *Behav Brain Res* 174: 143–150.

Csomor, P.A., Yee, B.K., Vollenweider, F.X., Feldon, J., Nicolet, T., and Quednow, B.B. (2008) On the influence of baseline startle reactivity on the indexation of prepulse inhibition. *Behav Neurosci* 122: 885–900.

Davis, M. (1997) Neurobiology of fear responses. *J Neuropsychiatry Clin Neurosci* 9: 382–402.

Dirks, A., de Jongh, R., Groenik, L., van der Gugten, J., Hijzen, T.H., and Olivier, B. (2001) Footshock-induced sensitization of the acoustic startle response in two strains of mice. *Behav Brain Res* 123: 17–21.

Falls, W.A., Carlson, S., Turner, J.G., and Willott, J.F. (1997) Fear-potentiated startle in two strains of inbred mice. *Behav Neurosci* 111: 855–861.

Fendt, M. and Fanselow, M.S. (1999) The neuroanatomical and neurochemical basis of conditioned fear. *Neurosci Biobehav Rev* 23: 743–760.

Fendt, M., Li, L., and Yeomans, J.S. (2001) Brain stem circuits mediating prepulse inhibition of the startle reflex. *Psychopharmacology* 156: 216–224.

Geyer, M.A., McIlwain, K.L., and Paylor, R. (2002) Mouse genetic models for prepulse inhibition: an early review. *Mol Psychiatry* 7: 1039–1053.

Graham, F.K., Putnam, L.E., and Leavitt, L.A. (1975) Lead-stimulation effects on human cardiac orienting and blink reflexes. *J Exp Psychol Hum Percept Perform* 104: 175–182.

Green, M.F., Butler, P.D., Chen, Y., Geyer, M., Silverstein, S., Wynn, J.K., et al. (2009) Perception measurement in clinical trials of schizophrenia: promising paradigms from CNTRICS. *Schizophr Bull* 35: 163–181.

Heldt, S., Sundin, V., Willott, J.F., and Falls, W.A. (2000) Posttraining lesions of the amygdala interfere with fear-potentiated startle to both visual and auditory conditioned stimuli in C57BL/6J mice. *Behav Neurosci* 114: 749–759.

Herry, C., Trifilieff, P., Micheau, J., Lüthi, A., and Mons, N. (2006) Extinction of auditory fear conditioning requires MAPK/ERK activation in the basolateral amygdala. *Eur J Neurosci* 24: 261–269.

Hitchcock, J.M., Sananes, C.B., and Davis, M. (1989) Sensitization of the startle reflex by footshock: blockade by lesions of the central nucleus of the amygdala or its efferent pathway to the brainstem. *Behav Neurosci* 103: 509–518.

Horlington, M. (1968) A method for measuring acoustic startle response latency and magnitude in rats: detection of a single stimulus effect using latency measurements. *Physiol Behav* 3: 839–844.

Ison, J.R. (2001) The acoustic startle response: reflex elicitation and reflex modification by preliminary stimuli. In Willot, J.F. (ed.), *Handbook of Mouse Auditory Research*. CRC Press, London, pp. 59–82.

Ison, J.R., Allen, P.D., and O'Neill, W.E. (2007) Age-related hearing loss in C57BL/6J mice has both frequency-specific and non-frequency-specific components that produce a hyperacusis-like exaggeration of the acoustic startle reflex. *J Assoc Res Otolaryngol* 8: 539–550.

Ison, J.R., Bowen, G.P., Pak, J., and Gutierrez, E. (1997a) Changes in the strength of prepulse inhibition with variation in the startle baseline associated with individual differences and with old age in rats and mice. *Psychobiology* 25: 266–274.

Ison, J.R., Taylor, M.K., Bowen, G.P., and Schwarzkopf, S.B. (1997b) Facilitation and inhibition of the acoustic startle reflex in the rat after a momentary increase in background noise level. *Behav Neurosci* 111: 1335–1352.

Johnston, A.L. and File, S.E. (1991) Sex differences in animal tests of anxiety. *Physiol Behav* 49: 245–250.

Jordan, W.P. and Leaton, R.N. (1983) Habituation of the acoustic startle response in rats after lesions in the mesencephalic reticular formation or in the inferior colliculus. *Behav Neurosci* 97: 710–724.

Kinkead., B., Yan, F., Owens, M.J., and Nemeroff, C.B. (2008) Endogenous neurotensin is involved in estrous cycle related alterations in prepulse inhibition of the acoustic startle reflex in female rats. *Psychoneuroendocrinology* 33: 178–187.

Koch, M. (1998) Sensorimotor gating changes across estrous cycle in female rats. *Physiol Behav* 64: 625–628.

Koch, M. (1999) The neurobiology of startle. *Prog Neurobiol* 59: 107–128.

Koch, M. and Fendt, M. (2003) Startle response modulation as a behavioral tool in neuropharmacology. *Curr Neuropharmacol* 1: 175–185.

Koch, M., Lingenhohl, K., and Pilz, P.K. (1992) Loss of the acoustic startle response following neurotoxic lesions of the caudal pontine reticular formation: possible role of giant neurons. *Neuroscience* 49: 617–625.

Landis, C. and Hunt, W. (1939) *The Startle Pattern*. Farrar and Rinehard, Oxford.

Leaton, R.N., Cassella, J.V., and Borszcz, G.S. (1985) Short-term and long-term habituation of the acoustic startle response in chronic decerebrate rats. *Behav Neurosci* 99: 901–912.

Leaton, R.N. and Supple, W.F., Jr. (1986) Cerebellar vermis: essential for long-term habituation of the acoustic startle response. *Science* 232: 513–515.

Li, L., Steidl, S., and Yeomans, J.S. (2001) Contributions of the vestibular nucleus and vestibulospinal tract to the startle reflex. *Neuroscience* 106: 811–821.

Li, L. and Yeomans, J.S. (1999) Summation between acoustic and trigeminal stimuli evoking startle. *Neuroscience* 90: 139–152.

Lingenhohl, K. and Friauf, E. (1992) Giant neurons in the caudal pontine reticular formation receive short latency acoustic input: an intracellular recording and HRP-study in the rat. *J Comp Neurol* 325: 473–492.

Logue, S.F., Owen, E.H., Rasmussen, D.L., and Wehner, J.M. (1997) Assessment of locomotor activity, acoustic and tactile startle, and prepulse inhibition of startle in inbred mouse strains and F1 hybrids: implications of genetic background for single gene and quantitative trait loci analysis. *Neuroscience* 80: 1075–1086.

Lopiano, L., de'Sperati, C., and Montarolo, P.G. (1990) Long-term habituation of the acoustic startle response: role of the cerebellar vermis. *Neuroscience* 35: 79–84.

Mansbach, R.S. and Geyer, M.A. (1991) Parametric determinants in pre-stimulus modification of acoustic startle: interaction with ketamine. *Psychopharmacology* 105: 162–168.

Marsh, R., Hoffman, H.S., and Stitt, C.L. (1973) Temporal integration in the acoustic startle reflex of the rat. *J Comp Physiol Psychol* 82: 507–511.

Meziane, H., Ouagazzal, A.M., Aubert, L., Wietrzych, M., and Krezel, W. (2007) Estrous cycle effects on behavior of C57BL/6J and BALB/cByJ female mice: implications for phenotyping strategies. *Genes Brain Behav* 6: 192–200.

Ouagazzal, A.M., Reiss, D., and Romand, R. (2006) Effects of age-related hearing loss on startle reflex and prepulse inhibition in mice on pure and mixed C57BL and 129 genetic background. *Behav Brain Res* 172: 307–315.

Palanza, P. (2001) Animal models of anxiety and depression: how are females different? *Neurosci Biobehav Rev* 25: 219–233.

Palmer, A.A., Dulawa, S.C., Mottiwala, A.A., Conti, L.H., Geyer, M.A., et al. (2000) Prepulse startle deficit in the Brown Norway rat: a potential genetic model. *Behav Neurosci* 114: 374–388.

Paylor, R. and Crawley, J.N. (1997) Inbred strain differences in prepulse inhibition of the mouse startle response. *Psychopharmacology* 132: 169–180.

Pilz, P.K., Caeser, M., and Ostwald, J. (1988) Comparative threshold studies of the acoustic pinna, jaw and startle reflex in the rat. *Physiol Behav* 43: 411–415.

Pilz, P.K., Carl, T.D., and Plappert, C.F. (2004) Habituation of the acoustic and the tactile startle responses in mice: two independent sensory processes. *Behav Neurosci* 118: 975–983.

Pilz, P.K., Ostwald, J., Kreiter, A., and Schnitzler, H.U. (1997) Effect of the middle ear reflex on sound transmission to the inner ear of rat. *Hear Res* 105: 171–182.

Pilz, P.K. and Schnitzler, H.-U. (1996) Habituation and sensitization of the acoustic startle response in rats: amplitude, threshold, and latency measures. *Neurobiol Learn Mem* 66: 67–79.

Plappert, C.F., Kuhn, S., Schnitzler, H.-U., and Pilz, P.K. (2006) Experience increases the prepulse inhibition of the acoustic startle response in mice. *Behav Neurosci* 120: 16–23.

Plappert, C.F. and Pilz, P.K. (2001) The acoustic startle response as an effective model for elucidating the effect of genes on the neural mechanism of behavior in mice. *Behav Brain Res* 125: 183–188.

Plappert, C.F. and Pilz, P.K. (2002) Difference in anxiety and sensitization of the acoustic startle response between the two inbred mouse strains BALB/cAN and DBA/2N. *Genes Brain Behav* 1: 178–186.

Plappert, C.F. and Pilz, P.K. (2005) Long-term habituation of the startle response in mice evoked by acoustic and tactile stimuli. *Behav Brain Res* 162: 307–310.

Plappert, C.F., Pilz, P.K., and Schnitzler, H.-U. (2004) Factors governing prepulse inhibition and prepulse facilitation of the acoustic startle response in mice. *Behav Brain Res* 152: 404–412.

Plappert, C.F., Rodenbücher, A.M., and Pilz, P.K. (2005) Effects of sex and estrous cycle on modulation of the acoustic startle response in mice. *Physiol Behav* 84: 585–594.

Popova, N.K., Naumenko, V.S., Tibeikina, M.A., and Kulikov, A.V. (2009) Serotonin transporter, 5-HT_{1A} receptor, and behavior in DBA/2J mice in comparison with four inbred mouse strains. *J Neurosci Res* 87: 3149–3217.

Ralph, R.J., Paulus, M.P., Fumagalli, F., Caron, M.G., and Geyer, M.A. (2001) Prepulse inhibition deficits and perseverative motor patterns in dopamine transporter knock-out mice: differential effects of D1 and D2 receptor antagonists. *J Neurosci* 21: 305–313.

Reijmers, L.G. and Peeters, B.W. (1994) Effects of acoustic prepulses on the startle reflex in rats: a parametric analysis. *Brain Res* 661: 174–180.

Simons-Weidenmaier, N.S., Weber, M., Plappert, C.F., Pilz, P.K., and Schmid, S. (2006) Synaptic depression and short-term habituation are located in the sensory part of the mammalian startle pathway. *BMC Neurosci* 7: 38.

Stevens, K.E., Freedman, R., Collins, A.C., Hall, M., Leonard, S., Marks, M.J., et al. (1996) Genetic correlation of inhibitory gating of hippocampal auditory evoked response and alpha-bungarotoxin-binding nicotinic cholinergic receptors in inbred mouse strains. *Neuropsychopharmacology* 15: 152–162.

Swerdlow, N.R., Auerbach, P., Monroe, S.M., Hartston, H., Geyer, M.A., and Braff, D.L. (1993) Men are more inhibited than women by weak prepulses. *Biol Psychiatry* 34: 253–260.

Swerdlow, N.R. and Geyer, M. (1998) Using an animal model of sensorimotor gating to study the pathophysiology and new treatments of schizophrenia. *Schizophr Bull* 24: 285–301.

Swerdlow, N.R., Geyer, M.A., and Braff, D.L. (2001) Neural circuit regulation of prepulse inhibition of startle in the rat: current knowledge and future challenges. *Psychopharmacology* 156: 194–215.

Swerdlow, N.R., Stephany, N.L., Talledo, J., Light, G., Braff, D.L., Baeyens, D., et al. (2005) Prepulse inhibition of perceived stimulus intensity: paradigm assessment. *Biol Psychol* 69: 133–147.

Takahashi, K., Nagai, T., Kamei, H., Maeda, K., Matsuya, T., Arai, S., et al. (2007) Neural circuits containing pallidotegmental GABAergic neurons are involved in the prepulse inhibition of the startle reflex in mice. *Biol Psychiatry* 62: 148–157.

Tarantino, L.M., Gould, T.J., Druhan, J.P., and Bucan, M. (2000) Behavior and mutagenesis screens: the importance of baseline analysis of inbred strains. *Mamm Genome* 11: 555–564.

Taylor, M.K., Ison, J.R., and Schwarzkopf, S.B. (1995) Effects of single and repeated exposure to apomorphine on the acoustic startle reflex and its inhibition by a visual prepulse. *Psychopharmacology* 120: 117–127.

Thinus-Blanc, C., Save, E., Rossi-Arnaud, C., Tozzi, A., and Ammassari-Teule, M. (1996) The differences shown by C57BL/6 and DBA/2 inbred mice in detecting spatial novelty are subserved by a different hippocampal and parietal cortex interplay. *Behav Brain Res* 80: 33–40.

Tovote, P., Meyer, M., Pilz, P.K., Ronnenberg, A., Ogren, S.O., Spiess, J., et al. (2005) Dissociation of temporal dynamics of heart rate and blood pressure responses elicited by conditioned fear but not acoustic startle. *Behav Neurosci* 119: 55–65.

van den Buuse, M., Simpson, E.R., and Jones, M.E. (2003) Prepulse inhibition of acoustic startle in aromatase knock-out mice: effects of age and gender. *Genes Brain Behav* 2: 93–102.

Wagner, T., Pilz, P.K., and Fendt, M. (2000) The superior olivary complex is necessary for the full expression of the acoustic but not tactile startle response in rats. *Behav Brain Res* 108: 181–188.

Walker, D.L. and Davis, M. (1997) Anxiogenic effects of high illumination levels assessed with the acoustic startle response in rats. *Biol Psychiatry* 42: 461–471.

Weber, M., Schnitzler, H.U., and Schmid, S. (2002) Synaptic plasticity in the acoustic startle pathway: the neuronal basis for short-term habituation? *Eur J Neurosci* 16: 1325–1332.

Willott, J.F. (2001) Animal models of presbycusis and the aging auditory system. In Hof, P.R. and Mobbs, C.V. (eds.), *Functional Neurobiology of Aging*. Academic Press, New York, pp. 605–621.

Willott, J.F. (2009) Effects of sex, gonadal hormones, and augmented acoustic environments on sensorineural hearing loss and the central auditory system: insights from research on C57BL/6J mice. *Hear Res* 252: 89–99.

Willott, J.F., Aitkin, L.M., and McFadden, S.L. (1993) Plasticity of auditory cortex associated with sensorineural hearing loss in adult C57BL/6J mice. *J Comp Neurol* 329: 402–411.

Willott, J.F., Demuth, R.M., Lu, S.M., and Van Bergem, P. (1982) Abnormal tonotopic organization in the ventral cochlear nucleus of the hearing-impaired DBA/2 mouse. *Neurosci Lett* 34: 13–17.

Willott, J.F., Shnerson, A., and Urban, G.P. (1979) Sensitivity of the acoustic startle response and neurons in subnuclei of the mouse inferior colliculus to stimulus parameters. *Exp Neurol* 65: 625–644.

Willott, J.F., Tanner, L., O'Steen, J., Johnson, K.R., Bogue, M.A., and Gagnon, L. (2003) Acoustic startle and prepulse inhibition in 40 inbred strains of mice. *Behav Neurosci* 117: 716–727.

Willott, J.F., Turner, J.G., Carlson, S., Ding, D., Seegers Bross, L., and Falls, W.A. (1998) The BALB/c mouse as an animal model for progressive sensorineural hearing loss. *Hear Res* 115: 162–174.

Yee, B.K., Chang, T., Pietropaolo, S., and Feldon, J. (2005) The expression of prepulse inhibition of the acoustic startle reflex as a function of three pulse stimulus intensities, three prepulse stimulus intensities, and three levels of startle responsiveness in C57BL6/J mice. *Behav Brain Res* 163: 265–276.

Yeomans, J.S. and Frankland, P.W. (1995) The acoustic startle reflex: neurons and connections. *Brain Res Brain Res Rev* 21: 301–314.

Yeomans, J.S., Lee, J., Yeomans, M.H., Steidl, S., and Li, L. (2006) Midbrain pathways for prepulse inhibition and startle activation in rat. *Neuroscience* 142: 921–929.

Yeomans, J.S., Li, L., Scott, B.W., and Frankland, P.W. (2002) Tactile, acoustic and vestibular systems sum to elicit the startle reflex. *Neurosci Biobehav Rev* 26: 1–11.

Chapter 19

Mouse models of stress-induced depression-like behavior

Stress vulnerability and antidepressant response as traits

Howard K. Gershenfeld

Introduction

Stress is a well-known common risk factor contributing to psychiatric, inflammatory, and cardiovascular disorders (Brown, 1993; Brown et al., 1996; Cohen, 2005; Das and O'Keefe, 2006; Kendler and Karkowski-Shuman, 1997; Kendler et al., 1993, 1995; Kiecolt-Glaser and Glaser, 1995; Kloner, 2006; Rosengren et al., 2004). Stress responses vary greatly among individuals. The relationships between "stress" and psychiatric illness is complex, multifactorial, and poorly understood with a multitude of mediators (Agid et al., 2000). From genetic studies on the comorbidity of anxiety and depression in twins, Kendler and colleagues concluded that genes "appeared to act by predisposing to general distress rather than specifically to symptoms of either anxiety or depression" (Kendler et al., 1992). We believe that the study of these "stress responses" in rodents may probe this genetic shared liability for "general distress," which is a risk factor for psychiatric disorders. Anxiety and depressive disorders are common psychiatric illnesses. From a recent US epidemiological study, the National Comorbidity Survey Replication ($n = 9090$), estimated lifetime prevalence rates for *Diagnostic and Statistical Manual of Mental Disorders, Fourth Edition* (DSM-IV, 1994) disorders as follows: anxiety disorders, 28.8%; mood disorders, 20.8%; impulse-control disorders, 24.8%; substance use disorders, 14.6%; any disorder, 46.4% (Kessler et al., 2003, 2005a, 2005b). From the Global Burden of Disease Study, depression was projected to be the second leading cause of total disease burden (disability-adjusted life years, DALYs), surpassed only by ischemic heart disease (Murray and Lopez, 1997). Given the magnitude of depressive disorders, an important gap is the limited response to antidepressant agents (ADA) for major depression with only a third to a half of patients experiencing a full clinical resolution (Fava, 2003b). Another issue is the chronic and relapsing nature of depression with a subpopulation of patients relapsing even while on maintenance doses of ADA, generally interpreted as a "break through," tolerance, tachyphylaxis, or "poop out" (Fava, 2003a; Solomon et al., 2005). Unfortunately, "etiology unknown" persists as a prominent feature of psychiatric disorders. While monoamine theories of depression remain salient (Charney, 1998; Delgado, 2000; Heninger et al., 1996; Leonard, 2000; Schildkraut, 1965), gaps remain in the pathogenesis of depression and antidepressants' mechanism of action (Berton and Nestler, 2006). A burgeoning perspective views the monoamine systems as important neuromodulators, while yet to be discovered neurobiological systems and adaptive changes ("neuronal plasticity") have a more central role (Manji et al., 2003; Yamada et al., 2005).

An understanding of the molecular pathophysiology of the mammalian stress response is of potential utility in psychiatry and general medicine, providing the motivation for these mouse models. Similarly, the neurochemistry, neuroendocrine changes, signaling pathways, and loci influencing these stress-induced behaviors may define new pathways of influence, suggesting novel targets for improved pharmacotherapy. While "stress" is a broad, poorly defined construct with individual perceptions, coping styles, and contexts ("controllability of stressor") influencing the stressful event, the construct has strong face validity. A voluminous literature measures stress in rodents using a variety of stressors (an evoking stimulus), including immobilization, inescapable shock, early separation, predator odors, and social defeat. These stressors lead to profound changes in neurotransmitters, cytokines, and the hypothalamic–pituitary–adrenocortical (HPA) axis (Fink, 2000; Kvetnansky et al., 1995). Likewise, a wealth of human clinical and temperament studies measure stress from various perspectives, assessing by self-reports of distress, daily hassles, major life events (such as early parental loss, poor social support, divorce, assault, serious housing problems, serious illness), hostility, neuroendocrine levels (adrenocorticotropic hormone (ACTH), cortisol), or physiological variables. A preponderance of epidemiological studies have indicted negative life events, negative affect, and exposure to stressful life events as risk factors in causing depression, anxiety disorders, and myocardial infarctions in predisposed individuals (Brown, 1993; Brown et al., 1996; Kendler, 1998; Kendler et al., 1992; Mittleman et al., 1997; Paykel, 1994, 2001; Rozanski et al., 2005). For example, in a large female twin study containing 316 episodes of depression, the increased relative risk, an odds ratio (OR), for the onset of depression in the month after a stressful event was 5.64. The highest risk stressful life event was

Behavioral Genetics of the Mouse: Volume I. *Genetics of Behavioral Phenotypes*, eds. Wim E. Crusio, Frans Sluyter, Robert T. Gerlai, and Susanna Pietropaolo. Published by Cambridge University Press.

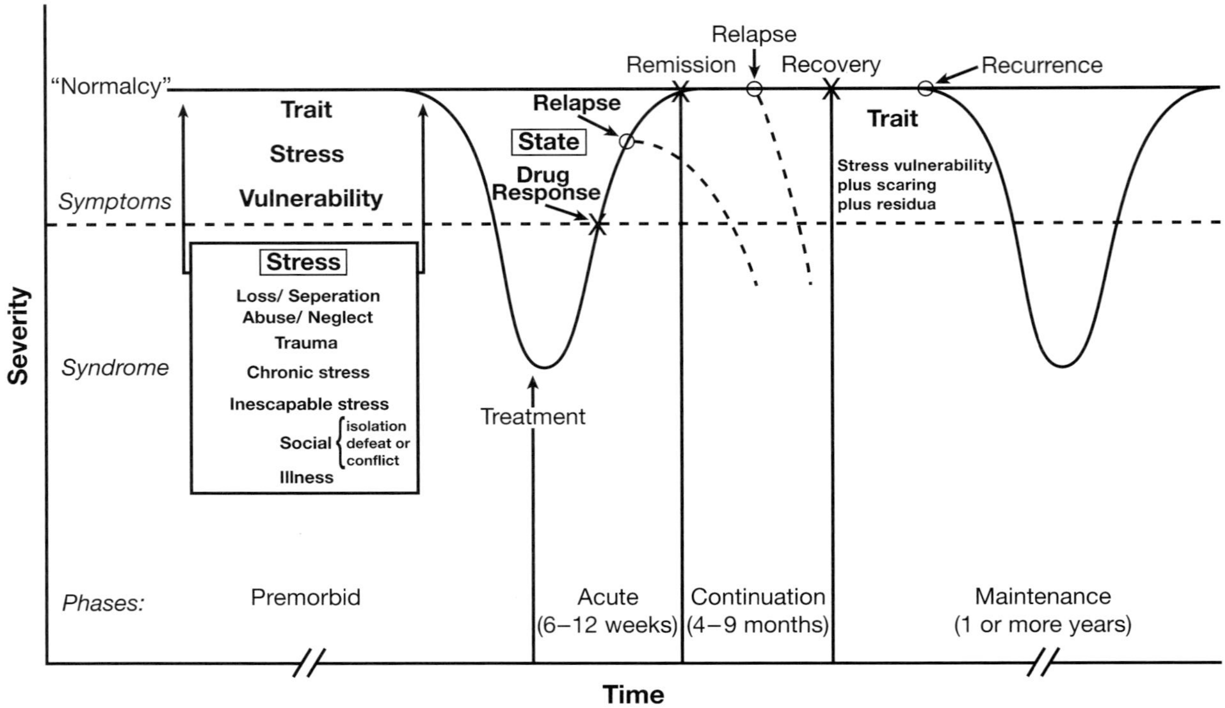

Figure 19.1 Overview of major depression with a longitudinal perspective, emphasizing distinct continuous traits for "stress vulnerability," response to treatment ("Drug response"), liability for relapse while on medication, and liability to recurrence. Severity of illness (number of symptoms, intensity of symptoms, and interference with functioning) is represented on the on the *y* axis and the *x* axis is time. (Adapted and modified from Kupfer, 1991, and Post, 1992.)

experiencing a personal assault with an OR = 25.4 (Kendler et al., 1999).

As a context for understanding these rodent models, a summary of the clinical syndrome of major depression is presented in Figure 19.1, highlighting a longitudinal, clinical perspective. Figure 19.1 emphasizes the role of "response to stressors" or "stress vulnerability" to major depression as a key continuous trait, which may vary among individuals and can be modeled by variation in mouse strains. Often depression occurs with an insidious onset and the accumulation of depressive symptoms over time (sadness, loss of interest, guilt, decreased energy, poor concentration, altered sleep, decreased appetite, psychomotor retardation). Depression becomes clinically significant when these symptoms interfere with functioning, reaching the level of a syndrome. Other important clinical features are the frequent initiation of treatment at the near nadir of the illness and the subsequent response to treatment (e.g., very much improved vs. much improved vs. better but not well vs. minimal or no response). This "response to antidepressant treatment" is viewed as another dimensional trait distinct from the response to a stressor. Finally, the time course of the antidepressant response and the chronicity of depression are highlighted along with their associated liabilities for relapse and recurrence adding further layers of complexity in modeling the illness. Rather than comprehensively modeling this disease, most mouse models provide "mini-models" selecting an aspect or component of the illness.

Given this body of literature on the phenomena of depression, the questions arise: How does one model components of depression? What kind of models in rodents conform to notions of "everyday, garden variety stress and distress" to model stress reactivity or stress vulnerability? Most stress-induced models involve the following chain of events: stressor, reaction to stressor *in a vulnerable individual*, neuroendocrine and neurochemical changes, immune response perturbations, adaptive changes in neural signaling and neurocircuitry with chronic stress, and the occurrence of behavioral, cognitive, vegetative, and emotional symptoms or illness behavior in a predisposed individual.

Why study mouse (and rat) model systems of these complex behaviors? Genetic studies towards understanding the pathophysiology of human depression, anxiety syndromes, and complex traits remain challenging with the following the inherent difficulties: (1) incomplete penetrance; (2) the presence of phenocopies (non-genetic causes in a normal genotype creating the disease phenotype); (3) genetic heterogeneity (several, different "genetic routes" giving rise to a single syndrome); (4) diagnostic uncertainties in classification without "gold standards"; (5) gene interactions (epistasis and gene × environment interactions; see, e.g., Caspi et al., 2003); and (6) the sample size limitations of existing pedigrees and population samples. In particular, the heterogeneity of depression, its comorbidity with anxiety disorders and stress response phenotypes, the relatively modest "effect" of any one of the polygenes, and incomplete penetrance make genetic dissection thorny with some noteworthy exceptions (Abkevich et al., 2003). Since a major obstacle to the genetic dissection of psychiatric disorders is their multifactorial etiology, a simpler genetic system in mice offers distinct advantages, biased by an evolutionary perspective. The neurobiological stress response system of lower animals provided a foundation for the later evolutionary developments in the adaptive stress response(s). As readers of this volume are aware, the advantages of using a mouse model include the existence of many outbred, inbred, recombinant inbred, and consomic lines, the ability to perform planned experiments and crosses, complete environmental control, a high-quality genetic map with an annotated genome and many markers, a refined map of conserved regions between mouse and man, the Mouse Phenome Project (Bogue and Grubb, 2004), good genome-wide expression tools (e.g., microarrays and the Allen Brain Bank), sophisticated transgenic and siRNA techniques, and access to brain tissue at any age or stage of development.

When selecting a mouse model of human psychopathology, legitimate concerns arise about any given model, its relevance, and its validity (Willner, 1984, 1991a; 1991b), discussed in greater detail in Volume II of this Handbook. Nonetheless, a major issue is the end user's ultimate question, goal, or purpose in modeling the behavior. A second key issue involves the appropriate "level" of analysis, from molecules to pharmacology to interacting neuronal circuits to whole brain imaging to an organism's emergent properties. Is one interested in the genetic level and molecular genetic correlates, testing new drugs, mechanisms of drug action, neurohormonal correlates, imaging, neural circuitry, or behavioral outputs integrating across levels? Briefly, we find the requirements of McKinney and Bunney for animal models a useful general framework (McKinney, 1984). These categories highlight the shared correspondence between a disease entity and an animal model in the domains of : (1) etiology; (2) symptoms; (3) biochemistry; and (4) response to treatment. We acknowledge the limitations of any animal model in that core mood and cognitive symptoms such as sadness, pessimism, guilt, and negative view of the self and the world are inherently impracticable to model in mute animals.

For psychiatric disorders, most animal paradigms have modeled the behavioral and physiological symptoms, emphasizing *response to treatment for validation*, namely "predictive validity." For example, if a model is specifically and dose-dependently responsive to various ADA and not anti-anxiety agents, it presumably models depressive-like behavior. However, others have privileged the criteria of face validity or construct validity or predictive validity with an appropriate time course of antidepressant response occurring only after chronic treatment (Harro, 2004; Mitchell and Redfern, 2005; Sarter and Bruno, 2002; Weiss and Kilts, 1998; Willner, 1984). While debates persist in weighing the relative merits of the many competing rodent models, we believe each model has certain advantages and disadvantages. As pragmatists, we privilege Porsolt's criteria for usefulness of a model valuing most: (1) non-dependence on a mechanism of action (i.e., atheoretical); (2) the existence of genetic determinants; (3) procedural simplicity; and (4) reproducibility (Porsolt, 2000). Hence, we briefly review learned helplessness (LH). Then, we selectively review "behavioral despair" paradigms within the "stress-depression topography." namely the tail suspension test (TST) and the forced swim test (FST). We focus more on the TST as it emerged as the most feasible, reliable, and validated experimental model for drug screening and genetic dissection based on the above criteria (Porsolt, 2000). In addition, this area overlaps our own laboratory's expertise. The interested reader is referred elsewhere (Cryan and Holmes, 2005; Harro, 2004; Malatynska and Knapp, 2005; Mitchell and Redfern, 2005; Sarter and Bruno, 2002; Weiss and Kilts, 1998) for other useful models such as: drug-induced (reserpine, clonidine, cytokine-induced, etc.) models, chronic mild stress paradigms, early maternal separation models, olfactory bulbectomy (Song and Leonard, 2005), and delayed reinforcement of low-response operant paradigms. Likewise, promising paradigms such as dominance-submission models (Malatynska and Knapp, 2005; Malatynska et al., 2005), recurrent social defeat (Avgustinovich et al., 2005), and improved chronic mild stress paradigms in mice (Strekalova et al., 2004) augur well for the field and are worthy of further exploration.

LH models

The LH paradigm has a long, distinguished, and contentious history with successful use in a variety of species (initially dogs, then rats and mice) (Overmier and Seligman, 1967; Seligman and Beagley, 1975; Seligman et al., 1968, 1980). The phenomenon centers on how an animal's exposure to intense, recurrent, unsignaled, inescapable shocks (footshocks or tail shocks), interferes 24 hours later with two-way shuttlebox avoidance–escape response learning, where escape from the shock is possible. This poor behavioral performance in a two-way shuttle box paradigm 24 hours after LH induction is measured as a delayed latency to escape or "escape failures." This deficit in acquiring shuttlebox escape–avoidance behavior was construed as learning "helplessness" and that "nothing I do matters" consistent with a cognitive theory of depression, interpreted as both a "motivational and emotional" deficit (Seligman et al., 1980). Historically, much of this LH work has been performed in rats and hence, we only briefly mention this rat work. The advantages of this paradigm include an extensive literature on the subject, the elegant use of a triadic design with escape ("controllable" shock), yoked (i.e., uncontrollable shock), and restrained (no shock) groups (Anisman and Merali, 2003; Drugan, 2003), evidence of strain variation among rat lines (Wieland et al., 1986), the generation of bidirectionally, selectively bred lines of rats for LH (Henn and Vollmayr, 2005), and adequate pharmacological validity for antidepressants given subchronically (Sherman et al., 1982). However, some serious concerns and controversies for this paradigm include the following: (a) interpretation of the procedure as inducing "conditioned inactivity" rather than LH; (b) the ephemeral duration of the escape deficit (being gone by 48–160 h); (c) the narrow "context specificity" of the LH-induced performance deficit (generally deficits only in tasks requiring sustained motor activity) (Francis et al., 1995); and (d) the LH behavioral deficit only develops in a proportion of the animals exposed to the inescapable stressor (0–80% among different rodent lines) (Chourbaji et al., 2005a; Wieland et al., 1986). However, this property of the LH paradigm where only a proportion of exposed animals develop LH permits interesting comparisons of non-helpless versus helpless rodents subjected to the same stressors, examining c-fos expression (Huang et al., 2004), neurochemistry, gene expression patterns (Kohen et al., 2005; Nakatani et al., 2004), and individual differences. In fact, using these differences in susceptibility to LH induction, bidirectionally selected lines have been developed by starting with a population of outbred rats (Henn and Vollmayr, 2005; Vollmayr and Henn, 2001).

The neurochemical changes induced by LH stress have been studied with changes in three main monoamines, namely dopamine, serotonin, and norepinephrine (NE). These changes have been shown to vary by brain region and strain. In the rat, the time course of the intense, uncontrollable shock-induced motor deficit correlated best with locus coeruleus (LC) region NE levels (Seligman et al., 1980; Weiss and Simson, 1988). Further studies suggest that the LH paradigm may be construed as a transitory, significant depletion (at least 20%) of NE levels in the LC region at 90 minutes or 48 hours due to excessive release of NE with stress compared to no shock controls (Weiss et al., 2005; Weiss and Kilts, 1998). Other experiments have emphasized more dopaminergic (Willner, 1983) or serotonergic mechanisms (Ronan et al., 2000; Willner, 1983). While NE clearly plays a contributory role, the concept of reciprocal, interacting networks of monoaminergic changes along with corticotropin-releasing hormone plus their corresponding receptors and signaling pathways should be appreciated (Maier and Watkins, 2005; Manji et al., 2003; Ridder et al., 2005; Taylor et al., 2005). For these stress paradigms, we encourage the movement away from a single neurotransmitter focus towards the complexity of "system biology," appreciating the emergent properties of behavior as well as its robustness to environmental perturbations (Aderem, 2005; Butcher et al., 2004). Finally, for the rat, Maier demonstrated that multiple re-exposures to the context of the inescapable shock can prolong the duration of the behavioral deficit (Maier, 2001), highlighting the similarity between LH behavioral depression models and post-traumatic stress disorder.

For mice, a common version of LH developed by H. Anisman's group exposes mice to anywhere from 60 to usually 360 inescapable footshocks (0.3 mA intensity) in a single session followed by single housing to induce various degrees of LH motor deficit (Anisman and Merali, 2003; Shanks and Anisman, 1988). For some experiments in CD-1 mice, looking at the effect of "chronic" stressors on NE and corticosterone levels, this 360 inescapable, footshocks procedure has been repeated up to 15 days (Irwin et al., 1986). The control groups were usually exposed to the same apparatus with no shock or escapable shock. Then, 24 hours later, the mice are tested in a shuttle-box escape task with best results obtained by requiring a 2 seconds delay of gate opening after the shock, thereby necessitating sustained motor activity. Using this protocol with outbred CD1 mice, an increased latency of response and more escape failures were found in the LH group compared to the no shock controls (Shanks and Anisman, 1988). As in rats, inbred mice strains demonstrate variability in their motor behavioral deficits from this procedure and in their neurochemical response (Shanks and Anisman, 1988, 1989; Shanks et al., 1990, 1994). In particular, using the 360 shock single session protocol, six inbred mouse strains (A/J, BALB/cByJ, C57BL/6J, DBA/2J, C3H/HeJ, and the outbred CD-1 as a control) were examined for their escape deficit response and trunk blood corticosterone response. BALB/cByJ and C3H/HeJ had the largest motor deficit, followed by C57BL/6J and the outbred CD-1, while DBA/2J seemed resistant to LH induction (see Figure 3 in Shanks and Anisman, 1988). The control (no shock) group of the A/J strain had a shuttlebox performance deficit with increased latency in some trial blocks, making this strain a less than ideal choice for LH. The plasma corticosterone response to shock varied by strain with BALB/cByJ showing the largest magnitude response, increasing approximately threefold in corticosterone levels compared to no shock controls immediately following both acute (single session of 360 shocks) and chronic (14 daily sessions) exposure to footshock. The time from the acute stressor-induced, peak plasma corticosterone levels to baseline also varied by strain. Most strains showed an immediate spike with return to basal levels in 1 hour, while the DBA/2J line had levels five times higher than non-stressed controls at this 1 hour time point, returning to baseline at 3 hours (Shanks et al., 1990). In general, the issue of adaptation (habituation or sensitization with time) to stress and the variation in adaptation, as measured by corticosterone and NE levels in this case, provides an important and recurrent theme in the literature. Basically, one can see an increase, decrease, or no change in a measured variable secondary to the extent of adaptation (see Figure 1 Anisman and Zacharko, 1986), depending upon the strain, prior exposure to stress, the controllability, context, duration, frequency, spacing, and intensity of the stressors (Anisman and Matheson, 2005). For mice, this paradigm has good pharmacological validity with amitriptyline reversing the behavioral deficit and lesser effects for desipramine and buproprion with only some strains showing a response (Caldarone et al., 2003; Shanks and Anisman, 1989). For readers interested in appropriate genetic backgrounds for this LH paradigm for transgenics, the C57BL/6J, 129/J, and their F1 hybrids have been usefully characterized for their LH responses with evidence of long escape latencies in control B6129F1s and 129/J mice (Caldarone et al., 2000). Hence, the C57BL/6J background seems more useful. Procedurally, a more reliable and consistent LH paradigm (2 consecutive days of 360 × 0.15 mA footshocks) for C57BL/6N mice with imipramine pharmacological validity provides a noteworthy entry point for future work. In this improved model, 80% of C57BL/6N mice remain helpless for 1 week and 22% at 2 weeks (Chourbaji et al., 2005a). In summary, LH paradigms offer a fruitful model of understanding behavioral deficits induced by intense stress simulating depression and adaptation to stress, with perhaps greater utility in rats than mice. In a way, this paradigm provided a foundation for the behavioral despair models and stimulated the generation of more refined models.

"Behavioral despair" models

Development, interpretation, methodology, and variability issues

"Behavioral despair" models involve the exposure of rodents to the milder inescapable stress of: (1) being forced to swim in a cylinder of water (FST) (Porsolt et al., 1977, 1978a, 1978b); or (2) being suspended from the tail (TST) for several minutes (Chermat et al., 1986; Porsolt et al., 1987; Steru et al., 1985, 1987). These paradigms work in both mice and rats, sharing the behavioral outcome measure of psychomotor immobility to an inescapable stressor. Antidepressants reduce this duration of immobility, namely an "anti-immobility" effect. For the FST, an interesting procedural difference exists in the rat versus the mouse paradigms. The rat protocol was developed first and the procedure was a 2-day test, with a day 1 "pretest" being a 15 minute session of immobility, followed by the test session on day 2 of 5 minutes. Searching for a "more rapid and less costly" primary screening test for antidepressants, Porsolt and colleagues subsequently developed a single day test in the mouse, usually measuring duration of immobility in the last 4 minutes of the 6 minute test (Porsolt et al., 1977). Using the 2-day rat FST paradigm, Porsolt et al. recognized the importance of natural strain variation early on and demonstrated similar basal rates of immobility duration for Sprague Dawley and Wistar rats, yet differences in response versus non-response to imipramine (2 i.p. injections; 15 min after trial 1 and 1 h before trial 2) attributable to different suppliers of Wistar rats (Porsolt et al., 1978b). In contrast, comparing three strains of outbred mice (CD, NMRI, and OF-1) in the mouse FST, they showed marked differences in basal durations of immobility varying from CD mice remaining immobile 80% of the time ($\sim$190 s) to NMRI and OF-1 being immobile roughly 40% of the time ($\sim$90 s). Nonetheless, all strains showed dose-dependent reductions in immobility with differences in the slope of dose response curves. Both the FST and the TST have been widely disseminated and utilized, providing rapid, reliable screens for antidepressant activity with manual versions being inexpensive to set up, yet labor intensive to score and complicated by subjectivity in scoring "immobility" with its associated variability. Automated versions of both paradigms are available and preferred (Crowley et al., 2004; Hedou et al., 2001; Kurtuncu et al., 2005), while validation is critical whatever the version. Specifically, investigators are encouraged to validate their particular version by checking for a dose-dependent reduction in immobility (an "anti-immobility effect") with an established antidepressant, determining at least a 20% reduction in immobility compared to vehicle controls for effective doses in the published range. For validation, the literature suggests the following drugs have the most reproducible data for activity: imipramine as a tricyclic ADA, citalopram as a selective serotonin reuptake inhibitor (SSRI) ADA, and buproprion as an atypical agent (reviewed by Cryan et al., 2005a). Venlafaxine, a newer, serotonin and norepinephrine reuptake inhibitor (SNRI) ADA, with an improved receptor binding profile, fewer side effects, and impressive TST efficacy provides another alternative for investigators wanting to validate their paradigm (Liu et al., 2003; Sanchez and Hyttel, 1999). In the automated version of the TST, a mouse is suspended by its tail for 6 minutes and a strain gauge measures the movements the mouse makes, calculating the duration of immobility below a given threshold (Porsolt et al., 1987; Steru et al., 1987). During a testing session, the mouse alternates between active attempts to escape and passive immobility. As with the FST, the duration of immobility has been inferred as an index of "behavioral despair," where longer durations of immobility imply a greater degree of "behavioral despair." The patterns of initial activity and struggling alternating with the ensuing, increasing amounts of immobility in later minutes of these paradigms have been variously interpreted as "behavioral despair," low mood, alternations of protest (activity) and despair (immobility), an adaptive response to a stressful situation (Hawkins et al., 1978), or even as a model of an evolutionary, adaptive "search-waiting" behavior with appropriate initial "searching for a solution to problem" via escape alternating with a later "waiting strategy" (immobility) (Thierry et al., 1984). Despite these debates, the predictive validity and utility of these paradigms remain notable. A wide variety of antidepressants have been validated with the FST protocol in mice and rats and this domain has been adroitly reviewed (Borsini and Meli, 1988). In addition, the paradigm has been useful in examining augmentation strategies (Redrobe and Bourin, 1999), dissecting pharmacological mechanisms of antidepressant action (Conti et al., 2004; Cryan et al., 2005b), and in testing for effects of knockout mice transgenic research (Urani et al., 2005).

Likewise, the TST has been extensively validated as a sensitive biobehavioral assay for antidepressant activity with an impressive diversity of antidepressants (tricyclic (TCAs), SSRIs, buproprion, and atypical antidepressants, monoamine oxidase inhibitors (MAOIs)), and even electroconvulsive shock (Perrault et al., 1992; Steru et al., 1985, 1987; Teste et al., 1990, 1993; Ukai et al., 1998). The TST procedure is simple and reproducible, avoiding the severe hypothermic effects of water immersion in the FST (Thierry et al., 1986). Usually, the drug treatment is a single acute treatment 30–60 minutes prior to the tail suspension and merely a "screen" for antidepressant activity with incomplete understanding of mechanism. However, chronic oral delivery of ADA is feasible (Caldarone et al., 2003; Dulawa et al., 2004). A recent review has

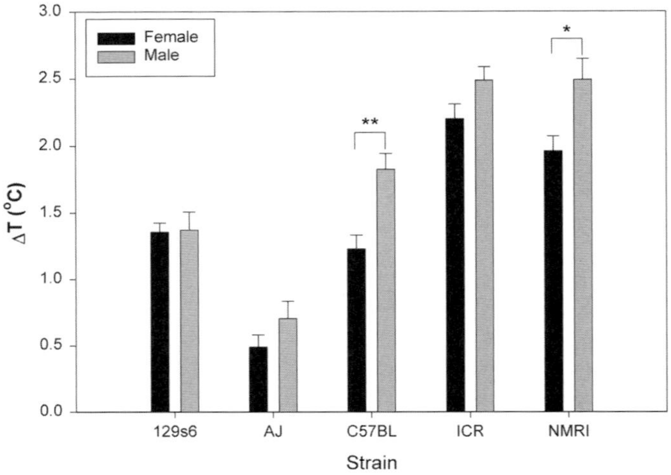

Figure 19.2 Strain and gender variations of tail suspension-induced hyperthermia (TSIH). ΔT was obtained by subtracting the mean temperature of the no tail suspension test (no-TST) group from the mean temperature of the TST mice. Asterisks indicate significant differences between the genders. ** $P < 0.01$, * $P < 0.05$.

comprehensively cataloged these pharmacological responses in the TST, further supporting the utility of this assay, and provides further explicit guidance on equipment, drugs, and doses (Cryan et al., 2005a). Overall, key advantages of this TST paradigm include: (1) higher sensitivity to ADA that were not detected in the original FST such as with SSRIs; (2) a high degree of pharmacological predictive validity; and (3) the absence of an *a priori* mechanism of antidepressant action. From our perspective, this atheoretical bias of the TST is beneficial. For example, the TST has demonstrated antidepressant activity for atypical agents such as σ1-receptor agonists (Ukai et al., 1998), St. John's wort (Butterweck et al., 1997; Kumar et al., 1999), a novel α-amino-3-hydroxy-5-methyl-4-isoxazolepropionic acid receptor (AMPA) receptor potentiator (Li et al., 2001), and metabotropic glutamate receptor (mGluR1 and mGluR5) antagonists (Belozertseva et al., 2007).

While the TST and FST paradigms seem similar paradigms on the surface, the FST and TST are also distinct paradigms as distinguished by profiles of outbred strains variation in basal immobility (Bai et al., 2001; Vaugeois et al., 1997). Also, the shapes of drug dose response curves for the response of C57BL/6 mice to imipramine gave a "U-shaped curve" in the FST, while the dose dependence was linear, monotonically decreasing in the TST (Bai et al., 2001). Finally, the inbred strain order and immobility times in the FST correlate non-significantly and weakly with the TST.

Beyond immobility in the TST, we have pharmacologically validated another dependent variable, namely tail suspension-induced hyperthermia (TSIH), as a robust increase in core temperature evoked by this paradigm (Figure 19.2) (Liu et al., 2003). Briefly, TSIH peaked after tail suspension and remained elevated at 120 minutes. Five (129/SvEvTac, A/J, C57BL/6J, NMRI, and ICR) strains were examined with or without TST. Significant strain variations were detected in TSIH, ranging from the inbred NMRI showing the highest temperature rise (2.3°C) to the A/J mice showing the lowest (0.6°C). Significant gender differences were also found for the C57BL/6J and NMRI strains on TSIH. Tail-suspension-induced hyperthermia and duration of immobility were not significantly correlated ($r = 0.22$, $P = 0.17$) in outbred mice. The benzodiazepine, diazepam dose-dependently increased immobility and decreased TSIH. Propranolol, a beta-blocker that crosses the blood-brain barrier, blocked TSIH. Nadolol, a lipophobic non-selective β-adrenergic blocker, which acts only peripherally, had no effect. Antidepressants showed more complex patterns of effects. Most importantly, from the data on propranolol, diazepam, and antidepressants, one can pharmacologically dissociate a distinct TSIH effect from the TST immobility effect. We believe this TSIH taps an interesting, unique aspect of "anxiety-like" behavior, providing a physiological measure to complement immobility. This TSIH trait models emotional hyperthermia (reviewed in Oka et al., 2001) seen across a variety of species and in humans.

Sources of consistent inconsistencies in the literature

Five broad technical issues regarding the TST and FST paradigms will be discussed to help explain the consistent inconsistencies in the literature. For this work, we believe the saying, "the experiment is only as good as the phenotyping." First, differences in methodology (different automated equipment vs. different settings vs. manual) of scoring the TST or FST can lead to differing results among laboratories. As one example for the TST, initially, we accepted the established and published rank order of nine inbred strains (Trullas et al., 1989) in the TST. For example, the BALB/cJ and DBA/2 mice differed the most in their mean duration of immobility (96 vs. 1 s, respectively) (Trullas et al., 1989). These TST results were obtained by manual observation of a mouse taped by its tail to a shelf, measuring the duration of immobility subjectively by eye. Although this method can be reliable, we wanted to avoid the subjectivity and variability of a manual testing, preferring an automated version of the TST paradigm to optimize reproducibility. Our goal was simply to validate the marked strain differences between BALB/cJ and DBA/2J and then select strains to perform quantitative trait loci (QTL) mapping. Surprisingly, we could not replicate these results in our laboratory with our automated version that measures immobility by a force transducer rather than visually. In fact, the strain order derived from the manual TST testing did not correlate with the automated TST rank order. Hence, though similar approaches to measuring TST immobility, these methods yield very different results for strain order, although similar reductions in immobility result for efficacy of antidepressants. Our settings were optimized for detection of ADA response. In contrast, a recent paper replicated the marked DBA/2J (~90 s) vs. C57BL/6J (~220 s) difference in basal immobility by using different gain settings which were specifically developed to give scores comparable with their prior manually scored data (Crowley et al., 2005). Hence, methodological differences may explain apparent discrepancies in the literature.

Second, as with many behavioral paradigms, variation in phenotyping protocols and differences in subtle nuance variables (housing (Chourbaji et al., 2005b), technician handling, lighting, seasons, suppliers) including nuances in prior stress exposure (e.g., rearing history, shipping, housing noise levels, cage density, handling, and trial effects) may explain differences among laboratories (Crabbe et al., 1999; Wahlsten et al., 2003). Perhaps relevant to the TST, an interesting effect of differential laboratory handling on aggressive behavior were found by J. Fuller and B. Ginsburg, explaining marked differences in male aggressive fighting behavior for C57BL/10 males (Ginsburg, 1967). In particular, the once per week handling at The Jackson Laboratories involved picking up mice by their tails with forceps (in a way, similar to a tail hang or perhaps perceived as an uncontrollable stressor or even a defeat), whereas the Ginsburg's laboratory mice were transferred by "being permitted to walk from one cage to another with an apposed entrance, or, if they balked, they were scooped up in a box." While the C3H agouti and a C-albino strains were unaffected in their aggressive behavior due to these handling differences, the forceps handled C57BL/10 males were significantly less aggressive than their non-handled counterparts. A similar tail suspension handling issue influenced the onset of seizures in susceptible EL mice (Leussis and Heinrichs, 2006). Hence, investigators should be mindful that such inadvertent "housekeeping" and handling differences in early learning histories may have genotype-specific, perturbing effects for adult behaviors, contributing to poor reliability among laboratories. Given the tradeoffs of competing particulars in phenotyping, we value simplicity, automation, and reliability. However, TST paradigms vary in procedure, testing naïve mice for 6 minutes in a single trial (Steru et al., 1985) to recording the last 2 minutes of a single 5 minute test trial to recording the last 6 minutes of a 7 minute third test trial, with two prior "training trials" of 7 minutes duration (Ukai et al., 1998). This last procedure may be especially advantageous in providing a more sensitive antidepressant activity assay. Presumably, the two "training trials" conditioned these outbred mice for longer immobility times and lowers within-group variability to provide a more stable and sensitive baseline, paralleling in a way the "pretest" used in rats for the FST.

Third, mice strains can differ in their behavior based on the supplier or substrain. For example, from a review of the literature among the European scientists using the ITEMATIC-TST device, the NMRI outbred mice from Centre d'Elevage Janvier, France, seem significantly more sensitive to imipramine (lowest effective dose = 2–4 mg/kg) (Porsolt et al., 1987; Ripoll et al., 2003) than the NMRI outbreds from other suppliers (lowest effective dose = 16–30 mg/kg) (Teste et al., 1993; van der Heyden et al., 1987) (Figure 19.3a), yet a reverse pattern of response between laboratories for desipramine (Figure 19.3b). Similarly, C57BL/6J mice from The Jackson Laboratories was one of the relatively less responsive strains to citalopram in the

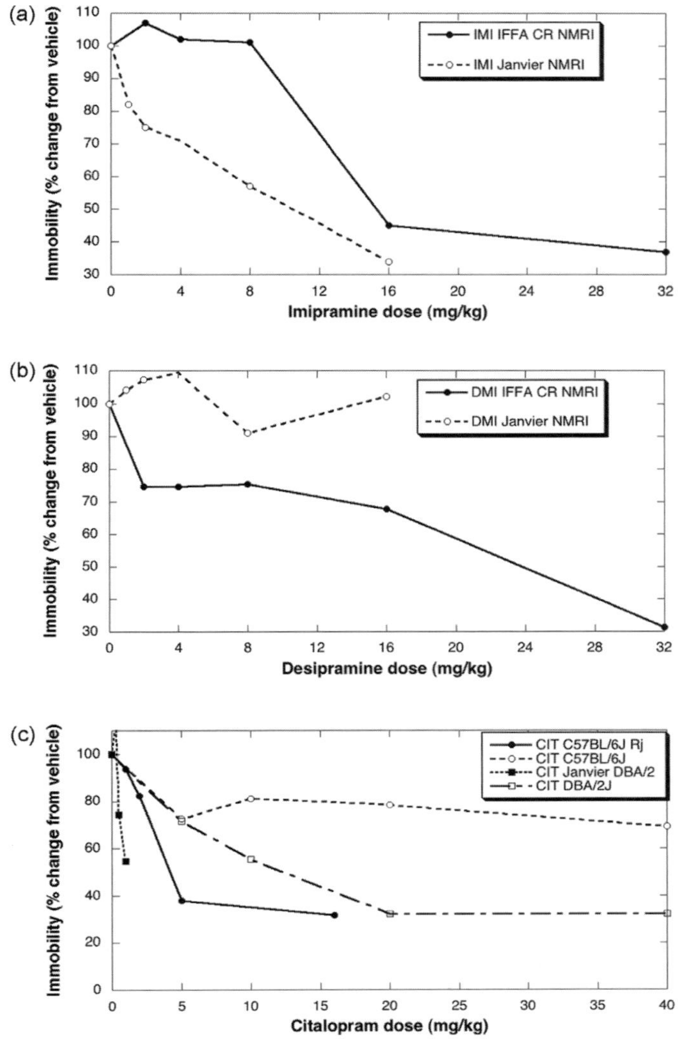

Figure 19.3 Comparison of published tail suspension test (TST) dose response curves for mice tested in different laboratories. The *y*-axis is the mean percentage change in immobility time compared to a vehicle control group, set at 100%. NMRI-outbred mice from two different suppliers (sourced from IFFA CR, as published in Teste et al., 1993, and from Janvier, as published in Ripoll et al., 2003) tested on the same Itematic-TST apparatus for response to (a) imipramine (IMI) and (b) desipramine (DMI). (c) Substrains of inbred DBA/2 and C57BL/6 are compared from two different sources (from The Jackson Laboratories, USA, as published in Crowley et al., 2005, and from France, as published in Ripoll et al., 2003) for response to citalopram (CIT) tested on different apparati.

TST (Crowley et al., 2005), while the C57BL/6J Rj subline in France was highly responsive to citalopram (Ripoll et al., 2003) (Figure 19.3c). Similarly, the dose response curves for citalopram in responsive DBA/2J strain (Figure 19.3c) show surprising differences in sensitivity. Hence, since colonies from different suppliers of these strains may vary, the helpful recommendation of M. Bourin to test ADA activity in screens using either the Swiss mice in the TST or the FST with Swiss and C57BL/6J Rj should be embraced in France (Bourin et al., 2005), yet the generalizability of this proposal seems poor.

Fourth, the importance of modeling the delayed onset (2–4 week "therapeutic lag") and time course of the therapeutic effect of antidepressant medications has been repeatedly identified as a key criterion and a weakness of these acute treatment models (Frazer and Morilak, 2005; Mitchell and Redfern, 2005). Often, clinicians and critics dismiss acute responses as irrelevant, privileging subchronic (days) and preferring chronic (weeks) dosing. While a focus on a delayed onset is useful, the relative merits of this perspective have been challenged and questioned by some recent findings. First, some patients have an early response to antidepressants even within the first week (Katz et al., 2004). Second, preliminary clinical trial data suggest the existence of rapidly acting antidepressants with onset of action at 48 hours and peak effect within 1 week (Feighner et al., 2001). Of note, this rapidly acting ADA, INN00835, was empirically developed using a modified rat FST paradigm (Hlavka et al., 1997). Third, an *inactive single* dose of the antidepressant fluoxetine (10 mg/kg, i.p.) in the TST can become *an effective dose* in the TST when given daily for 21 days, blurring the time course distinction (Naudon et al., 2002). Also, overvaluing this "lag period" for antidepressant response may contribute to misinterpretations and conflations of an SSRI's anti-anxiety activity with its antidepressant activity (Dulawa and Hen, 2005; Santarelli et al., 2003). In summary, while making a distinction between acute responses and chronic response to ADAs remains important, emphasizing this criterion for evaluating rodent models may have less persuasive force than in the past.

Finally, I. Lucki's group has reported the limitations of using the C57BL/6J strain of mice in the TST due to "tail climbing" in 70% of these mice. As many experiments are performed in this strain or on this background, two "workarounds" are worth noting. One approach has been to place the tape of the mouse's taped tail on a sharp hook preventing the climbing (Yoshikawa et al., 2002). Another simple approach we have reliably utilized places tape at the very base of the tail rather than the typical 2 cm from the distal tip of the tail, thereby reducing the tail climbing rate to 9% C57BL/6J (X. Liu, R. Reister, and H. Gershenfeld, unpublished observation).

FST – strain variation

The role of genotype was further demonstrated in the FST (3 min test; cylinder 12 cm diameter filled with 10 cm water) by testing nine strains of mice, showing marked variation in baseline immobility responses (Balb/c at about 105 s to C57BL/6 at 35 s and C3H/He at 5 s) (Nikulina et al., 1991). Six of these strains were further tested for their response to dopamine D_1 receptor agonists SCH23390 (0.5 mg/kg), which significantly reduced immobility in all strains. In fact, this response to D_1 agonist significantly correlated ($r = -0.92$) with their basal FST activity, yet the small sample size suggests caution. A more recent, noteworthy paper for FST behavior tested eight inbred strains and four outbred strains for: (1) basal immobility as well as; (2) antidepressant activity (desipramine and fluoxetine 0–20 mg/kg i.p.) in separate sets of naïve mice (Lucki et al., 2001). Again, significant and large variations in both immobility time and antidepressant response were found among strains. In this more standard, manual FST paradigm (6 min test; cylinders 46 cm high × 21 cm diameter filled with 15 cm depth of water), basal immobility ranged from 142 seconds in C57BL/6J mice to 13 seconds FVB/N/J with the decreasing rank order of immobility times being: C57BL/6J > CD-1 > BALB/cJ > DBA/2J > A/J > CF-1 > 129/SvImJ > Swiss Webster > C3H/HeJ > NIH Swiss > FVB/NJ. From the full desipramine dose response curves, the most sensitive strains were DBA/2J, C57BL6/J, and the CD-1 mice, showing antidepressant responses (reduced immobility times, namely an anti-immobility effect) at the 5–20 mg/kg doses. Less responsive were the BALB/cJ, 129/SvImj, and NIH Swiss mice showing responses only at the higher doses, while no effects were found in the C3H/HeJ, Swiss Webster, CF-1, and FVB/NJ lines. For the SSRI antidepressant fluoxetine, the most sensitive line was the DBA/2J strain responding at the 5–20 mg/kg doses, followed by the outbred Swiss Webster responding at 20 mg/kg. In general, the strains showed less responsiveness for fluoxetine than for desipramine. In fact, most strains showed no effect to fluoxetine and the 129/SvImj line even showed increased mobility. Finally, the basal immobility time in the FST was not correlated with response to antidepressant response. Further support for the FST uniqueness came from a study of outbred NIH Swiss mice, where the immobility time in the FST (10 min test; cylinder 17 cm high × 21 cm diameter filled with 8 cm of water) was shown not to significantly correlate with anxiety or locomotor tests as measured by the holeboard test or the elevated plus-maze test variables (Hilakivi and Lister, 1990).

TST – strain variation

Using the above-mentioned Porsolt criteria, the TST emerged as the most feasible, reliable, and validated experimental model for drug screening and genetic dissection. Initially outbred strain differences were found for both spontaneous, basal behavior, and imipramine response. For example, while NMRI strain showed large anti-immobility responses, the CD-1 mouse showed no response to imipramine (van der Heyden et al., 1987; Vaugeois et al., 1996). Later, spontaneous TST scores (duration of immobility; using either a manual visual method (Trullas et al., 1989) or an automated method (Crowley et al., 2005; Liu and Gershenfeld, 2001) for different inbred strains showed marked differences among strains. The TST immobility duration did not correlate significantly with open field activity or light⇔dark behavior among strains, suggesting a distinct set of loci affecting each behavior and distinct aspects of stressor response evoked by the TST (Liu and Gershenfeld, 2003; Trullas et al., 1989). An analysis of the spontaneous TST immobility among 60 NMRI versus 60 CD-1 mice showed substantial within-strain variation, ranging from 6 to 199 seconds with a mean of 79 seconds for CD-1 and ranging from 9 to 218 seconds with a mean of 111 seconds for NMRI (Vaugeois et al.,

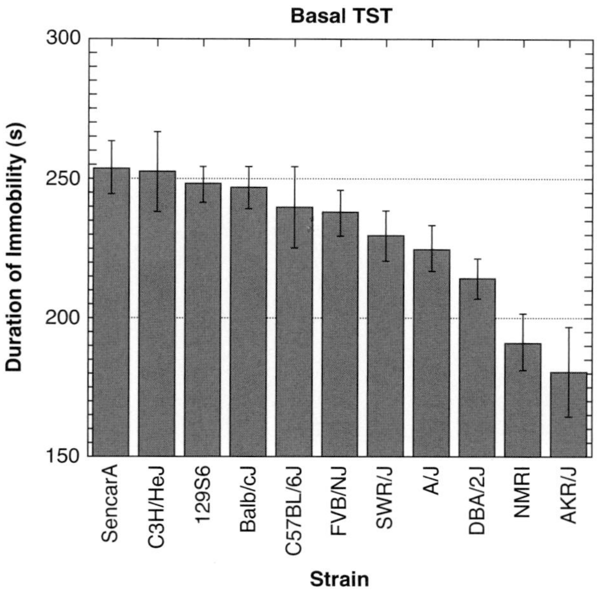

Figure 19.4 Strain distribution of basal immobility across 11 inbred strains, showing mean (±SEM) duration of immobility. Bars below the figure designate strains that significantly differ ($P < 0.05$). TST: tail suspension test.

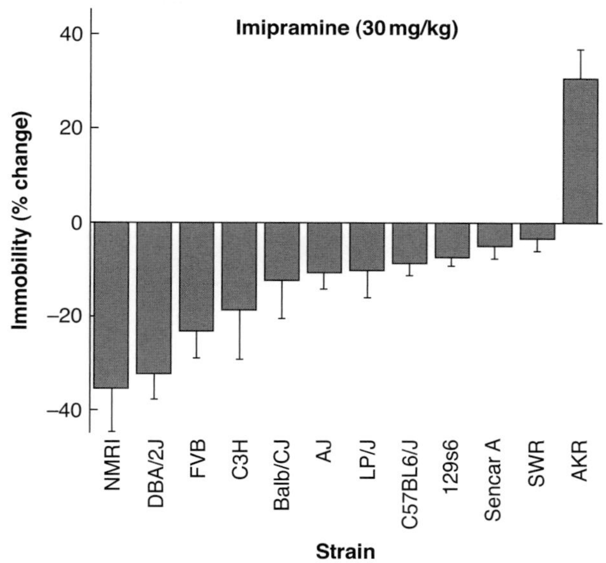

Figure 19.5 Strain distribution of imipramine response across 11 inbred strains, showing mean (± SEM) percentage change in duration of immobility. Percent change in duration of immobility was calculated as (TST immobility IMI – TST immobility saline) / TST saline) × 100%. TST: tail suspension test.

1997). Then, the differences in response to ADA was noted in a cohort of "high" versus "low" basal immobility CD1 mice (20% extremes of the distribution), leading to a genetic breeding program to develop selective bidirectional lines for this basal TST immobility (El Yacoubi et al., 2003; Vaugeois et al., 1996, 1997). In these experiments, only the "high" basal immobility group were responsive to ADA (imipramine, desipramine, and paroxetine). Subsequently, the Vaugeois' group has extensively phenotyped correlates of these "helpless" (HL; high immobility; >115 s) and non-learned helpless (NLH; low immobility; <3 s) lines, showing impressive differences consistent with a depressive phenotype. Specifically, the HL line had less wakefulness, shorter latency to REM sleep, higher basal corticosterone levels, higher serotonin (prefrontal cortex and hippocampal) levels, and lower basal temperatures compared to the NLH line. As these impressive findings come from a comparison of just two lines without a control line or replicate lines, one should be mindful of possible spurious findings (Henderson, 1997). In particular, the HL versus NLH line differences in antidepressant response may be an over-interpretation, since the low immobility groups provide a poor sensitivity bioassay for drug response rather than a simple direct relationship. Furthermore, the TST imipramine response is weakly correlated with basal immobility (Liu and Gershenfeld, 2003).

Our group has systematically screened 12 inbred strains in an automated TST paradigm with a distinct perspective influenced by our clinical knowledge of depression and its treatment (Figures 19.4 and 19.5; Liu and Gershenfeld, 2001). By judicious selection of mice from the major branches of the *Mus* genealogical tree (Atchley and Fitch, 1993), these inbred strains represent mousedom providing genetic diversity for heritability estimates. The narrow sense trait heritabilities for basal immobility and imipramine-induced reduction in immobility were estimated as 0.31 and 0.60, respectively. We also examined the dimensions of this paradigm by a factor analysis (Liu and Gershenfeld, 2003). In contrast to a recent strain distribution study of eight strains (Crowley et al., 2005), our phenotyping protocol involved initially a basal TST in naïve mice on day 1, followed by a response to imipramine at a full 30 mg/kg dose versus saline on day 3 in the same cohort. From our clinical perspective, treating naïve mice with antidepressants incompletely models the clinical situation of patients presenting near their nadir (Figure 19.1). Indeed antidepressants have no mood elevating effects in normal volunteers (Wilson et al., 2002). However, we believed previously stressed mice might more nearly simulate the stress of presenting patients. (An alternative perspective might view the injection and handling stress prior to the TST or FST as a sufficient stressor.) Likewise, we utilized the TCA imipramine at the "maximally tolerated dose," mirroring clinical practice. Imipramine is a "polyreceptor, dirty drug" thought to act predominantly through serotonin and norepinephrine reuptake inhibition. Imipramine was selected as the most validated drug with repeated demonstrations of a dose-dependent anti-immobility effect in the TST with maximal response 30 minutes following a single intraperitoneal injection at 30 mg/kg (Liu and Gershenfeld, 2001; Steru et al., 1987; Teste et al., 1993; van der Heyden et al., 1987). In the mouse, imipramine (20 mg/kg, i.v.) has a plasma elimination half-life of 50 minutes (Dingell et al., 1964). Through a factor analysis on both an F2 population and inbred strains using this protocol, three independent TST aspects of this paradigm were defined, namely: (1) spontaneous duration of immobility in the TST; (2) anti-immobility response to imipramine; and (3) TSIH (Liu and Gershenfeld, 2003).

Table 19.1 Summary of QTL mappings for behavioral despair paradigms.

Phenotype	Species	Strains crossed	Phenotyping method	Cross type	No.	Sex	QTL name	Chr.	Nearest marker	Pos (cM)	LOD	% Var	*Mus* chr. region	Phenotyping method
Basal TST	Mouse	DeFries High × DeFries Low selectively bred lines for Open field behavior derived from a founder F3 cross of BALB/c and C57BL/6	Manual scoring 3 min TST	F2	1636	M & F		3 5 11 19		34 14 50 40	6.7 10.9 6.9 6.2	2.4 3.9 2.5 2.2		Turri et al., 2001
Basal TST	Mouse	C57BL6 × C3H/He Source: Japan SLC Note: C3H mice are homozygous for the retinal degeneration (rd) allele of Pde6b, causing blindness by weaning age	Day 1, 10 min TST test automated, scoring immobility during the last 500 s	F2	560	M only	*Tst2* *Tst4* *Tst1* *Tst3* *Tst6*	4 8 11 14 6	D4Mit203 D8Mit242 D11Mit271 D14Mit257 D6Mit183	68 44 23 25 27	3.5 2.9 3.6 3.0 Interaction with D11Mit271 at 21 cM	3.8 2.5 3.2 2.3		Yoshikawa et al., 2002
Basal FST	Mouse		Day 2, given 15 min FST pretest, followed in 24 hours by Day 3, automated FST scoring immobility throughout 300 s session				*Fst5* *Fst1* *Fst3* *Fst2* *Fst3*	6 8 8 11 17	D6Mit289 D8Mit242 D8Mit93 D11Mit271 D17Mit185	48 51 66 20 35	2.8 4.2 3.0 3.7 2.9	3 5.4 4 2.9 2.2		Yoshikawa et al., 2002
TST citalopram response		BALB/cJ × A/J	Automated TST, 6 min; two tests with 1 week rest between TST; 1/2 mice i.p. Saline first and then 1/2 citalopram (i.p., 20 mg/kg); counterbalanced	F2	506	M & F		7 12 19		72 45 54	2.5 2.4 3.9	12 11 18		Crowley et al., 2006 only extremes geno-typed with overestimate of %var
Basal TST	Mouse	NMRI × 129S6	Automated TST, 6 min; two TST tests with basal TST on Day 1	F2	747	M only	*Tstb1* *Tstb2* *Tstb3* *Tst2*	5 12 18 4	D5Mit41 D12Mit158 D18Mit186 D4Mit204	61 43 51 62	5.7 5.2 3.0 3.1	8.3 7.4 2.3 6.8		Liu et al., 2007
TST imipramine response			Day 3 IMI (i.p., 30 mg/kg) 30 min prior to automated 6 min TST				*Tsti1* *Tsti2* *Tsti3*	5 1 4	D5Mit41 D1Mit410 D4Mit204	58 20 67	2.9 3.1 3.7	6 6.4 9.8		Liu et al., 2007
TSIH			Day 1 manual colonic temperature after TST				*Tsih1*	4 5	D4Mit142 D5Mit208	39 54	5.8 2.9	3.4 0.6		Liu et al., 2007
Basal FST	Rat	Wistar Kyoto(WKY) × Fisher 344	Day 1, Open field; Day 3, 15 min FST pretest; Day 4, manual 300 s FST scored by sampling every 5 s for immobility	F2	486	M & F	*Imm2* *Imm3* *Imm4*	3 5 8	D3Rat71 D5Rat40 D8Rat163	55 95 20	2.4 3.2 2.2	1.9 3.6 1.8	2 (57-87cM) 4 (43-84 cM) 9 (0-26 cM)	Solberg et al., 2004

FST: forced swim test; TST: tail suspension test.
Chromosome (Chr) and Position (cM) indicate the distance of the QTL from the centromere in centimorgans (cM) along with the nearest marker to the QTL peak.
LOD indicates logarithm of odds ratio score; %var, percent variance explained by the locus.
Mus chr. Region is the syntenic region of mouse estimated using the Rat-*Mus* syntenic map via the gene and position prediction (GAPP) map URL: http://gapp.gen.gu.se/

Table 19.1 summarizes the results of five rodent QTL mapping studies for animal behavioral despair paradigms (Crowley et al., 2006; Liu et al., 2007; Solberg et al., 2004; Turri et al., 2001; Yoshikawa et al., 2002). In general, the size of the QTLs defined comport with prior findings, where individual loci explain less than 10% of the variance (Valdar et al., 2006). Yoshikawa and colleagues found evidence for a shared, pleiotropic locus for both FST and TST behaviors, i.e., the chromosome 11 locus at about 20 cM (near D11Mit271). Of the QTLs mapped in a C57BL/6 × C3H/He cross (Yoshikawa et al., 2002), the suggestively significant chromosome 4 locus (*Tst2*) mapping at 68 cM (near D4Mit203; Lod 3.5) overlaps the Liu et al.'s suggestive QTL near D4Mit204 influencing both basal TST and TST imipramine response (Liu et al., 2007) as well as the rat *Imm3* locus influencing the FST (Solberg et al., 2003, 2004). The rat FST behaviors have been skillfully genetically dissected using a Wistar Kyoto (WKY; males with increased immobility) × Fisher 344 (F344) intercross population, studying both: (1) FST parameters (climbing, swimming, and immobility behaviors); and (2) HPA axis responses (basal and 10 min restraint-induced plasma corticosterone levels and adrenal weights). Of note, the HPA measures were independent of rat FST measures (Solberg et al., 2003) just as the basal TST is distinct from TSIH and TST imipramine response. None of the QTLs mapped for the SSRI citalopram TST response overlaps QTLs mapped for imipramine response, but their chromosome 12 locus overlaps the Liu et al. (2006) *Tstb2* locus for basal TST (Crowley et al., 2006). The reasons for such limited convergence among these QTL studies are probably multiple, including technical and procedural differences in measuring behaviors, genetic heterogeneity (different strains crossed and species differences in rat vs. mice), and different drugs used. Of note, the chromosome 5 locus defined by Liu et al. (2007) affected basal TST, antidepressant immobility response, and TSIH measures, suggesting the existence of a pleiotropic loci influencing both "stress vulnerability" and antidepressant responses. This chromosome region harbors a cluster of γ-amino butyric acid subunit A ($GABA_A$) receptor subunits and the human syntenic region of this 20 cM *Mus* chromosome 5 region, namely 9p21–23 and 1p31–35. Hence, the QTLs in Table 19.1 represent promising loci for testing in stress-diathesis models of human psychiatric illness and antidepressant responsiveness. In general, psychiatric illnesses probably arise from numerous susceptibility genes, each conferring a small odds ratio of increased relative risk. Given the heterogeneity of depressive and anxiety behaviors, the challenge of mapping these traits in humans with modest effect loci will continue. Hopefully, rodent models will provide regions of interest for integration with emerging human linkage, candidate gene association, and whole genome association studies.

Conclusions and future directions

In summary, the LH, FST, and TST paradigms provide foundations for understanding the pathophysiology of "stress vulnerability" and for screening ADA in rodent models at multiple levels, from molecules and drugs to the emergent properties of behavior. For these "stress vulnerability" and ADA responses, strain variation offers a model of human individual differences in temperament, vulnerability to major depression and anxiety disorders, and responses to treatment. The labels "stress response," depression, and anxiety are broad generic labels with many meanings at different levels from neuroendocrine responses to whole behavior. Factor analyses and correlational studies demonstrate independent aspects of a "stress response" and the variation in "adaptation to stress." The integration of approaches across levels through genome-wide studies, the study of multiple inbred lines via the Mouse Phenome Project (Bogue and Grubb, 2004), and complementary studies in renewable mouse populations like larger recombinant inbred (RI) strain panels or the Mouse Collaborative Cross Consortium (Churchill et al., 2004) may contribute to a systems approach. In some ways, future progress in developing improved antidepressants and the screening of atypical antidepressants depends on these simple "atheoretical models" as well as eventually understanding the pathophysiology. An appreciation of strain variation and sources of unwanted variation in the FST and TST paradigms provide guidance for optimizing bioassay sensitivity and the refinement of drug screening, minimizing false positives and negatives. Furthermore, preclinical drug development may eventually leverage this variation through selective breeding approaches (with or without the use of mapped QTL) to develop strains simulating "treatment resistance" to typical antidepressants, providing an assay for new, atypical agents. The neurochemical, cytokine, signaling, pharmacological, neurohormonal, neuroimaging, and gene expression changes along with genetic vulnerability loci correlating with variation in "stress vulnerability" or antidepressant response in these models provide testable hypotheses for validation in human depression. While the "therapeutic lag" of antidepressants can be modeled in TST using chronic low dosing, the refined chronic mild stress paradigm (Strekalova et al., 2004) incorporating repeated, daily TST seems particularly promising for modeling the chronic, neuroadaptive plasticity of major depression. Beyond better drugs for "treatment resistant" depression and dissecting the response versus non-response issue, future directions would target, develop, and model some neglected aspects of major depression highlighted in Figure 19.1, namely depression's chronicity with relapses, the issue of relapse while on medication ("maintenance phase"), the mechanisms perpetuating a depressive episode, the etiology of relapse, and the nature of the sequelae and residua of recurrent depressive episodes on the trait of "stress vulnerability." Among these challenges, simulating a relapse while on medication seems a feasible starting point, though continued research on all fronts provides hope for patients suffering with depressive disorders.

Chapter abbreviations

ADA, antidepressive agent
DALYs, disability-adjusted life years

FST, forced swim test
GABA, γ-amino butyric acid
HPA, hypothalamic–pituitary–adrenocortical
LC, locus coeruleus
LH, learned helplessness
LOD, likelihood of the odds ratio score
MAOIs, monoamine oxidase inhibitors
MDD, major depressive disorder
NE, norepinephrine
OR, odds ratio
QTL, quantitative trait loci
SNRI, serotonin and norepinephrine reuptake inhibitor
SSRI, selective serotonin reuptake inhibitor
TCAs, tricyclic antidepressants
TSIH, tail suspension induced hyperthermia
TST, tail suspension test

Acknowledgments

We appreciated helpful conversations with X. Liu, D. Lagace, D. Theobald, V. Krishnan, R. Reister, O. Berton, our University of Texas Southwestern Medical Center colleagues, S. Maxson for bringing B. Ginsburg's work to our attention, and our anonymous reviewers for their critiques. This project has been generously supported by a NARSAD Independent Investigators Award (H. G.) and NIH MH67211 (H. G.).

References

Abkevich, V., Camp, N.J., Hensel, C.H., Neff, C.D., Russell, D.L., Hughes, D.C., et al. (2003) Predisposition locus for major depression at chromosome 12q22–12q23.2. *Am J Hum Genet* 73: 1271–1281.

Aderem, A. (2005) Systems biology: its practice and challenges. *Cell* 121: 511–513.

Agid, O., Kohn, Y., and Lerer, B. (2000) Environmental stress and psychiatric illness. *Biomed Pharmacother* 54: 135–141.

Anisman, H. and Matheson, K. (2005) Stress, depression, and anhedonia: caveats concerning animal models. *Neurosci Biobehav Rev* 29: 525–546.

Anisman, H. and Merali, Z. (2003) Rodent models of depression: learned helplessness induced in mice. In Crawley, J., Gerfen, C., Rogawski, M., Sibley, D., Skolnick, P., and Wray, S. (eds.), *Current Protocols in Neuroscience*. John Wiley and Sons, Hoboken, NJ, USA, Unit 8.10C.

Anisman, H. and Zacharko, R.M. (1986) Behavioral and neurochemical consequences associated with stressors. *Ann N Y Acad Sci* 467: 205–225.

Atchley, W.R. and Fitch, W. (1993) Genetic affinities of inbred mouse strains of uncertain origin. *Mol Biol Evol* 10: 1150–1169.

Avgustinovich, D.F., Kovalenko, I.L., and Kudryavtseva, N.N. (2005) A model of anxious depression: persistence of behavioral pathology. *Neurosci Behav Physiol* 35: 917–924.

Bai, F., Li, X., Clay, M., Lindstrom, T., and Skolnick, P. (2001) Intra- and interstrain differences in models of "behavioral despair." *Pharmacol Biochem Behav* 70: 187–192.

Belozertseva, I.V., Kos, T., Popik, P., Danysz, W., and Bespalov, A.Y. (2007) Antidepressant-like effects of mGluR1 and mGluR5 antagonists in the rat forced swim and the mouse tail suspension tests. *Eur Neuropsychopharmacol* 17: 172–9.

Berton, O. and Nestler, E.J. (2006) New approaches to antidepressant drug discovery: beyond monoamines. *Nat Rev Neurosci* 7: 137–151.

Bogue, M.A. and Grubb, S.C. (2004) The Mouse Phenome Project. *Genetica* 122: 71–74.

Borsini, F. and Meli, A. (1988) Is the forced swimming test a suitable model for revealing antidepressant activity? *Psychopharmacology (Berl)* 94: 147–160.

Bourin, M., Chenu, F., Ripoll, N., and David, D.J. (2005) A proposal of decision tree to screen putative antidepressants using forced swim and tail suspension tests. *Behav Brain Res* 164: 266–269.

Brown, G.W. (1993) The role of life events in the aetiology of depressive and anxiety disorders. In Stanford, S.C. and Salmon, P. (eds.), *Stress: From Synapse to Syndrome*. Academic Press, San Diego, CA, USA, pp. 1–22.

Brown, G.W., Harris, T.O., and Eales, M.J. (1996) Social factors and comorbidity of depressive and anxiety disorders. *Br J Psychiatry Suppl* 30: 50–57.

Butcher, E.C., Berg, E.L., and Kunkel, E.J. (2004) Systems biology in drug discovery. *Nat Biotechnol* 22: 1253–1259.

Butterweck, V., Wall, A., Lieflander-Wulf, U., Winterhoff, H., and Nahrstedt, A. (1997) Effects of the total extract and fractions of *Hypericum perforatum* in animal assays for antidepressant activity. *Pharmacopsychiatry* 30 (Suppl 2): 117–124.

Caldarone, B.J., George, T.P., Zachariou, V., and Picciotto, M.R. (2000) Gender differences in learned helplessness behavior are influenced by genetic background. *Pharmacol Biochem Behav* 66: 811–817.

Caldarone, B.J., Karthigeyan, K., Harrist, A., Hunsberger, J.G., Wittmack, E., King, S.L., et al. (2003) Sex differences in response to oral amitriptyline in three animal models of depression in C57BL/6J mice. *Psychopharmacology (Berl)* 170: 94–101.

Caspi, A., Sugden, K., Moffitt, T.E., Taylor, A., Craig, I.W., Harrington, H., et al. (2003) Influence of life stress on depression: moderation by a polymorphism in the 5-HTT gene. *Science* 301: 386–389.

Charney, D.S. (1998) Monoamine dysfunction and the pathophysiology and treatment of depression. *J Clin Psychiatry* 59: 11–14.

Chermat, R., Thierry, B., Mico, J.A., Steru, L., and Simon, P. (1986) Adaptation of the tail suspension test to the rat. *J Pharmacol* 17: 348–350.

Chourbaji, S., Zacher, C., Sanchis-Segura, C., Dormann, C., Vollmayr, B., and Gass, P. (2005a) Learned helplessness: validity and reliability of depressive-like states in mice. *Brain Res Brain Res Protoc* 16: 70–78.

Chourbaji, S., Zacher, C., Sanchis-Segura, C., Spanagel, R., and Gass, P. (2005b) Social and structural housing conditions influence the development of a depressive-like phenotype in the learned helplessness paradigm in male mice. *Behav Brain Res* 164: 100–106.

Churchill, G.A., Airey, D.C., Allayee, H., Angel, J.M., Attie, A.D., Beatty, J., et al. (2004) The Collaborative Cross, a community resource for the genetic analysis of complex traits. *Nat Genet* 36: 1133–1137.

Cohen, S. (2005) Keynote presentation at the Eighth International Congress of Behavioral Medicine: the Pittsburgh common cold studies: psychosocial predictors of susceptibility to respiratory infectious illness. *Int J Behav Med* 12: 123–131.

Conti, A.C., Kuo, Y.C., Valentino, R.J., and Blendy, J.A. (2004) Inducible cAMP early repressor regulates corticosterone suppression after tricyclic antidepressant treatment. *J Neurosci* 24: 1967–1975.

Crabbe, J.C., Wahlsten, D., and Dudek, B.C. (1999) Genetics of mouse behavior: interactions with laboratory environment. *Science* 284: 1670–1672.

Crowley, J.J., Blendy, J.A., and Lucki, I. (2005) Strain-dependent antidepressant-like effects of citalopram in the mouse tail suspension test. *Psychopharmacology (Berl)* 183: 257–264.

Crowley, J.J., Brodkin, E.S., Blendy, J.A., Berrettini, W.H., and Lucki, I. (2006) Pharmacogenomic evaluation of the antidepressant citalopram in the mouse tail suspension test. *Neuropsychopharmacology* 31: 2433–2442.

Crowley, J.J., Jones, M.D., O'Leary, O.F., and Lucki, I. (2004) Automated tests for measuring the effects of antidepressants in mice. *Pharmacol Biochem Behav* 78: 269–274.

Cryan, J.F. and Holmes, A. (2005) The ascent of mouse: advances in modelling human depression and anxiety. *Nat Rev Drug Discov* 4: 775–790.

Cryan, J.F., Mombereau, C., and Vassout, A. (2005a) The tail suspension test as a model for assessing antidepressant activity: review of pharmacological and genetic studies in mice. *Neurosci Biobehav Rev* 29: 571–625.

Cryan, J.F., Valentino, R.J., and Lucki, I. (2005b) Assessing substrates underlying the behavioral effects of antidepressants using the modified rat forced swimming test. *Neurosci Biobehav Rev* 29: 547–569.

Das, S. and O'Keefe, J.H. (2006) Behavioral cardiology: recognizing and addressing the profound impact of psychosocial stress on cardiovascular health. *Curr Atheroscler Rep* 8: 111–118.

Delgado, P.L. (2000) Depression: the case for a monoamine deficiency. *J Clin Psychiatry* 61: 7–11.

Dingell, J.V., Sulser, F., and Gillette, J.R. (1964) Species differences in the metabolism of imipramine and desmethylimipramine. *J Pharmacol and Exp Ther* 143: 14–22.

Drugan, R.C. (2003) Rodent models of depression: learned helplessness using a triadic design in rats. In Crawley, J., Gerfen, C., Rogawski, M., Sibley, D., Skolnick, P., and Wray, S. (eds.), *Current Protocols in Neuroscience*. John Wiley and Sons, Hoboken, NJ, USA, Unit 8.10B.

DSM-IV (1994) *Diagnostic and Statistical Manual of Mental Disorders, Fourth Edition (DSM-IV)*. American Psychiatric Association, Washington, DC.

Dulawa, S.C. and Hen, R. (2005) Recent advances in animal models of chronic antidepressant effects: the novelty-induced hypophagia test. *Neurosci Biobehav Rev* 29: 771–783.

Dulawa, S.C., Holick, K.A., Gundersen, B., and Hen, R. (2004) Effects of chronic fluoxetine in animal models of anxiety and depression. *Neuropsychopharmacology* 29: 1321–1330.

El Yacoubi, M., Bouali, S., Popa, D., Naudon, L., Leroux-Nicollet, I., Hamon, M., et al. (2003) Behavioral, neurochemical, and electrophysiological characterization of a genetic mouse model of depression. *Proc Natl Acad Sci USA* 100: 6227–6232.

Fava, G.A. (2003a) Can long-term treatment with antidepressant drugs worsen the course of depression? *J Clin Psychiatry* 64: 123–133.

Fava, M. (2003b) Diagnosis and definition of treatment-resistant depression. *Biol Psychiatry* 53: 649–659.

Feighner, J.P., Ehrensing, R.H., Kastin, A.J., Patel, A., Sverdlov, L., Hlavka, J., et al. (2001) Double-blind, placebo-controlled study of INN 00835 (netamiftide) in the treatment of outpatients with major depression. *Int Clin Psychopharmacol* 16: 345–352.

Fink, G.E. (2000) *Encyclopedia of Stress* Academic Press, San Diego, CA, USA.

Francis, D.D., Zaharia, M.D., Shanks, N., and Anisman, H. (1995) Stress-induced disturbances in Morris water-maze performance: interstrain variability. *Physiol Behav* 58: 57–65.

Frazer, A. and Morilak, D.A. (2005) What should animal models of depression model? *Neurosci Biobehav Rev* 29: 515–523.

Ginsburg, B.E. (1967) Genetic parameters in behavioral research. In Hirsch, J. (ed.), *Behavior-Genetics Analysis*. McGraw-Hill Co., New York, pp. 146–151.

Harro, J. (2004) Animal models for better antidepressants: can pathogenetic approaches make a difference? *Preclinica* 2: 402–407.

Hawkins, J., Hicks, R.A., Phillips, N., and Moore, J.D. (1978) Swimming rats and human depression. *Nature* 274: 512–513.

Hedou, G., Pryce, C., Di Iorio, L., Heidbreder, C.A., and Feldon, J. (2001) An automated analysis of rat behavior in the forced swim test. *Pharmacol Biochem Behav* 70: 65–76.

Henderson, N.D. (1997) Spurious associations in unreplicated selected lines. *Behav Genet* 27: 145–154.

Heninger, G.R., Delgado, P.L., and Charney, D.S. (1996) The revised monoamine theory of depression: a modulatory role for monoamines, based on new findings from monoamine depletion experiments in humans. *Pharmacopsychiatry* 29: 2–11.

Henn, F.A. and Vollmayr, B. (2005) Stress models of depression: forming genetically vulnerable strains. *Neurosci Biobehav Rev* 29: 799–804.

Hilakivi, L.A. and Lister, R.G. (1990) Correlations between behavior of mice in Porsolt's swim test and in tests of anxiety, locomotion, and exploration. *Behav Neural Biol* 53: 153–159.

Hlavka, J., Nicolau, G., Noble, J.F., and Abajian, H. (1997) INN 00835. *Drug Fut* 22: 1314–1318.

Huang, Y.H., Cheng, C.Y., Hong, C.J., and Tsai, S.J. (2004) Expression of c-Fos-like immunoreactivity in the brain of mice with learned helplessness. *Neurosci Lett* 363: 280–283.

Irwin, J., Ahluwalia, P., Zacharko, R.M., and Anisman, H. (1986) Central norepinephrine and plasma corticosterone following acute and chronic stressors: influence of social isolation and handling. *Pharmacol Biochem Behav* 24: 1151–1154.

Katz, M.M., Tekell, J.L., Bowden, C.L., Brannan, S., Houston, J.P., Berman, N., et al. (2004) Onset and early behavioral effects of pharmacologically different antidepressants and placebo in depression. *Neuropsychopharmacology* 29: 566–579.

Kendler, K.S. (1998) Anna-Monika-Prize paper. Major depression and the environment: a psychiatric genetic perspective. *Pharmacopsychiatry* 31: 5–9.

Kendler, K.S., Karkowski, L.M., and Prescott, C.A. (1999) Causal relationship between stressful life events and the onset of major depression. *Am J Psychiatry* 156: 837–841.

Kendler, K.S. and Karkowski-Shuman, L. (1997) Stressful life events and genetic liability to major depression: genetic control of exposure to the environment? *Psychol Med* 27: 539–547.

Kendler, K.S., Kessler, R.C., Walters, E.E., MacLean, C., Neale, M.C., Heath, A.C., et al. (1995) Stressful life events, genetic liability, and onset of an episode of major depression in women. *Am J Psychiatry* 152: 833–842.

Kendler, K.S., Neale, M.C., Kessler, R.C., Heath, A.C., and Eaves, L.J. (1992) Major depression and generalized anxiety disorder. Same genes, (partly) different environments? *Arch Gen Psychiatry* 49: 716–722.

Kendler, K.S., Neale, M.C., Kessler, R.C., Heath, A.C., and Eaves, L.J. (1993) Major depression and phobias: the genetic and environmental sources of comorbidity. *Psychol Med* 23: 361–371.

Kessler, R.C., Berglund, P., Demler, O., Jin, R., Koretz, D., Merikangas, K.R., et al. (2003) The epidemiology of major depressive disorder: results from the National Comorbidity Survey Replication (NCS-R). *JAMA* 289: 3095–3105.

Kessler, R.C., Chiu, W.T., Demler, O., Merikangas, K.R., and Walters, E.E. (2005a) Prevalence, severity, and comorbidity of 12-month DSM-IV disorders in the National Comorbidity Survey Replication. *Arch Gen Psychiatry* 62: 617–627.

Kessler, R.C., Demler, O., Frank, R.G., Olfson, M., Pincus, H.A., Walters, E.E., et al. (2005b) Prevalence and treatment of mental disorders, 1990 to 2003. *N Engl J Med* 352: 2515–2523.

Kiecolt-Glaser, J.K. and Glaser, R. (1995) Psychoneuroimmunology and health consequences: data and shared mechanisms. *Psychosom Med* 57: 269–274.

Kloner, R.A. (2006) Natural and unnatural triggers of myocardial infarction. *Prog Cardiovasc Dis* 48: 285–300.

Kohen, R., Kirov, S., Navaja, G.P., Happe, H.K., Hamblin, M.W., Snoddy, J.R., et al. (2005) Gene expression profiling in the hippocampus of learned helpless and nonhelpless rats. *Pharmacogenomics J* 5: 278–291.

Kumar, V., Singh, P.N., Jaiswal, A.K., and Bhattacharya, S.K. (1999) Antidepressant activity of Indian *Hypericum perforatum* Linn in rodents. *Indian J Exp Biol* 37: 1171–1176.

Kupfer, D.J. (1991) Long-term treatment of depression. *J Clin Psychiatry* 52 Suppl: 28–34.

Kurtuncu, M., Luka, L.J., Dimitrijevic, N., Uz, T., and Manev, H. (2005) Reliability assessment of an automated forced swim test device using two mouse strains. *J Neurosci Methods* 149: 26–30.

Kvetnansky, R., Pacak, K., Fukuhara, K., Viskupic, E., Hiremagalur, B., Nankova, B., et al. (1995) Sympathoadrenal system in stress. Interaction with the hypothalamic–pituitary–adrenocortical system. *Ann N Y Acad Sci* 771: 131–158.

Leonard, B.E. (2000) Evidence for a biochemical lesion in depression. *J Clin Psychiatry* 61: 12–17.

Leussis, M.P. and Heinrichs, S.C. (2006) Routine tail suspension husbandry facilitates onset of seizure susceptibility in EL mice. *Epilepsia* 47: 801–804.

Li, X., Tizzano, J.P., Griffey, K., Clay, M., Lindstrom, T., and Skolnick, P. (2001) Antidepressant-like actions of an AMPA receptor potentiator (LY392098). *Neuropharmacology* 40: 1028–1033.

Liu, X. and Gershenfeld, H.K. (2001) Genetic differences in the tail-suspension test and its relationship to imipramine response among 11 inbred strains of mice. *Biol Psychiatry* 49: 575–581.

Liu, X. and Gershenfeld, H.K. (2003) An exploratory factor analysis of the Tail Suspension Test in 12 inbred strains of mice and an F2 intercross. *Brain Res Bull* 60: 223–231.

Liu, X., Peprah, D., and Gershenfeld, H.K. (2003) Tail-suspension induced hyperthermia: a new measure of stress reactivity. *J Psychiatr Res* 37: 249–259.

Liu, X., Stancliffe, D., Lee, S., Mathur, S., and Gershenfeld, H. (2007) Genetic dissection of the tail suspension test: a mouse model of stress vulnerability and antidepressant response. *Biol Psychiatry* 62: 81–91.

Lucki, I., Dalvi, A., and Mayorga, A.J. (2001) Sensitivity to the effects of pharmacologically selective antidepressants in different strains of mice. *Psychopharmacology (Berl)* 155: 315–322.

Maier, S.F. (2001) Exposure to the stressor environment prevents the temporal dissipation of behavioral depression/learned helplessness. *Biol Psychiatry* 49: 763–773.

Maier, S.F. and Watkins, L.R. (2005) Stressor controllability and learned helplessness: the roles of the dorsal raphe nucleus, serotonin, and corticotropin-releasing factor. *Neurosci Biobehav Rev* 29: 829–841.

Malatynska, E. and Knapp, R.J. (2005) Dominant-submissive behavior as models of mania and depression. *Neurosci Biobehav Rev* 29: 715–737.

Malatynska, E., Rapp, R., Harrawood, D., and Tunnicliff, G. (2005) Submissive behavior in mice as a test for antidepressant drug activity. *Pharmacol Biochem Behav* 82: 306–313.

Manji, H.K., Quiroz, J.A., Sporn, J., Payne, J.L., Denicoff, K., Gray, N., et al. (2003) Enhancing neuronal plasticity and cellular resilience to develop novel, improved therapeutics for difficult-to-treat depression. *Biol Psychiatry* 53: 707–742.

McKinney, W.T. (1984) Animal models of depression: an overview. *Psychiatr Dev* 2: 77–96.

Mitchell, P.J. and Redfern, P.H. (2005) Animal models of depressive illness: the importance of chronic drug treatment. *Curr Pharm Des* 11: 171–203.

Mittleman, M.A., Maclure, M., Nachnani, M., Sherwood, J.B., and Muller, J.E. (1997) Educational attainment, anger, and the risk of triggering myocardial infarction onset. The determinants of myocardial infarction onset study investigators. *Arch Intern Med* 157: 769–775.

Murray, C.J. and Lopez, A.D. (1997) Alternative projections of mortality and disability by cause 1990–2020: global burden of disease study. *Lancet* 349: 1498–1504.

Nakatani, N., Aburatani, H., Nishimura, K., Semba, J., and Yoshikawa, T. (2004) Comprehensive expression analysis of a rat depression model. *Pharmacogenomics J* 4: 114–126.

Naudon, L., El Yacoubi, M., Vaugeois, J.M., Leroux-Nicollet, I., and Costentin, J. (2002) A chronic treatment with fluoxetine decreases 5-HT1A receptors labeling in mice selected as a genetic model of helplessness. *Brain Res* 936: 68–75.

Nikulina, E.M., Skrinskaya, J.A., and Popova, N.K. (1991) Role of genotype and dopamine receptors in behaviour of inbred mice in a forced swimming test. *Psychopharmacology* 105: 525–529.

Oka, T., Oka, K., and Hori, T. (2001) Mechanisms and mediators of psychological stress-induced rise in core temperature. *Psychosom Med* 63: 476–486.

Overmier, J.B. and Seligman, M.E. (1967) Effects of inescapable shock upon subsequent escape and avoidance responding. *J Comp Physiol Psychol* 63: 28–33.

Paykel, E.S. (1994) Life events, social support and depression. *Acta Psychiatr Scand Suppl* 377: 50–58.

Paykel, E.S. (2001) The evolution of life events research in psychiatry. *J Affect Disord* 62: 141–149.

Perrault, G., Morel, E., Zivkovic, B., and Sanger, D.J. (1992) Activity of litoxetine and other serotonin uptake inhibitors in the tail suspension test in mice. *Pharmacol Biochem Behav* 42: 45–47.

Porsolt, R.D. (2000) Animal models of depression: utility for transgenic research. *Rev Neurosci* 11: 53–58.

Porsolt, R.D., Anton, G., Blavet, N., and Jalfre, M. (1978a) Behavioural despair in rats: a new model sensitive to antidepressant treatments. *Eur J Pharmacol* 47: 379–391.

Porsolt, R.D., Bertin, A., and Jalfre, M. (1977) Behavioral despair in mice: a primary screening test for antidepressants. *Arch Int Pharmacodyn Ther* 229: 327–336.

Porsolt, R.D., Bertin, A., and Jalfre, M. (1978b) "Behavioural despair" in rats and mice: strain differences and the effects of imipramine. *Eur J Pharmacol* 51: 291–294.

Porsolt, R.D., Chermat, R., Lenegre, A., Avril, I., Janvier, S., and Steru, L. (1987) Use of the automated tail suspension test for the primary screening of psychotropic agents. *Arch Int Pharmacodyn Ther* 288: 11–30.

Post, R.M. (1992) Transduction of psychosocial stress into the neurobiology of recurrent affective disorder. *Am J Psychiatry* 149: 999–1010.

Redrobe, J.P. and Bourin, M. (1999) Augmentation of antidepressant pharmacotherapy: a preclinical approach using the mouse forced swimming test. *CNS Spectrum* 4: 73–81.

Ridder, S., Chourbaji, S., Hellweg, R., Urani, A., Zacher, C., Schmid, W., et al. (2005) Mice with genetically altered glucocorticoid receptor expression show altered sensitivity for stress-induced depressive reactions. *J Neurosci* 25: 6243–6250.

Ripoll, N., David, D.J., Dailly, E., Hascoet, M., and Bourin, M. (2003) Antidepressant-like effects in various mice strains in the tail suspension test. *Behav Brain Res* 143: 193–200.

Ronan, P.J., Steciuk, M., Kramer, G.L., Kram, M., and Petty, F. (2000) Increased septal 5-HIAA efflux in rats that do not develop learned helplessness after inescapable stress. *J Neurosci Res* 61: 101–106.

Rosengren, A., Hawken, S., Ounpuu, S., Sliwa, K., Zubaid, M., Almahmeed, W.A., et al. (2004) Association of psychosocial risk factors with risk of acute myocardial infarction in 11119 cases and 13648 controls from 52 countries (the INTERHEART study): case-control study. *Lancet* 364: 953–962.

Rozanski, A., Blumenthal, J.A., Davidson, K.W., Saab, P.G., and Kubzansky, L. (2005) The epidemiology, pathophysiology, and management of psychosocial risk factors in cardiac practice: the emerging field of behavioral cardiology. *J Am Coll Cardiol* 45: 637–651.

Sanchez, C. and Hyttel, J. (1999) Comparison of the effects of antidepressants and their metabolites on reuptake of biogenic amines and on receptor binding. *Cell Mol Neurobiol* 19: 467–489.

Santarelli, L., Saxe, M., Gross, C., Surget, A., Battaglia, F., Dulawa, S., et al. (2003) Requirement of hippocampal neurogenesis for the behavioral effects of antidepressants. *Science* 301: 805–809.

Sarter, M. and Bruno, J.P. (2002) Animal models in biological psychiatry. In D'haenen, H., den Boer, H., and Willner, P. (eds.), *Biological Psychiatry*, Vol. 1. John Wiley and Sons, Chichester, UK, pp. 37–44.

Schildkraut, J.J. (1965) The catecholamine hypothesis of affective disorders: a review of supporting evidence. *Am J Psychiatry* 122: 509–522.

Seligman, M.E. and Beagley, G. (1975) Learned helplessness in the rat. *J Comp Physiol Psychol* 88: 534–541.

Seligman, M.E., Maier, S.F., and Geer, J.H. (1968) Alleviation of learned helplessness in the dog. *J Abnorm Psychol* 73: 256–262.

Seligman, M.E., Weiss, J., Weinraub, M., and Schulman, A. (1980) Coping behavior: learned helplessness, physiological change and learned inactivity. *Behav Res Ther* 18: 459–512.

Shanks, N. and Anisman, H. (1988) Stressor-provoked behavioral changes in six strains of mice. *Behav Neurosci* 102: 894–905.

Shanks, N. and Anisman, H. (1989) Strain-specific effects of antidepressants on escape deficits induced by inescapable shock. *Psychopharmacology (Berl)* 99: 122–128.

Shanks, N., Griffiths, J., and Anisman, H. (1994) Norepinephrine and serotonin alterations following chronic stressor exposure: mouse strain differences. *Pharmacol Biochem Behav* 49: 57–65.

Shanks, N., Griffiths, J., Zalcman, S., Zacharko, R.M., and Anisman, H. (1990) Mouse strain differences in plasma corticosterone following uncontrollable footshock. *Pharmacol Biochem Behav* 36: 515–519.

Sherman, A.D., Sacquitne, J.L., and Petty, F. (1982) Specificity of the learned helplessness model of depression. *Pharmacol Biochem Behav* 16: 449–454.

Solberg, L.C., Ahmadiyeh, N., Baum, A.E., Vitaterna, M.H., Takahashi, J.S., Turek, F.W., et al. (2003) Depressive-like behavior and stress reactivity are independent traits in a Wistar Kyoto × Fisher 344 cross. *Mol Psychiatry* 8: 423–433.

Solberg, L.C., Baum, A.E., Ahmadiyeh, N., Shimomura, K., Li, R., Turek, F.W., et al. (2004) Sex- and lineage-specific inheritance of depression-like behavior in the rat. *Mamm Genome* 15: 648–662.

Solomon, D.A., Leon, A.C., Mueller, T.I., Coryell, W., Teres, J.J., Posternak, M.A., et al. (2005) Tachyphylaxis in unipolar major depressive disorder. *J Clin Psychiatry* 66: 283–290.

Song, C. and Leonard, B.E. (2005) The olfactory bulbectomised rat as a model of depression. *Neurosci Biobehav Rev* 29: 627–647.

Steru, L., Chermat, R., Thierry, B., Mico, J.A., Lenegre, A., Steru, M., et al. (1987) The automated tail suspension test: a computerized device which differentiates psychotropic drugs. *Prog Neuropsychopharmacol Biol Psychiatry* 11: 659–671.

Steru, L., Chermat, R., Thierry, B., and Simon, P. (1985) The tail suspension test: a new method for screening antidepressants in mice. *Psychopharmacology (Berl)* 85: 367–370.

Strekalova, T., Spanagel, R., Bartsch, D., Henn, F.A., and Gass, P. (2004) Stress-induced anhedonia in mice is associated with deficits in forced swimming and exploration. *Neuropsychopharmacology* 29: 2007–2017.

Taylor, C., Fricker, A.D., Devi, L.A., and Gomes, I. (2005) Mechanisms of action of antidepressants: from neurotransmitter systems to signaling pathways. *Cell Signal* 17: 549–557.

Teste, J.F., Martin, I., and Rinjard, P. (1990) Electrotherapy in mice: dopaminergic and noradrenergic effects in the tail suspension test. *Fundam Clin Pharmacol* 4: 39–47.

Teste, J.F., Pelsy-Johann, I., Decelle, T., and Boulu, R.G. (1993) Anti-immobility activity of different antidepressant drugs using the tail suspension test in normal or reserpinized mice. *Fundam Clin Pharmacol* 7: 219–226.

Thierry, B., Steru, L., Chermat, R., and Simon, P. (1984) Searching-waiting strategy: a candidate for an evolutionary model of depression? *Behav Neural Biol* 41: 180–189.

Thierry, B., Steru, L., Simon, P., and Porsolt, R.D. (1986) The tail suspension test: ethical considerations. *Psychopharmacology (Berl)* 90: 284–285.

Trullas, R., Jackson, B., and Skolnick, P. (1989) Genetic differences in a tail suspension test for evaluating antidepressant activity. *Psychopharmacology (Berl)* 99: 287–288.

Turri, M.G., Datta, S.R., DeFries, J., Henderson, N.D., and Flint, J. (2001) QTL analysis identifies multiple behavioral dimensions in ethological tests of anxiety in laboratory mice. *Curr Biol* 11: 725–734.

Ukai, M., Maeda, H., Nanya, Y., Kameyama, T., and Matsuno, K. (1998) Beneficial effects of acute and repeated administrations of sigma receptor agonists on behavioral despair in mice exposed to tail suspension. *Pharmacol Biochem Behav* 61: 247–252.

Urani, A., Chourbaji, S., and Gass, P. (2005) Mutant mouse models of depression: candidate genes and current mouse lines. *Neurosci Biobehav Rev* 29: 805–828.

Valdar, W., Solberg, L.C., Gauguier, D., Burnett, S., Klenerman, P., Cookson, W.O., et al. (2006) Genome-wide genetic association of complex traits in heterogeneous stock mice. *Nat Genet* 38: 879–887.

van der Heyden, J.A., Molewijk, E., and Olivier, B. (1987) Strain differences in response to drugs in the tail suspension test for antidepressant activity. *Psychopharmacology (Berl)* 92: 127–130.

Vaugeois, J.M., Odievre, C., Loisel, L., and Costentin, J. (1996) A genetic mouse model of helplessness sensitive to imipramine. *Eur J Pharmacol* 316: R1–2.

Vaugeois, J.M., Passera, G., Zuccaro, F., and Costentin, J. (1997) Individual differences in response to imipramine in the mouse tail suspension test. *Psychopharmacology (Berl)* 134: 387–391.

Vollmayr, B. and Henn, F.A. (2001) Learned helplessness in the rat: improvements in validity and reliability. *Brain Res Brain Res Protoc* 8: 1–7.

Wahlsten, D., Metten, P., Phillips, T.J., Boehm, S.L., 2nd, Burkhart-Kasch, S., Dorow, J., et al. (2003) Different data from different labs: lessons from studies of gene–environment interaction. *J Neurobiol* 54: 283–311.

Weiss, J.M., Boss-Williams, K.A., Moore, J.P., Demetrikopoulos, M.K., Ritchie, J.C., and West, C.H. (2005) Testing the hypothesis that locus coeruleus hyperactivity produces depression-related changes via galanin. *Neuropeptides* 39: 281–287.

Weiss, J.M. and Kilts, C.D. (1998) Animal models of depression and schizophrenia. In Schatzberg, A.F. and Nemeroff, C.B. (eds.), *Textbook of Psychopharmacology*. American Psychiatric Press, Washington, D.C., pp. 89–131.

Weiss, J.M. and Simson, P.E. (1988) Neurochemical and electrophysiological events underlying stress-induced depression in an animal model. *Adv Exp Med Biol* 245: 425–440.

Wieland, S., Boren, J.L., Consroe, P.F., and Martin, A. (1986) Stock differences in the susceptibility of rats to learned helplessness training. *Life Sci* 39: 937–944.

Willner, P. (1983) Dopamine and depression: a review of recent evidence. II. Theoretical approaches. *Brain Res* 287: 225–236.

Willner, P. (1984) The validity of animal models of depression. *Psychopharmacology (Berl)* 83: 1–16.

Willner, P. (1991a) Animal models as simulations of depression. *Trends Pharmacol Sci* 12: 131–136.

Willner, P. (1991b) *Behavioral Models in Psychopharmacology: Theoretical, Industrial and Clinical Perspectives.* Cambridge University Press, Cambridge.

Wilson, S.J., Bailey, J.E., Alford, C., Weinstein, A., and Nutt, D.J. (2002) Effects of 5 weeks of administration of fluoxetine and dothiepin in normal volunteers on sleep, daytime sedation, psychomotor performance and mood. *J Psychopharmacol* 16: 321–331.

Yamada, M., Yamada, M., and Higuchi, T. (2005) Antidepressant-elicited changes in gene expression: remodeling of neuronal circuits as a new hypothesis for drug efficacy. *Prog Neuropsychopharmacol Biol Psychiatry* 29: 999–1009.

Yoshikawa, T., Watanabe, A., Ishitsuka, Y., Nakaya, A., and Nakatani, N. (2002) Identification of multiple genetic loci linked to the propensity for "behavioral despair" in mice. *Genome Res* 12: 357–366.

Chapter 20

Behavioral phenotyping of mouse grooming and barbering

Peter R. Canavello, Jonathan M. Cachat, Peter C. Hart, Dennis L. Murphy, and Allan V. Kalueff

Introduction

Grooming is an ancient behavior that is represented (with a striking homology) across most vertebrate species, including humans (Fentress, 1977a, 1977b; Sachs, 1988; Young and Thiessen, 1991). Its biological functions include care of body surface, removal of ectoparasites, thermoregulation, chemo-communication, sensory stimulation of skin, displacement, de-arousal, and maternal and social interaction (Ferkin and Leonard, 2005; Ferkin et al., 1996; Spruijt et al., 1992).

Both self- and hetero-grooming behaviors are commonly seen in wild and laboratory rodents (Berntson et al., 1988; Bolles, 1960; Dunn et al., 1987; Sachs, 1988; van Abeelen, 1966). Self-grooming is common in mice, representing greater than 15% of all waking time, and, after sleep, is the most frequently observed activity in their repertoire. It can be induced in animals by various factors, such as novelty, misting with water, swimming, exposure to predators or their odors, and pain (Boccalon et al., 2006; Burne et al., 2006; File et al., 1988; Kalueff and Tuohimaa, 2004a). It is also often seen during pregnancy (Harvey and Mann, 1987), as a "transitional element" between rest and exploration (Fentress, 1988), or as a part of sexual behavior (Sachs, 1988).

Many endogenous substances, including neuromediators such as dopamine, γ-aminobutyric acid (GABA), serotonin, and hormones (adrenocorticotropin, vasopressin, oxytocin, prolactin, substance P, somatostatin, bombesin, corticotropin-releasing hormone, opioids, cholecystokinin; Bertolini et al., 1988; Dunn et al., 1987, 1988; Gorris and van Abeelen, 1981; Spruijt et al., 1992; Ukai et al., 1989), as well as multiple brain regions (especially basal ganglia and hypothalamus) are involved in the regulation of rodent self-grooming (Aldridge et al., 2004; Berntson et al., 1988; Kruk et al., 1998; Roeling et al., 1993). Various psychotropic drugs (Audet et al., 2006; Barros et al., 1994; Dunn, 1988; File et al., 1988; Kalueff and Tuohimaa, 2005c; Yalcin et al., 2005) also affect rodent grooming, collectively making its analysis a useful tool in behavioral neuroscience (Kalueff et al., 2007).

In addition to grooming, behavior-associated hair loss (over-grooming, or barbering (in humans – trichotillomania, which afflicts approximately 1% of the population)) has long been observed in many species (Hauchka, 1952; Honess et al., 2005; Reinhardt, 2005; Sarna et al., 2000), suggesting that both behaviors may have a common evolutionary origin. Described in the rodent literature as whisker/hair or fur trimming, de-whiskering, or the Delilah effect, barbering is particularly common in laboratory mice (Garner et al., 2004a, 2004b; Van de Weerd et al., 1992). Mouse barbering includes plucking of fur or whiskers from cage-mates (hetero-barbering) or oneself (self-barbering) (Kurien et al., 2005; Long, 1972; Strozik and Festing, 1981).

Mounting data indicate that mouse barbering may be a form of dominant behavior and a strong indicator of social hierarchy (Long, 1972; Sarna et al., 2000). In groups of mice, there will often be one with unbarbered whiskers, playing a dominant role in the cage (Carruthers et al., 1997; Kalueff et al., 2006; Sarna et al., 2000; Strozik and Festing, 1981). Whisker removal occurs even if the mice are separated by wire mesh, indicating that both animals cooperate in this behavior (Van den Broek et al., 1993). Several husbandry factors have been reported to affect mouse barbering (e.g., diet, weaning age, and enrichment; De Luca, 1997; Myers, 1997), suggesting that it may represent a stress-evoked response (e.g., coping with inappropriate housing; Van den Broek et al., 1993), or a pathological, compulsive-like behavior (Garner et al., 2004a, 2004b). Given its frequent occurrence in mice, barbering appears to be an important part of their behavioral repertoire. This chapter will focus on behavioral phenotyping of mouse hetero-barbering and self-grooming, the most representative types of each of these behaviors.

Phenotyping of self-grooming

Mouse grooming appears in early ontogenesis, confirming the evolutionarily ancient age of this behavior (Golani and Fentress, 1985; Sachs, 1988; Sviderskaia and Dmitrieva, 1993). Usually more frequent in the home cage (Berridge et al., 2005), self-grooming is also seen in stressful situations, such as novelty-based anxiety tests (e.g., elevated plus-maze, open field, and holeboard tests), implying that this behavior is not only important to assess, but can also interfere with the primary

Behavioral Genetics of the Mouse: Volume I. *Genetics of Behavioral Phenotypes*, eds. Wim E. Crusio, Frans Sluyter, Robert T. Gerlai, and Susanna Pietropaolo. Published by Cambridge University Press.

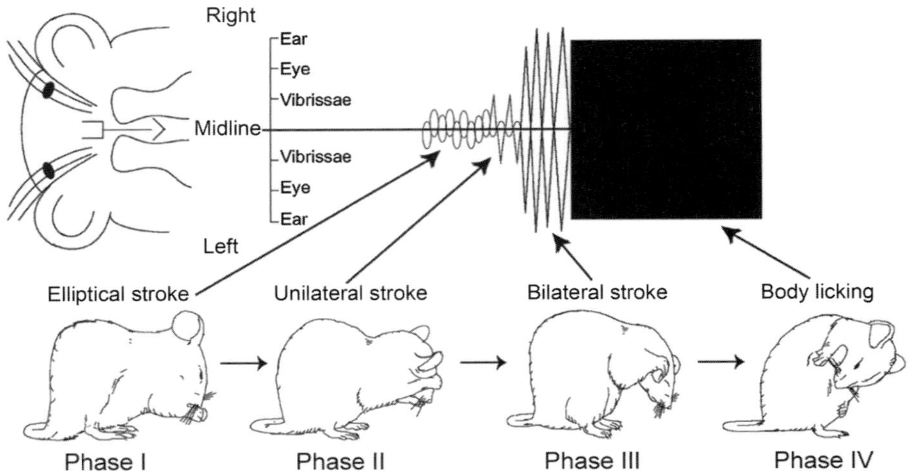

Figure 20.1 Prototypical syntactic grooming chain pattern in mice. *Phase I:* series of ellipse-shaped strokes tightly around the nose (paw, nose grooming). *Phase II:* series of unilateral strokes (each made by one paw) that reach up the mystacial vibrissae to below the eye (face grooming). *Phase III:* series of bilateral strokes made by both paws simultaneously. Paws reach back and upwards, ascending usually high enough to pass over the ears (head grooming). *Phase IV:* body licking, preceded by postural cephalocaudal transition from paw/head grooming to body grooming. (Adapted with permission from Berridge et al., 2005.)

behavior being assessed, thus becoming a confounding factor, if neglected.

Nevertheless, in many studies this behavior has received little attention. Surprisingly, there is little or no grooming data in the extensive Mouse Phenome Database (MPD), and in recent comprehensive studies (Rogers et al., 1999; Tang et al., 2002; Voikar et al., 2001, 2004). Some authors assessed several grooming patterns but reported only one global behavioral score (e.g., Dunn et al., 1987). Others compared different strains but analyzed only cumulative scores (Cabib and Bonaventura, 1997; Leppanen et al., 2006; Nyberg et al., 2003), or combined grooming data in "overall activity scores" with horizontal and vertical exploration (e.g., Wirth-Dzieciolowska et al., 2005). Still other studies assessed grooming indices in different strains cursorily among other measures, mostly including gross activity characteristics such as the latency, frequency, and duration of grooming (e.g., Dell'Omo et al., 1994; Rodgers et al., 2002; Voikar et al., 2005).

There is a growing understanding, however, that quantitative measures alone may be insufficient for correct interpretation and analysis of mouse grooming data (Fentress, 1988; Kalueff and Tuohimaa, 2004a). Rodent grooming is a complex, ethologically rich ritual, which normally proceeds in a cephalocaudal direction and consists of several stages: licking the paws, washing movements over the head, body fur licking (progressively shifting to a more posterior body focus), and tail/genital cleaning (Berridge et al., 2005; Spruijt et al., 1992) (Figure 20.1). Dissectible at a number of levels, ranging from individual organ kinematics (Kobayashi et al., 2003) to sequencing and habituation in grooming patterns and their association with non-grooming behaviors (Berridge et al., 2005; Cabib et al., 1990; Fentress, 1988), such complex patterned structure is particularly attractive for behavioral phenotyping. Importantly, this approach can be used not only to assess "normal" mouse behaviors and their patterning, but also as a source for ethologically oriented experimental models of human stereotypic behavioral disorders. For example, the utility of grooming assessment has recently been suggested for mouse models of obsessive-compulsive disorder (OCD) and Tourette syndrome (Albin and Mink, 2006; Berridge et al., 2005). Consistent with this, psychotropic drugs such as m-CPP, a serotonin receptor 2 agonist that exacerbates OCD in clinical patients (Pigott et al., 1993), have been reported to elicit robust rodent grooming (Graf et al., 2003). These findings suggest that grooming phenotyping may also hold potential for extensive use in behavioral pharmacology (also see Kalueff and Tuohimaa, 2005c for discussion). Table 20.1 summarizes methodological approaches to mouse grooming phenotyping.

Interesting data continue to accumulate on the behavioral genetics of mouse grooming. Van Abeelen (1963a, 1963b, 1963c) found that increased grooming is associated with the *pink-eyed dilution* (*p*) and *brown coat color* (*b*) loci on chromosomes 7 and 4, respectively.[1] Allelic variations in the *p* locus have been found in many strains of laboratory mice (Brilliant et al., 1994), and may be associated with some strain differences in self-grooming behavior. On the other hand, some data suggest that both *p* and *b* loci are associated with increased anxiety (Clement and Chapouthier, 1998; Clement et al., 1994), implying interaction between "grooming" and "anxiety" domains at both genetic and behavioral levels. In line with this, ABP/Le and C57BL/6 mice demonstrated association of increased novelty-induced grooming with the autosomal *short ear* (*se*) locus on chromosome 9 (Clement et al., 1994), also implicated in mouse avoidance behavior (see van Abeelen, 1966 for details). This study also confirmed the role of the *p* locus, suggesting multi-locus contributions to grooming phenotype in mice. Moreover, although

[1] The *p* locus (28.0 cM) on chromosome 7 is close to a cluster of GABAergic genes (28.2–28.6 cM) encoding $\gamma 3$, $\alpha 5$, and $\beta 3$ subunits of GABA-A receptors (Culiat et al., 1994; also MGI). The GABAergic system is known to regulate both grooming and emotional behaviours, and it is possible that these subunits play a role in grooming phenotypes (also see tendency to a lower grooming in mice selectively bred for sensitivity to β-carboline seizures in: Suaudeau et al., 2000).

Table 20.1 Methodological approaches to mouse grooming.

General cumulative measures

- The latency to onset, the duration, and the number of grooming episodes (bouts)
- The following patterns can be recorded for each bout: paw licking; nose/face grooming; head washing; body and leg grooming/scratching; tail/genitals grooming
- Additional cumulative indices: the average duration of a single grooming bout, total number of transitions between grooming stages, and average number of transitions per bout

Patterning

- The percentages of incorrect transitions (transitions against a cephalocaudal progression of grooming activity), as well as interrupted and incomplete grooming bouts (see details in Kalueff and Tuohimaa, 2004b, 2005a, 2005b, 2005c; Kalueff et al., 2005)

Regional distribution of grooming

- Can be assessed as directed to the following five anatomic areas: forepaws, head, body, hind-legs, and tail/genitals. Rostral grooming includes forepaw (preliminary rostral grooming) and head grooming. Body, legs, and tail/genital grooming can be considered as caudal grooming
- Each bout can be categorized as being directed to (1) multiple regions or (2) a single region, and the percentages of grooming bouts and of time spent grooming can be calculated for both categories (see details in Dunn et al., 1987; Kalueff and Tuohimaa, 2004b, 2005a, 2005b, 2005c)

Additional useful indices of grooming

- Probability of chain initiation (frequency of chain initiation per minute of grooming time)
- Probability of pattern completion once initiated (see Berridge and Aldridge, 2000, for the rat protocol, which can be easily adapted to use with mice)

Table 20.2 Examples of strain differences in mouse grooming behavior.

Strain differences	References
Frequency	
DBA/2J, F1 [C57BL/6J-DBA/2J] > C57BL/6J	Van Abeelen, 1966
A/Ibg, BALB/cIbg > DBA/2Ibg, C57BL/6Ibg	Streng, 1971
DBA/2J = CPB-K-Nmg = C3H/St = C57BL/6J	Crusio and van Abeelen, 1986
FVB/N = C57BL/6J	Mineur and Crusio, 2002
BALB/c = 129S1 = NMRI	Kalueff and Tuohimaa, 2005b
C57BL/6J > 129S1	Kalueff and Tuohimaa, 2004b
C57BL/6 = DBA/2	Cabib and Bonaventura, 1997
Duration	
DBA/2J > CPB-K-Nmg > C3H/St, C57BL/6J	Crusio and van Abeelen, 1986
C57BL/6J > FVB/N	Mineur and Crusio, 2002
C57BL/6J > BALB/cByJ	Ducottet and Belzung, 2004
C57BL/6J > 129S1	Kalueff and Tuohimaa, 2004b
BALB/c > 129S1, NMRI	Kalueff and Tuohimaa, 2005b

the *se* locus has also been linked to anxiety (Clement et al., 2002), it is crucial for normal growth and patterning of skeletal structures (Kingsley et al., 1992), implying a possible musculoskeletal component. Clearly, further studies are needed to map grooming to specific genes and chromosomes in mice and other rodents (e.g., Fernandez-Teruel et al., 2002; Flint, 2004).

Many neurobehavioral studies have considered strain differences in mouse grooming (Table 20.2). For example, Streng (1971) tested BALB/cIbg, C57BL/6Ibg, A/Ibg and DBA/2Ibg mice in the open field, analyzing their cumulative grooming scores and reporting significant time effect (grooming increase over a test period, observed in all strains) and time × strain effects. Schoots et al. (1978) assessed grooming frequency and duration in C57BL/6 and DBA/2 mice, reporting more grooming in DBA/2 strain regardless of treatment (i.e., after intranasal saline administration or zinc-induced peripheral anosmia). Clement et al. (1994) noted that ABP/Le mice (carrying homozygous recessive genes *se*, *p*, and *b*) groomed in the open field significantly more than less anxious C57BL/6, while F1 ABP/Le-C57BL/6 mice groomed more than F1 C57BL/6-ABP/Le (suggesting possible maternal effects). Another group analyzed genetic architectures of mouse novelty grooming, reporting ambidirectional dominance (for both frequency and duration), albeit rather weak, since it disappeared in the 5 × 5 diallel cross (Crusio and van Abeelen, 1986, 1987; Crusio et al., 1989).

There were also different baseline levels of sweet water spray-evoked grooming in C57BL/6J and BALB/cByJ mice markedly different in their emotionality (Table 20.2), whereas reduced grooming duration was seen in stressed (vs. non-stressed) BALB/cByJ mice, compared to "resistant" C57BL/6J grooming (Ducottet and Belzung, 2004). Voikar et al. (2005) examined C57BL/6JOlaHsd and DBA/2OlaHsd mice, reporting a significant strain effect on grooming in the elevated plus-maze for both socially and single-housed mice (B6: ↓ duration, ↑ latency), and in single-housed mice in the Y-maze (B6: ↑ latency). In contrast, they found no strain effect on grooming duration in the Y-maze in all four groups, and reported a significant housing effect (↑ latency) in group-housed mice of both strains.

Rodgers et al. (2002) compared C57BL/6 and some 129 substrains, reporting lower light–dark test grooming in 129SvEm (tendency) and 129SvHsD (significant decrease); similar results were seen for 129S1 mice in the social interaction (Hossain et al., 2004) and novelty- or water-evoked grooming tests (Kalueff and Tuohimaa, 2004a). Grooming scores rose in C57BL/6J mice during the dark (vs. the light) phase, but remained very low in 129S1 and F1 (C57BL/6J-129S1) mice (Hossain et al., 2004), negating possible maternal B6 influences. Mineur and Crusio (2002) compared open field behaviors of FVB/N and C57BL/6J strains, noting similar frequency but marked strain differences in the duration of grooming (Table 20.2). This again confirms that grooming indices, such as frequency and duration, may vary independently in different mouse strains, and therefore merit careful assessment in detail. Interestingly, grooming activity in this study did not correlate with open field horizontal and vertical activity (FVB/N > B6), suggesting that grooming *activity* represents a separate dimension in the structure of mouse behavior.

In a recent study (Kalueff and Tuohimaa, 2005b), the latency to start grooming was not statistically different in three mouse strains, although NMRI mice showed a clear tendency toward earlier onset of grooming, compared to both 129S1 and BALB/c strains (Table 20.2). This group also found no clear correlation between grooming activity measures and anxiety – with anxious strains showing both high (BALB/c) and low (129S1) grooming profiles, and non-anxious mice showing moderate-to-high grooming profiles (NMRI, C57BL/6) (Kalueff and Tuohimaa, 2004b, 2005b). Analyzing grooming in 129S1 mice and their F1 hybrids (129S1-C57BL/6, 129S1-BALB/c, 129S1-NMRI) in the open field test, they also (unpublished data; in line with Hossain et al., 2004) found more bouts in NMRI and BALB/c-derived, than in 129S1 and C57-derived F1 mice (both displaying similar low-grooming phenotype). In contrast, BALB/c and NMRI mice (regardless of 129 maternal influences) seem to increase grooming activity in F1 progeny, resembling high-grooming phenotypes of their parents (Kalueff et al., 2005).

More recently, Yalcin et al. (2008) performed a study combining the unpredictable chronic mild stress (UCMS) paradigm with administration of four different antidepressant drugs in both outbred Swiss and BALB/c mice. While BALB/c cohorts showed significantly reduced grooming under all treatments (including vehicle), Swiss mice under several treatments (vehicle, desipramine, maprotiline) groomed more frequently than non-stressed controls, though results in the latter strain were not significant. Another study (Avgustinovich et al., 2007), comparing open field behaviors in C57BL/6J and CBA/Lac mice, reported an increase in the number of grooming bouts in stressed subjects of B6 strain, but no such change in CBA/Lac groups.

Notably, while the literature on strain differences in grooming is vast, relatively few studies focused on its patterning aspect. Analysis of grooming behavioral microstructure in 129S1, C57BL/6, BALB/c, and NMRI strains (Kalueff and Tuohimaa, 2004b, 2005b) showed that 129S1 mice display significantly higher percentages of incorrect transitions and interrupted grooming bouts. However, these measures did not differ between 129S1 and F1 hybrid strain (unpublished data), suggesting that 129 genes, maternal influences (or both) may determine grooming patterning. Perhaps, the "quantity" and the "quality" (patterning) of mouse grooming may have different genetic determinants, underlining the need for further studies in behavioral genetics of mouse grooming and its sequencing.

Grooming responses vary across selectively bred mouse strains. For example, female mice of Turku Aggressive strain spent less time grooming during predatory aggression, compared to Turku Non-aggressive strain (Sandnabba, 1995). In contrast, anxious high-thigmotaxis strain (derived from Swiss albino outbred stock) produced fewer grooming bouts in the open field than did less anxious mice of the low-thigmotaxis strain (Leppanen and Ewalds-Kvist, 2005; Leppanen et al., 2006, 2008). There are also interesting cross-fostering data revealing similar grooming activity in both high- and low-thigmotaxis strains (Leppanen and Ewalds-Kvist, 2005; Leppanen et al., 2006), and implying that both genetic and epigenetic factors influence mouse grooming. Since grooming is regulated by numerous endogenous mediators and hormones, their strain-specific variation may lead to altered grooming phenotypes. In line with this notion, Krehbiel et al. (1986) studied the DeFries mice, and found strain differences in self-grooming (L1 > C1, H2) paralleled by altered catecholamine neurotransmission.

Robust alterations in grooming have also been reported in several mouse strains with spontaneously occurring mutations. For example, *Lurcher* mice display reduced grooming of forelimb, abdomen, back, and hind-limb, also showing less complex bouts (Strazielle and Lalonde, 1998). In contrast, *weaver* mice show a greater number, but shorter duration of grooming bouts, perform a higher proportion of smaller forelimb strokes and more incomplete grooming sequences (Bolivar et al., 1996; Coscia and Fentress, 1993). *Jimpy* mice (Bolivar and Brown, 1994) spend less time grooming, whereas *staggerer* mice show less grooming but unaltered patterning (Fentress, 1999).[2] Finally, aberrant grooming has been reported in many transgenic and knockout mice (e.g., Greer and Capecchi, 2002; Kalueff et al., 2005), which are not the scope of this volume (see, however, the Mouse Genome Informatics (MGI) database for details (http://www.informatics.jax.org)).

Mouse barbering phenotyping

The fact that mouse barbering profiles do not correlate with altered self-grooming (Sarna et al., 2000) suggests that they represent distinct behavioral domains (see, however, some overlap between barbering and hetero-grooming in the same study). In general, there are several reasons why assessment of mouse barbering may be important for neurobehavioral research. First, like grooming, it is an interesting behavior per se, representing an essential part of mouse activity (Sarna et al., 2000; Whishaw et al., 2001). Second, since barbering usually affects whiskers (regarded as a crucial source of sensory input in rodents), and whiskering represents an essential part of rodent behavioral repertoire, altered whisker status in mice due to barbering may affect other behaviors (including behavioral performance in various tasks, as well as eating, sexual behaviors, predator evasion, and various social interactions).

Finally, barbering is particularly common in some strains, especially C57BL/6, A2G, and NMRI mice (Kalueff et al., 2006;

[2] Interestingly, *staggerer* locus on mouse chromosome 9 (36.0 cM) is very close to *se* locus (42.0 cM), linked to mouse grooming in earlier studies (Clement et al., 1994), suggesting that this area may indeed be important for mouse grooming.

Table 20.3 Assessment of mouse barbering behavior.

- The following five-point scale can be used to assess mouse barbering: 0 – no barbering; 1 – whisker removal or shortening; 2 – snout/face denuding; 3 – individual bald patches on head and body; 4 – multiple alopecic areas on head and/or body; 5 – severe alopecia including complete snout denuding and large alopecic areas on head and body
- Hair loss shall be scored as barbering only if the hair lesion was non-puritic, there was no scarring or scabbing around the lesion, and the animal was otherwise in good health and the fur (where present) was in good conditions (Garner et al., 2004a, 2004b)
- The following parameters of barbering can be assessed: the number (%) of cages in which the barbering occurred; the average severity of barbering in each cage; and the percentages of barbers and barbered animals (of total animals of each strain)
- Barber animals can be easily identified as the single intact mouse in the cage (for details, see Garner et al., 2004a, 2004b; Sarna et al., 2000)
- If necessary, self-barbering may be assessed in mice housed individually (to prevent hetero-barbering) for 3–4 weeks (for details, see Kalueff et al., 2006)
- Sometimes, excessive grooming in mice (e.g., Greer and Capecchi, 2002) may lead to pronounced barbering-like alopecia (further home-cage observations may be needed in such cases to distinguish between the two behaviors)

Long, 1972; Sarna et al., 2000; Strozik and Festing, 1981; The Jackson Laboratory, 1987), suggesting a strong genetic component in this behavior (Hauchka, 1952; McElwee et al., 1999; Militzer and Wecker, 1986; Van den Broek et al., 1993). As it stands, further in-depth ethological analyses will be necessary to understand in detail the etiological and genetic nature of mouse barbering (Table 20.3). Furthermore, phenotyping of mouse barbering can also lead to ethologically oriented experimental models of human behavioral disorders. Indeed, the utility of barbering assessment has recently been suggested for mouse models of trichotillomania (Garner et al., 2004a, 2004b, Kurien et al., 2005), although it may bear relevance to other OCD and aggression-related behaviors as well.

Overall, there is limited available data on the behavioral genetics of murine barbering. In regard to genetic mapping of mouse barbering, this scarcity of data currently hinders comprehensive genotypic analysis of barbering phenotypes across strains. There have been surprisingly few data comparing barbering behaviors in different strains and their intercrosses. De Luca (1997) reported interesting effects of enrichment on mouse barbering but did not specify the strain (also see Kurien et al., 2005), while Sarna et al. (2000) examined barbering behavior in detail, but only in one (C57BL/6J) strain. In a series of recent experiments (Kalueff et al., 2006; also unpublished 2004–2005 data) we assessed barbering in several strains and their F1 progeny, revealing distinct domains, such as social dominance barbering in same-sex cages in C57BL/6J, A/J, 129S1, and NMRI (but not BALB/c) mice, sexual barbering (male by females) in breeding C57BL/6J, 129S1, and NMRI (but not BALB/c) mice, and maternal barbering (ventral fur removal by pups in lactating C57BL/6 and 129S1 dams), as well as whisker barbering by 129S1 mothers to their pups.

Figure 20.2 shows the variety of mouse barbering styles in different strains. Notably, the percentage of hair-barbering varies widely from strain to strain. For example, the BALB/c mice usually do not show barbering, in contrast to ~20% of C57BL/6 and A/J (Landau et al., 2001; Long, 1972), ~75% of A2G (Strozik and Festing, 1981), and ~80–100% of NMRI mice (Kalueff et al., 2006). Sarna et al. (2000) have recently reported that C57BL/6J mice may demonstrate individual "cutting styles." Our studies showed that mice also displayed consistent strain-specific cutting styles (Figure 20.2: snout + whiskers: NMRI; head and/or back: C57BL/6; whiskers: A/J and 129S1; face: F1 C57–129SvJ), further supporting the behaviorally complex and multifactorial nature of mouse barbering (also see similar data for C57BL/6J mice and whisker barbering in another 129 strain, 129S6; Holmes et al., 2002). Notably, several F1 hybrid strains – B6-129S1 or 129S1-C57, NMRI-129S1, or 129S1-NMRI – display barbering styles which appear to be a mixture of those of both parental strains, and do not seem to be affected by maternal influences (Kalueff et al., 2006; also unpublished data). In contrast, the lack of barbering in F1 hybrids derived from BALB/c strain suggests strong low-barbering effects in this strain, also independent of maternal influences. Finally, cross-fostering experiments with several strains showed that both genetic and epigenetic factors may contribute to mouse barbering, since it was observed in mice of non-barbering strains raised with barbering foster parents, as well as in barbering strains raised by non-barbering foster parents (Carruthers et al., 1998).

In general, barbering activity (NMRI, B6 > 129 >> BALB) negatively correlated with overall strain aggressiveness (BALB > 129, B6 > NMRI), supporting the notion that barbering might emerge in mice as a behavioral adaptation to minimize potential aggression (Kalueff et al., 2006). Alternatively, it might appear as a result of an absence of complex behavior to parry hair plucking (Whishaw et al., 2001), although this seems less plausible. In addition, further studies are needed to assess the potential relation of barbering to overall activity levels and to carry out in-depth ethological analyses in different mouse strains.

On a relevant note, an assessment of behavioral phenotypes of non-barbering (in addition to barbering) strains may be another useful approach to behavioral genetics of mouse barbering. For example, over 800 BALB/c mice used in Carruthers et al. (1997, 1998) studies did not barber (see similar results in Kalueff et al., 2006). Strain differences in sociability (Brodkin et al., 2004; Sankoorikal et al., 2006) have been recently suggested to underline barbering phenotypes (Brodkin, 2007). This interesting hypothesis, if confirmed, may explain the high intensity of barbering in "sociable" strains (such as C57Bl/6 mice) and the poor barbering in "autistic" mouse strains, such as BALB/c. Thus, barbering emerges as an essential part of mouse social behavior, and its strain differences may reflect (or underlie) different aspects and strategies of animal socialization. The latter, in turn, may confound all other behavioral domains, implying that in-depth analyses of strain barbering phenotypes may be even more important than has been previously recognized.

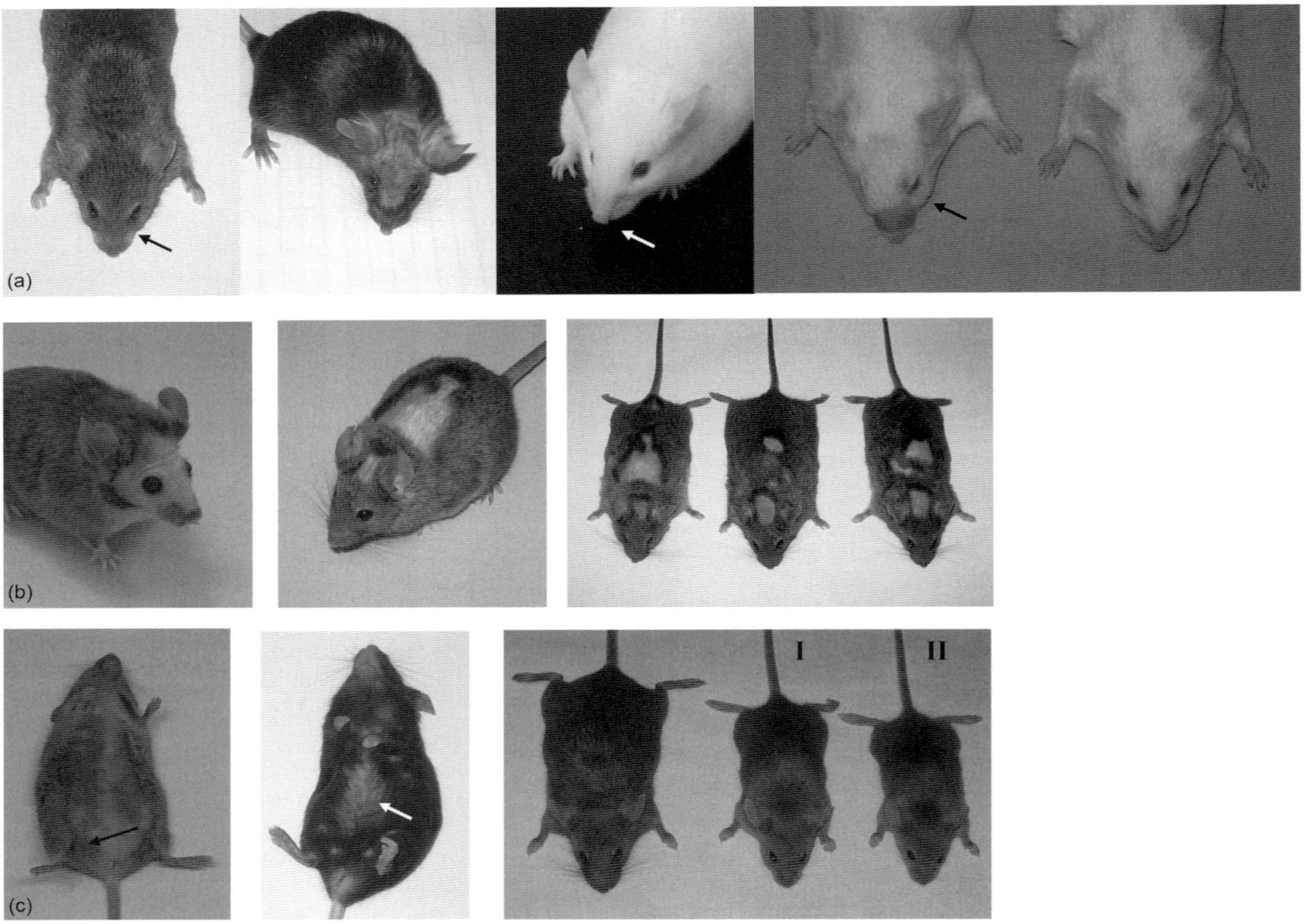

Figure 20.2 Typical barbering patterns in different mouse strains. (a) Social (dominant) barbering (from left to right) in inbred 129S1, C57BL/6J, A/J, and outbred NMRI male mice. Note missing whiskers in 129S1 and A/J mice, "haircut-style" in C57BL/6J, and snout denuding in one NMRI mouse (indicated by arrow; barber is a bigger mouse with normal whisker). (b) Social (dominant) barbering in several F1 hybrid strains: C57BL/6–129SvJ, C57BL/6–129S1, and 129S1-NMRI mice. (c) Maternal barbering: ventral hair loss due to barbering by suckling pups in 129S1 and C57BL/6 female mice (indicated by arrows), and barbering (whisker trimming) by lactating dams of their pups in 129S1 strain (note normal whiskers in the dam, half-short whiskers in pup I, and completely missing whiskers in pup II).

Finally, aberrant barbering is often seen in various mutant mice (e.g., Holmes et al., 2002; see MGI for details), collectively representing an important behavioral phenotypic effect with multiple mechanisms, genetic influences, neural substrates, forms, and contexts. It is of likely relevance, therefore, to human psychopathologies, like autism and OCD spectrum disorders, that are characterized by repetitive behavior and/or dysfunctions in social interaction.

Conclusions

Overall, there are clear potential benefits of in-depth mouse grooming and barbering phenotyping. First, it allows the assessment of strain differences in these behaviors per se. Second, as already mentioned, grooming activity, and especially its sequencing, may reflect strain differences in activity, anxiety, and motor patterning, in addition to the existing methods of emotionality phenotyping (discussed in other chapters of this volume). Third, given the sensitivity of mouse grooming and its sequencing to various pharmacological and physiological manipulations in different strains, ethologically oriented analysis of grooming may be useful in pharmacogenetics and neurophysiology (e.g., for dissection of brain substrates involved in the regulation of behaviors). Finally, altered grooming and barbering profiles may indicate behavioral stereotypy (Garner et al., 2004ab; Kalueff and Tuohimaa, 2004a). Indeed, it has been suggested that stereotypies in animals may originate from their natural behavioral repertoire of displacement activity (Low, 2003; Wurbel et al., 1998ab). Therefore, profiling of grooming and barbering phenotypes may allow researchers to indirectly assess potential strain differences in "stereotypicity."

Taken together, ethological analyses of mouse grooming and barbering can be a rich source of information in behavioral neurogenetics. Understanding how mice spend greater than 15% of their waking time will enable us to better understand their behavior in general. Clearly, more attention has to be paid to grooming and barbering indices, including standard mouse behavioral protocols (e.g., SHIRPA; Rogers et al., 1997, 1999) and behavioral analytical software (e.g., Ethovision, CleverSys, or any other program). In addition, since many published studies include data on C57BL/6J mice, and given its active high-grooming/high-barbering phenotype, this strain can be suggested as a reference strain for future studies focusing on different aspects of grooming and barbering behavior. The existing phenotype databases (e.g., MPD) may also benefit from the inclusion of systematic grooming and barbering profiles for different mouse strains. Providing more comprehensive coverage of mouse behavioral phenotypes and offering clever ideas on how known grooming/barbering peculiarities may affect other behaviors in different mouse strains will assist researchers in correct data interpretation and selection of appropriate mouse models for their tests.

Chapter abbreviations

GABA, γ-amino butyric acid
MGI, Mouse Genome Informatics
MPD, Mouse Phenome Database
OCD, obsessive-compulsive disorder

Acknowledgments

This work was supported by the NARSAD YI award (to A. V. K.), Georgetown University's Stress Physiology and Research Center (SPaRC) and Tulane University intramural research funds. The authors thank Marco Elegante for his help with this manuscript.

References

Albin, R.L. and Mink, J.W. (2006) Recent advances in Tourette syndrome research. *Trends Neurosci* 29: 175–182.

Aldridge, J.W., Berridge, K.C., and Rosen, A.R. (2004) Basal ganglia neural mechanisms of natural movement sequences. *Can J Physiol Pharmacol* 82: 732–739.

Audet, M.C., Goulet, S., and Dore, F.Y. (2006) Repeated subchronic exposure to phencyclidine elicits excessive atypical grooming in rats. *Behav Brain Res* 167: 103–110.

Avgustinovich, D.F., Kovalenko, I.L., and Koryakina, L.A. (2007) Effects of single episodes severe stress on the behavior of male and female CBA/Lac and C57BL/6J mice. *Neurosci Behav Physiol* 37(7): 731–737.

Barros, H.M., Tannhauser, S.L., Tannhauser, M.A., and Tannhauser, M. (1994) The effects of GABAergic drugs on grooming behaviour in the open field. *Pharmacol Toxicol* 74: 339–344.

Berntson, G.G., Jang, J.F., and Ronca, A.E. (1988) Brainstem systems and grooming behaviors. *Ann NY Acad Sci* 525: 350–361.

Berridge, K.C. and Aldridge, J.W. (2000) Super-stereotypy II: enhancement of a complex movement sequence by intraventricular dopamine D1 agonists. *Synapse* 37: 205–215.

Berridge, K.C., Aldridge, J.W., Houchard, K.R., and Zhuang, X. (2005) Sequential super-stereotypy of an instinctive fixed action pattern in hyper-dopaminergic mutant mice: a model of obsessive compulsive disorder and Tourette's. *BMC Biol* 3: 1–16.

Bertolini, A., Poggioli, R., and Vergoni, A.V. (1988) Cross-species comparison of the ACTH-induced behavioral syndrome. *Ann N Y Acad Sci* 525: 114–129.

Boccalon, S., Scaggiante, B., and Perissin, L. (2006) Anxiety stress and nociceptive responses in mice. *Life Sci* 78: 1225–1230.

Bolivar, V.J., and Brown, R.E., (1994) The ontogeny of ultrasonic vocalizations and other behaviors in male jimpy (jp/Y) mice and their normal male littermates. *Dev Psychobiol* 27(2): 101–110.

Bolivar, V.J., Danilchuk, W., and Fentress, J.C. (1996) Separation of activation and pattern in grooming development of weaver mice. *Behav Brain Res* 75: 49–58.

Bolles, R.C. (1960) Grooming behavior in the rat. *J Comp Physiol Psychol* 53: 306–310.

Brilliant, M.H., Ching, A., Nakatsu, Y., and Eicher, E.M. (1994) The original pink-eyed dilution mutation (p) arose in Asiatic mice: implications for the H4 minor histocompatibility antigen, Myod1 regulation and the origin of inbred strains. *Genetics* 138: 203–211.

Brodkin, E.S. (2007) BALB/c mice: low sociability and other phenotypes that may be relevant to autism. *Behav Brain Res* 176: 53–65.

Brodkin, E.S., Hagemann, A., Nemetski, S.M., and Silver, L.M. (2004) Social approach-avoidance behavior of inbred mouse strains towards DBA/2 mice. *Brain Res* 1002: 151–157.

Burne, T.H., Johnston, A.N.B., McGrath, J.J., and Mackay-Sim, A. (2006) Swimming behaviour and post-swimming activity in vitamin D receptor knockout mice. *Brain Res Bull* 69: 74–78.

Cabib, S., Algeri, S., Perego, C., and Puglisi-Allegra, S. (1990) Behavioral and biochemical changes monitored in two inbred strains of mice during exploration of an unfamiliar environment. *Physiol Behav* 47: 749–753.

Cabib, S. and Bonaventura, N. (1997) Parallel strain-dependent susceptibility to environmentally-induced stereotypies and stress-induced behavioral sensitization in mice. *Physiol Behav* 61: 499–506.

Carruthers, E.L., Halkin, S.L., and King, T.R. (1997) Are mouse "barbers" dominant to their cage mates? *Anim Behav Soc Abstracts.*

Carruthers, E.L., Halkin, S.L., and King, T.R. (1998) Mouse barbering: investigations of genetic and experiential control. *Anim Behav Soc Abstracts.*

Clement, Y., Adelbrecht, C., Martin, B., and Chapouthier, G. (1994) Association of autosomal loci with the grooming activity in mice observed in open-field. *Life Sci* 55: 1725–1734.

Clement, Y., Calatayud, F., and Belzung, C. (2002) Genetic basis of anxiety-like behaviour: a critical review. *Brain Res Bull* 57: 51–71.

Clement, Y. and Chapouthier, G. (1998) Biological bases of anxiety. *Neurosci Biobehav Res* 22: 623–633.

Coscia, E.M. and Fentress, J.C. (1993) Neurological dysfunction expressed in the grooming behavior of developing weaver mutant mice. *Behav Genet* 23: 533–541.

Crusio, W.E., Schwegler, H., and van Abeelen, J.H. (1989) Behavioral responses to novelty and structural variation of the hippocampus in mice. I. Quantitative-genetic analysis of behavior in the open-field. *Behav Brain Res* 32: 75–80.

Crusio, W.E. and van Abeelen, J.H. (1986) The genetic architecture of behavioural responses to novelty in mice. *Heredity* 56: 55–63.

Crusio, W.E. and van Abeelen, J.H. (1987) Zinc-induced peripheral anosmia and behavioral responses to novelty in mice: a quantitative-genetic analysis. *Behav Neural Biol* 48: 63–82.

Culiat, C.T., Stubbs, L.J., Montgomery, C.S., Russell, L.B., and Rinchik, E.M. (1994) Phenotypic consequences of deletion of the gamma 3, alpha 5, or beta 3 subunit of the type A gamma-aminobutyric acid receptor in mice. *Proc Natl Acad Sci USA* 91: 2815–2818.

Dell'Omo, G., Fiore, M., and Alleva, E. (1994) Strain differences in mouse response to odours of predators. *Behav Processes* 32: 105–116.

De Luca, A.M. (1997) Environmental enrichment: does it reduce barbering in mice? *AWIC Newslett* 8: 7–8.

Ducottet, C. and Belzung, C. (2004) Behaviour in the elevated plus-maze predicts coping after subchronic mild stress in mice. *Physiol Behav* 81: 417–426.

Dunn, A.J. (1988) Studies on the neurochemical mechanisms and significance of ACTH-induced grooming. *Ann N Y Acad Sci* 290: 150–166.

Dunn, A.J., Berridge, C.W., Lai, Y.I., and Yachabach, T.L. (1987) CRF-induced excessive grooming behavior in rats and mice. *Peptides* 8: 841–844.

Dunn, A.J., Berridge, C.W., Lai, Y.I., Yachabach, T.L., and File S.E. (1988) Excessive grooming behavior in rats and mice induced by corticotropin-releasing factor. *Ann N Y Acad Sci* 290: 391–393.

Fentress, J.C. (1977a) Opening remarks: constructing the potentialities of phenotype. *Ann N Y Acad Sci* 290: 220–225.

Fentress, J.C. (1977b) The tonic hypothesis and the patterning of behavior. *Ann N Y Acad Sci* 290: 370–395.

Fentress, J.C. (1988) Expressive contexts, fine structure, and central mediation of rodent grooming. *Ann N Y Acad Sci* 525: 18–26.

Fentress, J.C. (1999) Tracing behavioral phenotypes in neurological mutant mice. *IBANGS Conf Proc.*

Ferkin, M.H. and Leonard, S.T. (2005) Self-grooming by rodents in social and sexual contexts. *Acta Zool Sinica* 51: 772–779.

Ferkin, M.H., Sorokin, E.S., and Johnston, R.E. (1996) Self-grooming as a sexually dismorphic communicative behaviour in meadow voles, *Microtus pennsylvanicus. Anim Behav* 51: 801–810.

Fernandez-Teruel, A., Escorihuela, R.M., Gray, J.A., Aguilar, R., Gil, L., Gimenez-Llort, L., et al. (2002) A quantitative trait locus influencing anxiety in the laboratory rat. *Genome Res* 12: 618–626.

File, S.E., Mabbutt, P.S., and Walker, J.H. (1988) Comparison of adaptive responses in familiar and novel environments: modulatory factors. *Ann N Y Acad Sci* 525: 69–79.

Flint, J. (2004) The genetic basis of neuroticism. *Neurosci Biobehav Revs* 28: 307–316.

Garner, J.P., Dufour, B., Gregg, L.E., Weisker, S.M., and Mench, J.A. (2004a) Social and husbandry factors affecting the prevalence and severity of barbering ("whisker trimming") by laboratory mice. *Appl Anim Lab Sci* 89: 263–282.

Garner, J.P., Weisker, S.M., Dufour, B., and Mench, J.A. (2004b) Barbering (fur and whisker trimming) by laboratory mice as a model of human trichotillomania and obsessive-compulsive spectrum disorders. *Comp Med* 54: 216–224.

Golani, I. and Fentress, J.C. (1985) Early ontogeny of face grooming in mice. *Dev Psychobiol* 18: 529–544.

Gorris, L.G. and van Abeelen, J.H. (1981) Behavioural effects of (–) naloxone in mice from four inbred strains. *Psychopharmacology* 74: 355–359.

Graf, M., Kantor, S., Anheuer, Z.E., Modos, E.A., and Bagdy, G. (2003) m-CPP-induced self-grooming is mediated by 5-HT2C receptors. *Behav Brain Res* 142: 175–179.

Greer, J.M. and Capecchi, M.R. (2002) Hoxb8 is required for normal grooming behavior in mice. *Neuron* 33: 23–34.

Harvey, P.J. and Mann, M.A. (1987) Pregnancy-induced autogrooming in mice: the effects of nipple removal. *Dev Psychobiol* 20: 593–602.

Hauchka, T.S. (1952) Whisker-eating mice. *J Hered* 43: 77–80.

Holmes, A., Yang, R.J., Murphy, D.L., and Crawley, J.N. (2002) Evaluation of antidepressant-related behavioral responses in mice lacking the serotonin transporter. *Neuropsychopharmacology* 27: 914–923.

Honess, P., Gimpel, J., Wolfensohn, S., and Mason, G. (2005) Alopecia scoring: the quantitative assessment of hair loss in captive macaques. *Altern Lab Anim* 33: 193–206.

Hossain, S.M., Wong, B.K., and Simpson, E.M. (2004) The dark phase improves genetic discrimination for some high throughput mouse behavioral phenotyping. *Genes Brain Behav* 3: 167–177.

Jackson Laboratory (1987) Alopecia (loss of hair) in C57BL/6J and related strains. *JAX Notes* 431: 1–2.

Kalueff, A.V., Aldridge J.W., Laporte J.L., Murphy, D.L., and Tuohimaa, P. (2007) Analyzing grooming microstructure in neurobehavioral experiments. *Nat Protoc* 2: 2538–2544.

Kalueff, A.V., Lou, Y.R., Laaksi, I., and Tuohimaa, P. (2005) Abnormal behavioral organization of grooming in mice lacking the vitamin D receptor gene. *J Neurogenet* 19: 1–24.

Kalueff, A.V., Minasyan, A., Keisala, T., Shah, Z.H., and Tuohimaa, P. (2006) Hair barbering in mice: implications for neurobehavioural research. *Behav Processes* 71: 8–15.

Kalueff, A.V. and Tuohimaa, P. (2004a) Grooming analysis algorithm for neurobehavioural stress research. *Brain Res Protocols* 13: 151–158.

Kalueff, A.V. and Tuohimaa, P. (2004b) Contrasting grooming phenotypes in C57Bl/6 and 129S1/SvImJ mice. *Brain Res* 1028: 75–82.

Kalueff, A.V. and Tuohimaa, P. (2005a) The grooming analysis algorithm discriminates between different levels of anxiety in rats: potential utility for neurobehavioural stress research. *J Neurosci Methods* 43: 169–177.

Kalueff, A.V. and Tuohimaa, P. (2005b) Contrasting grooming phenotypes in three mouse strains markedly different in anxiety and activity (129S1, BALB/c and NMRI). *Behav Brain Res* 160: 1–10.

Kalueff, A.V. and Tuohimaa, P. (2005c) Mouse grooming microstructure is a reliable anxiety marker bidirectionally sensitive to GABAergic drugs. *Eur J Pharmacol* 508: 147–153.

Kingsley, D.M., Bland, A.E., Grubber, J.M., Marker, P.C., Russell, L.B., Copeland, N.G., et al. (1992) The mouse short ear skeletal morphogenesis locus is associated with defects in a bone morphogenetic member of the TGF beta superfamily. *Cell* 71: 399–410.

Kobayashi, M., Masuda, Y., Fujimoto, Y., Matsuya, T., Yamamura, K., Yamada, Y., et al. (2003) Electrophysiological analysis of rhythmic jaw movements in the freely moving mouse. *Physiol Behav* 75: 377–385.

Krehbiel, D., Bartel, B., Dirks, M., and Wiens, W. (1986) Behavior and brain neurotransmitters: correlations in different strains of mice. *Behav Neural Biol* 46: 30–45.

Kruk, M.R., Westphal, K.G., Van Erp, A.M., van Asperen, J., Cave, B.J., Slater, E., et al. (1998) The hypothalamus: cross-roads of endocrine and behavioural regulation in grooming and aggression. *Neurosci Biobehav Rev* 23: 163–177.

Kurien, B.T., Gross, T., and Scofield, R.H. (2005) Barbering in mice: a model for trichotillomania. *BMJ* 331: 1503–1505.

Landau, J.M., Wang, Z.-Y., Yang, G.-U., Ding W., and Yang, C. (2001) Inhibition of spontaneous formation of lung tumors and rhabdomyosarcomas in A/J mice by black and green tea. *Carcinogenesis* 19: 501–507.

Leppanen, P.K. and Ewalds-Kvist, S.B. (2005) Crossfostering in mice selectively bred for high and low levels of open-field thigmotaxis. *Scand J Psychol* 46: 21–29.

Leppanen, P.K., Ravaja, N., and Ewalds-Kvist, S.B. (2006) Twenty-three generations of mice bidirectionally selected for open-field thigmotaxis: selection response and repeated exposure to the open field. *Behav Processes* 72: 23–31.

Leppanen, P.K., Ravaja, N., and Ewalds-Kvist, S.B. (2008) Prepartum and postpartum open-field behavior and maternal responsiveness in mice bidirectionally selected for open-field thigmotaxis. *J Gen Psychol* 135: 37–53.

Long, S.Y. (1972) Hair-nibbling and whisker-trimming as indicators of social hierarchy in mice. *Anim Behav* 20: 10–12.

Low, M. (2003) Stereotypies and behavioural medicine: confusions in current thinking. *Austral Vet J* 81: 192–198.

McElwee, K.L., Boggess, D., Miller, J., King L.E., and Sundberg, J.P. (1999) Spontaneous alopecia areata-like hair loss in one congenic and seven inbred laboratory mouse strains. *J Investig Dermatol Symp Proc* 4: 202–206.

Militzer, K. and Wecker, E. (1986) Behavior-associated alopecia-areata in mice. *Lab Anim* 20: 9–13.

Mineur, Y.S. and Crusio, W.E. (2002) Behavioral and neuroanatomical characterization of FVB/N inbred mice. *Brain Res Bull* 57: 41–47.

Myers, D.D. (1997) C57BL/6J update – C57BL/6J skin lesion problem eliminated. *JAX Mice Anim Health Bull* 1–6.

Nyberg, J.M., Vekovischeva, O., and Sandnabba, N.K. (2003) Anxiety profiles of mice selectively bred for intermale aggression. *Behav Genet* 33: 503–511.

Pigott, T.A., Hill, J.L., Grady, T.A., L'Heureux, F., Bernstein, S., Rubenstein, C.S., et al. (1993) A comparison of the behavioral effects of oral versus intravenous mCPP administration in OCD patients and the effect of metergoline prior to i.v. mCPP. *Biol Psychiatry* 33: 3–14.

Reinhardt, V. (2005) Hair pulling: a review. *Lab Anim* 39: 361–369.

Rodgers, R.J., Boullier, E., Chatzimichalaki, P., Cooper, G.D., and Shorten, A. (2002) Contrasting phenotypes of C57BL/6JOlaHsd, 129S2/SvHsd and 129/SvEv mice in two exploration-based tests of anxiety-related behaviour. *Physiol Behav* 77: 301–310.

Roeling, T.A., Veening, J.G., Peters, J.P., Vermelis, M.E., and Nieuwenhuys, R. (1993) Efferent connections of the hypothalamic "grooming area" in the rat. *Neuroscience* 56: 199–225.

Rogers, D.C., Fisher, E.M., Brown, S.D., Peters, J., Hunter, A.J., and Martin, J.E. (1997) Behavioral and functional analysis of mouse phenotype: SHIRPA, a proposed protocol for comprehensive phenotype assessment. *Mamm Genome* 8: 711–713.

Rogers, D.C., Jones, D.N., Nelson, P.R., Jones, C.M., Quilter, C.A., Robinson, T.L., et al. (1999) Use of SHIRPA and discriminant analysis to characterise marked differences in the behavioural phenotype of six inbred mouse strains. *Behav Brain Res* 105: 207–217.

Sachs, B.D. (1988) The development of grooming and its expression in adult animals. *Ann N Y Acad Sci* 525: 1–17.

Sandnabba, N.K. (1995) Predatory aggression in male mice selectively bred for isolation-induced intermale aggression. *Behav Genet* 25: 361–366.

Sankoorikal, G.M., Kaercher, K.A., Boon, C.J., Lee, J.K., and Brodkin, E.S. (2006) A mouse model system for genetic analysis of sociability: C57BL/6J versus BALB/cJ inbred mouse strains. *Biol Psychiatry* 59: 415–423.

Sarna, J.R., Dyck R.H., and Whishaw, I.Q. (2000) The Dalila effects: C57BL6 mice barber whiskers by plucking. *Behav Brain Res* 108: 39–45.

Schoots, A.F., Crusio, W.E., and Van Abeelen, J.H. (1978) Zinc-induced peripheral anosmia and exploratory behavior in two inbred mouse strains. *Physiol Behav* 21: 779–784.

Spruijt, B.M., van Hooff, J.A., and Gispen, W.H. (1992) Ethology and neurobiology of grooming behavior. *Physiol Rev* 72: 825–852.

Strazielle, C. and Lalonde, R. (1998) Grooming in Lurcher mutant mice. *Physiol Behav* 64: 57–61.

Streng, J. (1971) Open-field behavior in four inbred mouse strains. *Can J Psychol* 25: 62–68.

Strozik, E. and Festing, M.F.W. (1981) Whisker trimming in mice. *Lab Anim* 15: 309–312.

Suaudeau, C., Rinaldi, D., Lepicard, E., Venault, P., Crusio, W.E., Costentin, J., et al. (2000) Divergent levels of anxiety in mice selected for differences in sensitivity to a convulsant agent. *Physiol Behav* 71: 517–523.

Sviderskaia, G.E. and Dmitrieva, L.E. (1993) The development of grooming in the ontogeny of rats and mice. *Zh Evol Biokhim Fiziol* 29: 379–386.

Tang, X., Orchard, S.M., and Sanford, L.D. (2002) Home cage activity and behavioral performance in inbred and hybrid mice. *Behav Brain Res* 136: 555–569.

Ukai, M., Toyoshi, T., and Kameyama, T. (1989) Multi-dimensional analysis of behavior in mice treated with the delta opioid agonists DADL (D-Ala2-D-Leu5-enkephalin) and DPLPE (D-Pen2-L-Pen5-enkephalin). *Neuropharmacology* 28: 1033–1039.

van Abeelen, J.H.F. (1963a) Mouse mutants studied by means of ethological methods. I. Ethogram. *Genetica* 34: 79–94.

van Abeelen, J.H.F. (1963b) II. Mutants and methods. *Genetica* 34: 95–101.

van Abeelen, J.H.F. (1963c) III. Results with yellow, pink-eyed dilution, brown and jerker. *Genetica* 34: 270–286.

van Abeelen, J.H.F. (1966) Effects of genotype on mouse behaviour. *Anim Behav* 14: 218–225.

Van den Broek, F.A.R., Omzight C.M., and Beynen, A.C. (1993) Whisker trimming behaviour in A2G mice is not prevented by offering means of withdrawal from it. *Lab Anim* 27: 270–272.

Van de Weerd, H.A., Van den Broek, F.A.R., and Beynen, A.C. (1992) Removal of the vibrissae in male mice does not influence social dominance. *Behav Processes* 27: 205–208.

Voikar, V., Polus, A., Vasar, E., and Rauvala, H. (2005) Long-term individual housing in C57BL/6J and DBA/2 mice: assessment of behavioral consequences. *Genes Brain Behav* 4: 240–252.

Voikar, V., Vasar, E., and Rauvala, H. (2001) Strain and gender differences in the behavior of mouse lines commonly used in transgenic studies. *Physiol Behav* 72: 271–281.

Voikar, V., Vasar, E., and Rauvala, H. (2004) Behavioral alterations induced by repeated testing in C57BL/6J and 129S2/Sv mice: implications for phenotyping screens. *Genes Brain Behav* 3: 27–38.

Whishaw, I.Q., Metz, G.A.S., Kolb B., and Pellis, S.M. (2001) Accelerated nervous system development contributes to behavioral efficiency in the laboratory mouse: a behavioral review and theoretical proposal. *Dev Psychobiol* 39: 151–170.

Wirth-Dzieciolowska, E., Lipska, A., and Wesierska, M. (2005) Selection for body weight induces differences in exploratory behavior and learning in mice. *Acta Neurobiol Experim* 65: 243–253.

Wurbel, H., Chapman, R., and Rutland, C. (1998a) Effect of feed and environmental enrichment on development of stereotypic wire-gnawing in laboratory mice. *Appl Anim Behav Sci* 60: 69–81.

Wurbel, H., Freire, R., and Nicol, C.J. (1998b) Prevention of stereotypic wire-gnawing in laboratory mice: effects on bevavious and implications for stereotypy as a coping response. *Behav Processes* 42: 61–72.

Yalcin, I., Aksu, F., and Belzung, C. (2005) Effects of desipramine and tramadol in a chronic mild stress model in mice are altered by yohimbine but not by pindolol. *Eur J Pharmacol* 514: 165–174.

Yalcin, I., Belzung, C., and Surget, A. (2008) Mouse strain differences in the unpredictable chronic mild stress: a four-antidepressant study. *Behav Brain Res* 193: 140–143.

Young, R.K. and Thiessen, D.D. (1991) Washing, drying, and anointing in adult humans (*Homo sapiens*): commonalities with grooming sequences in rodents. *J Comp Psychol* 105: 340–344.

Chapter 21

Social behaviors in wild and laboratory mice with a special emphasis on the C57BL/6J inbred strain

D. Caroline Blanchard, Jacqueline N. Crawley, Hiroyuki Arakawa, and Robert J. Blanchard

Mice have been, perhaps from the beginning of modern biology (Berry and Scriven, 2005), a premier subject species for biobehavioral research. In recent years knowledge of the mouse genome and ready availability of genetically manipulated mouse lines have further increased the focus of research on mice, increasing their use vis-à-vis rats, the other most widely used subject species. Genetic approaches have also emphasized the necessity for understanding how contemporary laboratory mice have evolved from their wild ancestors and the degree to which specific aspects of their physiology and behavior may or may not represent the characteristics of those ancestors.

History

Laboratory mice are the result of a process of breeding that began in the early days of the twentieth century (Berry, 1981). While the taxonomy of *Mus* continues to involve some controversy, it is clear that the two most widespread forms are *Mus musculus* and *Mus domesticus*, and that both (along with *Mus castaneus*: Wade et al., 2002) contributed extensively to the derivation of laboratory mice (Berry and Scriven, 2005; Wotjak, 2004). A striking feature of these, and other, *Mus* species (over 130 subspecies having been described from wild-caught animals: Berry, 1981) is their adaptability with reference to living in close association with humans, relying on the latter for shelter, and, especially, food. This commensal life style has characterized many mouse species for at least 6000 years (Brothwell, 1981) potentially selecting for different behavioral traits than are adaptive for mice that have lived without this association (Bronson, 1979; Gray and Hurst, 1997).

Human populations have been directly involved in the geographic distribution of house mice for at least 3000 years (Cucchi et al., 2005). Frynta et al. (2005) make the interesting case that house mouse populations in more recently colonized areas such as the United Kingdom, The Americas, Australia, and New Zealand, whether currently living as commensals or not, have likely passed through a commensal stage that was important in their dispersal. Moreover, house mice in such areas are strongly represented in the ancestry of laboratory mice, suggesting that the effects of long-term commensalism may have been a factor in the adaptability of house mice to even more focused human contact in the form of laboratory use. This suggests that laboratory-bred wild mice may already represent genetic selection for willingness to breed in close proximity to people, calling into question the view that laboratory-bred wild mice truly represent wild ancestral stock. The potential magnitude of this effect may be striking: Barnett and Stoddart (1969) reported "taming" of animals of a different wild rodent species following laboratory breeding for several generations without human selection. However, in addition to any possible genetic effects of commensalism, behavioral differences of current commensal and non-commensal mice may also reflect systematic differences in the immediate experience of opportunities and restrictions afforded by human-related versus wild habitats (Pocock et al., 2005).

The adaptability of house mice extends to geographical and habitat distribution (Lathan and Mason, 2004). They are found in both commensal and non-commensal situations over much of the world, on every continent except Antarctica (but in some subantarctic islands) and in climates from the tropics to frigid zones (Berry and Scriven, 2005). They can subsist on a wide variety of diets and go without free water as long as their food has a water content of 15%, and they can breed in temperatures as low as −30°C (Baumans, 1999; Randall, 1999; Southern, 1954). These habitat differences virtually ensure that the social behaviors of mice must also be capable of adapting to new and widely varying situations. This suggests that an analysis of normative wild mouse behavior must take into account important features of their habitat.

Dispersal of mice in natural habitats

House mice might well serve as ideal subjects for analysis of how mammalian sociality changes in different habitats. Unfortunately, due to the extreme difficulty of observing social behaviors of wild mice – small, often fossorial, rodents – in natural communities, studies of behavior that involve wild populations in natural habitats tend to be few, rather sketchy, and gathered with great effort. As an example of this last feature, Sutherland et al. (2005) trapped and examined wild mice in fields in Australia over 11 months. They took samples of each for DNA analysis, marked the animals with a Passive Integrated

Behavioral Genetics of the Mouse: Volume I. *Genetics of Behavioral Phenotypes*, eds. Wim E. Crusio, Frans Sluyter, Robert T. Gerlai, and Susanna Pietropaolo. Published by Cambridge University Press.

Transponder tag to enable monitoring, released them at the site of capture, and retrapped them; to evaluate abundance of mice over time and to analyze reproductive condition and relatedness among the animals, in terms of their distribution patterns. These rather heroic measures were aimed at analysis of "mouse plagues," a poorly understood "boom–bust" population growth phenomenon common in Australia but seen, albeit to a lesser degree and possibly with different underlying mechanisms, in other populations as well (see Singleton et al., 2005 for review).

Pocock et al. (2005) have recently reviewed field studies of mouse dispersal with a particular emphasis on commensal versus non-commensal populations, indicating that such studies tend to report lower rates of dispersal in commensals, and over shorter distances; a pattern associated with greater density of food resources, and also greater densities of mice, in human habitats. Dispersion of commensal mice is also more likely to involve males, particularly younger males, consonant with a view that aggression from dominant males is an important factor in dispersion of maturing male mice.

Van Oortmerssen and Busser (1989) used a natural, large (69 × 24 m) habitat with a fence that precluded mouse entry, to investigate population dynamics of wild-trapped wild mice over a longer period of time; about 2.5 years from beginning to population crash. The population boom–bust cycles in this area were dramatic. The first involved a decline from several hundred animals to less than 10; a second was a total population crash, with no mice left within the enclosure. Large numbers of dead and wounded animals were found at about the time of the population crashes, strongly implicating fighting as a factor. Also, at about the same time, the proportion of females in the population declined, and the number of pregnancies decreased to near zero. A number of measures suggested that gradually increasing levels of intolerance of conspecifics occurred over a period of more than a year prior to the final population crash, and might have been influential as a mediating factor.

The authors attribute the crash, not to changes in the environment for individual animals, but to changes in the animals themselves, with conditions in the mouse garden selecting for males that show more aggression and greater intolerance of conspecifics. A subsequent study in the same report (Van Oortmerssen and Busser, 1989) compared the descendants of wild mice trapped during an "increase phase" (of the population) with descendants of those trapped during a "decrease phase." In accord with predictions of enhanced aggression in the latter population, the proportions of contacts involving attack were higher for adult males on other adult males, on adult and on juvenile females; and by juvenile males on adult females. Adult females showed enhanced attack on males, but decreased attack on juveniles, male or female. This pattern, while suggesting higher aggression in males generally, additionally suggests a particular upsurge of male attacks on females, possibly in conjunction with enhanced sexual behavior, or as the result of a reduction in the normal inhibition of male attack on females. These findings were important to subsequent analyses of the short attack latency (SAL) and long attack latency (LAL) patterns that these authors and their colleagues have investigated in animals from this wild population (e.g., Van Oortmerssen and Bakker, 1981; Veenema et al., 2004). Krakow (2003) also reported a significant relationship of latency to disperse from an established group (into which the subject is placed) between fathers and sons among wild-caught and first-generation laboratory-bred wild mice. However, there was no significant relationship between dispersers and non-dispersers (over 14 days or more) for fathers and sons.

The relationship between the onset of agonistic behavior and tendencies for subordinate males to disperse has often been reported (e.g., Lidicker, 1976). Rusu and Krackow (2005) found that when the dominant member of a sibling male pair, and one member of an amicable sibling male pair (wild-caught to third-generation laboratory bred) were simultaneously introduced into an established colony of females, one male always became dominant, and dispersal from the group was particularly common among previously dominant males who were defeated.

Laboratory studies: habitats, sociality, and dispersal

Work with wild mice in laboratory settings indicates that sociality in mice is strongly influenced by habitat size and structure. Butler (1980) used first-generation offspring of wild house mice (*Mus musculus*) to form two males/one female (2M/1F) adult groups, housed in enclosures containing two (small enclosure) or four (larger enclosure) such groups, separated by walls. Six days later, after analysis of males within each group as dominant or subordinate, these internal walls were removed. Agonistic behaviors among dominant males or females of the now mixed areas involved long fights, but subordinates generally avoided strange animals. In the large enclosures, formation of territories was common for both males and females, whereas in the small enclosure, territories were rarely seen, suggesting that a dominance hierarchy rather than a territory had resulted from the mix when only four males were involved. This fits well with our personal observations of a number of both rat and mouse colonies, that dominance is difficult and stressful to achieve and maintain when there are more than four or five same-sex adults that must be subjugated: When more than this number is involved, a territory may be easier to defend than a dominant position in a male hierarchy.

In the Butler study (1980) "emigration" was possible, although difficult. About 20% did so, and 35 of the 36 emigrants were subordinates; 75% of them male. (As two-thirds of animals in the enclosures were male, this does not suggest a gender difference in emigration. However Gerlach (1996) observed twice as many males as females emigrating from cage systems containing a breeding pair and their offspring.) The single dominant (male) that emigrated did so immediately after losing a series of fights. These data are strongly supportive of a view that emigration of subordinates is an important factor in dispersal of wild mice, further indicating that subordinate females as well as males may emigrate. Finally, it is notable that in these situations, even though food and water were available *ad libitum* and animals were not crowded, agonistic interactions constituted over 80% of the interactions seen. Amicable interactions, even under these apparently favorable conditions, appear to be somewhat uncommon among adult house mice.

Lloyd (1975) investigated two substantial groups of house mice, originally (25 years previously) bred from wild populations. These groups were each maintained in two-level enclosures, with a total area of about 2.5 m, containing 10 nesting cages. As populations grew (to maxima of 38 and 60 animals in the two enclosures), they each experienced a relatively long (280 or 130 days, respectively) initial period in which maturing males set up and patrolled territories; females tended to live within a single male's territory; birth rates were high; and pup survival to weaning substantial (66 or 91% respectively). Following this initial phase, a great deal of fighting was associated with instability in territorial boundaries, and a precipitous decline in successful breeding, to less than 5% of pups being weaned. In a third phase (which occurred in the first colony only), births continued to be depressed, but about 50% of pups were weaned. This phase was characterized by a reduction in fighting among territorial males (except those initiated by a single male, whose territory now encompassed about half the enclosure), although attacks on non-territorial males remained high. These findings are in agreement with those of Butler (1980) and others in suggesting that when many males are present and space is sufficient, male mice tend to set up territories, and that such territories promote breeding success. It additionally suggests that crowding past an optimal level promotes fighting and reduces the ability of males to maintain territories, with disastrous consequences for reproductive success. A report by Anderson (1961) that territoriality is a major form of social organization for house mice at low densities, is in agreement with this view.

Resident status is another strong determinant of social systems in wild house mice (Gray and Hurst, 1997). Habitats containing resources such as food and water, or nest sites, are most strongly defended against intruders (Gray et al., 2002). Residents typically attacked same-sex intruders. When no resources were found in the defended habitat, resident males seldom attacked intruder females although continuing to attack intruder males. Gray et al. (2000) also reported that mice prefer complex habitats. They created habitat structures of differing complexity and captive-born male house mice were allowed to pass between pairs of such enclosures. The mice strongly preferred walled or complex enclosures over an open enclosure, suggesting a preference for potential concealment. Although residents chased and attacked intruders, many more residents in complex enclosures lost the intruder they were chasing, among the obstacles. Thus habitat complexity may increase the difficulty of residents (or predators) finding and attacking mice, enabling a higher population density to develop. In particular, this might increase the possibility for subordinate animals to live in the territory of a dominant mouse.

These outcomes may vary with the location and distribution of resources, and change over time. A series of studies (Gray et al., 2002; Jensen et al., 2003, 2005) investigated mouse pairs living in habitats varying in complexity, presence or absence of overhead cover, and clumping versus scattering of resources. When adjacent groups were allowed to interact, aggression was initially greater in open habitats, but eventually became most prolonged when scattered resources were to be found in complex habitats. Overhead cover was also important in providing shelter from predators, and aggression stopped immediately when investigators loomed. These studies were also notable for using mice initially trapped in two different geographic locations, with substantial differences between animals trapped in the two sites; in aggressiveness, use of cover, activity, and willingness to invade a neighbor's territory. However, the overall patterns of habitat invasion and territorial aggression were not systematically different for the two groups.

Habitat and sexual behavior

In an important review of the reproductive ecology of house mice, Bronson (1979) makes the point that female mouse urinary cues stimulate males to produce more reproductive pheromones. These, in turn, induce estrus in the females, forming a feedback loop that makes female reproductive readiness considerably independent of many of the environmental features that regulate reproduction in other species. Moreover, urinary cues from other females in the area can block this effect, particularly with reference to prepubertal females, effectively prolonging their non-reproductive state. These factors should reduce breeding of females remaining in the nest area, but promote rapid attainment of estrus when a female leaves the nest, a factor greatly potentiating the effective dispersal of mice.

There are vanishingly few studies involving direct observation of sexual behavior in free-ranging wild house mice. However, Hurst (1986) has described in detail a single mating sequence among wild house mice living in deep-pit poultry houses. This encompassed 31 mounts over a period of nearly an hour. Several features of this sequence were strikingly different from laboratory-based descriptions of wild mouse mating: many more mounts; little chasing by the male but instead mutual moving of the two animals to the specific mating site to copulate followed by independent withdrawal of the two afterwards; use of a site not often utilized by other mice, with defense of the site by the male for some time after the female had departed. The specifics of this pattern suggest substantial female choice in sexual activity, with as much female chasing of the male, as vice versa, and with repeated and apparently unforced approach by both to the copulatory site.

Relationships among females: communal nesting and success in rearing young

Given the opportunity at an early postpartum stage, *M. musculus* mothers consistently combine their litters and show communal nursing behaviors, a tendency that is stronger yet in mothers that have previously reared young communally (Sayler and Salmon, 1969). Mothers with older litters (>5 but <10 days) took longer to do so, and mothers whose litters were considerably younger than the communally nursed offspring were more aggressive toward the latter. This consistent response may reflect a substantial advantage of communal nesting, higher weight gain in the pups (see Branchi et al., 2006a, 2006b, for additional changes in communally reared pups); but with selective disadvantages to younger pups when litters of different ages are nursed communally.

König (1994) reported that, among first and second generation wild-caught house mice, nearly 90% of related and familiar (i.e., reared together) female pairs, housed in adjacent but communicating cages and bred simultaneously, successfully weaned young. For unfamiliar and unrelated females, or unfamiliar but related (cross-fostered) females, just over 50% did so, although one of the two females in each such pair did wean young, among an additional 25–35% of the pairs. For familiar but unrelated females success rates for both members of a pair were intermediate, about 70%. These differences were heightened for the female of each pair that delivered last, with nearly five times as many offspring weaned for familiar/related females that were the second dams to deliver in the pair, as for unfamiliar dams (regardless of relatedness) delivering pups after their pair-partners. This study suggests that both familiarity and relatedness of females promote reproductive success in communally living wild mouse females, but familiarity – having been reared together – appears to be more important.

Palanza et al. (2005) add male cohabitation as a factor in the role of familiarity versus relatedness in successful rearing of young. In laboratory-born second-generation offspring of wild house mice, *M. m. domesticus*, prior individual cohabitation with a male results in females becoming very aggressive to female intruders, even sisters with whom they had been raised. However, if females are introduced into the cohabitation situation *together*, whether before, after, or at the same time as the male, then the effects of prior familiarity and relatedness in reducing aggression becomes clearer; less aggression toward non-sibling females, and almost none to sisters. Moreover, most sibling female pairs introduced together delivered pups, and over 60% weaned at least one pup per female. For non-sibling female pairs, significantly fewer dams delivered pups and very few weaned them. When one female cohabited first, less than 10% of sibling pairs and no non-sibling pairs both successfully weaned pups, suggesting that even the potent combination of relatedness and prior familiarity is incapable of inducing tolerance in females that already cohabit with a male.

Female–female aggression is particularly high when several females (wild-caught to third laboratory generation) are housed with a single male, rather than several males (Rusu and Krackow, 2004). Again sibling females were less aggressive to each other, but sisters with a single male were almost as aggressive as non-sisters, while sisters with three males showed almost no aggression. Pregnancy rate was higher among sisters, and lower in the one-male enclosures. Dominant females within each group were reproductively more successful, due to a greater number of pregnancies and higher weaning rates for pups. In groups where female dominance was not seen, all females delivered and weaned some or all pups.

Under normal conditions, familiarity and relatedness would be expected to show strong covariance, and, a tendency for female siblings to remain together after weaning should promote contact prior to and at the time of breeding. Each dam in such a relationship would show an adaptive advantage through enhancement of offspring of her related pair-partner, as well as through advantages to her own offspring, consonant with the report that communal nesting preferences are highest among cagemate sisters (Manning et al., 1992).

These results, and others indicating communal nesting in wild-caught mice, or mice bred from wild-caught stock (e.g., Baker, 1981; Crowcroft and Rowe, 1963; Lloyd, 1975), were from studies done in laboratory situations. Wilkinson and Baker (1988) examined communal nest sites for wild mice in a Colorado feed shed, finding direct evidence of communal nesting (localization by fitted transmitters) in five of 36 nests found, with visual evidence for a sixth communal nest. This by no means implies that other nest sites were not communal, as these would not have been identified for transmitter fitted females cohabiting with non-fitted females. Blood and tissue samples of the five cohabiting (transmitter-fitted) female pairs indicated significantly similar genotypes. These pregnant or lactating females had smaller home ranges than those of adult males, and overlapped substantially with the ranges of only four of the 10 males whose home ranges could be estimated. The ranges of these four males showed reduced overlap with those of other adult males, suggesting that breeding males are more strongly territorial than non-breeding males.

Breeding success rate in females also depends on the level of aggression among males in the area. When females are housed with two males that are brothers, reared together, and less inclined to fight, their average number of pups reared is about twice as large as when the two cohabiting males are not brothers, and fight more (Ensminger and Meikle, 2005). Litter size was not different, but many females in the latter condition did not produce litters, suggesting that male fighting is stressful for females, possibly altering female cycling or inducing abortion. Females preferred areas containing the odors of males that did not fight, compared to areas with odors of males that fight among themselves, regardless of whether the male pairs were brothers or not; consonant with the view that the operative factor affecting female breeding success is the level of fighting, not whether her cohabiting males were brothers.

Finally, breeding success in females, as well as the extended quality of the progeny, may be enhanced by female mate choice. When females derived from wild stock were given a choice between males and then mated with the preferred or non-preferred male, females mated with a preferred male produced offspring that, when tested as adults, showed enhanced aggression and dominance; made better nests; and had higher rates of survival when released into a field enclosure for 60 days. Progeny from matings with preferred males also made better nests in the field enclosure, and, were significantly harder to trap (Drickamer et al., 2000). These finding suggest that female selection of mates is based on cues reflecting such aggression and survival qualities.

Scent marking

Scent marking with urine constitutes a major form of communication among house mice (Hurst, 1990a, 1990b, 1990c). Territorial and dominant males strongly countermark other males' urine traces, as well as those of females, while subordinates investigate but do not countermark the dominant's trace. Resident breeding females countermark scents of some non-resident males, but show strong countermarking of non-resident breeding females. Subadult females investigate, but generally do not mark, urine traces from both residents and non-residents. Breeding females, but not non-breeding females, are attracted to the marks of their own resident males. Non-breeding females also tend to ignore territorial boundaries. Such findings suggest that scent marks are used to advertise dominance and breeding status, especially to same-sex conspecifics (Hurst, 1990a, 1990b, 1990c). This hypothesis received considerable support from the finding that females in estrus are more strongly attracted to the odors of male holders of exclusively marked territories, in which that male's scent marks are not accompanied by marks of other males; and that they show more sexual behavior in interactions with these males (Rich and Hurst, 1998).

A number of recent developments in analysis of the olfactory mechanisms underlying recognition of individual conspecifics, assessment of their familiarity, and analysis of potential mate quality by female mice (Becker and Hurst, 2009; Hurst, 2009) suggests specific roles for both male scent marking patterns, and female behavior, in the latter process. Relevant olfactory information is provided through three separate sets of olfactory signals released by mice: Major urinary proteins (MUPs), detected through the vomeronasal system, provide information on genetic identity, including kin recognition and, potentially, genetic heterozygosity. Major histocompatibility complex (MHC) peptides, detected through the main olfactory system, provide volatile signatures capable of responding to a male mouse's androgenic status and other health related factors. Exocrine-gland secreting peptides (ESPs) are secreted by a number of orofacial glands, as well as the preputial gland, but their functions are less well understood.

Female mice respond to volatile male odors, by approaching and investigating the male or its scent marks. This process brings them into contact with non-volatile MUPs, providing an individual genetic identity signature that is experienced in tandem with the male's volatile peptides, leading to a learned association with the latter. Subsequently, the female can recognize the individual male through his airborne signature alone, which was not possible prior to this association (Becker and Hurst, 2009). Although much information relative to the male's health and status (diet, parasitism status, etc.) may be provided by the volatile odors, a major index of dominance, the patterning of scent marks and countermarks in the environment is detected through the age of scent marks deposited by the males involved: Females are more attracted to the males whose marks are freshest, indicating that they have countermarked scents from other males (Rich and Hurst, 1999). This system is facilitated by the relative lack of degradation of MUPs over time, and the fact that in wild mouse populations MUP patterns are so varied that males are highly unlikely to share the same pattern (Hurst, 2009). The effectiveness of this complex mechanism for recognition of mate quality also depends on differences in the patterning of marking by dominant and subordinate male mice, a well-established phenomenon (Desjardins et al., 1973).

In a program systematically describing scent marking in C57BL/6J males, to other male mice, Arakawa et al. (2007) reported that these males showed substantial marking toward outbred CD-1 males or to the odor alone of CD-1 males, but not toward C57BL/6J males, suggesting a lack of ability to discriminate between individual males of the highly inbred C57BL/6J strain. Such marking appeared at about 60 days of age, somewhat earlier than marking toward a C57BL/6J female. Arakawa et al. (2008a) reported that C57BL/6J (or CD-1) males showed rapid habituation of scent marking with repeated exposure at 24 hour intervals, to either stimulus males (CD-1 or C57BL/6J), or to an empty chamber, when no prior chamber familarization was given: Subsequent exposure to a genetically different novel mouse (CD-1 after CD-1 exposure, or CD-1 after C57BL/6J exposure) or to a novel context (different shaped chamber) produced recovery of marking, demonstrating that male mice differentiate familiar and novel conspecifics when they are genetically different, and also that scent marking can differentiate novel from familiar contexts. However, responses to a novel but genetically identical mouse (C57BL/6J after C57BL/6J exposure) did not show recovery, suggesting that initial habituation to the C57BL/6J mouse actually reflected habituation to the chamber where the stimulus mouse was presented. Scent marking was suppressed by presentation of a predator stimulus (cat fur/skin odor), both when the cat odor was presented in the test situation, and, when it was presented in the subject's home cage prior to testing. This suppression lasted for at least 7 days (Arakawa et al., 2008b).

Scent marking is modulated by subject and stimulus housing condition: Arakawa et al. (2009) found that in a novel situation without mouse odors, pair-housed male subjects deposited fewer scent marks than singly-housed males. Moreover, singly-housed males made fewer marks to pair-housed than to singly-housed stimulus males, while pair-housed males showed few marks to either type of stimulus. Reduced marking of singly-housed to pair-housed males can be mediated by the urinary odor, alone, of the pair-housed stimulus, and this odor is effective even when presented prior to testing. Singly-housed males also showed less marking to the scent marks of subordinate, compared to dominant or singly-housed males, an effect that did not depend on the quantity of the scent mark stimulus suggesting that it is a particular component in urinary odor rather than the magnitude of the odor, that plays a significant role in recipient behavior. Together, these experiments indicate that scent marking can serve as a sensitive index of a number of social relationships among male mice.

Summary

These studies suggest that mice are an extremely adaptable species, to the point that it may be more precise to describe what they are unlikely to do, rather than what they are likely to do in a range of situations. One likely reason for this great adaptability is the long history of mice living as human commensals, an environment in which ability to adjust to, and successfully breed in, rapidly altering conditions and in close proximity to potential predators is particularly crucial. This suggests that although laboratory breeding of mice may, in general, further reduce the animals' stress responses, it is less likely to produce major changes in sociality, the complex of social responses and relationships within and between groups of mice. This pattern appears to be one of limited but definite sociality in which mice are attracted to other mice, and spend more time investigating strangers compared to known animals, but are more tolerant of familiars, particularly those with whom they have been raised. This suggests that attraction to another mouse may involve mixed motives, with sexual and aggressive motivations in addition to prosocial or amicable attraction. There appears to be some ability to recognize kin, and relationships with kin are somewhat more peaceful than with non-kin, even with familiarity held constant. Adult female mice are more tolerant of other females than are males of other males, although this tolerance may be very limited when the female is cohabiting with a male. In wild populations, dyads of females sharing a nest or nest area appear to be common, and laboratory studies suggest this may be advantageous in terms of successful rearing of young. Dispersal is the normal outcome of postpubescent status for many male mice and some females, but the timing and immediate occasion for this may vary with food and space availability, crowding, and other environmental features. Within limits, aggression is a very adaptive behavior for individual male mice, but its limitations may include enhanced stress and reproductive failure of females exposed to such aggression, in addition to possible damage to other (potentially related) males, and stress to the aggressive male himself when a number of opponents are present.

C57BL/6J mice

These findings for wild mice provide a background for analysis of social behaviors in laboratory mouse strains. One particularly interesting strain is the C57BL/6J, used for sequencing the mouse genome (Mouse Genome Sequencing Consortium, 2002). This strain has served as background for a number of genetic manipulations (e.g., Crawley et al., 1997; Homanics et al., 1999; Logue et al., 1997; Silver, 1995). As specific behavioral or physiological parameters of a background strain may influence the expression or the salience of genetic manipulation effects (e.g., Le Roy et al., 2000; Mogil and Wilson, 1997; Young et al., 2002), a thorough description of relevant parameters for the strain may be important in interpreting such manipulations.

Anxiety in response to aggression

Because the topic of aggression is treated in a separate chapter in this volume, it will not be described here. However, a number of associated findings are consonant with a view that *either* C57BL/6J mice are highly aggressive, or, C57BL/6Js are more defensive/anxious in response to situations that involve some element of threat. In tests involving within-strain social behaviors, C57BL/6J females show anxiety-like changes on the elevated plus-maze and open field tests, along with reduced recognition of familiar conspecifics and a prolonged diestrus phase of the estrus cycle, after prolonged housing with an aggressive male (Avgustinovich and Kovalenko, 2004). Similarly, after intermittent social confrontations in a partition test, C57BL/6J male losers showed an anxiety-like pattern on the elevated plus-maze test, while winners also showed anxiety-like elevated plus-maze test behaviors after 20, but not 10, days of confrontations (Avgustinovich et al., 1997). Following 20 defeats, loser males of the C57BL/6J strain demonstrate withdrawal, passive defense postures, and new (i.e., not seen in early bouts) immobile postures instead of the active defense they had displayed in first encounters. Defeated C57BL/6J males also displayed a decrease of ambulation in the open field test and higher immobility times in the Porsolt test (Kudriavtseva and Bakshtanovskaia, 1991). Subordinate male C57BL/6J mice (from within-strain encounters) show reduced attack compared to BALB/cJ males, and enhanced submissive behaviors to dominant males (Mondragón et al., 1987). C57BL/6J losers, but not winners, showed enhanced ethanol intake (Kudriavtseva and Bakshtanovskaia, 1991), again a possible consequence of enhanced aggression by the attacking C57BL/6J victors, or, a reflection of enhanced emotional responsivity to threat in this strain.

With reference to a behavior that potentially interacts with both aggression and anxiety, C57BL/6J males have been described as more sexually active than DBA/2J (Ogawa et al., 1996). They show short latencies to mount, intromit, and ejaculate in comparison to other strains (D'Udine and Manning, 1983; McGill, 1962; McGill and Blight, 1963; McGill and Tucker, 1964). Following 10 daily bouts of fighting, victorious C57BL/6J males show enhanced sexual motivation, while this is reduced in males consistently losing the fights. After 20 days of fighting, however, victorious males also show reduced sexual motivation (Amikishieva and Ovsyukova, 2003; Dubrovina and Loskutova, 2003). Decreases in sexual motivation after prolonged fighting in C57BL/6J males have also been reported by Kudriavtseva et al. (1997). All of these findings suggest strong stress effects of prolonged fighting for C57BL/6J males, perhaps reflecting high levels of offensive aggression in males of this strain. An interpretation of enhanced responsivity to defeat is less clear here, as these males had not been defeated.

Maternal behavior

In an extensive review of behavioral tests involving comparison of C57BL/6J and C57BL/6JBy with as many as 12 other mouse strains (Le Roy et al., 1998), C57BL/6J mice had more pups living the day following birth, than eight strains, but less than four strains. Similarly, Table 4.1 of Silver (1995) compares C57BL/6J to 11 other strains on reproductive characteristics. C57BL/6Js are among the highest on litter size, number of litters, and relative fecundity. In a standardized test, C57BL/6J mothers had very low (i.e., good) scores on seconds to initial contact with pups (1st of 13); first retrieval of pups following contact (4th of 13), and time taken to place all pups in the nest (1st of 13) indicating rapid response to pups. They were the best of all 13 strains in terms of low rates of moving away from pups outside the nest without picking them up. However, they were intermediate (6 of 13 strains) on time from retrieval to placing the first pup in the nest, and among the lowest strains (3rd of 13) on time spent in the nest after placing the first pup in the nest. The size of the constructed nest was intermediate (5 of 7 strains evaluated).

Hennessy et al. (1980a) found no differences in maternal behavior between C57BL/6J and A/J dams responding to undisturbed infants. However, Hennessy et al. (1980b) reported that 19–20-day-old C57BL/6J but not A/J pups preferred the mother to a virgin female, a difference that might reflect either strain differences in pup response, or differential experience of maternal care. With reference to the latter possibility, C57BL/6J mothers have been extensively compared to BALB/cJ mothers in terms of licking/grooming their pups, and consistently shown to do this more than the latter (Anisman et al., 2001; Francis et al., 2003; Zaharia et al., 1996). Such experience may have enduring, but subtle, effects: Priebe et al. (2005) found that although C57BL/6J mothers showed more licking/grooming behaviors toward pups than BALB/cJ mothers, basal, but not stress-induced corticosterone levels increased for C57BL/6J adults that had been cross-fostered to BALB/cJ mothers, and anxiety-like behavior in the open field or elevated plus-maze did not change (see also Anisman et al., 1998).

Although the C57BL/6J mothers appear to show good – or at least rapid – response to pups (Le Roy et al., 1998) and more licking/grooming of pups than at least one other inbred strain (Priebe et al., 2005), a number of other studies comparing them to DBA/2Js suggest some problems. Female C57BL/6J mothers are less proficient nest builders (Broida and Svare, 1982). In agreement with the Le Roy's finding that C57BL/6J dams show little time on the nest after retrieving pups, Brown et al. (1999) reported that DBA/2J females spend more time with pups, performing maternal behaviors. Bond et al. (2002) summarize studies suggesting that C57BL/6J mothers have higher pup mortality in first litters (Brown et al., 1999; Nagasawa and Yani, 1973) and high rates of infanticide (Ghiraldi and Svare, 1986; Mann et al., 1983; Svare and Broida, 1982).

Social preferences

Comparison of inbred strains of mice on social approach demonstrated high levels of sociability in C57BL/6J. In an automated three-chambered choice test, C57BL/6J males and females, at ages from 6 weeks to 6 months, spent more time with a novel stranger mouse of either the same or a different strain than with a novel object (Bolivar et al., 2007; Moy et al., 2004, 2007, 2008; McFarlane et al., 2008; Nadler et al., 2004; Ryan et al., 2008; Yang et al., 2007a, 2007b, 2009). When given a choice between a familiar and a completely new stranger mouse, C57BL/6J showed a highly significant preference for social novelty (McFarlane et al., 2008; Moy et al. 2004, 2007; Nadler et al., 2004). Of the mouse strains in the top tier of the Mouse Phenome Project, C57BL/6J mice was one of only five strains that displayed both significant sociability and significant preference for social novelty, along with AKR/J, C57L/J, FVB/NJ, and PL/J (Moy et al., 2007, 2008). Lack of sociability and/or absence of preference for social novelty was reported for A/J, AKR/J, BALB/cByJ, BTBR T+tf/J, C3H/HeJ, C58/J, NOD/LtJ, NZB/BINJ, PERA/EiJ, SJL/J, SWR/J, and 129S1/SvImJ, and variable findings were obtained with DBA/2J (Moy et al., 2007, 2008).

Brodkin et al. (2004) reported that C57BL/6J mice showed the highest levels of social-approach behaviors of six mouse strains tested, in the three-chambered testing box situation. Female mice at ages 3 to 4 weeks were used in this study, to minimize sexual and aggressive motivations. In a subsequent study this preference was confirmed for 4 and 9-week-old male and female C57BL/6Js; these animals showing a stronger preference for chambers containing unfamiliar prepubescent mice of the same sex (Sankoorikal et al., 2006).

When tested in a novel chamber with brief (5 min) previous habituation, Pieper et al. (1997) found that in adult mice, housed together, DBA/2Js showed higher social preferences than C57BL/6Js. Avgustinovich et al. (2000) reported that BALB/cJ male pairs showed stronger approach tendencies toward each other, across a wire mesh barrier. Notably, both studies involving adult mice are more susceptible to approach based on sexual or aggressive motivations than are those with prepubescent females. C57BL/6J mice showed the highest level of social grooming of four strains investigated (Vale et al., 1971). However, such social grooming is also potentially related

Figure 21.1 A photograph of visible burrow system for mouse, which consists of the surface area (left side) and tunnels and three chambers (right side). The floor of surface area, tunnels, and chambers are covered by sawdust. Chambers are connected to the surface area via tunnels. Two chambers are connected to each other, and another chamber is unconnected to other chambers. Food hopper and water bottle are placed on the corner of the surface area.

to aggression, in that "aggressive grooming" or "barbering" is common in C57BL/6J mice (Kalueff et al., 2006).

Social relationships of C57BL/6J mice in the visible burrow system

Such comparisons of C57BL/6J mice with other strains, while relatively abundant in some areas, nonetheless fail to provide a full picture of social behavior in groups of these mice. In line with more ethological approaches, and ultimately to facilitate comparison with wild mouse data, we are examining behaviors of these mice in visible burrow systems (VBS) in which nest boxes are provided, leading to an open "surface" area containing food and water. The program will include both adult behaviors and developmental changes in behavior of mice born in the VBS, an environment that is enriched and provides a number of features of mouse demes. The following is a brief report of behaviors of adults, in 1M/3F groups, and of male–male groups as a function of age, from postnatal day (PND) 22 through PND 59.

Mixed-sex colonies

Subjects for the 1M/3F VBS study were four groups of C57BL/6J mice, 8 weeks of age at colony formation, each animal marked for identification with a commercial hair dye. This "single male" grouping was chosen because adult male mice, in contrast to rats, show lethal attacks on other adult males when females are present (Blanchard and Blanchard, 1989; Blanchard et al., 1995). Colonies (Figure 21.1) were housed in 86 × 21 × 26 cm rectangular metal bins, containing three, 12 × 7 × 6 cm, black Plexiglas chambers connected to the "surface" via clear Plexiglas tubes through the barrier wall dividing the chamber and "surface" areas. Two of the three chambers were connected to each other via a similar Plexiglas tube. The chambers and surface were covered by a clear Plexiglas sheet to permit videotaping, and the floor was covered by a 1 cm deep layer of sawdust. A video camera connected to a DVD recorder was mounted on the ceiling over the VBS.

All colonies were illuminated on a 12:12 light–dark cycle with lighting on between 07:00 and 19:00 hours. Infrared light illumination permitted videotaping during the dark period. Video-recordings were made for 4 hours per recording period, during five light and five dark periods at intervals over the 15 days of colony grouping. All findings noted below were significant for females at the $P < 0.05$ level. As many fewer males (four) were involved, these scores were generally not significant, although patterns were usually similar to those of the females. Where males and females showed different patterns these will be noted.

During the initial 4 hour (dark) period after colony formation, animals remained in the surface area during nearly 90% of observations. However, during the following light period, animals were usually found in chambers, and a preference (70% occupancy) for a particular chamber was evident in each colony at this time. The preferred chamber was always one of the two interconnected chambers. Given the tendency for many fossorial rodents to favor tunnel/chamber systems with multiple entry/exits to the surface, this preference appears to be based on the additional escape route possibility offered to both chambers by this dual tunnel system. The third chamber, with a single entry/exit tunnel, was occupied less than 5% of the time by any colony. This third, unconnected, chamber became the immediate choice of a latrine for each group.

Although a sleeping chamber was apparent during the first inactive (light) period, an increasing preference for location in the same preferred chamber over days was seen in the dark period as well. During the dark (active) phase, mice spent about 50–55% of their time in the open or surface area of the VBS, and virtually all the remaining time (about 40%) in the same favored chamber in which they slept. During the light phase, about 15% of time was spent on the surface and virtually all the rest (up to 90%) in their preferred chamber. Overall there were remarkably few sex differences in locations of animals within the VBS, during either the dark or the light cycle.

Huddling (lying in contact with another animal: Figure 21.2), a major behavior for both males and females, was higher during the light period compared to the dark period. Allogrooming (either given or received) was not different between lighting phases or over days in the habitat. Self-grooming was more common during the dark period, but did not change over days. "Alone" scores were more frequent during the dark periods.

Approaches by one mouse to another were a particular focus of attention. All approaches (i.e., approaches to back or front of the other animal, for both males and females) were very common during the dark phase of day 1, declining substantially thereafter. Approaches were an order of magnitude more common during the dark compared to the light phases. The responses of the approached animal differed by type of

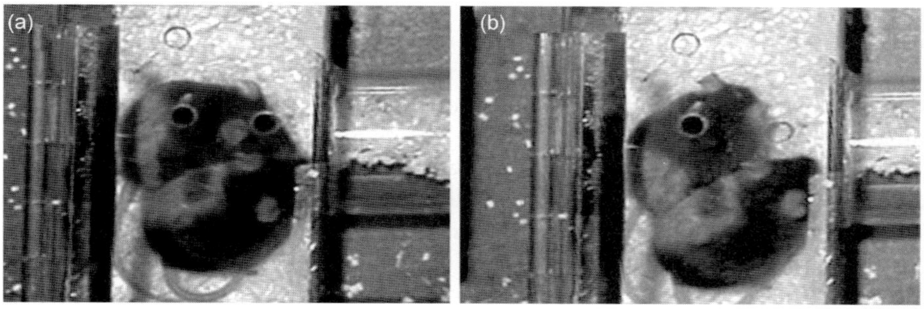

Figure 21.2 Two males of C57BL/6J mice huddled together in the preferred chamber (a and b).

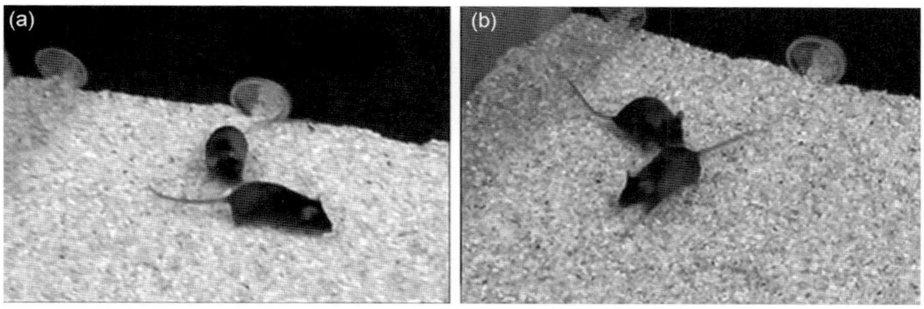

Figure 21.3 Photographs of approach behavior from the back of an approached animal in male C57BL/6J mice (a and b).

Figure 21.4 Photographs of approach behavior from the front of an approached animal in male C57BL/6J mice (a and b).

approach. Approaches to the back (Figure 21.3) were followed most commonly by flight of the approached animal (154 of 330 approaches on day 1), whereas flight was a much less common response (30 of 277 approaches) to approach from the front (Figure 21.4). Flight as a consequence of approach declined for both types of approach over time, but the difference in responsivity to front or back approaches continued through the last day of videotaping. On day 15, 14% of female approaches to the back of a female led to flight, while only 3% of approaches to the front produced this consequence. Although male–female differences were not analyzed statistically, due to the paucity of the former animals, the patterns for all approaches, i.e., male to female, female to male, female to female, appeared to have similar outcomes in terms of flight.

Approaches during the active phase, from the rear, by males, were particularly high on day 1 of colony formation. This is likely related to sexual behavior as mounting by males was seen only during this period. In agreement with VBS data on rats (Blanchard et al., 2001) this suggests that mixed-sex VBS grouping may bring female rodents into estrus. Most female mice in this study became pregnant during early stages of VBS housing, which may account for lack of observations of copulation during succeeding test days. However, females showed flight to female as well as male approaches from behind, suggesting that these flight responses may have been part of a defensive response to such approach, rather than a specific attempt to avoid copulation. In this context, the *lack* of female flight to frontal approach, by male or female conspecifics, suggests that such approaches are regarded by the approached animal as prosocial rather than threatening.

Social behaviors of VBS males over development

In the developmental study, subjects were male C57BL/6J mice. They had been born and reared in standard laboratory cages with their dams until 22 days of age, then removed and placed without their dams in VBS chambers identical to those of the mixed-sex groups. Each of 4 colony groups contained 3 sibling males, and these groups were maintained for 37 days, until PND 59 for the subjects. Videotaping schedules, measures taken, and data analyses were similar to those of the mixed-sex study, except that the videotaping/analysis was repeated at intervals throughout the period.

Overall use of space within the VBS was very similar to that of the mixed-sex groups. Preference for one of the two connected chambers was strong on day 1, and continued throughout development; during inactive (light) periods, about 90% of time was spent in this chamber, whereas the mice were on the surface about 50% of time during the dark period. Huddling was by far the most prevalent activity during the light period, and was strong also (about 40% of time) during the dark period. Huddling in the preferred chamber during the dark period of day 1 of colony grouping was especially strong. However, this may represent a response to the removal of the dam rather than a genuine age effect, as this declined by the following day (PND 23) to the levels at which it would be maintained throughout the grouping period. Feeding, allogrooming, and self-grouping showed no pronounced changes in frequency over time in the colony situation.

Except for the dark period of day 1 the number of approaches did not vary with time in the colony/age of animals. The day 1 dark period data, with reduced approaches, likely reflects that with 75% of time spent huddling, there was less time available for approaches to occur. Chase, flight, follow, and struggle showed no change over time.

As with the mixed-sex groups, the consequences of approach to the back, or approach to the front, were evaluated for each animal. Approaches to the front elicited no response on about 95% of occasions, with the approached animal fleeing on less than 5%. In striking contrast, only about 37% of approaches from the rear elicited no response, with flight the most common reaction of the approached animal (circa 60% of approaches led to flight).

This pattern, while very similar to that for the mixed-sex colonies, reflects an even greater difference in the effects of rear versus front approaches. Findings of very few struggles (<0.1% of approaches led to struggles on any day) suggest that relationships among these young sibling males were essentially amicable. It is notable that rear approaches (total of 112) were substantially fewer than front approaches (total of 446) throughout development, consonant with an interpretation that the front approaches are amicable or eusocial, while the rear approaches are more threatening. Rear approaches did show a (non-significant) tendency to be most frequent on PNDs 31 and 39, suggesting that some of these may have reflected play behavior. However, they declined only slightly after PNDs 47–59, ages when play fighting is sharply lower in mice (Pellis and Pasztor, 1999; Walker and Byers, 1991). These remaining approaches, like those in adult male and female mice, strongly suggest that play is not the only motivation for approach to the back; while the flight responses seen in the approached animals, both young and adult, suggest that these animals interpret back approaches as potentially threatening. Conversely, the overwhelming lack of flight responses to frontal approaches indicate that these are interpreted by the approached animal as amicable.

Discussion

In studies of the social behaviors of mixed sex groups of adult C57BL/6J, or in developmental studies of male C57BL/6J mice reared from PND 22 to PND 59 in a VBS (Arakawa et al., unpublished observations), all groups tended to show early and strong cohesiveness in terms of huddling in a specific preferred chamber during the inactive (lights on) period. Huddling and another behavior, front approach to a conspecific, show promise as indices of amicable sociality. The behavior of the mouse approached from the front suggests that this animal regards such approaches as non-threatening, whereas approaches from the rear tend to produce flight in the approached mouse. These and other in-progress VBS studies of C57BL/6J, including developmental studies, may provide additional means of analyzing amicable behaviors in this important mouse strain.

References

Amikishieva, A.V. and Ovsyukova, M.V. (2003) Effects of alternative social experience on the sexual function of male mice. *Bull Exp Biol Med* 136: 607–610.

Anderson, P.K. (1961) Density, social structure, and nonsocial environment in house-mouse populations and the implications for regulation of numbers. *Trans N Y Acad Sci* 23: 447–451.

Anisman, H., Hayley, S., Kelly, O., Borowski, T., and Merali, Z. (2001) Psychogenic, neurogenic, and systemic stressor effects on plasma corticosterone and behavior: mouse strain-dependent outcomes. *Behav Neurosci* 115: 443–454.

Anisman, H., Zaharia, M.D., Meaney, M.J., and Merali, Z. (1998) Do early-life events permanently alter behavioral and hormonal responses to stressors? *Int J Dev Neurosci* 16: 149–164.

Arakawa, H., Arakawa, K., Blanchard, D.C., and Blanchard, R.J. (2007) Scent marking behavior in male C57BL/6J mice: sexual and developmental determination. *Behav Brain Res* 182: 73–79.

Arakawa, H., Arakawa, K., Blanchard, D.C., and Blanchard, R.J. (2008a) A new test paradigm for social recognition evidenced by urinary scent marking behavior in C57BL/6J mice. *Behav Brain Res* 190: 97–104.

Arakawa, H., Arakawa, K., Blanchard, D.C., and Blanchard, R.J. (2009) Social features of scent-donor mice modulate scent marking of C57BL/6J recipient males. *Behav Brain Res* 205: 138–145.

Arakawa, H., Blanchard, D.C., Arakawa, K., Dunlap, C., and Blanchard, R.J. (2008b) Scent marking behavior as an odorant communication in mice. *Neurosci Biobehav Rev* 7: 1236–1248.

Avgustinovich, D.F., Gorbach, O.V., and Kudryavtseva, N.N. (1997) Comparative analysis of anxiety-like behavior in partition and plus-maze tests after agonistic interactions in mice. *Physiol Behav* 61: 37–43.

Avgustinovich, D.F. and Kovalenko, I.L. (2004) Development of pathological behavior in C57BL/6J female mice under the long-lasting psychoemotional influence. *Ross Fiziol Zh Im I M Sechenova* 90: 1324–1336.

Avgustinovich, D.F., Lipina, T.V., Bondar, N.P., Alekseyenko, O.V., and Kudryavtseva, N.N. (2000) Features of the genetically defined anxiety in mice. *Behav Genet* 30: 101–109.

Baker, A.E.M. (1981) Gene flow in house mice: introduction of a new allele into free-living population. *Evolution* 35: 243–258.

Barnett, S.A. and Stoddart, R.C. (1969) Effects of breeding in captivity on conflict among wild rats. *J Mammal* 50, 321–325.

Baumans, V. (1999) The laboratory mouse. In Poole, T. (ed.), *The UFAW Handbook on the Care and Management of Laboratory Animals*, Vol. 1, 7th edn. Blackwell Publishing, London, pp. 282–312.

Becker, S.D. and Hurst, J.L. (2009) Female behaviour plays a critical role in controlling murine pregnancy block. *Proc Biol Sci* 276: 1723–1729.

Berry, R.J. (ed.) (1981) *Biology of the House Mouse*. Symposium of the Zoological Society of London, No. 47. Academic Press, London.

Berry, R.J. and Scriven, P.N. (2005) The house mouse: a model and motor for evolutionary understanding. *Biol J Linn Soc* 84: 335–347.

Blanchard, D.C., Spencer, R.L., Weiss, S.M., Blanchard, R.J., McEwen, B., and Sakai, R.R. (1995) Visible burrow system as a model of chronic social stress: behavioral and neuroendocrine correlates. *Psychoneuroendocrinol* 20: 117–134.

Blanchard, R.J. and Blanchard, D.C. (1989) Attack and defense in rodents as ethoexperimental models for the study of emotion. *Prog Neuropsychopharmacol Biol Psychiatry* 13 (Suppl): S3–14.

Blanchard, R.J., Dulloog, L., Markham, C., Nishimura, O., Compton, J.N., Jun, A., et al. (2001) Sexual and aggressive interactions in a visible burrow system with provisioned burrows. *Physiol Behav* 72: 245–254.

Bolivar, V.J., Walters, S.R., and Phoenix, J.L. (2007) Assessing autism-like behavior in mice: variations in social interactions among inbred strains. *Behav Brain Res* 176: 21–26.

Bond, T.L., Neumann, P.E., Mathieson, W.B., and Brown, R.E. (2002) Nest building in nulligravid, primigravid and primiparous C57BL/6J and DBA/2J mice (*Mus musculus*). *Physiol Behav* 75: 551–555.

Branchi, I., D'Andrea, I., Flore, M., Di Fausto, V., Aloe, L., et al. (2006b) Early social enrichment shapes social behavior and nerve growth factor and brain-derived neurotrophic factor levels in the adult mouse brain. *Biol Psychiatry* 60: 690–696.

Branchi, I., D'Andrea, I., Sietzema, J., Fiore, M., Di Fausto, V., Aloe, L., et al. (2006a) Early social enrichment augments adult hippocampal BDNF levels and survival of BRDU-positive cells while increasing anxiety- and "depression"-like behavior. *J Neurosci Res* 83: 965–973.

Brodkin, E.S., Hagemann, A., Nemetski, S.M., and Silver, L.M. (2004) Social approach-avoidance behavior of inbred mouse strains towards DBA/2 mice. *Brain Res* 1002: 151–157.

Broida, J. and Svare, B. (1982) Strain-typical patterns of pregnancy-induced nestbuilding in mice: maternal and experiential influences. *Physiol Behav* 25: 153–157.

Bronson, F.H. (1979) The reproductive ecology of the house mouse. *Q Rev Biol* 54: 265–299.

Brothwell, D.R. (1981) The Pleistocene and Holocene ecology of the house mouse and related species. *Symp Zool Soc London* 47: 1–13.

Brown, R.E., Mathieson, W.B., Stapleton, J., and Neumann, P.E. (1999) Maternal behavior in female C57BL/6J and DBA/2J inbred mice. *Physiol Behav* 67: 599–605.

Butler, R.G. (1980) Population size, social behaviour, and dispersal in house mice: a quantitative investigation. *Anim Behav* 28: 78–85.

Crawley, J.N., Belknap, J.K., Collins, A., Crabbe, J.C., Frankel, W., Henderson, N., et al. (1997) Behavioral phenotypes of inbred mouse strains: Implications and recommendations for molecular studies. *Psychopharmacol* 132: 107–124.

Crowcroft, P. and Rowe, F.P. (1963) Social organisation and territorial behaviour in the wildhouse mouse (*Mus musculus L.*). *Proc Zool Soc Lond B* 140: 517–531.

Cucchi, T., Vigne, J.-D., and Auffray, J.-C. (2005) First occurrence of the house mouse (*Mus musculus domesticus* Schwarz and Schwarz, 1943) in the western Mediterranean: a zooarchaeological revision of sub-fossil house mouse occurrences. *Biol J Linn Soc* 84: 429–445.

Desjardins, C., Maruniak, J.A., and Bronson, F.H. (1973) Social rank in the house mouse: differentiation revealed by ultraviolet visualization of urinary marking patterns. *Science* 182: 939–941.

Drickamer, L.C., Gowaty, P.A., and Holmes, C.M. (2000) Free female mate choice in house mice affects reproductive success and offspring viability and performance. *Anim Behav* 59: 371–378.

Dubrovina, N.I. and Loskutova, L.V. (2003) The effects of novelty and behavioral stereotype on the development of amnesia in mice. *Neurosci Behav Physiol* 33: 789–793.

D'Udine, B. and Manning, A. (1983) Rooting and the development of male sexual behavior in two inbred mouse strains. *Dev Psychobiol* 16: 311–322.

Ensminger, A.L. and Meikle, D.B. (2005) Effects of male kinship and agonistic behaviour on reproduction and odour preferences of female house mice, *Mus domesticus*. *Anim Behav* 69: 1147–1155.

Francis, D.D., Szegda, K., Campbell, G., Martin, W.D., and Insel, T.R. (2003) Epigenetic sources of behavioral differences in mice. *Nat Neurosci* 6: 445–446.

Frynta, D., Munclinger, P., Slábová, M., Volfová, R., and Třešťíková, H. (2005) Aggression and commensalism in house mouse: a comparative study across Europe and Near East. *Aggress Behav* 31: 283–293.

Gerlach, G. (1996) Emigration mechanisms in feral house mice: a laboratory investigation of the influence of social structure, population density, and aggression. *Behav Ecol Sociobiol* 39: 159–170.

Ghiraldi, L.L. and Svare, B. (1986) Postpubertal isolation decreases infanticide and elevates parental care in C57BL/6J male mice. *Physiol Behav* 36: 59–62.

Gray, S.J. and Hurst, J.L. (1997) Behavioural mechanisms underlying the spatial dispersion of commensal *Mus domesticus* and grassland *Mus spretus*. *Anim Behav* 53: 511–524.

Gray, S.J., Jensen, S.P., and Hurst, J.L. (2000) Structural complexity of territories: preference, use of space and defence in commensal house mice, *Mus domesticus*. *Anim Behav* 60: 765–772.

Gray, S.J., Jensen, S.P., and Hurst, J.L. (2002) Effects of resource distribution on activity and territory defense in house mice, *Mus domesticus. Anim Behav* 63: 531–539.

Hennessy, M.B., Coyle, S., and Levine, S. (1980a) Altered plasma corticosterone responses during lactation in mice of two inbred strains. *Neuroendocrinology* 31: 369–374.

Hennessy, M.B., Li, J., Lowe, E.L., and Levine, S. (1980b) Maternal behavior, pup vocalizations, and pup temperature changes following handling in mice of two inbred strains. *Dev Psychobiol* 13: 573–584.

Homanics, G.E., Quinlan, J.J., and Firestone, L.L. (1999) Pharmacologic and behavioral responses of inbred C57BL/6J and strain 129/SvJ mouse lines. *Pharmacol Biochem Behav* 63: 21–26.

Hurst, J.L. (1986) Mating in free-living wild house mice (*Mus. domesticus*). *J Zool Ser* 210: 623–628.

Hurst, J.L. (1990a) Urine marking in populations of wild house mice *Mus domesticus Rutty*. I. Communication between males. *Anim Behav* 40: 209–222.

Hurst, J.L. (1990b) Urine marking in populations of wild house mice *Mus domesticus Rutty*. II. Communication between females. *Anim Behav* 40: 223–232.

Hurst, J.L. (1990c) Urine marking in populations of wild house mice *Mus domesticus Rutty*. III. Communication between the sexes. *Anim Behav* 40: 233–243.

Hurst, J.L. (2009) Female recognition and assessment of males through scent. *Behav Brain Res* 200: 295–303.

Jensen, S.P., Gray, S.J., and Hurst, J.L. (2003) How does habitat structure affect activity and use of space among house mice? *Anim Behav* 66: 239–250.

Jensen, S.P., Gray, S.J., and Hurst, J.L. (2005) Excluding neighbours from territories: effects of habitat structure and resource distribution. *Anim Behav* 69: 785–795.

Kalueff, A.V., Minasyan, A., Keisala, T., Shah, Z.H., and Tuohimaa, P. (2006) Hair barbering in mice: implications for neurobehavioural research. *Behav Processes* 71: 8–15.

König, B. (1994) Fitness effects of communal rearing in house mice: the role of relatedness versus familiarity. *Anim Behav* 48: 1449–1457.

Krackow, S. (2003) Motivational and heritable determinants of dispersal latency in wild male house mice (*Mus musculus musculus*). *Ethology* 109: 671–689.

Kudriavtseva, N.N. and Bakshtanovskaia, I.V. (1991) The neurochemical control of aggression and submission. *Zh Vyssh Nerv Deiat Im I P Pavlova* 41: 459–466.

Kudriavtseva, N.N., Bakshtanovskaia, I.V., and Avgustinovich, D.F. (1997) The effect of the repeated experience of aggression in daily confrontations on the individual and social behavior of male mice. *Zh Vyssh Nerv Deiat Im I P Pavlova* 47: 86–97.

Latham, N. and Mason, G. (2004) From house mouse to mouse house: The behavioural biology of free-living *Mus musculus* and its implications in the laboratory. *Appl Anim Behav Sci* 86: 261–289.

Le Roy, I., Pothion, S., Mortaud, S., Chabert, C., Nicolas, L., Cherfouh, A., et al. (2000) Loss of aggression, after transfer onto a C57BL/6J background, in mice carrying a targeted disruption of the neuronal nitric oxide synthase gene. *Behav Genet* 30: 367–373.

Le Roy, I., Roubertoux, P.L., Jamot, L., Maarouf, F., Tordjman, S., Mortaud, S., et al. (1998) Neuronal and behavioral differences between *Mus musculus domesticus* (C57BL/6JBy) and *Mus musculus castaneus* (CAST/Ei). *Behav Brain Res* 95: 135–142.

Lidicker, W.Z. (1976) Social behaviour and density regulation in house mice living in large enclosures. *J Anim Ecol* 45: 677–697.

Lloyd, J. (1975) Social structure and reproduction in two freely-growing populations of house mice. (*Mus musculus L.*). *Anim Behav* 23: 413–420.

Logue, S.F., Owen, E.H., Rasmussen, D.L., and Wehner, J.M. (1997) Assessment of locomotor activity, acoustic and tactile startle, and prepulse inhibition of startle in inbred mouse strains and F1 hybrids: Implications of genetic background for single gene and quantitative trait loci analyses. *Neurosci* 80: 1075–1086.

Mann, M.A., Kinsley, C., Broida, J., and Svare, B. (1983) Infanticide exhibited by female mice: genetic, developmental and hormonal influences. *Physiol Behav* 30: 697–702.

Manning, C.J., Wakeland, E.K., and Potts, W.K. (1992) Communal nesting patterns in mice implicate MHC genes in kin recognition. *Nature* 360: 581–583.

McFarlane, H.G., Kusek, G.K., Yang, M., Phoenix, J.L., Bolivar, V.J., and Crawley, J.N. (2008) Autism-like behavioral phenotypes in BTBR T+tf/J mice. *Genes Brain Behav* 7: 152–163.

McGill, T.E. (1962) Sexual behavior in three inbred strains of mice. *Behaviour* 19: 341–350.

McGill, T.E. and Blight, W.C. (1963) Effects of genotype on the recovery of sex drive in the male mouse. *J Comp Physiol Psychol* 56: 887–888.

McGill, T.E. and Tucker, G.R. (1964) Genotype and sex drive in intact and in castrated male mice. *Science* 145: 514–515.

Mogil, J.S. and Wilson, S.G. (1997) Nociceptive and morphine antinociceptive sensitivity of 129 and C57BL/6 inbred mouse strains: Implications for transgenic knock-out studies. *Eur J Pain* 1: 293–297.

Mondragón, R., Mayagoitia, L., López-Luján, A., and Díaz, J.-L. (1987) Social structure features in three inbred strains of mice, C57BL/6J, BALB/cJ, and NIH: a comparative study. *Behav Neural Biol* 47: 384–391.

Mouse Genome Sequencing Consortium (2002) Initial sequencing and comparative analysis of the mouse genome. *Nature* 420: 520–562.

Moy, S.S., Nadler, J.J., Perez, A., Barbaro, R.P., Johns, J.M., Magnuson, T.R., et al. (2004) Sociability and preference for social novelty in five inbred strains: an approach to assess autistic-like behavior in mice. *Genes Brain Behav* 3: 287–302.

Moy, S.S., Nadler, J.J., Young, N.B., Nonneman, R.J., Segall, S.K., Andrade, G.M., et al. (2008) Social approach and repetitive behavior in eleven inbred mouse strains. *Behav Brain Res* 191: 118–129.

Moy, S.S., Nadler, J.J., Young, N.B., Perez, A., Holloway, P., Barbaro, R.P., et al. (2007) Mouse behavioral tasks relevant to autism: phenotypes of ten inbred strains. *Behav Brain Res* 176: 4–20.

Nadler, J.J., Moy, S.S., Dold, G., Trang, D., Simmons, N., Perez, A., et al. (2004) Automated apparatus for quantitation of social approach behaviors in mice. *Genes Brain Behav* 3: 303–314.

Nagasawa, H. and Yani, R. (1973) Effects of adrenalectomy and-or deficiency of pituitary prolactin secretion on initiation and maintenance of lactation in mice. *J Endocrinol* 58: 67–73.

Ogawa, S., Robbins, A., Kumar, N., Pfaff, D.W., Sundaram, K., and Bardin, C.W. (1996) Effects of testosterone and 7α-methyl-10-nortestosterone (MENT) on sexual and aggressive behaviors in two inbred strains of male mice. *Horm Behav* 30: 74–84.

Palanza, P., Della Seta, D., Ferrari, P.F., and Parmigiani, S. (2005) Female competition in wild house mice depends upon timing of female/male settlement and kinship between females. *Anim Behav* 69: 1259–1271.

Pellis, S.M. and Pasztor, T.J. (1999) The developmental onset of a rudimentary form of play fighting in C57 mice. *Dev Psychobiol* 34: 175–182.

Pieper, J.O., Forester, D.C., and Elmer, G.I. (1997) Mice show strain differences in social affiliation: implication for open field behavior. *Ann N Y Acad Sci* 807: 552–555.

Pocock, M.J.O., Hauffe, H.C., and Searle, J.B. (2005) Dispersal in house mice. *Biol J Linn Soc* 84: 565–583.

Priebe, K., Brake, W.G., Romeo, R.D., Sisti, H.M., Mueller, A., McEwen, B.S., et al. (2005) Maternal influences on adult stress and anxiety-like behavior in C57BL/6J and BALB/CJ mice: a cross-fostering study. *Dev Psychobiol* 47: 398–407.

Randall, C. (ed.) (1999) *Vertebrate Pest Management: A Guide for Commercial Applicators (Extension Bulletin E-2050).* Michigan State University, East Lansing, MI, USA.

Rich, T.J. and Hurst, J.L. (1998) Scent marks as reliable signals of the competitive ability of mates. *Anim Behav* 56: 727–735.

Rich, T.J. and Hurst, J.L. (1999) The competing countermarks hypothesis: reliable assessment of competitive ability by potential mates. *Anim Behav* 58: 1027–1037.

Rusu, A.S. and Krackow, S. (2004) Kin-preferential cooperation, dominance-dependent reproductive skew, and competition for males in communally nesting female house mice. *Behav Ecol Sociobiol* 56: 298–305.

Rusu, A.S. and Krackow, S. (2005) Agonistic onset marks emotional changes and dispersal propensity in wild house mouse males (*Mus domesticus*). *J Comp Psychol* 119: 58–66.

Ryan, B.C., Young, N.B., Moy, S.S., and Crawley, J.N. (2008) Olfactory cues are sufficient to elicit social approach behaviors but not social transmission of food preference in C57BL/6J mice. *Behav Brain Res* 193: 235–242

Sankoorikal, G.M., Kaercher, K.A., Boon, C.J., Lee, J.K., and Brodkin, E.S. (2006) A mouse model system for genetic analysis of sociability: C57BL/6J versus BALB/cJ inbred mouse strains. *Biol Psychiatry* 59: 415–423.

Sayler, A. and Salmon, M. (1969) Communal nursing in mice: influence of multiple mothers on the growth of the young. *Science* 164: 1309–1310.

Silver, L.M. (1995) *Mouse Genetics*. Oxford University Press, Oxford.

Singleton, G.R., Brown, P.R., Pech, R.P., Jacob, J., Mutze, G.J., and Krebs, C.J. (2005) One hundred years of eruptions of house mice in Australia: a natural biological curio. *Biol J Linn Soc* 84: 617–627.

Southern, H.N. (1954) *Control of Rats and Mice*. Clarendon Press, Oxford.

Sutherland, D.R., Spencer, P.B., Singleton, G.R., and Taylor, A.C. (2005) Kin interactions and changing social structure during a population outbreak of feral house mice. *Mol Ecol* 14: 2803–2814.

Svare, B. and Broida, J. (1982) Genotypic influences on infanticide in mice: Environmental, situational and experiential determinants. *Physiol Behav* 28: 171–175.

Vale, J.R., Vale, C.A., and Harley, J.P. (1971) Interaction of genotype and population number with regard to aggressive behavior, social grooming, and adrenal and gonadal weight in male mice. *Commun Behav Biol* 6: 209–221.

Van Oortmerssen, G.A. and Bakker, T.C. (1981) Artificial selection for short and long attack latencies in wild *Mus musculus domesticus*. *Behav Genet* 11: 115–126.

Van Oortmerssen, G.A. and Busser, J. (1989) Studies in wild house mice: disruptive selection on aggression as a possible force in evolution. In Brain, P.F., Mainardi, D., and Parmigiani, S. (eds.), *House Mouse Aggression: A Model for Understanding the Evolution of Social Behaviour*. Harwood Academic, Chur, Switzerland, pp. 87–117.

Veenema, A.H., Koolhaas, J.M., and de Kloet, E.R. (2004) Basal and stress-induced differences in HPA axis, 5-HT responsiveness, and hippocampal cell proliferation in two mouse lines. *Ann N Y Acad Sci* 1018: 255–265.

Wade, C.M., Kulbokas, E.J., III, Kirby, A.W., Zody, M.C., Mullikin, J.C., Lander, E.S., et al. (2002) The mosaic structure of variation in the laboratory mouse genome. *Nature* 420: 574–578.

Walker, C. and Byers, J.A. (1991) Heritability of locomotor play in house mice, *Mus domesticus*. *Anim Behav* 42: 891–897.

Wilkinson, G.S. and Baker, A.E.M. (1988) Communal nesting among genetically similar house mice. *Ethology* 77: 103–114.

Wotjak, C.T. (2004) Of mice and men: potentials and caveats of behavioural experiments with mice. *B I F Futura* 19: 158–169.

Yang, M., Clarke, A., and Crawley, J.N. (2009) Postnatal lesion evidence against a primary role for the corpus callosum in mouse sociability. *Eur J Neurosci* 29: 1663–1677.

Yang, M., Scattoni, M.L., McFarlane, H.G., Zhodzishsky, V., Chen, T., and Crawley, J.N. (2007a) Social approach behaviors are similar on conventional versus reverse lighting cycles, and in replications across cohorts, in BTBR T+ tf/J, C57BL/6J, and vasopressin receptor 1B mutant mice. *Front Behav Neurosci* 1: 1–9.

Yang, M., Zhodzishsky, V., and Crawley, J.N. (2007b) Social deficits in BTBR T+tf/J mice are unchanged by cross-fostering with C57BL/6J mothers. *Int J Dev Neurosci* 25: 515–521.

Young, L.J., Pitkow, L.J., and Ferguson, J.N. (2002) Neuropeptides and social behavior: animal models relevant to autism. *Mol Psychiatry* 7: S38–39.

Zaharia, M.D., Kulczycki, J., Shanks, N., Meaney, M.J., and Anisman, H. (1996) The effects of early postnatal stimulation on Morris water-maze acquisition in adult mice: genetic and maternal factors. *Psychopharmacol* 128: 227–239.

Chapter 22

Mouse sex: sexual differentiation and sexual behavior in *Mus musculus*

Lee Niel and Douglas A. Monks

My brain? That's my second favorite organ.
Woody Allen

Introduction

Why study mouse sex? For some of us, the answer is obvious; sex is inherently interesting and entertaining to study! Even for those who don't find sex particularly fascinating, a minimal working knowledge of mouse sex is useful in propagating mouse strains. This chapter has two main aims: first, to convince investigators of the importance of including a characterization of sex in standard behavioral phenotyping of mouse strains; and second, to encourage investigators to consider the ecological validity of the studies they conduct on mouse sex. This chapter is intended to provide an overview of the theoretical considerations in this field of study, and to consider the available information concerning the naturalistic aspects of mouse sex and the underlying genetic diversity. As we shall see, there have been lamentably few studies characterizing sex in inbred strains or using population genetics. Similarly, certain areas of mouse sex have received insufficient attention (e.g., feminine sexual behaviors). Nonetheless, there is sufficient information to warrant a review, and missing information can be supplemented where necessary by studies in rats, in which sex has been more intensively studied. Studies of specific genetic models (transgenics and gene-targeted animals) and methodological considerations will be covered in subsequent chapters in Volumes II and III of this Handbook and so these omissions are intentional.

Sex has a dual meaning: referring both to sexual behavior and to the categorization of males and females. The intimate association of these two concepts is warranted, as sexual behavior is ultimately a manifestation of sexual reproduction,[1] which in mice occurs between male and female individuals, whose sex is determined genetically. The study of sex therefore includes both sexual behavior and sexual differentiation (i.e., the developmental processes that result in male and female individuals). Sex deserves the attention of behavior geneticists as it provides an example of specific and biologically relevant genetic contributions to individual differences in the nervous system and behavior. Indeed, one could argue that a full characterization of a mouse strain requires consideration of both types of individuals that are required for its propagation and how they interact with each other.

One of the powerful advantages of studying sex is that we understand and can readily manipulate a major physiological variable that influences its expression. Steroid hormones produced by the gonads (ovaries in females or testes in males) are necessary for (i.e., activate) normal sexual behavior. Additionally, production of gonadal steroid hormones around the time of birth determines to what extent the capacity for male-typical or female-typical sexual behavior is retained in adulthood (i.e., perinatal hormones organize adult behavior). Any study of mouse sexual behavior must consider the relationships between perinatal gonadal steroid hormones and sexual differentiation of brain and behavior and also between adult gonadal steroid hormone production and sexual behavior. We will begin by considering the development of sex differences (i.e., sexual differentiation) in special reference to the nervous system and behavior.

Sexual differentiation

The study of mammalian sexual differentiation has a sufficiently long history to have developed a central dogma. This dogma holds that individuals begin as bipotential embryos that become differentiated along male or female lines according to a hierarchical process which is initiated by a genetic sex determining event (see Morris et al., 2004, for a recent review). According to this model, sex is determined by the presence or absence of a Y chromosome, which contains genetic information necessary and sufficient for the formation of a testis from a bipotential gonad. Recent evidence has indicated that the

[1] Which is not to say that mice (or other animals) engage in sexual behavior with the express purpose of reproduction, but rather there is every reason to believe that mice have sex because this activity is rewarding to individuals, as are many other behaviors that promote inclusive fitness (see Hull et al., 2002, for further discussion).

Behavioral Genetics of the Mouse: Volume I. *Genetics of Behavioral Phenotypes*, eds. Wim E. Crusio, Frans Sluyter, Robert T. Gerlai, and Susanna Pietropaolo. Published by Cambridge University Press.

Y chromosome gene *Sry* is both necessary and largely sufficient for the formation of testes, although additional sex chromosome genes appear to be necessary for normal gonadal differentiation (Koopman et al., 1991). In the absence of a Y chromosome (or *Sry*), ovaries are formed.

As a rule, testes secrete more testosterone into circulation than do ovaries[2]. Males therefore have relatively high levels of blood testosterone during development and females relatively low levels. This sexually differentiated pattern of testosterone production resulting from gonadal differentiation largely accounts for known sexual dimorphisms. The timing of testosterone exposure is also important, and there is a so-called "critical period" during which testosterone can act to promote male-typical developmental outcomes. It is important to acknowledge that different critical periods exist for different sexual dimorphisms, but the majority of these fall within the period immediately preceding and immediately following birth (perinatal period).

Males have two pronounced perinatal peaks in testosterone production, one several days prior to birth and another on the first day of life. Testosterone produced by males *in utero* influences not only their own development, but also influences the development of their female littermates, who are exposed to biologically significant doses of their brothers' testosterone (Gandelman et al., 1977). These perinatal testosterone surges in males are of central importance to the sexual differentiation of the nervous system and behavior, although somewhat surprisingly, testosterone influences brain sex primarily via aromatization to an estrogen (Fisher et al., 1998; Matsumoto et al., 2003; Ogawa et al., 1998). In contrast, androgenic actions of testosterone or its metabolites are essential for masculine genital differentiation and for several spinal cord dimorphisms (Morris et al., 2004). It is unclear whether there are additional actions of androgens per se (rather than estrogenic metabolites of testosterone) in brain differentiation, although there is some evidence to suggest this may be the case (Burns-Cusato et al., 2004; Sato et al., 2004).

As a consequence of sexual differentiation of the nervous system, males retain the capacity to express male typical sexual behavior in adulthood (i.e., they are masculinized) and lose the capacity to express female-typical sexual behaviors in adulthood (they are defeminized). Removing testes (and therefore reducing blood testosterone) from neonatal males results in more female-typical behavior (demasculinzation and feminization), and administering exogenous testosterone to females during this perinatal critical period results in more male-typical behavior (masculinization and defeminization). It is worth mentioning that these experimental treatments are rarely successful in completely sex-reversing behavior, which likely reflects insufficient knowledge of the required dosing regimens (Burns-Cusato et al., 2004). These masculine and feminine dimensions have been defined because they are dissociable within individuals, as masculinization and feminization result from different endocrine actions of testosterone during the perinatal critical period. For example, genetic females may be experimentally masculinized (i.e., they will mount a receptive stimulus female) but not defeminized (i.e., they will show feminine levels of receptivity when paired with a stimulus male) with regard to adult sexual behavior (Kudwa et al., 2005).

The masculinization/feminization terminology is somewhat confusing as it refers to the *potential* for behavior in adult animals, and indeed the expression of sexual behaviors in adulthood requires activation by sex-typical gonadal steroid hormones (i.e., testosterone for male-typical behaviors and estrogen and progesterone for female-typical behaviors). The classic assay for masculinization and feminization of the nervous system is therefore to gonadectomize adult animals, treat them with sex-typical hormones (either testosterone or estrogen and progesterone), and then measure masculine or feminine sexual behavior. Masculine and feminine sexual behaviors are described more fully in the sexual behavior section.

The central dogma of sexual differentiation described above, wherein genetic sex results in gonadal sex, which results in hormonal sex, which results in neural and behavioral sex, has recently been questioned in light of evidence for effects of the sex chromosomes that do not obviously result from alterations in gonadal steroid hormones (see (Arnold and Chen, 2009) for a comprehensive review). For example, sex differences in gene expression and phenotype of hypothalamic neurons can be demonstrated prior to obvious sex differences in circulating gonadal hormones (Carruth et al., 2002; Dewing et al., 2003), and some sexually differentiated features of mice whose gonadal sex is dissociated from their genetic sex via an autosomal Sry transgene correspond to genetic rather than gonadal sex (De Vries et al., 2002). Perhaps most compelling are observed differences in X-chromosome gene expression that exist as a consequence of X chromosome inactivation (see Lyon, 2002 for review). Whereas heterozygosity of X chromosomes and consequent variability due to tissue mosaicism of X-chromosome gene expression between XX individuals is not a factor in inbred strains, this is potentially a very significant factor in natural populations. Furthermore, there is good reason to believe that dosage compensation resulting from X-chromosome inactivation (XCI) is imperfect, and there is significant evidence indicating sex differences in the expression of X-chromosome genes in the nervous systems of mice (Xu et al., 2005).

Despite this new-found appreciation for the possibility of direct genetic effects, it is practically difficult to experimentally disentangle gonadal hormonal effects from sex chromosome effects with any confidence. For example, even

[2] It is a common misconception that testosterone is therefore an exclusively male hormone, which is not the case. Rather, the ovaries produce biologically significant amounts of testosterone, which aside from being the precursor to estradiol, is also released by the ovaries into circulation. Similarly, estrogens and progestins contribute significantly to male endocrinology.

if perinatal and adult androgens are controlled for, pubertal androgens are also important for male-typical development and may explain differences in aggression originating from Y chromosome loci (Shrenker et al., 1985). Nonetheless, it is clear that the influence of the sex chromosomes on sex is not limited to sex determination but extends into the process of sexual differentiation and adult behavior as well.

Sexual dimorphisms in the nervous system

There is widespread sexual dimorphism in the nervous system, and it is believed that these structural differences between the sexes are important for sex differences in behavior (Morris et al., 2004). At least some of these sexually dimorphic structures are important for sexual behavior, including the anteroventral periventricular nucleus (AVPV), the bed nucleus of the stria terminalis (BST), the medial amygdala (MEA), the medial preoptic area (MPOA), the ventromedial nucleus of the hypothalamus (VMH), and the spinal nucleus of the bulbocavernosus (SNB). Most of these (e.g., the SNB) can be visualized histologically on the basis of simple Nissl stain (Monks et al., 2003; Park et al., 2002; Wee and Clemens, 1987), although others (e.g., the MPOA) require the use of phenotypic markers such as androgen receptor expression (Shah et al., 2004). There is good reason to believe that some of these dimorphic structures form interconnected "circuits" that mediate sex differences in behavior, such as differences in responses to olfactory/pheromonal cues and copulation (e.g., Boehm et al., 2005; Simerly, 2002).

The available evidence concerning the mechanisms responsible for these dimorphisms is largely consistent with the central dogma of sexual differentiation outlined above (i.e., that testosterone produced perinatally by males organizes the developing nervous system). Furthermore, whereas androgenic mechanisms account for the SNB, estrogenic mechanisms largely account for the rest. It is additionally clear that regulation of cell death by gonadal steroid hormone production is of central importance to these sexual dimorphisms (Forger, 2006). As with all aspects of mouse sex, there is little information available about strain differences in sexually dimorphic structures, although there is some indication that these differences exist. For example, differences in the morphology and androgen responsiveness of SNB motoneurons have been reported between C57BL/6J and DBA/2J strains and the BDF1 hybrid between these (Wee and Clemens, 1987). Activational (i.e., adult) actions of gonadal hormones also contribute to sex differences in neural structures such as SNB motoneurons (Park et al., 2002; Zuloaga et al., 2007). Furthermore, sexual dimorphisms in the nervous system can be modulated by the performance of relevant behaviors and by social interactions with conspecifics, and these effects do not appear to be dependent on changes in gonadal hormones (Breedlove et al., 1999 and references therein). For example, male rats with sexual experience have smaller SNB soma and nuclei than inexperienced males, even when constant testosterone levels are maintained via implants (Breedlove, 1997). In contrast to the above described organizational influences of gonadal hormones, these activational effects on morphology result primarily from changes in the size of neurons and their connectivity rather than in cell number.

One general consideration that must always be borne in mind when evaluating sex differences in behavior is that these differences may not necessarily result directly from differences in the nervous system, but may result indirectly from differences in the periphery (Beach, 1971). For example, some male-typical and female-typical sexual behaviors require sex-typical genitals. Furthermore, peripheral structures may determine not only the sensory quality of experiences, but also the manner in which conspecifics interact with each other. For example, glands associated with the penis of male rat pups produce pheromones that stimulate maternal licking, and this licking in turn influences future sexual dimorphism and sexual behavior in these developing males (Moore, 1992). Finally, peripheral structures may directly contribute to nervous system development. For example, sexually dimorphic development of SNB motoneurons appears to be a function of testosterone action on their target muscles rather than on the motoneurons themselves (Morris et al., 2004).

Adult hormone effects

Relatively high production of gonadal steroid hormones by both sexes is observed beginning at puberty. There is a growing appreciation that the pubertal period provides another critical period for the organization of brain and behavior (Romeo et al., 2002; Schulz et al., 2004; Shrenker et al., 1985). After puberty, gonadal hormone production becomes more regular, although rhythms in hormone production are observed, especially in females. In males, testicular production of testosterone remains high throughout adult life, although testosterone production fluctuates considerably within individuals over the course of a day and declines gradually with age (see the 'Masculine copulatory behaviors' section below for the activational effects of testosterone during adulthood). Interestingly, there is an acute surge in testosterone following male exposure to estrous females in males of some inbred and outbred strains (Batty, 1978a; James and Nyby, 2002; James et al., 2006). The function of this surge is uncertain, although immediate benefits may occur as relatively rapid facilitation of testosterone on the initiation of copulation by males has been described (James and Nyby, 2002). Alternatively, these testosterone surges may serve to enhance future reproductive opportunities (Nyby, 2008).

Relative to males, females have a complex pattern of gonadal endocrinology. Adult females have a surge in ovarian production of estradiol and progesterone triggered by and participating in ovulation, which usually occurs every 4–5 days in mice that are not pregnant or lactating. Yet, ovarian function is powerfully influenced by the sociosexual milieu of the female (see below) and considerable variation in this "spontaneous" cycle is observed depending on housing conditions. The ovulatory pattern of estradiol followed by progesterone is the major endocrine event facilitating sexual receptivity in females. Gestational endocrinology is also complex and generally inhibitory to female sexual behavior, although a remarkable postpartum estrus occurs in which ovulation and a period of sexual receptivity lasting several hours occurs in dams that have just delivered a litter. The result of this is a remarkable efficiency in reproduction in which females may produce litters roughly every three weeks if a sire is available postpartum.

Sexual behavior

Reproductive strategy

In naturalistic and semi-naturalistic conditions, mice live in small, stable social groups within territories held by a dominant male, who has relatively exclusive sexual access to several females. Mice breed year-round, given adequate food and shelter. The social organization and reproductive strategy of mice therefore can be described as opportunistic harem polygyny. Subordinate adult males may coexist with the dominant, but appear to have little reproductive success (Wolff, 1985). It is not clear whether a further hierarchy exists within subordinates, but subordinate males differ significantly from dominants in several respects. Subordinate males have distinct behavioral patterns (Desjardins et al., 1973) and presumably differences in physiology and endocrinology as well. It is unlikely that subordination results simply from lesser testosterone production, as equalizing testosterone between dominants and subordinates does not change their relationship significantly (Maruniak et al., 1977). Nonetheless, in view of the dramatically reduced sexual performance of subordinates in semi-naturalistic and naturalistic settings, it may be appropriate to consider dominance of adult males living in social groups as a potential factor in their sexual behavior. As we shall see, courtship displays and to some extent copulation itself likely revolves around male advertisement of dominance.

It is a truism that individuals of sexually reproducing species invest far more energy in the pursuit of sex than in copulation itself. Mice are no exception to this, spending much of their time establishing and defending territory in the case of males and acquiring desirable nest sites and fertilization from desirable males in the case of females. The importance of territory for mouse sexual behavior is illustrated by studies of Wolff (1985), who trapped feral mice (*Mus musculus domesticus*) and observed their behavior over the course of several days in a semi-naturalistic setting. Wolff found that the introduced male mice established non-overlapping territories, each with a single dominant male. The vast majority (80%) of copulations (and presumably of pups) were obtained by dominant males in territories with walls and internal structural features. Relatively little study has been made of female choice and possible inter-female competition in this context, but it is clear that females actively solicit copulations from dominant males and actively avoid copulation with subordinates. Furthermore, in carefully controlled laboratory studies female mice have been shown to prefer dominant males to subordinate males, visit dominants more often and show greater receptivity during copulation with dominants (Tomihara, 2005). It is likely that cryptic female choice for dominant males has led to rather elaborate copulatory behavior necessary for fertilization (described in detail below), which doubtless also contributes to low fertilization rates of females by subordinate males.

There is additional evidence that agonistic encounters between males are a principal behavioral determinant of their reproductive fitness. Horn (1974) studied reproductive success of inbred mouse strains with different levels of aggressivity in semi-naturalistic environments and found that the strain with the greatest degree of intermale aggression (RF/J) won almost all (93%) of the territorial challenges and obtained a striking proportion (96%) of the sired pups in the generation following introduction. Of course these results should be interpreted with caution due to better fertilization rate of the female strain (C57BL/6J) by the RF/J males relative to the other strains used (Horn, 1974). More recent work on the neurobiological and genetic basis of aggression in mice provide promising avenues for research in elucidating the factors that contribute to individual and strain differences in aggression (Ogawa et al., 2004).

In the final analysis, it remains uncertain to what extent these results are applicable to natural settings with varying population densities, food availability, predation, intrinsic features and so on. Nonetheless, it seems clear that dominance and the quality of territory that males hold is an important aspect of female choice of sexual partner in semi-naturalistic conditions. This raises the question of whether sexual selection occurs purely on these bases or whether females also choose males on the basis of other intrinsic characteristics.

Courtship behaviors

Molecular genetic study of paternity in outbred strains in semi-naturalistic settings indicates that approximately one third of female matings occurred outside of their home territory and about half of all litters had major histocompatibility complex (MHC) alleles of extra-territorial sires (Potts et al., 1991). Additionally, females appear to avoid mating with males with a high parasite load and with their litter mates (Potts et al., 1994). These studies suggest that females do indeed use criteria other than the nearest dominant male in mate selection, and that these extra-territorial matings are of major importance for male reproductive success. It is therefore important to consider characteristics of males that serve to attract females aside from their dominance. Courtship behaviors are usually defined as those behaviors that appear solely to attract prospective mates, with a familiar example being birdsong. Both vocalizations and scent marking have been implicated as potential courtship behaviors.

Chemosensory

Perhaps the best characterized male chemosensory courtship behavior is the scent marking behavior of dominant males.

There are at least two important aspects of urinary scent marking that differ between dominant and subordinate mice and that presumably contribute to male attractivity. The first difference in urine marking is dramatic and qualitative: dominant males maintaining a territory urinate relatively uniformly throughout that area whereas subordinate males urinate in large pools at the boundaries of the territory. The second major difference relates to androgen-dependent, sexually dimorphic urinary proteins. Dominant males tend to have higher circulating androgens, and females display a clear preference for urine from intact males over that from castrates. Destruction of the olfactory epithelium in females results in elimination of female preference for intact males and reduced receptivity during copulation (Keller et al., 2006a). While the presence and variability of particular urinary proteins is known to vary between inbred and outbred strains of different lineages (Cheetham et al., 2009), the relevance of these findings to attractivity has not yet been explored.

The accessory olfactory system also plays a critical role in female sexual behavior. For example, vomeronasal organ removal abolishes receptive behavior altogether in female mice, likely via a reduction in gonadotropin production in response to male cues (Keller et al., 2006b). Housing conditions, and especially housing density and the presence or absence of males, are important variables that determine the reproductive status in female mice. For example, puberty is delayed (Vandenbergh, 1967) and the timings between ovulations are longer (van der Lee and Boot, 1955) or ovulation is suspended altogether (Whitten, 1957) in all-female groups and the magnitude of this effect depends on population density. In contrast, male exposure to group-housed females speeds puberty and induces ovulation (Whitten, 1956). Finally, introduction of a "strange" male can result in termination of pregnancy in dams, termed the "Bruce effect" (Bruce, 1959, 1960).

Auditory

Like other rodent species, male mice produce ultrasonic vocalizations preceding and during copulation (Holy and Guo, 2005; Pomerantz et al., 1983; Sales, 1972; White et al., 1998; Whitney et al., 1973). These vocalizations range from 30 to 70 kHz. While most work has focused on those centered around 70 kHz as potential courtship behaviors, other categories of vocalizations, such as the high intensity 40 kHz vocalization, have also been examined but are less reliably associated with copulation and courtship (White et al., 1998). Furthermore, a recent fine-grained analysis of male vocalizations has revealed unsuspected complexity and interindividual variability (Holy and Guo, 2005), raising the possibility that these vocalizations might constitute courtship behaviors analogous to birdsong, and be subjected to sexual selection.

Studies of vocalization during male–female interactions in which one of the partners is devocalized have indicated that 70 kHz calls are produced primarily by males (White et al., 1998). It further appears that these vocalizations are attractive to females as they spend more time with males that are able to vocalize than those that are not, and this preference is restored by playback of recordings of 70 kHz male vocalizations (Pomerantz et al., 1983). It is also likely that vocalizations participate in copulation as well, as male vocalizations differ systematically with different stages of copulation (discussed below).

Female courtship behaviors

Female courtship behaviors are not usually defined, although there is probably a stronger case for female behaviors that appear to be used exclusively in the context of acquiring copulation with a specific individual. Females do actively solicit copulation from males, primarily via approach and remaining immobile near the stimulus male. This proceptive posture is dependent on estrogenic action in the hypothalamus (Musatov et al., 2006). A more dramatic form of behavior that might be considered solicitation occurs in the case of unresponsive males, when females will mount males in a display of masculine sexual behavior. Whereas males vary in their apparent sexual motivation with different females, it is unclear to what extent male choice contributes to mating patterns in naturalistic settings. It may be the case that female solicitation behaviors serve to initiate and optimize the timing of copulation (see below), rather than in competitive mate acquisition as is implied in the term courtship. Nonetheless, if one defines courtship behaviors as those that facilitate the initiation of copulation, then female solicitations qualify.

Copulation

Species and sex-typical sexual behavior tends to be rather stereotyped and does not vary extensively in qualitative aspects between wild and captive populations and naturalistic, semi-naturalistic, and laboratory settings (Estep et al., 1975; McGill, 1965; Wolff, 1985). Nonetheless, considerable individual variation is observed and there are striking quantitative differences in various aspects of copulatory behavior between inbred strains (McGill, 1965), and copulatory behaviors do become more efficient with experience.

The full expression of mouse copulation is normally only observed during estrus in females, which is an increased period of sexual activity that occurs around the time of ovulation. Sequential surges of ovarian estradiol then progesterone associated with ovulation initiate alterations in the nervous system of females that permit the expression of female-typical sexual behavior, including increased locomotion and seeking of sexual partners. Females will only solicit males or permit copulation to occur during this period. In the interest of convenience and reducing variability in ovarian hormones between females, laboratory tests of copulation are usually performed with females that are ovariectomized and injected with estrogen and progesterone.

The initial phases of copulation include male vocalizations and olfactory investigation of the female's anogenital region. Estrous females may also solicit males during this phase by

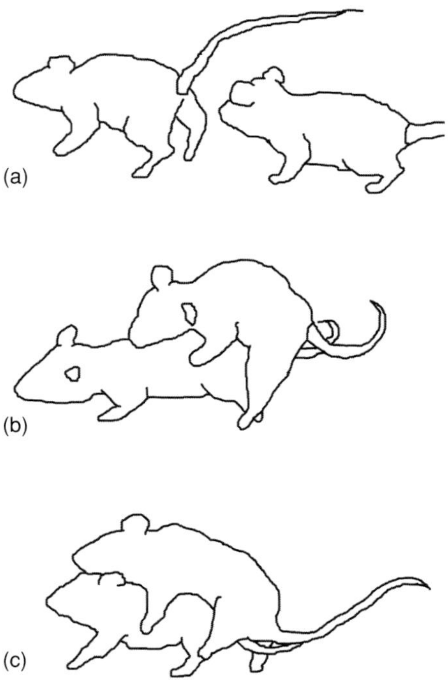

Figure 22.1 Copulation. (a) *Anogenital investigation and following:* during this phase a female initially approaches a male, who then investigates the anogenital area of the female with his snout, often lifting her hindquarters off the substrate. Following this, the male will follow the female (depicted, the female is on the left and the male on the right). (b) *Mount:* the male grasps the female flank with its forepaws and engages in rapid, shallow pelvic thrusting. The male will then either successfully intromit or dismount. The female may remain immobile and enter lordosis or continue walking. (c) *Intromission/lordosis:* the mounting male successfully intromits his penis and begins thrusting more deeply and slowly. The male may also grasp the female with one hind-paw. The female is in lordosis, with back arched.

approaching and remaining proximate to males. This is usually followed by male anogenital investigation and following of females (Figure 22.1a). It is believed that this initial phase serves to "prime" copulation by increasing male arousal to the point where mounting is initiated. Indeed the latency for males to begin copulatory attempts is often used as an index of male libido. It also appears that chemosensory priming is important for the expression of female sexual behavior (see above).

Copulation itself is arguably the most intensively studied aspect of mouse sexual behavior. In describing copulatory behaviors we will follow the convention of Blaustein and Erskine (2002), and adopt the terms masculine and feminine behaviors in recognition that both males and females normally retain some capacity for copulatory behaviors typically associated with the opposite sex. These surprising aptitudes can be observed with the appropriate adult hormone priming. Nonetheless, in standard copulation tests between males and females, masculine sexual behaviors are normally performed by males and feminine behaviors by females.

Feminine copulatory behaviors

Two main types of feminine sexual behaviors are usually studied: receptive behaviors and proceptive behaviors. Receptive behavior consists simply of a reflexive postural change in response to male mounts, termed lordosis. Lordosis results from an arching of the back and reflection of the tail, which allows for penile intromission, and is dependent on ovarian hormones for its normal expression. The usual measure of feminine receptive behavior is lordosis quotient, which is simply the percentage of male mounts that result in the female displaying lordosis. In rats, it is also possible to subjectively measure the intensity of the lordosis reflex (i.e., the degree of dorsal flexion), but the magnitude of this reflex is relatively smaller in mice than rats, and so the range of lordotic response may be more difficult to distinguish.

Proceptive behaviors are those that serve to increase the likelihood of male copulatory behaviors. As described above, feminine proceptive behaviors appear to be important for initiating copulation, but they are also displayed during the course of the copulatory series. It is likely that these behaviors serve to modulate the pace of copulation (Blaustein and Erskine, 2002). In rats, optimal pacing of copulation provides sensory information necessary for the initiation of pregnancy and is usually studied by providing females a means of avoiding males. Female pacing of copulation is not usually studied in mice, although recently, methods for studying this aspect of feminine sexual behavior have been described (Johansen et al., 2008).

Masculine copulatory behaviors

Multiple units of masculine rodent sexual behavior are typically defined (Mosig and Dewsbury, 1976). A copulatory series consists of several bouts of activity. In each bout, a male will mount an estrous female, grasping her flanks with his forepaws and engaging in relatively rapid and shallow pelvic thrusting during these mounts (Figure 22.1b). During some of the mounts, the male will insert his penis into the female's vagina (termed intromission), which is signalled by deeper and slower pelvic thrusting (Figure 22.1c). The male may grasp the female with forepaws only or additionally with one hind-paw. Following each mount or intromission, males will engage in genital grooming and several seconds or minutes may pass before the next mount or intromission. After several intromissions (potentially dozens), each accompanied by several thrusts (potentially hundreds across an ejaculatory series), males will ejaculate, ending the copulatory series. Ejaculation is quite dramatic in mice and the pair often will fall over together and remain immobile for several seconds. The units of masculine sexual behavior that contribute to a copulatory series are therefore mounts, intromissions, and ejaculations. One may study the number and frequency of mounts and intromissions as well as the number of intromissions and/or pelvic thrusts preceding ejaculation. Of additional interest, following ejaculation, there is a refractory period during which no male sexual behavior is observed. Many of these common measures of male copulatory behavior vary markedly within the small number of inbred strains in which copulation has been compared. For example, the post-ejaculatory refractory period of males is extremely variable between inbred mouse strains, ranging from minutes to days (McGill, 1965).

In addition to the above-described motor aspects of male copulation, vocalization by males changes according to the nature of the copulatory activity. As mentioned previously, males produce 70 kHz vocalizations in the presence of females, and males vocalize to the same extent towards estrous and non-estrous females (Pomerantz et al., 1983). However, the rate of 70 kHz vocalizations increases preceding and during mounts, and decreases during intromissions (White et al., 1998). Interestingly, the vocalizations increase during the ejaculatory intromission and are depressed during the post-ejaculatory interval (White et al., 1998).

Both appetitive and consummatory aspects of male sexual behavior, including associated 70 kHz vocalizations, depend on androgens. All of these behaviors decrease markedly in males that are castrated (although some inbred strains have a remarkably persistent ability to copulate following castration; McGill, 1965 and see below), and maintenance or replacement of testosterone prevents these behavioral deficits. As with many of the actions of testosterone on the rodent brain, testosterone has its effects on these behaviors primarily via conversion to estradiol. Furthermore, intracranial testosterone implants into the preoptic area are sufficient to maintain these behaviors in castrated males (Nyby et al., 1992).

Cryptic female choice and copulation

In common with many other rodent species, female mice spontaneously ovulate, but require coital stimulation to produce ovarian hormones necessary for pregnancy. As previously mentioned, there has been some investigation of the aspects of coitus that initiate the endocrine reflex that supports pregnancy (commonly termed pseudopregnancy, as female mice that mate with sterile males will appear pregnant for several weeks). For mice, in contrast to other rodent species, the essential stimulation for pseudopregnancy is ejaculation itself. Several aspects of ejaculation might conceivably contribute to female fertility. For example, semen coagulates to form a plug that completely fills the vagina, and contractions of perineal striated muscles associated with the penis result in dramatic alterations in the morphology of the glans penis. It is likely that the ejaculatory penile reflexes rather than the copulatory plug are responsible for pseudopregnancy as surgical removal of the glands responsible for seminal emission does not prevent pseudopregnancy in females receiving ejaculations (McGill, 1972; McGill and Coughlin, 1970; McGill et al., 1968).

Erectile function

Erectile function is essential for male copulatory performance, and may be studied in the context of copulation or *ex copula*. In general, more information is available concerning the latter type of test. The exception to this is provided by observations of glans morphology at ejaculation, and its role in pseudopregnancy. The ejaculatory reflex is caused by the contraction of striated perineal muscles associated with the bulb of the penis (bulbospongiosus muscles), which are innervated by the sexually dimorphic SNB motoneurons described above. It is possible to study these reflexes in mice, although it appears to require removing tonic inhibition by supraspinal afferents (Elmore and Sachs, 1988; Sachs, 1980). It is important to bear in mind that these reflexive erections are not so much a measure of sexual arousal, as might be imagined, but rather are spinal reflexes, although sexual stimulation can facilitate their expression (Sachs, 1995). Nonetheless, erections can be used as an index of sexual arousal as more traditionally defined (Sachs, 1995). For example, rats manifest non-contact erection (NCEs) in response to odors from estrous females (Sachs et al., 1994), which are closer to the concept of sexual arousal. There have been few studies reporting tests of NCEs in mice (Matsumoto et al., 2003; Soukhoval-O'Hare et al., 2007) and studies to date have not examined strain differences, therefore it is unclear whether NCEs are easily observed in all strains of mice.

Role of learning

Mouse copulatory behavior is relatively stereotyped and mice that have never witnessed or experienced copulation do so in a species-typical fashion. Furthermore, wild mice have qualitatively similar copulatory behavior to inbred and outbred laboratory strains. Nonetheless, there are clear experience-dependent effects on sexual behavior in this species. One of the most obvious is an increase in efficiency of copulation. In comparison with virgins, male mice that have experience with limited time copulation tests have shorter intromission and ejaculation latencies and achieve more ejaculations. Similarly, as mentioned above, priming of ovariectomized females with ovarian hormones is more effective following sexual experience. It is unclear to what extent this reflects improved sexual function or training of the mice to the rather artificial constraints (e.g., time, social stimuli, spatial) imposed by commonly used testing conditions.

Differences between inbred strains in sexual differentiation and behavior

Differences between inbred strains of mice in sexual differentiation and behavior are the subject of a recent, excellent review (Burns-Cusato et al., 2004), and we concur with the authors that there has been insufficient study of strain differences in these areas. Interested readers are directed to this review, and we will briefly review some of the major work on this topic and limit any comments to a few salient points relevant to the current review. Because of methodological differences among the studies, it is difficult if not impossible to directly compare results. Because of this, we will consider each group of studies separately and a summary of the rankings of strains compared within studies can be found in Table 22.1.

Levine and colleagues (Levine, 1958; Levine et al., 1966), compared male copulatory behavior and competitive mating in ST (albino) and CBA (pigmented) strains. Male mice were

Table 22.1 Rankings of strains within studies on measures of male copulation.

Study	Measures	Ranking
Batty, 1978a, b	Copulation success, mount latency and frequency	BDF1,C57BL/6J>DBA/2J> BALB/c,CBA/H
Canastar and Maxson, 2003	Sexual aggression (male directed to female)	FVB>C57BL/6J
	Mounts, intromissions, thrusts	C57BL/6J > FVB
Levine, 1958, 1966	Copulation success, paternity	ST>CBA
McGill, 1965	Copulation success, intromissive behaviors	C57BL/6~BDF1 >DBA/2J
Ogawa et al., 1996	Copulation success	C57BL/6J>DBA/2J

either tested either alone or during co-habitation with a male of the alternate strain, and females of both strains were used as stimulus animals. In general, and contrary to a general pattern of pigmented superiority reported by Batty (1978a, 1978b), the ST males outperformed the CBA males on measures of copulatory success and paternity. It is worth noting that complex interactions with social condition and sexual partner were observed, indicating that these variables are indeed important to consider.

Perhaps the most comprehensive series of studies comparing strains on male copulation was undertaken by McGill (reviewed in McGill, 1965), whose conventions in strain choice and methodology have been influential. These early studies identified a dozen strain differences in aspects of male copulatory behavior between the C57BL/6J and DBA/2J strains when tested with BDF1 hybrid stimulus females. The source of these differences was then investigated by testing F1 crosses between two of these strains (BDF1 and DBF1). The results of these crosses suggest that some strain differences can be attributed to one parent, indicating a dominant pattern of inheritance, including both more masculine patterns (i.e., the shorter mount latency, greater number of thrusts to ejaculation, and number of intromissions of the B6) and less masculine patterns (i.e., the greater number of inappropriately oriented mounts by DBA males). Other traits were intermediate between the parents, indicating additive effects (e.g., interintromission interval). Finally, some traits were more extreme than either parent, indicating a heterotic pattern of inheritance (e.g., number of mounts, thrusts per mount and intromission interval). The authors of these studies were appropriately cautious concerning the inferences concerning mode of inheritance drawn from the F1 crosses as these studies are in themselves insufficient to draw such conclusions. Furthermore, as detailed above, diverse physiological mechanisms might underlie these strain differences, including differences in endocrinology and pup–dam interactions.

Batty (1978a, 1978b) examined the relationship between adult androgens levels and male copulation in four inbred strains (BALB/c, CBA/H, DBA/2J, C57BL/6Fa) and in some hybrids between these (BDF1, DBF1, CBF1). She observed strain differences in both copulation and androgens, but a surprising inverse relationship between these. She categorized the mice as low-frequency mounters (BALB/c, CBA/H), medium-frequency mounters (DBA/2J), and high-frequency mounters (C57BL/6Fa and BDF1). She also described a greater sexual vigor in the pigmented strains in relation to the albino strains. It is worth noting that, as in the tests by McGill and colleagues, all tests were performed with BDF1 stimulus females, and so the generalizability of these results to other female partners may be limited.

Canastar and Maxson (2003) investigated differences between the FVB/NtacfBR and C57BL/6J strains in male sexual aggression and copulation towards FVB/NtacfBR stimulus females. When paired with estrous females, FVB/NtacfBR males exhibited more aggression and fewer copulatory behaviors than C57BL/6J mice. These results are particularly interesting as they illustrate that differences between inbred strains on measures of copulation may well be secondary to incompatible behaviors, such as propensity to aggression.

Strain differences in androgen response

There has been little study of strain differences in female-typical endocrinology. One early study examined the ability of progestins to activate receptive behavior in the CD-1 and SW lines and found that whereas lordosis in both lines can be primed by progesterone, only the CD-1 line responds to dihydroprogesterone (Gorzalka and Whalen, 1974), and subsequent crosses of these lines indicated that hybrids respond to the dihydroprogesterone as do the CD-1 (Gorzalka and Whalen, 1976). It is worth noting that individual differences in progesterone response were more pronounced than strain differences in this measure.

In contrast, several groups have reported strain differences in androgen response; the results of these studies are presented in Table 22.2. In an early report, McGill and Manning (1976) noted a remarkable persistence of masculine sexual behavior following castration in BDF1 hybrids, with these males continuing to copulate for many months following castration (usually males cease copulating within weeks of castration). This remarkable persistence is also observed in recombinant mice, indicating an allelic effect rather than a linkage effect (Coquelin, 1991). The physiological correlates of this relatively androgen independent copulatory behavior have not been identified, despite investigation of several likely physiological mechanisms, such as differences in androgen endocrinology, brain aromatase, and brain estrogen receptors (Sinchak et al., 1996).

As previously mentioned, early studies indicate strain differences in androgen response of males to sexual contact (Batty, 1978a). In further investigation of this effect, James et al. (2006) compared the CF-1, CK, and DBA/2J strains in their response to female sexual cues (urine from an estrous female). In the CF-1 and DBA/2J males, this exposure triggered both increases in

Table 22.2 Ranking of strains on measures of androgen endocrinology.

Study	Measure	Ranking
Batty, 1978a, b	Adult basal testosterone	BALB/c,CBA/H>CBF1, C57BL/6>BDF1>DBF1
Batty, 1978a, b	Testosterone response to female cues	DBA/2J,C57BL/6>BALB /c,CBA/H,BDF1
Dominguez-Salazar et al., 2004	Copulation in ERKO crosses	C57Bl/6JXDBA,C57BL/6JX BALB/c>C57BL/6JXAKR/J
James et al., 2006	Testosterone response to female cues	DBA/2J,CF-1>CK
McGill and Manning, 1976	Continued copulation after castration	BDF1>C57BL/6>BALB /c, DBA/2
Selmanoff et al., 1977	Peripubertal testosterone	DBA/1/Bg>C57BL/10/Bg (hybrids indicate DBA dominance)
Shrenker and Maxson, 1986	Testis and seminal vesicle weight	C57BL/10>DBA/1J (backcrosses indicate Y-chromosome locus)

testosterone and vocalizations, but the same was not observed in the CK strain.

Luttge and Hall (1973) compared the efficacy of testosterone or its androgenic metabolite DHT in the activation of copulation in castrated CD-1 and SW mice. These mice were castrated peripubertally (32 days) and the hormone treatments initiated at 68 days of age. Interestingly, whereas testosterone activated mounting in both strains, DHT, which activates androgen receptors but is not readily converted to estradiol, maintained copulation in SW but not the CD-1 strain. These authors then reported that cyproterone acetate (CA), which has mixed androgenic and anti-androgenic action, was also efficacious in activating copulation in similarly castrated SW mice (Hall and Luttge, 1975). Ogawa et al. (1996), compared androgen maintenance of copulatory and aggressive behaviors in castrated C57BL/6J and DBA/2J mice using either testosterone or MENT, an androgen which is not readily 5-α-reduced to DHT and is therefore a weaker activator of the androgen receptor. As had been found in previous studies, intact B6 males showed higher levels of intromissions than DBA males. Interestingly, while testosterone and MENT restored normal intromission rates in castrated males of both strains, mounting rates were only restored to intact levels in C57 males and not DBA males. Finally, differences have been observed in the effects of an estrogen receptor alpha null mutation on sex according to the background strain used (Dominguez-Salazar et al., 2004). These differences are presumably due to indirect differences in androgen endocrinology relating to estrogenic metabolism of testosterone and further illustrate the genetic diversity contributing to endocrine control of mouse sex.

In summary, strain differences can readily be observed on several measures of sexual differentiation, male copulation, and androgen endocrinology in response to sexual cues. Preliminary studies of hybrid strains suggest distinct patterns of inheritance for several of these traits. To the extent to which it has been studied, differences in androgen endocrinology appear to account for at least some of these differences and some remarkable differences suggest considerable variability in alleles contributing to androgen endocrinology in mice (e.g., the resistance of copulation to castration in BDF1 hybrids). In none of these cases has the genetic or physiological basis of these differences been elucidated. Another striking gap in our knowledge is found in possible strain differences in feminine sexual behavior, with the exception of the contribution of female preference for MHC allele polymorphisms. Needless to say, a systematic characterization of mouse sex within inbred strains would be of significant value to this field.

References

Arnold, A.P. and Chen, X. (2009) What does the "four core genotypes" mouse model tell us about sex differences in the brain and other tissues? *Front Neuroendocrinol* 30: 1–9.

Batty, J. (1978a) Acute changes in plasma testosterone levels and their relation to measures of sexual behaviour in the male house mouse (*Mus musculus*). *Anim Behav* 26: 349–357.

Batty, J. (1978b) Plasma levels of testosterone and male sexual behaviour in strains of the house mouse (*Mus musculus*). *Anim Behav* 26: 339–348.

Beach, F.A. (1971) Hormonal factors controlling the differentiation, development, and display of copulatory behavior in the ramstergig and related species. In Tobach, E., Aronson, L.R., and Shaw, E. *The Biopsychology of Development*. Academic Press, New York, pp. 249–296.

Blaustein, J.D. and Erskine, M.S. (2002) Feminine sexual behavior: Cellular integration of hormonal and afferent information in the rodent forebrain. In Pfaff, D.W., Arnold, A.P., Etgen, A.M., Fahrbach, S.E., and Rubin, R.T. (eds.), *Hormones, Brain and Behavior*, Vol. 1. Academic Press, New York, pp. 139–214.

Boehm, U., Zou, Z., and Buck, L.B. (2005) Feedback loops link odor and pheromone signaling with reproduction. *Cell* 123: 683–695.

Breedlove, S.M. (1997) Sex on the brain. *Nature* 389: 801–801.

Breedlove, S.M., Cooke, B.M., and Jordan, C.L. (1999) The orthodox view of brain sexual differentiation. *Brain Behav Evol* 54: 8–14.

Bruce, H.M. (1959) An exteroceptive block to pregnancy in the mouse. *Nature* 184: 105.

Bruce, H.M. (1960) A block to pregnancy in the mouse caused by proximity of strange males. *J Reprod Fertil* 1: 96–103.

Burns-Cusato, M., Scordalakes, E.M., and Rissman, E.F. (2004) Of mice and missing data: what we know (and need to learn) about male sexual behavior. *Physiol Behav* 83: 217–232.

Canastar, A. and Maxson, S.C. (2003) Sexual aggression in mice: effects of male strain and of female estrous state. *Behav Genet* 33: 521–528.

Carruth, L.L., Reisert, I., and Arnold, A.P. (2002) Sex chromosome genes directly affect brain sexual differentiation. *Nat Neurosci* 5: 933–934.

Cheetham, S.A., Smith, A.L., Armstrong, S.D., Beynon, R.J., and Hurst, J.L. (2009) Limited variation in the major urinary proteins of laboratory mice. *Physiol Behav* 96: 253–261.

Coquelin, A. (1991) Persistent sexual behavior in castrated, recombinant inbred mice. *Biol Reprod* 45: 680–684.

Desjardins, C., Maruniak, J.A., and Bronson, F.H. (1973) Social rank in house mice: differentiation revealed by ultraviolet visualization of urinary marking patterns. *Science* 182: 939–941.

De Vries, G.J., Rissman, E.F., Simerly, R.B., Yang, L.Y., Scordalakes, E.M., Auger, C.J., et al. (2002) A model system for study of sex chromosome effects on sexually dimorphic neural and behavioral traits. *J Neurosci* 22: 9005–9014.

Dewing, P., Shi, T., Horvath, S., and Vilain, E. (2003) Sexually dimorphic gene expression in mouse brain precedes gonadal differentiation. *Brain Res Mol Brain Res* 118: 82–90.

Dominguez-Salazar, E., Bateman, H.L., and Rissman, E.F. (2004) Background matters: the effects of estrogen receptor alpha gene disruption on male sexual behavior are modified by background strain. *Horm Behav* 46: 482–490.

Elmore, L.A. and Sachs, B.D. (1988) Role of the bulbospongiosus muscles in sexual behavior and fertility in the house mouse. *Physiol Behav* 44: 125–129.

Estep, D.Q., Lanier, D.L., and Dewsbury, D.A. (1975) Copulatory behavior and nest building behavior of wild house mice (*Mus musculus*). *Anim Learn Behav* 3: 329–336.

Fisher, C.R., Graves, K.H., Parlow, A.F., and Simpson, E.R. (1998) Characterization of mice deficient in aromatase (ArKO) because of targeted disruption of the cyp19 gene. *Proc Natl Acad Sci USA* 95: 6965–6970.

Forger, N.G. (2006) Cell death and sexual differentiation of the nervous system. *Neuroscience* 138: 929–938.

Gandelman, R., vom Saal, F.S., and Reinisch, J.M. (1977) Contiguity to male foetuses affects morphology and behaviour of female mice. *Nature* 266: 722–724.

Gorzalka, B.B. and Whalen, R.E. (1974) Genetic regulation of hormone action: selective effects of progesterone and dihydroprogesterone (5alpha-pregnane-3,20-dione) on sexual receptivity in mice. *Steroids* 23: 499–505.

Gorzalka, B.B. and Whalen, R.E. (1976) Effects of genotype on differential behavioral responsiveness to progesterone and 5-a-dihydroprogesterone in mice. *Behav Genet* 6: 7–15.

Hall, N.R. and Luttge, W.G. (1975) Maintenance of sexual behavior in castrate male SW mice using the anti-androgen, cyproterone acetate. *Pharmacol Biochem Behav* 3: 551–555.

Holy, T.E. and Guo, Z. (2005) Ultrasonic songs of male mice. *PLoS Biol* 3: e386.

Horn, J.M. (1974) Aggression as a component of relative fitness in four inbred strains of mice. *Behav Genet* 4: 373–381.

Hull, E.M., Meisel, R.L., and Sachs, B.D. (2002) Male sexual behavior. In Pfaff, D., Arnold, A.P., Etgen, A.M., Fahrbach, S.E., and Rubin, R.T. (eds.), *Hormones, Brain and Behavior*, Vol. 1. Elsevier, New York, pp. 3–137.

James, P.J. and Nyby, J.G. (2002) Testosterone rapidly affects the expression of copulatory behavior in house mice (*Mus musculus*). *Physiol Behav* 75: 287–294.

James, P.J., Nyby, J.G., and Saviolakis, G.A. (2006) Sexually stimulated testosterone release in male mice (*Mus musculus*): roles of genotype and sexual arousal. *Horm Behav* 50: 424–431.

Johansen, J.A., Clemens, L.G., and Nunez, A.A. (2008) Characterization of copulatory behavior in female mice: evidence for paced mating. *Physiol Behav* 95: 425–429.

Keller, M., Douhard, Q., Baum, M.J., and Bakker, J. (2006a) Destruction of the main olfactory epithelium reduces female sexual behavior and olfactory investigation in female mice. *Chem Senses* 31: 315–323.

Keller, M., Pierman, S., Douhard, Q., Baum, M.J., and Bakker, J. (2006b) The vomeronasal organ is required for the expression of lordosis behaviour, but not sex discrimination in female mice. *Eur J Neurosci* 23: 521–530.

Koopman, P., Gubbay, J., Vivian, N., Goodfellow, P., and Lovell-Badge, R. (1991) Male development of chromosomally female mice transgenic for Sry. *Nature* 351: 117–121.

Kudwa, A.E., Bodo, C., Gustafsson, J.A., and Rissman, E.F. (2005) A previously uncharacterized role for estrogen receptor beta: defeminization of male brain and behavior. *Proc Natl Acad Sci USA* 102: 4608–4612.

Levine, L. (1958) Studies on sexual selection in mice. I. Reproductive competition between albino and black-agouti males. *Am Nat* 92: 21–26.

Levine, L., Barsel, G.E., and Diakow, C.A. (1966) Mating behaviour of two inbred strains of mice. *Anim Behav* 14: 1–6.

Luttge, W.G. and Hall, N.R. (1973) Differential effectiveness of testosterone and its metabolites in the induction of male sexual behaviour in two strains of albino mice. *Horm Behav* 4: 31–43.

Lyon, M.F. (2002) X-chromosome inactivation and human genetic disease. *Acta Paediatr Suppl* 91: 107–112.

Maruniak, J.A., Desjardins, C., and Bronson, F.H. (1977) Dominant-subordinate relationships in castrated male mice bearing testosterone implants. *Am J Physiol* 233: E495–499.

Matsumoto, T., Honda, S., and Harada, N. (2003) Alteration in sex-specific behaviors in male mice lacking the aromatase gene. *Neuroendocrinology* 77: 416–424.

McGill, T.E. (1965) Studies of the sexual behavior of male laboratory mice: effects of genotype, recovery of sex drive, and theory. In Beach, F.A. (ed.), *Sex and Behavior*. Wiley, New York, pp. 76–88.

McGill, T.E. (1972) Preejaculatory stimulation does not induce luteal activity in the mouse *Mus musculus*. *Horm Behav* 3: 83–85.

McGill, T.E., Corwin, D.M., and Harrison, D.T. (1968) Copulatory plug does not induce luteal activity in the mouse *Mus musculus*. *J Reprod Fertil* 15: 149–151.

McGill, T.E. and Coughlin, R.C. (1970) Ejaculatory reflex and luteal activity induction in *Mus musculus*. *J Reprod Fertil* 21: 215–220.

McGill, T.E. and Manning, A. (1976) Genotype and retention of the ejaculatory reflex in castrated male mice. *Anim Behav* 24: 507–518.

Monks, D.A., Xu, J., O'Malley, B.W., and Jordan, C.L. (2003) Steroid receptor coactivator-1 is not required for androgen-mediated sexual differentiation of spinal motoneurons. *Neuroendocrinology* 78: 45–51.

Moore, C.L. (1992) The role of maternal stimulation in the development of sexual behavior and its neural basis. *Ann N Y Acad Sci* 662: 160–177.

Morris, J.A., Jordan, C.L., and Breedlove, S.M. (2004) Sexual differentiation of the vertebrate nervous system. *Nat Neurosci* 7: 1034–1039.

Mosig, D.W. and Dewsbury, D.A. (1976) Studies of the copulatory behavior of house mice (*Mus musculus*). *Behav Biol* 16: 463–473.

Musatov, S., Chen, W., Pfaff, D.W., Kaplitt, M.G., and Ogawa, S. (2006) RNAi-mediated silencing of estrogen receptor-alpha in the ventromedial nucleus of hypothalamus abolishes female sexual behaviors. *Proc Natl Acad Sci USA* 103: 10456–10460.

Nyby, J.G. (2008) Reflexive testosterone release: a model system for studying the nongenomic effects of testosterone upon male behavior. *Front Neuroendocrinol* 29: 199–210.

Nyby, J.G., Matochik, J.A., and Barfield, R.J. (1992) Intracranial androgenic and estrogenic stimulation of male-typical behaviors in house mice (*Mus domesticus*). *Horm Behav* 26: 24–45.

Ogawa, S., Choleris, E., and Pfaff, D. (2004) Genetic influences on aggressive behaviors and arousability in animals. *Ann N Y Acad Sci* 1036: 257–266.

Ogawa, S., Robbins, A., Kumar, N., Pfaff, D.W., Sundaram, K., and Bardin, C.W. (1996) Effects of testosterone and 7 alpha-methyl-19-nortestosterone (MENT) on sexual and aggressive behaviors in two inbred strains of male mice. *Horm Behav* 30: 74–84.

Ogawa, S., Washburn, T.F., Taylor, J., Lubahn, D.B., Korach, K.S., and Pfaff, D.W. (1998) Modifications of testosterone-dependent behaviors by estrogen receptor-alpha gene disruption in male mice. *Endocrinology* 139: 5058–5069.

Park, J.J., Zup, S.L., Verhovshek, T., Sengelaub, D.R., and Forger, N.G. (2002) Castration reduces motoneuron soma size but not dendritic length in the spinal nucleus of the bulbocavernosus of wild-type and BCL-2 overexpressing mice. *J Neurobiol* 53: 403–412.

Pomerantz, S.M., Nunez, A.A., and Bean, N.J. (1983) Female behavior is affected by male ultrasonic vocalizations in house mice. *Physiol Behav* 31: 91–96.

Potts, W.K., Manning, C.J., and Wakeland, E.K. (1991) Mating patterns in seminatural populations of mice influenced by MHC genotype. *Nature* 352: 619–621.

Potts, W.K., Manning, C.J., and Wakeland, E.K. (1994) The role of infectious disease, inbreeding and mating preferences in maintaining MHC genetic diversity: an experimental test. *Philos Trans R Soc Lond B Biol Sci* 346: 369–378.

Romeo, R.D., Richardson, H.N., and Sisk, C.L. (2002) Puberty and the maturation of the male brain and sexual behavior: recasting a behavioral potential. *Neurosci Biobehav Rev* 26: 381–391.

Sachs, B.D. (1980) Sexual reflexes of spinal male house mice. *Physiol Behav* 24: 489–492.

Sachs, B.D. (1995) Placing erection in context: the reflexogenic-psychogenic dichotomy reconsidered. *Neurosci Biobehav Rev* 19: 211–224.

Sachs, B.D., Akasofu, K., Citron, J.H., Daniels, S.B., and Natoli, J.H. (1994) Noncontact stimulation from estrous females evokes penile erection in rats. *Physiol Behav* 55: 1073–1079.

Sales, G.D. (1972) Ultrasound and aggressive behaviour in rats and other small mammals. *Anim Behav* 20: 88–100.

Sato, T., Matsumoto, T., Kawano, H., Watanabe, T., Uematsu, Y., Sekine, K., et al. (2004) Brain masculinization requires androgen receptor function. *Proc Natl Acad Sci USA* 101: 1673–1678.

Schulz, K.M., Richardson, H.N., Zehr, J.L., Osetek, A.J., Menard, T.A., et al. (2004) Gonadal hormones masculinize and defeminize reproductive behaviors during puberty in the male Syrian hamster. *Horm Behav* 45: 242–249.

Selmanoff, M.K., Goldman, B.D., and Ginsburg, B.E. (1977) Serum testosterone, agonistic behavior, and dominance in inbred strains of mice. *Horm Behav* 8: 107–119.

Shah, N.M., Pisapia, D.J., Maniatis, S., Mendelsohn, M.M., Nemes, A., and Axel, R. (2004) Visualizing sexual dimorphism in the brain. *Neuron* 43: 313–319.

Shrenker, P. and Maxson, S.C. (1986) Effects of the DBA/Bg Y chromosome on testis weight and aggression. *Behav Genet* 16: 263–270.

Shrenker, P., Maxson, S.C., and Ginsburg, B.E. (1985) The role of postnatal testosterone in the development of sexually dimorphic behaviors in DBA/1Bg mice. *Physiol Behav* 35: 757–762.

Simerly, R.B. (2002) Wired for reproduction: organization and development of sexually dimorphic circuits in the mammalian forebrain. *Annu Rev Neurosci* 25: 507–536.

Sinchak, K., Roselli, C.E., and Clemens, L.G. (1996) Levels of serum steroids, aromatase activity, and estrogen receptors in preoptic area, hypothalamus, and amygdala of B6D2F1 male house mice that differ in the display of copulatory behavior after castration. *Behav Neurosci* 110: 593–602.

Soukhoval-O'Hare, G.K., Schmidt, M.H., Nozdrachev, A.D., and Gozal, D. (2007) A novel mouse model for assessment of male sexual function. *Physiol Behav* 91: 535–543.

Tomihara, K. (2005) Selective approach to a male and subsequent receptivity to mounting comprise mate-choice behavior of female mice. *Jpn Psychol Res* 47: 22–30.

Vandenbergh, J.G. (1967) Effect of the presence of a male on the sexual maturation of female mice. *Endocrinology* 81: 345–349.

Van der Lee, S. and Boot, L.M. (1955) Spontaneous pseudopregnancy in mice. *Acta Physiol Pharmacol Neerl* 4: 442–444.

Wee, B.E. and Clemens, L.G. (1987) Characteristics of the spinal nucleus of the bulbocavernosus are influenced by genotype in the house mouse. *Brain Res* 424: 305–310.

White, N.R., Prasad, M., Barfield, R.J., and Nyby, J.G. (1998) 40- and 70-kHz vocalizations of mice (*Mus musculus*) during copulation. *Physiol Behav* 63: 467–473.

Whitney, G., Coble, J.R., Stockton, M.D., and Tilson, E.F. (1973) Ultrasonic emissions: do they facilitate courtship of mice? *J Comp Physiol Psychol* 84: 445–452.

Whitten, W.K. (1956) Modification of the oestrous cycle of the mouse by external stimuli associated with the male. *J Endocrinol* 13: 399–404.

Whitten, W.K. (1957) Effect of exteroceptive factors on the oestrous cycle of mice. *Nature* 180: 1436.

Wolff, R.J. (1985) Mating behaviour and female choice: their relation to social structure in wild-caught house mice (*Mus musculus*) housed in a semi-natural environment. *J Zool Lond (A)* 207: 43–51.

Xu, J., Taya, S., Kaibuchi, K., and Arnold, A.P. (2005) Sexually dimorphic expression of Usp9x is related to sex chromosome complement in adult mouse brain. *Eur J Neurosci* 21: 3017–3022.

Zuloaga, D.G., Morris, J.A., Monks, D.A., Breedlove, S.M., and Jordan, C.L. (2007) Androgen-sensitivity of somata and dendrites of spinal nucleus of the bulbocavernosus (SNB) motoneurons in male C57BL6J mice. *Horm Behav* 51: 207–212.

Chapter 23

Thermoregulatory behavior and the genetic correlation structure of adaptive phenotypes in house mice

Abel Bult-Ito

Introduction

Thermoregulatory nest-building behavior is a characteristic of an adaptive phenotype in populations of wild house mice, *Mus domesticus*. On a latitudinal cline along the east coast of the USA, wild-caught mice from Maine (north) build larger nests than mice from Florida (south) after one to three generations of breeding in the laboratory (Lynch, 1992). A similar latitudinal cline was found in mice from Colorado and Texas in the USA (Plomin and Manosevitz, 1974). Consequently, nest-building behavior appears to be under natural selection. Taken together with results from studies on laboratory house mice, to be discussed in this chapter, nest-building behavior can be considered an adaptation to cold ambient temperatures.

More than 30 years of behavioral genetics analyses of thermoregulatory nest-building behavior have contributed considerably to the understanding of the genetic and environmental contributions to the control of this behavior in the laboratory house mouse. In this chapter, I will review the results of replicated bidirectional selection for big and small nest-building in mice and extensive investigations into the correlated responses to this selection. These responses include nest-building behavior at 4–5°C, maternal nest-building behavior, reproductive measures, body weight, food consumption, brown adipose tissue, aggression, and circadian organization of behavior and physiology. I will conclude that the behavioral genetic analysis of nest-building behavior has contributed importantly to a better understanding of the genetic correlation structure of behavioral, physiological, and neurobiological traits of phenotypes adapted to cold and high latitude environments in the house mouse. Lastly, I will briefly discuss unresolved issues.

Replicated bidirectional artificial selection for thermoregulatory nest-building behavior at room temperature

Original bidirectional selection for thermoregulatory nest-building behavior

Lynch (1980) reported on the original bidirectional within-family selection experiment for high and low levels of thermoregulatory nest-building behavior. Lynch used the total nesting score as the selection criterion. The total nesting score is the total amount of cotton used during a 4-day testing period. A cotton roll of known weight is placed on the cage top and 24 hours later the roll is weighed again and the old nest removed. Additional cotton is added on the cage top if necessary. This procedure is repeated for 4 consecutive days. All the studies discussed in this chapter have used the same or very similar methods to quantify nest-building behavior.

After about three to four generations of selection, the replicate big and replicate small nest-building lines started to diverge from the randomly bred replicate control lines. By generation 15 of selection, the big and small nest-builders showed an eight-fold difference in the total nesting score. The narrow-sense heritability estimate of nest-building behavior, measured as the total nesting score, pooled across the entire experiment was 0.28 ± 0.05. This is similar to the heritability estimate of 0.33 obtained from parent–offspring regression of the founding population (Lynch, 1980). Narrow-sense heritability of a trait refers to the additive genetic variance contributing to the expression of this trait that is transferred from parent to offspring. Parent–offspring regression, using the slope of the linear regression between parent trait values on the x-axis and offspring trait values on the y-axis, is one way of estimating narrow-sense heritability (Bult and Lynch, 2000; Falconer, 1973). Cross-fostering studies have shown that maternal effects do not account for the response to bidirectional selection. Surprisingly, big nest-builder females depress the total nesting score of their offspring (Laffan, 1989).

About 70% of the variation among individuals was probably mainly due to environmental factors and some potential non-additive genetic effects, such as dominance, epistasis, and genotype × environment interactions (Falconer, 1981; Lynch et al., 1988). The environmental effects probably included ambient temperature differences in the animal room or between the top and the bottom of cage racks. Animals were tested for nest-building behavior in a period of 3 to 4 weeks with an ambient temperature of generally 22 ± 1°C. A variation of up to 2°C may have occurred in average ambient temperature from week to week, which would have led to increased nesting scores during weeks with slightly cooler temperatures.

Behavioral Genetics of the Mouse: Volume I. *Genetics of Behavioral Phenotypes*, eds. Wim E. Crusio, Frans Sluyter, Robert T. Gerlai, and Susanna Pietropaolo. Published by Cambridge University Press.

Table 23.1 Total nesting scores (grams of cotton per 4-day testing period) of females and males of seven inbred mouse lines, consisting of a total of 15 sub-lines, and one heterogeneous line (HS/Ibg) at different ambient temperatures.

<table>
<tr><th rowspan="2">Inbred line</th><th colspan="2">Sex</th></tr>
<tr><th>Females</th><th>Males</th></tr>
<tr><td>A/J</td><td>–
–
4.1 (24°C)[b]
–</td><td>42.0 (15°C)[a]
32.8 (20°C)[a]
5.5 (24°C)[b]
19.2 (25°C)[a]</td></tr>
<tr><td>BALB/cBYJ</td><td>14.0 (25–27°C)[c]</td><td>–</td></tr>
<tr><td>BALB/cIbg</td><td colspan="2">30.2 (4°C)[d]
12.5 (21°C)[d]
12.5 (25–27°C)[c]</td></tr>
<tr><td>BALB/cJ</td><td>–
–
29.2–31.2 (21.4°C)[e]
9.0 (24°C)[b]
–</td><td>72.0 (15°C)[a]
64.8 (20°C)[a]
35.6–36.8 (21.4°C)[e]
9.5 (24°C)[b]
48.0 (25°C)[a]</td></tr>
<tr><td>C3H/2Ibg</td><td colspan="2">16.0 (4°C)[d]
3.3 (21°C)[d]
1.6 (25–27°C)[c]</td></tr>
<tr><td>C3H/HeJ</td><td>1.9–2.1 (24°C)[b]
4.0 (25–27°C)[c]</td><td>2.1–2.5 (24°C)[b]
–</td></tr>
<tr><td>C3HeB/FeJ</td><td>8.8 (21.4°C)[e]</td><td>12.0 (21.4°C)[e]</td></tr>
<tr><td rowspan="2">C57BL/6J</td><td>15.2–16.0 (21.4°C)[e]
1.7–2.5 (24°C)[b]
3.6 (25–27°C)[c]</td><td>24–24.4 (21.4°C)[e]
4.7–5.6 (24°C)[b]
–</td></tr>
<tr><td colspan="2">23.3 (4°C)[d]
10.0 (21°C)[d]</td></tr>
<tr><td>C57BL/10J</td><td>–
–
–</td><td>40.8 (15°C)[a]
32.0 (20°C)[a]
18.4 (25°C)[a]</td></tr>
<tr><td>C57BL/10Sn</td><td>4.4 (25–27°C)[c]</td><td>–</td></tr>
<tr><td>CBA/J</td><td>1.2–4.7 (24°C)[b]</td><td>1.5–5.0 (24°C)[b]</td></tr>
<tr><td>DBA/1BG</td><td colspan="2">10.8 (4°C)[d]
1.9 (21°C)[d]
3.6 (25–27°C)[c]</td></tr>
<tr><td>DBA/1J</td><td>1.9 (24°C)[b]2.0
(25–27°C)[c]</td><td>2.5 (24°C)[b]
–</td></tr>
<tr><td>DBA/2J</td><td>–
–
–</td><td>26.0 (15°C)[a]
24.0 (20°C)[a]
16.8 (25°C)[a]</td></tr>
<tr><td>HS/Ibg</td><td>16.8–19.9 (21°C)[f]</td><td>21.1–22.7 (21°C)[f]</td></tr>
<tr><td>SJL/J</td><td>34.0 (21.4°C)[e]</td><td>32.8 (21.4°C)[e]</td></tr>
</table>

[a] Lee and Wong (1970); average daily amount of cotton used estimated from a figure and multiplied by four to obtain the total nesting score, no light–dark cycle indicated.
[b] Lynch and Hegmann (1972); total amount of cotton used during the first 4 days of a 5 day testing period to obtain the total nesting score, 16–8 hour light–dark cycle.
[c] Schneider et al. (1982); average daily amount of cotton used estimated from figures and multiplied by four to obtain the total nesting score, 16–8 hour light–dark cycle.
[d] Lynch (1992); total nesting scores, averaged for females and males, estimated from a figure, 16–8 hour light–dark cycle.
[e] Lee (1973); average daily amount of cotton used estimated from a figure and a table and multiplied by four to obtain the total nesting score, no light–dark cycle indicated.
[f] Lacy and Lynch (1979); total nesting scores from a table, 16–8 hour light–dark cycle.

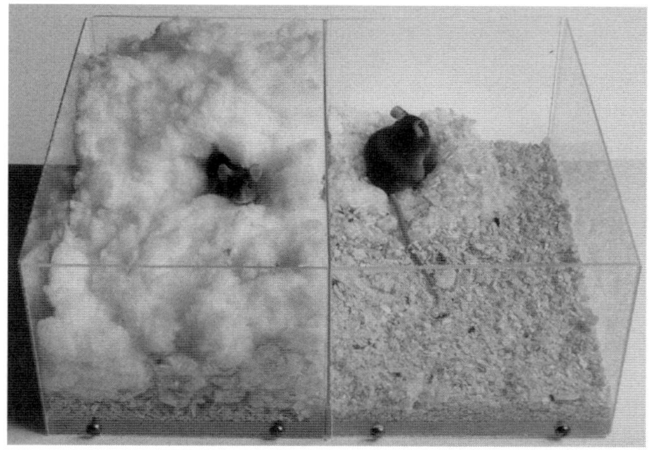

Figure 23.1 Representative examples of nests build by a big (left) and small (right) nest-builder male mouse. The mice had access to cotton for 24 hours.

Mice nest more at colder ambient temperatures (as discussed in the "Nest-building behavior at 4–5°C" section, later). Humidity was not controlled in the animal facilities and depending on the season and local weather conditions, it may have been substantially different from day to day and week to week. This could have affected the animal's perception of ambient temperature as well as the physical characteristics of the cotton used to measure nest-building behavior. Differences in litter size would have also affected the local ambient temperature environment of the pups, which could have affected their response to ambient temperatures as adults during the testing period (Lynch et al., 1976). Furthermore, maternal effects may have affected the nesting scores of the offspring, in which females with high nesting scores may have depressed the nesting scores of their offspring (Laffan, 1989), thus introducing additional variation among families. Significant family effects were always observed (Bult and Lynch, 1996, 1997, 2000; Lynch, 1980). Family effects constitute an important environmental factor affecting variation in nest-building behavior.

Males on average build larger nests than females. This sex difference is consistent from generation to generation of selection and at different ambient temperatures (Bult and Lynch, 1996, 2000; Laffan, 1989; Lynch, 1980). Others have also confirmed this sex difference (Lee, 1973; Plomin and Manosevitz, 1974; Table 23.1).

The selected lines reached apparent selection plateaus after approximately 15 (big nest-builder lines) and 25 (small nest-builder lines) generations of selection (Laffan, 1989; Lynch, 1994). By this time, the big and small nest-builders showed about a 30-fold difference in the total nesting score (Figures 23.1 and 23.2). At generation 36 of selection, Laffan (1989) performed reverse selection of the big and small nest-builder lines. Both big nest-builder lines and one small nest-builder line responded to reverse selection, which revealed that genetic variance for nest-building behavior remained in three of the four selected lines after reaching apparent selection limits.

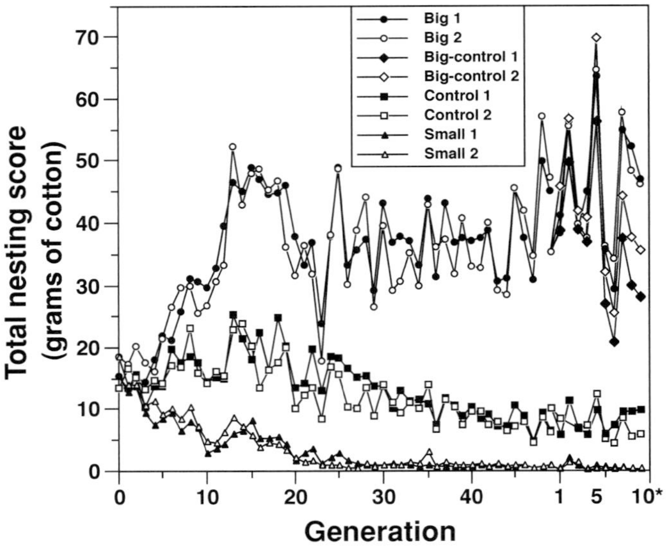

Figure 23.2 The average total nesting scores of the least-square means of the females and males of the replicate big-selected, big-control, control, and the small-selected nest-builder lines after 49 generations of original selection and subsequent 10 generations of renewed selection. For renewed selection, generation 49 is generation 0 of renewed selection and represents the F3 generations of the crosses of the replicate big lines, the replicate control lines, and replicate small lines after 46 generations of selection in the original selection experiment (Bult and Lynch, 1996). *Generation of renewed selection. For clarity, the small-control lines are shown in Figure 23.3. (Figure adapted from data published in Lynch, 1994, and Bult and Lynch, 2000.)

Interestingly, the genetic factors responsible for the response to reverse selection or the lack thereof were likely different for each line (Laffan, 1989). In the big nest-builder lines, one replicate line reached a new selection limit within a few generations of reversed selection. This indicated that the remaining genetic variability was probably due to rare unfavorable recessive alleles that were quickly fixed when selected for in the reversed selection experiment. The other big nest-builder replicate line continued to respond for the 10 generations of reversed selection. This finding is consistent with overdominance being responsible for the apparent selection limit of this line in the original selection experiment. Although this big nest-builder line continued to respond to reversed selection, it would not have been able to be selected towards the levels of the small nest-builder lines because most likely several alleles for building big nests had already been fixed in this line. In the small nest-builder lines, one replicate line appeared to be fixed for the genes influencing nest-building behavior because it did not respond to reversed selection. The other replicate line responded to 10 generations of reversed selection, which indicated that this line might have reached a selection limit due to opposing natural selection or a floor effect (Laffan, 1989). For example, some alleles contributing to building small nests may also have had pleiotropic effects on reproduction, e.g., resulting in infertility or very small litter size. Therefore, fixing these alleles in this line would have been opposed and some genetic variability would have been maintained.

To confirm that genetic variance remained in the selected lines and to test for changes in the genetic correlation structure, I crossed the replicate big nest-builder lines, the replicate randomly bred control lines, and the replicate small nest-builder lines at generation 46 of selection, i.e., big × big, control × control, and low × low, respectively, and produced F1, F2, and F3 generations for each of the big, control, and low crosses (Bult and Lynch, 1996). The parent lines were also maintained for the same number of generations to make direct comparisons possible. All the F1 crosses resulted in significant heterosis for nest-building behavior compared to the corresponding parent line generations, i.e., the offspring of the crosses built bigger nests on average than the offspring of the corresponding generation within each of the two parent lines. This finding of heterosis in the F1 crosses indicated that the replicate lines within each nest-building level, i.e., big, control, small, were genetically different, although they were phenotypically very similar. Heterosis also confirmed that dominance is in the big nest building direction (Lacy and Lynch, 1979; Laffan, 1989; Lee, 1973; Lynch, 1977; Lynch and Hegmann, 1972; Lynch and Sulzbach, 1984; Lynch et al., 1986).

Parent–offspring heritability estimates of the F3 on the F2 generations revealed remaining genetic variance for nest-building behavior with the highest level in the randomly bred control lines (0.27±0.13), followed by the big (0.16±0.10) and small nest-builder (0.07±0.10) lines. This finding revealed a potential to respond to renewed selection and break through the selection limits (Bult and Lynch, 2000).

Realized heritability estimates in the renewed selection experiment (Bult and Lynch, 2000) and the original selection experiment (Lynch, 1980) are bigger than those obtained in crosses of inbred lines, 0.29 to 0.30 and 0.28, respectively. Realized heritability is an estimate of the additive genetic variance that was actually used to change the trait under directional selection, and is another way of estimating narrow-sense heritability. Realized heritability is estimated by graphing cumulative selection differentials of successive generations of selection on the *x*-axis and response to selection on the *y*-axis at the corresponding generation. The slope of the linear regression is an estimate of realized heritability. The selection differential refers to the difference in the trait under selection between the animals successfully producing the next generation of offspring and the average of all the animals in that generation. The response to selection at a particular generation of selection refers to the difference between the average of the selected line and the randomly bred control line(s) (Bult and Lynch, 2000; Falconer, 1973). Using crosses between C57BL/6J and BALB/cJ, and among BALB/cJ, SJL/J, C57BL/6J, and C3HeB/FeJ inbred lines resulted in narrow-sense heritability estimates of 0.07–0.21 (Lee, 1973). Crosses between C3H/HeJ and DBA/1J and between BALB/cJ and C57BL/6J revealed very low (0.001–0.092) to negative narrow-sense heritability estimates, respectively (Lynch and Hegmann, 1972). A full diallel cross, using BALB/cIbg, C57BL/6J, C3H/2Ibg, and DBA/1BG inbred lines, revealed a narrow-sense heritability estimate of 0.23–0.32 (Lynch and Sulzbach, 1984). Lynch (1980) employed the HS/Ibg heterogeneous line that was the result of crosses among eight inbred lines (McClearn et al., 1970) for starting the original selection experiment. This resulted in larger genetic variance in this selection experiment than the crosses used in the Lee and Lynch and the Hegmann studies. At generations 22 and 23, the HS/Ibg line still showed robust heritability estimates for nest-building behavior of 0.31–0.32 (parent–offspring) and 0.23 to 0.29 (intraclass correlation of full sibs) (Lacy and Lynch, 1979). This may explain some of the differences in narrow-sense heritability estimated among these studies.

The parent–offspring heritability estimates using the F2 and F3 generations of crosses between replicate lines were lower (0.07 and 0.16) than the realized heritability estimates (0.29–0.30) after renewed selection (Bult and Lynch, 2000). Incidentally, these parent–offspring heritability estimates are very similar to the narrow-sense heritability estimates reported by Lee (1973) and Lynch and Hegmann (1972) using crosses between inbred lines. Selection experiments may sometimes provide higher estimates of the heritability of nest-building behavior in mice because they are obtained from measurements over several generations and several years. Therefore, the overall effects of environmental variation within a particular period, such as experienced during relatively short-term experiments using crosses, are potentially reduced by averaging out some of the extreme environmental variation over successive generations of selection. In addition, heritability estimates may have been underestimated when using crosses because of linkage disequilibrium among loci controlling nest-building behavior (Bult and Lynch, 2000). The linkage disequilibrium might take several generations to break up, which would have contributed additional additive genetic variation that bidirectional selection could act upon, and thereby increasing the realized heritability estimate.

Renewed bidirectional selection for thermoregulatory nest-building behavior

The F3 generation was used as the founding population for renewed replicated selection in the big and small nest-builder lines. Randomly bred replicate control lines were maintained at each nest-building level, i.e., big-control and small-control lines. Ten generations of renewed within-family selection revealed clear responses to selection in the big and small nest-builder lines compared to the randomly bred big-control and small-control lines, respectively. The difference in nest-building behavior between the big and small nest-builders increased to about 40-fold as a result of renewed selection (Figures 23.1–23.3).

Realized heritabilities were 0.29±002 and 0.30±0.004 in the big and small nest-builder lines, respectively (Bult and Lynch, 2000). These realized heritabilities were very similar to those in the original selection experiment (Lynch, 1980) and combined,

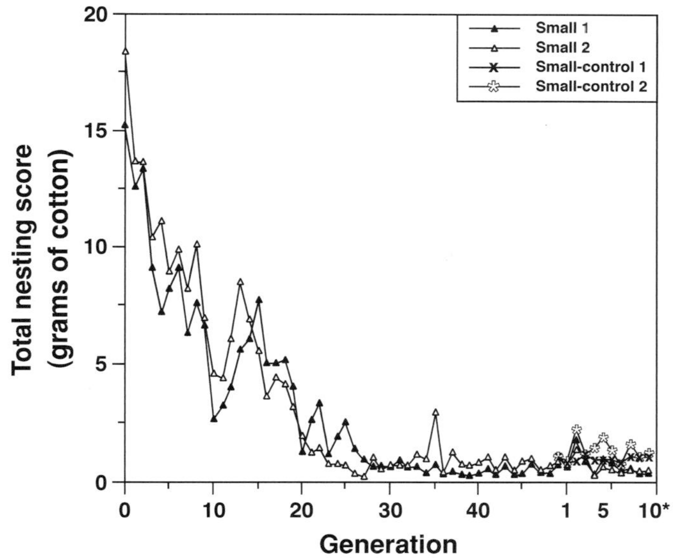

Figure 23.3 The average total nesting scores of the least-square means of the females and males of the replicate small-selected and small-selected nest-builder lines after 49 generations of original selection and subsequent 10 generations of renewed selection. For renewed selection, generation 49 is generation 0 of renewed selection and represents the F3 generations of the crosses of the replicate small lines after 46 generations of selection in the original selection experiment (Bult and Lynch, 1996). *Generation of renewed selection. (Figure adapted from data published in Lynch, 1994, and Bult and Lynch, 2000.)

these results show that many genes contribute to the control of nest-building behavior.

In summary, about 30% of the variance in nest-building behavior among individuals can be explained by additive genetic components, while about 70% is due to environmental and non-additive genetic components. In addition, nest-building behavior of house mice is a truly polygenic behavioral trait. Correlated responses to selection and their consequences for adaptive phenotypes are discussed in the next section.

Inbred lines and the genetics of nest-building behavior

Variation among inbred strains has also been important in revealing genetic variation that contributes to the regulation of nest-building behavior in mice. Table 23.1 shows the total nesting scores of seven inbred mouse strains, consisting of a total of 15 sub-lines, and one heterogeneous line (HS/Ibg) at different ambient temperatures. The total nesting score is quite variable among research groups and among different studies, but some general trends are clear. Different BALB substrains and the SJL/J strain use the largest amount of cotton to build a nest, while C3H and DBA substrains, and CBA/J mice build the smallest nests. C57BL substrains, A/J mice, and the HS/Ibg heterogeneous line have intermediate values (Table 23.1). Factors that account for the variation among the studies summarized in Table 23.1 include different ambient temperatures, different cage sizes, different light–dark cycles, different ages of the study animals, and different consistencies of the cotton used.

Genetic correlation structure in the original and renewed replicated selection experiment

Nest-building behavior at 4–5°C

Big, randomly bred control, and small nest-builder mice build larger nests at cold ambient temperatures of 4–5°C (total nesting score of 74, 24, and 3 g of cotton, respectively) than at room temperature around 21–22°C (46, 8, and <1 g, respectively; Bult and Lynch, 2000). Nest-building behavior at room temperature is highly correlated with nesting in the cold and, therefore, the differences between the big and small nest-builder lines and randomly bred control mice were maintained in the cold (Bult and Lynch, 2000; Lynch, 1994; Lynch et al., 1988).

This genetic correlation structure was confirmed in wild house mice along a latitudinal cline. All the mice nested more at 4–5°C than at room temperature and the Maine mice from the north build larger nests than those from Florida in the south at both temperatures. This genotype by environment effect is confirmed in several studies using inbred lines (Table 23.1).

In the renewed selection experiment, the tendency was for the two big renewed selected lines to nest more than the two big-control lines at 4–5°C, while this tended to be less in the two small renewed selected lines compared to the two small-control lines. However, the differences were not significant, probably because of relatively large replicate effects (Bult and Lynch, 2000). These results indicate no change in correlation structure for nest-building behavior at 22°C and 4–5°C as a result of the original selection experiment.

In summary, mice nest more at colder ambient temperatures and line differences are maintained or amplified. This is consistent with thermoregulatory nest-building behavior being an adaptation to cold ambient temperatures that is expected to increase individual survival and potential fitness.

Maternal nest-building behavior

A positive correlation is expected between thermoregulatory nest-building behavior and maternal nesting because larger maternal nests are expected to increase the survival of offspring in the nest and therefore contribute to fitness (see next section). Indeed, the correlation coefficient for thermoregulatory and maternal nesting ranges from 0.37 to 0.49 (Lynch and Possidente, 1978), 0.44 to 0.58 (Lynch, 1981), and 0.59 to 1.00 (Schneider et al., 1982). Although genetic factors contribute to differences in maternal and progesterone-induced nest-building behavior (Schneider et al., 1982), these differences do not appear to be due to plasma progesterone differences (Schneider and Lynch, 1984). Male mice build bigger thermoregulatory nests than female mice (Bult and Lynch, 1996, 1997, 2000; Lynch, 1980), and this result is, therefore, not unexpected. Nevertheless, this is interesting because other studies have shown that progesterone is a required hormone in mice for the elicitation of maternal nest-building behavior (Lisk, 1971; Lisk et al., 1969, 1973). Consistent with big nest-builders having larger litters (see next section), Gandelman (1975) revealed that the level of maternal nest-building behavior is positively correlated with the number of fetuses from day 4 through day 18 of gestation in Rockland-Swiss albino mice.

Reproductive measures

Percent infertile matings

In the original selection experiment, the number of infertile matings was significantly higher in the control and small nest-builder lines than in the big nest-builder lines (Lynch, 1980). This result shows a positive correlation between fertility and building big nests.

Litter size

In the original experiment, the big nest-builder lines had larger litter sizes at birth, weaning (20 days of age) and at 40 days of age than the small nest-builder lines, with the randomly bred control lines generally at intermediate levels (Bult and Lynch 1996, 1997, 2000; Lynch, 1980; Marsteller and Lynch, 1983).

In the renewed selection experiment, the two big renewed selected lines had larger litter sizes at birth and at weaning compared to the two randomly bred big-control lines. No correlated response was observed in the small selected and small-control lines (Bult and Lynch, 2000).

These results confirm that building big nests is associated with higher fitness, both at room temperature (Bult and Lynch 1996, 1997, 2000; Falconer, 1973; Lynch, 1980; Marsteller and Lynch, 1983) and at 4–5°C (Bult and Lynch, 1997; Lynch and Possidente, 1978).

Lifetime reproductive success

Lifetime reproductive success, measured as the number of offspring surviving to 40 days of age during a 6 month period, was higher in big nest-builder lines compared to control and small nest-builder lines at room temperature and at 4–5°C (Bult and Lynch, 1997). These findings are in agreement with results on litter sizes at birth, weaning, and at 40 days of age presented above, and again reveal a positive correlation between fitness and building big nests. So, why do all mice not build big nests and produce large litters? As shown by Lynch (1992), mice from Florida build smaller nests than mice in Maine, but unfortunately litter size has not been reported for these mice. In hot southern climates, selection for small body size is expected so that mice are able to lose heat more effectively and increase survival (Mayr, 1963). This may confer more fitness benefits in hot environments than the larger fitness effects of larger litter size that is genetically correlated with larger body size and the building of larger nests.

Body weight

Body weight is positively correlated with nest-building behavior in the original selection experiment (Bult and Lynch, 2000, Lynch, 1980). No change was observed after 10 generations of renewed selection, probably because the nest-building genes having pleiotropic effects on body weight were fixed in the original selection experiment (Bult and Lynch, 2000). In another study, positive genetic correlations between nest-building behavior and body weight were also found (Lynch et al., 1988). Interestingly, mice selected for large and small size also showed a positive genetic correlation between these two traits (Lynch and Roberts, 1984; Lynch et al., 1986).

This relationship was confirmed in wild house mice along the east coast of the USA. Animals from the north are heavier than mice from the south, while they also nest more in the north compared to the south. The intermediate locations follow the latitudinal cline for both traits (Lynch, 1992). In addition, this finding shows that body weight of wild house mice follows Bergmann's rule (Bergmann, 1847), i.e., larger body size at higher latitudes, and probably represents an adaptation to cold ambient temperatures by reducing heat loss as the surface to volume ratio declines with larger size (Mayr, 1963).

In summary, the genetic correlation structure for thermoregulatory nest-building behavior and body weight is quite robust as these traits show consistent positive correlations in mouse populations from diverse genetic backgrounds.

Food consumption

Food consumption, measured as grams of food eaten per gram body weight, both at room temperature and after cold acclimation is negatively correlated with thermoregulatory nest-building behavior in the original selection experiment (Lacy et al., 1978; Laffan, 1989), in mice selected for big and small size (Lynch and Roberts, 1984), and in crosses of four inbred lines (Lynch et al., 1988). This genetic correlation structure is considered to be an adaptation to cold ambient temperatures (Bult and Lynch, 2000). In addition, mice selected for large size showed lower food consumption than mice selected for small size, while randomly bred controls are intermediate (Roberts, 1981). This finding is consistent with the big nest-builders being larger than the small nest-builders, while the randomly bred controls are intermediate (Bult and Lynch, 2000; Lynch, 1980). The larger body size reduces heat loss, and consequently energy loss, as the surface to volume ratio declines with larger size (Mayr, 1963), which decreases energy intake requirements per body volume unit and, therefore, decreases food intake per gram body weight.

Surprisingly, in the renewed selection experiment the correlation structure reversed and the randomly bred big-control mice showed lower food consumption than the big nest-builder mice. This indicates that selection for big nests in the original selection experiment probably resulted in fixing genes that increased nest-building behavior while decreasing food consumption. The remaining genetic variance involved genes with a positive correlation between the two traits (Bult and Lynch, 2000). This remaining genetic correlation structure could therefore indicate a genetic constraint that would oppose further adaptive evolution of nest-building behavior in the cold if these genes were to be prominent in natural mouse populations.

Interscapular brown adipose tissue

Interscapular brown adipose tissue plays an important role in body temperature regulation and temperature acclimation (Chaffee and Roberts, 1971). However, this trait was not affected by 11 generations (Lacy et al., 1978) and 48 generations of selection for big and small nest-builders in the original selection experiment. Although replicates nested within each nest-building phenotype were significantly different after 48 generations of selection ($F_{3,69} = 25.75$, $P < 0.0001$), no overall selection line effects were found for the amount of fat-free dry weight of interscapular brown adipose tissue per gram body weight ($F_{2,3} = 0.66$, NS). The least-square mean dry weight values ($\pm$SE) of the replicate big, control, and small nest-building females were 0.900 ± 0.011 and 0.761 ± 0.008, 0.764 ± 0.010 and 0.650 ± 0.010, and 0.713 ± 0.009 and 0.928 ± 0.017, respectively. The least-square mean dry weight values ($\pm$SE) of the replicate big, control, and small nest-building males were 0.826 ± 0.009 and 0.703 ± 0.008, 0.724 ± 0.009 and 0.591 ± 0.012, and 0.645 ± 0.009 and 0.794 ± 0.012, respectively. The family effect nested within line and replicate ($F_{69,427} = 6.76$, $P < 0.0001$), sex effect ($F_{1,427} = 106.45$, $P < 0.0001$), and age effect ($F_{1,427} = 3.67$, $P < 0.001$) were significant. The line by sex ($F_{2,427} = 1.88$, NS) and line by age ($F_{1,427} = 2.24$, NS) interaction effects were not significant. For age, the selected line ($F_{2,3} = 0.40$, NS) and replicate nested within line ($F_{2,3} = 1.16$, NS) effects were not significant. The mice were exposed to 21–25 days of cold acclimation at 4°C before they were killed at 102–145 days of age by cervical dislocation, weighed, and the interscapular brown adipose tissue removed. The tissue was prepared as described previously and for statistical analysis the brown adipose tissue values were square root transformed (Lynch and Sulzbach, 1984).

These results are consistent with the absence of a cline in interscapular brown adipose tissue in wild house mouse populations along the east coast of the USA (Lynch, 1992). In addition, heritability estimates for interscapular brown adipose tissue in mice are generally low (Lacy and Lynch, 1979; Lynch and Sulzbach, 1984; Lynch et al., 1988), and genetic correlations between thermoregulatory nest-building behavior and interscapular brown adipose tissue are also very low (Lynch et al., 1988).

Aggression

Big nest-builders are more aggressive, i.e., shorter attack latency, than small nest-builders. In addition, mice selected for high levels of aggression build bigger nests than mice selected for low aggression (Sluyter et al., 1995). Furthermore, mice selected for high levels of aggression produce larger numbers of offspring than mice selected for low aggression (van Oortmerssen and Bakker, 1981), which are in the same direction as the mice bi-directionally selected for thermoregulatory nest-building behavior (Bult and Lynch, 1997, 2000). These findings in mice of distinct genetic background indicate a robust genetic correlation structure for aggression, number of offspring produced, and nest-building behavior.

Although the Y chromosome plays an important role in the regulation of aggression in mice (Carlier et al., 1990; Maxson, 1992; Sluyter et al., 1996), the Y chromosome does not appear to contribute to differences in nest-building behavior and aggressions in the nest-builder lines (Sluyter et al., 1997). This finding indicates that the genes with pleiotropic effects on nest-building behavior and aggression are not located on the Y chromosome.

Activity level

Small nest-builders are about 1.5–2.0 times more active while using a running wheel than the big nest-builders, while the activity level of randomly bred control mice was similar to that of the big nest-builders (Bult et al., 1993). Interestingly, mice selected for high levels of voluntary wheel running also nested less when body weight difference between the high-activity lines and randomly bred control lines were ignored (Carter et al., 2000). In these mice, higher levels of activity were correlated with smaller size on average (Koteja et al., 1999), which is in the same direction as in the mice selected for big and small nests (Bult and Lynch, 2000; Bult et al., 1993). I would like to note that in the nest-building lines activity level appears to refer to distance run (Bult et al., 1993), while in the voluntary running lines it refers to running speed (Carter et al., 2000). In mice selected for high and low levels of aggression, the more aggressive mice were less active during the dark period of a 12:12 hour light–dark cycle (Benus et al., 1988), while these mice nested more (Sluyter et al., 1995). In a recent study (Castillo et al., 2005), we did not find a difference in wheel-running activity level between big and small nest-builders. This may have been due to the use of a different type of wheel or a change from a 16:8 hour light–dark cycle to a 12:12 hour light–dark cycle.

Lower activity levels reduce energy expenditure. Therefore, the negative genetic correlation between activity level and thermoregulatory nest-building behavior in mice of different genetic backgrounds would be adaptive in colder environments by reducing metabolic demands due to heat loss and activity level resulting in higher survival rates.

Circadian organization of behavior and physiology

Behavioral rhythms

Nest-building behavior and wheel-running activity in mice are expressed in clear circadian (about 24 hours) cycles with high levels during the night and low levels during the day (Possidente et al., 1979). Circadian locomotor activity rhythms are very distinct between the big and small nest-builders. Small nest-builders have very robust and precise wheel-running activity patterns, while the big nest-builders display very fragmented lower amplitude activity rhythms that are less precisely entrained to 16:8 hour and 12:12 hour light–dark cycles (Amy et al., 2000; Bult et al., 1993, 2001; Yan et al., 2003).

One of the characteristics of the circadian system is its response to light in constant dark conditions. The master circadian clock in mice, the suprachiasmatic nucleus (SCN) (Weaver, 1998), has an endogenous rhythm that is a little less than 24 hours (Amy et al., 2000; Bult et al., 1993; Yan et al., 2003), and therefore this clock needs to shift its phase every day to stay entrained to the 24 hour light–dark cycle. The light-induced phase shifts accomplish this entrainment.

The big nest-builders phase delay their endogenous master circadian clock about 2–3 times more than the small nest-builders in response to the same 15 minute light pulse in constant dark. The control mice have intermediate values (Amy et al., 2000; Bult et al., 1993; Yan et al., 2003). These behavioral differences in response to a phase shifting light pulse are also reflected in higher light-induced immediate early gene FOS protein expression in the SCN, while control mice have intermediate values (Amy et al., 2000). In addition, light-induced clock gene mouse Period2 (*mPer2*) mRNA expression in the SCN is higher in big compared to small nest-builders, while *mPer1* mRNA expression does not differ (Yan et al., 2003).

We reported originally that the endogenous free-running period of wheel-running activity of small nest-builders was shorter than those of the randomly bred controls and big-nest-builders (Bult et al., 1993). However, we have not been able to confirm these differences in a recent study for wheel-running activity and body temperature rhythms (Castillo et al., 2005).

In summary, the big nest-builders have less robust behavioral circadian organization than the small nest-builders, which is also reflected at the level of the master circadian clock. These differences are characterized by larger fragmentation, lower amplitude rhythms, less precise entrainment to the light–dark cycle, and larger light-induced phase delays in big compared to small nest-builders. A very robust master circadian clock is assumed to be resistant to light (small phase shifts), while a weakly organized master circadian clock would be more easily shifted by light (larger phase shifts).

Body temperature

Mouse body temperature rhythms in a light–dark cycle peak during the night and are at their lowest levels during the day (Castillo et al., 2005; Connolly and Lynch, 1983; Hotz et al., 1987; Schneider and Lynch, 1984). Less stable circadian organization in the big compared to the small nest-builders is also confirmed with circadian body temperature rhythms. This is exemplified by less robust circadian rhythms of body temperature and less control of the rhythm amplitude in big compared to small nest-builders (Castillo et al., 2005).

Interestingly, the mean daily body temperature is not different between the big and small nest-builders (Castillo et al., 2005), which is in general agreement with earlier findings (Lacy et al., 1978; Lynch et al., 1988). Mice selected for different sizes that also differ in thermoregulatory nest-building behavior also do not differ in body temperature (Lynch and Roberts, 1984). In addition, no latitudinal cline was observed for body temperature in wild house mice, while nest-building behavior increased with more northern location (Lynch, 1992). Furthermore, the heritability estimate of main daily body temperature from a cross between C57BL/6J and C3H/2Ibg inbred mouse lines is near zero (Connolly and Lynch, 1983) and, therefore, no correlated response to selection was expected.

Although heritability estimates are population and environment specific, these findings are consistent with very low heritability estimates for body temperature in laboratory house mouse lines, which indicates that mean daily body temperature in mice was under strong selection in the past that eliminated most genetic variability (Lacy and Lynch, 1979; Lynch and Sulzbach, 1984). In addition, these combined results support Scholander et al.'s (1950) studies on arctic and tropical species, in which they found that mammalian mean daily body temperature appears not to be changed by natural selection.

Neuroanatomy of the master circadian clock

Small nest-builders were found to have a larger number of arginine-vasopressin (AVP) peptide containing neurons in the SCN compared to big nest-builders and randomly bred controls (Bult et al., 1992). Originally we thought that this difference might be a reflection of the difference in the circadian activity rhythms (Bult et al., 1993). However, a similar analysis on mice selected for high levels of voluntary wheel running did not support this hypothesis (Hochstetler et al., 2004). We have subsequently demonstrated that the difference in the number of AVP neurons in the SCN is not due to an absolute difference in the number of AVP expressing neurons, but rather due to a difference in release or transport of AVP (Van der Veen et al., 2005). The functional relevance of the AVP differences in the SCN remains to be determined, although I would like to propose a hypothesis. We know that AVP plays a role in body temperature regulation (Epstein et al., 1990; Kasting, 1989; Pittman et al., 1998) while the SCN controls most behavioral and physiological rhythms of the organism (Weaver, 1998). Consequently, I hypothesize that AVP in the SCN modulates the amplitude of the circadian body temperature rhythm. The larger differences in the amplitude of the circadian body temperature rhythm may be due to a more dynamic AVP regulatory system in the big nest-builders, while the amplitude in the small nest-builders may be less variable due to a more static AVP regulatory system in the SCN.

The second messenger protein kinase CβI (PKCβI) has a higher expression level in the SCN of the small nest-builders compared to big nest-builders, with the control mice having intermediate levels. The protein levels in the SCN of PKCα, PKCβII, and PKCγ did not differ among the nest-building lines (Bult et al., 2001). Whether the differences in PKCβI in the SCN are functionally linked to some of the circadian rhythm differences among the lines, e.g., phase shifting responses or robustness of the activity and body temperature rhythms, or to nest-building behavior remains to be elucidated.

Non-photic entrainment

Although photic signals, such as sunlight, are the most effective timing cues for the circadian system (Amy et al., 2000; Bult et al., 1993; Yan et al., 2003), non-photic cues, such as scheduled feeding, can shift circadian behavioral and physiological rhythms and entrain the SCN in the absence of light cues (Abe et al., 2007; Caldelas et al., 2005; Castillo et al., 2004; Lamont et al., 2005; Mendoza et al., 2005).

The small nest-builders overall are more likely to entrain to scheduled feeding in constant darkness than the big nest-builders, but the replicate lines differ significantly. In one small nest-builder replicate line, the mice always entrain to the scheduled feeding paradigm (49 of 49), while in the other replicate line this is 47.4% (nine of 19). In one big nest-builder replicate line, only 13.8% (four of 29) of the mice entrained to scheduled feeding, while in the other this was 46.7% (14 of 30) (Castillo et al., 2004; Hochstetler and Bult-Ito, unpublished). The large differences between the big replicate lines and between the small replicate lines, and the similarity of one big and one small replicate line, does not support a genetic correlation between nest-building behavior and entrainment to scheduled feeding.

Adaptive value of circadian organization

Most organisms on earth have endogenous control of circadian rhythm expression, which suggests that maintenance of circadian organization has adaptive value. However, the evolutionary significance of circadian rhythmicity in mammals remains a largely unexplored field (DeCoursey and Krulas, 1998; DeCoursey et al., 1997, 2000; Horton, 2001).

Why would mice that build bigger nests have less stable circadian organization? As previously discussed, building big nests is an adaptation to cold ambient temperatures and therefore to northern latitudes, as was shown in wild house mice (Lynch, 1992). The characteristics of the ambient light–dark cycle differ substantially along a latitudinal cline from the equator to the North Pole. While the light–dark cycle varies very little from a 12 hour dark and 12 hour light period throughout the year at the equator, in the Arctic the light–dark cycle changes from near constant light in summer to near constant dark in winter. These extreme changes in day length may require special circadian rhythm adaptations (Tavernier et al., 2004).

We have postulated two hypotheses that could explain the adaptive value of a less stable circadian system at higher latitudes (Tavernier et al., 2004). Firstly, the ability of a small mammal in the arctic to keep up with relatively large changes in day length, especially in spring and fall, is possibly one of the most important features of the master circadian clock that confers fitness benefits, e.g., predator avoidance (DeCoursey and Krulas, 1998; DeCoursey et al., 1997, 2000; Horton, 2001). This response would only be possible if the circadian system is less stable and more sensitive to light, resulting in larger light-induced phase shifts and therefore an increased ability to keep up with larger changes in day length. Secondly, a less stable master circadian clock at higher latitudes might be adaptive by allowing mice to be active during longer periods of time by extending their activity into the light period, or even to become ultradian or arrhythmic under constant environmental lighting conditions. This would permit the utilization of (limited) resources for longer periods of time or around the clock, respectively.

Some evidence exists that mammals at northern latitudes have weak circadian organization. For example, 50% of wild-caught and laboratory bred northern red-backed voles, *Clethrionomys rutilus*, from Fairbanks, Alaska, loose their circadian organization in constant dark, while this is 70% in constant light (Tavernier et al., 2004). In addition, two subspecies of reindeer, *Rangifer tarandus platyrhynchus* and *R. t. tarandus*, living in Svalbard and northern Norway, respectively, lost circadian organization in mid winter and mid summer, while they tracked the light–dark cycle in spring and fall (Van Oort et al., 2005). An earlier study had reported a loss of circadian rhythms of plasma melatonin rhythms in female reindeer in Finland (Eloranta et al., 1992). Further research on mice and other species will be necessary to test the two hypotheses.

Not all mammals lose their circadian organization under constant lighting conditions in the Arctic. For example, circadian aboveground activity rhythms of the Arctic black-capped marmots were observed throughout the Arctic summer (Semenov et al., 2001). In addition, captive Arctic foxes, wolves, and wolverines in Barrow, Alaska, showed circadian heart rate, EKG, and body temperature rhythms throughout the year (Folk, 1964). The maintenance of circadian rhythms during summer and winter in these arctic predators raises interesting questions. What role does predator avoidance play in relation to the two proposed hypotheses? Could selection for predator avoidance and consequently more stable circadian organization balance selection for weaker organization and explain the heterogeneity of circadian phenotypes in Alaskan voles (Tavernier et al., 2004). Is weaker circadian organization under stronger selective pressure than predator avoidance because of a need to be able to shift the endogenous clock during spring and summer or to exploit limited resources? Or, is selection for predator avoidance consistent with the ability to shift the circadian system quickly to be able to avoid predators during the spring and fall seasons?

Conclusions

The behavioral genetic analysis of nest-building behavior has contributed significantly to a better understanding of the genetic correlation structure of behavioral, physiological, and neurobiological traits of phenotypes adapted to cold and high-latitude environments in the house mouse. Bidirectional selection for thermoregulatory nest-building behavior in the mouse has revealed a very robust genetic correlation structure that is supported by independent studies on other mouse lines and wild house mice of diverse genetic backgrounds. A phenotype adapted to cold and high-latitude natural environments appears

Table 23.2 Correlated responses to bidirectional selection for nest-building behavior.

	Direction of selection	
Trait	**Big**	**Small**
Nest-building behavior at 4–5°C	+	–
Maternal nest-building behavior	+	–
Percent infertile matings	–	0
Litter size at 4–5°C and 22°C	+	–
Number of young weaned at 4–5°C and 22°C	+	–
Lifetime reproductive success at 4–5°C and 22°C	+	0
Body weight	+	–
Food consumption at 4–5°C and 22°C	– (+*)	+
Interscapular brown adipose tissue at 4°C	0	0
Wheel-running activity level	0	+ (0**)
Robustness of the circadian wheel-running activity rhythm	–	+
Amplitude of the circadian wheel-running activity rhythm	0	+
Light-induced phase shift in constant dark	+	–
Light-induced FOS in the SCN in constant dark	+	–
Free-running period of circadian wheel-running activity rhythm	0	– (0**)
Mean daily body temperature	0	0
Number of AVP neurons in the SCN	0	+
PKCα protein expression pattern in the SCN	0	0
PKCβI protein expression pattern in the SCN	–	+
PKCβII protein expression pattern in the SCN	0	0
PKCγ protein expression pattern in the SCN	0	0
Comparing big to small nest-builders (randomly bred controls not measured)		
Aggression	+	
Light-induced *mPer1* in the SCN in constant dark	0	
Light-induced *mPer2* in the SCN in constant dark	+	
Control of mean amplitude of daily body temperature rhythm	–	
Free-running period of circadian body temperature rhythm	0	
Robustness of circadian body temperature rhythm	–	
Release or transport of AVP in the SCN	+	
Entrainment to scheduled feeding in constant dark	0	

AVP: arginine-vasopressin; PKC, protein kinase C; SCN: suprachiasmatic nucleus.

Note: In both directions of selection for nest-building behavior, the +, –, and 0 refer to a significant increase, decrease, or no change, respectively, compared to the control lines. When comparing big to small nest-builders, the +, –, and 0 refer to a significant increase, decrease, or no change, respectively, of the big compared to the small nest-builders.

*Correlated response in renewed selection experiment. Data compiled from Amy et al. (2000), Bult and Lynch (1996, 1997, 2000), Bult et al. (1992, 1993, 2001), Castillo et al. (2004, **2005), Lacy et al. (1978), Laffan (1989), Lynch (1980, 1994), Lynch and Possidente (1978), Lynch and Sulzbach (1984), Lynch et al. (1988), Marsteller and Lynch (1983), Schneider et al. (1982), Sluyter et al. (1995), Van der Veen et al. (2005), and Yan et al. (2003).

to consist of a mouse that builds big nests while alone or while raising a litter, produces more and better quality offspring, is larger, eats less food per gram body weight, is more aggressive, is less active, and has weak circadian organization of physiological and behavioral traits (Table 23.2). The validity of this adaptive phenotype can now be tested by elucidating clinal variation in wild house mouse populations of the traits discussed in this chapter. A crucial component of such a study would be to replicate it across the globe to confirm this proposed adaptive phenotype independently.

I would not be surprised if such a phenotype adapted to cold ambient temperatures would also be applicable to other small, and perhaps larger, non-hibernating mammals. For example, different species of the genus *Peromyscus* reveal a cline in nest-building behavior, in which northern species build larger nests than southern species (King et al., 1964).

Unresolved issues

Testing of hypotheses related to adaptive phenotypes in the house mouse

Testing the hypotheses discussed in this chapter related to phenotypes adapted to cold and high latitude environments will be challenging and will require the efforts of several laboratories. Not only should latitudinal clines be replicated at different locations on earth, the laboratory experiments should go hand in hand with field experiments to confirm that the behaviors and physiological traits that are measured in the laboratory actually occur as expected in populations in their natural environment.

Identification of genes that contribute to the control of thermoregulatory nest-building behavior

The large and consistent differences in behavioral and physiological traits between the big and small nest-builders will allow us to start identifying genes that contribute to the control of nest-building behavior and correlated traits by using modern genomics and proteomics approaches. Because of the size of data sets produced by these techniques, the identification of candidate gene groups will be important to make these analyses manageable and relevant to the phenotype.

Identification of neural pathways controlling thermoregulatory nest-building behavior

The large phenotypic differences between the big and small nest-builders will also allow us to start identifying candidate neural pathways controlling nest-building behavior and associated correlated traits. The neuroanatomical differences in the SCN and circadian organization among the big and small nest-builder and control mice are good examples, and recently published data support an important role for serotonergic pathways in the regulation of thermoregulatory nest-building behavior (Greene-Schloessen et al., 2011).

References

Abe, H., Honma, S., and Honma, K. (2007) Daily restricted feeding resets the circadian clock in the suprachiasmatic nucleus of CS mice. *Am J Physiol Regul Integr Comp Physiol* 292: 607–615.

Amy, S.P., Chari, R., and Bult, A. (2000) Fos in the suprachiasmatic nucleus of house mouse lines that reveal a different phase-delay response to the same light pulse. *J Biol Rhythms* 15: 95–102.

Benus, R.F., Koolhaas, J.M., and van Oortmerssen, G.A. (1988) Aggression and adaptation to the light–dark cycle: role of intrinsic and extrinsic control. *Physiol Behav* 43: 131–137.

Bergmann, C. (1847) Uber die verhalnisse der warmeeokonomie der thiere zu inrer grosse. *Gottinger Studien* 3: 595–708.

Bult, A., Hiestand, L., VanderZee, E.A., and Lynch, C.B. (1993) Circadian rhythms differ between selected mouse lines: a model to study the role of vasopressin neurons in the suprachiasmatic nuclei. *Brain Res Bull* 32: 623–627.

Bult, A., Kobylk, M.E., and Van der Zee, E.A. (2001) Differential expression of protein kinase C βI (PKCβI) but not PKCα and PKCβII in the suprachiasmatic nucleus of selected house mouse lines, and the relationship to arginine-vasopressin. *Brain Res* 914: 123–133.

Bult, A. and Lynch, C.B. (1996) Multiple selection responses in house mice bidirectionally selected for thermoregulatory nest-building behavior: crosses of replicate lines. *Behav Genet* 26: 439–446.

Bult, A. and Lynch, C.B. (1997) Nesting and fitness: lifetime reproductive success in house mice bidirectionally selected for thermoregulatory nest-building behavior. *Behav Genet* 27: 231–240.

Bult, A. and Lynch, C.B. (2000) Breaking through artificial selection limits of an adaptive behavior in mice and the consequences for correlated responses. *Behav Genet* 30: 193–206.

Bult, A., Van der Zee, E.A., Compaan, J.C., and Lynch, C.B. (1992) Differences in the number of arginine-vasopressin-immunoreactive neurons exist in the suprachiasmatic nuclei of house mice selected for differences in nest-building behavior. *Brain Res* 578: 335–338.

Caldelas, I., Feillet, C.A., Dardente, H., Eclancher, F., Malan, A., Gourmelen, S., et al. (2005) Timed hypocaloric feeding and melatonin synchronize the suprachiasmatic clockwork in rats, but with opposite timing of behavioral output. *Eur J Neurosci* 22: 921–929.

Carlier, M., Roubertoux, P.L., Kottler, M.L., and Degrelle, H. (1990) Y chromosome and aggression in strains of laboratory mice. *Behav Genet* 20: 137–156.

Carter, P.A., Swallow, J.G., Davis, S.J., and Garland, T., Jr. (2000) Nesting behavior of house mice (*Mus domesticus*) selected for increased wheel-running activity. *Behav Genet* 30: 85–94.

Castillo, M.R., Hochstetler, K.J., Greene, D.M., Firmin, S.I., Tavernier, R.J., Raap, D.K., et al. (2005) Circadian rhythm of core body temperature in two laboratory mouse lines. *Physiol Behav* 86: 538–545.

Castillo, M.R., Hochstetler, K.J., Tavernier, R.J., Jr., Greene D.M., and Bult-Ito, A. (2004) Entrainment of the master circadian clock by scheduled feeding. *Am J Physio : Regul Integr Comp Physiol* 287: R551–R555.

Chaffee, R.R.J. and Roberts, J.C. (1971) Temperature acclimation in birds and mammals. *Ann Rev Physiol* 33: 155–197.

Connolly, M.S. and Lynch, C.B. (1983) Classical genetic analysis of body temperature rhythms in mice. *Behav Genet* 13: 491–500.

Decoursey, P.J. and Krulas, J.R. (1998) Behavior of SCN-lesioned chipmunks in natural habitat: a pilot study. *J Biol Rhythms* 13: 229–244.

Decoursey, P.J., Krulas, J.R., Mele, G., and Holley, D.C. (1997) Circadian performance of suprachiasmatic nuclei (SCN)-lesioned antelope ground squirrels in a desert enclosure. *Physiol Behav* 62: 1099–1108.

Decoursey, P.J., Walker, J.K., and Smith, S.A. (2000) A circadian pacemaker in free-living chipmunks: essential for survival? *J Comp Physiol* 186: 169–180.

Eloranta, E., Timisjarvi, J., Nieminen, M., Ojutkangas, V., Leppaluoto, J., and Vakkuri, O. (1992) Seasonal and daily patterns in melatonin secretion in female reindeer and their calves. *Endocrinol* 130: 1645–1652.

Epstein, Y., Horowitz, M., and Shapiro, Y. (1990) Hypothalamic and extrahypothalamic-limbic system vasopressin concentration under exposure to hyperosmolarity, hypovolemia and heat stress. *J Therm Biol* 15: 177–181.

Falconer, D.S. (1973) Replicated selection for body weight in mice. *Genet Res* 22: 291–321.

Falconer, D.S. (1981) *Introduction to Quantitative Genetics*, 2nd edn. Longman, New York.

Folk, G.E., Jr. (1964) Daily physiological rhythms of carnivores exposed to extreme changes in Arctic daylight. *Fed Proc* 23: 1221–1228.

Gandelman, R. (1975) Maternal nest-building performance and fetal number in Rockland-Swiss albino mice. *J Reprod Fert* 44: 551–554.

Greene-Schloessen, D.M., Van der Zee, E.A., Sheppard, D.K., et al. (2011) Predictive validity of a non-induced mouse model of compulsive-like behavior. *Behav Brain Res* 221: 55–62.

Hochstetler, K.J., Garland, T., Jr., Swallow, J.G., Carter, P.A., and Bult-Ito, A. (2004) Number of arginine vasopressin neurons in the suprachiasmatic nuclei is not related to level or circadian characteristics of wheel-running activity in house mice. *Behav Genet* 34: 131–136.

Horton, T.H. (2001) Conceptual issues in the ecology and evolution of circadian rhythms. In Norman, T.A. (ed.), *Handbook of Behavioral Neurobiology*, Vol. 12. Takahashi, J.S., Turek, F.W., and Moore, R.Y. (eds.), *Circadian Rhythms*. Kluwer Academic/Plenum Publishers, New York, pp. 45–57.

Hotz, M.M., Connolly, M.S., and Lynch, C.B. (1987) Adaptation to daily meal-timing and its effect on circadian temperature rhythms in two inbred strains of mice. *Behav Genet* 17: 37–51.

Kasting, N.W. (1989) Criteria for establishing a physiological-role for brain peptides – a case in point – the role of vasopressin in thermoregulation during fever and antipyresis. *Brain Res Rev* 14: 143–153.

King, J.A., Maas, D., and Weisman, R.G. (1964) Geographic variation in nest size among species of *Peromyscus*. *Evolution* 18: 230–234.

Koteja, P., Garland, T., Jr., Sax, J.K., Swallow, J.G., and Carter, P.A. (1999) Behaviour of house mice artificially selected for high levels of voluntary wheel running. *Animal Behav* 58: 1307–1318.

Lacy, R.C. and Lynch, C.B. (1979) Quantitative genetic analysis of temperature regulation in *Mus musculus*. I. Partitioning of variance. *Genetics* 91: 743–753.

Lacy, R.C., Lynch, C.B., and Lynch, G.R. (1978) Development and adult acclimation effects of ambient temperature on temperature regulation of mice selected for high and low levels of nest-building. *J Comp Physiol B* 123: 185–192.

Laffan, E.A. (1989) *Artificial Selection for Thermoregulatory Nest-Building in the House Mouse: Analysis of the Lines at their Limits*. Ph.D. thesis, Wesleyan University, Middletown, CT, USA.

Lamont, E.W., Diaz, L.R., Barry-Shaw, J., Stewart, J., and Amir, S. (2005) Daily restricted feeding rescues a rhythm of Period2 expression in the arrhythmic suprachiasmatic nucleus. *Neurosci* 132: 245–248.

Lee, C.T. (1973) Genetic analyses of nest-building behavior in laboratory mice (*Mus musculus*). *Behav Genet* 3: 247–256.

Lee, C.T. and Wong, P.T.P. (1970) Temperature effect and strain differences in the nest-building behavior of inbred mice. *Psychon Sci* 20: 9–10.

Lisk, R.D. (1971) Oestrogen and progesterone synergism and elicitation of maternal nest-building in the mouse (*Mus musculus*). *Anim Behav* 19: 606–610.

Lisk, R.D., Pretlow, R.A., 3rd, and Friedman, S.M. (1969) Hormonal stimulation necessary for elicitation of maternal nest-building in the mouse (*Mus musculus*). *Anim Behav* 17: 730–737.

Lisk, R.D., Russell, J.A., Kahler, S.G., and Hanks, J.B. (1973) Regulation of hormonally mediated maternal nest structure in the mouse (*Mus musculus*) as a function of neonatal hormone manipulation. *Anim Behav* 21: 296–301.

Lynch, C.B. (1977) Inbreeding effects upon animals derived from a wild population of *Mus musculus*. *Evolution* 31: 526–537.

Lynch, C.B. (1980) Response to divergent selection for nesting behavior in *Mus musculus*. *Genetics* 96: 757–765.

Lynch, C.B. (1981) Genetic correlation between two types of nesting in *Mus musculus*: direct and indirect selection. *Behav Genet* 11: 267–272.

Lynch, C.B. (1992) Clinal variation in cold adaptation in *Mus domesticus*: verification of predictions from laboratory populations. *Am Nat* 139: 1219–1236.

Lynch, C.B. (1994) Evolutionary inferences from genetic analyses of cold adaptation in laboratory and wild populations in the house mouse. In Boake C.R.B. (ed.), *Quantitative Genetic Studies of Behavioral Evolution*. University of Chicago Press, Chicago, pp. 278–301.

Lynch, C.B. and Hegmann, J.P. (1972) Genetic differences influencing behavioral temperature regulation in small mammals. I. Nesting by *Mus musculus*. *Behav Genet* 2: 43–53.

Lynch, C.B. and Possidente, B.P. (1978) Relationship of maternal nesting to thermoregulatory nesting in house mice (*Mus musculus*) at warm and cold temperatures. *Anim Behav* 26: 1136–1143.

Lynch, C.B. and Roberts, R.C. (1984) Aspects of temperature regulation in mice selected for large and small size. *Genet Res Camb* 43: 299–306.

Lynch, C.B., Roberts, R.C., and Hill, W.G. (1986) Heterosis among lines of mice selected for large and small size. 3. Thermoregulation. *Genet Res* 48: 95–100.

Lynch, C.B. and Sulzbach, D.S. (1984) Quantitative genetic analysis of temperature regulation in *Mus domesticus*. II. Diallel analysis of individual traits. *Evolution* 38: 527–540.

Lynch, C.B., Sulzbach, D.S., and Connolly (1988) Quantitative genetic analysis of temperature regulation in *Mus domesticus*. IV. Pleiotropy and genotype-by-environment interactions. *Am Nat* 132: 521–537.

Lynch G.R., Lynch, C.B., Dube, M., and Allen, C. (1976) Early cold exposure: effects on behavioral and physiological thermoregulation in the house mouse, *Mus musculus. Physiol Zool* 49: 191–199.

Marsteller F.A. and Lynch C.B. (1983) Reproductive consequences of food restriction at low temperature in lines of mice divergently selected for thermoregulatory nesting. *Behav Genet* 13: 397–410.

Maxson, S.C. (1992) Methodological issues in genetic analyses of an agonistic behavior (offense) in male mice. In Goldowitz, D., Wahlsten, D., and Wimer, R. (eds.), *Techniques in the Analysis of Brain and Behavior: Focus on the Mouse, Techniques in the Behavioral and Neural Sciences*, Vol. 8. Elsevier, Amsterdam, pp. 349–373.

Mayr, E. (1963) *Animal Species and Evolution*. Belknap, Cambridge, MA, USA.

McClearn, G.E., Wilson, J.R., and Meredith, W. (1970) The use of isogenic and heterogenic mouse stocks in behavioral research. In Lindzey, G. and Theissen, D.D. (eds.), *Contributions to Behavior-Genetic Analysis: The Mouse as a Prototype*. Appleton-Century-Crofts, New York, pp. 3–22.

Mendoza, J., Angeles-Castellanos, M., and Escobar, C. (2005) A daily palatable meal without food deprivation entrains the suprachiasmatic nucleus of rats. *Eur J Neurosci* 22: 2855–2862.

Pittman, Q.J., Chen, X.H., Mouihate, A., Hirasawa, M., and Martin, S. (1998) Arginine vasopressin, fever and temperature regulation. *Adv Brain Vasopressin Prog Brain Res* 119: 383–392.

Plomin, R.J. and Manosevitz, M. (1974) Behavioral polytypism in wild *Mus musculus. Behav Genet* 4: 145–157.

Possidente, B., Hegmann, J.P., O'Rourke, S., and Birnbaum, S. (1979) Control of behavioral circadian rhythms for nesting and wheel running in mice. *Physiol Behav* 23: 1141–1146.

Roberts, R.C. (1981) The growth of mice selected for large and small size in relation to food intake and the efficiency of conversion. *Genet Res* 38: 9–24.

Schneider, J.E. and Lynch, C.B. (1984) Investigation of a common physiological mechanism underlying progesterone-induced and maternal nesting in mice, *Mus musculus. J Comp Psychol* 98: 165–176.

Schneider, J.E., Lynch, C.B., Possidente B., and Hegmann, J.P. (1982) Genetic association between progesterone-induced and maternal nesting in mice. *Physiol Behav* 29: 97–105.

Scholander, P.F., Hock, R., Walters, V., and Irving, L. (1950) Adaptation to cold in arctic and tropical birds and mammals in relation to body temperature, insulation, and basal metabolic rate. *Biol Bull* 99: 259–271.

Semenov, Y., Ramousse, R., Le Berre, M., Vassiliev, V., and Solomonov, N. (2001) Above ground activity rhythm in Arctic black-capped marmot (*Marmota camtschatica bungei Katschenko 1991*) under polar day conditions. *Acta Oecol Intl J Ecol* 22: 99–107.

Sluyter, F., Bult, A., Lynch, C.B., van Oortmerssen, G.A., and Koolhaas J.M. (1995) A comparison between house mouse lines selected for attack latency or nest-building: evidence for a genetic basis for alternative behavioral strategies. *Behav Genet* 25: 247–252.

Sluyter, F., Bult, A. Meeter, F., Van Oortmerssen, G.A., and Lynch C.B. (1997) A comparison between F1 reciprocal hybrids of house mouse lines bidirectionally selected for attack latency or nest-building behavior: no Y chromosomal effects on alternative behavioral strategies. *Behav Genet* 27: 477–482.

Sluyter, F., van Oortmerssen, G.A., De Ruiter, A.J.H., and Koolhaas J.M. (1996) Aggression in wild house mice current state of affairs. *Behav Genet* 26: 489–496.

Tavernier, R.J., Largen, A.L., and Bult-Ito, A. (2004) Circadian organization of a subarctic rodent, the northern red-backed vole (*Clethrionomys rutilus*). *J Biol Rhythms* 19: 238–247.

Van der Veen, D.R., Castillo, M.R., Van der Zee, E.A., Jansen, K., Gerkema, M.P., and Bult-Ito A. (2005) Circadian dynamics of vasopressin in mouse selection lines: translation and release in the SCN. *Brain Res* 1060: 16–25

Van Oort, B.E.H., Tyler, N.J.C., Gerkema, M.P., Folkow, L., Blix, A.S., and Stokkan, K.-A (2005) Circadian organization in reindeer – these arctic animals abandon their daily rhythms when it is dark all day or light all night. *Nature* 438: 1095–1096.

Van Oortmerssen, G.A. and Bakker, T.C.M. (1981) Artificial selection for short and long attack latencies in wild *Mus musculus domesticus. Behav Genet* 11: 115–126.

Weaver, D.R. (1998) The suprachiasmatic nucleus: a 25-year retrospective. *J Biol Rhythms* 13: 100–112.

Yan, L., Hochstetler, K.J., Silver, R., and Bult-Ito, A. (2003) Relationship between phase shifts and *per* gene expression in mouse suprachiasmatic nucleus. *Neuroreport* 14: 1247–1251.

Chapter 24

Aggression

Stephen C. Maxson, Sietse F. de Boer, and Frans Sluyter

It is commonly accepted in biology that, throughout the animal kingdom, aggression is a highly functional form of social behavior that ultimately contributes to fitness and survival of individuals. Animals and humans fight each other for territory, status, mates, food, and other vital resources. Although most individuals engage in these social conflicts with appropriate and well-controlled forms of aggressive behavior, some individuals become extremely violent. Much of the scientific interest in aggression is motivated by these violent and presumably less adaptive forms aggression in humans, Regardless, a considerable part of our current knowledge on the ethology, neurobiology, and genetics of this behavior is based on experimental and laboratory studies of aggressive behavior in animals. Most (neuro)genetic work on aggressive behavior is performed in murine species that show high levels of territorial aggression characteristic of their dispersive social structure.

In this chapter, we will consider agonistic behaviors that are related to physical attacks of one mouse against another. As defined by Scott (1966), agonistic behaviors are adaptations for situations involving physical conflict or contests between members of the same species. There are two types of agonistic behaviors in both males and females related to attacks (Adams, 1980): offense and defense. These differ in motor patterns, bite or attack targets, adaptive function, and proximate neurobiological mechanisms. For offense, the motor patterns are chase, offensive upright posture, offensive sideways posture, and attacks (simple bites or bite and kick). The bite targets are primarily the hindquarters, back, and base of tail. The function is to obtain and retain resources such as space, food, and mates. For defense, the motor patterns are flight, defensive upright posture, defensive sideways posture, and attacks (lunge and bite). The bite targets are primarily the face and shoulders. The function is to defend one's self, mates, and progeny from attacks of other mice. In many – but certainly not all – agonistic encounters, both mice show offense and defense.

In the laboratory, agonistic encounters are staged either on familiar or unfamiliar ground. The former is often referred to as a resident–intruder test, and the latter is often referred to as a neutral-cage test. There are four types of encounters: male versus male opponent, male versus female opponent, female versus male opponent, and female versus female opponent. The opponents may be of the same genotypes as the subject, of a standard genotype for all subjects, or be of several genotypes. These are respectively known as homogenous set tests, standard opponent tests, and round-robin tests. We will consider separately what is known about some of the genetics of each type of encounter. We will review the literature on selected strains, inbred strains, and genetic crosses to identify contributions of chromosomes and genes to individual differences, and associations of phenotype with gene expression. The literature on transgenics, knockouts, and aggression will be reviewed in Volume II of this Handbook.

Selection studies

Selective breeding is a powerful tool to gain insight in the neurobiological and genetic underpinnings of aggressive behavior. It is based on the fact that the offspring of animals with a desired quality are more likely to demonstrate that quality than the progeny of random individuals. To put it more technically: the response to selective breeding R (the difference between offspring mean and mean of the unselected population) is equal to narrow-sense heritability (h^2) multiplied by the selection differential (S) (the difference between the mean of the selected parents and the unselected population mean, or $R = h^2S$) (see McClearn and DeFries, 1973).

Males attacking males

To our knowledge there are currently three pairs of mouse lines bidirectionally selected for intermale aggression: (1) the Finnish Turku Aggressive and Non Aggressive lines (TA and TNA); (2) the Dutch Short and Long Attack Latency lines (SAL and LAL); and (3) the American North Carolina 900 and 100 lines (NC900 and NC100).

The TA and TNA lines were initiated by Lagerspetz, in 1961 in Turku, Finland. They originate from an outbred colony of Swiss albino mice and have been selected on the basis of a rating score. TA males score high on a seven-point aggression

Behavioral Genetics of the Mouse: Volume I. *Genetics of Behavioral Phenotypes*, eds. Wim E. Crusio, Frans Sluyter, Robert T. Gerlai, and Susanna Pietropaolo. Published by Cambridge University Press.

scale and TNA males score low on this scale during a 7 minute neutral-cage test with a non-aggressive male mouse from the control line. For more details about the selection procedure, see Lagerspetz (1964); for more details about the behavioral structure of aggression and levels of aggression in different paradigms, see Nyberg et al. (2003, 2004). Significant differences between the breeding lines were already observed in the second generation. Narrow-sense heritability was estimated at 0.36 for the TA line.

The SAL and LAL lines originated from wild house mice caught in the vicinity of Groningen, The Netherlands, in 1971. The realized heritability was estimated at 0.31 for the SAL line. Aggressive behavior towards a standard opponent is tested in a cage, which combines very familiar and less familiar territory (Sluyter et al., 2003; Van Oortmerssen and Bakker, 1981). Males were tested on 3 consecutive days and selection was based on attack latency scores (averaged over 3 days). Significant differences between the lines were observed immediately, although selection for long attack latencies proved to be difficult (hence, the difference in generation between SAL and LAL). SAL males also appear to display abnormal forms of attack. For instance, they attack vulnerable body parts, females, immobilized intruders, and they disregard submissive signals from their opponents (Caramaschi et al., 2007; Natarajan et al., 2009a, 2009b; Sluyter and de Boer, unpublished findings).

The NC900 and NC100 lines were developed in North Carolina (USA). They were derived from an Institute of Cancer Research (ICR) Swiss Webster-type outbred stock. Selection is based on "attack frequency/attack latency" scores towards a standard opponent from the control line during a 10 minute test in a neutral cage. Selection resulted in a high-aggressive line (NC900) and a low-aggressive line (NC100). Line differences in intermale aggressive behavior were first obtained in the third generation (Cairns et al., 1983).

Although these lines come from different genetic backgrounds and different laboratories (with their own housing idiosyncrasies) and although selection criteria have been different (testing environment, type of test, type of opponent), the aggression levels of these pairs are qualitatively, but not quantitatively, comparable. In recent experiments, Caramaschi et al. (2007) and Natarajan et al. (2009a, 2009c) tested males of all three pairs in the same laboratory against the same type of opponent and found that the time spent on aggressive behavior was roughly similar across the different pairs. The aggressive lines spent on average 30–40% of their time on attack, threat, and chase, whereas the non-aggressive lines – with the exception of the TNA line – showed very few aggressive acts. In contrast to the TA and NC900 animals, SAL males confronted with a male opponent, will attack this animal within seconds and, if not stopped, will continue to fight, often seriously wounding or even killing the opponent. They are not inhibited by signals of submission and will even attack females, showing persistent and indiscriminate forms of aggression. Such excessive, maladaptive forms of aggression – also known as pathological aggression – are very relevant for the study of excessive and deviant human aggressive behavior that results in harm and injury.

Obviously the genotype plays a prominent role in the aggressive phenotype of these selection lines, but this does not mean that environmental factors, such as housing, are trivial in the development of aggression. For instance, prolonged isolation (>5 weeks) seems to increase aggressive behavior in these lines and push already aggressive animals so that they reach extremely high levels of aggression (Nyberg et al., 2004). In fact, isolation after weaning is essential for the development of differences between NC900 and NC100 males (Hood and Cairns, 1989).

However, variations in the maternal environment – either prenatal or postnatal – do not, by and large, seem to affect the line difference in aggressive behavior. Sluyter et al. (1996a) transferred embryos of SAL and LAL at the blastocyst stage to NMRI females and found that sharing the same maternal environment does not affect aggression differences in these lines. Also, when SAL newborns are fostered by LAL mothers and LAL newborns by SAL mothers, the line differences are maintained. However, there are small and test and genotype-specific effects, i.e., only attack latencies for the first day of testing are influenced by the maternal environment and only in reciprocal crosses between SAL and LAL (Sluyter et al., 1995; Van Oortmerssen et al., 1985; Van Zegeren, 1980). Moreover, maternal factors do not seem to be crucial for the development of aggressive behavior in the other two pairs. In TA and TNA, reciprocal cross-fostering only marginally affected aggressive behavior (Lagerspetz and Wuorinen, 1965). Also, neither cross-fostering of NC900 and NC100 pups to a dam of the control line (Cairns et al., 1983) nor rearing these lines in different groups (Hood and Cairns, 1989) changed their difference in aggression. However, it should be noted that variation in maternal behavior exists in at least two of these three pairs of selection lines (Benus and Röndigs, 1996; Gariépy et al., 2001).

The three pairs of selected strains have been investigated extensively for behavioral, physiological, and neuronal markers associated with differences in aggression. Not surprisingly, each pair is characterized by its own, typical set of data, due mainly to the specific research interests of the different laboratories. For instance, the population dynamics of SAL and LAL have been thoroughly investigated in the 1970s (reviewed by van Oortmerssen and Busser, 1989) while research in the 1980s focused on individual variation and behavioral strategies (summarized in Benus et al., 1991; Bohus et al., 1987). The 1990s were characterized by a more genetic approach to aggression and related behaviors in SAL and LAL (Sluyter et al., 1996b). Similar stories exist for the other two pairs. For example, the Finnish pair is the only pair in which the effects of post-weaning exposure to aggression and sexual behavior has been investigated (Sandnabba and Korpela, 1994; Sandnabba et al., 1994), while the American pair has also been studied from an immunological angle (e.g., Hood et al., 2003).

Despite the variety in research aims and findings these studies converge on the hypothesis that aggression is part of a more fundamental behavioral strategy of how to cope with environmental challenges. Aggressive animals are characterized by an active behavioral strategy, while low-aggressive animals usually cope passively with environmental demands, a phenomenon also observed in other species (Koolhaas et al., 1999). These fundamentally different behavioral strategies are a reflection of distinct neuroanatomical features and neurophysiological processes. For instance, SAL and LAL differ with regard to their stress reactivity (measured by corticosterone levels), distribution of vasopressin neurons, and response to serotonin in the hippocampus (Veenema et al., 2004). Results on neurophysiological differences in TA and TNA are limited, but serotonin contents of the forebrain seem to be lower in TA males than in TNA males (Lagerspetz et al., 1968). The NC lines are distinct with regard to γ-amino butyric acid (GABA) system functioning: the low-aggressive line produces high levels of endogenous GABA, while the high-aggressive line has low levels of endogenous GABA. The lines also differ in the levels of dopamine in nucleus accumbens, and serotonin in specific brain areas (Weerts et al., 1992). Taken together these studies suggest that variation in serotonergic, dopapminergic, and GABAergic neurotransmission have roles in individual differences in coping strategy and levels of aggression. Also, a neural model based on studies with SAL and LAL mice suggests that the attack latency of the SAL mice is due to increase in facilitatory input from the central amygdala to the hypothalamic attack area and decrease in inhibitatory input from the lateral septum to the hypothalamic attack area (Veenema and Neuman, 2007). Genetic effect in the same brain areas may underlie the coping strategy and aggressive behavior differences between the other two selected strains.

At this point it should be emphasized that the aggression levels of SAL, TA, and NC900 are much higher than those of most aggressive inbred strains and genetically modified animals. For instance, Miczek (1999) showed that the number of attacks per minute is four times higher in NC900 males as compared to 5-HT$_{1B}$ knockouts. However, the pathway to the heightened aggression in SAL, TA, and NC900 mice is not the same (Natarajan et al., 2009a). The sequence of offensive behaviors is different in the three strains. Threat (tail rattle, offensive upright posture, offensive sideways postures, and aggressive groom) are more likely to precede attacks in TA and NC900 mice than in SAL mice.

We have focused exclusively on territorial offense in these lines. We are not aware of any studies on competitive offense or defense.

Males attacking females

There is a conflicting literature on adult males attacking adult females. On the one hand, many adult females make a pheromone that inhibits males from attacking them at least in laboratory tests (Dixon and MacIntosh, 1971, 1975). On the other, in a semi-natural deme and territory, males attacked both intruder males and females (Rowe and Redfern, 1969). Regardless, there are strain differences for males attacking females for the TA and TNA selected strains and the SAL and LAL selected strains.

In a resident–intruder test, for instance, all isolated TA males attacked an intruder female while very few female-paired TA males or isolated TNA males did (Nyberg et al., 2004). There is also a study of male attacks on females in SAL and LAL (Benus et al., 1990). Male mice of each strain had zero, two, or nine confrontations with a male intruder of the Mas-Gro albino strain. After the last test with a male, they encountered on the next day a female of the same strain. With no prior history of male encounters, very few of the SALs or LALs attacked the females. With 2 or 9 days of encounters, more SALs than LALs attacked the females and the SALs attacked the females more after 9 than after 2 days of male encounters with males. Recently, Caramaschi et al. (2008) replicated and extended these findings by demonstrating that SAL males, but not TA and NC900 male mice, attacked female intruders in both the home-cage and in a novel-cage test after repeated aggressive (winning) experiences.

Female aggression: some general remarks

In general, females may be aggressive when non-pregnant and non-lactating (NPNL), when pregnant (P), when lactating (L), and when both pregnant and lactating (PL). Although in demes most females are pregnant and lactating, aggression of females in either NPNL, P, or L is usually studied in the laboratory. For any of the reproductive states, female aggression may be assessed with male or female (usually NPNL) opponents. It has been suggested that females show offense usually toward non-infanticidal females and defense usually to infanticidal males (Parmigiani et al., 1998, 1999).

Females attacking males

Isolated TA and TNA females (NPNL or P) do not show any aggressive behavior towards a male intruder in a neutral-cage test (Lagerspetz and Lagerspetz, 1983). The same is true for SAL and LAL females when non-pregnant and lactating (NPL) in a modified resident–intruder test (Compaan et al., 1993) and for NC900 and NC100 females when NPNL in a neutral-cage test (Hood and Cairns, 1988). However, when females of the latter pair are lactating, there is a difference in aggression. Lactating (i.e., 3–9 days postpartum) NC900 females are more aggressive towards an ICR male than lactating NC100 females (Hood and Cairns, 1988). Similar results have been reported for the TA and TNA pair of selected strains (Sandnabba, 1992). On days 3 and 9 postpartum, lactating TA females were more aggressive than TNA females in a resident–intruder test with a male standard opponent.

There are also lines of female mice bidirectionally selected for high and low aggression towards an intruder female (Ebert 1983, and see below). When NPL, females of these lines also differ in aggression against intruder males.

Females attacking females

There is a single selection study on female offensive agonistic behaviors. Wild male and female mice were trapped in Ohio (USA) during the fall of 1972 and transported to the laboratory. After weaning the offspring at 28 days of age, all females were isolated for 28 days. There were two resident–intruder tests. NPNL females were the residents and previously pair-housed C57BL/6 females the intruders. During the agonistic encounter, the female resident was rated on a seven-point scale, and an aggression score was based on this rating. Variation in offensive aggression was observed in the wild base population. From the base population, replicate lines of female mice differing in offensive aggression were selectively bred (Ebert, 1983). Inbreeding was avoided. Hence, for the high lines (H1, H2), high-scoring females were bred to the brothers of high-scoring females of another litter, while for the low lines (L1, L2), low-scoring females were bred to the brothers of low-scoring females from another litter. Randomly bred control lines were also maintained. The realized heritability at the eighth generation of selection was 0.12 for H1, 0.14 for H2, 0.34 for L1, and 0.46 for L2. The difference between H and L lines in realized heritability reflects an asymmetry in response to selection which has many possible explanations including scalar asymmetry and genetic asymmetry. (Falconer and McKay, 1996). In the H1 and H2 lines, aggression scores were highest in proestrus and metestrus and lowest in estrus and diestrus.

Several studies have attempted to correlate these line differences with other types of aggression. Thus, lactating H1 and H2 females show higher maternal aggression against a C57BL6 male intruder than L1 and L2. However, in a resident–intruder test, isolated males from the H and L lines do not differ in aggression against C57BL6 male intruder. Also, the latency to attack a cricket was similar across the six lines. Hence, in these lines there appears to be a (genetic) correlation between aggression of non-pregnant, non-lactating females and aggression of lactating females, but no genetic correlation of these types of female aggression with intermale aggression or predatory attacks. A similar lack of genetic correlation between male and female aggression has been observed in SAL and LAL (van Oortmerssen and Bakker, 1981). Here, males were tested against male intruders, and females against female intruders. Intruders were from the Mas-Gro albino strain.

There is an interesting story with regard to the NC900 and NC100 females (Hood and Cairns, 1988). When group-housed female mice were tested at 30 days of age, no line differences were observed. However, at 210 days of age NC900 females were more aggressive towards female intruders than NC100 females. In this selection pair, similar findings were reported for 180-day-old lactating females. Hood and Cairns conclude that when both sexes are tested in conditions that have been shown previously to be linked to the expression of aggression (the postpartum period in females; isolation in males), strong within-line similarities between males and females are found. They propose that these cross-sex similarities are mediated by a common genetic and central nervous system pathway.

Selection studies and genetic correlations

The results of these selection studies have been used to argue that the same (or different) genes are involved in intermale aggression and other behaviors, such as aggression towards a female, or biological traits, such as quantitative differences in hippocampal structures. There are, however, methodological and conceptual limits to these conclusions.

The TA and TNA, the SAL and LAL, and the NC900 and NC100 selection studies do not have replicate lines. Also, the limited size of the mouse colonies would have resulted in some inbreeding with random fixation of genes in each line. Thus, the associations between male and female aggression or between a type of aggression and a biological trait in these selected strains may not be due to genetic correlations. This issue is fully discussed by Henderson (1997). However, the selection study of Ebert (1983) had replicate lines, which means that between-line correlations for this selection study are more likely to be genetic. One should also take into account that the base populations of these four selection studies are not the same (SAL and LAL: wild mice trapped in The Netherlands; TA and TNA: Swiss mice; NC900 and NC100: ICR mice; H and L: wild mice from two locations in the USA). It is therefore very likely that the variant genes were not the same in each of these base populations, and that this may explain some of the observed differences in behavioral and biological correlations across selection studies.

Regardless, the three selection studies can be considered as being three replicate high and three replicate low lines, and the same association across the three of them may be evidence of genetic correlations. For each pair of selected strains there are similar line differences in attack latency and duration of offense behaviors in a resident–intruder test (Caramaschi et al., 2007). For all three selection pairs the prefrontal cortex levels of serotonin were lowest in the aggressive line than in the non-aggressive line. Also, for the three selection studies, there are similar line differences in 5-HT_{1A} receptor sensitivity as measured by the effects of a 5-HT_{1A} agonist on hypothermia. There are also differences among the three selection studies. Physiological hypoarousal is associated with aggression in the SAL versus LAL lines but not in the TN and TNA lines or NC 900 and NC 100 lines (Caramaschi et al., 2008a). Also, SAL males, but not TA or NC900 males attack females without any prior resident–intruder testing (Caramaschi et al., 2008b).

Inbred strain studies

Males attacking males

Territorial offense

The genetic analysis of aggressive behavior started more than 60 years ago when two papers appeared at approximately the same time (Ginsburg and Allee, 1942; Scott, 1942). Both studies investigated the aggressive behavior of the same three inbred strains. Although there were strain differences for aggression in each study, the rank order of the strains for aggression was not the same for each study. This was later found to be due to different handling procedures during testing.

Since then many studies have been carried out using inbred strains for the genetic analysis of aggressive behavior in male mice. The largest collection of data on inbred strain differences in male mouse aggression has been assembled by Roubertoux and colleagues in France. Their paper (Roubertoux et al., 2005) summarizes the data – collected over years – on the attack behavior of 12 inbred strains in different paradigms, against different opponents, and exposed to different housing conditions. Not surprisingly, and in agreement with the findings in selection lines, offensive intermale aggression of inbred strains depends on various factors. Social isolation is again an important factor. For instance, 13 days of isolation is sufficient to increase aggression in most inbred strains. There are strains, such as NZB (and to a lesser extent CAST, FVB, and SJL), which readily attack without isolation. These strains are the exception as in most inbred strains less than 50% of the animals attack without isolation. There are also strains which rarely attack even with isolation. A/J males, for instance, are famous for their placid nature and therefore often used as standard opponents, for which docility is the *conditio sine qua non*. Males from the 129 substrains seem relatively reluctant to attack as well, though stress regimes, such as exposure to uncontrolled chronic mild stress for several weeks, can increase their aggressive behavior (Sluyter and de Boer, unpublished findings). The aggressive behavior of C57BL/6J and DBA/2J, arguably the two most widely used strains in mouse research and progenitors of the BXD recombinant inbred strains, is rather unpredictable. B6 males, for instance, have been reported to be relatively non-aggressive by Roubertoux and Maxson in various paradigms, whereas Kulikov et al. (2005) found more than half of their B6 males to behave aggressively. The same is true for D2 males which have been found to be aggressive in some situations, but not in others. In general, D2 males rarely show attack behavior when they are not familiar with the test arena, whereas isolation and subsequent testing in their home cage nearly always result in aggression towards a standard opponent.

Similar to findings in selection lines, changes in the maternal environment do not lead to considerable changes in aggression. From the 10 studies that investigated the effects of cross-fostering on aggression in inbred strains (see Maxson, 1992, for a review of most studies), seven studies reported no effects of cross-fostering. However, mice from reciprocal crosses of inbred strains appear to be more vulnerable to effects of the maternal environment. Thus, using both reciprocal F1 crosses and backcrosses between the unrelated CBA/H and NZB strains, as well as ovarian grafting and cross-fostering to disentangle the effects of the prenatal and postnatal environment, Carlier et al. (1991) found an interaction between the effects of the Y chromosome and the postnatal maternal environments on aggression. In the postnatal maternal environment of their own inbred strain, males with the NZB Y chromosome were more aggressive than those with the CBA/H Y chromosome, whereas in the postnatal maternal environment of F1s, males with the CBA/H Y chromosome were more aggressive than males with the NZB Y chromosome.

Few studies have systematically investigated the neuronal correlates of aggressive behavior across a reasonable (>4) number of strains. Roubertoux et al. (2005) carried out a factor analysis, which included intermale aggressive behavior against a standard opponent (A/J) in different paradigms (neutral cage and resident–intruder) following different housing conditions (e.g., isolation or not), as well as a number of physiological variables, including testes weight, plasma testosterone levels, 5-HT, steroid sulfatase (STS), Met-enkaphalin, dynorphin, β-endorphin, and adrenocorticotropic hormone (ACTH). They found three factors, which explained 71% of the variation. The first factor was characterized by positive loading on aggression in the neutral cage with or without isolation and positive loading for testes weight, testosterone levels, and STS concentration and negative loadings with brain 5-HT levels, β-endorphin levels, and ACTH levels. The second factor was characterized by positive loadings from isolation-induced aggression in the resident–intruder test, paired testes weight, and negative loadings from brain 5-HT levels; while the third contained a mixture of physiological variables and relatively low (positive) loadings from aggression in the non-isolated neutral cage test. Two important conclusions can be drawn from this analysis. First, similar to aggression in selection lines, isolation is a crucial factor in the sense that it leads to both qualitatively (different factor) and quantitatively (higher scores) different levels of aggression. Second, aggression in the neutral cage without prior isolation, which Roubertoux et al. call "spontaneous" aggression, and aggression in the resident–intruder test after isolation are characterized by different physiological variables (and, accordingly, different sets of genes). For instance, whereas spontaneous aggression specifically covaries with steroid metabolism, isolation-induced aggression does not appear to be set apart by most of the physiological measures in this (meta-) analysis.

Unfortunately, the sizes of the intra/infrapyramidal mossy fiber (IIP-MF) terminal fields were not taken into account in this analysis. These hippocampal structures have been shown to correlate genetically with aggression in studies using selection lines (SAL and LAL, see Sluyter et al., 1994a), inbred strains (Guillot et al., 1994), and mutants (Sluyter et al., 1999). In general, smaller sizes of the IIP-MF terminal fields are associated with higher levels of aggressive behavior against standard opponents. This negative genetic correlation between the IIP-MF terminal fields and attack behavior seems to be specific for isolation-induced (2 weeks) aggression in the resident–intruder test as no association was found for spontaneous aggression in the neutral cage test across strains.

Competitive offense

Offense also occurs in the context of competition for food and water (Adams, 1980). Both males and females show competitive fighting, and, unlike territorial offense, it is not abolished by olfactory bulbectomy. Strain differences occur in competitive fighting. In one study (Hahn, 1983), male mice of four inbred strains (C57BL/6, DBA/2, BALB/cJ, and SJL/J) and their reciprocal F1 hybrids were isolated for 24 days and then were food deprived for 24 hours. After the 24 hours of food deprivation, an F1 male (C57BL/6 × SJL/J and food pellet were placed in the cage. The latency to attack and percentage of time the subject possessed the food were recorded. There were strain and F1 differences in both measures. The two measures were correlated across the strains. However, there was little or no correlation for latency to attack in this competitive test with that of the same strains and F1s in a territorial test. Both tests were in a neutral cage but opponent type was not the same for both tests. The low or absent correlation between attack latency in the territorial and in the competitive encounters suggests that different genes are involved in attack behavior in competitive versus territorial fighting.

Defense

As far as we know, strain variation in defense per se has not been investigated. However, it has been looked at in two strains (DBA/2 and C57BL/6) after low levels of conflict and pain (10 bites to the subject) or after high levels of conflict and pain (30–50 bites to the subject, Siegfried et al., 1990). At both levels, DBA/2 males show more escape than C57BL/6 males, whereas C57BL/6 show more immobility, defensive upright postures, and defensive sidewise postures than DBA/2. Also, shock-induced fighting has been assessed in several inbred strains and in F2s of C57BL/6 and BALB/c (Popova et al., 1993). Shock-induced fighting may be defensive. In this study there was no correlation across strains or in the F2 for offensive, shock-induced, or predatory attacks.

Males attacking females

Males of some inbred strains attack females. One of these strains is FVB/NtacfBr. Another is the closely related SJL/J strain. For FVB but not C57BL/6, males have been observed to attack females in the home cage and in mating tests (Canastar and Maxson, 2003). The females in the mating tests were FVBs either in estrus or diestrus as assessed by vaginal smears. Regardless of estrous state, FVB males had higher frequency of chase, upright offensive posture, sidewise offensive posture, and offensive attacks than C57BL/6 males. FVBs also had shorter latencies to chase, upright offensive posture, sidewise offensive posture, and offensive attacks than C57BL/6 males. There are identical strain differences for these agonistic behaviors in resident–intruder tests of male versus male encounters.

Females attacking males

When non-pregnant and non-lactating, the females of most inbred strains do not attack male opponents in a resident–intruder test. Females of the DBA/2J, C57BL/6, C3H/He, BALB/c, and ICR/JCL strains are not aggressive in this situation, whereas AKR/J females are (Ogawa and Makino, 1981). They have high levels of attacks and bites. The opponent in these tests was an ICR/JCL male.

When pregnant or lactating, females of the same strains were assessed for resident–intruder aggression with a male opponent. There were different strain patterns for attacks, bites, and chases. While pregnant and lactating, BALB/c showed very low levels of these three variables, AKR displayed high levels during pregnancy and even higher ones during lactation. During pregnancy, DBA/2 had high levels of attacks and bites, whereas C57BL/6 had high levels of chases. C3H showed essentially no aggression. During lactation, DBA/2s and C3Hs had more attacks, whereas C57BL/6 had more chases. ICRs had high levels of attack during both pregnancy and lactation. The findings for pregnant females were replicated in Ogawa and Makino (1984).

Subsequently, Svare examined the aggression of pregnant or lactating DBA/2 females and of C57BL/6 males in resident–intruder tests. The opponents were outbred R-S males. In these studies, both pregnant and lactating DBA/2 were more aggressive than C57BL/6 females in the same conditions (Svare, 1989). There were no differences in aggression of pregnant or lactating females for reciprocal F1 hybrids of these two strains. Also when cross-fostered, the levels of aggression of pregnant and lactating females resemble that of the biological and not the fostered parent.

In an earlier study, St. John and Corning (1973) compared the aggression of males attacking males and of lactating females towards females in four inbred strains. In their tests DBA/2 and BALB/c males and lactating females were more aggressive than C3H and C57BL/6 males and lactating females. Similarly, Jones and Brain (1987) found a strain correlation for aggression of isolated or female-paired males and aggression of lactating females. The strains were Tuck Ordinary (TO), Swiss Webster, NZW/Ola, BALB/c, C57BL/10, DBA/2, CBA/Ca, and C3H/He.

Across many but not all studies, it would appear that DBA/2 pregnant or lactating females are more aggressive toward males than are C57BL/6 females in the same conditions. It is not clear whether this variation in attacks directed toward males is due to differences in offense, defense, both, or another kind of aggression.

Females attacking females

In small groups, females will attack lactating females. There is strain variation in this behavior (Haug et al., 1992). In a resident–intruder test, the mean number of attacks is higher for Swiss and CFW than for TO females. Similarly, the mean number of attacks by C57BL/6 females is higher in a resident–intruder test than by C3H/He females. Also, the latency to first attack is lower for C57BL/6 than C3H females. Reciprocal F1 hybrids of these strains do not attack lactating females. The results were the same for C57BL/6 and C3H lactating female intruders.

This behavior is hormonally dependent. Intact males do not attack lactating females, whereas they do after gonadectomy. Both testosterone and estrogen treatments reverse this effect of gonadectomy. Conversely, treating neonatal females with the testosterone precursor or treating gonadectomized adult females with testosterone reduces the resident females attacks on lactating intruders. There may also be a role of neurosteroids in this type of female aggression. Long-term treatment with the neurosteroid dehydroepiandrosterone (DHEA) attenuates the resident's attacks on lactating females.

There are also studies of strain differences in so-called competitive fighting in food-deprived male–male and female–female pairs. In one study (Fredericson and Birnbaum, 1954), both male and female C57BL/10 mice fight vigorously to maintain possession of a piece of chow. In contrast, both male and female BALB/c mice share the food without fights. In general, although there are strain differences for this type of aggression, there are no within-strain sex differences.

Inbred strains and genetic correlation

There are two issues in using inbred strains to assess genetic correlations among behaviors or between behaviors and other biological traits. First, a large set of unrelated strains should be used. This attenuates the effects of correlations due to random fixation of genes during inbreeding. With rare exceptions, the strain sets employed in most existing studies have not been large enough to do this. Second, strain correlations due to direct genetic effects should be separated from indirect ones due to maternal environments. This identifies the mother or the individual as the site of gene action.

Regardless, there appears to be between-strain correlations for a single nucleotide polymorphism of the tryptophan hydroxylase 2 enzyme and aggression in both resident–intruder test without isolation of the subject and neutral-cage test with isolation of the subject (Kulikov et al., 2005), for a spontaneous stop codon mutant in exon eight of the monoamine oxidase A genes and aggression in a resident–intruder test with isolation (Scott et al., 2008). Tryptophan hydroxylase 2 is the rate-limiting enzyme in the postsynaptic synthesis of serotonin, and monoamine oxidase A is an enzyme in dopaminergic, adrenergic, and serotonergic neurons that degrades their respective neurotransmitter after reuptake from the synapse. There is also a spontaneous mutant in the C57BL/6 strain with effects on the size of the hippocampal mossy fibers and aggression in a modified resident–intruder test with isolation of the subject (Sluyter et al., 1999).

Chromosomes and genes

Here we review the research that seeks with breeding studies to identify the naturally-occurring individual chromosomal, quantitative trait loci (QTLs), and gene variants that cause individual differences in mouse aggression. All of this research has focused on male–male encounters. There is none on male–female encounters, female–male encounters, or female–female encounters. All have identified chromosome regions containing gene variants with effects on male–male offense. We will consider the following: (1) the male-specific part of the Y chromosome; (2) the recombining part of the heterosomes and the gene for STS; (3) QTLs and candidate genes from a study of the F2s of NZB × AJ; and (4) QTLs and candidate genes from a study of the F2s of NZB and C57BL/6.

Male-specific part of the Y chromosome

One part of the mouse Y chromosome is strictly passed from father to son: the non-recombining or non-pseudoautosomal part. To our knowledge there are five studies that related different non-recombining parts of the Y chromosome with differences in offensive behavior. These studies compared the following Y chromosomes: DBA1 versus C57BL/10, CBA/H versus NZB, SAL versus LAL, CBA/Fa versus C57BL6, and PHH versus PHL (see reviews by Roubertoux and Carlier, 2003; Maxson, 1996). The effects of each of these pairs of Y chromosomes have been shown to depend on the genetic background. Also, most if not all of the effects of the DBA1 and C57BL/10 Y-chromosome pair and of the CBA/H and NZB Y-chromosome pair on offense depend on the genotype of the opponent and thereby on the opponent's chemosignals.

There are approximately 13 protein-coding genes on the male-specific part of the Y chromosome (Mitchell, 2000). Seven of these are expressed in the brain at one or more stages of development (Lahr et al., 1995; Mayer et al., 2000; Xu et al., 2002). They are candidates for the effects on offense of gene variants in the male-specific part of the Y chromosome.

Recombining part of the heterosomes and STS

There is a single gene in the region of the Y chromosome that pairs and recombines with the homologous region of the X chromosome. This is the gene for the enzyme steroid sulfatase. Steroid sulfatase is found in liver and brain. In the brain, it is expressed in glial cells but not in neurons. In glial cells, STS regulates whether neurosteroids are sulfated or not. Free or sulfated neurosteroids appear to have opposite effects on neurotransmitters systems. For example, DHEAS and pregnalone sulfate act as antagonists of the GABA receptor, whereas allopregnanolone acts as an agonist of the GABA receptor.

Initially, it was shown that for crosses of CBA/H and NZB strains, this region of the heterosomes cosegregates with the proportion of males attacking offensively (Roubertoux et al., 1994). The association between brain STS and offense has also been shown across 11 inbred strains (Le Roy et al. 1999), and, as will be described below, in F2s derived from CBA/H and NZB strains. This association only occurs for male mice that have been housed with a female and were tested in a neutral cage. It does not occur for male mice that have been isolated or that were evaluated in a resident–intruder test.

The pairing region of the heterosomes of SAL and LAL mice also differs in its effect on attack latency (Sluyter et al., 1994b). Presumably, this is an effect of variants of the STS gene. The mice in this study were also housed with a female prior to testing.

QTLs and offense

There have been two studies using DNA markers to map QTLs associated with male offense in crosses of inbred strains.

The first study (Brodkin et al., 2002) tested F2 males from reciprocal crosses of NZB/B1NJ and A/J inbred strains. The encounters occurred in the resident's cage and the resident had been individually housed from weaning. The opponents were males of the 129T2/SvEmsJ strain. The opponent was dangled into the resident's cage by having its tail taped to the plastic cage top so that its front paws rested on the floor of the resident's cage. Each resident mouse had a test once a day for 3 days. Bites and lunges were scored as attacks by the resident. If a resident attacked an intruder on two of the three tests, it was classified as aggressive, whereas if the resident attacked an intruder on one or no days, it was classified as non-aggressive. The cosegregation of the phenotype was assessed for microsatellite DNA markers spaced about 20cM apart across the mouse genome. Two QTLs were identified. One was on chromosome 10 at 14.5 cM and one was on the X chromosome at 17.2 cM. These QTLs were confirmed with a backcross. The authors suggested that the gene for the QTLs on chromosome 10 might be that for diaglycerol kinase α subunit (*Dagk1*) and on the X chromosome might be that for the glutamate receptor subunit AMPA3 (*Gria3*). These suggestions need to be confirmed (Brodkin, 2005).

The second study (Roubertoux et al., 2005) tested F2 males from reciprocal crosses of NZB/BINJ and C57BL/6 inbred strains. There were two contrasting life history and testing conditions. For one group (Group 1), there were resident–intruder tests with the resident isolated from weaning. For the other group (Group 2), there were neutral-cage tests with the subjects cohabiting with a female from weaning. For both groups, the opponent was an A/J male. The phenotypes were latency to first tail rattle, frequency of tail rattles, latency to first attack, and frequency of attacks. In each group, the cosegregation of these phenotypes was assessed for simple sequence length polymorphisms (SSLP) DNA markers spaced about 22.5cM apart across the mouse genome. (Maxson, 2009, with kind permission from Springer Science of Business Media B.V.)

For Group 1, QTLs were reported on chromosomes 8 and 9 for tail rattling latency, on chromosomes 11 and 12 for tail rattling frequency, for attack latency on chromosome 12 and the X chromosome, and on chromosome 11 and 12 for attack frequency. For Group 2, QTLs were reported on chromosome 8, 9, and X for tail rattling latency, on chromosome 9 and X for attack latency, and on chromosomes 8, 11, 12, and X chromosomes for attack frequency. There are three points of note in these findings. First, some but not all QTLs are the same between groups. Genes in these QTLS would affect offense regardless of test environment or pretest housing. Second, within a group, some but not all QTLs are the same for each measure of aggression. Genes within these QTLs have effects on all measures of offensive aggression assessed in this study. Third, these QTLs accounted for the same amount of phenotypic variance for latency to tail rattle and latency to attack for both groups but very different amounts for attack frequency in each group (28% in Group 1; 87% in Group 2).

Candidate genes have been proposed for these QTLs. For Group 1 these are: *Got1* (glutamic oxaloacetate transaminase) for latency to tail rattling; *Gabra1* (gamma-aminobutyric acid A receptor, subunit $\alpha 1$), *Gria1* (glutamate receptor, ionotrophic AMPA1) for frequency of tail rattling; *Ar* (androgen receptor) and *Gabra3* (gamma-aminobutyric acid A receptor, subunit $\alpha 3$) for latency to attack; and *Gria1*, *Esr2* (β-estrogen receptor) for frequency of attacks. For Group 2 these are: *Got1*, *Htr1b* (5-hydroxytraptamine receptor, 1b), *Sts* (steroid sulfatase) for latency to tail rattling; *Sts* for latency to attack; and *Gabra3*, *Esr2*, *Sts* for frequency of attack. There were two reasons for selecting these as candidate genes: (1) each gene is within one of the QTL regions; and (2) other evidence is consistent with a role of this gene and its biology regarding offense. The role of all of these, except *Sts*, needs further confirmation.

There are several reasons for differences in QTLs for male offense in the resident–intruder tests between the studies of Brodkin et al. (2002) and Roubertoux et al. (2005). These include the use of: (1) F2s descended from different inbred strains; (2) intruders of different strains; (3) intruders tethered or free; and (4) different measures of offense.

Gene expression studies

To date few studies have used gene expression profiling as a tool to identify genes affecting variation in aggressive behavior. Applying SAGE, microarray analysis (Affymetrix) as well as *in situ* hybridization to hippocampal samples, Feldker et al. (2003a, 2003b) showed that non-aggressive LAL males, as compared to aggressive SAL males, display a higher expression of numerous cytoskeleton genes, including cofilin and several tubulin isotypes, suggesting that LAL males have longer hippocampal axonal and dendritic projections. LAL males also display higher expression of several calmodulin-related genes and genes encoding components of a MAPK cascade (raf-related oncogene and ERK2). In fact, the Affymetrix data show that nearly all differences in gene expression fall in this category, i.e., LAL exhibiting higher expression levels than SAL. The only exception is *gas5*, which does not encode a protein but several small RNAs. Although these results are promising, it cannot be excluded that at least some of these differences are the result of genetic drift and have nothing to do with the development of individual differences in aggression. If similar clusters of alterations in hippocampal gene expression profiles would emerge in the Finnish pair and/or the American pair, the evidence for the involvement of these genes in aggression would be much stronger. The hippocampus was chosen for study with the SAL and LAL mice because these strains differ in size of the projection of the hippocampal mossy fibers and because there is a strain correlation between size of the hippocampal mossy fibers and offensive aggression in males.

A similar approach has been used for comparing gene expression in the hypothalamus and preoptic region in a mouse line selected for high levels of maternal aggression and its control line (Gammie et al., 2006a, 2006b). These areas were chosen because of their involvement in maternal and aggressive behaviors. Maternal aggression was tested in the female's home cage on the 3rd, 4th, and 5th days postpartum against a male intruder. Pups were removed just before the test. The selected line had shorter attack latency, performed more attacks, and spent more time fighting than the control line. The realized heritability in the selected line was 0.403. A microarray representing 40 000 genes or expressed sequences was used. After confirmation with reverse-transcription polymerase chain reaction (RT-PCR), there were significant decreases in the selected line for gene expression of neurotensin and neuropeptide Y receptor 2, and there were also significant increases in the expression of a potassium channel subunit, *Kcna1*. There was also higher expression in the selected line of the genes for corticotrophin releasing factor binding protein, GABA receptor subunit alpha 1, adenosine 1A receptor, the transcription factors c-Fos and Erg-1, and 24 metabolic proteins.

Another approach to identify genes using microarrays is expression correlation. For instance, Fernandes et al. (2004) have carefully constructed a data set of baseline hippocampal gene expression profiles in a large panel of inbred mice and correlated gene expression data with intermale offensive aggressive behavior. Using this technique two genes were detected: *Comt* and *Fgf*. In particular, catechol-O-methyl transferase (COMT) (an enzyme involved in the metabolism of dopamine and noradrenaline), is a compelling candidate that merits further investigation. Thus, this finding agrees with other studies reporting that male mice having only one functional COMT gene were much more aggressive than males having two functional copies of the gene (Gogos et al., 1998). Catechol-O-methyl transferase also seems to be important in human aggression where individuals homozygous for one COMT allele (leading to lower activity) are generally more aggressive than those homozygous for the other variant (high activity) (Lachman et al., 1998). Hence, in humans the direction of the COMT effect on aggression is similar to that found in the Fernandes study: the lower the level and/or activity of COMT, the more aggressive the individual.

Summary and conclusions

We believe that the material here reviewed is consistent with five conclusions: (1) Offense (and perhaps defense) are heritable in both male and female mice. (2) Some but not all the genes with effects on male–male, male–female, female–male, and female–female offense may be the same. (3) The effects of genetic variants on offense depend on life history and test situation parameters. (4) Some naturally-occurring variants of chromosome regions and known genes have been tentatively or firmly shown to effect offense. (5) As we will consider in Volume II of this Handbook, there are more than 90 knockout mutants with effects on offense (for recent reviews, see Maxson, 2009; Maxson and Canastar, 2007). It has not been determined whether there are naturally-occurring variants of these genes, nor have the effects of any such variants on agonistic behaviors in either males or female mice been studied.

References

Adams, D.B. (1980) Motivational systems of agonistic behavior: a comparative review and neural model. *Aggress Behav* 4: 295–346.

Benus, R.F., Bohus, B., Koolhaas, J.M., and Van Oortmerssen, G.A. (1991) Heritable variation for aggression as a reflection of individual coping strategies. *Experientia* 47: 1008–1019.

Benus, R.F., Den Daas, S., Koolhaas, J.M., and Van Oortmerssen, G.A. (1990) Routine formation and flexibility in social and non-social behaviour of aggressive and non-aggressive male mice. *Behaviour* 112: 176–193.

Benus, R.F. and Röndigs, M. (1996) Patterns of maternal effort in mouse lines bidirectionally selected for aggression. *Anim Behav* 51: 67–75.

Bohus, B., Benus, R.F., Fokkema, D.S., Koolhaas, J.M., Nyakas, C., Van Oortmerssen, G.A., et al. (1987) Neuroendocrine states and behavioural and physiological responses. *Prog Brain Res* 72: 57–71.

Brodkin, E.S. (2005) Quantitative trait locus analysis of aggressive behaviours in mice. *Novartis Found Symp* 268: 57–69.

Brodkin, E.S., Goforth S.A., Keene, A.H., Fossella, J.A., and Silver, L.M. (2002) Identification of quantitative trait loci that affect aggressive behavior in mice. *J Neurosci* 22: 1165–1170.

Cairns, R.B., MacCombie, D.J., and Hood, K.E. (1983) A developmental-genetic analysis of aggressive behavior in mice I: behavioral outcomes. *J Comp Psychol* 97: 69–89.

Canastar, A. and Maxson, S.C. (2003) Sexual aggression in mice: effects of strain and estrous state. *Behav Genet* 33: 521–528.

Caramaschi, D., de Boer S.F., de Vries, H., and Koolhaas, J.M. (2008b) Development of violence in mice through repeated victory along with changes in prefrontal cortex neurochemistry. *Behav Brain Res* 189: 263–272.

Caramaschi, D., de Boer, S.F., and Koolhaas, J.M. (2007) Differential role of the 5-HT1A receptor in aggressive and non-aggressive mice: an across-strain comparison. *Physiol Behav* 90: 590–601.

Caramaschi, D., de Boer S.F., and Koolhouse, J.M. (2008a) Is hyper-aggressiveness associated with physiological hypoarousal? A comparative study on mouse lines selected for high and low aggressiveness. *Physiol Behav* 95: 591–598.

Carlier, M., Roubertoux, PL., and Pastoret, C. (1991) The Y chromosome effect on intermale aggression in mice depends on the maternal environment. *Genetics* 129: 231–236.

Compaan, J.C., van Wattum, G., de Ruiter, A.J., van Oortmerssen, G.A., Koolhaas, J.M., and Bohus, B. (1993) Genetic differences in female house mice in aggressive response to sex steroid hormone treatment. *Physiol Behav* 54: 899–902.

Dixon, A.K. and MacIntosh, J.H. (1971) Effects of female urine upon social behaviour in adult male mice. *Anim Behav* 19: 138–140.

Dixon, A.K. and MacIntosh, J.H. (1975) The relationship between physiological condition of female mice and effects of their urine on the social behaviour of adult males. *Anim Behav* 23: 513–520.

Ebert, P.D. (1983) Selection for aggression in a natural population. In Simmel, E.C., Hahn, M.E., and Walters, J.K. (eds.), *Aggressive Behavior: Genetic and Neural Approaches*. Lawrence Erlbaum, Hillsdale, NJ, USA, pp. 103–127.

Falconer, D.S. and Mackay, T.F.C. (1996) *Introduction to Quantitative Genetics*, 4th edn. Addison Wesley Longman, Harlow, Essex, UK.

Feldker, D.E., Datson N.A., Veenema, A.H., Meulmeester, E., de Kloet E.R., and Vreugdenhil, E. (2003a) Serial analysis of gene expression predicts structural differences in hippocampus of long attack latency and short attack latency mice. *Eur J Neurosci* 17: 379–387.

Feldker, D.E., Datson, N.A., Veenema, A.H., Proutski, V., Lathouwers, D., de Kloet, E.R., et al. (2003b) Gene chip analysis of hippocampal gene expression profiles of short- and long-attack-latency mice: technical and biological implications. *J Neurosci Res* 74: 701–716.

Fernandes, C., Paya-Cano, J.L., Sluyter, F., D'Souza, U., Plomin, R., and Schalkwyk, L.C. (2004) Hippocampal gene expression profiling across eight mouse inbred strains: towards understanding the molecular basis for behaviour. *Eur J Neurosci* 19: 2576–2582.

Fredericson, E. and Birnbaum, E.A. (1954) Competitive fighting between mice with different hereditary backgrounds. *J Genet Psychol* 85: 271–280.

Gammie, S.C., Auger, A.P., Jessen, H.M., Vanto, R.J., Awad, T.A., and Stevenson, S.A. (2006b) Altered gene expression in mice selected for high maternal aggression. *Genes Brain Behav* 6: 432–443.

Gammie, S.C., Garland, Jr., T., and Stevenson, S.A. (2006a) Artificial selection for increased maternal defense behavior. *Behav Genet* 36: 713–722.

Gariépy, J.L., Nehrenberg, D., and Mills-Koonce, R. (2001) Maternal care and separation stress in high- and low-aggressive line mice. *Dev Psychobiol* 38: 203.

Ginsburg, B.G. and Allee, W.C. (1942) Some effects of conditioning on social dominance and subordination in inbred strains of mice. *Physiol Zool* 15: 485–506.

Gogos, J.A., Morgan, M., Luine, V., Santha, M., Ogawa, S., Pfaff, D., et al. (1998) Catechol-*O*-methyltransferase-deficient mice exhibit sexually dimorphic changes in catecholamine levels and behavior. *Proc Natl Acad Sci USA* 95: 9991–9996.

Guillot, P.V., Roubertoux, P.L., and Crusio, W.E. (1994) Hippocampal mossy fiber distributions and intermale aggression in seven inbred mouse strains. *Brain Res* 660: 67–69.

Hahn, M.E. (1983) Genetic "artifacts" and aggressive behavior. In Simmel, E.C., Hahn, M.E., and Walters, J.K. (eds.), *Aggressive Behavior: Genetic and Neural Approaches*. Lawrence Erlbaum, Hillsdale, NJ, USA, pp. 67–88.

Haug, M., Johnson, F.J., and Brain, P.F. (1992) Biological correlates of attack on lactating intruders by female mice: a topical review. In Bjorkqvist, K. and Pirkko, P. (eds.), *Of Mice and Women: Aspects of Female Aggression*. Academic Press, New York, pp. 381–393.

Henderson, N.D. (1997) Spurious associations in unreplicated selection lines *Behav Genet* 27: 145–154.

Hood, K.E. and Cairns, R.B. (1988) A developmental analysis of aggressive behavior in mice. II. Cross-sex inheritance. *Behav Genet* 18: 605–619.

Hood, K.E. and Cairns, R.B. (1989) A developmental-genetic analysis of aggressive behavior in mice: IV. Genotype–environment interaction. *Aggress Behav* 15: 361–380.

Hood, K.E., Dreschel, N.A., and Granger, D.A. (2003) Maternal behavior changes after immune challenge of neonates with developmental effects on adult social behavior. *Dev Psychobiol* 42: 17–34.

Jones, S.E. and Brain, P.F. (1987) Performances of inbred and outbred laboratory mice in putative tests of aggression. *Behav Genet* 17: 87–96.

Koolhaas, J.M., Korte, S.M., de Boer, S.F., van der Vegt, B.J., van Reenen, C.G., Hopster, H., et al. (1999) Coping styles in animals: current status in behavior and stress-physiology. *Neurosci Biobehav Rev* 23: 925–936.

Kulikov, A.V., Osipova, D.V., Naumenko V.S., and Popova, N.K. (2005) Association between Tph2 gene polymorphism, brain tryptophan hydroxylase activity and aggressiveness in mouse strains. *Genes Brain Behav* 4: 482–485.

Lachman, H.M., Nolan, K.A., Mohr P., Saito, T., and Volavka, J. (1998) Association between catechol *O*-methyltransferase genotype and violence in schizophrenia and schizoaffective disorder. *Am J Psychiatry* 155: 835–837.

Lagerspetz, K.M.J. (1964) Studies on the aggressive behavior of mice. *Ann Finnish Acad Sci Series B* 131: 1–131.

Lagerspetz, K.M.J. and Lagerspetz, K.Y.H. (1983) Genes and aggression. In Simmel, E.C., Hahn, M.E., and Walters, J.K. (eds.), *Aggressive Behavior: Genetic and Neural Approaches*. Lawrence Erlbaum, Hillsdale, NJ, USA, pp. 89–101.

Lagerspetz, K.M.J., Tirri, R., and Lagerspetz, K.Y.H. (1968) Neurochemical and endocrinological studies in mice selectively bred for aggressiveness. *Scand J Psychol* 9: 157–160.

Lagerspetz, K.M.J. and Wuorinen, K. (1965) A cross-fostering experiment with mice selectively bred for aggressiveness and non-aggressiveness. *Rep Inst Psychol Univ Turku* 17: 1–6.

Lahr, G., Maxson, S.C., Mayer, A., Just, W., Pilgrim, C., and Reisert, I. (1995) Transcription of the Y chromosomal gene, Sry, in adult mouse brain. *Mol Brain Res* 33: 179–182.

Le Roy, I., Mortaud, S., Tordjman, S., Donsez-Darcel, E., Carlier, M., Degrelle, H., et al. (1999) Genetic correlation between steroid sulfatase concentration and initiation of attack behavior in mice. *Behav Genet* 29: 131–136.

Maxson, S.C. (1992) Methodological issues in genetic analyses of an agonistic behavior (offense) in male mice. In Goldowitz, D., Wahlsten, D., and Wimer, R. (eds.), *Techniques for the Genetic Analysis of Brain and Behavior: Focus on the Mouse, Techniques in the Behavioral and Neural Sciences*, Vol. 8. Elsevier, Amsterdam, pp. 349–373.

Maxson, S.C. (1996) Searching for candidate genes with effects on an agonistic behavior, offense, in mice. *Behav Genet* 26: 471–476.

Maxson, S.C. (2009) The genetics of offensive aggression in mice. In Kim, Y-K. (ed.), *Handbook of Behavior Genetics*. Springer, New York, pp. 301–316.

Maxson, S.C. and Canastar, A. (2007) The genetics of aggression in mice. In Flannery, D.J., Vazony, A.T., and Waldman, I.D. (eds.), *The Cambridge Handbook of Violent Behavior and Aggression*. Cambridge University Press, New York, pp. 91–110.

Mayer, A., Mosler, G., Just, W., Pilgrim, C., and Reisert, I. (2000) Developmental profile of Sry transcripts in mouse brain. *Neurogenetics* 3: 25–30.

McClearn, G.E. and DeFries, J.C. (1973) *Introduction to Behavioral Genetics*. W. H. Freeman, San Francisco, CA, USA.

Miczek, K.A. (1999) Aggressive and social stress responses in genetically modified mice: from horizontal to vertical strategy. *Psychopharmacology* 147: 17–19.

Mitchell, M.J. (2000) Spermatogenesis and the mouse Y chromosome: specialization out of decay. *Results Probl Cell Differ* 28: 233–270.

Natarajan, D., de Boer, S.F., and Koolhaas, J.M. (2009c) Lack of differential serotonin biosynthesis capacity in genetically selected low and high aggressive mice. *Physiol Behav* 98: 411–415.

Natarajan, D., de Vries, H., de Boer, S.F., and Koolhaas, J.M. (2009b) Violent phenotype in SAL mice is inflexible and fixed in adulthood. *Aggress Behav* 35: 430–436.

Natarajan, D., de Vries, H., Saaltink, D.J., de Boer, S.F., and Koolhaas, J.M. (2009a) Delineation of violence from functional aggression in mice: an ethological approach. *Behav Genet* 39: 73–90.

Nyberg, J.M., Sandnabba, N.K., Schalkwijk, L., and Sluyter, F. (2004) Genetic and environmental (inter)actions in male mouse lines selected for aggressive and nonaggressive behavior. *Genes Brain Behav* 3: 101–109.

Nyberg, J.M., Vekovischeva, O., and Sandnabba, N.K. (2003) Anxiety profiles of mice selectively bred for intermale aggression. *Behav Genet* 33: 503–511.

Ogawa, S. and Makino, J. (1981) Maternal aggression in inbred strains of mice: effects of reproductive state. *Jpn J Psychol* 52: 78–84.

Ogawa, S. and Makino, J. (1984) Aggressive behavior in inbred strains of mice during pregnancy. *Behav Neural Biol* 40: 195–204.

Parmigiani, S., Ferrari, P.F., and Palanza, P. (1998) An evolutionary approach to behavioral pharmacology: using drugs to understand proximate and ultimate mechanisms of different forms of aggression. *Neurosci Biobehav Rev* 23: 143–153.

Parmigiani, S., Palanza, P.S., Rodgers, J., and Ferrari, P.F. (1999) Selection, evolution of behavior and animal models in behavioral neuroscience. *Neurosci Biobehav Rev* 23: 957–970.

Popova, N.K., Nikulina, E.M., and Kulikov, A.V. (1993) Genetic analysis of different kinds of aggressive behavior. *Behav Genet* 23: 491–497.

Roubertoux, P.L. and Carlier, M. (2003) Y chromosome and antisocial behavior. In Mattson, M.P. (ed.), *Neurobiology of Aggression: Understanding and Preventing Violence*. Humana, Totowa, NJ, USA, pp. 119–134.

Roubertoux, P.L., Carlier, M., Degrelle, H., Phillips, J., Tordjamn, S., Dupertuis-Haas, M.C., et al. (1994) Co-segregation of intermale aggression with the pseudoautosomal region of the Y chromosome in mice. *Genetics* 135: 225–230.

Roubertoux, P.L., Guillot, P.V., Mortaud, S., Pratte, M., Jamon, M., Cohen-Salmon, C., et al. (2005) Attack behaviors in mice: from factorial structure to quantitative trait loci mapping. *Eur J Pharmacol*. 526: 172–185.

Rowe, F.P. and Redfern, R. (1969) Aggressive behaviour in related and unrelated wild house mice (*Mus musculus* L.). *Ann Appl Biol* 64: 425–431.

Sandnabba, N.K. (1992) Aggressive behavior in female mice as a correlated characteristic in selection for aggressiveness in male mice. In Bjorkqvist, K. and Pirkko, P. (eds.), *Of Mice and Women: Aspects of Female Aggression*. Academic Press, New York, pp. 367–379.

Sandnabba, N.K. and Korpela, S.R. (1994) Effects of early exposure to mating on adult sexual behavior in male varying in their genetic disposition for aggressive behavior. *Aggress Behav* 20: 429–439.

Sandnabba, N.K., Lagerspetz, K.M.J., and Jensen, E. (1994) Effects of testosterone exposure and fighting experience on the aggressive behavior of female and male mice selectively bred for intermale aggression. *Horm Behav* 28: 219–231.

Scott, A.L., Bortolato, M., Chen, K. and Shih, J.C. (2008) Novel monoamine A knock out mice with human-like spontaneous mutation. *Neuroreport* 19: 739–743.

Scott, J.P. (1942) Genetic differences in the social behavior on inbred strains of mice. *J Hered* 33: 11–15.

Scott, J.P. (1966) Agonistic behavior in mice and rats: a review. *Am Zool* 6: 683–701.

Siegfried, B., Frischknecht, H.R., and Nunes de Souza, R.L. (1990) An ethological model for the study of activation and interaction of pain, memory and defensive systems in the attacked mouse. Role of endogenous opioids. *Neurosci Biobehav Rev* 14: 481–490.

Sluyter, F., Arseneault, L., Moffitt, T.E., Veenema, A.H., De Boer, S.F., and Koolhaas, J.M. (2003) Towards an animal model for antisocial behavior: parallels between mice and men. *Behav Genet* 33: 563–574.

Sluyter, F., Marican, C.C., and Crusio, W.E. (1999) Further phenotypical characterisation of two substrains of C57BL/6J inbred mice differing by a spontaneous single-gene mutation. *Behav Brain Res* 98: 39–43.

Sluyter, F., Meijeringh, B.J., Van Oortmerssen, G.A., and Koolhaas, J.M. (1995) Studies on wild house mice. VIII. Postnatal maternal influence on intermale aggression in reciprocal F1s. *Behav Genet* 25: 367–370.

Sluyter, F., Van der Vlugt, J.J., Van Oortmerssen, G.A., Koolhaas, J.M., Van der Hoeven, F., and De Boer, P. (1996a)

Studies on wild house mice. VII. The prenatal maternal environment and aggression. *Behav Genet* 26: 513–518.

Sluyter, F., Van Oortmerssen, G.A., De Ruiter, A.J.H., and Koolhaas, J.M. (1996b) Aggression in wild house mice. Current state of affairs. *Behav Genet* 5: 489–496.

Sluyter, F., Van Oortmerssen, G.A., and Koolhouse, J.P. (1994b) Studies on wild house mice. VI. Differential effects of the Y chromosome on intermale aggression. *Aggress Behav* 20: 379–386.

St. John, R.D. and Corning, P.A. (1973) Maternal aggression in mice. *Behav Biol* 9: 635–639.

Svare, B. (1989) Recent advances in the study of female aggressive behavior in mice. In Brain, P.F., Mainardi, D., and Parmigiani, S. (eds.), *House Mouse Aggression*. Academic Press, New York, pp. 135–159.

Van Oortmerssen, G.A. and Bakker, T.C.M. (1981) Artificial selection for short and long attack latencies in wild *Mus musculus domesticus*. *Behav Genet* 11: 115–126.

Van Oortmerssen, G.A., Benus, I., and Dijk, D.J. (1985) Studies in wild house mice: genotype–environment interactions for attack latency. *Neth J Zool* 35: 155–169.

Van Oortmerssen, G.A. and Busser, J. (1989) Studies in wild house mice 3: disruptive selection on aggression as a possible force in evolution. In Brain, P.F., Mainardi, D., and Parmigiani, S. (eds.), *House Mouse Aggression*. Academic Press, New York, pp. 87–117.

Van Zegeren, K. (1980) Variation in aggressiveness and the regulation of numbers in house mouse populations. *Neth J Zool* 30: 635–770.

Veenema, A.H., Koolhaas, J.M., and de Kloet, E.R. (2004) Basal and stress-induced differences in HPA axis, 5-HT responsiveness, and hippocampal cell proliferation in two mouse lines. *Ann N Y Acad Sci* 1018: 255–265.

Veenema, A.H. and Neuman, I.D. (2007) Neurobiological mechanisms of aggression and stress coping: a comparative study in mouse and rat selection lines *Brain Behav Evol* 70: 274–285.

Weerts, E.M., Miller, L.G., Hood, K.E, and Miczek, K.A (1992) Increased GABAA-dependent chloride uptake in mice selectively bred for low aggressive behavior. *Psychopharmacology* 108: 196–204.

Xu, J., Burgoyne, P.S., and Arnold, A.P. (2002) Sex differences in sex chromosome gene expression in mouse brain. *HumMol Genet* 11: 1409–1419.

Chapter 25 Latent inhibition

Thomas J. Gould and Sheree F. Logue

Overview

The ability to adapt to environmental change is essential for the survival of any organism. Adaptations to environmental change include the ability to learn associations and the ability to modify these associations through processes such as attention, sensory gating, extinction, and response inhibition. Studies of these cognitive processes will further the understanding of the neurobiology and genetics of learning and of mental illnesses such as Alzheimer's disease, attention deficit hyperactivity disorder, autism, bipolar disorder, obsessive-compulsive disorder, and schizophrenia that are associated with altered attention and/or response inhibition (Amieva et al., 2004; Baron-Cohen and Belmonte, 2005; Baruch et al., 1988; Clark and Goodwin, 2004; Gray et al., 1995; Kaplan et al., 2005; Lubow and Gewirtz, 1995; Lubow and Josman, 1993; Muller and Roberts, 2005; Pantelis et al., 2004; Swerdlow et al., 1999; Tyson et al., 2004; Vaitl and Lipp, 1997; Weiner, 2003; Willcutt et al., 2005). This chapter focuses on one process that modifies associations, latent inhibition. We will discuss the properties of latent inhibition, the theories of latent inhibition, the behavioral genetics of latent inhibition, and the neurobiology of latent inhibition including the neural substrates, pharmacology, and cellular substrates.

Introduction to latent inhibition

Latent inhibition is the behavioral phenomenon in which the pre-exposure to a conditioned stimulus, prior to the pairing of this conditioned stimulus with an unconditioned stimulus, decreases the subsequent conditioned responses. In other words, repeated experience with a stimulus such as a tone before learning to associate the tone with another stimulus decreases the level of learning observed. Two conceptual difficulties are often associated with this phenomenon: (1) understanding the meaning of the phrase latent inhibition; and (2) understanding what this behavioral phenomenon models (e.g., attention, sensory filtering/gating, or response inhibition). The former is easier to explain and thus we will start there. The first latent inhibition experiment, published in 1959 by Lubow and Moore, was conducted during a period when there was great interest in latent learning or learning that was not immediately observed but only later expressed when conditions were appropriate. As described by Lubow (1989), the learning field in the late 1950s was embroiled in debate as to whether reinforcement was necessary for the formation of associations. Thus, experiments were designed to demonstrate that latent learning could occur. The majority of these experiments focused on demonstrating latent learning in instrumental paradigms, paradigms in which voluntary behaviors are conditioned through the association with reinforcements or punishments. Lubow and Moore (1959), however, were interested in demonstrating that latent learning could occur in classical conditioning paradigms, that is paradigms in which innate responses to unconditioned stimuli come to be elicited by neutral conditioned stimuli after repeated pairings of the conditioned and unconditioned stimuli. If latent learning could occur in classical conditioning paradigms, it was predicted that conditioned responding would be facilitated by pre-exposure to a conditioned stimulus prior to conditioning; interestingly, the opposite effect occurred. Thus, the name latent inhibition emerged because the experiments were intended to demonstrate latent learning but instead demonstrated an inhibition of subsequent conditioned responses.

Understanding what the behavioral phenomenon of latent inhibition models is a conceptual difficulty whose explanation will not be as brief as the explanation of the term latent inhibition. It is our hope that by the end of this chapter the reader will have a basic understanding of the behavioral phenomenon of latent inhibition as well as a solid overview of the neurobiology of latent inhibition as it is understood to date.

Behavioral properties of latent inhibition

Latent inhibition can be demonstrated in a variety of behavioral paradigms including cued fear conditioning, conditioned taste aversion, avoidance conditioning, appetitive conditioning, and eye blink conditioning, to name but a few (Gould et al., 2001; Han et al., 1995; Lubow, 1973; Mcintosh and Tarpy, 1977; Nicholson and Freeman, Jr., 2002). Regardless of the behavioral paradigm used for the learning component, assessment of latent inhibition is a three-phase procedure. The three phases of latent inhibition are pre-exposure to the conditioned stimulus, training (or conditioning), and testing. Latent inhibition is

Behavioral Genetics of the Mouse: Volume I. *Genetics of Behavioral Phenotypes*, eds. Wim E. Crusio, Frans Sluyter, Robert T. Gerlai, and Susanna Pietropaolo. Published by Cambridge University Press.

identified by comparing the degree of conditioned responding during the testing phase between two groups of subjects; one group that experienced the pre-exposure to the conditioned stimulus during the pre-exposure phase and one group that only experienced a visit to the context for the pre-exposure phase. Procedural manipulations in one or more of the three phases can influence the development of latent inhibition.

In the pre-exposure phases various properties of the pre-exposed stimulus modulate the strength of the latent inhibition. These properties include the number of pre-exposures to the conditioned stimulus either within a single session or across multiple sessions, the duration of the conditioned stimulus (e.g., a 10 s vs. a 30 min conditioned stimulus), and its intensity (e.g., 60 dB vs. 100 dB). Manipulations of these parameters, individually or in combination, influence the strength of the resultant latent inhibition. Latent inhibition is stronger as the number of pre-exposures to the conditioned stimulus increases and is reduced or absent when the number of pre-exposures is low (Lubow, 1973). Stronger latent inhibition also results when a more intense conditioned stimulus is used during pre-exposure (Crowell and Anderson, 1972; Schnur and Lubow, 1976). Just as the intensity of the conditioned stimulus at pre-exposure correlates with the strength of latent inhibition, so does the duration of the conditioned stimulus. Latent inhibition for conditioned taste aversion develops with just one conditioned stimulus pre-exposure if a 20 minute pre-exposure is used (Kalat and Rozin, 1973). Thus, an increase in the number of pre-exposures to the conditioned stimulus, the intensity of the conditioned stimulus, or the duration of the conditioned stimulus will increase the strength of latent inhibition. Additionally, these individual properties of the pre-exposed conditioned stimulus can interact to determine the strength of the latent inhibition. Disrupting latent inhibition by decreasing the number of pre-exposures can be overcome by an increase in the intensity of the pre-exposed conditioned stimulus (Crowell and Anderson, 1972).

Other factors that influence latent inhibition include the training protocol and the duration of the intervals between phases. For instance, the number of training trials impacts the strength of the latent inhibition such that increasing the number of training trials will decrease latent inhibition (Gaisler-Salomon and Weiner, 2003; Lipina et al., 2005). In addition, an increase in the interval between pre-exposure and training can disrupt latent inhibition (see review by Lubow (1989)). For example, McIntosh and Tarpy (1977) demonstrated latent inhibition of conditioned taste aversion with a one-day interval between the pre-exposure and train phases but not with a 24-day interval. The duration of the conditioned stimulus pre-exposure, however, interacts with the interval between pre-exposure and training to influence the expression of latent inhibition. Lubow (1989) reviewed 11 studies and concluded that when the interval between pre-exposure and training is 3.5 hours or less then both short and long conditioned stimulus pre-exposures result in latent inhibition. On the other hand, if a long interval between pre-exposure and training is used then only long conditioned stimulus pre-exposures will produce latent inhibition. Because methodological variations greatly impact latent inhibition, familiarity with the effects of procedural variations on latent inhibition aids in understanding differences when comparing the results across studies.

Theories of latent inhibition

Whereas the behavioral properties of latent inhibition are well studied and fairly straightforward, the cognitive process that latent inhibition may reflect is still debated. There are multiple, and sometimes conflicting, theories of latent inhibition with varying levels of experimental support. Generally, the theories of latent inhibition can be divided into two classes: traditional theories proposing that latent inhibition results from altering the learning of associations and more recent theories proposing that latent inhibition results from altering the expression of the learned association during the testing phase. The traditional theories that propose latent inhibition reduces the associability of the conditioned stimulus suggest that the brain acts in a pre-emptive manner to reduce the impact of the flood of information that bombards the senses. For example, one such theory proposes that pre-exposure to a conditioned stimulus may modify the salience of the conditioned stimulus such that during training, the conditioned stimulus would have a low level of salience. The low salience of the conditioned stimulus would prevent a strong conditioned stimulus–unconditioned stimulus association from forming and thus disrupt the formation of conditioned responses; which would be seen as latent inhibition (Rescorla, 1971). Studies have shown that changing the salience of the conditioned stimulus to a higher level, achieved by changing the context between the pre-exposure and training phases, will disrupt latent inhibition (Hall and Channell, 1985; Hall and Minor, 1985; Lovibond et al., 1984; Swartzentruber and Bouton, 1986; Wright et al., 1986). If the theories proposing that latent inhibition decreases the ability to form a learned association are correct then conditioned responses should not be seen under any testing conditions in subjects pre-exposed to the conditioned stimulus.

The second class of latent inhibition theories put forward by Kasprow et al. (1984) and further elaborated on by Bouton (1993) propose that latent inhibition results from the modification of expression of the learned association. That is, conditioned stimulus pre-exposure does not disrupt formation of a conditioned stimulus–unconditioned stimulus association but instead alters the ability to express conditioned responses. How the ability to express the conditioned response is altered is unknown and could be due to an inability to access or successfully retrieve a learned conditioned stimulus–unconditioned stimulus association or a deficit in the ability to express conditioned responses. These theories suggest that the brain may make irrelevant or unnecessary associations to stimuli that bombard the senses but then the brain modulates the expression of these associations in response to environmental changes. If this class of theory is correct, then it should be possible to reveal the learned association in conditioned stimulus pre-exposed subjects via manipulation of experimental conditions, assuming the conditions in the training phase were sufficient to produce a learned association.

Experimental manipulations following the training phase that result in the expression of the conditioned response provide strong support for the contention that latent inhibition involves a deficit in the retrieval of the learned association or in expression of the conditioned response and not a modification of the learning. For instance, Holt and Maren (1999) report the emergence of conditioned responses when the testing context differed from the pre-exposure context. In addition to physically changing the testing context, contextual extinction after training, achieved by exposing the subject to the context without exposure to the unconditioned stimulus, resulted in recovery of the conditioned response (Talk et al., 2005). These results showing that latent inhibition was reversed and conditioned responses were recovered after manipulating the context strongly suggest that latent inhibition does not disrupt learning but instead alters the expression of the learned responses.

As with manipulation of context, changes in the interval between training and testing phases also impact latent inhibition. Latent inhibition develops with a short interval between training and testing, but latent inhibition is disrupted and conditioned responses are expressed when the training-testing interval is lengthened (Bakner et al., 1991). The ability of pre-exposed subjects to show conditioned responses after an extended train-test interval is difficult to reconcile with the view that conditioned stimulus pre-exposure disrupts the formation of conditioned stimulus–unconditioned stimulus associations. Instead, the presence of conditioned responses suggests that conditioned stimulus pre-exposure produced a change in neural function that gates the expression of the learned response and that the influence of this neural change dissipates with time.

Further evidence that learning does occur in pre-exposed animals comes from a study that examined the effects of a reminder trial on latent inhibition. Kasprow et al. (1984) demonstrated that in pre-exposed animals two presentations of the unconditioned stimulus after training in a novel environment disrupted latent inhibition and conditioned responding was seen. Furthermore, in addition to behavioral manipulations resulting in expression of conditioned responses after latent inhibition, neural manipulations can also result in the expression of conditioned responses in pre-exposed animals. We have found that inactivation of the entorhinal cortex in the testing phase restores conditioned responses in pre-exposed mice (Lewis and Gould, 2007a). Thus, latent inhibition does not block learning but modulates the expression of the learned response. The data demonstrating the ability of pre-exposed subjects to switch between showing latent inhibition and expressing the conditioned responses suggests that latent inhibition and learning are indeed two separate and distinct processes.

Although learning and latent inhibition are separate processes, they do appear to have some commonalities. For instance, latent inhibition is also sensitive to reinstatement as has been demonstrated for learning (Bouton, 1993; Bouton, 2004). Latent inhibition was reinstated by a single conditioned stimulus exposure during a 10-day train–test interval, an interval shown to disrupt latent inhibition (Ackil et al., 1992). This single conditioned stimulus reminder treatment was insufficient to produce latent inhibition in non pre-exposed controls.

In order to adapt to ever-changing environmental stimuli and associations, the nervous system may need to both filter incoming information as well as modulate expression of associations in subsequent situations. As proposed by Lubow and Kaplan (2005), changes in both attention and retrieval processes may account for latent inhibition. This issue requires further study. The application of behavioral genetics to neural and behavioral studies of latent inhibition will strengthen the understanding of the cognitive processes involved in latent inhibition.

Behavioral genetics of latent inhibition

Clearly many important issues surrounding latent inhibition remain unresolved. Increased application of behavioral genetics will aid in the clarification of these issues. This section will review the current behavioral genetic analyses of latent inhibition that are shedding light on the cognitive processes and neural substrates involved in latent inhibition.

One of the first behavioral genetic studies of latent inhibition was a strain survey comparing latent inhibition of cued fear conditioning in nine inbred mouse lines (Gould and Wehner, 1999). Mice received either 40 pre-exposures to the conditioned stimulus or no pre-exposures to the conditioned stimulus before training. The degree of latent inhibition varied across strains with the rank order from the strongest latent inhibition to absent latent inhibition being: 129S6/SvevTac > C57BL/6J > BALB/cByJ > AKR/J > DBA/2J > 129X1/SvJ = A/J = C3H/Ibg. Of the eight strains that were able to learn, all but three developed latent inhibition; the 129X1/SvJ, A, and C3H strains failed to show latent inhibition. The CBA strain was also evaluated but did not learn cued fear conditioning and thus it was impossible to assess if latent inhibition developed in this strain. This study clearly demonstrated that genetic variation contributes to variation in latent inhibition, and that there is an underlying genetic basis of latent inhibition.

In addition to Gould and Wehner (1999), other studies have examined strain differences in latent inhibition. Baarendse et al. (2008) compared the latent inhibition of passive avoidance between C57BL/6J and DBA/2J mice and found that the C57BL/6 mice showed superior latent inhibition. In fact, the DBA/2 mice had poorer baseline performance but showed improvement with pre-exposure, an effect opposite of latent inhibition. De Bruin et al. (2006), however, compared C57BL/6Crl, 129X1/SvCrl, and F2 mice from a B6X129 cross in latent inhibition of conditioned taste aversion and found no difference between genotypes. Whereas the Baarendse et al. (2008) results were similar to the results of Gould and Wehner (1999) in that the C57BL/6 mice performed better than the DBA/2 mice, the de Bruin et al. (2006) results differed because

Table 25.1 The influence of genetics and nucleus accumbens lesions on latent inhibition.

	C57BL/6 mice		DBA/2 mice	
Pre-exposure	NAc intact	NAc lesion	NAc intact	NAc lesion
4 sessions	No LI	LI	LI	No LI
7 sessions	LI	LI	LI	LI

LI: latent inhibition.
Reproduced with permission from Restivo et al. (2002).

Gould and Wehner (1999) did not see strong latent inhibition in 129X1/Sv mice. These findings that 129X1/Sv mice did not show strong latent inhibition of fear conditioning but did show strong latent inhibition of conditioned taste aversion suggest that the genetics of latent inhibition may be influenced by procedural factors such as the base task that latent inhibition is performed on. Differences in the source of the mice, Charles River versus The Jackson Laboratories, or in baseline learning, as 129X1/Sv did not learn fear conditioning as well as C57BL/6 mice but showed equal levels of taste conditioning, could also contribute to the differences in latent inhibition.

Another study further examined strain differences in latent inhibition by testing if the number of conditioned stimulus pre-exposure sessions and lesions of the nucleus accumbens interact to differentially affect latent inhibition of cued fear conditioning in C57BL/6Crl and DBA/2Crl mice (Restivo et al., 2002). In the pre-exposure phase, C57BL/6 and DBA/2 mice with intact or lesioned nucleus accumbens received four 30 second pre-exposures to the conditioned stimulus for either 4 days or 7 days (see Table 25.1 for study summary). Similar to previous results (Gould and Wehner, 1999), both C57BL/6 mice and DBA/2 mice, with intact nucleus accumbens, developed latent inhibition. However, the C57BL/6 mice only showed latent inhibition following the 7-day pre-exposure while the DBA/2 mice developed latent inhibition under both the 4-day and 7-day pre-exposure conditions. The nucleus accumbens lesions had no impact on the development of latent inhibition under the 7-day pre-exposure condition for either strain but produced effects in both strains under the 4-day pre-exposure condition. Nucleus accumbens lesions in C57BL/6 mice resulted in the development of latent with 4-day pre-exposure whereas this lesion in DBA/2 mice prevented the development of latent inhibition. The authors interpret the data as reflecting differences in the strains' use of configural information; that is the use of complex cues such as a context rather than single cues such as an auditory tone. DBA/2 mice have deficits in tasks requiring configural processing compared to C57BL/6 mice (Logue et al., 1997; Paylor et al., 1994, 1996; Rossi-Arnaud et al., 1991). Thus, Restivo et al. (2002) propose that the deficit in latent inhibition in C57BL/6 mice when the 4-day pre-exposure protocol was used is due to their use of a configural strategy in which discrete elements of the environment are not processed individually but instead processed as a whole. In C57BL/6 mice, the use of configural strategy could essentially diminish the saliency of the auditory conditioned stimulus thereby preventing development of latent inhibition. DBA/2 mice using a cue-based strategy, rather than configural, would give greater weight to strong individual cues such as the auditory conditioned stimulus and thus latent inhibition would develop when pre-exposure sessions were used. Nucleus accumbens lesions may alter the way an animal processes individual cues as well as how individual cues are integrated into complex cues. The role of the nucleus accumbens in latent inhibition is discussed further in the neural substrates section, below. In configural-biased C57BL/6 mice, the lesion may shift them to an individual cue-based processing strategy thereby promoting the development of latent inhibition. In contrast, in cue-biased DBA/2 mice lesions of the nucleus accumbens may affect processing of individual cues in a manner that prevents the development of latent inhibition when few pre-exposure sessions are given. Therefore, this study demonstrates a genetic difference between the strains in how they process cues, which impacts the development of latent inhibition. In addition, the results suggest that these genes may modulate nucleus accumbens function and impact the function of the nucleus accumbens in cognitive processes.

Deficits in latent inhibition and other forms of information processing such as prepulse inhibition of the acoustic startle reflex are associated with schizophrenia. To examine if similar genetic factors were involved in responsiveness to neuroleptics and latent inhibition, Kline et al. (1998) compared latent inhibition of conditioned avoidance in mice selected from a heterogeneous stock for responsiveness to neuroleptics. Mice from an eight-way cross of A/J, AKR/J, BALB/cJ, CBA/J, C3H/HeJ, C57BL/6J, DBA/2J, and LP/J mice were selected into two lines that either showed haloperidol-induced catalepsy or did not. Lines were inbred for 12–16 generations. The non-responsive neuroleptic line showed a deficit in latent inhibition and also in prepulse inhibition of the acoustic startle response. These results demonstrating genetic overlap for responsiveness to haloperidol, prepulse inhibition, and latent inhibition have implications for understanding schizophrenia.

Strain differences in latent inhibition have been demonstrated in rats and these results further suggest a link between the genetics of schizophrenia and latent inhibition. Inbred BN and WKY rats were tested in latent inhibition of conditioned taste aversion and prepulse inhibition of the acoustic startle reflex (Conti et al., 2001). BN rats showed deficits in both latent inhibition and prepulse inhibition; they further demonstrated shared genetics for latent inhibition and prepulse inhibition. In another experiment investigating the potential link between schizophrenia and performance in prepulse inhibition and latent inhibition, outbred Wistar rats selected for responsiveness to the dopaminergic agonist apomorphine were behaviorally tested in the seventeenth and eighteenth generations (Ellenbroek et al., 1995). Rats with increased sensitivity to apomorphine had deficits in prepulse inhibition of the acoustic startle reflex and latent inhibition of conditioned taste aversion. Thus, the Ellenbroek et al. (1995), Conti et al. (2001), and Kline et al. (1998) studies all suggest that genes involved in pharmacological phenotypes and behavioral phenotypes of schizophrenia may overlap. However, environmental manipulations such as maternal separation and social isolation have dissociable effects on prepulse inhibition and latent inhibition (Weiss et al., 2001), suggesting the tasks may have different underlying neural substrates.

Future behavior genetics studies may shed more light on how genetic variation contributes to sensitivity to procedural manipulations that modulate latent inhibition and the relationship between latent inhibition and mental disorders. The interesting results with the nucleus accumbens lesions and genetic background and with sensitivity to drugs associated with schizophrenia and changes in latent inhibition demonstrate the power of applying behavioral genetics techniques to the evaluation of the neurobiology of latent inhibition and disease states. The following sections will review the neurobiology of latent inhibition starting with neural substrates, then pharmacology and ending with potential cell signaling cascades of latent inhibition because identifying these substrates, especially the cell signaling substrates, may provide targets for analysis of candidate genes involved in latent inhibition.

Latent inhibition: neural substrates

Studies exploring the neural substrates underlying latent inhibition have demonstrated that multiple brain regions are involved in latent inhibition. Several brain regions, the hippocampus, amygdala, and orbitofrontal cortex, appear to play a modulatory role in latent inhibition as indicated by their involvement in latent inhibition primarily under non-optimal conditions. On the other hand, two brain areas, the nucleus accumbens and entorhinal cortex, may be critical sites for latent inhibition.

Based on multiple lines of evidence, the nucleus accumbens is thought to play an important role in the control of latent inhibition. Similar to the effects of nucleus accumbens lesions in DBA/2 mice (Restivo et al., 2002), electrolytic and excitotoxic lesions of the nucleus accumbens disrupt latent inhibition in rats (Tai et al., 1995). However, Gal et al. (2005) found that nucleus accumbens shell lesions disrupted latent inhibition while lesions of the entire nucleus accumbens resulted in stronger latent inhibition; neither type of lesion disrupted learning. Based on the differential effects of full and partial nucleus accumbens lesions on latent inhibition, Gal et al. (2005) suggest that the shell and the core of the nucleus accumbens interact to regulate the expression of latent inhibition via modulation of dopamine levels during latent inhibition. This supposition is supported by results of an earlier study in which latent inhibition was associated with differential levels of dopamine in the nucleus accumbens shell and core. During expression of latent inhibition, the conditioned release of dopamine in the shell of the nucleus accumbens was reduced in animals pre-exposed to the conditioned stimulus but no change was seen in the core (Murphy et al., 2000).

The second brain area critically involved in latent inhibition is the entorhinal cortex. Irreversible lesions of the entorhinal cortex disrupted latent inhibition in pre-exposed subjects but learning of the conditioned stimulus–unconditioned stimulus association was unaffected in non pre-exposed animals (Coutureau et al., 1999, 2002; Oswald et al., 2002; Shohamy et al., 2000; Yee et al., 1997). Recently, we found that reversible inactivation of the entorhinal cortex in either the pre-exposure phase or testing phase disrupted latent inhibition in C57BL/6 mice (Lewis and Gould, 2007a). Reversible inactivation of the entorhinal cortex during the training phase had no effect on learning in non pre-exposed mice. In a similar study in rats, inactivation of the entorhinal cortex at pre-exposure but not training altered latent inhibition but not learning (Seillier et al., 2007). These data suggest that the entorhinal cortex is critically involved in latent inhibition but not learning the association between conditioned and unconditioned stimulus. While it is unclear what the exact role of the entorhinal cortex is in latent inhibition, current data suggest that the entorhinal cortex gates retrieval of conditioned stimulus–unconditioned stimulus associations or expression of the conditioned response.

The hippocampus has been shown to play critical and modulatory roles in many types of cognitive processes (Bakshi and Geyer, 1998; Eichenbaum, 1999; Rudy and Sutherland, 1995) and the majority of the current data indicates a modulatory role for the hippocampus in latent inhibition. Multiple conflicting reports indicate that hippocampal lesions either disrupt (Han et al., 1995; Kaye and Pearce, 1987; Oswald et al., 2002; Schmajuk et al., 1994; Solomon and Moore, 1975) or have no effect on latent inhibition (Clark et al., 1992; Coutureau et al., 1999; Gallo and Candido, 1995; Shohamy et al., 2000). Differences in lesion techniques and differences in the behavioral procedures, which as reviewed earlier can greatly impact the strength of latent inhibition, may contribute to these conflicting results. Thus, the hippocampus may play a greater or lesser role in latent inhibition in different circumstances. In support, Honey and Good (1993), and Holt and Maren (1999) found that hippocampal lesions, which failed to disrupt latent inhibition, did disrupt the context specificity of latent inhibition. That is the disruption of latent inhibition caused by a shift in context failed to occur in hippocampus-lesioned animals. Finally, although neurotoxic lesions of the ventral hippocampus failed to disrupt latent inhibition, activation of the ventral hippocampus did disrupt latent inhibition (Pouzet et al., 2004). Collectively, these data suggest that the hippocampus is not critically involved but may modulate latent inhibition and provide information for the context specificity associated with latent inhibition.

The amygdala and orbitofrontal cortex may also modulate latent inhibition. Initial studies that examined the role of the amygdala in latent inhibition under standard parameters produced conflicting results. One study found disrupted latent inhibition following lesions of the basolateral nucleus (Coutureau et al., 2001) whereas other studies found no impact on latent inhibition following lesions of either the basolateral (Weiner et al., 1996b) or central nucleus (Holland and

Gallagher, 1993) of the amygdala. In contrast, when latent inhibition is produced under suboptimal parameters, the involvement of the basolateral nucleus of the amygdala becomes clear. Normally, latent inhibition is not produced when too few pre-exposures to the conditioned stimulus or too many conditioning trials are given; however, latent inhibition was produced under these conditions in rats with basolateral amygdala lesions (Schiller and Weiner, 2004, 2005; Schiller et al., 2006). Similarly, lesions of the orbitofrontal cortex also resulted in expression of latent inhibition when suboptimal latent inhibition procedures were used (Schiller and Weiner, 2004; Schiller et al., 2006). These cortical effects appear to be specific to the orbitofrontal region of prefrontal cortex because multiple studies have shown that lesions of the medial prefrontal cortex do not alter latent inhibition (Joel et al., 1997; Lacroix et al., 1998, 2000; Schiller and Weiner, 2004). Thus, the basolateral nucleus of the amygdala and orbitofrontal cortex may inhibit the expression of latent inhibition when conditions exist that normally favor the expression of conditioned responses; but when basolateral amygdala function or orbitofrontal cortical function is disrupted, the system may shift, giving greater weight to the pre-exposure effect and thus producing latent inhibition.

It is clear that multiple neural areas are involved in latent inhibition and the role of these areas in latent inhibition may be determined by behavioral parameters. For instance, the results from lesion studies of the hippocampus appear equivocal but the degree of engagement of the hippocampus may depend on task parameters and on the role of contextual information in the latent inhibition procedure. The basolateral nucleus of the amygdala and the orbitofrontal cortex may not be critically involved in latent inhibition but may modulate the expression of latent inhibition. In contrast, the entorhinal cortex appears critically involved in the development and expression of latent inhibition. The nucleus accumbens may also regulate the expression of latent inhibition through modulating dopamine levels. These results emphasize how a greater understanding of latent inhibition can be gained through studying both the neural substrates and the neurochemical substrates of latent inhibition.

Latent inhibition: pharmacology

The role of dopamine, as well as other monoamine neurotransmitters such as serotonin, in the control of latent inhibition has been well studied. In addition to monoaminergic neurotransmitters, evidence indicates a role for the cholinergic and glutamatergic neurotransmitter systems in latent inhibition. The involvement of these four neurotransmitters in latent inhibition will be reviewed briefly.

Studies investigating the role of the dopaminergic system in latent inhibition have been motivated in part by the potential of latent inhibition to model deficits associated with schizophrenia; patients with schizophrenia show disrupted latent inhibition (Baruch et al., 1988; Gray et al., 1995; Lubow and Gewirtz, 1995; Vaitl and Lipp, 1997; Weiner, 2003). In both humans and rodents, amphetamine, which increases synaptic levels of dopamine, disrupts latent inhibition (Meyer et al., 2004; Swerdlow et al., 2003; Thornton et al., 1996). In fact, direct infusion of amphetamine into the nucleus accumbens disrupted latent inhibition (Solomon and Staton, 1982). The amphetamine-induced disruption of latent inhibition can be reversed by administration of both typical and atypical antipsychotics, which work primarily through antagonism of dopamine D_2 receptors in the nucleus accumbens and striatum (Russig et al., 2003; Warburton et al., 1994; Weiner et al., 1996a). Furthermore, both typical and atypical antipsychotics, which decrease the level of synaptic dopamine in the nucleus accumbens, can enhance latent inhibition in normal animals (Feldon and Weiner, 1991; Ruob et al., 1998; Trimble et al., 1997). These results further support the hypothesis that latent inhibition is regulated by dopamine signaling in the nucleus accumbens. Weiner (2003) proposed that during latent inhibition two associations are learned and the expression of these associations at testing is determined by dopamine levels. During pre-exposure, an association between the conditioned stimulus and no stimulus (or no event) is formed, and during training, an association between the conditioned stimulus and the unconditioned stimulus is formed. In this model, if dopamine levels are high in the testing phases then the conditioned stimulus–unconditioned stimulus association is favored and conditioned responses are expressed, but if dopamine levels are low then the conditioned stimulus–no stimulus association is favored and no conditioned responses are expressed (Weiner, 1990, 2003).

Serotonin is another monoamine neurotransmitter shown to be involved in latent inhibition. Depletion of serotonin disrupts latent inhibition (Lorden et al., 1983; Loskutova, 2001; Loskutova et al., 1990; Solomon et al., 1978). Studies investigating the involvement of serotonin in latent inhibition demonstrate that different subregions of the raphe nucleus are involved in latent inhibition; lesions of the medial raphe nucleus, but not the dorsal raphe, disrupted latent inhibition (Solomon et al., 1980). Interestingly, the 5-$HT_{2A/2C}$ agonist 1-(2,5-dimethoxy-4-iodophenyl)-2-aminopropane disrupted latent inhibition (Hitchcock et al., 1997), while both 5-HT_{1A} (Killcross et al., 1997) and 5-HT_{2A} antagonists (McDonald et al., 2003) all enhanced latent inhibition. However, it should be noted that 1-(2,5-dimethoxy-4-iodophenyl)-2-aminopropane has been shown to be a potent inhibitor of serotonin neuronal firing (Wright et al., 1990). Furthermore, 5-HT_{1A} presynaptic autoreceptors can exert an inhibitory control over the firing rate of serotonergic neurons (Evrard et al., 1999). Thus, it is possible that agonism of 5-$HT_{2A/2C}$ or 5-HT_{1A} receptors could produce an effect similar to that seen with serotonin depletion. Whereas the monoamines dopamine and serotonin are involved in latent inhibition, not all monoamines appear to be critically involved in latent inhibition. Disruption of noradrenergic signaling had no detrimental effect on latent inhibition (Archer et al., 1983; Lorden et al., 1983; Tsaltas et al., 1984).

For the cholinergic neurotransmitter system, the overall evidence suggests a modulatory rather than critical role in latent inhibition. Lesions of the nucleus basalis magnocellularis, the source of cortical cholinergic afferents, did not disrupt latent inhibition (Chiba et al., 1995; Schauz and Koch, 1999). In another study, nucleus basalis magnocellularis lesions appeared to alter learning (Rochford et al., 1996b), which is a result that precludes clear evaluation of latent inhibition. Whether the medial septum/diagonal band, the source of cholinergic input into the hippocampus, has a critical role in latent inhibition is less clear. In one study, 192 IgG-saporin lesions of the medial septum/diagonal band did not disrupt latent inhibition of conditioned taste aversion (Dougherty et al., 1996); however, another study found that similar 192 IgG-saporin lesions disrupted latent inhibition of an appetitive conditioning task (Baxter et al., 1997). Baxter et al. (1997) suggest that different neural mechanisms may support latent inhibition depending on whether the task is aversive or appetitive. This hypothesis requires further study.

In further support of the contention that the cholinergic system may not be required for latent inhibition, knockout mice lacking the β_2-nicotinic acetylcholinergic receptor subunit (Caldarone et al., 2000) and mice treated with the broad-spectrum nicotinic receptor antagonist mecamylamine (Gould and Lewis, 2005; Gould et al., 2001) both showed intact latent inhibition. However, ligand-mediated effects at nicotinic acetylcholinergic receptors may modulate latent inhibition. The results on the effects of nicotine on latent inhibition have been mixed. One study in mice found that nicotine enhanced latent inhibition (Gould et al., 2001), while the opposite effect was described in rats (Joseph et al., 1993). An additional rat study found that the direction of the effect of nicotine on latent inhibition depends on the number of pre-exposures to the conditioned stimulus (Rochford et al., 1996a). Thus, the effects of nicotine on latent inhibition may depend on the species and behavioral procedures used. In summary, activation of nicotinic acetylcholinergic receptors may be sufficient to modulate latent inhibition but may not be critically involved in latent inhibition; though the involvement of the cholinergic system in latent inhibition may depend on whether the task is appetitive or aversive.

Studies of synaptic plasticity have suggested that the calcium influx gated by *N*-methyl-D-aspartate (NMDA) glutamate receptors may be critically involved in many cognitive processes (for review see Lisman, 2003; Riedel et al., 2003). For latent inhibition, there are conflicting results using a variety of NMDA antagonists and treatment schedules, but overall when reviewed as a whole there is a clear role for the NMDA receptor in latent inhibition. Some studies using phencyclidine and ketamine, both of which are less potent and less selective NMDA antagonists relative to the non-competitive antagonists MK-801 and DL-2-amino-5-phosphonopentanoic acid (Ellison, 1995), have failed to find effects on latent inhibition. For instance, phencyclidine (1 and 5 mg/kg) had no effect on latent inhibition of cued fear conditioning (Weiner and Feldon, 1992). In another study, ketamine (25 mg/kg) did not alter latent inhibition of conditioned taste aversion when ketamine was given at both the pre-exposure phase and the training phase (Aguado et al., 1994). Although this result suggests no role for NMDA receptors in latent inhibition there is a confound due to giving an NMDA antagonist at the training phase. An NMDA antagonist administered at training could disrupt learning; and if learning is disrupted in the training phase, the resultant lack of conditioned responses at testing could be misinterpreted as latent inhibition. Furthermore, the low dose of ketamine in the Aguado et al. (1994) study did disrupt latent inhibition when ketamine was administered only in the pre-exposure phase. Additionally, another study found that 50 mg/kg of ketamine disrupted latent inhibition of conditioned taste aversion (Gallo et al., 1998). Overall, these results suggest that NMDA receptors may be involved in latent inhibition. Studies using more potent and selective non-competitive NMDA receptor antagonists clearly demonstrate the involvement of NMDA receptors in latent inhibition. Schauz and Koch (2000) found that direct infusion of the NMDA receptor antagonist DL-2-amino-5-phosphonopentanoic acid into the basolateral amygdala blocked latent inhibition of fear-potentiated startle, and Traverso et al. (2003) found that the NMDA receptor antagonist MK-801 disrupted latent inhibition of conditioned taste aversion. Similarly, we found that doses of 0.5 and 1.0 mg/kg of MK-801 disrupted latent inhibition of cued fear conditioning (Davis and Gould, 2005; Lewis and Gould, 2004). These results strongly suggest that NMDA receptors are involved in latent inhibition.

Discrepancies between the recent studies that used the NMDA receptor antagonists DL-2-amino-5-phosphonopentanoic acid and MK-801 and the earlier studies that used ketamine and phencyclidine could be due to differences in the effective doses of the drugs or to compensatory mechanisms that may be able to overcome effects of low doses of NMDA antagonists. Recently, we reported that nicotinic acetylcholinergic receptors and NMDA receptors could mediate similar cellular processes during latent inhibition (Gould and Lewis, 2005). Thus, latent inhibition may not be disrupted when a low or moderate dose of a NMDA receptor antagonist is used because compensation for the effect of the NMDA inhibitor may occur through nicotinic acetylcholinergic receptor-mediated processes. However, when higher doses of NMDA receptor antagonists are used, processes mediated through nicotinic acetylcholinergic receptors may not be sufficient to compensate for a greater level of NMDA receptor inhibition.

In summary, multiple neurotransmitter systems are involved in latent inhibition. The dopaminergic system may modulate the expression of latent inhibition; with high levels of dopamine decreasing latent inhibition. Involvement of the acetylcholinergic system in latent inhibition may depend upon behavioral procedures. In contrast, depletion of serotonin disrupted latent inhibition but the function of serotonin in latent inhibition needs further examination. Finally, NMDA receptors appear to play an important role in latent inhibition. Inhibition of glutamate NMDA receptors disrupted latent inhibition and thus cellular processes mediated by NMDA receptor activation may be critically involved in establishing latent inhibition.

Latent inhibition: cellular substrates

We have proposed that activation of NMDA receptors during pre-exposure to the conditioned stimulus initiates cell signaling cascades that mediate latent inhibition in a manner similar to the NMDA receptor-mediated cell signaling cascades that underlie various other learning processes. As established for other learning processes, the activation of NMDA receptors can initiate the cyclic AMP/protein kinase A (cAMP/PKA) second messenger signaling pathways (Abel and Kandel, 1998; Abel et al., 1997; Chetkovich et al., 1991; Selcher et al., 2002; Szapiro et al., 2003) with PKA then activating mitogen-activated protein kinase (i.e., MAPK) (Berman et al., 2000; Blum et al., 1999; Roberson et al., 1999; Schafe et al., 2000; Selcher et al., 1999; Vossler et al., 1997; Waltereit, 2003; Walz et al., 1999) and leading to activation of the downstream gene transcription factor cAMP response element binding protein (i.e., CREB) (Davis et al., 2000; Perkinton et al., 1999). We conducted a series of experiments to identify the cell signaling cascades involved in latent inhibition by targeting different points within the described NMDA receptor cell-signaling cascade.

If NMDA-receptor antagonism interferes with latent inhibition by decreasing cAMP/PKA signaling, then direct activation of cAMP/PKA signaling should offset NMDA antagonist-induced disruption of latent inhibition. To test this, we examined if amplification of the cAMP signaling pathway by rolipram, a selective type 4 cAMP phosphodiesterase inhibitor, would reverse MK-801 induced impairments in latent inhibition of cued fear conditioning. Because phosphodiesterases inactivate cAMP, phosphodiesterase inhibition should increase cAMP levels, which may counter the effects of NMDA receptor inhibition by MK-801. Mice treated with the NMDA receptor antagonist MK-801 at the pre-exposure phase showed disrupted latent inhibition but mice treated with rolipram and MK-801 did not show deficits in latent inhibition (Davis and Gould, 2005). These results support the hypothesis that NMDA receptors and cAMP/PKA signaling are involved in latent inhibition.

Upon showing that modulation of cAMP/PKA signaling could affect latent inhibition, we sought to determine if modulating the downstream targets of PKA would impact latent inhibition. Knowing that MAPK is activated by PKA (Roberson et al., 1999; Vossler et al., 1997; Waltereit, 2003) and that numerous studies have demonstrated a role of MAPK in multiple types of learning (Berman et al., 2000; Blum et al., 1999; Schafe et al., 2000; Selcher et al., 1999; Walz et al., 1999), we treated mice with an inhibitor of MAPK signaling and successfully disrupted latent inhibition of cued fear conditioning (Lewis et al., 2004). In addition, direct infusion of the NMDA receptor antagonist DL-2-amino-5-phosphonopentanoic acid, or the cAMP inhibitor Rp-cAMPS, or the MAPK inhibitor U0126 into the entorhinal cortex at CS pre-exposure all disrupted latent inhibition (Lewis and Gould, 2007b), suggesting that a NMDA receptor → cAMP/PKA → MAPK cell signaling cascade in the entorhinal cortex may be involved in latent inhibition.

The mechanisms through which MAPK may be exerting its affects on latent inhibition of cued fear conditioning remain unknown but potentially include two downstream targets of MAPK; CREB (Davis et al., 2000; Perkinton et al., 1999) and microtubule-associated proteins (i.e., MAP2) (Quinlan and Halpain, 1996). If MAPK activation supports latent inhibition through CREB activation of Cre-mediated genes, then latent inhibition would be dependent on protein synthesis. Our recent data, however, suggest that latent inhibition can occur during protein synthesis inhibition.

We found that latent inhibition of cued fear conditioning can be established in the presence of the protein synthesis inhibitor anisomycin at the pre-exposure phase and when anisomycin is administered immediately after the pre-exposure phase (Lewis and Gould, 2004). Our findings suggest that, while MAPK can activate CREB and thus modulate transcription, MAPK may support latent inhibition of cued fear conditioning through an alternative mechanism. One potential mechanism could involve changes in MAP2. Lynch and Baudry proposed that cytoskeletal changes that occur during learning could increase the number of receptors available at the synapse and result in an increase in post-synaptic responses (Baudry and Lynch, 2001; Lynch and Baudry, 1984). Thus, during latent inhibition, MAPK could phosphorylate MAP2 producing an increase in activity-dependent cytoskeletal rearrangement that would support latent inhibition. This hypothesis needs to be tested.

Conclusions

In summary, the development of latent inhibition depends on a balance of multiple factors within and between the phases of pre-exposure, training, and testing. Neurotransmitters such as dopamine may modulate latent inhibition by regulating the influence of the pre-exposure condition. Neurotransmitters such as glutamate may mediate the cellular processes involved in establishing latent inhibition. Neural areas such as the nucleus accumbens, amygdala, and orbitofrontal cortex may modulate the expression of latent inhibition, and the entorhinal cortex may be involved in the establishment of latent inhibition. However, many important issues need further resolution. For instance, exactly how pre-exposure to a conditioned stimulus disrupts future conditioned responding is unknown. In addition, the underlying cellular and genetic substrates of latent inhibition are only beginning to be understood. Furthermore, it is unknown if the neural, cellular, and genetic substrates of latent inhibition are a constant across tasks or vary depending on properties of each task such as whether the task is aversive or appetitive. Neural and genetic studies of latent inhibition will aid in answering these questions and will further understanding of cognitive processes and disease states.

References

Abel, T. and Kandel, E. (1998) Positive and negative regulatory mechanisms that mediate long-term memory storage. *Brain Res Brain Res Rev* 26: 360–378.

Abel, T., Nguyen, P.V., Barad, M., Deuel, T.A., Kandel, E.R., and Bourtchouladze, R. (1997) Genetic demonstration of a role for PKA in the late phase of LTP and in hippocampus-based long-term memory. *Cell* 88: 615–626.

Ackil, J.K., Carman, H.M., Bakner, L., and Riccio, D.C. (1992) Reinstatement of latent inhibition following a reminder treatment in a conditioned taste aversion paradigm. *Behav Neural Biol* 58: 232–235.

Aguado, L., San Antonio, A., Perez, L., del Valle, R., and Gomez, J. (1994) Effects of the NMDA receptor antagonist ketamine on flavor memory: conditioned aversion, latent inhibition, and habituation of neophobia. *Behav Neural Biol* 61: 271–281.

Amieva, H., Phillips, L.H., Della, S.S., and Henry, J.D. (2004) Inhibitory functioning in Alzheimer's disease. *Brain* 127: 949–964.

Archer, T., Mohammed, A.K., and Jarbe, T.U. (1983) Latent inhibition following systemic DSP4: effects due to presence and absence of contextual cues in taste-aversion learning. *Behav Neural Biol* 38: 287–306.

Baarendse, P.J.J., van Grootheest, G., Jansen, R.F., Pieneman, A.W., Ögren, S.O., Verhage, M., et al. (2008) Differential involvement of the dorsal hippocampus in passive avoidance in C57bl/6J and DBA/2J mice. *Hippocampus* 18: 11–19.

Bakner, L., Strohen, K., Nordeen, M., and Riccio, D.C. (1991) Postconditioning recovery from the latent inhibition effect in conditioned taste aversion. *Physiol and Behav* 50: 1269–1272.

Bakshi, V.P. and Geyer, M.A. (1998) Multiple limbic regions mediate the disruption of prepulse inhibition produced in rats by the noncompetitive NMDA antagonist dizocilpine. *J Neurosci* 18: 8394–8401.

Baron-Cohen, S. and Belmonte, M.K. (2005) Autism: a window onto the development of the social and the analytic brain. *Annu Rev Neurosci* 28: 109–126.

Baruch, I., Hemsley, D.R., and Gray, J.A. (1988) Differential performance of acute and chronic schizophrenics in a latent inhibition task. *J Nerv Ment Dis* 176: 598–606.

Baudry, M. and Lynch, G. (2001) Remembrance of arguments past: how well is the glutamate receptor hypothesis of LTP holding up after 20 years? *Neurobiol Learn Mem* 76: 284–297.

Baxter, M.G., Holland, P.C., and Gallagher, M. (1997) Disruption of decrements in conditioned stimulus processing by selective removal of hippocampal cholinergic input. *J Neurosci* 17: 5230–5236.

Berman, D.E., Hazvi, S., Neduva, V., and Dudai, Y. (2000) The role of identified neurotransmitter systems in the response of insular cortex to unfamiliar taste: activation of ERK1–2 and formation of a memory trace. *J Neurosci* 20: 7017–7023.

Blum, S., Moore, A.N., Adams, F., and Dash, P.K. (1999) A mitogen-activated protein kinase cascade in the CA1/CA2 subfield of the dorsal hippocampus is essential for long-term spatial memory. *J Neurosci* 19: 3535–3544.

Bouton, M.E. (1993) Context, time, and memory retrieval in the interference paradigms of Pavlovian learning. *Psychol Bull* 114: 80–99.

Bouton, M.E. (2004) Context and behavioral processes in extinction. *Learn Mem* 11: 485–494.

Caldarone, B.J., Duman, C.H., and Picciotto, M.R. (2000) Fear conditioning and latent inhibition in mice lacking the high affinity subclass of nicotinic acetylcholine receptors in the brain. *Neuropharm* 39: 2779–2784.

Chetkovich, D.M., Gray, R., Johnston, D., and Sweatt, J.D. (1991) N-methyl-D-aspartate receptor activation increases cAMP levels and voltage-gated Ca^{2+} channel activity in area CA1 of hippocampus. *Proc Natl Acad Sci USA* 88: 6467–6471.

Chiba, A.A., Bucci, D.J., Holland, P.C., and Gallagher, M. (1995) Basal forebrain cholinergic lesions disrupt increments but not decrements in conditioned stimulus processing. *J Neurosci* 15: 7315–7322.

Clark, A.J., Feldon, J., and Rawlins, J.N. (1992) Aspiration lesions of rat ventral hippocampus disinhibit responding in conditioned suppression or extinction, but spare latent inhibition and the partial reinforcement extinction effect. *Neurosci* 48: 821–829.

Clark, L. and Goodwin, G.M. (2004) State- and trait-related deficits in sustained attention in bipolar disorder. *Eur Arch Psychiatry Clin Neurosci* 254: 61–68.

Conti, L.H., Palmer, A.A., Vanella, J.J., and Printz, M.P. (2001) Latent inhibition and conditioning in rat strains which show differential prepulse inhibition. *Behav Genet* 31: 325–333.

Coutureau, E., Blundell, P.J., and Killcross, S. (2001) Basolateral amygdala lesions disrupt latent inhibitionin rats. *Brain Res Bull* 56: 49–53.

Coutureau, E., Galani, R., Gosselin, O., Majchrzak, M., and Di Scala, G. (1999) Entorhinal but not hippocampal or subicular lesions disrupt latent inhibition in rats. *Neurobiol Learn Mem* 72: 143–157.

Coutureau, E., Lena, I., Dauge, V., and Di, S.G. (2002) The entorhinal cortex-nucleus accumbens pathway and latent inhibition: a behavioral and neurochemical study in rats. *Behav Neurosci* 116: 95–104.

Crowell, C.R. and Anderson, D.C. (1972) Variations in intensity, interstimulus interval, and interval between preconditioning CS exposures and conditioning with rats. *J Comp Physiol Psychol* 79: 291–298.

Davis, J.A. and Gould, T.J. (2005) Rolipram attenuates MK-801-induced deficits in latent inhibition. *Behav Neurosci* 119: 595–602.

Davis, S., Vanhoutte, P., Pages, C., Caboche, J., and Laroche, S. (2000) The MAPK/ERK cascade targets both Elk-1 and cAMP response element-binding protein to control long-term potentiation-dependent gene expression in the dentate gyrus *in vivo*. *J Neurosci* 20: 4563–4572.

de Bruin, N., Mahieu, M., Patel, T., Willems, R., Lesage, A., and Megens, A. (2006) Performance of F2 B6x129 hybrid mice in the Morris water maze, latent inhibition and prepulse inhibition paradigms: comparison with C57Bl/6J and 129sv inbred mice. *Behav Brain Res* 172: 122–134.

Dougherty, K.D., Salat, D., and Walsh, T.J. (1996) Intraseptal injection of the cholinergic immunotoxin 192-IgG saporin fails to disrupt latent inhibition in a conditioned taste aversion paradigm. *Brain Res* 736: 260–269.

Eichenbaum, H. (1999) The hippocampus and mechanisms of declarative memory. *Behav Brain Res* 103: 123–133.

Ellenbroek, B.A., Geyer, M.A., and Cools, A.R. (1995) The behavior of APO-SUS rats in animal models with construct validity for schizophrenia. *J Neurosci* 15: 7604–7611.

Ellison, G. (1995) The N-methyl-D-aspartate antagonists phencyclidine, ketamine and dizocilpine as both behavioral and anatomical models of the dementias. *Brain Res Rev* 20: 250–267.

Evrard, A., Laporte, A.M., Chastanet, M., Hen, R., Hamon, M., and Adrien, J. (1999) 5-HT1A and 5-HT1B receptors control the firing of serotoninergic neurons in the dorsal raphe nucleus of the mouse: studies in 5-HT1B knock-out mice. *Eur J Neurosci* 11: 3823–3831.

Feldon, J. and Weiner, I. (1991) The latent inhibition model of schizophrenic attention disorder. Haloperidol and sulpiride enhance rats' ability to ignore irrelevant stimuli. *Biol Psychiatry* 29: 635–646.

Gaisler-Salomon, I. and Weiner, I. (2003) Systemic administration of MK-801 produces an abnormally persistent latent inhibition which is reversed by clozapine but not haloperidol. *Psychopharm* 166: 333–342.

Gal, G., Schiller, D., and Weiner, I. (2005) Latent inhibition is disrupted by nucleus accumbens shell lesion but is abnormally persistent following entire nucleus accumbens lesion: The neural site controlling the expression and disruption of the stimulus preexposure effect. *Behav Brain Res* 162: 246–255.

Gallo, M., Bielavska, E., Roldan, G., and Bures, J. (1998) Tetrodotoxin inactivation of the gustatory cortex disrupts the effect of the N-methyl-D-aspartate antagonist ketamine on latent inhibition of conditioned taste aversion in rats. *Neurosci Lett* 240: 61–64.

Gallo, M. and Candido, A. (1995) Dorsal hippocampal lesions impair blocking but not latent inhibition of taste aversion learning in rats. *Behav Neurosci* 109: 413–425.

Gould, T.J., Collins, A.C., and Wehner, J.M. (2001) Nicotine enhances latent inhibition and ameliorates ethanol-induced deficits in latent inhibition. *Nicotine Tob Res* 3: 17–24.

Gould, T.J. and Lewis, M.C. (2005) Coantagonism of glutamate receptors and nicotinic acetylcholinergic receptors disrupts fear conditioning and latent inhibition of fear conditioning. *Learn Mem* 12: 389–398.

Gould, T.J. and Wehner, J.M. (1999) Genetic influences on latent inhibition in mice. *Behav Neurosci* 113: 1291–1296.

Gray, N.S., Pilowsky, L.S., Gray, J.A., and Kerwin, R.W. (1995) Latent inhibition in drug naive schizophrenics: relationship to duration of illness and dopamine D2 binding using SPET. *Schizophr Res* 17: 95–107.

Hall, G. and Channell, S. (1985) Differential effects of contextual change on latent inhibition and on the habituation of an orienting response. *J Exper Psychol Anim Behav Process* 11: 470–481.

Hall, G. and Minor, H. (1985) A search for context-stimulus associations in latent inhibition. *Q J Exp Psychol B* 36: 145–169.

Han, J.S., Gallagher, M., and Holland, P. (1995) Hippocampal lesions disrupt decrements but not increments in conditioned stimulus processing. *J Neurosci* 15: 7323–7329.

Hitchcock, J.M., Lister, S., Fischer, T.R., and Wettstein, J.G. (1997) Disruption of latent inhibition in the rat by the 5-HT2 agonist DOI: effects of MDL 100,907, clozapine, risperidone and haloperidol. *Behav Brain Res* 88: 43–49.

Holland, P.C. and Gallagher, M. (1993) Amygdala central nucleus lesions disrupt increments, but not decrements, in conditioned stimulus processing. *Behav Neurosci* 107: 246–253.

Holt, W. and Maren, S. (1999) Muscimol inactivation of the dorsal hippocampus impairs contextual retrieval of fear memory. *J Neurosci* 19: 9054–9062.

Honey, R.C. and Good, M. (1993) Selective hippocampal lesions abolish the contextual specificity of latent inhibition and conditioning. *Behav Neurosci* 107: 23–33.

Joel, D., Weiner, I., and Feldon, J. (1997) Electrolytic lesions of the medial prefrontal cortex in rats disrupt performance on an analog of the Wisconsin card sorting test, but do not disrupt latent inhibition: implications for animal models of schizophrenia. *Behav Brain Res* 85: 187–201.

Joseph, M.H., Peters, S.L., and Gray, J.A. (1993) Nicotine blocks latent inhibition in rats – evidence for a critical role of increased functional-activity of dopamine in the mesollimbic system at conditioning rather than preexposure. *Psychopharm* 110: 187–192.

Kalat, J.W. and Rozin, P. (1973) "Learned safety" as a mechanism in long-delay taste-aversion learning in rats. *J Comp Physiol Psychol* 83: 198–207.

Kaplan, O., Dar, R., Rosenthal, L., Hermesh, H., Fux, M., and Lubow, R.E. (2006) Obsessive-compulsive disorder patients display enhanced latent inhibition on a visual search task. *Behav Res Ther* 44: 1137–1145.

Kasprow, W.J., Catterson, D., Schachtman, T.R., and Miller, R.R. (1984) Attenuation of latent inhibition by post-acquisition reminder. *Q J Exp Psychol B* 36: 53–63.

Kaye, H. and Pearce, J.M. (1987) Hippocampal lesions attenuate latent inhibition of CS and a neutral stimulus. *Psychobiol* 15: 293–299.

Killcross, A.S., Stanhope, K.J., Dourish, C.T., and Piras, G. (1997) WAY100635 and latent inhibition in the rat: selective effects at preexposure. *Behav Brain Res* 88: 51–57.

Kline, L., Decena, E., Hitzemann, R., and McCaughran, J., Jr. (1998) Acoustic startle, prepulse inhibition, locomotion, and latent inhibition in the neuroleptic-responsive (NR) and neuroleptic-nonresponsive (NNR) lines of mice. *Psychopharm* 139: 322–331.

Lacroix, L., Broersen, L.M., Weiner, I., and Feldon, J. (1998) The effects of excitotoxic lesion of the medial prefrontal cortex on latent inhibition, prepulse inhibition, food hoarding, elevated plus maze, active avoidance and locomotor activity in the rat. *Neuroscience* 84: 431–442.

Lacroix, L., Spinelli, S., White, W., and Feldon, J. (2000) The effects of ibotenic acid lesions of the medial and lateral prefrontal cortex on latent inhibition, prepulse inhibition and amphetamine-induced hyperlocomotion. *Neurosci* 97: 459–468.

Lewis, M.C., Davis, J.A., and Gould, T.J. (2004) Inhibition of mitogen-activated protein kinase/extracellular signal-regulated kinase disrupts latent inhibition of cued fear conditioning in C57BL/6 mice. *Behav Neurosci* 118: 1444–1449.

Lewis, M.C. and Gould, T.J. (2004) Latent inhibition of cued fear conditioning: an NMDA receptor-dependent process that can be established in the presence of anisomycin. *Eur J Neurosci* 20: 818–826.

Lewis, M.C. and Gould, T.J. (2007a) Reversible inactivation of the entorhinal cortex disrupts the establishment and expression of latent inhibition of cued fear conditioning in C57BL/6 mice. *Hippocampus* 17: 462–470.

Lewis, M.C. and Gould, T.J. (2007b) Signal transduction mechanisms within the entorhinal cortex that support latent inhibition of cued fear conditioning. *Neurobiol Learn Mem* 88: 359–368.

Lipina, T., Labrie, V., Weiner, I., and Roder, J. (2005) Modulators of the glycine site on NMDA receptors, D-serine and ALX 5407, display similar beneficial effects to clozapine in mouse models of schizophrenia. *Psychopharm* 179: 54–67.

Lisman, J. (2003) Long-term potentiation: outstanding questions and attempted synthesis. *Philos Trans R Soc Lond B Biol Sci* 358: 829–842.

Logue, S.F., Paylor R., and Wehner J.M. (1997) Hippocampal lesions cause learning deficits in inbred mice in the Morris water maze and conditioned-fear task. *Behav Neurosci* 111: 104–113.

Lorden, J.F., Rickert, E.J., and Berry, D.W. (1983) Forebrain monoamines and associative learning: I. Latent inhibition and conditioned inhibition. *Behav Brain Res* 9: 181–199.

Loskutova, L.V. (2001) The effects of a serotoninergic substrate of the nucleus accumbens on latent inhibition. *Neurosci Behav Physiol* 31: 15–20.

Loskutova, L.V., Luk'yanenko, F.Y., and Il'yuchenok, R.Y. (1990) Interaction of serotonin- and dopaminergic systems of the brain in mechanisms of latent inhibition in rats. *Neurosci Behav Physiol* 20: 500–505.

Lovibond, P.F., Preston, G.C., and Mackintosh, N.J. (1984) Context specificity of conditioning, extinction, and latent inhibition. *J Exp Psychol Anim Behav Process* 10: 360–375.

Lubow, R.E. (1973) Latent inhibition. *Psychol Bull* 79: 398–407.

Lubow, R.E. (1989) *Latent Inhibition and Conditional Attention Theory*. Cambridge University Press, Cambridge.

Lubow, R.E. and Gewirtz, J.C. (1995) Latent inhibition in humans: data, theory, and implications for schizophrenia. *Psychol Bull* 117: 87–103.

Lubow, R.E. and Josman, Z.E. (1993) Latent inhibition deficits in hyperactive children. *J Child Psychol Psychiatry* 34: 959–973.

Lubow, R.E. and Kaplan, O. (2005) The visual search analogue of latent inhibition: implications for theories of irrelevant stimulus processing in normal and schizophrenic groups. *Psychon Bull Rev* 12: 224–243.

Lubow, R.E. and Moore, A. (1959) Latent inhibition: the effect of nonreinforced pre-exposure to the conditional stimulus. *J Comp Physiol Psychol* 52: 415–419.

Lynch, G. and Baudry, M. (1984) The biochemistry of memory: a new and specific hypothesis. *Science* 224: 1057–1063.

McDonald, L.M., Moran, P.M., Vythelingum, G.N., Joseph, M.H., Stephenson, J.D., and Gray, J.A. (2003) Enhancement of latent inhibition by two 5-HT2A receptor antagonists only when given at both pre-exposure and conditioning. *Psychopharm* 169: 321–331.

Mcintosh, S.M. and Tarpy, R.M. (1977) Retention of latent inhibition in a taste-aversion paradigm. *Bull Psychon Soc* 9: 411–412.

Meyer, U., Chang, D.L., Feldon, J., and Yee, B.K. (2004) Expression of the CS- and US-pre-exposure effects in the conditioned taste aversion paradigm and their abolition following systemic amphetamine treatment in C57BL6/J mice. *Neuropsychopharm* 29: 2140–2148.

Muller, J. and Roberts, J.E. (2005) Memory and attention in obsessive-compulsive disorder: a review. *J Anxiety Disord* 19: 1–28.

Murphy, C.A., Pezze, M., Feldon, J., and Heidbreder, C. (2000) Differential involvement of dopamine in the shell and core of the nucleus accumbens in the expression of latent inhibition to an aversively conditioned stimulus. *Neurosci* 97: 469–477.

Nicholson, D.A. and Freeman, J.H., Jr. (2002) Medial dorsal thalamic lesions impair blocking and latent inhibition of the conditioned eyeblink response in rats. *Behav Neurosci* 116: 276–285.

Oswald, C.J., Yee, B.K., Rawlins, J.N., Bannerman, D.B., Good, M., and Honey, R.C. (2002) The influence of selective lesions to components of the hippocampal system on the orientating response, habituation and latent inhibition. *Eur J Neurosci* 15: 1983–1990.

Pantelis, C., Harvey, C.A., Plant, G., Fossey, E., Maruff, P., Stuart, G.W., et al. (2004) Relationship of behavioural and symptomatic syndromes in schizophrenia to spatial working memory and attentional set-shifting ability. *Psychol Med* 34: 693–703.

Paylor, R., Baskall-Baldini, L., Yuva, L., and Wehner, J.M. (1996) Developmental differences in place-learning performance between C57BL/6 and DBA/2 mice parallel the ontogeny of hippocampal protein kinase C. *Behav Neurosci* 110: 1415–1425.

Paylor, R., Tracy, R., Wehner, J.M., and Rudy, J.W. (1994) DBA/2 and C57BL/6 mice differ in contextual fear but not auditory fear conditioning. *Behav Neurosci* 108: 810–817.

Perkinton, M.S., Sihra, T.S., and Williams, R.J. (1999) Ca^{2+}-permeable AMPA receptors induce phosphorylation of cAMP response element-binding protein through a phosphatidylinositol 3-kinase-dependent stimulation of the mitogen-activated protein kinase signaling cascade in neurons. *J Neurosci* 19: 5861–5874.

Pouzet, B., Zhang, W.N., Weiner, I., Feldon, J., and Yee, B.K. (2004) Latent inhibition is spared by n-methyl–aspartate (nmda)-induced ventral hippocampal lesions, but is attenuated following local activation of the ventral hippocampus by intracerebral nmda infusion. *Neurosci* 124: 183–194.

Quinlan, E.M. and Halpain, S. (1996) Emergence of activity-dependent, bidirectional control of microtubule-associated protein MAP2 phosphorylation during postnatal development. *J Neurosci* 16: 7627–7637.

Rescorla, R.A. (1971) Summation and retardation tests of latent inhibition. *J Comp Physiol Psychol* 75: 77–81.

Restivo, L., Passino, E., Middei, S., and Ammassari-Teule, M. (2002) The strain-specific involvement of nucleus accumbens in latent inhibition might depend on differences in processing configural- and cue-based information between C57BL/6 and DBA mice. *Brain Res Bull* 57: 35–39.

Riedel, G., Platt, B., and Micheau, J. (2003) Glutamate receptor function in learning and memory. *Behav Brain Res* 140: 1–47.

Roberson, E.D., English, J.D., Adams, J.P., Selcher, J.C., Kondratick, C., and Sweatt, J.D. (1999) The mitogen-activated protein kinase cascade couples PKA and PKC to cAMP response element binding protein phosphorylation in area CA1 of hippocampus. *J Neurosci* 19: 4337–4348.

Rochford, J., Sen, A.P., and Quirion, R. (1996a) Effect of nicotine and nicotinic receptor agonists on latent inhibition in the rat. *J Pharmacol Exp Ther* 277: 1267–1275.

Rochford, J., Sen, A.P., Rousse, I., and Welner, S.A. (1996b) The effect of quisqualic acid-induced lesions of the nucleus basalis magnocellularis on latent inhibition. *Brain Res Bull* 41: 313–317.

Rossi-Arnaud, C., Fagioli, S., and Ammassari-Teule, M. (1991) Spatial learning in two inbred strains of mice: genotype-dependent effect of amygdaloid and hippocampal lesions. *Behav Brain Res* 45: 9–16.

Rudy, J.W. and Sutherland, R.J. (1995) Configural association theory and the hippocampal formation: an appraisal and reconfiguration. *Hippocampus* 5: 375–389.

Ruob, C., Weiner, I., and Feldon, J. (1998) Haloperidol-induced potentiation of latent inhibition: interaction with parameters of conditioning. *Behav Pharm* 9: 245–253.

Russig, H., Kovacevic, A., Murphy, C.A., and Feldon, J. (2003) Haloperidol and clozapine antagonise amphetamine-induced disruption of latent inhibition of conditioned taste aversion. *Psychopharm* 170: 263–270.

Schafe, G.E., Atkins, C.M., Swank, M.W., Bauer, E.P., Sweatt, J.D., and Ledoux, J.E. (2000) Activation of ERK/MAP kinase in the amygdala is required for memory consolidation of Pavlovian fear conditioning. *J Neurosci* 20: 8177–8187.

Schauz, C. and Koch, M. (1999) Lesions of the nucleus basalis magnocellularis do not impair prepulse inhibition and latent inhibition of fear-potentiated startle in the rat. *Brain Res* 815: 98–105.

Schauz, C. and Koch, M. (2000) Blockade of NMDA receptors in the amygdala prevents latent inhibition of fear-conditioning. *Learn Mem* 7: 393–399.

Schiller, D. and Weiner, I. (2004) Lesions to the basolateral amygdala and the orbitofrontal cortex but not to the medial prefrontal cortex produce an abnormally persistent latent inhibition in rats. *Neurosci* 128: 15–25.

Schiller, D. and Weiner, I. (2005) Basolateral amygdala lesions in the rat produce an abnormally persistent latent inhibition with weak preexposure but not with context shift. *Behav Brain Res* 163: 115–121.

Schiller, D., Zuckerman, L., and Weiner, I. (2006) Abnormally persistent latent inhibition induced by lesions to the nucleus accumbens core, basolateral amygdala and orbitofrontal cortex is reversed by clozapine but not by haloperidol. *J Psychiatric Res* 40: 167–177.

Schmajuk, N.A., Lam, Y.W., and Christiansen, B.A. (1994) Latent inhibition of the rat eyeblink response: effect of hippocampal aspiration lesions. *Physiol Behav* 55: 597–601.

Schnur, P. and Lubow, R.E. (1976) Latent inhibition: the effects of ITI and CS intensity during preexposure. *Learn Motiv* 7: 540–550.

Seillier, A., Dieu, Y., Herbeaux, K., Di Scala, G., Will, B., and Majchrzak, M. (2007) Evidence for a critical role of entorhinal cortex at pre-exposure for latent inhibition disruption in rats. *Hippocampus* 17: 220–226.

Selcher, J.C., Atkins, C.M., Trzaskos, J.M., Paylor, R., and Sweatt, J.D. (1999) A necessity for MAP kinase activation in mammalian spatial learning. *Learn Mem* 6: 478–490.

Selcher, J.C., Weeber, E.J., Varga, A.W., Sweatt, J.D., and Swank, M. (2002) Protein kinase signal transduction cascades in mammalian associative conditioning. *Neuroscientist* 8: 122–131.

Shohamy, D., Allen, M.T., and Gluck, M.A. (2000) Dissociating entorhinal and hippocampal involvement in latent inhibition. *Behav Neurosci* 114: 867–874.

Solomon, P.R., Kiney, C.A., and Scott, D.R. (1978) Disruption of latent inhibition following systemic administration of parachlorophenylalanine (PCPA). *Physiol Behav* 20: 265–271.

Solomon, P.R. and Moore, J.W. (1975) Latent inhibition and stimulus generalization of the classically conditioned nictitating membrane response in rabbits (*Orcytolagus cuniculus*) following dorsal hippocampal ablation. *J Comp Physiol Psychol* 89: 1192–1203.

Solomon, P.R., Nichols, G.L., Kiernan, J.M., III, Kamer, R.S., and Kaplan, L.J. (1980) Differential effects of lesions in medial and dorsal raphe of the rat: latent inhibition and septohippocampal serotonin levels. *J Comp Physiol Psychol* 94: 145–154.

Solomon, P.R. and Staton, D.M. (1982) Differential effects of microinjections of d-amphetamine into the nucleus accumbens or the caudate putamen on the rat's ability to ignore an irrelevant stimulus. *Biol Psychiatry* 17: 743–756.

Swartzentruber, D. and Bouton, M.E. (1986) Contextual control of negative transfer produced by prior CS-US pairings. *Learn Motiv* 17: 366–385.

Swerdlow, N.R., Hartston, H.J., and Hartman, P.L. (1999) Enhanced visual latent inhibition in obsessive-compulsive disorder. *Biol Psychiatry* 45: 482–488.

Swerdlow, N.R., Stephany, N., Wasserman, L.C., Talledo, J., Sharp, R., and Auerbach, P.P. (2003) Dopamine agonists disrupt visual latent inhibition in normal males using a within-subject paradigm. *Psychopharm* 169: 314–320.

Szapiro, G., Vianna, M.R., McGaugh, J.L., Medina, J.H., and Izquierdo, I. (2003) The role of NMDA glutamate receptors, PKA, MAPK, and CAMKII in the hippocampus in extinction of conditioned fear. *Hippocampus* 13: 53–58.

Tai, C.T., Cassaday, H., Feldon, J., and Rawlins, J.N.P. (1995) Both electrolytic and excitotoxic lesions of nucleus accumbens disrupt latent inhibition of learning in rats. *Neurobiol Learn Mem* 64: 36–48.

Talk, A., Stoll, E., and Gabriel, M. (2005) Cingulate cortical coding of context-dependent latent inhibition. *Behav Neurosci* 119: 1524–1532.

Thornton, J.C., Dawe, S., Lee, C., Capstick, C., Corr, P.J., Cotter, P., et al. (1996) Effects of nicotine and amphetamine on latent inhibition in human subjects. *Psychopharm* 127: 164–173.

Traverso, L.M., Ruiz, G., and De la Casa, L.G. (2003) Latent inhibition disruption by MK-801 in a conditioned taste-aversion paradigm. *Neurobiol Learn Mem* 80: 140–146.

Trimble, K.M., Bell, R., and King, D.J. (1997) Enhancement of latent inhibition in the rat by the atypical antipsychotic agent remoxipride. *Pharmacol Biochem Behav* 56: 809–816.

Tsaltas, E., Preston, G.C., Rawlins, J.N., Winocur, G., and Gray, J.A. (1984) Dorsal bundle lesions do not affect latent inhibition of conditioned suppression. *Psychopharm* 84: 549–555.

Tyson, P.J., Laws, K.R., Roberts, K.H., and Mortimer, A.M. (2004) Stability of set-shifting and planning abilities in patients with schizophrenia. *Psychiatry Res* 129: 229–239.

Vaitl, D. and Lipp, O.V. (1997) Latent inhibition and autonomic responses: a psychophysiological approach. *Behav Brain Res* 88: 85–93.

Vossler, M.R., Yao, H., York, R.D., Pan, M.G., Rim, C.S., and Stork, P.J. (1997) cAMP activates MAP kinase and Elk-1 through a B-Raf- and Rap1-dependent pathway. *Cell* 89: 73–82.

Waltereit, R. (2003) Signaling from cAMP/PKA to MAPK and synaptic plasticity. *Mol Neurobiol* 27: 99–106.

Walz, R., Rockenbach, I.C., Amaral, O.B., Quevedo, J., and Roesler, R. (1999) MAPK and memory. *Trends Neurosci* 22: 495.

Warburton, E.C., Joseph, M.H., Feldon, J., Weiner, I., and Gray, J.A. (1994) Antagonism of amphetamine-induced disruption of latent inhibition in rats by haloperidol and ondansetron – implications for a possible antipsychotic action of ondansetron. *Psychopharm* 114: 657–664.

Weiner, I. (1990) Neural substrates of latent inhibition: the switching model. *Psychol Bull* 108: 442–461.

Weiner, I. (2003) The "two-headed" latent inhibition model of schizophrenia: modeling positive and negative symptoms and their treatment. *Psychopharm* 169: 257–297.

Weiner, I. and Feldon, J. (1992) Phencyclidine does not disrupt latent inhibition in rats: implications for animal models of schizophrenia. *Pharmacol Biochem Behav* 42: 625–631.

Weiner, I., Shadach, E., Tarrasch, R., Kidron, R., and Feldon, J. (1996a) The latent inhibition model of schizophrenia: further validation using the atypical neuroleptic, clozapine. *Biol Psychiatry* 40: 834–843.

Weiner, I., Tarrasch, R., and Feldon, J. (1996b) Basolateral amygdala lesions do not disrupt latent inhibition. *Behav Brain Res* 72: 73–81.

Weiss, I.C., Domeney, A.M., Moreau, J.L., Russig, H., and Feldon J. (2001) Dissociation between the effects of pre-weaning and/or post-weaning social isolation on prepulse inhibition and latent inhibition in adult Sprague–Dawley rats. *Behav Brain Res* 121: 207–218.

Willcutt, E.G., Doyle, A.E., Nigg, J.T., Faraone, S.V., and Pennington, B.F. (2005) Validity of the executive function theory of attention-deficit/hyperactivity disorder: a meta-analytic review. *Biol Psychiatry* 57: 1336–1346.

Wright, D.C., Skala, K.D., and Peuser, K.A. (1986) Latent inhibition from context-dependent retrieval of conflicting information. *Bull Psychon Soc* 24: 152–154.

Wright, I.K., Garratt, J.C., and Marsden, C.A. (1990) Effects of a selective 5-HT2 agonist, DOI, on 5-HT neuronal firing in the dorsal raphe nucleus and 5-HT release and metabolism in the frontal cortex. *Br J Pharmacol* 99: 221–222.

Yee, B.K., Feldon, J., and Rawlins, J.N. (1997) Cytotoxic lesions of the retrohippocampal region attenuate latent inhibition but spare the partial reinforcement extinction effect. *Exp Brain Res* 115: 247–256.

Chapter 26

Executive function
Response inhibition, attention, and cognitive flexibility

Sheree F. Logue and Thomas J. Gould

Overview

In order to survive, all individuals must be able to adapt to ever-changing environments. Executive function is thought to mediate the ability to adapt by regulating reflexive reactions to current salient stimuli so that goals requiring complex behaviors can be attained. It is theorized that executive function is a heterogeneous group of higher order cognitive abilities that enable individuals to orient towards the future, demonstrate self-control, and successfully complete goal-directed behavior (Baddeley, 1998; Robbins, 1996; Stuss and Alexander, 2000). In humans and other animals executive function is disrupted following brain injury involving the frontal cortical regions (Bechara and Van Der Linden, 2005; Dalley et al., 2004; Schoenbaum et al., 2006) and there are numerous psychiatric disorders which are characterized by disruptions in executive function; schizophrenia, attention deficit hyperactivity disorder, bipolar disorder, substance abuse, antisocial behavior, and obsessive-compulsive disorder (Brower and Price, 2001; Cavedini et al., 2006; Daban et al., 2006; Schoenbaum et al., 2006; Willcutt et al., 2005). Thus, further study of these higher order cognitive processes will not only advance the understanding of the cognitive processes and neurobiology of executive function but also aid in the discovery of new treatments for patients with brain damage and mental illnesses. This chapter will briefly outline executive function in humans, discuss the animal models of executive function that are analogous to human tests, and review the current behavioral genetic data in these models.

Executive function: humans

As highlighted by the well-known case of Phineas Gage – the railroad foreman who survived an accident in which a tamping iron passed through his skull and frontal lobe – people with damage to the frontal lobes demonstrate poor control and regulation over their behavior and usually have problems in their everyday lives. Extensive research on frontal lobe patients using neuropsychological test batteries (see Table 26.1 for a partial listing) has consistently demonstrated impairment on complex executive tasks (e.g., the Wisconsin card sorting test and the Tower of Hanoi or Tower of London tasks), even though the patients may demonstrate intact performance on other cognitive tasks such as IQ tests (Damasio, 1994). Poor performance on complex executive tasks such as the Wisconsin card sorting test can be due to deficits in one or more of the higher order cognitive processes that may be involved in complex executive function. The higher order cognitive processes that have been theorized to comprise executive function include response inhibition, attention, working memory, cognitive flexibility, planning, judgment, and decision-making (Baddeley, 1998; Robbins, 1996; Stuss and Alexander, 2000). Most current researchers consider executive function to reflect multiple individual processes that are inter-related and research focuses on how these inter-related process function in order to complete a complex task.

Table 26.1 Tasks used to assess executive function in humans.

Attention and cognitive flexibility	Response inhibition
Subtask switch (number–letter; color–shape category switch)	Antisaccade task Go/no-go task
Wisconsin card sorting test	Competing motor task
Trail making test	Stop-signal task
Tower of London	Stroop task
Tower of Hanoi	Continuous performance test
Continuous performance test	

Studies of executive function in normal subjects have provided insight into the relationships between the multiple processes involved in executive function and the relationships between these processes and general intelligence, as well as lower level cognitive processes such as reaction time (Cheung et al., 2004; Friedman et al., 2006; Luciano et al., 2001; Miyake et al., 2000). When evaluated at the level of latent variables (i.e., focusing on the common features shared among multiple tests measuring each executive function), response inhibition, cognitive flexibility, and working memory executive function processes were demonstrated to be moderately correlated; yet they were clearly separate processes (Miyake et al., 2000). Additionally, the processes of response inhibition, cognitive flexibility, and working memory executive function differentially contribute to performance on complex tasks of executive function (Miyake et al., 2000) and to fluid and crystallized intelligence

Behavioral Genetics of the Mouse: Volume I. *Genetics of Behavioral Phenotypes*, eds. Wim E. Crusio, Frans Sluyter, Robert T. Gerlai, and Susanna Pietropaolo. Published by Cambridge University Press.

Table 26.2 Tasks used to assess executive function in rodents.

Attention and cognitive flexibility	Response inhibition/impulsivity	
	Impulsive choice	Impulsive section
Five-choice serial reaction time		
Attentional set-shifting task	Delay discounting	Go/no-go task
Reversal learning	Delayed reinforcement	Stop-signal reaction time
		Five-choice serial reaction time
		Differential reinforcement of low rates of responding
		Signaled nose poke task

(Friedman et al., 2006). For example, by using individual differences in a genetically diverse population, Miyake et al. (2000) found that performance on the Wisconsin card sorting test was more strongly related to cognitive flexibility than to either response inhibition or working memory while performance on the Tower of Hanoi was related most strongly to response inhibition even though both tests are considered tests of executive function. Thus, studies in normal subjects can elucidate subtle differences in the relationships between the different cognitive processes of executive function and this information will facilitate continued refinement of theories on executive function. Also important is the information gained about the assays used to study executive function. Even though one can assume that several different assays all measure a single cognitive process of executive function, the data described above clearly indicates that this is not the case and this information can be used to refine the tools used to study executive function.

Although there is some disagreement about the relative validity of comparing complex cognitive processes between humans and non-primate research species, rodent studies do provide crucial information toward understanding the processes involved in executive function. Table 26.2 contains a list of assays used in rodents to evaluate executive function. The majority of the assays were developed to be analogous to the tests of executive function in humans. Using these assays, researchers have explored the neurobiology of executive function in the rat and the adaptation of these assays to mice provides an opportunity to further the understanding of executive function through the application of behavioral genetic approaches. As researchers continue to use the natural genetic variability in humans and animals to probe complex cognitive processes the question of how executive functions and other cognitive processes interact to successfully complete complex cognitive tasks will be answered.

Executive function: rodents

There are a growing number of studies using inbred and genetically modified mice in tasks that measure response inhibition, attention, and cognitive flexibility processes in executive function. The remainder of this chapter will focus on studies comparing inbred mice to highlight the contribution of genetic variability to performance within individual tasks and across multiple tasks measuring executive function. Inbred strains and selective crosses have typically been compared in a single assay designed to measure one process of executive function and these studies demonstrate a clear impact of genetic variability on performance. This approach is highly suited to the study of the genes and neurobiology underlying a selected measure of executive function. In order to gain the most information on complex cognitive function by utilizing natural genetic variability, mice will need to be evaluated in multiple assays measuring response inhibition, attention, and cognitive flexibility as has been done in humans. To date, no study has utilized this approach in an effort to understand executive function as has been done for other cognitive processes like hippocampal-dependent learning (Owen et al., 1997a). But the insights into the role of different cognitive processes in executive function gained from the current overview may spark interest in research focused on understanding the functional relationships required for executive function.

There are two caveats relevant to the interpretation of the relationships mentioned in the subsequent sections. The first caveat is that the inbred strains from different suppliers are considered to be a single inbred strain. There is a potential problem for the genetic interpretations made herein if there has been enough genetic drift to result in several substrains with different behavioral characteristic rather than a single strain and if the environmental differences between suppliers are significant enough to impact the behavioral performances. Secondly, the impact of differences in methodology across the studies is only given minor consideration and such differences in method could have a very large impact on the performance of different strains in the assays for executive function as has been demonstrated for other behavioral measures (Wahlsten et al., 2003).

Attention and cognitive flexibility

Five-choice serial reaction time task

The most well-studied rodent model of attentional processes is the five-choice serial reaction time (5CSRT) task developed by Robbins and colleagues (see review, Robbins, 2002). The 5CSRT is analogous to the continuous performance test in humans. In the basic 5CSRT assay, the rodent scans an array of apertures for a brief 500 ms light stimulus that can be presented in any one of five spatial locations in the array; a nose poke in the illuminated aperture results in a food reinforcement. The behavioral measures obtained in the 5CSRT allow for determination of discrete changes in attention by ruling out changes in sensory and motivational factors. These measures in the 5CSRT include percent correct responding (i.e., accuracy), percent omissions (i.e., the failure to respond to the stimulus), response latency, latency to collect the reinforcement, and premature responding (i.e., nose pokes during the intertrial interval). Premature responding is a measure of response inhibition and will be discussed in more detail in that section. Manipulations that differentially affect aspects of 5CSRT are informative for understanding executive function. For instance, an increase in response latency without an increase in latency to collect the reward would be interpreted as an alteration in decision processing rather than a change in motivation. For attention per se, a decrease in percent correct responding, without an increase in omissions, and an increase in latency to collect the reward may reflect a very selective change in attention while a decrease in accuracy concomitant with an increase in omissions but not accompanied by a change in magazine latency may be due to gross impairments in attention.

There are several parametric manipulations that can be used in the 5CSRT to increase the attentional load or to tax the ability to inhibit responding. Attentional load can be increased by decreasing the duration or the intensity of the visual stimulus, and by adding an auditory stimulus to challenge the subject's ability to focus on the visual stimulus while ignoring the auditory stimulus. If the occurrence of the visual stimulus is made unpredictable by varying the length of the intertrial interval, then premature responding increases suggest a decrease in the ability to withhold pre-potent responses. This point will be addressed more fully in the response inhibition section.

Following the initial report of successful adaptation of the 5CSRT to mice (Humby et al., 1999) researchers began using the task to evaluate inbred strains and crosses (de Bruin et al., 2006; Greco et al., 2005; Marston et al., 2001; Patel et al., 2006) and to phenotype genetically modified mice (Davies et al., 2007; Greco and Carli, 2006; Higgins et al., 2003; Hoyle et al., 2006; Kerr et al., 2004; Wrenn et al., 2006; Young et al., 2004, 2007). Overall, the performance of mice on the 5CSRT task was similar to the performance of rats and was affected by cholinergic manipulations as has been shown in rats, monkeys, and humans (Humby et al., 1999), thus validating the use of the 5CSRT test in mice to study attentional processes.

Under baseline training conditions (1 second stimulus duration, 5 second limited hold, and a stable intertrial interval), the response latency on correct trials was approximately 1 second in all studies for all strains evaluated (de Bruin et al., 2006; Greco et al., 2005; Humby et al., 1999; Marston et al., 2001; Patel et al., 2006). This was the only measure in the 5CSRT that did not appear to be affected by either genetic variability or differences in methodology. Percent omissions, on the other hand, had the greatest degree of variability across studies even when the same strain was compared. Based on the current data, it is unclear to what degree the percent omissions measure is impacted by genetic variability versus differences in methodology. For the measure of percent correct responses, the attentional accuracy demonstrated by the C57BL/6 mice (Greco et al., 2005; Marston et al., 2001; Patel et al., 2006) was equivalent to the accuracy of the129P2/OlaHsd mice (Marston et al., 2001), but greater than the DBA/2 mice (Greco et al., 2005; Patel et al., 2006). The accuracy of the C57BL/6xDBA/2 F1 and C57BL/6x129/Sv F1 crosses was quite high (Humby et al., 1999) and the accuracy of the C57BL/6xDBA/2 F1 mice was more similar to the level demonstrated by the C57BL/6 strain than the DBA/2 stain. In contrast, the accuracy of a C57BL/6Jx129P2/SvPas F2 cross was lower than the reported percent correct responses for either a C57BL/6x129/Sv F1 cross or the C57BL/6 strain but given that there was no strain or cross for comparison with the C57BL/6Jx129P2/SvPas F2 cross, the differences in attention could reflect methodological differences.

The performance of C57BL/6, 129P2/OlaHsd, DBA/2, C57BL/6xDBA/2 F1, C57BL/6x129/Sv F1, and C57BL/6Jx129P2/SvPas F2 mice was also evaluated in the 5CSRT under increased attentional load. Decreasing the duration of the visual stimulus, decreasing the intensity of the visual stimulus, and adding a distracting auditory stimulus increased the load on attention in the mice as had been demonstrated in the rat. When the duration of the visual stimulus was decreased, the accuracy of performance for all strains was disrupted at short durations and some of the strains were more severely affected than others. C57BL/6 and 129P2/OlaHsd mice were similarly affected by the decrease in stimulus duration (de Bruin et al., 2006; Marston et al., 2001), but the DBA/2 mice were more affected than the C57BL/6 (Patel et al., 2006) mice. The decrease in accuracy in the C57BL/6 and DBA/2 mice at the short duration stimulus was not associated with a change in latency to make a correct response or in percent omissions, suggesting that the effect was very selective to attention. The accuracy of the C57BL/6xDBA/2 F1 mice was less affected at short durations than the C57BL/6x129/Sv F1 mice (Humby et al., 1999). However, the decreased accuracy in the C57BL/6x129/Sv F1 mice was paralleled by an increase in percent omissions; a profile of effects suggesting broad impact on attentional processes. A similar profile of decreased accuracy and increased omissions when the stimulus duration was shortened was also seen in C57BL/6Jx129P2/SvPas F2 mice (de Bruin et al., 2006). The ability of the C57BL/6xDBA/2 F1 mice to maintain attentional accuracy when the stimulus duration was short appeared to be due to the ability to slow their response speed. This utilization of an error-speed trade-off may reflect greater cognitive flexibility in these mice.

When the intensity of the visual stimulus was decreased there was an overall decrease in accuracy for all strains and again there were strain differences in sensitivity to the attentional load. The 129P2/OlaHsd mice were more sensitive to the dimming of the stimulus than the C57BL/6 mice (de Bruin et al., 2006; Marston et al., 2001) and the C57BL/6x129/Sv F1 were greatly affected, while the C57BL/6xDBA/2 F1 mice showed equivalent accuracy at all intensities (Humby et al., 1999). As with the stimulus duration manipulation, the C57BL/6x129/Sv F1 showed a concomitant increase in percent omissions while the C57BL/6xDBA/2 F1 mice appeared to maintain accuracy by slowing latency to respond.

The final manipulation used to increase attentional load was the introduction of an auditory stimulus as a distractor.

There was no decrease in accuracy when exposed to a distracting auditory stimulus in either the C57BL/6x129/Sv F1 or C57BL/6xDBA/2 F1 mice (Humby et al., 1999), but an increase in percent omissions was observed in the C57BL/6x129/Sv F1 mice as well as a slowing of the response latency on correct trials in the C57BL/6xDBA/2 F1 mice specifically when the distracting stimulus was presented immediately prior to the onset of the visual stimulus. In a C57BL/6Jx129P2/SvPas F2 cross, accuracy was not affected when a distractor stimulus was interpolated into the intertrial interval and there was a slowing of response latency similar to the C57BL/6xDBA/2 F1 mice but there was also a decrease in omissions which was contrary to the response in the C57BL/6x129/Sv F1 mice (de Bruin et al., 2006).

A variation of the 5CSRT that includes direct assessment of correct rejections or responses on non-signal or no/go trials, the five-choice continuous performance test (5C-CPT) has been developed in mice (Young et al., 2009). The measurement of correct rejections allows this assay to assess vigilance using signal detection calculations as is done in human versions of CPT. In the 5C-CPT, 83% of the trials were identical to the trials of 5CSRT requiring the mouse to nose poke in the lit aperture while the remaining 17% of trials were no/go trials during which all five apertures were lit and the mouse had to refrain from nose poking in any aperture. C57BL/6J and DBA/2J mice were trained in the 5C-CPT and demonstrated similar attentional accuracy and premature responding but differed on several important measures. Relative to DBA/2J mice, C57BL/6J mice had lower percent omissions, a higher sensitivity index score indicating greater ability to discriminate between go and no/go trials, and a more liberal response index bias. The strain difference was retained when the length of the session was extended from 120 to 250 trials in order to increase the attentional load.

The strain differences in accuracy on the baseline task of the 5CSRT and on sensitivity index and response bias index in the 5C-CPT indicate sufficient genetic variability to make further studies using behavioral genetics approaches very interesting. The genetics underlying the difference in attention between the C57BL/6 and DBA/2 strains could be evaluated using the BxD recombinant inbred lines and comparisons with the known behavioral and neurological profile of these strains in other cognitive assays could be made (Martin et al., 2006; Owen et al., 1997a; Risinger and Cunningham, 1998). Using the three inbred and two F1 crosses as reviewed, it appears that there are interesting genetic interactions underlying the differences in accuracy under baseline conditions as compared to the response of the subject to increased attentional load. Baseline attention was equivalent between the 129P2/OlaHsd, C57BL/6, C57BL/6x129/Sv F1, and C57BL/6xDBA/2 F1 mice and all were better than the DBA/2. The lower accuracy at baseline in the DBA/2 corresponded to an increased sensitivity to disruption by increased attentional load. In the F1 crosses the sensitivity to attentional load was not predicted by baseline performance given that the C57BL/6x129/Sv F1 and C57BL/6xDBA/2 F1 mice were similar at baseline but only the C57BL/6xDBA/2 F1 mice seemed capable of adjusting their behavior to accommodate for the increased load while the attention of the C57BL/6x129/Sv F1 mice seemed to be disrupted at a more gross level (e.g., the decrease in accuracy along with the increase in percent omissions). Further exploration of the genetic interactions underlying the differential response of the F1 crosses under increased attentional load would facilitate understanding of the genetic contributions to disrupted attentional processing.

Attentional set-shifting

Attentional set-shifting is another assay measuring attention and cognitive flexibility that was adapted from a non-human primate test for use in rats and then mice. Attentional set-shifting was developed to evaluate the individual cognitive components involved in the Wisconsin card sorting test. Roberts et al. (1988) developed the attentional set-shifting model in non-human primates to evaluate the ability of the animal to shift attention within a single dimension (e.g., shape) and between two dimensions of a compound stimulus (e.g., shape to color) based on changing feedback/reinforcement. An intra-dimensional shift, switching from a circle shape to a square shape, should be less demanding with respect to attention and cognitive flexibility than an extra-dimensional shift, switching from a shape to a color. The increased difficulty with which the extra-dimensional shift is acquired indicates that an attentional set (i.e., knowing that paying attention to shape is critical for solving the problem) was acquired.

The rat version of attentional set-shifting (Birrell and Brown, 2000) was adapted to mice by Colacicco et al. (2002) using C57BL/6J, 129S6/SvEvTac, and C57BL/6x129S6/SvEv F1 mice to assess performance on intra-dimensional and extra-dimensional attention set-shifting using odor and digging medium as the dimensions. In order to establish an attention set, the mouse must progress from learning and using simple constant rules to guide behavior (e.g., a simple discrimination in which an odor is always reinforced and everything else is irrelevant) to learning how to figure out complex variable rules in order to guide behavior. The training phases move from a simple discrimination to a compound discrimination (e.g., odor 1 reinforced and odor 2 not reinforced and ignore what you are digging in to get the reinforcer) and then to a reversal of the compound discrimination (e.g., odor 1 now not reinforced and odor 2 is reinforced and ignore the digging medium). At this point the mouse has learned that paying attention to odor is important for accessing the reinforcer but which odor is important can change. In the intra-dimensional shift phase, the mouse is presented with two novel odors and two novel digging media but the rule for guiding behavior is the same as it was for the compound discrimination (e.g., figure out which odor to choose to get access to the reinforcer and ignore digging medium). In the extra-dimensional shift phase, the mouse is again presented with two novel odors and two novel digging media, but in this phase the rule regarding which cue should be attended to and which cue should be ignored has also changed. The mouse now has the added difficulty of finding out the rule has changed in addition to shifting attention from the odor medium to the digging medium. The learning in the extra-dimensional shift phase should take longer than the intra-dimensional shift because of the need to figure out there was a rule change in addition to solving the discrimination. In the rat model (Birrell and Brown, 2000), all phases of training were completed in one test session and the compound discrimination, intra-dimensional shift, and extra-dimensional shifts were followed by reversal training before proceeding to the next phase. In the mouse (Colacicco et al., 2002), each phase occurred on a separate day and only the compound discrimination was followed by reversal learning with no reversals of the intra- or extra-dimensional shifts.

C57BL/6 mice required significantly more trials to reach criterion on the simple discrimination than the 129S6/SvEvTac mice, while the F1 mice performed more similarly to the C57BL/6 parental strain (Colacicco et al., 2002). Trials to reach criterion on the conditional discrimination, conditional discrimination reversal, compound reversal repetition, intra-dimensional shift and intra-dimensional repetition were equivalent between the C57BL/6, 129S6/SvEvTac, and F1 mice. On the extra-dimensional shift, the C57BL/6 mice again required more trials to reach criterion than the 129S6/SvEvTac mice and in this phase the performance of the F1 mice matched the 129S6/SvEvTac parental strain, but there was no difference in trials to criterion between the intra- and extra-dimensional shift phases. Although there was no demonstration of attention sets, there may be a hint of an interesting relationship in the genetic contributions to the learning in the simple discrimination and extra-dimensional shift phases. The C57BL/6 mice required more trials in each of these phases than did the 129S6/SvEvTac mice but there was divergence in the performance of the F1 cross between these phases. On the simple discrimination the trials to criterion in the F1 mice was similar to the C57BL/6, while in the extra-dimensional phase the F1 mice performed identically to the 129S6/SvEvTac parental strain thus suggesting a difference in learning requirements for the two phases. Since neither the C57BL/6 nor the 129S6/SvEvTac mice demonstrated the expected increase in difficulty when shifting between the intra- and extra-dimensional phases, there is no evidence of the mice forming an attentional set.

Colacicco et al. (2002) discussed that the failure of the mice to demonstrate formation of an attentional set was possibly due to methodological differences between their study and the study in rats (Birrell and Brown, 2000). For example, the training phases did not include an intra-dimensional shift reversal, which in rats has been shown to be required to produce a difference between the intra- and extra-dimensional shifts indicative of the formation of an attentional set. Furthermore, the separation of the training phases across days may have impacted the ability to form an attentional set. Similar to Colacicco et al. (2002), two other studies testing C57BL/6 mice also failed to show evidence of acquisition of an attentional set using an odor and digging medium procedure (Glickstein et al., 2005) and a novel procedure using visual stimuli on touch screen highly similar to the non-human primate model (Brigman et al., 2005). Given that Glickstein et al. (2005) included reversal of the intra-dimensional shift and performed all phases in a single session, the question is raised as to what the specific procedural requirements would be to produce formation of attentional sets in mice. The question of the specific procedural requirements needed to achieve and demonstrate attentional set-shifting in mice was answered by Garner et al. (2006). Several changes in procedure including time of day for testing, over training, and the use of an eight out of 10 consecutive correct trials as criterion were made but the major change was making the digging bowls the compound stimuli with the use of outer texture as well as the standard digging medium and using odor as the new dimension during extra-dimensional shift testing. This protocol change makes the task more similar to a refined intra-dimensional and extra-dimensional task (Owen et al., 1993) used in humans utilizing the "extra-dimensional stuck-in-set" shift rather than the original extra-dimensional shift. Using this refined procedure for mice, C57BL/6 mice required significantly more trials to reach criterion on the extra-dimensional shift than had been required for the intra-dimensional shift (Garner et al., 2006).

Using a procedure that truly demonstrates attentional set-shifting in mice, the role of the medial prefrontal cortex in this executive function assay has been verified in mice (Bissonette et al., 2008). The contribution of specific genes to attention and cognitive flexibility is being explored by using this assay to evaluate genetically modified mice (Papaleo et al., 2008). Not only does the attention set shift assay provide information on the development of attentional sets but the reversal phases included in the assay can provide useful information on the role of reversal learning as a measure of cognitive flexibility. For example, the roles of the medial and orbital prefrontal cortical areas in attention and reversal learning were dissociated (Bissonette et al., 2008).

Reversal learning

Cognitive flexibility as measured by reversal learning in mice has been evaluated using different behavioral paradigms from simple discriminations using visual or odor cues, to avoidance Y-maze discrimination, and to spatial learning in a water maze. In non-water maze procedures, DBA/2 mice were consistently better at reversal learning than the C57BL/6 mice (Gasbarri et al., 1997; Izquierdo et al., 2006; Mihalick et al., 2000). Needing fewer trials to reach criterion on discrimination reversal may indicate better cognitive flexibility in the DBA/2 mice. When evaluated on reversal of spatial learning in the water maze, DBA/2 mice were slower to learn the reversal than C57BL/6 mice (Francis et al., 1995). The disparate performance of the DBA/2 on reversal learning between non-water maze and water maze based tasks may be related to known differences between the strains on spatial (Ammassari-Teule et al., 1993; Logue et al., 1997; Owen et al., 1997b; Thinus-Blanc et al., 1996) versus non-spatial tasks (Anisman, 1975; Cabib et al., 2002; Weinberger et al., 1992) and begs the question as to what assays should be used to measure cognitive flexibility. The test-dependency of strain differeneces also suggests that there may be distinct subtypes of reversal learning.

The C57BL/6 strain has also been compared to a variety of 129 substrains on reversal learning in water maze tasks. Based on escape latency over reversal training, the C57BL/6 mice were better than 129S2/SvHsd mice (Voikar et al., 2004), and equal to 129T2/Sv mice (Wolff et al., 2002), 129S1/SvImJ, 129S2/SvPasIsoCrlBR, and 129S6/SvEvTac, but slower than the B6x129S1 F1 mice (Clapcote and Roder, 2004). However, during the probe trial test following reversal training the 129S6/SvEvTac mice showed more marked preference for the trained location. Based on the search-time preference score, the 129S6/SvEvTac showed the greatest preference for the trained location compared to all other strains except for the B6x129S1 F1 mice while based on the crossing index measure the 129S6/SvEvTac were only superior to the C57BL/6J and C57BL/6N (Clapcote and Roder, 2004).

Response inhibition

The response inhibition associated with executive function, herein considered synonymous of impulsivity, is a multifactorial phenomenon which includes, as delineated in Table 26.2, direct control of actions as well as of choices so that the most beneficial outcome for the individual is achieved (Evenden, 1999). The direct control over responses or impulsive action covers a range of response inhibition from withholding a prepotent response (e.g., go/no-go task) to stopping a response that has already been initiated (e.g., stop-signal task) to inhibiting a response during a delay when the response will preclude acquisition of a reward (e.g., differential reinforcement of low rates of responding task, the premature responding measure in the 5 CSRT test, and the signaled nose poke task). Impulsive choice is typically defined as a preference for small, immediate rewards over large delayed rewards and depends more on the ability to assess the outcome of a behavioral response than to control the motor aspects of a response (Evenden, 1999). In the following section, we will review the studies that have compared inbred mouse strains on impulsive choice (Helms et al., 2006; Isles et al., 2004; Otobe and Makino, 2004) and on impulsive action (Abraham et al., 2004; Logue et al., 1998; Pattij et al., 2003, 2004; Ripley et al., 2001).

Impulsive choice

Several inbred strains have been evaluated for impulsive choice using operant based, nose poke or lever press response tasks of delayed reinforcement or delay discounting. In delayed reinforcement tasks, a choice is made between a small immediate reinforcer and a large delayed reinforcer (Isles et al., 2004; Otobe and Makino, 2004). All subjects should show a choice bias for the larger reinforcer when there is no delay but the choice bias should shift to the smaller reinforcer as the delay to obtaining the larger reinforcer is increased. An impulsive mouse should show the shift in choice bias to the smaller reinforcer at shorter delay times relative to normal mice. Isles et al. (2004) used a delayed reinforcement procedure that required a choice between 25 μl of immediate reinforcer and 50 μl of reinforcer delayed from 2 to 8 seconds with the mice experiencing all delays within each session. In this study, the four strains, C57BL/6, BALB/c, CBA/Ca, and 129S2/SvHsd, showed the expected bias toward the large reinforcer at the zero delay and the expected shift in bias toward the small immediate reinforcer when the length of delay to the large reinforcer increased. However, the C57BL/6 and BALB/c strains shifted bias to the small immediate reward at shorter delay times than did the CBA/Ca and the 129S2/SvHsd mice, and there was no difference between the C57BL/6 and BALB/c mice in the timing of the shift in bias. Thus, the C57BL/6 and BALB/c mice were more impulsive than the CBA/Ca and 129S2/SvHsd strains (Isles et al., 2004). Otobe and Makino (2004) used a similar procedure where a single pellet reinforcer was always available within 6 seconds while a larger two pellet reinforcer could be delayed up to 30 seconds. In this protocol the C57BL/6, BALB/c, and DBA/2 mice were tested with one delay length within a session but the delay increased in length across sessions. Similar to the report by Isles et al. (2004), all three strains showed a choice bias for the large reinforcer at the shortest delay and a shift in bias toward the small reinforcer as the delay to the large reinforcer increased to 30 seconds. The C57BL/6 and DBA/2 mice demonstrated an equivalent shift in bias to the small reinforcer at all delays, but in contrast to the Isles et al. (2004) study, the BALB/c mice were significantly more impulsive than the C57BL/6 mice.

In delay discounting tasks, the choice is between a large delayed reinforcer and access to an immediate reinforcer of variable size with the aim of determining the degree to which an individual will devalue or "discount" the large reinforcer as the delay to the large reinforcer increases (Green and Myerson, 2004). As the delay increases, the value of the large reinforcer to the subject decreases, thus increasingly smaller immediate reinforcers will be accepted instead. C57BL/6 and DBA/2 mice were compared on impulsive choice using an adjusting amount procedure (Helms et al., 2006). The adjusting amount procedure in mice is analogous to the one used in human research (Green and Myerson, 2004) and thus allows for direct comparisons between mouse studies and human studies. In the mouse adjusting amount procedure, the delayed reinforcer is a fixed amount (9.76 μl) and is contrasted with an immediate reinforcer that varied in amount by 10% across trials from a starting amount of 4.88 μl. The amount of the immediate variable reinforcer is increased or decreased depending on the choice made on the previous trial such that choice of the large delayed reinforcer results in an increase in the immediate variable reinforcer whereas sequential choice of the immediate variable reinforcer results in a decrease in the immediate variable reinforcer. Impulsive choice in the mouse adjusting amount procedure is characterized by the "indifference point" which reflected the amount of the immediate variable reinforcer required for the subject to choose equally between the immediate and delayed reinforcer. When no delay is interposed, a normal subject should choose equally between the fixed and variable reinforcers when the amount of the reinforcers is roughly equal. Furthermore, as the delay to the large reinforcer increases, the "indifference point" would require progressively smaller amounts of the immediate variable reinforcer as the subject devalues the delayed reinforcer. At zero delay, the C57BL/6 mice accepted a significantly lower amount of the variable reinforcer at their "indifference point" than did the DBA/2 mice. In fact, the amount of the variable reinforcer accepted at the "indifference point" at the zero delay for the C57BL/6 mice was significantly lower than the fixed reinforcer. The authors (Helms et al., 2006) attributed this strain difference to a "side bias" in responding for the C57BL/6 mice and controlled for this difference in order to selectively assess impulsive choice. To control for the divergent "indifference points" at baseline, the "indifference points" across the delays were analyzed as a proportion of the baseline and this analysis determined the DBA/2 mice to be more impulsive than C57BL/6 mice because they had significantly lower "indifference points" across all delays indicating a greater propensity to devalue the large reward (Helms et al., 2006).

With respect to impulsive choice, the primary observation is the variability in the performance of strains across the limited number of studies. The difference in the level of impulsive choice displayed by C57BL/6 mice in relation to the BALB/c mice (Isles et al., 2004; Otobe and Makino, 2004) could be due to differences in methodology while the reversed degree of impulsivity between the C57BL/6 and DBA/2 mice (Helms et al., 2006; Otobe and Makino, 2004) could be an indication that the delayed reinforcement tests and the delay discounting task are measuring different components of impulsive choice that are genetically dissociable. Additional studies are required in order to determine to what degree impulsive choice is influenced by genetic variation, differences in methodology, or an interaction between the two.

Impulsive action

As with assays for impulsive choice, relatively few studies have evaluated mice in assays that measure impulsive action. The stop-signal reaction time test developed to measure the ability of a rat to inhibit a response that has been initiated (Eagle and Robbins, 2003a, 2003b) has to date not been adapted to mice. A go/no-go odor discrimination task has been developed in C57BL/6 mice to examine information processing in the olfactory system but the procedure has not been used to compare impulsive action in inbred mice (Abraham et al., 2004). The assays measuring the ability to withhold pre-potent response that have been adapted to mice include the differential reinforcement of low rates of responding task, the premature response measure in the 5CSRT test, and the signaled nose poke task.

The differential reinforcement of low rates of responding task (DRL) requires the mouse to withhold responding for a fixed interval of time after which a response is reinforced. Successful performance on a DRL task requires the mouse to be able to accurately assess the passage of time and to inhibit responding during the specified time interval in order to optimize access to reinforcers. During successful acquisition of a DRL task, the number of reinforcers obtained typically increases, the relative frequency of inter-response times becomes normally distributed with the peak equal to the specified interval and the incidence of burst responding (i.e., a high number of responses clustered immediately after the reinforced response) decreases. C57BL/6 and 129/Sv mice demonstrated similar acquisition profiles in a DRL-15 second task with the expected increase in the number of reinforcers obtained and an equivalent level of burst responding over training (Ripley et al., 2001). Thus, for the two strains examined there appears to be no genetic variability in the ability to withhold responding under the constraints of a DRL-15 task.

As stated previously, the 5CSRT test provides multiple behavioral measures that are relevant to several of the executive functions, one of which is response inhibition. The measure of premature responding in the 5CSRT test is considered a measure of the ability to withhold responding during the constraints of this attention test (Robbins, 2002). In the 5CSRT test, a nose poke response during the intertrial interval delays the onset of stimulus trial until a limited-hold period is successfully completed; this is considered a premature response. During baseline training of rats in the 5CSRT test, premature responding decreases over training and, once rats are well trained, premature responding can be increased by increasing the length of the intertrial interval (Robbins, 2002). As seen with rats, premature responding by the mice decreased over training but the affect of increasing the length of the intertrial interval on rate of premature responding was variable. In two direct comparisons between C57BL/6 and DBA/2 mice (Greco et al., 2005; Patel et al., 2006), the DBA/2 mice showed more premature responding than the C57BL/6 mice at the completion of baseline training, indicating a higher level of impulsive responding or more difficulty in withholding pre-potent responses. Marston et al. (2001) reported no difference in premature responding between C57BL/6 and 129P2/OlaHsd mice although the data was not shown and the statistical analyses were not cited. C57BL/6x129/Sv F1 and C57BL/6xDBA/2 F1 mice demonstrated very low premature responding at completion of baseline training and did not show an increase in premature responding (Humby et al., 1999) as seen in rats (Robbins, 2002). C57BL/6Jx129P2/SvPas F2 mice, on the other hand, also had low premature response rates at the end of baseline training but did show the expected increase in premature responding in response to manipulating the intertrial interval (de Bruin et al., 2006).

The final assay measuring ability to inhibit pre-potent responses, the signaled nose poke task, has been used to evaluate the greatest number of inbred strains of mice. In the signaled nose poke task, adapted from rats (Steinmetz et al., 1993), mice are trained to nose poke in response to a tone in order to obtain food reinforcement (Logue et al., 1998). As with the 5CSRT test, a limited hold varying between 1 and 8 seconds requires the mouse to withhold responding in order to gain the opportunity for a reinforced signaled response. Thirteen inbred strains were assessed for degree of impulsive action by determining their efficiency (i.e., the greatest number of reinforcements with the fewest nose poke responses) in obtaining signaled reinforcers. DBA/2 mice were not included in this study due to inability to acquire the basic signaled nose poke response. At the end of 10 training sessions, BALB/cBy and 129S6/SvEvTac were the least impulsive making only two responses per signaled reinforcer whereas CBA, C57BL/6, SJL, and 129X1/SvJ were more impulsive making approximately four responses per signaled reinforcer. In the assays of impulsive action discussed herein, the greater degree of premature responding by the DBA/2 mice relative to the C57BL/6 mice was consistent across two studies but there is not sufficient overlap of strains evaluated across the studies in order to make any preliminary assessment of whether the three assays all measure the same aspects of behavioral inhibition.

When considering the performance of the various strains in the assays of impulsive choice (e.g., delayed reinforcement and delay discounting) and impulsive action (e.g., DRL, premature responding in 5CSRT, and the signaled nose poke task), there appears to be an interesting disconnect between the two components of impulsivity. BALB/c mice were consistently impulsive across two studies of impulsive choice (Isles et al., 2004; Otobe and Makino, 2004), but this strain was one of the least impulsive in the signaled nose poke task (Logue et al., 1998) purported to measure impulsive action. In contrast, CBA mice showed less impulsive choice relative to BALB mice, but a greater propensity for impulsive action. The results of these early studies raise questions about the relationship between impulsive choice and impulsive action. Additional behavioral genetics studies using a variety of assays for impulsive choice and action would provide insights into the relationship between these aspects of impulsivity.

Conclusions

This review highlights the problems of the limited number of strains which have been compared within a test either by the same researcher or using the same method and of the lack of a sufficient number of assays adapted to mice to allow for a strong behavioral genetic evaluation of the cognitive processes of executive function. No general conclusion can be made about relationships between the individual assays thought to measure each cognitive process of executive function even though there are hints that some of the assays may not be measuring the same process (e.g., delayed reinforcement task and delay discounting task as measures of impulsive choice and the spatial versus non-spatial tests of reversal learning as a measure of cognitive flexibility).

References

Abraham, N.M., Spors, H., Carleton, A., Margrie, T.W., Kuner, T., and Schaefer, A.T. (2004) Maintaining accuracy at the expense of speed: stimulus similarity defines odor discrimination time in mice [see comment]. *Neuron* 44: 865–876.

Ammassari-Teule, M., Hoffmann, H.J., and Rossi-Arnaud, C. (1993) Learning in inbred mice: strain-specific abilities across three radial maze problems. *Behav Genet* 23: 405–412.

Anisman, H. (1975) Acquisition and reversal learning of an active avoidance reponse in three strains of mice. *Behav Biol* 14: 51–58.

Baddeley, A. (1998) The central executive: a concept and some misconceptions. *J Int Neuropsychol Soc* 4: 523–526.

Bechara, A. and Van Der Linden, M. (2005) Decision-making and impulse control after frontal lobe injuries. *Curr Opin Neurol* 18: 734–739.

Birrell, J.M. and Brown, V.J. (2000) Medial frontal cortex mediates perceptual attentional set shifting in the rat. *J Neurosci* 20: 4320–4324.

Bissonette, G.B., Martins, G.J., Franz, T.M., Harper, E.S., Schoenbaum, G., and Powell, E.M. (2008) Double dissociation of the effects of medial and orbital prefrontal cortical lesions on attentional and affective shifts in mice. *J Neurosci* 28: 11124–11130.

Brigman, J.L., Bussey, T.J., Saksida, L.M., and Rothblat, L.A. (2005) Discrimination of multidimensional visual stimuli by mice: intra- and extradimensional shifts. *Behav Neurosci* 119: 839–842.

Brower, M.C. and Price, B.H. (2001) Neuropsychiatry of frontal lobe dysfunction in violent and criminal behaviour: a critical review. *J Neurol Neurosurg Psychiatry* 71: 720–726.

Cabib, S., Puglisi-Allegra, S., and Ventura, R. (2002) The contribution of comparative studies in inbred strains of mice to the understanding of the hyperactive phenotype. *Behav Brain Res* 130: 103–109.

Cavedini, P., Gorini, A., and Bellodi, L. (2006) Understanding obsessive-compulsive disorder: focus on decision making. *Neuropsychol Rev* 16: 3–15.

Cheung, A.M., Mitsis, E.M., and Halperin, J.M. (2004) The relationship of behavioral inhibition to executive functions in young adults. *J Clin Exp Neuropsychol* 26: 393–404.

Clapcote, S.J. and Roder, J.C. (2004) Survey of embryonic stem cell line source strains in the water maze reveals superior reversal learning of 129S6/SvEvTac mice. *Behav Brain Res* 152: 35–48.

Colacicco, G., Welzl, H., Lipp, H.P., and Wurbel, H. (2002) Attentional set-shifting in mice: modification of a rat paradigm, and evidence for strain-dependent variation. *Behav Brain Res* 132: 95–102.

Daban, C., Martinez-Aran, A., Torrent, C., Tabares-Seisdedos, R., Balanza-Martinez, V., Salazar-Fraile, J., et al. (2006) Specificity of cognitive deficits in bipolar disorder versus schizophrenia. A systematic review. *Psychother Psychosom* 75: 72–84.

Dalley, J.W., Cardinal, R.N., and Robbins, T.W. (2004) Prefrontal executive and cognitive functions in rodents: neural and neurochemical substrates. *Neurosci Biobehav Rev* 28: 771–784.

Damasio, A.R. (1994) *DescartesÕ error: Emotion, reason, and the human brain.* Avon books, New York, USA.

Davies, W., Humby, T., Isles, A.R., Burgoyne, P.S., and Wilkinson, L.S. (2007) X-monosomy effects on visuospatial attention in mice: a candidate gene and implications for Turner syndrome and attention deficit hyperactivity disorder. *Biol Psychiatry* 61: 1351–1360.

De Bruin, N.M.W.J., Fransen, F., Duytschaever, H., Grantham, C., and Megens, A.A.H.P. (2006) Attentional performance of (C57BL/6Jx129Sv) F2 mice in the five-choice serial reaction time task. *Physiol Behav* 89: 692–703.

Eagle, D.M. and Robbins, T.W. (2003a) Lesions of the medial prefrontal cortex or nucleus accumbens core do not impair inhibitory control in rats performing a stop-signal reaction time task. *Behav Brain Res* 146: 131–144.

Eagle, D.M. and Robbins, T.W. (2003b) Inhibitory control in rats performing a stop-signal reaction-time task: effects of lesions of the medial striatum and d-amphetamine. *Behav Neurosci* 117: 1302–1317.

Evenden, J.L. (1999) Varieties of impulsivity. *Psychopharmacology (Berl)* 146: 348–361.

Francis, D.D., Zaharia, M.D., Shanks, N., and Anisman, H. (1995) Stress-induced disturbances in Morris water-maze performance: interstrain variability. *Physiol Behav* 58: 57–65.

Friedman, N.P., Miyake, A., Corley, R.P., Young, S.E., Defries, J.C., and Hewitt, J.K. (2006) Not all executive functions are related to intelligence. *Psychol Sci* 17: 172–179.

Garner, J.P., Thogerson, C.M., Wurbel, H., Murray, J.D., and Mench, J.A. (2006) Animal neuropsychology: validation of the intra-dimensional extra-dimensional set shifting task for mice. *Behav Brain Res* 173: 53–61.

Gasbarri, A., Sulli, A., Pacitti, C., Puglisi-Allegra, S., Cabib, S., Castellano, C., et al. (1997) Strain-dependent effects of D2 dopaminergic and muscarinic-cholinergic agonists and antagonists on memory consolidation processes in mice. *Behav Brain Res* 86: 97–104.

Glickstein, S.B., Desteno, D.A., Hof, P.R., and Schmauss, C. (2005) Mice lacking dopamine D2 and D3 receptors exhibit differential activation of prefrontal cortical neurons during tasks requiring attention. *Cereb Cortex* 15: 1016–1024.

Greco, B. and Carli, M. (2006) Reduced attention and increased impulsivity in mice lacking NPY Y2 receptors: relation to anxiolytic-like phenotype. *Behav Brain Res* 169: 325–334.

Greco, B., Invernizzi, R.W., and Carli, M. (2005) Phencyclidine-induced impairment in attention and response control depends on the background genotype of mice: reversal by the mGLU(2/3) receptor agonist LY379268. *Psychopharmacology (Berl)* 179: 68–76.

Green, L. and Myerson, J. (2004) A discounting framework for choice with delayed and probabilistic rewards. *Psychol Bull* 130: 769–792.

Helms, C.M., Reeves, J.M., and Mitchell, S.H. (2006) Impact of strain and d-amphetamine on impulsivity (delay discounting) in inbred mice. *Psychopharmacology (Berl)* 188: 144–151.

Higgins, G.A., Ballard, T.M., Huwyler, J., Kemp, J.A., and Gill, R. (2003) Evaluation of the NR2B-selective NMDA receptor antagonist Ro 63-1908 on rodent behaviour: evidence for an involvement of NR2B NMDA receptors in response inhibition. *Neuropharmacology* 44: 324–341.

Hoyle, E., Genn, F.R., Fernandes, C., and Stolerman, I.P. (2006) Impaired performance of $\alpha 7$ nicotinic receptor knockout mice in the five-choice serial reaction time task. *Psychopharmacology (Berl)* 189: 211–223.

Humby, T., Laird, F.M., Davies, W., and Wilkinson, L.S. (1999) Visuospatial attentional functioning in mice: interactions between cholinergic manipulations and genotype. *Eur J Neurosci* 11: 2813–2823.

Isles, A.R., Humby, T., Walters, E., and Wilkinson, L.S. (2004) Common genetic effects on variation in impulsivity and activity in mice. *J Neurosci* 24: 6733–6740.

Izquierdo, A., Wiedholz, L.M., Millstein, R.A., Yang, R.J., Bussey, T.J., Saksida, L.M., et al. (2006) Genetic and dopaminergic modulation of reversal learning in a touchscreen-based operant procedure for mice. *Behav Brain Res* 171: 181–188.

Kerr, L.E., Mcgregor, A.L., Amet, L.E., Asada, T., Spratt, C., Allsopp, T.E., et al. (2004) Mice overexpressing human caspase 3 appear phenotypically normal but exhibit increased apoptosis and larger lesion volumes in response to transient focal cerebral ischaemia. *Cell Death Differ* 11: 1102–1111.

Logue, S.F., Paylor, R., and Wehner, J.M. (1997) Hippocampal lesions cause learning deficits in inbred mice in the Morris water maze and conditioned-fear task. *Behav Neurosci* 111: 104–113.

Logue, S.F., Swartz, R.J., and Wehner, J.M. (1998) Genetic correlation between performance on an appetitive-signaled nosepoke task and voluntary ethanol consumption. *Alcohol Clin Exp Res* 22: 1912–1920.

Luciano, M., Wright, M., Smith, G.A., Geffen, G.M., Geffen, L.B., and Martin, N.G. (2001) Genetic covariance among measures of information processing speed, working memory, and IQ. *Behav Genet* 31: 581–592.

Marston, H.M., Spratt, C., and Kelly, J.S. (2001) Phenotyping complex behaviours: assessment of circadian control and five-choice serial reaction learning in the mouse. *Behav Brain Res* 125: 189–193.

Martin, M.V., Dong, H., Vallera, D., Lee, D., Lu, L., Williams, R.W., et al. (2006) Independent quantitative trait loci influence ventral and dorsal hippocampal volume in recombinant inbred strains of mice. *Genes Brain Behav* 5: 614–623.

Mihalick, S.M., Langlois, J.C., and Krienke, J.D. (2000) Strain and sex differences on olfactory discrimination learning in C57BL/6J and DBA/2J inbred mice (*Mus musculus*). *J Comp Psychol* 114: 365–370.

Miyake, A., Friedman, N.P., Emerson, M.J., Witzki, A.H., Howerter, A., and Wager, T.D. (2000) The unity and diversity of executive functions and their contributions to complex "frontal lobe" tasks: a latent variable analysis. *Cognit Psychol* 41: 49–100.

Otobe, T. and Makino, J. (2004) Impulsive choice in inbred strains of mice. *Behav Process* 67: 19–26.

Owen, A.M., Roberts, A.C., Hodges, J.R., Summers, B.A., Polkey, C.E., and Robbins, T.W. (1993) Contrasting mechanisms of impaired attentional set-shifting in patients with frontal lobe damage or Parkinson's disease. *Brain* 116: 1159–1175.

Owen, E.H., Christensen, S.C., Paylor, R., and Wehner, J.M. (1997a) Identification of quantitative trait loci involved in contextual and auditory-cued fear conditioning in BxD recombinant inbred strains. *Behav Neurosci* 111: 292–300.

Owen, E.H., Logue, S.F., Rasmussen, D.L., and Wehner, J.M. (1997b) Assessment of learning by the Morris water task and fear conditioning in inbred mouse strains and F1 hybrids: implications of genetic background for single gene mutations and quantitative trait loci analyses. *Neuroscience* 80: 1087–1099.

Papaleo, F., Crawley, J.N., Song, J., Lipska, B.K., Pickel, J., Weinberger, D.R., et al. (2008) Genetic dissection of the role of catechol-*O*-methyltransferase in cognition and stress reactivity in mice. *J Neurosci* 28: 8709–8723.

Patel, S., Stolerman, I.P., Asherson, P., and Sluyter, F. (2006) Attentional performance of C57BL/6 and DBA/2 mice in the five-choice serial reaction time task. *Behav Brain Res* 170: 197–203.

Pattij, T., Broersen, L.M., Peter, S., and Olivier, B. (2004) Impulsive-like behavior in differential-reinforcement-of-low-rate 36-s responding in mice depends on training history. *Neurosci Lett* 354: 169–171.

Pattij, T., Broersen, L.M., Van Der Linde, J., Groenink, L., Van Der Gugten, J., Maes, R.A., et al. (2003) Operant learning and differential-reinforcement-of-low-rate 36-s responding in 5-HT1A and 5-HT1B receptor knockout mice. *Behav Brain Res* 141: 137–145.

Ripley, T.L., Horwood, J.M., and Stephens, D.N. (2001) Evidence for impairment of behavioural inhibition in performance of operant tasks in $TPA^{-/-}$ mice. *Behav Brain Res* 125: 215–227.

Risinger, F.O. and Cunningham, C.L. (1998) Ethanol-induced conditioned taste aversion in BXD recombinant inbred mice. *Alcohol Clin Exp Res* 22: 1234–1244.

Robbins, T.W. (1996) Dissociating executive functions of the prefrontal cortex. *Philos Trans R Soc Lond B Biol Sci* 351: 1463–1470; discussion 1470–1461.

Robbins, T.W. (2002) The five-choice serial reaction time task: behavioural pharmacology and functional neurochemistry. *Psychopharmacology (Berl)* 163: 362–380.

Roberts, A.C., Robbins, T.W., and Everitt, B.J. (1988) The effects of intradimensional and extradimensional shifts on visual discrimination learning in humans and non-human primates. *Q J Exp Psychol B* 40: 321–341.

Schoenbaum, G., Roesch, M.R., and Stalnaker, T.A. (2006) Orbitofrontal cortex, decision-making and drug addiction. *Trends Neurosci* 29: 116–124.

Steinmetz, J.E., Logue, S.F., and Miller, D.P. (1993) Using signaled barpressing tasks to study the neural substrates of appetitive and aversive learning in rats: behavioral manipulations and cerebellar lesions. *Behav Neurosci* 107: 941–954.

Stuss, D.T. and Alexander, M.P. (2000) Executive functions and the frontal lobes: a conceptual view. *Psychol Res* 63: 289–298.

Thinus-Blanc, C., Save, E., Rossi-Arnaud, C., Tozzi, A., and Ammassari-Teule, M. (1996) The differences shown by C57BL/6 and DBA/2 inbred mice in detecting spatial novelty are subserved by a different hippocampal and parietal cortex interplay. *Behav Brain Res* 80: 33–40.

Voikar, V., Vasar, E., and Rauvala, H. (2004) Behavioral alterations induced by repeated testing in C57BL/6J and 129S2/Sv mice: Implications for phenotyping screens. *Genes Brain Behav* 3: 27–38.

Wahlsten, D., Metten, P., Phillips, T.J., Boehm, S.L., 2nd, Burkhart-Kasch, S., Dorow, J., et al. (2003) Different data from different labs: lessons from studies of gene–environment interaction. *J Neurobiol* 54: 283–311.

Weinberger, S.B., Koob, G.F., and Martinez, J.L., Jr. (1992) Differences in one-way active avoidance learning in mice of three inbred strains. *Behav Genet* 22: 177–188.

Willcutt, E.G., Doyle, A.E., Nigg, J.T., Faraone, S.V., and Pennington, B.F. (2005) Validity of the executive function theory of attention-deficit/hyperactivity disorder: a meta-analytic review. *Biol Psychiatry* 57: 1336–1346.

Wolff, M., Savova, M., Malleret, G., Segu, L., and Buhot, M.C. (2002) Differential learning abilities of 129T2/Sv and c57BL/6J mice as assessed in three water maze protocols. *Behav Brain Res* 136: 463–474.

Wrenn, C.C., Turchi, J.N., Schlosser, S., Dreiling, J.L., Stephenson, D.A., and Crawley, J.N. (2006) Performance of galanin transgenic mice in the five-choice serial reaction time attentional task. *Pharmacol Biochem Behav* 83: 428–440.

Young, J.W., Crawford, N., Kelly, J.S., Kerr, L.E., Marston, H.M., Spratt, C., et al. (2007) Impaired attention is central to the cognitive deficits observed in alpha 7 deficient mice. *Eur Neuropsychopharmacol* 17: 145–155.

Young, J.W., Finlayson, K., Spratt, C., Marston, H.M., Crawford, N., Kelly, J.S., et al. (2004) Nicotine improves sustained attention in mice: Evidence for involvement of the alpha7 nicotinic acetylcholine receptor. *Neuropsychopharmacol* 29: 891–900.

Young, J.W., Light, G.A., Marston, H.M., Sharp, R., and Geyer, M.A. (2009) The five-choice continuous performance test: evidence for a translational test of vigilance for mice. *PLoS One* 4: e4227.

Chapter 27

Water navigation tasks

David P. Wolfer, Giovanni Colacicco, and Hans Welzl

The circular water maze permits one to implement a large variety of tasks and to probe different aspects of brain function. This versatility, its technical simplicity, and the large amount of information that can be collected during experiments by means of video-tracking are the main reasons for the ongoing popularity of the water maze in behavioral research with both mice and rats. This chapter begins with a summary of the basic concepts and most important procedures for rats and mice (Vorhees and Williams, 2006). After a brief discussion of the relevant differences between the two rodent species and some important technical aspects, an overview is given on the natural variation of performance displayed by mice in the most frequently used water-maze procedure, the place navigation task.

Origin and history of the water maze

A new procedure to study spatial reference memory in rats

In response to controversies about the neural basis of spatial and working memory (O'Keefe and Nadel, 1978; Olton et al., 1979) and ideas that learning about spatial relations may be fundamentally different from other forms of associative learning, Richard G. M. Morris introduced a water-maze navigation task to study spatial localization in the rat (Morris, 1981). Rats were placed into a large circular pool of water from which they could escape onto a platform. To make sure that the platform offered no local cues to guide escape behavior, the water was rendered opaque and the platform was hidden beneath the water surface. Rats very quickly learned to swim directly towards the hidden platform from any starting position at the circumference of the pool. The accurate directionality of their escape behavior provided evidence that they learned the spatial location of the platform relative to distal extra-maze cues. Because accurate directionality required continual monitoring of the animal's position in relation to extra-maze cues, it was expected that, unlike in T-maze tasks (e.g., spontaneous alternation) which require binary decisions and in which responses are either correct or false, performance in the water-maze navigation task could reflect spatial memory and its disruption by drugs or lesions in a quantitative manner.

After two pre-training swims without opportunity to escape, the rats were given five blocks of four training trials distributed over 3 days. Within each block the rats were released at different locations and spent the 60 seconds intertrial interval sitting on the platform. To avoid false conclusions, subgroups of rats were trained on four variants of the task. In the *place* (Figure 27.1a) and *place-random* condition, animals were trained with a hidden platform, held at a constant location for any given animal, or moving at random between quadrants, respectively. The *cue + place* and *cue only* (Figure 27.1b) tasks were run in the same ways, but using a visible platform which was painted black and had a border protruding 1 cm above the water surface. Near optimal escape latencies were reached by Trial 6 in all groups except in the place-random group, whose purpose was to establish that no local cues were associated with the hidden platform. The place-random group showed only a small improvement of performance reflecting their learning to swim away from the sidewalls, thereby increasing their likelihood of contacting the platform by chance. In order to assess the acquired spatial bias, the last block of training trials was immediately followed by a 1 minute *probe trial* in which the platform was absent or in a new location. Focused searching was observed in the place and to a lesser degree in the cue + place groups, but not in the other two groups, establishing that rats were able to learn the spatial location of an invisible platform relative to distal cues. The pool had a diameter of 1.32 m with a platform size of 11 cm. Later a new pool was constructed with a diameter of 2.14 m (Morris et al., 1982). Small pools may be less effective in discriminating normal from impaired animals because of increased probability to hit the platform by chance. In addition, one would expect that a very small pool prevents the subjects from seeing distant cues from sufficiently different angles.

Working memory, reversal, procedural, and discrimination learning

In subsequent methodological publications (Morris, 1984; Morris and Seifert, 1983), modifications of the place navigation task as well as new tasks were introduced. The place

Behavioral Genetics of the Mouse: Volume I. *Genetics of Behavioral Phenotypes*, eds. Wim E. Crusio, Frans Sluyter, Robert T. Gerlai, and Susanna Pietropaolo. Published by Cambridge University Press.

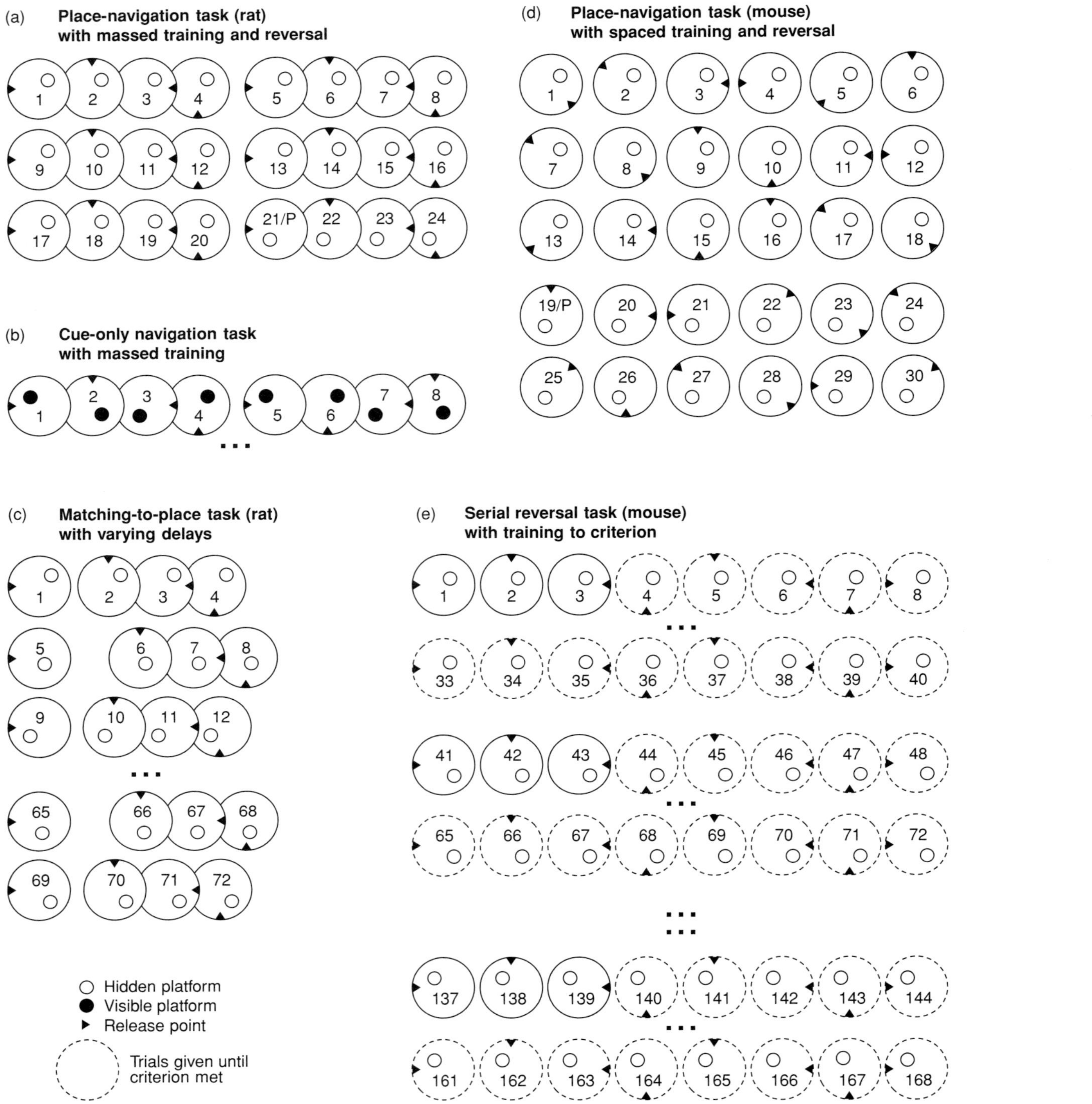

Figure 27.1 Examples of various types of water-maze tasks. Training trials are numbered consecutively. Probe trials are labeled as "P." Platform position and release point are shown for each trial. Overlapping trial representations indicate massed training where the intertrial interval (ITI) of circa 1 minute is spent on the goal platform. Non-overlapping representations indicate spaced training with longer ITI during which animals are allowed to dry off in their home cage or in a waiting cage. Trials of the same day are shown in one line. (a) Place-navigation task as introduced by Richard Morris to assess spatial reference memory in rats (Morris, 1981, 1984; Morris et al., 1982). Similar protocols are popular in mouse studies, but usually more trials are needed to acquire the task (e.g., three blocks of four trials for 3 days, Logue et al., 1997). If there is no reversal training, probe trials are typically run without goal platform. (b) Cue-only navigation task (Morris, 1981, 1984) as it can be used in rats or mice. The platform is visible and changes position across trials. (c) Delayed matching-to-place task implemented to study spatial working memory in rats (Panakhova et al., 1984; Steele and Morris, 1999). The goal position changes every day. The first trial serves as instruction trial. One to four test trials are given after intervals of typically 30 seconds to 2 hours. Unlike rats, mice do not show clear one-trial learning in this protocol (Whishaw, 1995). (d) Place-navigation task with reversal for mice (Mohajeri et al., 2004), which exploits the higher training efficiency that results from spaced training with an ITI of 30–45 minutes. This protocol has been used to collect the data presented in Figures 27.2 and 27.3. (e) Serial reversal task (Chen et al., 2000) introduced as a model of episodic-like memory. Mice are trained for five consecutive place tasks with different platform positions. As soon as they meet the criterion of three consecutive trials with escape latencies less than 20 seconds, animals proceed to the next task. The learning criterion was determined empirically and may need to be adjusted when pool configuration changes. Performance can be measured as trials to criterion or learning capacity, that is the number of tasks mastered within a given number of training days.

navigation task was extended by a second phase in which the animals were trained with a new platform location. This is usually referred to as a "*reversal*" procedure even though this term is inexact because the allocentric spatial relationship of the distal room cues remains invariant. The animal needs only to learn the new position of the platform within that array of cues. Further, delays of 4–336 hours were introduced to measure the decline of spatial bias due to *forgetting* (loss of the acquired information during long retention intervals) or *extinction* (learning during repeated probe trials of the new information that there is no longer a platform in the expected position). It was found that spatial bias was relatively insensitive to extinction effects and that the rate of forgetting strongly depended on the intensity of prior training, that is the number of trials and their spacing.

In order to prevent chance finding of the hidden target, Buresova and colleagues designed a place-navigation task with a platform that is available only after the rat has stayed in the target area for a predetermined amount of time (Buresova et al., 1985). This "on-demand" or "Atlantis" platform (Spooner et al., 1994), was later used to study the interaction of spatial and *procedural memory* systems, encoding the position of the platform and mediating the procedure of localized searching, respectively (Micheau et al., 2004; Riedel et al., 1999). As a complement to the cue navigation task, the water maze was also soon used to assess egocentric *position-response navigation* by releasing rats always at the same point relative to a hidden platform that was either held in a constant position or moved between trials (Whishaw, 1985a).

While the first place-navigation task was designed to measure spatial *reference memory*, a new matching to place procedure was implemented for the water maze as a test of spatial *working memory*. Each day, an instruction trial, during which the platform is always placed in a new position, is followed by one (Morris, 1984; Morris and Seifert, 1983) or several (Steele and Morris, 1999) test trials, in which the hidden platform is kept in the same position. Rats learn to benefit from information about the platform's position gained during the single instruction trial and show significantly shorter escape latencies during the test trials. This improvement is only observed if the platform is kept in the same position during both instruction and test trials, proving that rats show a true spatial win-stay tendency in this water-maze task and that the improvement is not due to non-specific or motivational differences. A *delayed matching-to-place* (DMP) task (Figure 27.1c) introduces delays between instruction and test trials. A rapid decay of working memory occurs in normal rats during the first hour and an exponential slow down of the decline is observed with longer intervals. However, significant time savings occur even with intervals of 24 hours (Panakhova et al., 1984). Whishaw published a similar procedure and called it *place learning-set* task (Whishaw, 1985c). This type of continued one-trial spatial learning seems best suited for eliminating non-spatial strategies and permits continuous monitoring of spatial memory over extended periods of time (Whishaw, 1985c).

Because hidden platform procedures do not provide any index of spatial discriminability, a two-platform procedure was developed which can be used to study both spatial and non-spatial *discrimination learning* (Morris, 1984). Both platforms are visible. One is rigid, similar to the one used in the cue-navigation protocols. The other floats on the water surface and is anchored by means of nylon threads to a metal base on the bottom of the pool. As the floating platform sinks beneath the animal, it soon learns to approach the rigid one preferentially. Training can be in two forms. In the spatial discrimination task, the platforms look alike and can only be discriminated by their position in space relative to distal room cues. In the visual discrimination task, room cues are hidden behind curtains and the platforms are defined by brightness or pattern. Curtains may also be used to test to which degree animals trained in a spatial procedure rely on extramaze cues to localize the platform.

Using water-maze tasks to study the neural basis of spatial memory

After its introduction the water maze rapidly gained wide popularity. A large number of studies investigated the effect of surgical and pharmacological manipulations on performance in the above-described and similar tasks in order to study the neural basis of spatial memory (for review see Brandeis et al., 1989). Pioneering publications demonstrated that place-navigation was specifically abolished by complete bilateral aspiration lesions of the hippocampus (Morris et al., 1982), but also impaired by pharmacological cholinergic receptor blockade (Sutherland et al., 1982b), medial frontal cortical damage (Sutherland et al., 1982a), or trans-section of the fimbria/fornix (Sutherland and Rodriguez, 1989). Rats with hippocampal damage failed to show any place learning. They performed in the place-navigation task as if it were a place-random task, but were normal in the cue navigation task. By contrast, catecholamine depletion by neurotoxins impaired more procedural aspects of learning (Hagan et al., 1983). Sutherland et al. were the first to document place-navigation deficits using neurotoxin-induced lesions (Sutherland et al., 1983). Morris and colleagues provided the first direct evidence for the involvement of *N*-methyl-D-aspartate (NMDA) receptors in spatial learning (Morris et al., 1986). Steele et al. used a DMP task with one instruction trial and three test trials following after delays of 15 seconds to 2 hours to demonstrate that this task is hippocampal-dependent and that inactivation of NMDA receptors by 2-amino-5-phosphonovalerate (AP5) causes a delay-dependent impairment (Steele and Morris, 1999).

Water-maze tasks for mice

Laboratory mice

Because rats and mice are two species of different behavioral ecology it was of interest to see how laboratory mice responded to the procedural and spatial challenges of water-maze procedures. The first report that mice are able to learn the water-maze place navigation task was already published in 1987. C57BL/6 mice were trained for up to 36 trials in a pool of 1.22 m diameter with two blocks of six or three blocks of four trials per day (Upchurch and Wehner, 1987). A more systematic comparison between C57BL/6 mice and Long Evans rats was later undertaken by Whishaw (1995). Independently of platform size, mice had longer escape latencies and higher error probabilities in the place navigation task. In contrast to rats, mice also failed to display the one-trial learning required in the delayed matching to place task. Extension of the comparison to dry-land spatial tasks on a radial maze (Whishaw and Tomie, 1996) suggested that the advantage of rats over mice in the water maze was mainly a consequence of better adaptation to water rather than superior spatial abilities per se. Both species inhabit burrow systems which provides them with competency in mazes, but rats are also adapted to water. This should be kept in mind when interpreting data from experimental manipulations that lead to decreased escape performance. In mice (Lipp and Wolfer, 1998; Paul et al., 2009), but also in rats (Harris et al., 2009), reduced training performance especially when associated with increased passive floating or a stronger tendency to follow the side wall of the pool (thigmotaxis) may reflect impaired adaptation to the unnatural environment rather than a deficit in spatial orientation. Ideally, any suspected spatial deficits should be confirmed with a second task, for example a dry-land maze (Patil et al., 2009). In fact, training of C57BL/6 mice on a dry-land maze such as the Barnes maze induces smaller increases in plasma corticosterone than water-maze training and, unlike in the water maze levels of corticosterone are not inversely correlated with performance (Harrison et al., 2009). Introducing longer intervals instead of giving the next trial immediately after a short interval on the platform and returning mice rapidly to their waiting cage so they can dry off (spaced instead of massed training) greatly improves the efficiency of learning (Kogan et al., 1997). With the standard protocol of our laboratory (Figure 27.1d), which gives six trials per day with intertrial intervals of 30–45 minutes, C57BL/6 mice and most genetically mixed populations can easily learn two different platform positions in 5 days. Recent attempts to adapt the water-maze apparatus to the specific difficulties and needs of mice (Schmitt et al., 2004; Wahlsten et al., 2005) in order to rescue the poor performance of some strains and to extend the validity of the test have been only partially successful, and lead to the conclusion that no single test can reveal the full richness of spatially guided behavior in a wide range of mouse genotypes (Wahlsten et al., 2005). Mice have occasionally been used in DMP or place learning-set tasks (e.g., Fordyce and Wehner, 1993b) despite their failure to show clear one-trial learning in such settings (Whishaw, 1995). Because mice take several trials to learn each new platform position this test situation is best described as a *serial reversal task.* Chen and colleagues have recently used a similar protocol as a model of episodic-like memory. Each time the mice had learned to escape reliably and quickly to the hidden platform at one location, it was moved to a new location. This was repeated until the mice had learned five subsequent spatial problems (Chen et al., 2000) (Figure 27.1e). In such a procedure, memory retrieval must be selective for the most recently encoded location even though earlier locations may still be stored in long-term memory, potentially causing interference.

As in the rat, water-maze procedures were soon used in conjunction with various experimental manipulations to investigate the neural basis of spatial learning and navigation. Already the first water-maze experiment with mice demonstrated that spatial learning was abolished after chronic treatment with the organophosphate DFP (Upchurch and Wehner, 1987). Stereotaxic lesions of the nucleus basalis of Meynert impaired place-navigation learning of BALB/cByJ mice (Sweeney et al., 1988). Hippocampal lesions disrupted place-navigation learning in C57BL/6 and DBA/2 mice, as well as in a F1 cross between these strains (Logue et al., 1997). However, cue navigation was not as completely preserved as in hippocampal rats, indicating that cue learning may not be completely independent of the hippocampus in mice. Nevertheless, combinations of place and cue navigation protocols have been used in mice with the aim to dissociate hippocampal and striatal contributions to navigation (Lee et al., 2008; Sung et al., 2008). Because several of the most poorly performing strains (see below) have malformations of the corpus callosum, it was attempted to correlate place-learning performance deficits with callosal agenesis. No individual relationship could be established, however (Balogh et al., 1999). The first paper applying the water-maze place-navigation task to genetically modified mice was published in 1991. It demonstrated delayed learning but also reduced swim speed in transgenic mice overexpressing the amyloid precursor protein 695 isoform (Yamaguchi et al., 1991). The first studies using gene targeted mice were published 1 year later. One showed that targeted deletion of the α-calcium-calmodulin kinase II gene impaired both hippocampal long-term potentiation and place navigation in the water maze (Silva et al., 1992a; 1992b). The other showed similar consequences in conjunction with morphological abnormalities in the hippocampus in fyn null mice (Grant et al., 1992). These pioneering papers have established the water-maze place-navigation task as one of the most popular tests of cognitive function in genetically modified mice (Grant and Silva, 1994; Lee and Silva, 2009; Silva and Giese, 1994).

Wild-living mouse species

Several groups have also studied place navigation in wild-living mouse species. Galea et al. (1994a) used two subspecies of deer mice (*Peromyscus maniculatus*) and found superior hidden platform acquisition performance in animals from an island population (*P. maniculatus angustus*) compared to those from a mainland population (*P. maniculatus artemisiae*). Kavaliers, Galea, and others (Galea et al., 1994a, 1994b; Kavaliers and Galea, 1994) reported that meadow voles (*Microtus pennsylvanicus*) learned well in a place-navigation task when they could use the sun and associated celestial cues, but not when it was overcast. This prompted the speculation that the cues generally available to the mice in the laboratory are adequate for dry-land spatial learning, but in some way inadequate for swimming-navigation learning. In a direct comparison, deer mice performed better than laboratory mice of the Swiss Webster or DBA strains (Petrie, 1995). Wood mice (*Apodemus sylvaticus*) outperformed CD1 as well as C57BL/6J (Patil et al., 2008). Good learning abilities were also found in meadow voles (*Microtus pennsylvanicus*) (Galea et al., 1996). CAST/Ei mice of the castaneus subspecies (*Mus musculus castaneus*) showed learning performance similar to C57BL/6JBy laboratory mice (*Mus musculus domesticus*) (Le Roy et al., 1998). Both bank voles (*Clethrionomys glareolus*) which are adapted to a wide range of different habitats, and root voles (*Microtus oeconomus*) living in homogenous grassland habitats with small home ranges learned faster than laboratory mice as judged from escape latencies (Pleskacheva et al., 2000). However, search patterns were more goal-oriented only in *C. glareolus*. *Microtus oeconomus* mainly excelled by faster swimming speed, showing strong wall hugging, circular swimming, and chance level performance in probe trials reminiscent of many mutant knockout mice considered to be deficient in spatial memory. The remarkable differences in performance and preferred strategy in the water maze that were observed between different mouse species illustrate the strong influence of evolutionary constraints on learning caused by differences in behavioral ecology: what would be considered impaired spatial learning in one species may be completely normal behavior in another.

Data collection and quantification

In their first studies (Morris, 1981; Morris et al., 1982), Richard Morris and colleagues measured escape latency to monitor training performance. Because interpretation of water-maze experiments is impossible using escape latencies alone, particularly important swims such as final training and probe trials were recorded on videotape and transcribed to charts. This permitted assessment of spatial bias by measuring the time spent in each pool quadrant (Morris, 1981) or the number of crossings over annuli indicating the surface and former positions of the platforms (Morris et al., 1982). Path length was measured as well as directionality, defined as deviation from the correct heading angle towards the platform at various distances along the path. The first video-tracking device was published in 1984 (Morris, 1984). It identified the animal in the video image automatically by analyzing contrast in real-time and transmitted *x*- and *y*- position coordinates to a computer 10 times per second. Special software permitted display of the paths taken by the animals and computation of swim speed and the above measures in a time-saving and objective way. Other laboratories used manual tracing on a digitizing tablet in order to trace swim paths and to convert them into streams of *xy* coordinates for quantitative analysis (Denenberg et al., 1990; Whishaw, 1985a). Modern video-tracking systems determine the animal's position by real-time digital image processing, and are very reliable and easy to use.

Electronic tracing techniques allowed new measures of spatial accuracy to be devised, such as Whishaw's error (Whishaw, 1985a, 1985b). It counts the trials in which the animal swims outside a straight corridor that connects the release point with the goal platform. The width of this corridor depends on the geometry of the pool and was 18 cm in Whishaw's pool of 146 cm diameter. Gallagher and coworkers introduced new measures that allowed detection of age-related impairments of spatial ability with high sensitivity (Gallagher et al., 1993). A goal proximity measure was obtained by calculating the distance of the subject from the escape platform at 1-second intervals. Cumulative distance was used as a search error measure on training trials and average distance from the goal as a measure of spatial selectivity on probe trials. Bias resulting from variation of starting points relative to the goal was removed by a correction procedure. Based on the average swimming speed for each trial, the minimal amount of time required to swim to the goal from the start location was estimated and removed from the record prior to computing trial performance. The output of modern tracking systems permits computation of a large variety of geometric measures which can help to monitor the learning progress during place-navigation training, discrimination of spatial from non-spatial strategies, and recognition of aberrant swim patterns that are due to motor impairments (Balschun et al., 2003; Janus, 2004; Mohajeri et al., 2004; Wolfer et al., 2001).

Natural variation of place learning in mice

Strain differences

A large body of literature documents that performance in water-maze place-navigation tasks varies strongly between different inbred strains of mice (for reviews see e.g., Crawley et al., 1997; D'Hooge and De Deyn, 2001; Schimanski and Nguyen, 2004) with some strains showing rapid learning and good performance in both cue and place-navigation tasks and others failing completely even in cue navigation. As already mentioned, the first replication of a place-navigation task in mice (Upchurch and Wehner, 1987) used the C57BL/6 strain. A year later the same laboratory published the first comparison of different inbred strains in the place- and cue-navigation task (Upchurch and Wehner, 1988a). It marked the beginning of a long and still growing series of studies examining and comparing the spatial learning abilities of inbred strains, of crosses thereof, and of outbred strains. The findings of a selection of these papers are summarized in Table 27.1. C57BL/6 mice have been used particularly often and have gained the reputation of being the most suitable strain for place-navigation tests in the water maze even though they are not the only inbred strain to perform reasonably well.

Visual ability of mouse strains is a critical determinant of performance in the water maze (Brown and Wong, 2007; Wong

Table 27.1 Place-navigation performance of inbred, hybrid, and outbred laboratory mouse strains. Learning was rated based on published acquisition performance and probe trial results.

Strain	Learning	Selected references	Notes
C57BL/6Ibg	Good	Upchurch and Wehner, 1987, 1988a	
C57BL/10J	Good	Owen et al., 1997	Fast swimmers
SM/J	Good	Clapcote and Roder, 2004	Good probe trial
NZB	Intermediate	Wright et al., 2004	
DBA/2Ibg	Intermediate	Owen et al., 1997; Upchurch and Wehner, 1988a	Good cue learning
DBA/2J	Intermediate	Nguyen et al., 2000	Good cue learning
DBA/2J	Good	Francis et al., 1995	Cue learning not tested
DBA/2J	Good	Holmes et al., 2002; Owen et al., 1997	Good cue learning
DBA/2HeJ	Good	Schopke et al., 1991	Inferior to C57 only during reversal
DBA/2OlaHsd	Good	Brooks et al., 2005	
BALB/cByJ*	Poor	Brooks et al., 2005; Upchurch and Wehner, 1988b	Poor cue learning
BALB/cByJ*	Poor	Owen et al., 1997	Good cue learning
BALB/cByJ*	Poor/good	Francis et al., 1995	20% good place learning
BALB/cOlaHsd*	Intermediate	Klapdor and van der Staay, 1996	Slow swimmers, good by path length
BALB/c*	Good	Royle et al., 1999; Wahlsten et al., 2005	After extensive pre-training
BALB/cByJ*	Good	Zaharia et al., 1996	Cross-fostered to C57BL/6 dam
BALB/cJ*	Poor	Francis et al., 2003	Cross-fostered to BALB/cJ dam
C57BL/6J	Poor		
LP/J	Poor	Leil et al., 2002; Owen et al., 1997	Good cue learning
LP/J	Poor	Clapcote and Roder, 2004	Floating, slow swimmers
129X1/SvJ*	Poor	Leil et al., 2002; Owen et al., 1997	Good cue learning
129P3/J*	Poor	Montkowski et al., 1997	Poor cue learning, pink-eyed dilution
129S5/SvEvBrd	Poor	Van Dam et al., 2006	Learn only in small pool
129S3/SvImJ	Excellent	Clapcote and Roder, 2004	Floating
129T1/Sv-Ter+	Good	Montkowski et al., 1997	Slow swimmers
129P2/Ola*	Good	Montkowski et al., 1997; Royle et al., 1999	Sham-operated controls, reluctant to swim
129S6/SvEvTac	Good	Clapcote and Roder, 2004; Owen et al., 1997	
129S6/SvEvTac	Intermediate	Holmes et al., 2002	
129S6/SvEvTac	Intermediate	Balogh et al., 1999; Gerlai, 2002	
129S6/SvEvTac	Poor	Wolfer et al., 1997	
129T2/SvEms	Good	Nguyen et al., 2000	
129T2/SvEms	Intermediate	Wolff et al., 2002	
129T2/SvEms	Poor	Fox et al., 1999	
129S1/Sv	Good	Moy et al., 2007; Rogers et al., 1999	Passive floating
129S1/Sv	Poor	Wahlsten et al., 2005	Delayed acquisition, good probe trial
129S2/SvPas	Excellent	Clapcote and Roder, 2004	Poor probe trial
129S2/SvPas	Good	Contet et al., 2001	
129S2/SvHsd	Intermediate	Voikar et al., 2001	
129S2/SvPas	Poor	Brooks et al., 2005	
A/J*	Poor	Moy et al., 2007; Owen et al., 1997	Poor cue learning
C3H/Ibg	Poor	Owen et al., 1997; Upchurch and Wehner, 1988b	Poor cue learning, homozygous for *Pde6b*^{rd1}
SJL/J*, FVB/NJ*	Poor	Owen et al., 1997	Poor cue learning, homozygous for *Pde6b*^{rd1}
Bub/BnJ*	Poor	Owen et al., 1997	Poor cue learning, homozygous for *Pde6b*^{rd1}
CBA/J	Poor	Nguyen et al., 2000	Poor cue learning, homozygous for *Pde6b*^{rd1}
CBALac-Sto	Good	Leitinger et al., 1994	No retinal degeneration
CBA/CaOlaHsd	Good	Brooks et al., 2005	No retinal degeneration
B6D2F1	excellent	(Owen et al., 1997; Upchurch and Wehner, 1988a)	
FVBB6F1	excellent	(Mohajeri et al., 2004; Owen et al., 1997)	heterozygous for *rd* mutation
FVB129F1, SLJB6F1	excellent	(Owen et al., 1997)	heterozygous for *rd* mutation
C3B6F1	excellent	(Owen et al., 1997)	heterozygous for *rd* mutation
129B6F1	excellent	(Owen et al., 1997; Wolfer et al., 1997)	
cBy129F1	Intermediate	Owen et al., 1997	Good cue navigation
NMRI*, CFW1*	Intermediate	Klapdor and van der Staay, 1996	Outbred, good probe trial
CD-1*	Good	Adams et al., 2002; Francis et al., 1995	Outbred
ICR*	Poor	Adams et al., 2002	Outbred, poor cue learning, visual impairment
Swiss Webster*	Intermediate	Petrie, 1995; Wright et al., 2004	Outbred
OF1	Intermediate	Carrie et al., 1999	Outbred
NIH Swiss*, Black Swiss	Poor	Clapcote et al., 2005b	Outbred, homozygous for *Pde6b*^{rd1}

* Albino strains.
Good = performance is comparable to C57BL/6 mice tested in the same or a similar protocol and pool.
Excellent = better than C57BL/6.
Intermediate = strain learns but worse than C57BL/6.
Poor = little or no evidence of spatial learning.

and Brown, 2007). While albinism does not prevent good learning of the place-navigation task, the low end of the performance spectrum is formed by those strains that are homozygous for the retinal degeneration allele *Pde6b*^{rd1} (Bowes et al., 1993). This mutation leads to blindness at wean age in homozygous animals, making them unsuitable for any kind of visually guided water-maze task (Brown and Wong, 2007; Clapcote et al., 2005a; Moy et al., 2007). However, for some of these strains congenic lines are available which do not carry the mutation (Brooks et al., 2005; Clapcote et al., 2005a; Errijgers et al., 2007; Leitinger et al., 1994).

DBA/2 mice are capable of cue learning, but several laboratories reported impaired place learning relative to C57BL/6 mice (Nguyen et al., 2000; Owen et al., 1997; Upchurch and Wehner, 1988a, 1988b), prompting the hypothesis that abnormal hippocampal function may exist under baseline conditions in this strain (Paylor et al., 1993, 1996; Wehner et al., 1990). However, being rather mild, the impairment of DBA/2 mice relative to C57BL/6 has not been consistently observed across laboratories and protocols (Sung et al., 2008). The albino strain BALB/c shows extremely variable and often very poor performance with a strong tendency to display passive floating. Hyperreactivity to stressors (Francis et al., 1995) and insufficient adaptation to the test situation (Yoshida et al., 2001) seem to play important roles in this behavioral profile which is largely determined by maternal factors (Francis et al., 2003; Zaharia et al., 1996).

129 strains are the primary source for the ES cells used in gene-targeting studies. Due to genetic linkage, alleles derived from the ES cell donor strain may contribute to the phenotypic profile of the derived mutant animals even after extensive backcrossing to another strain (Gerlai, 1996; Wolfer et al., 2002). Because water-maze learning is frequently used to examine gene-targeted mice, knowing the behavioral profile of the 129 strain in water-maze tasks is of particular interest. Unfortunately, deliberate and accidental outcrossing has lead to extensive genetic variability among 129 substrains and the ES cells derived from them (Simpson et al., 1997; Threadgill et al., 1997). The comparison of published studies is further complicated by the fact that the nomenclature of 129 substrains has been revised a few years ago (Festing et al., 1999). The available literature shows that performance in the water-maze place navigation task is extremely heterogeneous among 129 substrains with a spectrum ranging from complete non-performers to apparently excellent learners. There is good agreement that 129X1/SvJ and 129P3/J are poor performers whereas 129P2/Ola, 129S/SvImJ, and 129T1/Sv–Ter+ perform as well or better than C57BL/6. 129S5/SvEvBrd were only able to acquire the task in a very small pool of 75 cm diameter (Van Dam et al., 2006). Very heterogeneous results have been obtained with 129S6/SvEvTac, 129T2/SvEms, 129S1/Sv, and 129S2/SvPas substrains, suggesting that their performance may strongly depend on the experimental conditions.

Already in 1988, Upchurch and colleagues tested first-generation (F1) hybrids between DBA/2Ibg and C57BL/6Ibg mice for spatial learning in the place navigation task (Upchurch et al., 1988). The hybrids performed better than either parental strain, suggesting a heterosis effect with both parents contributing genes for spatial learning ability. Anticholinergic treatment, which abolished spatial learning ability in C57BL/6Ibg mice, produced only minor impairments in the hybrids. In a subsequent more extensive genetic analysis (Upchurch and Wehner, 1989), at least part of this heterosis effect was found to be preserved in F2 as well as in first and second generations of a backcross to C57BL/6Ibg. Owen and colleagues tested seven F1 crosses which performed well with the only exception of the cross between BALB/cByJ and 129X1/SvJ (Owen et al., 1997). Because the mutation is expressed recessively, good performance is also seen in F1 crosses if one of the parental strains carries the retinal degeneration allele *Pde6b*^{rd1} (Gass et al., 1998; Mohajeri et al., 2004; Owen et al., 1997; Voikar et al., 2001). However, if such hybrids are further crossed *Pde6b*^{rd1} will again segregate and blind animals will reappear in the population (Wolfer et al., 2001). Genotyping for *Pde6b*^{rd1} permits elimination of the allele from the population. Among outbred strains, NMRI, CFW1, CD1, Swiss Webster, and OF1 were successfully used in place-navigation tasks, whereas ICR, NIH Swiss, and Black Swiss seem to be unsuitable, with the latter two suffering from retinal degeneration (Clapcote et al., 2005b).

It is generally recommended that behavioral studies, in particular water-maze experiments, should be conducted in an inbred C57BL/6 background, combining good performance with low variability due to the genetic homogeneity of the population. A retrospective survey of place-navigation data collected in our own laboratory during the period 1987–2005 shows that, in line with the literature summarized above, genetically mixed populations are just as suitable and that there is no absolute need to establish a congenic population before investigating the effects of a mutation on water-maze learning. In this context it should be noted that standard backcrossing programs do not necessarily solve the flanking gene problem (Wolfer et al., 2002). Among 109 experimental groups consisting of C57BL/6, 129 substrains or crosses thereof, the parental strains C57BL/6 and 129 were outperformed not only by F1 hybrids, but also by F2/F3 generations and by partial backcrosses to C57BL/6. This difference was found both for training and probe trial measures (Figure 27.2a and b), confirming the importance of the heterosis effect. Our retrospective analysis also indicates that the beneficial effect of genetic homogeneity on behavioral variability should not be overestimated (Figure 27.2c,d). While as expected experimental groups with C57BL/6 or F1 background showed somewhat less variability than F2/F3 populations during acquisition, variability was clearly larger in groups with an inbred 129 background. Spatial selectivity during the probe trial was similarly consistent in C57BL/6, F1, and F2/F3 groups, but again more variable in groups with an inbred 129 background. For both measures, variability of partial backcrosses was indistinguishable from pure C57BL/6 groups.

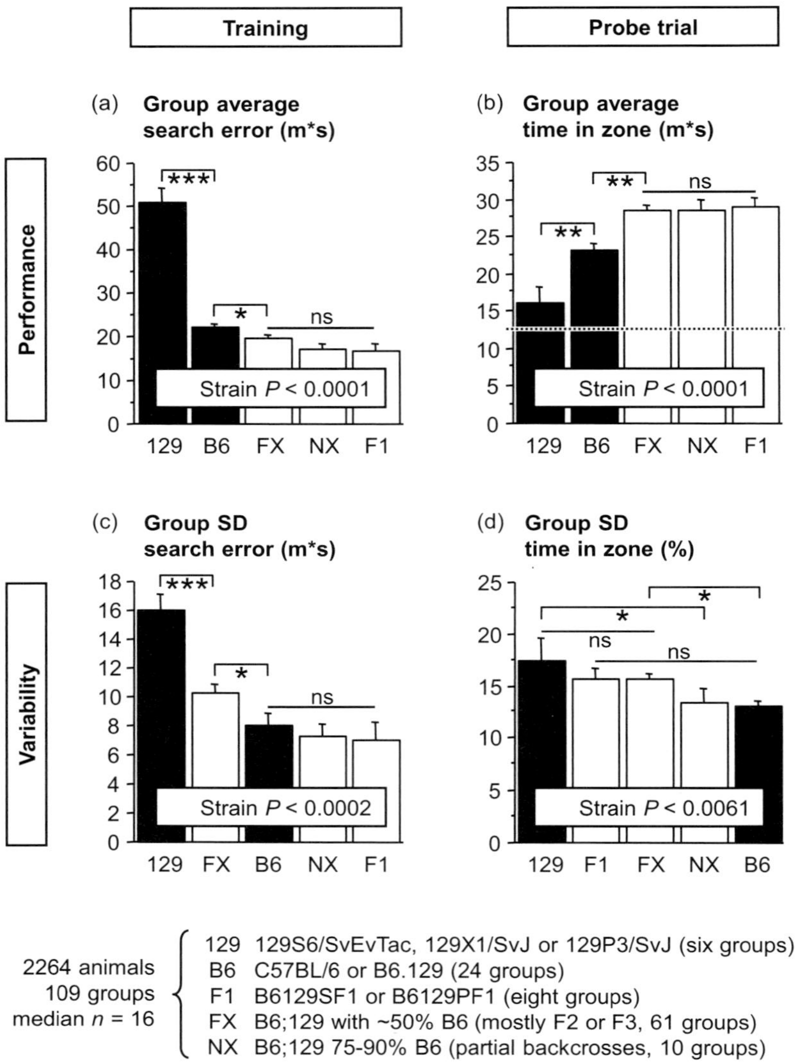

Figure 27.2 Strain effects on place navigation. Retrospective analysis of training (a and c) and probe trial (b and d) data collected 1987–2005 using the protocol shown in Figure 27.1d. To permit comparison of both performance level (a and c) and variability (b and d), 109 experimental groups were chosen as units of observation, all controls or experimental groups without treatment effect. C57BL/6 mice and 129 substrains (filled bars) were included as parental strains, as well as three different populations derived from crosses between them (empty bars). The three 129 substrains did not differ in any of the measures shown and are represented as one category. Category B6 includes inbred as well as congenic lines. Significant effects of strain revealed by one-way factorial analysis of variance (ANOVA) were further explored using Fisher's protected least significant difference (PLSD) post-hoc test. (a) Cumulative search error (see main text) served as a measure of training performance and was averaged across training trials and group members. Bars show mean and standard error (SE) of the group averages in each category. (b) Time spent in the target zone was used as measure of spatial retention during the probe trial and averaged across group members. (c) Standard deviation (SD) of cumulative search error was calculated for each group as measure of training performance variability. Bars show mean and SE of group SD values in each category. (d) Similarly, SD of time in target zone was calculated as measure of probe trial performance variability.

Sex differences

Assuming that performance of female mice in spatial water-maze tasks is inferior compared with males and more variable due to the estrous cycle, it is often recommended to use only male mice. Effects of hormonal and reproductive status on water-maze place navigation are well documented in wild-living mouse species. Male deer mice (*Peromyscus maniculatus*) showed superior performance during the breading season, but compared to females no difference was found outside the breading season (Galea et al., 1994a). Female meadow voles (*Microtus pennsylvanicus*) with high estrogen levels performed more poorly than males, whereas performance of females with low estrogen levels was indistinguishable from males (Galea et al., 1995). Laboratory mice are less well studied in this respect and, perhaps due to differences in estrous cycle, there is no unequivocal experimental evidence for female inferiority with respect to place-navigation performance. While Upchurch and colleagues reported superior performance of male C57BL/6 mice (Upchurch et al., 1988), inconsistent or opposite trends were found in a later study with the same and other strains (Upchurch and Wehner, 1989). Frick and colleagues found no sex differences in 5- or 22-month-old C57BL/6 mice in a water-maze place-navigation task (Frick et al., 2000). A meta-analysis of data from published and unpublished sources found overall a small female advantage for mice in the water maze, whereas a large reliable male advantage was evident in rats (Jonasson, 2005). Retrospective analysis of our own water-maze data (Figure 27.3a and c) revealed a small overall trend toward better performance in males which accounts for only about 1% of the variance in the data. But we found no evidence for increased variability in female mice (Figure 27.3b and d). Probe trial scores were even significantly more stable in females. These data representing 3273 mice suggest that a priori exclusion of female mice from water-maze studies is not justified.

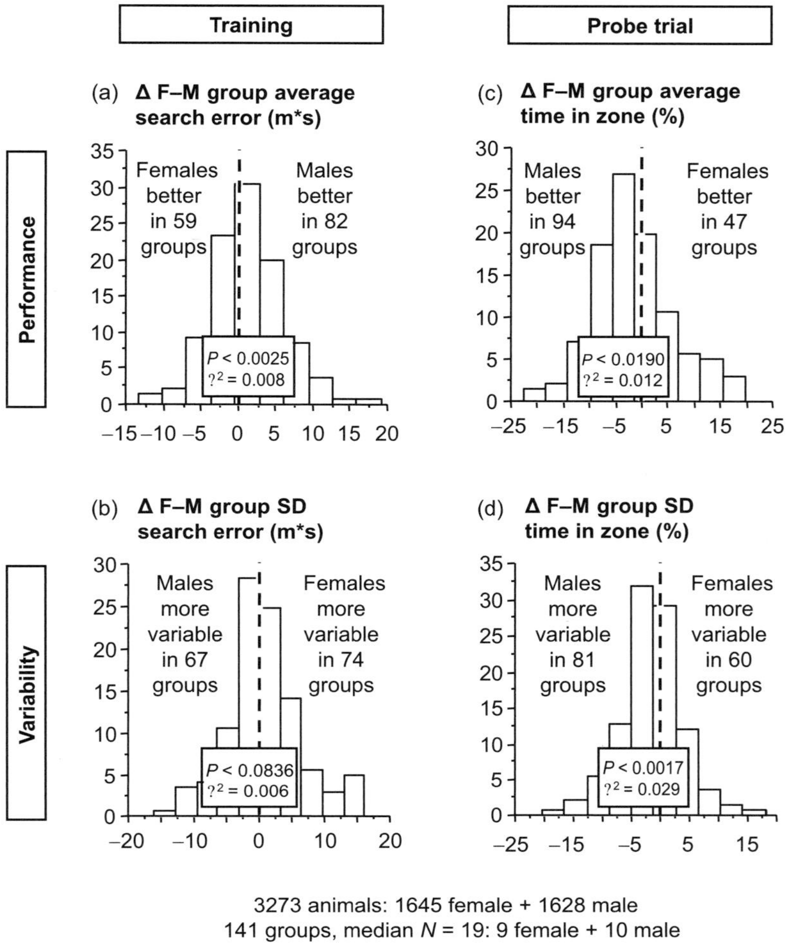

Figure 27.3 Sex effects on place navigation. Retrospective analysis of training (a and b) and probe trial (c and d) data collected 1987–2005 using the protocol shown in Figure 27.1d. To permit comparison of both performance level (a and c) and variability (b and d), 141 experimental groups including mice of both sexes were chosen as units of observation, all controls or experimental groups without treatment effect. Female–male differences were calculated in each group, plotted as histograms, and tested against zero using one-sample t-tests. (a) Female–male subgroup differences of cumulative search error (see main text) as a measure of training performance, averaged across training trials. (b) Female–male subgroup differences of time spent in target zone as measure of spatial retention during the probe trial. (c) Standard deviation (SD) differences between female and male subgroups in cumulative search error as measure of training performance variability. (d) The SD differences between female and male subgroups in time spent in the target zone as a measure of probe trial performance variability.

Age effects

It is well known that performance of rats in spatial water-maze tasks declines with age and age-related performance deficits are also well documented in several mouse strains. In female NMRI mice, performance in the water-maze place-navigation task began to decline at 9 months of age (Lamberty and Gower, 1990) and they were already strongly impaired at 11 months (Gower and Lamberty, 1993; Lamberty and Gower, 1991, 1992). Female mice of the outbred strain OF1 showed a place learning deficit at the age of 17 months (Carrie et al., 1999). Wong and Brown found faster age-related decline of place-navigation performance in DBA/2J mice than in C57BL/6J and attributed this to age-related changes in visual acuity (Wong and Brown, 2007). In male C57BL/6Nia mice, Fordyce and Wehner (1993a) found a mild impairment at 14 months which became marked at the age of 25 months and correlated with a reduction of membrane-bound protein kinase C activity. In the same study, B6D2F1/Nia mice showed greater protein kinase C activity than C57BL/6Nia and no significant decline of place-navigation performance until 25 months of age. A more recent study correlated the impairment of 22-month-old C57BL/6Nia mice to oxidative protein damage in the brain (Forster et al., 1996). Calhoun and coworkers found largely preserved place learning as well as hippocampal cell and synapse numbers in C57BL/6 mice up to the age of 31 months, but found that individual performance was correlated positively with neuron and synapse numbers (Calhoun et al., 1998). Using probe trial performance as criterion, Magnusson found that male C57BL/6Nia mice were mildly impaired at 10 months when fed ad libitum, but not after dietary restriction. At 26 months all tested animals were impaired, but the impairment was stronger in mice that were fed at libitum (Magnusson, 1997). Age-related performance decline in C57BL/6 mice can also be slowed by physical activity (van Praag et al., 2005) and enriched housing (Frick et al., 2003). Based on their observation that 17-month-old C57BL/6Nia females performed similarly to 25-month-old females, whereas 17-month-old males performed more like 5-month-old males, Frick and colleagues concluded that the age-related decline may begin at an earlier age in females than in males (Frick et al., 2000). Similarly, Benice et al. found stronger place-navigation deficits in 18–20-month-old female C57BL/6J mice than in males of the same age (Benice et al., 2006). Taken together, these data indicate that the age-related decline is strongly strain dependent and also varies with sex and environmental factors.

The ontogeny of place learning in mice is much less well studied. Meadow voles (*Microtus pennsylvanicus*) acquired a place navigation task already at the preweaning age of 10 days, but reached adult performance levels only after weaning at the age of 20 days (Galea et al., 1994b). In a study with C57BL/6 mice, 22-day-old individuals showed adult-like performance in various water-maze tasks including place navigation, but were more easily distracted by irrelevant proximal cues (Chapillon and Roullet, 1996). Overall it appears that no age-related confounds should be expected if mice are tested in the water-maze place-navigation task between 2 and 6 months of age.

Housing conditions

Place-navigation performance is also influenced by environmental effects. For example, physical activity enhances place-learning performance in C57BL/6 (Fordyce and Wehner, 1993b; van Praag et al., 1999, 2005), DBA/2Ibg (Fordyce and Wehner, 1993b), and ICR (Rhodes et al., 2003) mice. This effect has been linked to a stimulation of protein kinase C (Fordyce and Wehner, 1993b), as well as adult neurogenesis (van Praag et al., 1999). Apart from physical activity, place-learning performance in adult healthy mice seems to be rather insensitive to changes of housing conditions. Voikar and colleagues reported negative effects of long-lasting individual versus group housing in male C57BL/6 and DBA/2 mice in several learning tasks, but failed to see a change in the water-maze place-navigation task (Voikar et al., 2005), confirming an earlier study which also failed to demonstrate such effects (Moragrega et al., 2003). Further, grouped housing in enriched versus standard cages increased swimming speed but had no significant effect on place navigation acquisition or retention in C57BL/6, DBA/2, and B6D2F1 mice even though the study was replicated in three different laboratories (Wolfer et al., 2004). However, in contrast to the above papers a more recent study which tested the effects of physical activity and enriched housing on C57BL/6 mice in a 2 × 2 factorial design found that enrichment but not exercise improved acquisition and reversal in a water-maze place-navigation task (Pietropaolo et al., 2006).

References

Adams, B., Fitch, T., Chaney, S., and Gerlai, R. (2002) Altered performance characteristics in cognitive tasks: comparison of the albino ICR and CD1 mouse strains. *Behav Brain Res* 133: 351–361.

Balogh, S.A., McDowell, C.S., Stavnezer, A.J., and Denenberg, V.H. (1999) A behavioral and neuroanatomical assessment of an inbred substrain of 129 mice with behavioral comparisons to C57BL/6J mice. *Brain Res* 836: 38–48.

Balschun, D., Wolfer, D.P., Gass, P., Mantamadiotis, T., Welzl, H., Schutz, G., et al. (2003) Does cAMP response element-binding protein have a pivotal role in hippocampal synaptic plasticity and hippocampus-dependent memory? *J Neurosci* 23: 6304–6314.

Benice, T.S., Rizk, A., Kohama, S., Pfankuch, T., and Raber, J. (2006) Sex-differences in age-related cognitive decline in C57BL/6J mice associated with increased brain microtubule-associated protein 2 and synaptophysin immunoreactivity. *Neuroscience* 137: 413–423.

Bowes, C., Li, T., Frankel, W.N., Danciger, M., Coffin, J.M., Applebury, M.L., et al. (1993) Localization of a retroviral element within the rd gene coding for the beta subunit of cGMP phosphodiesterase. *Proc Natl Acad Sci USA* 90: 2955–2959.

Brandeis, R., Brandys, Y., and Yehuda, S. (1989) The use of the Morris water maze in the study of memory and learning. *Int J Neurosci* 48: 29–69.

Brooks, S.P., Pask, T., Jones, L., and Dunnett, S.B. (2005) Behavioural profiles of inbred mouse strains used as transgenic backgrounds. II: cognitive tests. *Genes Brain Behav* 4: 307–317.

Brown, R.E. and Wong, A.A. (2007) The influence of visual ability on learning and memory performance in 13 strains of mice. *Learn Mem* 14: 134–144.

Buresova, O., Krekule, I., Zahalka, A., and Bures, J. (1985) On-demand platform improves accuracy of the Morris water maze procedure. *J Neurosci Methods* 15: 63–72.

Calhoun, M.E., Kurth, D., Phinney, A.L., Long, J.M., Hengemihle, J., Mouton, P.R., et al. (1998) Hippocampal neuron and synaptophysin-positive bouton number in aging C57BL/6 mice. *Neurobiol Aging* 19: 599–606.

Carrie, I., Debray, M., Bourre, J.M., and Frances, H. (1999) Age-induced cognitive alterations in OF1 mice. *Physiol Behav* 66: 651–656.

Chapillon, P. and Roullet, P. (1996) Use of proximal and distal cues in place navigation by mice changes during ontogeny. *Dev Psychobiol* 29: 529–545.

Chen, G., Chen, K.S., Knox, J., Inglis, J., Bernard, A., Martin, S.J., et al. (2000) A learning deficit related to age and beta-amyloid plaques in a mouse model of Alzheimer's disease. *Nature* 408: 975–979.

Clapcote, S.J., Lazar, N.L., Bechard, A.R., and Roder, J.C. (2005a) Effects of the rd1 mutation and host strain on hippocampal learning in mice. *Behav Genet* 35: 591–601.

Clapcote, S.J., Lazar, N.L., Bechard, A.R., Wood, G.A., and Roder, J.C. (2005b) NIH Swiss and Black Swiss mice have retinal degeneration and performance deficits in cognitive tests. *Comp Med* 55: 310–316.

Clapcote, S.J. and Roder, J.C. (2004) Survey of embryonic stem cell line source strains in the water maze reveals superior reversal learning of 129S6/SvEvTac mice. *Behav Brain Res* 152: 35–48.

Contet, C., Rawlins, J.N., and Bannerman, D.M. (2001) Faster is not surer – a comparison of C57BL/6J and 129S2/Sv mouse strains in the water maze. *Behav Brain Res* 125: 261–267.

Crawley, J.N., Belknap, J.K., Collins, A., Crabbe, J.C., Frankel, W., Henderson, N., et al. (1997) Behavioral phenotypes of inbred mouse strains: implications and recommendations for molecular studies. *Psychopharmacology (Berl)* 132: 107–124.

Denenberg, V.H., Talgo, N.W., Waters, N.S., and Kenner, G.H. (1990) A computer-aided procedure for measuring morris maze performance. *Physiol Behav* 47: 1027–1029.

D'Hooge, R. and De Deyn, P.P. (2001) Applications of the Morris water maze in the study of learning and memory. *Brain Res Rev* 36: 60–90.

Errijgers, V., Van Dam, D., Gantois, I., Van Ginneken, C.J., Grossman, A.W., D'Hooge, R., et al. (2007) FVB.129P2-Pde6b(+) Tyr(c-ch)/Ant, a sighted variant of the FVB/N mouse strain suitable for behavioral analysis. *Genes Brain Behav* 6: 552–557.

Festing, M.F., Simpson, E.M., Davisson, M.T., and Mobraaten, L.E. (1999) Revised nomenclature for strain 129 mice. *Mamm Genome* 10: 836.

Fordyce, D.E. and Wehner, J.M. (1993a) Effects of aging on spatial learning and hippocampal protein kinase C in mice. *Neurobiol Aging* 14: 309–317.

Fordyce, D.E. and Wehner, J.M. (1993b) Physical activity enhances spatial learning performance with an associated alteration in hippocampal protein kinase C activity in C57BL/6 and DBA/2 mice. *Brain Res* 619: 111–119.

Forster, M.J., Dubey, A., Dawson, K.M., Stutts, W.A., Lal, H., and Sohal, R.S. (1996) Age-related losses of cognitive function and motor skills in mice are associated with oxidative protein damage in the brain. *Proc Natl Acad Sci USA* 93: 4765–4769.

Fox, G.B., LeVasseur, R.A., and Faden, A.I. (1999) Behavioral responses of C57BL/6, FVB/N, and 129/SvEMS mouse strains to traumatic brain injury: implications for gene targeting approaches to neurotrauma. *J Neurotrauma* 16: 377–389.

Francis, D.D., Szegda, K., Campbell, G., Martin, W.D., and Insel, T.R. (2003) Epigenetic sources of behavioral differences in mice. *Nat Neurosci* 6: 445–446.

Francis, D.D., Zaharia, M.D., Shanks, N., and Anisman, H. (1995) Stress-induced disturbances in Morris water-maze performance: interstrain variability. *Physiol Behav* 58: 57–65.

Frick, K.M., Burlingame, L.A., Arters, J.A., and Berger-Sweeney, J. (2000) Reference memory, anxiety and estrous cyclicity in C57BL/6NIA mice are affected by age and sex. *Neuroscience* 95: 293–307.

Frick, K.M., Stearns, N.A., Pan, J.Y., and Berger-Sweeney, J. (2003) Effects of environmental enrichment on spatial memory and neurochemistry in middle-aged mice. *Learn Mem* 10: 187–198.

Galea, L.A., Kavaliers, M., and Ossenkopp, K.P. (1996) Sexually dimorphic spatial learning in meadow voles *Microtus pennsylvanicus* and deer mice *Peromyscus maniculatus*. *J Exp Biol* 199: 195–200.

Galea, L.A., Kavaliers, M., Ossenkopp, K.P., and Hampson, E. (1995) Gonadal hormone levels and spatial learning performance in the Morris water maze in male and female meadow voles, *Microtus pennsylvanicus*. *Horm Behav* 29: 106–125.

Galea, L.A., Kavaliers, M., Ossenkopp, K.P., Innes, D., and Hargreaves, E.L. (1994a) Sexually dimorphic spatial learning varies seasonally in two populations of deer mice. *Brain Res* 635: 18–26.

Galea, L.A., Ossenkopp, K.P., and Kavaliers, M. (1994b) Developmental changes in spatial learning in the Morris water-maze in young meadow voles, *Microtus pennsylvanicus*. *Behav Brain Res* 60: 43–50.

Gallagher, M., Burwell, R., and Burchinal, M. (1993) Severity of spatial learning impairment in aging: development of a learning index for performance in the Morris water maze. *Behav Neurosci* 107: 618–626.

Gass, P., Wolfer, D.P., Balschun, D., Rudolph, D., Frey, J.U., Lipp, H.P., et al. (1998) Deficits in memory tasks of mice with CREB mutations depend on gene dosage. *Learn Mem* 5: 274–288.

Gerlai, R. (1996) Gene-targeting studies of mammalian behavior: is it the mutation or the background genotype. *Trends Neurosci* 19: 177–181.

Gerlai, R. (2002) Hippocampal LTP and memory in mouse strains: is there evidence for a causal relationship? *Hippocampus* 12: 657–666.

Gower, A.J. and Lamberty, Y. (1993) The aged mouse as a model of cognitive decline with special emphasis on studies in NMRI mice. *Behav Brain Res* 57: 163–173.

Grant, S.G.N., O'Dell, T.J., Karl, K.A., Stein, P.L., Soriano, P., and Kandel, E.R. (1992) Impaired long-term potentiation, spatial learning, and hippocampal development in fyn mutant mice. *Science* 258: 1903–1910.

Grant, S.G.N. and Silva, A.J. (1994) Targeting learning. *Trends Neurosci* 17: 71–75.

Hagan, J.J., Alpert, J.E., Morris, R.G., and Iversen, S.D. (1983) The effects of central catecholamine depletions on spatial learning in rats. *Behav Brain Res* 9: 83–104.

Harris, A.P., D'Eath, R.B., and Healy, S.D. (2009) Environmental enrichment enhances spatial cognition in rats by reducing thigmotaxis (wall hugging) during testing. *Anim Behav* 77: 1459–1464.

Harrison, F.E., Hosseini, A.H., and McDonald, M.P. (2009) Endogenous anxiety and stress responses in water maze and Barnes maze spatial memory tasks. *Behav Brain Res* 198: 247–251.

Holmes, A., Wrenn, C.C., Harris, A.P., Thayer, K.E., and Crawley, J.N. (2002) Behavioral profiles of inbred strains on novel olfactory, spatial and emotional tests for reference memory in mice. *Genes Brain Behav* 1: 55–69.

Janus, C. (2004) Search strategies used by APP transgenic mice during navigation in the Morris water maze. *Learn Mem* 11: 337–346.

Jonasson, Z. (2005) Meta-analysis of sex differences in rodent models of learning and memory: a review of behavioral and biological data. *Neurosci Biobehav Rev* 28: 811–825.

Kavaliers, M. and Galea, L.A. (1994) Spatial water maze learning using celestial cues by the meadow vole, *Microtus pennsylvanicus*. *Behav Brain Res* 61: 97–100.

Klapdor, K. and van der Staay, F.J. (1996) The Morris water-escape task in mice: strain differences and effects of intra-maze contrast and brightness. *Physiol Behav* 60: 1247–1254.

Kogan, J.H., Frankland, P.W., Blendy, J.A., Coblentz, J., Marowitz, Z., Schultz, G., et al. (1997) Spaced training induces normal long-term memory in CREB mutant mice. *Curr Biol* 7: 1–11.

Lamberty, Y. and Gower, A.J. (1990) Age-related changes in spontaneous behavior and learning in NMRI mice from maturity to middle age. *Physiol Behav* 47: 1137–1144.

Lamberty, Y. and Gower, A.J. (1991) Simplifying environmental cues in a Morris-type water maze improves place learning in old NMRI mice. *Behav Neural Biol* 56: 89–100.

Lamberty, Y. and Gower, A.J. (1992) Age-related changes in spontaneous behavior and learning in NMRI mice from middle to old age. *Physiol Behav* 51: 81–88.

Lee, A.S., Duman, R.S., and Pittenger, C. (2008) A double dissociation revealing bidirectional competition between striatum and hippocampus during learning. *Proc Natl Acad Sci USA* 105: 17163–17168.

Lee, Y.S. and Silva, A.J. (2009) The molecular and cellular biology of enhanced cognition. *Nat Rev Neurosci* 10: 126–140.

Leil, T.A., Ossadtchi, A., Cortes, J.S., Leahy, R.M., and Smith, D.J. (2002) Finding new candidate genes for learning and memory. *J Neurosci Res* 68: 127–137.

Leitinger, B., Poletaeva, I.I., Wolfer, D.P., and Lipp, H.P. (1994) Swimming navigation, open-field activity, and extrapolation behavior of two inbred mouse strains with Robertsonian translocation of chromosomes 8 and 17. *Behav Genet* 24: 273–284.

Le Roy, I., Roubertoux, P.L., Jamot, L., Maarouf, F., Tordjman, S., Mortaud, S., et al. (1998) Neuronal and behavioral differences between *Mus musculus domesticus* (C57BL/6JBy) and *Mus musculus castaneus* (CAST/Ei). *Behav Brain Res* 95: 135–142.

Lipp, H.P. and Wolfer, D.P. (1998) Genetically modified mice and cognition. *Curr Opin Neurobiol* 8: 272–280.

Logue, S.F., Paylor, R., and Wehner, J.M. (1997) Hippocampal lesions cause learning deficits in inbred mice in the Morris water maze and conditioned-fear task. *Behav Neurosci* 111: 104–113.

Magnusson, K.R. (1997) Influence of dietary restriction on ionotropic glutamate receptors during aging in C57B1 mice. *Mech Ageing Dev* 95: 187–202.

Micheau, J., Riedel, G., Roloff, E.L., Inglis, J., and Morris, R.G. (2004) Reversible hippocampal inactivation partially dissociates how and where to search in the water maze. *Behav Neurosci* 118: 1022–1032.

Micheau, J., Riedel, G., Roloff, E.L., Inglis, J., and Morris, R.G. (2004) Reversible hippocampal inactivation partially dissociates how and where to search in the water maze. *Behav Neurosci* 118: 1022–1032.

Mohajeri, M.H., Madani, R., Saini, K., Lipp, H.P., Nitsch, R.M., and Wolfer, D.P. (2004) The impact of genetic background on neurodegeneration and behavior in seizured mice. *Genes Brain Behav* 3: 228–239.

Montkowski, A., Poettig, M., Mederer, A., and Holsboer, F. (1997) Behavioural performance in three substrains of mouse strain 129. *Brain Res* 762: 12–18.

Moragrega, I., Carrasco, M.C., Vicens, P., and Redolat, R. (2003) Spatial learning in male mice with different levels of aggressiveness: effects of housing conditions and nicotine administration. *Behav Brain Res* 147: 1–8.

Morris, R.G. (1981) Spatial localization does not require the presence of local cues. *Learn Motiv* 12: 239–260.

Morris, R.G. (1984) Developments of a water-maze procedure for studying spatial learning in the rat. *J Neurosci Methods* 11: 47–60.

Morris, R.G., Anderson, E., Lynch, G., and Baudry, M. (1986) Selective impairment of learning and blockade of long-term potentiation by an N-methyl-D-aspartate antagonist, AP5. *Nature* 319: 774–776.

Morris, R.G., Garrud, P., Rawlins, J.N.P., and O'Keefe, J. (1982) Place navigation impaired in rats with hippocampal lesions. *Nature* 297: 681–683.

Morris, R.G. and Seifert, W. (1983) An attempt to dissociate "spatial-mapping" and "working-memory" theories of hippocampal function. In Siefert, W. (ed.), *Neurobiology of the Hippocampus*. Academic Press, London, pp. 405–432.

Moy, S.S., Nadler, J.J., Young, N.B., Perez, A., Holloway, L.P., Barbaro, R.P., et al. (2007) Mouse behavioral tasks relevant to autism: phenotypes of 10 inbred strains. *Behav Brain Res* 176: 4–20.

Nguyen, P.V., Abel, T., Kandel, E.R., and Bourtchouladze, R. (2000) Strain-dependent differences in LTP and hippocampus-dependent memory in inbred mice. *Learn Mem* 7: 170–179.

O'Keefe, J. and Nadel, L. (1978) *The Hippocampus as a Cognitive Map*. Clarendon Press, Oxford.

Olton, D.S., Becker, J.T., and Handelmann, G.E. (1979) Hippocampus, space, and memory. *Behav Brain Sci* 2: 313–365.

Owen, E.H., Logue, S.F., Rasmussen, D.L., and Wehner, J.M. (1997) Assessment of learning by the Morris water task and fear conditioning in inbred mouse strains and F1 hybrids: implications of genetic background for single gene mutations and quantitative trait loci. *Neuroscience* 80: 1087–1099.

Panakhova, E., Buresova, O., and Bures, J. (1984) Persistence of spatial memory in the Morris water tank task. *Int J Psychophysiol* 2: 5–10.

Patil, S.S., Sunyer, B., Hoger, H., and Lubec, G. (2008) Apodemus sylvaticus (LOXT) is a suitable mouse strain for testing spatial memory retention in the Morris water maze. *Neurobiol Learn Mem* 89: 552–559.

Patil, S.S., Sunyer, B., Hoger, H., and Lubec, G. (2009) Evaluation of spatial memory of C57BL/6J and CD1 mice in the Barnes maze, the multiple T-maze and in the Morris water maze. *Behav Brain Res* 198: 58–68.

Paul, G.M., Magda, G., and Abel, S. (2009) Spatial memory: theoretical basis and comparative review on experimental methods in rodents. *Behav Brain Res* 203: 151–164.

Paylor, R., Baskall, L., and Wehner, J.M. (1993) Bahavioral dissociations between C57BL/6 and DBA/2 mice on learning and memory tasks: a hippocampal dysfunction hypothesis. *Psychobiology* 21: 11–26.

Paylor, R., Baskall-Baldini, L., Yuva, L., and Wehner, J.M. (1996) Developmental differences in place-learning performance between C57BL/6 and DBA/2 mice parallel the ontogeny of hippocampal protein kinase C. *Behav Neurosci* 110: 1415–1425.

Petrie, B.F. (1995) Learning set spatial navigation performance in three mouse strains. *Psychol Rep* 77: 1339–1342.

Pietropaolo, S., Feldon, J., Alleva, E., Cirulli, F., and Yee, B.K. (2006) The role of voluntary exercise in enriched rearing: a behavioral analysis. *Behav Neurosci* 120: 787–803.

Pleskacheva, M.G., Wolfer, D.P., Kupriyanova, I.E., Nikolenko, D.L., Scheffrahn, H., Dell'Omo, G., et al. (2000) Hippocampal mossy fibers and swimming navigation learning in two vole species occupying different habitats. *Hippocampus* 10: 17–30.

Rhodes, J.S., van Praag, H., Jeffrey, S., Girard, I., Mitchell, G.S., Garland, T., Jr., et al. (2003) Exercise increases hippocampal neurogenesis to high levels but does not improve spatial learning in mice bred for increased voluntary wheel running. *Behav Neurosci* 117: 1006–1016.

Riedel, G., Micheau, J., Lam, A.G., Roloff, E., Martin, S.J., Bridge, H., et al. (1999) Reversible neural inactivation reveals hippocampal participation in several memory processes. *Nat Neurosci* 2: 898–905.

Rogers, D.C., Jones, D.N., Nelson, P.R., Jones, C.M., Quilter, C.A., Robinson, T.L., et al. (1999) Use of SHIRPA and discriminant analysis to characterise marked differences in the behavioral phenotype of six inbred mouse strains. *Behav Brain Res* 105: 207–217.

Royle, S.J., Collins, F.C., Rupniak, H.T., Barnes, J.C., and Anderson, R. (1999) Behavioural analysis and susceptibility to CNS injury of four inbred strains of mice. *Brain Res* 816: 337–349.

Schimanski, L.A. and Nguyen, P.V. (2004) Multidisciplinary approaches for investigating the mechanisms of hippocampus-dependent memory: a focus on inbred mouse strains. *Neurosci Biobehav Rev* 28: 463–483.

Schmitt, W.B., Deacon, R.M., Reisel, D., Sprengel, R., Seeburg, P.H., Rawlins, J.N., et al. (2004) Spatial reference memory in GluR-A-deficient mice using a novel hippocampal-dependent paddling pool escape task. *Hippocampus* 14: 216–223.

Schopke, R., Wolfer, D.P., Lipp, H.P., and Leisinger-Trigona, M.C. (1991) Swimming navigation and structural variations of the infrapyramidal mossy fibers in the hippocampus of the mouse. *Hippocampus* 1: 315–328.

Silva, A.J. and Giese, K.P. (1994) Platic genes are in! *Curr Opin Neurobiol* 4: 413–420.

Silva, A.J., Paylor, R., Wehner, J.M., and Tonegawa, S. (1992a) Impaired spatial learning in alpha-calcium-calmodulin kinase II mutant mice. *Science* 257: 206–211.

Silva, A.J., Stevens, C.F., Tonegawa, S., and Wang, Y. (1992b) Deficient hippocampal long-term potentiation in alpha-calcium-calmodulin kinase II mutant mice. *Science* 257: 201–206.

Simpson, E.M., Linder, C.C., Sargent, E.E., Davisson, M.T., Mobraaten, L.E., and Sharp, J.J. (1997) Genetic variation among 129 substrains and its importance for targeted mutagenesis in mice. *Nat Genet* 16: 19–27.

Spooner, R.I., Thomson, A., Hall, J., Morris, R.G., and Salter, S.H. (1994) The Atlantis platform: a new design and further developments of Buresova's on-demand platform for the water maze. *Learn Mem* 1: 203–211.

Steele, R.J. and Morris, R.G. (1999) Delay-dependent impairment of a matching-to-place task with chronic and intrahippocampal infusion of the NMDA-antagonist D-AP5. *Hippocampus* 9: 118–136.

Sung, J.Y., Goo, J.S., Lee, D.E., Jin, D.Q., Bizon, J.L., Gallagher, M., et al. (2008) Learning strategy selection in the water maze and hippocampal CREB phosphorylation differ in two inbred strains of mice. *Learn Mem* 15: 183–188.

Sutherland, R.J., Kolb, B., and Whishaw, I.Q. (1982a) Spatial mapping: definitive disruption by hippocampal or medial frontal cortical damage in the rat. *Neurosci Lett* 31: 271–276.

Sutherland, R.J. and Rodriguez, A.J. (1989) The role of the fornix/fimbria and some related subcortical structures in place learning and memory. *Behav Brain Res* 32: 265–277.

Sutherland, R.J., Whishaw, I.Q., and Kolb, B. (1983) A behavioural analysis of spatial localization following electrolytic, kainate- or colchicine-induced damage to the hippocampal formation in the rat. *Behav Brain Res* 7: 133–153.

Sutherland, R.J., Whishaw, I.Q., and Regehr, J.C. (1982b) Cholinergic receptor blockade impairs spatial localization by use of distal cues in the rat. *J Comp Physiol Psychol* 96: 563–573.

Sweeney, J.E., Hohmann, C.F., Moran, T.H., and Coyle, J.T. (1988) A long-acting cholinesterase inhibitor reverses spatial memory deficits in mice. *Pharmacol Biochem Behav* 31: 141–147.

Threadgill, D.W., Yee, D., Matin, A., Nadeau, J.H., and Magnuson, T. (1997) Genealogy of the 129 inbred strains: 129/SvJ is a contaminated inbred strain. *Mamm Genome* 8: 390–393.

Upchurch, M., Pounder, J.I., and Wehner, J.M. (1988) Heterosis and resistance to DFP effects on spatial learning in C57BL × DBA hybrids. *Brain Res Bull* 21: 499–503.

Upchurch, M. and Wehner, J.M. (1987) Effects of chronic diisopropylfluorophosphate treatment on spatial learning in mice. *Pharmacol Biochem Behav* 27: 143–151.

Upchurch, M. and Wehner, J.M. (1988a) DBA/2Ibg mice are incapable of cholinergically-based learning in the Morris water task. *Pharmacol Biochem Behav* 29: 325–329.

Upchurch, M. and Wehner, J.M. (1988b) Differences between inbred strains of mice in Morris water maze performance. *Behav Genet* 18: 55–68.

Upchurch, M. and Wehner, J.M. (1989) Inheritance of spatial learning ability in inbred mice: a classical genetic analyis. *Behav Neurosci* 103: 1251–1258.

Van Dam, D., Lenders, G., and De Deyn, P.P. (2006) Effect of Morris water maze diameter on visual-spatial learning in different mouse strains. *Neurobiol Learn Mem* 85: 164–172.

van Praag, H., Christie, B.R., Sejnowski, T.J., and Gage, F.H. (1999) Running enhances neurogenesis, learning, and long-term potentiation in mice. *Proc Natl Acad Sci USA* 96: 13427–13431.

van Praag, H., Shubert, T., Zhao, C., and Gage, F.H. (2005) Exercise enhances learning and hippocampal neurogenesis in aged mice. *J Neurosci* 25: 8680–8685.

Voikar, V., Koks, S., Vasar, E., and Rauvala, H. (2001) Strain and gender differences in the behavior of mouse lines commonly used in transgenic studies. *Physiol Behav* 72: 271–281.

Voikar, V., Polus, A., Vasar, E., and Rauvala, H. (2005) Long-term individual housing in C57BL/6J and DBA/2 mice: assessment of behavioral consequences. *Genes Brain Behav* 4: 240–252.

Vorhees, C.V. and Williams, M.T. (2006) Morris water maze: procedures for assessing spatial and related forms of learning and memory. *Nat Protoc* 1: 848–858.

Wahlsten, D., Cooper, S.F., and Crabbe, J.C. (2005) Different rankings of inbred mouse strains on the Morris maze and a refined 4-arm water escape task. *Behav Brain Res* 165: 36–51.

Wehner, J.M., Sleight, S., and Upchurch, M. (1990) Hippocampal protein kinase C activity is reduced in poor spatial learners. *Brain Res* 523: 181–187.

Whishaw, I.Q. (1985a) Cholinergic receptor blockade in the rat impairs local but not taxon strategies for place navigation in a swimming pool. *Behav Neurosci* 99: 979–1005.

Whishaw, I.Q. (1985b) Evidence for two types of place navigation in the rat. In Buzsaki, G. (ed.), *Electrical Activity of the Archicortex*. Hungarian Academy of Sciences, Budapest, pp. 233–253.

Whishaw, I.Q. (1985c) Formation of a place learning-set by the rat: a new paradigm for neurobehavioral studies. *Physiol Behav* 35: 139–143.

Whishaw, I.Q. (1995) A comparison of rats and mice in a swimming pool place task and matching to place task: some surprising differences. *Physiol Behav* 58: 687–693.

Whishaw, I.Q. and Tomie, J.A. (1996) Of mice and mazes: similarities between mice and rats on dry land but not water mazes. *Physiol Behav* 60: 1191–1197.

Wolfer, D.P., Crusio, W.E., and Lipp, H.P. (2002) Knockout mice: simple solutions to the problems of genetic background and flanking genes. *Trends Neurosci* 25: 336–340.

Wolfer, D.P., Litvin, O., Morf, S., Nitsch, R.M., Lipp, H.P., and Wurbel, H. (2004) Laboratory animal welfare: cage enrichment and mouse behaviour. *Nature* 432: 821–822.

Wolfer, D.P., Madani, R., Valenti, P., and Lipp, H.P. (2001) Extended analysis of path data from mutant mice using the public domain software Wintrack. *Physiol Behav* 73: 745–753.

Wolfer, D.P., Muller, U., Stagliar-Bozizevic, M., and Lipp, H.P. (1997) Assessing the effects of the 129/Sv genetic background on swimming navigation learning in transgenic mutants: a study using mice with a modified beta-amyloid precursor protein gene. *Brain Res* 771: 1–13.

Wolff, M., Savova, M., Malleret, G., Segu, L., and Buhot, M.C. (2002) Differential learning abilities of 129T2/Sv and C57BL/6J mice as assessed in three water maze protocols. *Behav Brain Res* 136: 463–474.

Wong, A.A. and Brown, R.E. (2007) Age-related changes in visual acuity, learning and memory in C57BL/6J and DBA/2J mice. Neurobiol. *Aging* 28: 1577–1593.

Wright, J.W., Alt, J.A., Turner, G.D., and Krueger, J.M. (2004) Differences in spatial learning comparing transgenic p75 knockout, New Zealand Black, C57BL/6, and Swiss Webster mice. *Behav Brain Res* 153: 453–458.

Yamaguchi, F., Richards, S.J., Beyreuther, K., Salbaum, M., Carlson, G.A., and Dunnett, S.B. (1991) Transgenic mice for the amyloid precursor protein 695 isoform have impaired spatial memory. *Neuroreport* 2: 781–784.

Yoshida, M., Goto, K., and Watanabe, S. (2001) Task-dependent strain difference of spatial learning in C57BL/6N and BALB/c mice. *Physiol Behav* 73: 37–42.

Zaharia, M.D., Kulczycki, J., Shanks, N., Meaney, M.J., and Anisman, H. (1996) The effects of early postnatal stimulation on Morris water-maze acquisition in adult mice: genetic and maternal factors. *Psychopharmacology (Berl)* 128: 227–239.

Chapter 28

Active and passive avoidance

Igor Branchi and Laura Ricceri

Introduction

Avoidance conditioning involves the acquisition of a response that serves to prevent the occurrence of an aversive (punishing) event. Avoidance tests have been used extensively for many years in the assessment of various types of treatment effects on behavior (for drug effects see reviews in Bammer, 1982; Myhrer, 2003; Sarter et al., 1992). They are also used in screening procedures, as shown for example by the use of active and passive avoidance tests in research on learning and memory abilities in genetically modified mice (e.g., Brambilla et al., 1997; Huang et al., 1996; King et al., 2003; Martin et al., 2002; Minichiello et al., 1999; Montag-Sallaz and Montag, 2003; Tzavara et al., 2003).

These tests are widely used also because they have several practical advantages: for example, two-way (bidirectional) active avoidance procedures allow an easy automation of the test. One-way active and passive avoidance learning tests are procedures that can be carried out quickly, requiring only a limited amount of time. In particular, in the case of passive avoidance, a two-trial protocol is enough to assess learning abilities in adult rodents, whereas a multi-trial protocol is usually employed for young, developing rats and mice (Ricceri, 2003). Finally, avoidance procedures do not require continuous monitoring of food or fluid intake aimed at maintaining appropriate motivation levels, as is the case with reward-based learning tasks.

Both active and passive avoidance tests quantify the strength of the association between a neutral stimulus (conditioned stimulus, CS) and an aversive stimulus (unconditioned stimulus, US), and allow assessment of the capacity of learning a response contingency by letting the subjects escape and avoid the US.

Active avoidance

By definition, active avoidance tests are procedures in which the animal, in order to avoid punishment (US, which is usually an electric shock), is required to respond to a discrete signal (CS), or within a specified amount of time.

Several types of active avoidance procedures have been studied, of which only two need to be illustrated for the purposes of the present review; namely, one-way active avoidance and two-way or bidirectional active avoidance.

One-way (unidirectional) active avoidance

In one-way active avoidance, the apparatus more commonly employed consists of two similar compartments (Figure 28.1). In the first one, the unsafe compartment, the US, e.g., an electric shock, can be delivered to the animal through a grid floor. The second compartment is always safe. When the animal is placed in the first compartment, if it does not move to the safe compartment within a given time (e.g., 5 s) or within a specified period of presentation of the CS (generally a visual or acoustic stimulus) the shock is delivered. In subsequent trials, the animal is returned to the unsafe compartment and acquisition is measured as a decreased latency (i.e., the period of time between time stimulus presentation and response) to move to the safe compartment. Good acquisition of the task entails decreasing escape latencies after the shock, followed by responding before punishment (successful avoidance).

It is important to mention that active avoidance has long been considered as a type of instrumental learning, in which the acquisition of a novel response or a novel response-reinforcement contingency is required. However, Bolles (1970) argued against this view on the basis of evidence indicating that the acquisition of active avoidance is due mainly to Pavlovian (classical) conditioning of a highly prepotent, species-specific defensive unconditioned response running or jumping away from a place where painful stimulation is experienced. The latter view has now become mainly accepted.

Two-way (bidirectional) active avoidance

Two-way active avoidance in a shuttle box (Figure 28.1) is a frequently utilized test procedure. The apparatus consists of two identical compartments separated either by a partition with a hole or by a hurdle. The animal is initially placed in one of the two compartments and when the discrete CS is presented, the subject must run over or jump into the other compartment before the delivery of the US. After a predetermined period of time (intertrial interval, ITI), which can be fixed or randomly varied, the trial is repeated and now in order for the

Behavioral Genetics of the Mouse: Volume I. *Genetics of Behavioral Phenotypes*, eds. Wim E. Crusio, Frans Sluyter, Robert T. Gerlai, and Susanna Pietropaolo. Published by Cambridge University Press.

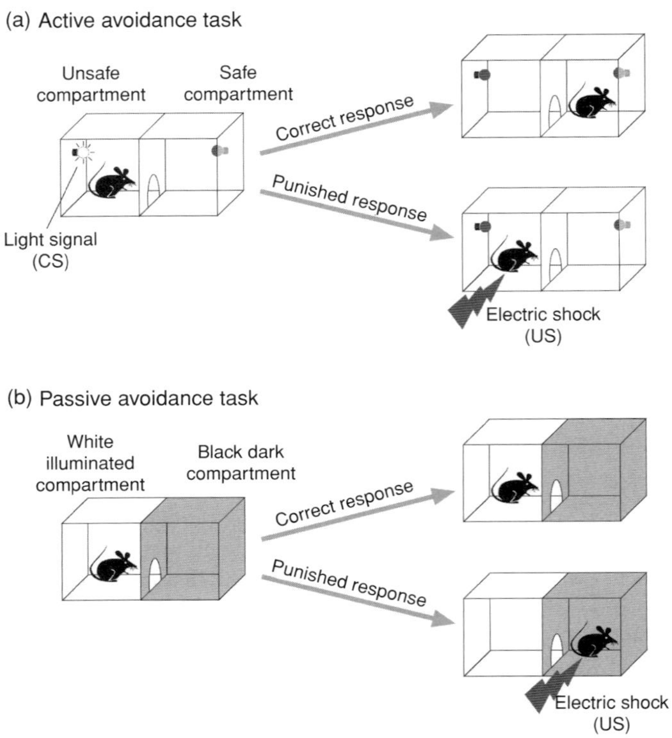

Figure 28.1 (a) Active avoidance learning. The apparatus consists of two compartments separated by a partition with a hole allowing the animal to move from one compartment to the other. The animal is placed in the first compartment and when a conditioned stimulus (CS), usually a light or a sound signal, is presented, the animal must run into the other compartment before the delivery of an electric foot-shock (unconditioned stimulus, US). In the one-way active avoidance procedure, one compartment is always the unsafe one where the CS is presented and the electric shock delivered, while the second compartment, different from the former (e.g., by the absence of grid floor), is always the safe one. In the two-way avoidance procedure, the unsafe and safe compartments are exchanged at each trial and must be identical. Thus, the CS is presented and the electric shock delivered in the compartment that, in the trial immediately before, was safe. (b) Passive avoidance learning: step-through passive avoidance test. One compartment is white, brightly illuminated and the other is black and dark. The animal is placed in the white compartment and learns to withhold the spontaneous response of entering the dark compartment in order to avoid a mild electric foot-shock.

subject to avoid the shock, it must move back to the original compartment.

Developmental changes in active avoidance performance of the mouse have been only partially investigated. Mice can consistently acquire avoidance in a jump-up task by 31–36 days of age (Stavnes and Sprott, 1975), though it is plausible that they may also learn the task earlier, if the protocol is adapted to pup characteristics (Branchi and Ricceri, 2002). In rats, a modification of the one-way active avoidance protocol, rendering the escape response not directional (Misanin et al., 1985), revealed that rats as young as 9 days old acquired foot-shock avoidance by climbing in any direction after a vibrotactile warning signal (Spear and Smith, 1978).

Numerous brain structures are involved in controlling the acquisition and performance of two-way avoidance. These are only briefly discussed here. Acquisition is disrupted by complete lesion of the amygdala (Eclancher and Karli, 1980), while it is facilitated by hippocampal lesions in rats (Jarrard, 1976). Since lesions of the limbic structures are known to produce disinhibition, their facilitating effect constitutes indirect additional evidence that bidirectional avoidance is normally retarded by response impairing mechanisms activated by conflict. In other words, the animal must cope with the need to negotiate between opposite tendencies, that is, leaving and re-entering places where shock has been previously experienced. The same applies to the "paradoxical" facilitating effects of cholinergic blockers of the anti-muscarinic type, which usually produce amnesia in a variety of other learning tests (Bignami, 1964, 1976). Moreover, muscarinic blockers have opposite effects on one-way and two-way avoidance acquisition, the former being impaired and the latter facilitated (Suits and Isaacson, 1968). This is what one would expect if amnesic effects were more than counteracted by response disinhibition effects in the latter task.

In good agreement with these lesion and pharmacological results, bidirectional avoidance performance correlates negatively with the size of the hippocampal intra/infrapyramidal mossy fiber (IIP-MF) projections (Schwegler et al., 1981). This correlation is observed both when changes in the size of the hippocampal IIP-MF projection are the result of genetic differences (Schwegler and Lipp, 1995; Schwegler et al., 1981) and when they are produced by experimental manipulations conducted in early life (Lipp et al., 1988).

Passive avoidance

In passive avoidance procedures, a specified motor sequence spontaneously emitted by the experimental subject – usually, a highly "prepotent" response – is punished and the resulting reluctance to respond is measured in a subsequent test: increased response latencies reflect acquisition of the task. In addition to drug and lesion studies (for reviews see Myhrer, 2003 and Bignami, 1976), passive avoidance procedures have also been extensively used in the analysis of genetically modified mice (e.g., Jeong et al., 2006; Kobayashi and Sano, 2000; Ma et al., 2008; Middei et al., 2004; Minichiello et al., 1999; Shobe, 2002) often as a non-spatial alternative to the water-maze spatial learning test (Minichiello et al., 1999). Two types of passive avoidance tests are used often in laboratory rodents: "step-through" and "step-down."

Step-through passive avoidance

The apparatus consists of a cage divided by a sliding door into two compartments, one white and brightly illuminated, the other one black and dark (Figure 28.1). The animal is placed in the white compartment and must learn to suppress the response of escaping from the lighted to the dark compartment in order to avoid an electric shock. Retention of the response is usually assessed 24 hours later simply by placing the animal in the white compartment and measuring the latency to step through to the dark compartment.

Step-down passive avoidance

The apparatus usually consists of a cage with a central, vibrating platform over a grid floor. Animals are placed on the platform and receive an electric shock if they display the spontaneous response of stepping down onto the grid floor. Acquisition of the task is usually assessed 24 hours later by placing the animal on the platform and measuring the latency to step down. This task differs from the step-through protocol because it requires the experimental subject to withhold responding in any direction, thus it does not require good vision and it is suitable to assess passive avoidance acquisition and retention capabilities in developing pre-weaning rodents.

Passive avoidance tasks have been proven to be a reliable test to assess acquisition and retention capabilities in adult and also in developing rodents. Indeed, the ability to withhold a highly prepotent response to avoid an aversive stimulus is already present in rodents in the second postnatal week (Alleva and D'Udine, 1988; Branchi et al., 2004; Calamandrei et al., 1996a; Ricceri et al., 1997). In the past, young rodents were deemed to be unable to acquire passive avoidance, but later it was found that this deficit was related to an inappropriateness of test features for pup competencies (Spear et al., 1990). Indeed, if the response requirement consists of withholding movement in any direction, as in step-down procedures, and not just in one direction, as in step-through procedures, mice as young as postnatal day 7 and rats as young as postnatal day 10 appear to be able to acquire the response (Nagy et al., 1977; Ray and Nagy, 1978; Stehouwer and Campbell, 1980). Furthermore, when age-related locomotor deficiencies were circumvented by the use of an air-puff-reinforced procedure (neonates move when exposed to the puff), even 2–3-hour-old rats were able to acquire passive avoidance (Myslivecek, 1997). It is important to underline that non-monotonic trends in maturation of acquisition and retention capabilities may occur in altricial rodents, and "matured" responses may disappear at critical (mainly, post-weaning) developmental transitions (Spear, 1979). Therefore, age-dependent characteristics need to be checked accurately before designing an avoidance task (Spear, 1979).

Passive avoidance tasks are quite sensitive to several pharmacological treatments. Agents affecting cholinergic, glutamatergic, GABAergic, and also serotoninergic neurotransmission have been widely studied (Bammer, 1982; Bignami, 1976; Bignami and Michalek, 1978; Myhrer, 2003). In particular, *N*-methyl-D-aspartate (NMDA) antagonists administered before or immediately after passive avoidance training impair retention in mice, indicating that the glutamatergic system plays a role in acquisition and performance of a passive avoidance task (Adriani et al., 1998; Myhrer, 2003; Venable and Kelly, 1990). Moreover, a crucial role for central cholinergic systems is strongly suggested by many pharmacological experiments showing impairments in passive avoidance acquisition and retention after treatments with muscarinic and nicotinic antagonists (for a review see Myhrer, 2003; Wilson and Cook, 1994). More recently, studies using a lesion of both medial septum and nucleus basalis magnocellularis have shown that the activity of basal forebrain cholinergic nuclei markedly affects passive avoidance performance (Leanza et al., 1995; for a review see Everitt and Robbins, 1997).

Pharmacological effects have also been studied in developing rodents. These studies showed that, in 2–3-week-old mice, passive avoidance performance is extremely sensitive to agents affecting central cholinergic function, such as nerve growth factor, anti-nerve growth factor, or choline (Calamandrei et al., 1996b; Ricceri and Berger-Sweeney, 1998; Ricceri et al., 1994). Furthermore, selective neonatal basal forebrain cholinergic lesions significantly impair passive avoidance acquisition by 15-day-old rat pups (Ricceri et al., 1997), confirming previous pharmacological ontogenetic evidence (Dumery et al., 1988; Ray and Nagy, 1978).

Finally, it should be emphasized that the differences in the mechanisms responsible for acquisition and retention of active and passive avoidance are further supported by the different, or even opposite, effects of drugs and lesion treatments in the two types of tests. For example, hippocampal lesions improve two-way active avoidance (see above) while impairing passive avoidance (Barros et al., 2001; Martinez et al., 2002). Similar dissociation between passive and one-way active avoidance is obtained after pharmacological manipulations of other brain regions (Prado-Alcala et al., 1980) or in mutant mice (Minichiello et al., 1999).

Species and strain differences

Comparison of learning performance of several rodent species in avoidance tasks has highlighted some of the mechanisms underlying behavior regulation. A simple example is the strong locomotory tendency of gerbils that translates into poor performance in passive avoidance and an efficient performance in two-way active avoidance tasks as compared with that of rats (Ashe and McCain, 1972; Galvani et al., 1975; Walters and Abel, 1971). With regard to differences between mice and rats, the smaller body size of mice may allow a faster initial acceleration upon exposure to shock in two-way active avoidance. This may explain the relative insensitivity of mouse bidirectional avoidance to the impairing effects of strong punishment compared to rats (Bignami et al., 1985; Levine, 1966). Conversely, mice and rats respond similarly to changes in the length of intertrial interval and also show similar effects of stimulus modality (Bignami et al., 1985).

Important differences in avoidance learning have been reported among mouse strains. These differences are due to random fixation of alleles. The following strains differ in avoidance performance with the first being best and last the worst: DBA, BALB/c, NMRI, C57BL/6, SM/J, ICR, and C3H.

The first paper illustrating differences in behavioral responses among mouse strains was published in 1969 by Daniel Bovet and colleagues (Bovet et al., 1969). In this seminal work, the authors provided the first description of the learning and memory performances in active avoidance of nine different strains, showing that the performances vary notably because of the genetic background, though further evidence pointed out that the genes are not only the cause of the behavioral differences described.

Subsequent studies ascribed the mouse strain differences in active avoidance behavior to a single gene (Buselmaier et al., 1978; Oliverio and Messeri, 1973; Simmel and Eleftheriou, 1977). By contrast, others studies supported a polygenic control system (Oliverio et al., 1972; Royce et al., 1971). In the 1980s, Dudley F. Peeler performed a number of experiments to definitively examine whether a single or multiple genes control avoidance behavior (Peeler, 1987). In particular, the performance of adult male mice of the inbred progenitor strains C57BL/6ByJ and BALB/cByJ, known to have divergent learning abilities, and their seven recombinant inbred strains were tested for active avoidance learning. The results clearly showed that the single gene hypothesis had to be rejected since only a polygenic control of avoidance learning was compatible with the behavioral data (Peeler, 1987). Following works, exploring whether the genetic contribution varied according to specific environmental constrains and determinants, confirmed that more than one gene control this behavioral response (Peeler, 1995). However, the latter work clearly pointed out that the active avoidance performance, and thus the mouse strain differences, are highly dependent on the environmental features of the context in which the test is performed.

When inheritance of avoidance behavior was analyzed in the two sexes in different inbred strains and F1 and F2 generations of their crossings, some differences emerged in inheritance mode: in males inheritance was transmitted predominantly as an overdominant trait, whereas in females both overdominance and partial dominance was observed (Holmes et al., 1974).

Since in an artificial selection for shuttle avoidance in a random-bred population of wild *Mus musculus*, heritability estimates were markedly lower than those extrapolated for random-bred domestic mice and crosses of inbred mice (0.23 vs. 0.45/0.48 respectively), some caveats should always be considered before generalizing heritability estimates to the mouse species or speculating about the adaptive value of avoidance in unspecified feral mouse populations (Spear and Smith, 1978).

As for neuroanatomy, strain differences in the sizes of the hippocampal IIP-MF projections (from granule cells terminating upon the dendrites of hippocampal pyramidal neurons) correlate with avoidance performances: the more IIP-MF, the poorer the learning (Schwegler and Lipp, 1983). The analysis of hippocampal traits functionally relevant for avoidance behavior in rodents benefited from a series of studies reviewed by Lipp et al. (1989). Less invasively than in classical lesion studies, the relationship between two-way avoidance performance and IIP-MF size has been evaluated in a series of experiments, ranging from individual and genetic correlational studies, to Mendelian crossings of mouse strains with extreme IIP-MF traits, and to IIP-MF size variations induced by developmental pharmacological manipulations. As a whole, this experimental strategy verified that the IIP-MF projection is indeed relevant for the behavioral differences found in two-way avoidance performance in rodents; variations in the mossy fibers, however, may not be directly related to "genuine learning (associative and/or cognitive) capacity," they may rather correlate with behavioral predictability, spatial processing, or capacity of coping with conflict or difficult tasks (Lipp et al., 1989). Interestingly the same authors concluded with a statement that also maintains intact its value nowadays: "contemporary brain research has amassed an impressive battery of analytical techniques. There is no corresponding sophistication in formulating experimental questions for employing those tools for the analysis of brain and behaviour."

Strain differences in avoidance learning are also influenced by differences in basal activity levels. C57BL/6 mice, which notably display high activity levels and a high number of shuttle-box crossings before CS onset, also show an impairment in the extinction of non-reinforced responses (Bovet et al., 1969; Cabib et al., 2002). The relationship between activity levels and avoidance capability is not a simple one, as suggested by the fact that in a passive avoidance task C57BL/6 mice can be as good or even better than DBA mice, depending on age and shock intensity (Sprott, 1972). However, C57BL/6 mice have been also reported to show no passive avoidance retention, with US ranging from 0.2 to 0.4 mA (Mathis et al., 1994). Differences in emotional reactivity (and also sensitivity to anxiolitic compounds) are well documented across inbred and outbred mouse strains (Griebel et al., 2000) and could certainly account for some of the differences in passive avoidance step-through paradigms (Mathis et al., 1994) in which the emotional reactivity to dark and light environments plays a substantial role. However, the use of complete test batteries (including emotionality, learning and memory, and activity) can provide useful information concerning emotionality and activity roles on avoidance performance (Brown et al., 2000; Crawley and Paylor, 1997). As an example, applying a battery of tests and subsequent covariate analyses, differences between C57BL/6 and DBA/2 mice in passive avoidance learning have been shown not to depend on well evident differences in emotionality and activity between the two strains (Podhorna and Brown, 2002). Finally, avoidance responding in various strains differs also according to the nature of the aversive stimulus used. In the two-way active avoidance task for example, an air pulse has been shown to be as effective as an electric shock to train C57BL/6, but not 129 mice (Clark et al., 2003).

A different approach to investigate the genetic bases of avoidance learning exploited mouse lines artificially selected for a given behavioral response. This allowed assessment of whether different behavioral responses were under control of the same genes. In particular, two mouse lines selected for aggressiveness – the short attack latency (SAL) and long attack latency (LAL) mice – have been tested in the active avoidance. The results showed that latency to attack and avoidance behavior are not associated, excluding a common genetic control. In particular, while all SAL mice, which show a proactive coping style, adopted an active strategy and showed high learning scores, not all LAL mice, which show a reactive coping style, adopted a passive strategy showing low learning scores, but some displayed an avoidance behavior similar to SAL mice (Benus et al., 1990).

Avoidance learning in an ethological perspective

In order to reliably assess behavior, experimental protocols should be considered from an ethological perspective. Animals have a species-specific behavioral repertoire that includes defensive behaviors, such as freezing, fleeing, and fighting to mention but a few types of responses. As a consequence, some specific avoidance responses are rapidly acquired because they are based on behaviors belonging to the animal species-specific repertoire (Bolles, 1970); by contrast, other responses may be difficult or even impossible to be acquired (an extreme case is the failure to learn to suppress tongue darting towards a fly by frogs and toads when the fly is impaled on a pin which injures the animal's tongue). In the case of avoidance acquisition, in order to avoid the US, the animal is favored when the response required is part of the species-specific defensive repertoire, including fleeing (favoring active avoidance) and freezing (favoring passive avoidance). The stimuli paired to punishing events employed in the avoidance procedures represent a learned threat that is aversive through prior association with pain or some other punishing event. Other stimuli having an eco-ethological relevance in the natural habitat for the species under investigation may allow further investigation of avoidance acquisition (Capone et al., 2002; Kamil and Mauldin, 1988). In fact, it has been shown that in classical (Pavlovian) defensive conditioning procedures stimuli of some modalities can be promptly paired, whereas stimuli of other modalities need a much more extended training. For example, rats can learn in a single trial to associate food with a nausea-inducing substance, but require many trials to avoid food associated with an electric foot-shock. The relevance of this propensity ("preparedness") appears evident when considering the possible adaptive value of this learning response, since it allows rats to avoid eating again food that previously made them sick (Garcia and Koelling, 1966).

When animals are tested in controlled conditions, as in the laboratory, arguments such as those just mentioned may appear to be merely theoretical, of little or no practical relevance. The ethological perspective, however, considerably increases data quality in passive avoidance studies (Gerlai and Clayton, 1999). Specifically, it can prevent undesirable interferences or even costly failures, due for example to animals' reactions to apparently irrelevant stimuli and/or neglect of stimuli deemed by the experimenter to be appropriate for solving the task.

More subtly, the learning performance may be affected by the fact that the mouse perceives differences in the location of the experimenter, who is usually misidentified as a predator by the experimental subject. Specifically, a location closer to the lighted or the dark compartment would lead, respectively, to a shorter or a longer latency to step-through. As a second example, data collected ignoring the behavioral ecology of the species may be biased by environmental variation, leading to less effective detection of behavioral differences among experimental groups. For instance, odors present in the experimental room may affect the passive avoidance performance. An estrous female odor may enhance the animal drive to explore the environment, decreasing the latency to step-through, while the odor of a rat, a potential predator, may trigger freezing response, increasing this latency.

Conclusions

In conclusion, the evidence analyzed here shows that avoidance tasks are widely used in experimental analyses of the behavior and brain function of the house mouse. Avoidance paradigms have revealed a range of alterations due to naturally-occurring genetic differences, induced mutations, lesions and pharmacological manipulations. These tests are used less often than other paradigms including the popular Morris water maze, but have turned out to provide additional results complementary to those obtained with commoner paradigms. Thus they may allow one to obtain a more complete picture of cognitive function/dysfunction in the mouse model. Finally, testing animals for both passive and active avoidance has been proven to be highly informative because it allows one to control for performance factors such as perception and motor function.

Acknowledgments

We wish to gratefully acknowledge the many useful comments on this chapter by Giorgio Bignami, former head of our Section, now retired, based on his extensive experience with avoidance methods. His passionate and rigorous approach to research on animal behavior has been for us an invaluable example. Supported by ISS-NIH 5301/530H "Neurobehavioral phenotyping of genetically-modified mouse models of mental retardation: from gene alteration to cognitive and social impairment" to L. R.

References

Adriani, W., Felici, A., Sargolini, F., Roullet, P., Usiello, A., Oliverio, A., et al. (1998) *N*-methyl-D-aspartate and dopamine receptor involvement in the modulation of locomotor activity and memory processes. *Exp Brain Res* 123: 52–59.

Alleva, E. and D'Udine, B. (1988) Early learning capability in rodents: a review. *Int J CompPsychol* Winter Issue: 107–125.

Ashe, V.M. and McCain, G. (1972) Comparison of one-way and shuttle-avoidance performance of gerbils and rats. *J Comp Physiol Psychol* 80: 293–296.

Bammer, G. (1982) Pharmacological investigations of neurotransmitter involvement in passive avoidance responding: a review and some new results. *Neurosci Biobehav Rev* 6: 247–296.

Barros, D.M., Mello, E., Souza, T., de Souza, M.M., Choi, H., DeDavid, E, et al. (2001) LY294002, an inhibitor of phosphoinositide 3-kinase given into rat hippocampus impairs acquisition, consolidation and retrieval of memory for one-trial step-down inhibitory avoidance. *Behav Pharmacol* 12: 629–634.

Benus, R.F., Bohus, B., Koolhaas, J.M., and van Oortmerssen, G.A. (1990) Behavioural strategies of aggressive and non-aggressive male mice in response to inescapable shock. *Behav Process* 21: 127–141.

Bignami, G. (1964) Effects of benactyzine and adiphenine on instrumental avoidance conditioning in a shuttle-box. *Psychopharmacologia* 5: 264–279.

Bignami, G. (1976) Nonassociative explanations of behavioral changes induced by central cholinergic drugs. *Acta Neurobiol Exp (Wars)* 36: 5–90.

Bignami, G., Alleva, E., Amorico, L., De Acetis, L., and Giardini, V. (1985) Bidirectional avoidance by mice as a function of CS, US, and apparatus variables. *Anim Learn Behav* 13: 439–450.

Bignami, G. and Michalek, H. (1978) Cholinergic mechanisms and aversively motivated behaviors. In Anisman, H. and Bignami, G. (eds.), *Psychopharmacology of Aversively Motivated Behavior*. Plenum Press, New York, pp. 173–255.

Bolles, R.C. (1970) Species-specific defense reactions and avoidance learning. *Psychol Rev* 71: 32–48.

Bovet, D., Bovet-Nitti, F., and Oliverio, A. (1969) Genetic aspects of learning and memory in mice. *Science* 163: 139–149.

Brambilla, R., Gnesutta, N., Minichiello, L., White, G., Roylance, A.J., Herron, C.E., et al. (1997) A role for the Ras signalling pathway in synaptic transmission and long-term memory. *Nature* 390: 281–286.

Branchi, I., Bichler, Z., Minghetti, L., Delabar, J.M., Malchiodi-Albedi, F., Gonzalez, M.C., et al. (2004) Transgenic mouse *in vivo* library of human Down syndrome critical region 1: association between DYRK1A overexpression, brain development abnormalities, and cell cycle protein alteration. *J Neuropathol Exp Neurol* 63: 429–440.

Branchi, I. and Ricceri, L. (2002) Transgenic and knock-out mouse pups: the growing need for behavioral analysis. *Genes Brain Behav* 1: 135–141.

Brown, R.E., Stanford, L., and Schellinnck, H.M. (2000) Developing standardized behavioral tests for knockout and mutant mice. *ILAR J* 41: 163–173.

Buselmaier, W., Geiger, S., and Reichert, W. (1978) Monogene inheritance of learning speed in DBA and C3H mice. A behavioral genetic study in the shuttle-box. *Hum Genet* 40: 209–214.

Cabib, S., Puglisi-Allegra, S., and Ventura, R. (2002) The contribution of comparative studies in inbred strains of mice to the understanding of the hyperactive phenotype. *Behav Brain Res* 130: 103–109.

Calamandrei, G., Pennazza, S., Ricceri, L., and Valanzano, A. (1996a) Neonatal exposure to anti-nerve growth factor antibodies affects exploratory behavior of developing mice in the hole board. *Neurotoxicol Teratol* 18: 141–146.

Calamandrei, G., Ricceri, L., and Valanzano, A. (1996b) Systemic administration of anti-NGF antibodies to neonatal mice impairs 24-h retention of an inhibitory avoidance task while increasing ChAT immunoreactivity in the medial septum. *Behav Brain Res* 78: 81–91.

Capone, F., Puopolo, M., Branchi, I., and Alleva, E. (2002) A new easy accessible and low-cost method for screening olfactory sensitivity in mice: behavioural and nociceptive response in male and female CD-1 mice upon exposure to millipede aversive odour. *Brain Res Bull* 58: 193–202.

Clark, M.G., Vasilevsky, S., and Myers, T.M. (2003) Air and shock two-way shuttlebox avoidance in C57BL/6J and 129X1/SvJ mice. *Physiol Behav* 78: 117–123.

Crawley, J.N. and Paylor, R. (1997) A proposed test battery and constellations of specific behavioral paradigms to investigate the behavioral phenotypes of transgenic and knockout mice. *Horm Behav* 31: 197–211.

Dumery, V., Derer, P., and Blozovski, D. (1988) Enhancement of passive avoidance learning through small doses of intra-amygdaloid physostigmine in the young rat. Its relation to the development of acetylcholinesterase. *Dev Psychobiol* 21: 553–565.

Eclancher, F. and Karli, P. (1980) Effects of infant and adult amygdaloid lesions upon acquisition of two-way active avoidance by the adult rat: influence of rearing conditions. *Physiol Behav* 24: 887–893.

Everitt, B.J. and Robbins, T.W. (1997) Central cholinergic systems and cognition. *Annu Rev Psychol* 48: 649–684.

Galvani, P.F., Riddell, W.I., and Foster, K.M. (1975) Passive avoidance in rats and gerbils as a function of species-specific exploratory tendencies. *Behav Biol* 13: 277–290.

Garcia, J. and Koelling, R.A. (1966) Relation of cue to consequence in avoidance learning. *Psychonomic Sci* 4: 123–124.

Gerlai, R. and Clayton, N.S. (1999) Analysing hippocampal function in transgenic mice: an ethological perspective. *Trends Neurosci* 22: 47–51.

Griebel, G., Belzung, C., Perrault, G., and Sanger, D.J. (2000) Differences in anxiety-related behaviours and in sensitivity to diazepam in inbred and outbred strains of mice. *Psychopharmacology (Berl)* 148: 164–170.

Holmes, T.M., Aksel, R., and Royce, J.R. (1974) Inheritance of avoidance behavior in *Mus musculus*. *Behav Genet* 4: 357–371.

Huang, Y.Y., Bach, M.E., Lipp, H.P., Zhuo, M., Wolfer, D.P., Hawkins, R.D., et al. (1996) Mice lacking the gene encoding tissue-type plasminogen activator show a selective interference with late-phase long-term potentiation in both Schaffer collateral and mossy fiber pathways. *Proc Natl Acad Sci USA* 93: 8699–8704.

Jarrard, L.E. (1976) Anatomical and behavioral analysis of hippocampal cell fields in rats. *J Comp Physiol Psychol* 90: 1035–1050.

Jeong, Y.H., Park, C.H., Yoo, J., Shin, K.Y., Ahn, S.M., Kim, H.S., et al. (2006) Chronic stress accelerates learning and memory impairments and increases amyloid deposition in APPV717I-CT100 transgenic mice, an Alzheimer's disease model. *FASEB J* 20: 729–731.

Kamil, A.C. and Mauldin, J.E. (1988) A comparative-ecological approach to the study of learning. In Bolles, R.C. and Beecher, M.D. (eds.), *Evolution and Learning*. Lawrence Erlbaum Associates, Hillsdale, NJ, USA, pp. 117–133.

King, S.L., Marks, M.J., Grady, S.R., Caldarone, B.J., Koren, A.O., Mukhin, A.G., et al. (2003) Conditional expression in corticothalamic efferents reveals a developmental role for nicotinic acetylcholine receptors in modulation of passive avoidance behavior. *J Neurosci* 23: 3837–3843.

Kobayashi, K. and Sano, H. (2000) Dopamine deficiency in mice. *Brain Dev* 22 (Suppl 1): S54–60.

Leanza, G., Nilsson, O.G., Wiley, R.G., and Bjorklund, A. (1995) Selective lesioning of the basal forebrain cholinergic system by intraventricular 192 IgG-saporin: behavioural, biochemical and stereological studies in the rat. *Eur J Neurosci* 7: 329–343.

Levine, S. (1966) UCS intensity and avoidance learning. *J Exp Psychol* 71: 163–164.

Lipp, H.P., Schwegler, H., Crusio, W.E., Wolfer, D.P., Heimrich, B., Driscoll, P., et al. (1989) Using genetically-defined rodent strains for the identification of hippocampal traits relevant for two-way avoidance learning: a non-invasive approach. *Experientia* 45: 845–859.

Lipp, H.P., Schwegler, H., Heimrich, B., and Driscoll, P. (1988) Infrapyramidal mossy fibers and two-way avoidance learning: developmental modification of hippocampal circuitry and adult behavior of rats and mice. *J Neurosci* 8: 1905–1921.

Ma, X.M., Kiraly, D.D., Gaier, E.D., Wang, Y., Kim, E.J., Levine, E.S., et al. (2008) Kalirin-7 is required for synaptic structure and function. *J Neurosci* 28: 12368–12382.

Martin, M., Ledent, C., Parmentier, M., Maldonado, R., and Valverde, O. (2002) Involvement of CB1 cannabinoid receptors in emotional behaviour. *Psychopharmacology (Berl)* 159: 379–387.

Martinez, I., Quirarte, G.L., Diaz-Cintra, S., Quiroz, C., and Prado-Alcala, R.A. (2002) Effects of lesions of hippocampal fields CA1 and CA3 on acquisition of inhibitory avoidance. *Neuropsychobiology* 46: 97–103.

Mathis, C., Paul, S.M., and Crawley, J.N. (1994) Characterization of benzodiazepine-sensitive behaviors in the A/J and C57BL/6J inbred strains of mice. *Behav Genet* 24: 171–180.

Middei, S., Geracitano, R., Caprioli, A., Mercuri, N., and Ammassari-Teule, M. (2004) Preserved fronto-striatal plasticity and enhanced procedural learning in a transgenic mouse model of Alzheimer's disease overexpressing mutant hAPPswe. *Learn Mem* 11: 447–452.

Minichiello, L., Korte, M., Wolfer, D., Kuhn, R., Unsicker, K., Cestari, V., et al. (1999) Essential role for TrkB receptors in hippocampus-mediated learning. *Neuron* 24: 401–414.

Misanin, J.R., Turns, L.E., and Hinderliter, C.F. (1985) Acquisition and retention of active avoidance behavior in previsual rats. *Am J Psychol* 98: 485–501.

Montag-Sallaz, M. and Montag, D. (2003) Severe cognitive and motor coordination deficits in tenascin-R-deficient mice. *Genes Brain Behav* 2: 20–31.

Myhrer, T. (2003) Neurotransmitter systems involved in learning and memory in the rat: a meta-analysis based on studies of four behavioral tasks. *Brain Res Brain Res Rev* 41: 268–287.

Myslivecek, J. (1997) Inhibitory learning and memory in newborn rats. *Prog Neurobiol* 53: 399–430.

Nagy, Z.M., Thaller, K., and Mazzaferri, T.A. (1977) Acquisition and retention of a passive-avoidance task as a function of age in mice. *Dev Psychobiol* 10: 563–573.

Oliverio, A., Castellano, C., and Messeri, P. (1972) Genetic analysis of avoidance, maze, and wheel-running behaviors in the mouse. *J Comp Physiol Psychol* 79: 459–473.

Oliverio, A. and Messeri, P. (1973) An analysis of single-gene effects on avoidance, maze, wheel running, and exploratory behavior in the mouse. *Behav Biol* 8: 771–783.

Peeler, D.F. (1987) Active avoidance performance in genetically defined mice. *Behav Neural Biol* 48: 83–89.

Peeler, D.F. (1995) Shuttlebox performance in BALB/cByJ, C57BL/6ByJ, and CXB recombinant inbred mice – environmental and genetic-determinants and constraints. *Psychobiology* 23: 161–170.

Podhorna, J. and Brown, R.E. (2002) Strain differences in activity and emotionality do not account for differences in learning and memory performance between C57BL/6 and DBA/2 mice. *Genes Brain Behav* 1: 96–110.

Prado-Alcala, R.A., Cruz-Morales, S.E., and Lopez-Miro, F.A. (1980) Differential effects of cholinergic blockade of anterior and posterior caudate nucleus on avoidance behaviors. *Neurosci Lett* 18: 339–345.

Ray, D. and Nagy, Z.M. (1978) Emerging cholinergic mechanisms and ontogeny of response inhibition in the mouse. *J Comp Physiol Psychol* 92: 335–349.

Ricceri, L. (2003) Behavioral patterns under cholinergic control during development: lessons learned from the selective immunotoxin 192 IgG saporin. *Neurosci Biobehav Rev* 27: 377–384.

Ricceri, L., Alleva, E., and Calamandrei, G. (1994) Impairment of passive avoidance learning following repeated administrations of antibodies against nerve growth factor in neonatal mice. *Neuroreport* 5: 1401–1404.

Ricceri, L. and Berger-Sweeney, J. (1998) Postnatal choline supplementation in preweanling mice: sexually dimorphic behavioral and neurochemical effects. *Behav Neurosci* 112: 1387–1392.

Ricceri, L., Calamandrei, G., and Berger-Sweeney, J. (1997) Different effects of postnatal day 1 versus 7 192 immunoglobulin G-saporin lesions on learning, exploratory behaviors, and neurochemistry in juvenile rats. *Behav Neurosci* 111: 1292–1302.

Royce, J.R., Yeudall, L.T., and Poley, W. (1971) Diallel analysis of avoidance conditioning in inbred strains of mice. *J Comp Physiol Psychol* 76: 353–358.

Sarter, M., Hagan, J., and Dudchenko, P. (1992) Behavioral screening for cognition enhancers: from indiscriminate to valid testing: part II. *Psychopharmacology (Berl)* 107: 461–473.

Schwegler, H. and Lipp, H.P. (1983) Hereditary covariations of neuronal circuitry and behavior: correlations between the proportions of hippocampal synaptic fields in the regio inferior and two-way avoidance in mice and rats. *Behav Brain Res* 7: 1–38.

Schwegler, H. and Lipp, H.P. (1995) Variations in the morphology of the septo-hippocampal complex and maze learning in rodents: correlation between morphology and behaviour. In Alleva, E., Fasolo, A., Lipp, H.P., Nadel, L. and Ricceri, L. (eds.), *Behavioural Brain Research in Naturalistic and Semi-Naturalistic Settings*, Vol. 82. Kluwer, Dodrecht, The Netherlands, pp. 259–276.

Schwegler, H., Lipp, H.P., Van der Loos, H., and Buselmaier, W. (1981) Individual hippocampal mossy fiber distribution in mice correlates with two-way avoidance performance. *Science* 214: 817–819.

Shobe, J. (2002) The role of PKA, CaMKII, and PKC in avoidance conditioning: permissive or instructive? *Neurobiol Learn Mem* 77: 291–312.

Simmel, E.C. and Eleftheriou, B.E. (1977) Multivariate and behavior genetic analysis of avoidance of complex visual stimuli and activity in recombinant inbred strains of mice. *Behav Genet* 7: 239–250.

Spear, L.P. (1979) The use of psychopharmacological procedures to analyze the ontogeny of learning and retention: issues and concerns. In Spear, N.E. and Campbell, B.A. (eds.), *Ontogeny of Learning and Memory*. Lawrence Erlbaum Associates, Hillsdale, NJ, USA, pp. 135–156.

Spear, N.E., Miller, J.S., and Jagielo, J.A. (1990) Animal memory and learning. *Annu Rev Psychol* 41: 169–211.

Spear, N.E. and Smith, G.J. (1978) Alleviation of forgetting in preweanling rats. *Dev Psychobiol* 11: 513–529.

Sprott, R.L. (1972) Passive-avoidance conditioning in inbred mice: effects of shock intensity, age, and genotype. *J Comp Physiol Psychol* 80: 327–334.

Stavnes, K. and Sprott, R.L. (1975) Effects of age and genotype on acquisition of an active avoidance response in mice. *Dev Psychobiol* 8: 437–445.

Stehouwer, D.J. and Campbell, B.A. (1980) Ontogeny of passive avoidance: role of task demands and development of species-typical behaviors. *Dev Psychobiol* 13: 385–398.

Suits, E. and Isaacson, R.L. (1968) The effects of scopolamine hydrobromide on one-way and two-way avoidance learning in rats. *Int J Neuropharmacol* 7: 441–446.

Tzavara, E.T., Bymaster, F.P., Felder, C.C., Wade, M., Gomeza, J., Wess, J., et al. (2003) Dysregulated hippocampal acetylcholine neurotransmission and impaired cognition in M2, M4 and M2/M4 muscarinic receptor knockout mice. *Mol Psychiatry* 8: 673–679.

Venable, N. and Kelly, P.H. (1990) Effects of NMDA receptor antagonists on passive avoidance learning and retrieval in rats and mice. *Psychopharmacology (Berl)* 100: 215–221.

Walters, G.C. and Abel, E.L. (1971) Passive avoidance learning in rats, mice, gerbils and hamsters. *Psychonomic Sci* 22: 269–270.

Wilson, W.J. and Cook, J.A. (1994) Cholinergic manipulations and passive avoidance in the rat: effects on acquisition and recall. *Acta Neurobiol Exp (Wars)* 54: 377–391.

Chapter 29

Radial maze

Wim E. Crusio and Herbert Schwegler

Origins of the radial maze

The radial maze was first proposed some 30 years ago by Olton and Samuelson (1976) to study spatial working memory in rats. Many different apparatuses have been used over the years, but the general principle is the same: a radial maze consists of a central platform from which a number of arms radiate outwards. The number of arms can vary from as few as four or five to up to 16. For mice, the most frequently used apparatus has eight arms (this is the number of arms used in all experiments discussed below, unless mentioned otherwise). At the end of each arm a reward may be present. In the case of dry mazes this may be a food or liquid reward, but an aquatic version of the task has also been developed (Hyde et al., 1998) in which the reward consists of an opportunity to escape the water. Most (but certainly not all) dry mazes are elevated above the floor of the experimental room, but obviously water mazes have to be placed on the floor or a sturdy table. Olton's rationale for his new task was stated as follows:

> The present experiment was designed to permit place learning and to utilize this learning to assess the capacity of rats to discriminate, remember, and process information derived from place learning in searching for food . . . a new experimental testing paradigm, one which requires sampling with replacement from a known set of items until the entire set is sampled (Olton and Samuelson, 1976, p. 97).

In the years that followed, the radial maze played an important role in the development of theories on hippocampal functions. Olton and colleagues (Olton et al., 1979) proposed, based mainly on their hippocampal lesion studies in rats (Olton, 1977; Olton and Papas, 1979; Olton and Schlosberg, 1978; Olton et al., 1980), that the hippocampus was predominantly involved in working memory (WM), a concept formulated by Honig (1978). Working memory was defined as "a short-term memory that requires flexible stimulus–response associations and is highly susceptible to interference" (Olton et al., 1979, p. 313) and it was proposed to process information that "is useful for one trial of an experiment, but not for subsequent trials" (Olton et al., 1979, p. 314). Operationally, WM errors were defined as repeated entries into an arm where the animal, during the same trial, had already eaten the reward. It was proposed that the hippocampus was involved in the processing of WM information, but not of its counterpart, reference memory (RM). The latter was defined as processing information that is useful for many trials and usually for the entire experiment.

The main competing theory on hippocampal functioning at the time was O'Keefe and Nadel's cognitive mapping theory (O'Keefe and Nadel, 1978, 1979), which proposed that the hippocampus is the locus of a mental representation of the space around an animal. Nadel and MacDonald used a lesion study in the radial maze to provide support for the cognitive mapping over the WM theory (Nadel and MacDonald, 1980). They used a modified radial-maze task to test both WM and RM in the same procedure by designating some arms as the "reward arms" and some as the "non-reward arms." Entries into non-reward arms would be RM errors, whereas repeated entries into rewarded arms were counted as WM errors. By using a spatial and a cued version of this task, they showed that hippocampally lesioned rats performed well on both WM and RM aspects of the cued, but not the spatial task, thereby providing supporting evidence for the cognitive mapping theory of hippocampal function.

Can mice learn a radial maze task?

Initially it was believed that radial maze tasks would be too difficult for mice and, indeed, Mizumori et al. (1982) reported a failure of male CD-1 mice to demonstrate spatial memory in this maze. Pretty soon thereafter, however, the first reports of successful radial-maze learning in mice surfaced (Reinstein et al., 1983), several of them also using male CD-1 mice (Levy et al., 1983; Pico and Davis, 1984), and since then it has become clear that the abilities of most mice to learn this task are at least equal to those of rats (Whishaw and Tomie, 1996). It is difficult to explain why the mice used by Mizumori et al. (1982) did not succeed in learning the eight-arm maze task. Although elevated, the maze was placed on a round Masonite board and height-induced anxiety might not have been a confounding factor in this experiment (Crusio, 1999). To facilitate learning, specific black–white patterns were provided as visual cues at the end of each arm, so the test used was as simple as possible. Given

Behavioral Genetics of the Mouse: Volume I. *Genetics of Behavioral Phenotypes*, eds. Wim E. Crusio, Frans Sluyter, Robert T. Gerlai, and Susanna Pietropaolo. Published by Cambridge University Press.

the existence of large strain differences in radial-maze learning ability (see below), one possibility is that the CD-1 mice used by Mizumori et al. (1982) carried more alleles unfavorable to good learning performance than those used by Pico and Davis (1984) or Levy et al. (1983), as the CD-1 line is a non-inbred and therefore non-standardized “strain” of mice, whose genetic composition may differ from vendor to vendor (and also between different locations or even breeding rooms from the same vendor: it should be noted that Mizumori et al. (1982) and Levy et al. (1983) obtained their animals from the same vendor). Another possible explanation derives from the fact that Mizumori et al. compared the performance of their mice to that of juvenile rats using the same apparatus. If these tests took place concurrently, then the presence of rats could have interfered with acquisition of the task in mice (Grootendorst et al., 2001).

Strain differences

Reinstein et al. (1983) were not only the first to successfully demonstrate significant radial-maze learning in mice (of undetermined sex), but also the first who reported that different inbred strains show significantly different learning abilities. These authors used three different inbred strains: C57BR/cdJ, BALB/cJ, and C57BL/6J, in decreasing order of learning ability. They related the strain differences found to differences in hippocampal choline acetyltransferase activity. Fimbria/fornix lesions disrupted learning in the first two strains, whereas even control C57BL/6J animals never performed above chance levels. Fairly soon thereafter, though, the first reports of good radial-maze learning ability in this strain surfaced (Ammassari-Teule and Caprioli, 1985; Bernstein et al., 1985). Although this is not mentioned in the article, the maze Reinstein et al. (1983) used was most likely elevated above the floor. In addition, the arms of the maze were rather short (19 × 6 cm). After an animal returned to the central platform, access to all arms was blocked by clear plastic doors and it was confined to the central platform for 5 seconds before it could make another choice. In contrast, Bernstein et al. (1985) used a rather huge elevated maze, with arms that were 667 cm long, whereas Ammassari-Teule and Caprioli (1985) used a moderately sized elevated maze (arms 35 cm long) and were the only ones to use a six-arm configuration and cardboard walls to minimize extra-maze cues. As is so often the case for behavioral tests, training procedures were also very different between the three studies: Reinstein et al. (1983) applied 1 trial per day (5 min maximum) in three 5-day blocks separated by 2-day recesses, with animals food-deprived to 85% of free-feeding body weight and using small food pellets as reward. The procedure used by Ammassari-Teule and Caprioli (1985) was similar, except that they gave 10 daily trials. Finally, Bernstein et al. (1985) applied three trials (10 min maximum) per day over a total of 36 days, with animals water-deprived during the course of the experiment with the only source of liquid being the rewards obtained.

Our group was the next to report on strain differences in radial maze learning (Crusio et al., 1987). The maze used had opaque walls and was placed on the floor. Arms were longer than in the previous study (25 × 6 × 6 cm) and no confinement procedure was used between subsequent arm choices. One trial was administered per day (maximum 15 min on first day, no time limit after that) on 3 subsequent days. In all experiments from our group, animals were food-deprived to 85% of their free-feeding body weight and small (10–20 mg) pieces of food pellets were used as rewards. In this first experiment we surveyed (to avoid the modern buzz word “phenotyped”) male mice from eight different inbred strains; ranked from good to bad learners these were: C3H/HeJ > BALB/cJ > C57BL/6J = C57BR/cdJ > BA > DBA/2J = NZB/B1NJ > CPB/K. Performance showed an almost perfect correlation with the extent of the hippocampal intra- and infrapyramidal mossy fibers. As correlations between these fibers and radial-maze learning have recently been treated exhaustively elsewhere (Crusio and Schwegler, 2005), they will not be further discussed here.

However, it was noted in the above-described experiment that better learners tended to use a simple kinesthetic strategy, by visiting arms adjacent to the previous ones in a clockwise or counter-clockwise manner. Therefore, it was not clear whether the animals had used a spatial, allocentric searching strategy or a non-spatial, egocentric one, and two follow-up experiments were carried out (Schwegler et al., 1990). In both cases, a confinement procedure was used, as Bolhuis et al. (1986) had shown in rats that such a procedure interrupts a kinesthetic strategy. In the spatial test, clear Plexiglas® arms were used, in combination with a 5-second confinement to the central platform. In the non-spatial test, arms made of opaque polyvinyl chloride (PVC) were closed by doors that the animals could push open and the small delay needed for this functioned as confinement in this procedure. As these tasks were more complicated than the previously used one, mice were now given five daily trials. This time, no obvious chaining (kinesthetic) strategies were encountered. Male mice from nine different strains were used and the rank order obtained in the spatial task was: C3H/HeJ > C57BL/6J > C57BR/cdJ >BALB/cJ > NMRI > CPB-K > BA > DBA/2J > NZB/B1NJ. This closely resembled the one found previously (Crusio et al., 1987), with a Spearman rank correlation (Siegel, 1956) of 0.77 ($df = 6$, $P < 0.05$) between the results of the common strains in the two experiments, suggesting that similar capabilities were used to solve the two different problems. The situation was drastically different for the non-spatial task, where the strain rank order obtained was: DBA/2J > C57BL/6J = C57BR > CPB-K > C3H/HeJ > BALB/cJ > NMRI > BA > NZB/B1NJ. These results did not correlate with those from the spatial task ($r_S = 0.34$, $df = 7$, ns) or with those from the earlier experiment ($r_S = -0.05$, $df = 6$, ns). To the best of our knowledge, this was also the first demonstration of the huge effect that procedural variations can have on strain rank orders obtained in radial-maze tasks. At the same time, these data demonstrated the robustness of these rank orders with respect to procedural variations that apparently are less important for the performance of the animals (see also below).

In contrast to all of the above-presented studies, Roullet and Lassalle (1992) used female mice in their experiments. In addition, they were the first to study F1 crosses between several inbred strains (BALB/cBy × C57BL/6J and C57BL/6J × DBA/2J). The apparatus they used was an ingeniously constructed elevated maze, allowing the lowering of parts of the arms to confine mice to the central platform. Two different procedures were used, one entailing a confinement procedure but the other not. Here, too, the use of a confinement procedure interrupted the use of kinesthetic strategies reported by Bolhuis et al. (1985) in rats and by Schwegler et al. (1990) in mice. After five daily training trials, the extra-maze spatial cues were removed and the possible perturbing effects of this on performance were studied in two additional "probe trials." In addition, they found that several groups that performed well in both tasks differed in their use of kinesthetic strategies (termed "radial strategy"), employing these more frequently in the non-confined task. Interestingly, after removing the extra-maze cues even strains using a kinesthetic strategy to solve the non-confined task showed impaired performance and it was concluded that "the use of a radial strategy does not preclude the processing of spatial information." This parallels the correlation between performances in confined and non-confined tasks reported by our group (Crusio et al., 1987; Schwegler et al., 1990). Good performers in their tasks were C57BL/6J, DBA/2J, and both F1 groups. Poor performers were C3H/He, NZB/B1N, and CBA/J. The fact that both F1s were as good as the best parental strains, but not better, suggests (but does not prove, see Crusio, 2007) the presence of dominance, but not heterosis, for higher learning performances. The strain rank order obtained in their study differed considerably from the one reported by Schwegler et al. (1990). Possible explanations, besides some procedural differences, might be sex differences or the fact that Roullet and Lassalle (1992) used an elevated maze, possibly inducing a confounding factor caused by anxiety (Crusio, 1999). In a later study, Roullet et al. (1993) modified their apparatus and procedure to investigate the influence of intra-maze cues on performance. To this end, they rotated the lowered arm parts used to confine animals to the central platform in between different arm choices. Their results suggested that BALB/cJ animals learned the task using intra-maze cues, as they were impaired after rotation. C57BL/6J animals, however, appeared to use spatial cues exclusively.

Several groups have also tested mice in the more complicated radial maze task used in rats to distinguish WM and RM in the same test (Nadel and MacDonald, 1980). Our group trained mice from nine different strains in 10 daily trials in a spatial and a non-spatial version of the radial maze in which only four out of eight arms were baited (Crusio et al., 1993). The set of rewarded arms was always the same for a particular mouse, but differed between individuals. In the spatial version of the task, clear Plexiglas® arms were used, with several extra-maze cues placed close to the apparatus. In addition, the maze was rotated 45° between daily trials to dissociate intra- and extra-maze cues, the position in the testing room determining the baited arm. In the non-spatial version of the task, opaque PVC arms were used with different black/white patterns painted on the floor. Here, rewards were always associated with particular patterns, regardless of position. Large strain differences were found for both WM and RM errors in the spatial task, but all animals showed very good performances in the non-spatial task and strain differences were not significant, probably due to a floor effect. In both tasks, RM and WM errors were correlated almost perfectly. The strain rank order obtained in the spatial task again closely resembled those obtained in earlier tasks with Spearman correlations between WM/RM and performance in the previously reported two tasks (Crusio et al., 1987; Schwegler et al., 1990) ranging from 0.71 to 0.90. Ammassari-Teule et al. (1993) used similar procedures, except for the absence of a confinement procedure and the facts that their maze was elevated and not rotated between trials. Their results agreed in smaller strain differences in the visually cued task and better performance of C57BL/6J animals compared to DBA/2J animals in the spatial tasks. Here, too, WM and RM did not appear to be dissociable in either the spatial or the non-spatial task. Their results differed in that the C3H strain was reported to perform at lower levels than both C57BL/6 and DBA/2, whereas this strain was always among the better performing strains in our hands. A possible explanation for this discrepancy may be that the C3H strain carries the *rd* mutation causing retinal degeneration. Although severely visually impaired, animals carrying this mutation are not yet completely blind around 2–3 months (Dräger and Hubel, 1978; Mrosovsky, 1994; Mrosovsky and Hampton, 1997; Nagy and Misanin, 1970), the age at which animals were tested in the experiments described above. The explanation for the relatively good performance of C3H mice observed by our group compared to others may therefore lie in the fact that we were the only ones to provide salient visual cues in close proximity to the arms of the maze, so that mice did not have to rely on relatively distal cues attached to the walls of the testing room, for example.

Conclusions

The fully baited eight-arm radial maze test has been shown to be a sensitive tool to detect not only subtle strain differences, but also the effects of mutations. For example, a spontaneous mutation that arose in the C57BL/6J strain maintained in the laboratory of Hans van Abeelen (Crusio et al., 1991) was shown to entail behavioral differences in the open field test and hippocampal neuroanatomy. This led to the hypothesis that the mutated substrain, C57BL/6JNmg, would show decreased radial-maze learning abilities, a prediction that was subsequently confirmed (Jamot et al., 1994). As described above, procedural changes may lead to very different results in different laboratories. It should be stressed, however, that the opposite is also possible: employing similar procedures and apparatuses can lead to highly replicable results, even in the hands of different experimenters in different laboratories (Crusio and Schwegler, 2005). Unfortunately, although many robust strain differences have been described, the exact loci and biochemical mechanisms underlying these differences are not well understood. We have still much to explore with regard to the genetic underpinnings of maze learning in mice.

Acknowledgments

We would like to thank Robert T. Gerlai (Toronto, Canada) for critically reading the manuscript. W. E. C was supported by the Centre National de la Recherche Scientifique (UMR 5106) and grants from the Conseil Régional d'Aquitaine, CNRS, the University of Bordeaux, and the National Institute of Mental Health (MH072920).

References

Ammassari-Teule, M. and Caprioli, A. (1985) Spatial learning and memory, maze running strategies and cholinergic mechanisms in two inbred strains of mice. *Behav Brain Res* 17: 9–16.

Ammassari-Teule, M., Hoffmann, H.J., and Rossi-Arnaud, C. (1993) Learning in inbred mice: strain-specific abilities across three radial maze problems. *Behav Genet* 23: 405–412.

Bernstein, D., Olton, D.S., Ingram, D.K., Waller, S.B., Reynolds, M.A., and London, E.D. (1985) Radial maze performance in young and aged mice: neurochemical correlates. *Pharmacol Biochem Behav* 22: 301–307.

Bolhuis, J.J., Bijlsma, S., and Ansmink, P. (1986) Exponential decay of spatial memory of rats in a radial maze. *Behav Neural Biol* 46: 115–122.

Bolhuis, J.J., Buresova, O., and Bures, J. (1985) Persistence of working memory of rats in an aversively motivated radial maze task. *Behav Brain Res* 15: 43–49.

Crusio, W.E. (1999) Methodological considerations for testing learning in mice. In Crusio, W.E. and Gerlai, R.T. (eds.), *Handbook of Molecular-Genetic Techniques for Brain and Behavior Research, Techniques in the Behavioral and Neural Sciences*, Vol. 13. Elsevier, Amsterdam, pp. 638–651.

Crusio, W.E. (2007) An introduction to quantitative genetics. In Jones, B.C. and Mormède, P. (eds.), *Neurobehavioral Genetics: Methods and Applications*, 2nd revised edn. CRC Press, Boca Raton, FL, USA, pp. 37–54.

Crusio, W.E. and Schwegler, H. (2005) Learning spatial orientation tasks in the radial-maze and structural variation in the hippocampus in inbred mice. *Behav Brain Func* 1: 3.

Crusio, W.E., Schwegler, H., and Brust, I. (1993) Covariations between hippocampal mossy fibres and working and reference memory in spatial and non-spatial radial maze tasks in mice. *Eur J Neurosci* 5: 1413–1420.

Crusio, W.E., Schwegler, H., and Lipp, H.-P. (1987) Radial-maze performance and structural variation of the hippocampus in mice: a correlation with mossy fibre distribution. *Brain Res* 425: 182–185.

Crusio, W.E., Schwegler, H., and van Abeelen, J.H.F. (1991) Behavioural and neuroanatomical divergence between two sublines of C57BL/6J inbred mice. *Behav Brain Res* 42: 93–97.

Dräger, U.C. and Hubel, D.H. (1978) Studies of visual function and its decay in mice with hereditary retinal degeneration. *J Comp Neurol* 180: 85–114.

Grootendorst, J., de Kloet, E.R., Vossen, C., Dalm, S., and Oitzl, M.S. (2001) Repeated exposure to rats has persistent genotype-dependent effects on learning and locomotor activity of apolipoprotein E knockout and C57Bl/6 mice. *Behav Brain Res* 125: 249–259.

Honig, W.K. (1978) Studies of working memory in the pigeon. In Hulse, S.H., Fowler, H. and Honig, W.K. (eds.), *Cognitive Processes in Animal Behavior*. Lawrence Erlbaum Associates, Hillsdale, NJ, USA, pp. 211–248.

Hyde, L.A., Hoplight, B.J., and Denenberg, V.H. (1998) Water version of the radial-arm maze: learning in three inbred strains of mice. *Brain Res* 785: 236–244.

Jamot, L., Bertholet, J.-Y., and Crusio, W.E. (1994) Neuroanatomical divergence between two substrains of C57BL/6J inbred mice entails differential radial-maze learning. *Brain Res* 644: 352–356.

Levy, A., Kluge, P.B., and Elsmore, T.F. (1983) Radial arm maze performance of mice: acquistion and atropine effects. *Behav Neural Biol* 39: 229–240.

Mizumori, S.J., Rosenzweig, M.R., and Kermisch, M.G. (1982) Failure of mice to demonstrate spatial memory in the radial maze. *Behav Neural Biol* 35: 33–45.

Mrosovsky, N. (1994) In praise of masking: behavioural responses of retinally degenerate mice to dim light. *Chronobiol Int* 11: 343–348.

Mrosovsky, N. and Hampton, R.R. (1997) Spatial responses to light in mice with severe retinal degeneration. *Neurosci Lett* 222: 204–206.

Nadel, L. and MacDonald, L. (1980) Hippocampus: cognitive map or working memory? *Behav Neural Biol* 29: 405–409.

Nagy, Z.M. and Misanin, J.R. (1970) Visual perception in the retinal degenerate C3H mouse. *J Comp Physiol Psychol* 72: 306–310.

O'Keefe, J.O. and Nadel, L. (1978) *The Hippocampus as a Cognitive Map*. Clarendon Press, Oxford.

O'Keefe, J.O. and Nadel, L. (1979) Precis of O'Keefe and Nadel's, *The Hippocampus as a Cognitive Map*. *Behav Brain Sci* 2: 487–533.

Olton, D.S. (1977) Spatial memory. *Sci Am* 236: 82–98.

Olton, D.S., Becker, J.T., and Handelmann, G.E. (1979) Hippocampus, space, and memory. *Behav Brain Sci* 2: 313–365.

Olton, D.S., Becker, J.T., and Handelmann, G.E. (1980) Hippocampal function: working memory or cognitive mapping? *Physiol Psychol* 8: 239–246.

Olton, D.S. and Papas, B.C. (1979) Spatial memory and hippocampal function. *Neuropsychologia* 17: 669–682.

Olton, D.S. and Samuelson, R.J. (1976) Remembrance of places passed: spatial memory in rats. *J Exp Psychol Anim Behav Process* 2: 97–117.

Olton, D.S. and Schlosberg, P. (1978) Food-searching strategies in young rats: win-shift predominates over win-stay. *J Comp Physiol Psychol* 92: 609–618.

Pico, R.M. and Davis, J.L. (1984) The radial maze performance of mice: assessing the dimensional requirements for serial order memory in animals. *Behav Neural Biol* 40: 5–26.

Reinstein, D.K., DeBoissere, T., Robinson, N., and Wurtman, R.J. (1983) Radial maze performance in three strains of mice: role of the fimbria/fornix. *Brain Res* 263: 172–176.

Roullet, P. and Lassalle, J.-M. (1992) Behavioural strategies, sensorial processes and hippocampal mossy fibre distribution in radial maze performance in mice. *Behav Brain Res* 48: 77–85.

Roullet, P., Lassalle, J.-M., and Jegat, R. (1993) A study of behavioral and sensorial bases of radial maze learning in mice. *Behav Neural Biol* 59: 173–179.

Schwegler, H., Crusio, W.E., and Brust, I. (1990) Hippocampal mossy fibers and radial-maze learning in the mouse: a correlation with spatial working memory but not with non-spatial reference memory. *Neuroscience* 34: 293–298.

Siegel, S. (1956) *Nonparametric Statistics for the Behavioral Sciences*. McGraw-Hill, New York.

Whishaw, I.Q. and Tomie, J.A. (1996) Of mice and mazes: similarities between mice and rats on dry land but not water mazes. *Physiol Behav* 60: 1191–1197.

Chapter 30 Other mazes

Timothy P. O'Leary and Richard E. Brown

Introduction

Mazes are important tools for studying learning and memory as they provide a quantifiable method for scoring performance, can be replicated in different laboratories, and provide a semi-natural environment for the mouse to explore, much like the runways and burrows in the mouse's natural habitat. Mazes generally consist of an open field or a system of alleyways through which the mouse must transverse to obtain food reward or escape from an aversive stimulus (Olton, 1979).

In this chapter, we review the use of six different mazes for the study of strain differences in learning and memory in mice: the Barnes maze, the holeboard, the T-maze, the Y-maze, the Lashley type III, and Hebb–Williams mazes (Figure 30.1). Although these mazes are less frequently used than others, such as the Morris water maze and radial arm maze (see Chapters 27 and 29), they are inexpensive to make and can assess the same forms of learning and memory as more commonly used tests. For example, depending on the test procedure, these mazes can be used to study spatial, cued, and response learning, as well as reference and working memory. The T- and Y-mazes can also be used to assess working memory through the measure of spontaneous alternation (Table 30.1). Although this chapter focuses on strain differences in measures of learning and memory in these mazes, sex differences are often found in studies of maze learning (Jonasson, 2005), and are discussed here when they have been reported.

In studies of maze learning, reference or long-term memory is the storage of information about which spatial locations, discrete cues, or behavioral responses reliably predict reinforcement (i.e., food reward or escape). Thus, in reference memory tasks the same spatial location, cue or response will reliably predict reinforcement on each of the training trials, which are usually completed over several days. Working or short-term memory is the storage of information over a short duration (Olton, 1979), usually within the time period of a single trial (e.g., spontaneous alternation), over a delay interval between a pair of trials (e.g., delayed non-matching to sample), or over a series of trials (e.g., Hebb–Williams maze). In working memory tasks, spatial locations, cues or responses do not reliably predict reward. Instead, animals must remember what predicted reward previously, and depending on the task, learn to use the same or a different spatial location, cue, or response to obtain reinforcement within the current trial or in a subsequent trial.

Barnes circular maze

The Barnes maze was originally designed by Carol Barnes to investigate the age-related decline of visuo-spatial learning and memory in rats. The original maze (Barnes, 1979) consisted of an elevated circular open field (122 cm in diameter) with 18 holes located at equal distances around the edge (Figure 30.1a). In the Barnes maze, rats or mice are placed into the center of the maze and are motivated to escape from bright light and the open space of the maze floor by entering a dark box that is placed beneath one of the holes (the escape hole). With training, animals learn where the escape hole is located relative to visual cues in the extra-maze environment, as performance deficits are found when extra-maze visual cues are not present (Barnes et al., 1980), or when the location of the escape hole is moved relative to the extra-maze visual cues (Barnes, 1979).

The original Barnes maze design used with rats is often adapted for use with mice by reducing the diameter and the number of holes in the maze (Holmes et al., 2002; Pompl et al., 1999). Bach et al. (1995), however, tested mice in a maze with the same diameter as the original Barnes design for rats, and increased the number of holes to 40. Some designs also include a wall around the edge, upon which intra-maze visual cues can be placed (Pompl et al., 1999). In addition to the bright light, buzzers (Bach et al., 1995) or fans (Pompl et al., 1999) have been used to provide additional aversive stimulation on the Barnes maze, as mice may be more reluctant to descend into the escape hole than rats. To reduce this problem, one design placed the holes in a wall surrounding the maze rather than in the maze floor (Koopmans et al., 2003).

Learning in the Barnes maze can be assessed with multiple behavioral measures including: (1) latency to enter the escape hole; (2) distance traveled; (3) errors (lowering the head into incorrect holes); and (4) hole deviation (number of holes that the first hole investigated was away from the escape hole; Barnes, 1979; McNaughton et al., 1986; O'Leary and Brown,

Behavioral Genetics of the Mouse: Volume I. *Genetics of Behavioral Phenotypes*, eds. Wim E. Crusio, Frans Sluyter, Robert T. Gerlai, and Susanna Pietropaolo. Published by Cambridge University Press.

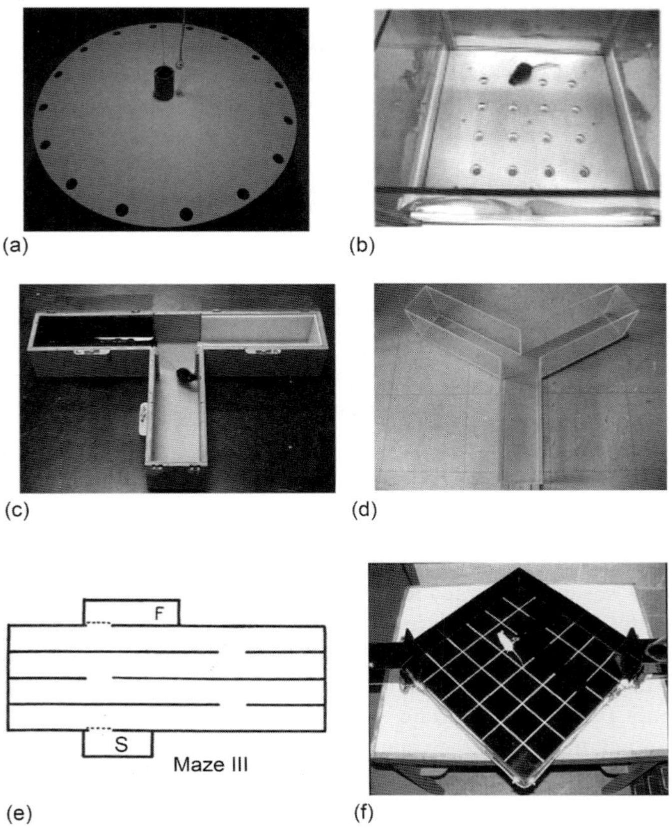

Figure 30.1 Six mazes used to assess learning and memory in mice. (a) The Barnes maze (as used in our laboratory). In this maze mice are motivated by bright light and noise to locate the hole that has a dark escape box beneath it. The animal uses extra-maze cues to learn where the escape hole is located. (b) The holeboard maze. In this maze mice remember the location of holes baited with food reward using extra-maze visual cues. (Reproduced with permission from Kuc et al., 2006. © Blackwell Publishing.) (c) The T-maze as used during cued training with visual cues (as used in our laboratory). On this task, mice discriminate between the black and white cues, one of which predicts the presence of food reward. (d) The Y-maze (as used in our laboratory). This maze is similar in design to the T-maze, but the arms are in the shape of a "Y" rather than a "T." (e) The Lashley type III maze. In this maze, mice learn the correct sequence of left and right turns to navigate from the start box (S), through a series of four alleyways, to locate the goal box that contains food (F) reward. (Reproduced with permission from Lashley, 1929. © Dover Publishing.) (f) The Hebb–Williams maze (as used by Stanford and Brown, 2003). In this maze mice navigate around barriers to locate the goal box. The spatial arrangement of barriers is altered to create six practice and 12 test problems of varying difficulty. Mice can be trained with food reward or water escape. The barrier in this figure is positioned as in practice problem "c" of Rabinovitch and Rosvold (1951).

2009). The search strategy that animals use to locate the escape hole can also be used as an index of learning. Rats and mice begin using a random search strategy, where holes are investigated in an unsystematic fashion with multiple entries into the center of the maze. They then progress to a serial search strategy, where adjacent holes are investigated in a clockwise or counter-clockwise fashion. Finally, many animals adopt the most efficient spatial search strategy, where the animal moves directly from the center of the maze toward the escape hole (Bach et al., 1995; Barnes, 1979). Memory for the location of the escape hole may be assessed with reversal learning procedures, where the escape box is moved to a new location, or with probe trials, where no escape box is present. Duration of time spent near, and head dips into the previously correct escape hole are then used as an index of visuo-spatial memory (O'Leary and Brown, 2009).

The Barnes maze has advantages over other mazes more commonly used to assess visuo-spatial learning and memory in mice, as food restriction is not required, and the stress responses to bright light, loud noise, and/or tactile stimulation (blowing air) may be less than to other types of aversive stimulation, such as immersion in water or foot shock (Pompl et al., 1999). For example, the corticosterone levels of mice after training on the Barnes maze with bright lights are lower than after training on the Morris water maze (Harrison et al., 2009). The Barnes maze may also be used as an alternative to the Morris water maze when mice exhibit poor or abnormal swimming behavior (Paylor et al., 2001). The performance of some mouse strains (BALB/cA), however, may differ on water-based and land-based visuo-spatial learning and memory tasks (Yoshida et al., 2001). Furthermore, to ensure that mice are motivated to enter the escape hole, they may be tested on the light–dark transition test (see Chapter 16) to determine their preference for dark over light regions prior to testing on the Barnes maze (Bach et al., 1995).

The size and design of the Barnes maze, however, can affect the performance of mice. For example, when the maze is surrounded by a wall and its size reduced, mice show increased use of the non-spatial serial search strategy (O'Leary and Brown, 2012), and blind mice can perform as well as sighted mice (Garcia et al., 2004), which suggests that mice do not use extra-maze visual cues to locate the escape hole, and that this maze design may not be a valid visuo-spatial learning and memory test (O'Leary et al., 2011).

As with the Morris water maze, the Barnes maze may be used to test spatial or cued learning and reference memory depending on the location of the visual cues (Bach et al., 1995). In spatial training, visual cues are distributed throughout the extra-maze environment, whereas in cued training a unique salient intra-maze visual cue is located near the escape hole. The cued training procedure may be used to complement spatial training, by providing general measures of visual ability and motivation to locate the escape hole (Bach et al., 1995).

Strain differences in learning and memory

Performance differences between inbred mouse strains in the Barnes maze have been attributed to differences in visuo-spatial learning and memory, as well as non-cognitive performance factors such as visual ability. Male C57BL/6J mice show superior learning compared to CBA/J, DBA/2J, and 129T2/SvEmsJ (129T2) mice (formerly 129/SvEms-+$^{Ter?}$/J or 129/SvEmsJ; Festing et al., 1999), based on the measure of deviation from the escape hole (Nguyen et al., 2000). Male C57BL/6J also find the escape hole faster, travel a shorter distance, and make fewer errors than male CD-1 mice (Patil et al., 2009).

Table 30.1 Types of learning and memory that have been assessed in mice using the six mazes discussed in the chapter.

Apparatus	Spatial learning[a]	Cued learning[b]	Response learning[c]	Reference memory[d]	Working memory[e]	Spontaneous alternation[f]
Barnes maze	+	+		+		
Holeboard	+			+	+	
T-maze	+	+	+	+	+	+
Y-maze	+	+	+	+	+	+
Lashley type III maze	+		+	+		
Hebb–Williams maze	+		+	+	+	

[a] In spatial learning, animals learn where a goal is located within a maze relative to multiple extra-maze visual cues.
[b] In cued learning, animals learn to associate a single intra-maze cue with a reward.
[c] In response learning, animals learn to use a motor response (turn left vs. turn right) to obtain a reward.
[d] Reference or long-term memory is the storage of information over a long time period, such as across training trials given over several days.
[e] Working or short-term memory is the storage of information over a short time period, such as a delay interval.
[f] Spontaneous alternation is a method used to assess working memory, and measures an animal's tendency to explore novel areas over areas recently visited.

Placing a curtain around the Barnes maze to block extra-maze visual cues impaired performance for the C57BL/6J and 129T2 strains, but not for the CBA/J and DBA/2J strains, suggesting that the latter strains use a non-spatial strategy to locate the escape hole. The poor performance of the DBA/2 strain in the Barnes maze agrees with their poor performance in the Morris water maze (Logue et al., 1997; Nguyen et al., 2000; Paylor et al., 1996) which suggests a deficit in visuo-spatial learning and memory. This deficit may be due to differences in hippocampal morphology (Crusio et al., 1990) and neurochemistry (Paylor et al., 1996) between DBA/2 and C57BL/6 mice. The poor performance of CBA/J mice in the Barnes maze and other visuo-spatial learning and memory tasks (Nguyen et al., 2000) may be due to their poor visual ability as these mice have retinal degeneration due to the *Pde6b*^{rd1} mutation (Chang et al., 2002), and mice with retinal degeneration are unable to solve visual discrimination tasks (Wong and Brown, 2006).

Strain differences in learning and memory in the Barnes maze may depend on the behavioral measure used. Male Swiss Webster, 129S2/SvPas (129S2) (formerly 129/SvPas; Festing et al., 1999), BALB/cBy, and C57BL/6 mice were tested in a modified Barnes maze, with holes in the wall around the edge of the maze rather than in the floor (Koopmans et al., 2003). Although the C57BL/6 mice showed robust learning on the Barnes maze, the Swiss Webster, 129S2, and BALB/cBy strains did not show reliable evidence of learning using the measures of distance traveled, errors, and latency to find the escape hole. Furthermore, moving the escape hole to the opposite side of the maze did not disrupt performance for the Swiss Webster, 129S2, or BALB/cBy mice, suggesting that they used a non-spatial search strategy to locate the escape hole.

Holmes et al. (2002) found no strain difference in learning between male and female C57BL/6J, 129S6/SvEvTac (129S6) (formerly 129/SvEvTac; Festing et al., 1999), and DBA/2J mice using the measure of errors. Strain differences in memory were found during a probe trial, however, as C57BL/6J mice searched more accurately for the escape hole than 129S6 mice. DBA/2J mice demonstrated an intermediate level of performance. The poor performance of the 129S6 strain on the probe trial was not attributed to deficits in visuo-spatial memory, but to decreased levels of exploratory behavior. Likewise, Fox et al. (1999) tested male C57BL/6J, FVB/N, and 129T2 mice in the Barnes maze with a 4-day training protocol, and found that only the C57BL/6J strain showed evidence of learning when latency to enter the escape hole was used to measure performance. The poor performance of the 129T2 and FVB/N strains was attributed to elevated anxiety during testing which resulted in little exploratory activity. Performance deficits in the FVB/N strain, however, may also be due to their poor visual ability (Wong and Brown, 2006), as this strain has retinal degeneration (Chang et al., 2002).

The 129S6 and other 129 mouse substrains often show low levels of exploratory behavior in the open field (Bothe et al., 2005; Holmes et al., 2002), and as a result measures of memory in the Barnes maze, such as head dips into the escape hole, may be artificially low. Therefore, the Barnes maze may not be suitable for assessing visuo-spatial memory in mice that show low levels of exploratory behavior. Indeed, some of the 129 substrains show good spatial memory in the Morris water maze (Clapcote and Roder, 2004; Võikar et al., 2001), but poor spatial memory in the Barnes maze (Holmes et al., 2002; Koopmans et al., 2003).

Comparisons of spatial learning and memory between mouse strains in the Barnes maze demonstrate that C57BL/6J mice consistently perform well in both acquisition and retention of the task. Strain differences in learning and memory in the Barnes maze may depend on differences in the hippocampus, as DBA/2J and C57BL/6J mice differ both in performance in this maze (Nguyen et al., 2000), and in hippocampal morphology (Crusio et al., 1990) and neurochemistry (Paylor et al., 1996). However, strain differences may also depend on the behavioral measure used (Koopmans et al., 2003) and the maze design (O'Leary and Brown, 2012), underlying the importance of using multiple measures to fully characterize learning and memory performance in the Barnes maze. The poor performance of mice with retinal degeneration (CBA/J, FVB/N) or low exploratory behavior (129S6) indicates that non-cognitive factors influence performance on the Barnes maze, and that these factors should be controlled before conclusions about strain differences in visuo-spatial learning and memory are made.

Holeboard spatial memory test

The *holeboard* consists of a circular or square open field with multiple holes arranged symmetrically in the floor (Figure 30.1b). It can be used to measure exploratory behavior as well as spatial learning and memory. In the holeboard test, mice are food restricted and motivated to find the one or more holes with food reward. To find the baited hole(s), mice learn the spatial location of the hole(s) relative to extra-maze visual cues (Yoshida et al., 2001). Measures used to assess learning and reference memory in the holeboard task include latency to locate the baited hole(s), errors (searching into unbaited holes), distance traveled, and time spent near the baited hole(s) during probe trials. Working memory can be assessed by the number of errors made at holes that were previously searched within a single trial (Kuc et al., 2006).

Strain differences in learning and memory

Male C57BL/6N mice have a shorter latency and distance traveled to reach the baited hole than BALB/cA mice during training in the holeboard (Yoshida et al., 2001). However, during a probe trial both the C57BL/6N and BALB/cA mice spent a similar amount of time in the correct quadrant, indicating no strain differences in memory for the spatial location of the baited hole. These results disagree with results from the Morris water maze in the same study, which suggest a memory deficit in BALB/cA mice. Differences in motivation (aversive vs. appetitive) may account for the different findings, and suggest that performance on spatial tasks for some inbred strains may depend on the type of task and motivation used (Yoshida et al., 2001).

Kuc et al. (2006) found strain differences in the overall number of working and reference memory errors made in the holeboard by male C57BL/6J, Swiss Webster, CD-1, and DBA/2J mice. DBA/2J mice made fewer reference memory errors than CD-1 mice, while C57BL/6J and Swiss Webster mice did not differ from CD-1 or DBA/2J mice. C57BL/6J and DBA/2J mice made fewer working memory errors than CD-1 mice, with Swiss Webster mice falling between these extremes, having more working memory errors than DBA/2J mice but not differing from C57BL/6J or CD-1 mice.

C57BL/6J mice often perform better then DBA/2J mice on tests of spatial learning and memory (Logue et al., 1997; Nguyen et al., 2000; Paylor et al., 1996), and Kuc et al. (2006) suggest that the lack of a difference in reference memory errors between DBA/2J and C57BL/6J mice on the holeboard may occur because this test is less demanding in terms of spatial memory than other tests. Kuc et al. (2006) propose that DBA/2J mice may use a non-spatial search strategy as efficiently as C57BL/6J mice use a spatial search strategy in the holeboard.

T- and Y-mazes

The T-maze consists of a straight alleyway or stem that meets two other alleyways or choice arms that run perpendicular to the stem in the shape of the letter "T" (Gerlai, 1998; Figure 30.1c). A start-box is located in the stem of the maze and holds animals before each trial. Similarly, the Y-maze consists of three straight alleyways that meet, and are positioned at angles greater than 90° relative to each other in the shape of the letter "Y" (Figure 30.1d; Crusio, 1999, p. 643). Guillotine doors are placed in each arm of the T- and Y-mazes, and may be lowered to prevent mice from entering or leaving an arm. T- and Y-mazes can be used with food reward or escape from water as motivation.

The structure of the T- and Y-mazes is dependent on the type of learning and memory being assessed, as spatial learning and memory requires low, or transparent, walls so that extra-maze cues are visible to mice (Deacon et al., 2002), while cued learning and memory requires high and opaque walls to block extra-maze visual cues. In T-mazes that use escape from water as motivation, the choice arms may be curved towards the stem so that mice cannot see escape ladders until they completely enter a choice arm (Boehm et al., 1998).

Reference memory

The T- and Y-mazes are used to assess both reference and working memory. In a reference memory task, a mouse is placed into one arm (the start box), and must choose between the two remaining arms, one of which is baited with food reward or provides escape from water. Different training procedures are used to assess cued, spatial, and response-based learning and reference memory. In cued training, a unique auditory or visual cue predicts which of the two choice arms provides reward. The visual stimuli in cued training are typically located within the choice arms of the maze (as in Figure 30.1c) and require the animal to discriminate between black and white (Denenberg et al., 1990) or horizontal and vertical lines (Balogh et al., 1999). Spatial learning and reference memory may be assessed in a similar manner, if multiple extra-maze cues are present with no unique cue indicating which of the two arms is correct, and if the spatial relationship between the location of the correct arm and extra-maze visual cues remains constant (Deacon et al., 2002). Finally, response learning and reference memory are assessed when no intra- or extra-maze cues predict which arm is correct, and the animal is required to make the same turn (left or right) to obtain reward (Crusio et al., 1990).

Working memory

Working memory is assessed in T- and Y-mazes using a delayed non-matching to sample (DNMS) procedure. The DNMS task involves a sample and a test trial, with an intervening delay period. In the sample trial, the mouse is placed in the start box and one choice arm is baited with food while the other choice arm is blocked. The animal is allowed to find the food and is then returned to the start box for a delay interval. Following the delay, there is a test trial in which the animal can choose either of the arms, but only the arm that was not baited in the sample trial contains a reward (Deacon et al., 2002). Therefore, the animal must use cued, spatial, or response-based information to learn which arm is baited on the sample trial, remember this during the delay interval, and choose the other arm on the test trial. By varying the duration of the interval between the sample and test trials, the duration of working memory can be assessed.

Two confounds which can occur when testing reference and working memory on the T- or Y-mazes are the development of response biases (side preferences) and the use of uncontrolled extra-maze cues. In cued training, response biases may be discouraged by varying the position of the correct choice arm (left or right) according to a pseudo-randomized sequence (Denenberg et al., 1990). In spatial training, the development of a response bias may be prevented by rotating the maze 180° relative to extra-maze cues between trials, and by baiting different arms of the maze before and after rotation. In this case, the correct arm is always in the same spatial location, but different responses are required to locate the correct arm. Conversely, the use of extra-maze cues may be prevented in response training by using high opaque walls to block extra-maze cues, or by rotating the maze between trials, and baiting the same arm of the maze before and after rotation. In this case, the correct arm will be in a different spatial location before and after rotation of the maze, but the same response may be used on both trials to correctly choose the baited arm (Crusio, 1999, p. 644).

Strain differences in DNMS and reference memory

Spatial learning and reference memory performance in the Y-maze and DNMS performance in the T-maze are dependant on the hippocampus, as female C57BL/6J mice with hippocampal lesions show decreased choice accuracy in both tasks (Deacon et al., 2002). There is no significant correlation between the size of intra- and infra-pyramidal mossy fiber terminal fields within the CA3 and CA4 regions of the hippocampus in inbred mice and performance on spatial or response-based tasks in the T-maze, suggesting that the size of intra- and infra-pyramidal mossy fiber terminal fields may not influence spatial or response-based learning and reference memory in the T-maze (Crusio et al., 1990).

Strain differences in a cued visual discrimination task using horizontal and vertical black and white stripes have been found in the T-maze, as male and female C57BL/6J mice made more correct choices during the acquisition phase than male 129S6 mice (Balogh et al., 1999). On a black/white visual discrimination in a water escape version of the T-maze, male and female DBA/2, NMRI, and C57BL/6 mice had shorter escape latencies and made fewer errors than C3H/HeJ, BALB/c, and BALB/cN mice (Schwegler and Buselmaier, 1981). No sex differences were found in black/white discrimination performance in the T-maze in these strains. Strain differences may be due to differences in visual ability rather than differences in learning, as C3H/HeJ mice have retinal degeneration. Crusio et al. (1990) tested spatial and response-based learning and memory in male A/J, CBA/H, NZB/B1NJ, XLII, BALB/cByJ, C57BL/6J, DBA/2J, BA, and CPB-K mice in a T-maze using food reward. Of these strains, BALB/cBy and NZB mice showed the best learning performance, as measured by the number of correct choices on the last 2 days of training, whereas BA mice performed the worst on both the spatial and response-based tasks. With response learning on the T-maze, SJL/J mice acquired the task with fewer trials than C58/J, PL/J, and SWR/J mice (Moy et al., 2008). Also, a low percentage of C57BL/6J, AKR/J, and C3H/HeJ mice reached criterion (80% correct per daily 10 trial block for 3 days) in response-based learning in the T-maze compared to DBA/2J, FVB/NJ, and A/J mice (Moy et al., 2007).

Spontaneous alternation

The T- and Y-mazes can be used to assess working memory using a test of spontaneous alternation, which makes use of the finding that during exploration mice prefer novel over familiar areas. Spontaneous alternation does not require food reward, and may be assessed with either continuous or discrete trial procedures in the T- and Y-mazes. In the discrete trial procedure, trials are administered in pairs with a forced trial given before each free trial (Bertholet and Crusio, 1991). In the forced trial procedure, the mouse is placed in the start arm and allowed to explore only one of the choice arms while the other is blocked. On the subsequent free trial, both arms are open and the mouse shows spontaneous alternation if it enters the arm that was blocked during the forced trial. The demand on working memory can be increased with the delayed alternation procedure, where a delay is introduced between the forced and choice trials (Jones et al., 2001). A drawback of the discrete trials procedure is that it involves extensive handling of the mice as they are removed from the choice arm by the experimenter and placed back into the start arm after every trial (Gerlai, 1998). Stress induced by handling is problematic as it reduces spontaneous alternation rates in mice. Indeed, C57BL/6J mice that are exposed to an inescapable stress (forced exploration of an open field) immediately before the free trial show lower rates of spontaneous alternation in a T-maze than mice which are unstressed or exposed to an escapable stressor (Bats et al., 2001).

Continuous alternation in T- or Y-mazes has an advantage over the discrete trial procedure as the handling of animals is minimized. In the T-maze, continuous alternation consists of one forced trial followed by a series of free trials (Gerlai, 1998). In the forced trial, mice are confined in the start arm for a brief period, and are then allowed to explore one of the choice arms of the T-maze while the other choice arm is blocked. Upon returning to the start arm on their own, mice complete a series of free trials where they have a choice of entering either of the choice arms. Upon entering a choice arm in a free trial, access to the other choice arm is blocked, and the trial ends when mice re-enter the start arm. A spontaneous alternation occurs when mice choose the arm that was not visited in a previous forced or free trial. Gerlai (1998) has demonstrated that C57BL/6J mice will use extra-maze visual cues while alternating, as reducing illumination so that visual cues were not visible reduced alternation rates to chance levels.

In the Y-maze, spontaneous alternation is typically assessed with the continuous procedure in which mice freely explore the maze for a set time without any forced trials (Contet et al., 2001; Heyser et al., 1999). An alternation occurs when a mouse enters all three arms of the Y-maze in a sequence that does not contain repeat visits to one arm (i.e., ABC rather than ABA or ABB). The spontaneous alternation rate is then calculated as the ratio of alternations performed, out of the maximum number of alternations possible, given the total number of arm entries minus one. As such, the Y-maze can provide concurrent measures of alternation rate and locomotor activity as measured by arm entries. Continuous spontaneous alternation in the Y-maze, however, is influenced by response biases, as mice that continually turn in the same direction will show high alternation rates (Deacon et al., 2002).

Spontaneous alternation is thought to involve working memory as the mouse must remember which arm was most recently visited using spatial or response-based information. Discrete trial and continuous spontaneous alternation in the T-maze are both dependent on the hippocampus, as male and female C57BL/6J mice with hippocampal lesions alternate less than sham control mice (Deacon et al., 2002; Gerlai, 1998). There are no significant correlations between the rate of discrete trial spontaneous alternation and the size of intra- and infrapyramidal mossy fiber terminal fields within the CA3 and CA4 regions of the hippocampus, suggesting the size of intra- and infrapyramidal mossy fiber terminal fields may not influence spontaneous alternation performance in the T-maze (Bertholet and Crusio, 1991).

Strain differences in spontaneous alternation

Bertholet and Crusio (1991) tested males of nine strains of inbred mice in discrete trial spontaneous alternation in the T-maze, and found that XLII, CBA/H, C57BL/6, CPB-K, BALB/c, A/J, and BA mice alternated at above chance levels, whereas DBA/2J and NZB/B1NJ mice did not alternate above chance. In this study, response and spatial strategies were dissociated by rotating the maze 180° between forced and choice trials. Of the strains that alternated above chance, the BA and XLII mice used a response strategy, while the CBA/H, C57BL/6, CPB-K, BALB/c, and A/J mice had a non-significant tendency to use a spatial strategy. With delayed alternation, C57BL/6 mice alternated more than DBA/2 mice when a 2 minute delay period was used (Jones et al., 2001).

On the continuous alternation task in the T-maze, male C57BL/6 and CD-1 mice displayed a higher rate of alternation than 129/Sv, 129/SvEv, and DBA/2 mice (Gerlai, 1998). Spowart-Manning and van der Staay (2004) found that male C57BL/6J and B6D2F1 mice alternated above chance on the continuous alternation task in the T-maze, while 129S6, C57BL/6Ntac, and HsDWin:CFW1 mice did not alternate above chance. C57BL/6J, B6D2F1, 129S6, and C57BL/6Ntac mice did not differ in rates of alternation, while HsDWin:CFW1 mice alternated less than any of the other strains. Male and female heterogeneous stock (HS) mice alternate above chance in continuous alternation in the T-maze, and no difference between sexes was found (Galsworthy et al., 2002, 2005).

In the continuous Y-maze task male C57BL/6J, BALB/cByJ, and B6 × SJL/J mice displayed a higher rate of spontaneous alternation than DBA/2J, SJL/J, and CD1 mice (Heyser et al., 1999). Contet et al. (2001) did not find a difference in alternation rates between C57BL/6J and 129S2 mice on a test of continuous alternation in the Y-maze, and both strains alternated only slightly above chance.

Lashley type III maze

The Lashley type III maze was developed by Karl Lashley to investigate the effects of electrolytic lesions on learning and memory in rats (Lashley, 1929, p. 31). The maze consists of a start box, and a series of alleyways through which a mouse must navigate in order to locate a goal box (Figure 30.1e). Originally animals were motivated to find the goal box by food reward, although modified versions of the maze use escape from water to motivate animals (Denenberg et al., 1991). The opportunity for mice to return to their home cages upon reaching the goal box has also been used as a reward in this maze (Blizard et al., 2003). Learning in the Lashley type III maze is assessed using latency to reach the goal box or number of errors, which are defined as the number of cul-de-sac entries, number of wrong turns at T-choices, or backtracking towards the start box (Denenberg et al., 1991).

In the Lashley type III maze, mice learn the sequence of left or right turns required to efficiently navigate from the start box to the goal box, avoiding the cul-de-sacs. Mice may also use spatial reference memory to locate the goal box if visual cues are present in the extra-maze environment (Denenberg et al., 1991). To determine if mice are using spatial reference memory, probe trials may be given in which the maze is rotated relative to extra-maze cues. Performance of NZB mice was impaired when the Lashley type III maze was rotated relative to extra-maze cues, suggesting that NZB mice can use spatial reference memory to locate the goal box (Hyde, 1998). Performance of female C57BL/6J mice, however, did not change when the Lashley type III maze was rotated relative to extra-maze cues, or when lesions were made to the hippocampus, suggesting that female C57BL/6J mice do not use spatial reference memory to locate the goal box (Deacon et al., 2002).

Strain differences in performance

Male HS mice make fewer errors in learning the Lashley type III maze than Swiss Webster mice (Blizard et al., 2003). Female C57BL/6J mice are able to learn this maze (Deacon et al., 2002), but male 129S6 mice are unable to learn this maze (Balogh et al., 1999). Poor performance has been found in male and female NZB mice relative to RF/B and BXSB mice on a water-escape version of the Lashley type III maze (Schrott et al., 1992). The performance deficit in NZB mice was not eliminated by extending the length of the training period or by including additional intra-maze visual cues within the Lashley type III maze. The performance deficits in NZB mice may be partially due to the development of neocortical ectopias (regions of abnormally developed neurons) within layer 1 of the somatosensory cortex (Sherman et al., 1985). Indeed, NZB mice that develop neocortical ectopias perform poorly on the Lashley type III maze relative to NZB mice that do not develop ectopias (Hyde et al., 2000).

Hebb–Williams maze

The Hebb–Williams maze (HWM) was designed by Donald O. Hebb and Kenneth Williams (Hebb and Williams, 1946) as an intelligence test for rats and standardized by Rabinovitch and Rosvold (1951). The maze consists of a square closed-field surrounded by an outer wall, with a goal box and start box located at opposite corners of the maze (Figure 30.1f). Removable barriers can be placed within the maze in different spatial arrangements to alter the maze configuration. In the HWM, the mouse is placed in the start box and must navigate around the barriers to locate the goal box, which contains food reward. Other designs use water escape to motivate animals with the goal box providing an escape platform (Hoplight et al., 1996; Rosvold and Mirsky, 1954). Mouse-adapted versions of the HWM are similar to the original rat design but are decreased in size (Meunier et al., 1986; Stanford and Brown, 2003).

Rabinovitch and Rosvold (1951) developed standardized spatial arrangements of barriers for six practice and 12 test problems of varying difficulty and defined the error zones for each of the test problems. The test procedure in the HWM allows for the assessment of spatial and response learning along with reference and working memory. Mice first complete a training phase consisting of a series of six practice problems. During training the animal learns to travel from the start box to the goal box, and habituates to the maze arena so that fear and exploratory responses are reduced. The training criterion is met when mice complete the six practice problems in less than 60 seconds on two consecutive training sessions (Rabinovitch and Rosvold, 1951). Spatial learning and reference memory are assessed during the training phase, as the location of the goal box remains in a fixed location relative to the extra-maze cues throughout training.

After reaching criterion on the practice problems, animals complete 12 test problems of varying difficulty, with one or two problems completed each day. During the test phase animals know the spatial location of the goal box based on reference memory established during the training phase, but must negotiate different sets of barriers to reach it. Therefore, during the test phase mice must learn the new route around the barriers to locate the goal box. The arrangement of barriers differs with each test problem, so response-based working memory is assessed by examining how performance improves over trials (five or six) for each test problem (Hoplight et al., 1996). Learning in the HWM is measured by the latency to locate the goal box, as well as number of entries into error zones (Rabinovitch and Rosvold, 1951).

An advantage of the HWM is that the test problems are of varying difficulty. Meunier et al. (1986) divided the 12 test problems into three classes of difficulty (low, moderate, and high) with ratio scores that used the number of errors made by mice, given the number of error zones previously defined for a test problem by Rabinovitch and Rosvold (1951). Subsets of test problems, which each comprise easy, medium, and hard test problems can be given at different time points to assess changes in performance, without having to present the same test problem at two different time points. Using this method, Meunier et al. (1986) found that BALB/c mice with electrolytic lesions of the cingulate cortex showed facilitation of performance by making fewer errors than sham controls at 22 and 35 days after surgery, but were impaired and made more errors at 44 days following surgery.

Strain differences in performance

Male and female C57BL/6J and B6-H-2K mice (congenic strains which differ in all regions of the major histocompatability complex) differ in performance on a food reward paradigm on the HWM. B6-H-2K mice located the goal box faster than C57BL/6J mice, but no differences were found in errors, suggesting that although the strains differed in performance the cognitive abilities of these two strains are similar (Stanford and Brown, 2003). Furthermore, no differences in performance were found between these strains in a water-escape version of the same HWM. Taken together, these results underscore the importance of using both latency and errors to assess performance in the HWM, and, once again, suggest that strain differences in performance may depend on the type of motivation used or on other non-cognitive factors.

Sex differences were found between C57BL/6J and B6-H-2K mice in the food-reward version of the HWM, with females requiring more trials to reach criterion on the practice problems during training than males (Stanford and Brown, 2003). Females also made more errors on the first trial of each test problem, although on subsequent trials the performance did not differ between the sexes. Similar findings have been found with HS mice, as males locate the goal box faster and make fewer errors than females on the HWM (Galsworthy et al., 2002, 2005). No sex difference in errors was found in CD-1 mice, however, suggesting that sex differences on the HWM may be strain dependent (Locurto and Scanlon, 1998).

Summary

This chapter has reviewed the use of six different mazes to test strain differences in mice in cued, spatial, and response learning, as well as reference and working memory (Table 30.2). More extensive use of these mazes might be made to further

Table 30.2 Summary of strain differences in performance in the six mazes discussed in this chapter.

Maze and source	Type of learning and memory	Measure(s)	Strain differences in performance
Barnes maze			
Nguyen et al., 2000	Spatial learning	Angular deviation	C57BL/6 > 129T2, CBA, DBA/2
Patil et al., 2009	Spatial learning	Latency, distance and errors	C57BL/6 > CD-1
Koopmans et al., 2003	Spatial learning	Latency, distance and errors	C57BL/6 reliably improved performance on all measures 129S2, BALB/cBy, and Swiss did not reliably improve performance based on all measures
Holmes et al., 2002	Spatial reference memory	Escape hole visits during probe trial	C57BL/6 > DBA/2 >129S6
Fox et al., 1999	Spatial learning	Latency	C57BL/6 improved performance FVB/N and 129T2 did not improve performance
Holeboard			
Yoshida et al., 2001	Spatial learning	Latency and distance	C57BL/6N > BALB/cA
	Spatial reference memory	Time in correct quadrant	C57BL/6N, BALB/cA
Kuc et al., 2006	Spatial reference and working memory	Reference and working memory errors	Reference: DBA/2 > CD-1 Working: DBA/2, C57BL/6J > CD-1 DBA/2 > Swiss Webster
T-maze			
Balogh et al., 1999	Cued learning	Correct choices	C57BL/6 > 129S6
Schwegler and Buselmaier, 1981	Cued learning	Latency and correct choices	DBA/2, NMRI, C57BL/6 > C3H/HeJ, BALB/c, BALB/cN
Crusio et al., 1990	Response and spatial learning	Correct choices	Best performance: BALB/cByJ, NZB/B1NJ Intermediate performance: A/J, CBA/H, NZB/B1NJ, XLII, C57BL/6J, DBA/2J, CPB-K Worst performance: BA
Moy et al., 2008	Response learning	Correct trials	SJL/J > C58/J, PL/J, SWR/J
Moy et al., 2007	Response learning	% of mice reaching criterion	DBA/2J, FVB/NJ, A/J >C57BL/6J, AKR/J, C3H/HeJ
Bertholet and Crusio, 1991	Working memory	SA rate	Above chance: XLII, CBA/H, C57BL/6, CPB-K, BALB/c, A/J, and BA At/below chance: DBA/2J and NZB/B1NJ
Jones et al., 2001	Working memory	Delayed SA rate	C57BL/6 > DBA/2
Gerlai, 1998	Working memory	SA rate	C57BL/6, CD-1 > 129/Sv, 129/SvEv, and DBA/2 mice
Spowart-Manning and van der Staay, 2004	Working memory	SA rate	Above chance: C57BL/6J and B6D2F1 At/below chance: 129/Sv, C57BL/6Ntac, and HsDWin:CFW1
Y-maze			
Heyser et al., 1999	Working memory	SA rate	C57BL/6J, BALB/cByJ, C57 × SJL/J > DBA/2J, SJL/J, CD-1
Contet et al., 2001	Working memory	SA rate	C57BL/6J, 129S2
Lashley type III			
Blizard et al., 2003	Response learning	Errors	HS > Swiss Webster
Deacon et al., 2002	Response learning	Errors	C57BL/6 mice improved performance with training
Balogh et al., 1999	Response learning	Errors	129S6 mice did not improve performance with training
Schrott et al., 1992	Response learning	Errors	RF/B, BXSB > NZB
Hebb–Williams maze			
Stanford and Brown, 2003	Spatial reference and working memory	Latency and errors	Latency: B6-H-2K > C57BL/6J Errors: B6-H-2K, C57BL/6J

A>B indicates that strain A performed better than B.
A,B indicates no significant strain difference in performance.
SA: spontaneous alternation.

assess learning and memory in inbred mouse strains. However, many of these mazes are designed to study visuo-spatial learning and memory in mice, and thus strains with poor vision (i.e., albino mice), and strains with retinal degeneration, which are blind, may perform poorly, or use non-visual strategies (Brown and Wong, 2007). In addition, differences in apparatus design, test procedure, and strain differences in anxiety may influence the pattern of strain differences in learning and memory (O'Leary and Brown, 2011). Thus, to understand the cognitive processes underlying learning and memory in these mazes it is necessary to use measures of learning (i.e., errors) and to analyze search strategies or response biases to dissociate cognitive from non-cognitive factors, which underlie measures of performance (i.e., latency) in these mazes.

References

Bach, M.E., Hawkins, R.D., Osman, M., Kandel, E.R., and Mayford, M. (1995) Impairment of spatial but not contextual memory in CaMKII mutant mice with a selective loss of hippocampal LTP in the range of the θ frequency. *Cell* 81: 905–915.

Balogh, S.A., McDowell, C.S., Stavnezer, A.J., and Denenberg, V.H. (1999) A behavioral and neuroanatomical assessment of an inbred substrain of 129 mice with behavioral comparisons to C57BL/6J mice. *Brain Res* 836: 38–48.

Barnes, C.A. (1979) Memory deficits associated with senescence: a neurophysiological and behavioral study in the rat. *J Comp Physiol Psychol* 93: 74–104.

Barnes, C.A., Nadel, L., and Honig, W.K. (1980) Spatial memory deficit in senescent rats. *Can J Psychol* 34: 29–39.

Bats, S., Thoumas, J.L., Lordi, B., Tonon, M.C., Lalonde, R., and Caston, J. (2001) The effects of a mild stressor on spontaneous alternation in mice. *Behav Brain Res* 118: 11–15.

Bertholet, J-V. and Crusio, W.E. (1991) Spatial and non-spatial spontaneous alternation and hippocampal mossy fibre distribution in nine inbred mouse strains. *Behav Brain Res* 43: 197–202.

Blizard, D.A., Klein, L.C., Cohen, R., and McClearn, G.E. (2003) A novel mouse-friendly cognitive task suitable for use in aging studies. *Behav Genet* 33: 181–189.

Boehm, G.W., Sherman, G.F., Hoplight, B.J., Hyde, L.A., Bradway, D.M., Galaburda, A.M., et al. (1998) Learning in year-old female autoimmune BXSB mice. *Physiol Behav* 64: 75–82.

Bothe, G.W.M., Bolivar, V.J., Vedder, M.J., and Geistfeld, J.G. (2005) Behavioral differences among fourteen inbred mouse strains commonly used as disease models. *Comp Med* 55: 326–334.

Brown, R.E. and Wong, A.A. (2007) The influence of visual ability on learning and memory performance in 13 strains of mice. *Learn Mem* 14: 134–144.

Chang, B., Hawes, N.L., Hurd, R.E., Davisson, M.T., Nusinowitz, S., and Heckenlively, J.R. (2002) Retinal degeneration mutants in the mouse. *Vision Res* 42: 517–525.

Clapcote, S.J. and Roder, J.C. (2004) Survey of embryonic stem cell line source strains in the water maze reveals superior reversal learning of 129S6/SvEvTac mice. *Behav Brain Res* 152: 35–48.

Contet, C., Rawlins, J.N.P., and Deacon, R.M.J. (2001) A comparison of 129S2/SvHsd and C57BL/6JOlaHsd mice on a test battery assessing sensorimotor, affective and cognitive behaviours: implications for the study of genetically modified mice. *Behav Brain Res* 124: 33–46.

Crusio, W.E. (1999) Methodological considerations for testing learning in mice. In Crusio, W.E. and Gerlai, R.T. (eds.), *Handbook of Molecular-Genetic Techniques for Brain and Behavior Research*, Vol. 13. Elsevier, Amsterdam, pp. 638–651.

Crusio, W.E., Bertholet, J.-Y., and Schwegler, H. (1990) No correlations between spatial and non-spatial reference memory in a T-maze task and hippocampal mossy fibre distribution in the mouse. *Behav Brain Res* 41: 251–259.

Deacon, R.M.J., Bannerman D.M., Kirby, B.P., Croucher, A., and Rawlins, J.N.P. (2002) Effects of cytotoxic hippocampal lesions in mice on a cognitive test battery. *Behav Brain Res* 133: 57–68.

Denenberg, V.H., Talgo, N.W., Carroll, D.A., Freter, S., and Deni, R. (1991) A computer-aided procedure for measuring Lashley III maze performance. *Physiol Behav* 50: 857–861.

Denenberg, V.H., Talgo, N.W., Schrott, L.M., and Kenner, G.H. (1990) A computer-aided procedure for measuring discrimination learning. *Physiol Behav* 47: 1031–1034.

Festing, M.F.W., Simpson, E.M., Davisson, M.T., and Mobraaten, L.E. (1999) Revised nomenclature for strain 129 mice. *Mamm Genome* 10: 836.

Fox, G.B., LeVasseur, R.A., and Faden, A.I. (1999) Behavioral responses of C57BL/6, FVB/N, and 129/SvEMS mouse strains to traumatic brain injury: implications for gene targeting approaches to neurotrauma. *J Neurotrauma* 16: 377–389.

Galsworthy, M.J., Paya-Cano, J.L., Liu, L., Monleón, S., Gregoryan, G., Fernandes, C., et al. (2005) Assessing reliability, heritability and general cognitive ability in a battery of cognitive tasks for laboratory mice. *Behav Genet* 35: 675–692.

Galsworthy, M.J., Paya-Cano, J.L., Monleón, S., and Plomin, R. (2002) Evidence for general cognitive ability (g) in heterogeneous stock mice and an analysis of potential confounds. *Genes Brain Behav* 1: 88–95.

Garcia, M.F., Gordon, M.N., Hutton, M., Lewis, J., McGowan, E., Dickey, C.A., et al. (2004) The retinal degeneration (rd) gene seriously impairs spatial cognitive performance in normal and Alzheimer's transgenic mice. *Neuroreport* 15: 73–77.

Gerlai, R. (1998) A new continuous alternation task in T-maze detects hippocampal dysfunction in mice. A strain comparison and lesion study. *Behav Brain Res* 95: 91–101.

Harrison, F.E, Hosseini, A.H., and McDonald, M.P. (2009) Endogenous anxiety and stress responses in water maze and Barnes maze spatial learning tasks. *Behav Brain Res* 198: 247–251.

Hebb, D.O. and Williams, K. (1946) A method of rating animal intelligence. *J Gen Psychol* 34: 59–65.

Heyser, C.J., McDonald, J.S., Polis, I.Y., and Gold, L.H. (1999) Strain distribution of mice in discriminated Y-maze avoidance learning: genetic and procedural differences. *Behav Neurosci* 113: 91–102.

Holmes, A., Wrenn, C.C., Harris, A.P., Thayer, K.E., and Crawley J.N. (2002) Behavioral profiles of inbred strains on novel olfactory, spatial and emotional tests for reference memory in mice. *Genes Brain Behav* 1: 55–69.

Hoplight, B.J., Boehm, G.W., Hyde, L.A., Deni, R., and Denenberg, V.H. (1996) A computer-aided procedure for measuring Hebb–Williams maze performance. *Physiol Behav* 60: 1171–1176.

Hyde, L.A. (1998) *The Effects of Neocortical Ectopias on Learning and Memory in the BXSB Mouse; and Massed vs. Spaced Learning in the NZB Mouse*. Doctoral dissertation, The University of Connecticut, CT, USA.

Hyde, L.A., Sherman, G.F., Stavnezer, A.J., and Denenberg, V.H. (2000) The effects of neocortical ectopias on Lashley III water maze learning in New Zealand Black mice. *Brain Res* 887: 482–483.

Jonasson, Z. (2005) Meta-analysis of sex differences in rodent models of learning and memory: a review of behavioral and biological data. *Neurosci Biobehav Rev* 28: 811–825.

Jones, M.W., Peckham, H.M., Errington, M.L., Bliss, T.V.P., and Routtenberg, A. (2001) Synaptic plasticity in the hippocampus of awake C57BL/6 and DBA/2 mice: interstrain differences and parallels with behavior. *Hippocampus* 11: 391–396.

Koopmans, G., Blokland, A., van Nieuwenhuijzen, P., and Prickaerts, J. (2003) Assessment of spatial learning abilities of mice in a new circular maze. *Physiol Behav* 79: 683–693.

Kuc, K.A., Gregersen, K.S., Gannon, K.S., and Dodart, J.-C. (2006) Holeboard discrimination learning in mice. *Genes Brain Behav* 5: 355–363.

Lashley, K.S. (1929) *Brain Mechanisms and Intelligence. A Quantitative Study of Injuries to the Brain* (Rev. edn. 1963). Dover Publishing, New York.

Locurto, C. and Scanlon, C. (1998) Individual differences and a spatial learning factor in two strains of mice (*Mus musculus*). *J Comp Psychol* 112: 344–352.

Logue, S.F., Paylor, R., and Wehner, J.M. (1997) Hippocampal lesions cause learning deficits in inbred mice in the Morris water maze and conditioned-fear task. *Behav Neurosci* 111: 104–113.

McNaughton, B.L., Barnes, C.A., Rao, G., Baldwin, J., and Rasmussen, M. (1986) Long-term enhancement of hippocampal synaptic transmission and the acquisition of spatial information. *J Neurosci* 6: 563–571.

Meunier, M., Saint-Marc, M., and Destrade, C. (1986) The Hebb–Williams test to assess recovery of learning after limbic lesions in mice. *Physiol Behav* 37: 909–913.

Moy, S.S., Nadler, J.J., Young, N.B., Nonneman, R.J., Segall, S.K., Andrade, G.M., et al. (2008) Social approach and repetitive behavior in 11 inbred mouse strains. *Behav Brain Res* 191: 118–129.

Moy, S.S., Nadler, J.J., Young, N.B., Perez, A., Holloway, L.P., Barbaro, R.P., et al. (2007) Mouse behavioral tasks relevant to autism: phenotypes of 10 inbred strains. *Behav Brain Res* 176: 4–20.

Nguyen, P.V., Abel, T., Kandel, E.R., and Bourtchouladze, R. (2000) Strain-dependant differences in LTP and hippocampus-dependant memory in inbred mice. *Learn Mem* 7: 170–179.

O'Leary, T.P. and Brown, R.E. (2012) The effects of apparatus design and test procedure on learning and memory performance of C57BL/6J mice on the Barnes maze. *J Neurosci Meth* 203: 315–324.

O'Leary, T.P. and Brown, R.E. (2009) Visuo-spatial learning and memory deficits on the Barnes maze in the 16-month-old APPswe/PS1dE9 mouse model of Alzheimer's disease. *Behav Brain Res* 201: 120–127.

Olton, D.S. (1979) Mazes, maps, and memory. *Am Psychologist* 34: 583–596.

Patil, S.S., Sunyer, B., Höger, H., and Lubec, G. (2009) Evaluation of spatial memory of C57BL/6J and CD1 mice in the Barnes maze, the multiple T-maze and in the Morris water maze. *Behav Brain Res* 189: 58–68.

Paylor, R., Baskall-Baldini, L., Yuva, L., and Wehner, J.M. (1996) Developmental differences in place-learning performance between C57BL/6 and DBA/2 mice parallel the ontogeny of hippocampal protein kinase C. *Behav Neurosci* 110: 1415–1425.

Paylor, R., Zhao, Y., Libbey, M., Westphal, H., and Crawley J.N. (2001) Learning impairments and motor dysfunctions in adult *Lhx5*-deficient mice displaying hippocampal disorganization. *Physiol Behav* 73: 781–792.

Pompl, P.N., Mullan, M.J., Bjugstad, K., and Arendash, G.W. (1999) Adaptation of the circular platform spatial memory task for mice: use in detecting cognitive impairment in the APPsw transgenic mouse model for Alzheimer's disease. *J Neurosci Methods* 87: 87–95.

Rabinovitch, M.S. and Rosvold, H.E. (1951) A closed-field intelligence test for rats. *Can J Psychol* 5: 122–128.

Rosvold, H.E. and Mirsky, A.F. (1954) The closed–field intelligence test for rats adapted for water-escape motivation. *Can J Psychol* 8: 10–16.

Schrott, L.M., Denenberg, V.H., Sherman, G.F., Rosen, G.D., and Galaburda, A.M. (1992) Lashley maze learning deficits in NZB mice. *Physiol Behav* 52: 1085–1089.

Schwegler, H. and Buselmaier, W. (1981) Behavior genetic analysis of water-T-maze learning in inbred strains of mice, their hybrids, and selected second generation crosses. *Psychol Res* 43: 335–345.

Sherman, G.F., Galaburda, A.M., and Geschwind, N. (1985) Cortical anomalies in brains of New Zealand mice: a neuropathologic model of dyslexia? *Proc Natl Acad Sci USA* 82: 8072–8074.

Spowart-Manning, L. and van der Staay, F.J. (2004) The T-maze continuous alternation task for assessing the effects of putative cognitive enhancers in the mouse. *Behav Brain Res* 151: 37–46.

Stanford, L.E. and Brown, R.E. (2003) MHC-congenic mice (C57BL/6J and B6-H-2K) show differences in speed but not accuracy in learning in the Hebb–Williams maze. *Behav Brain Res* 144: 187–197.

Võikar, V., Kõks, S., Vasar, E., and Rauvala, H. (2001) Strain and gender differences in the behavior of mouse lines commonly used in transgenic studies. *Physiol Behav* 72: 271–281.

Wong, A.A. and Brown, R.E. (2006) Visual detection, pattern detection and visual acuity in 14 strains of mice. *Genes Brain Behav* 5: 389–403.

Yoshida, M., Goto, K., and Watanabe, S. (2001) Task-dependent strain difference of spatial learning in C57BL/6N and BALB/c mice. *Physiol Behav* 73: 37–42.

Chapter 31

Cued and contextual fear conditioning

Robert T. Gerlai

Introduction

One of the most difficult questions concerning brain function is learning and memory. Not only are these phenomena extremely complex but also they cannot be studied directly in animals. While in human memory research certain forms of memories (e.g., those called declarative or episodic) can be explicitly remembered and declared, i.e., verbally stated by the subject (Zola-Morgan and Squire, 1993), in animals a conclusion whether the subject remembers or not can only be deduced from its performance in behavioral tests. Why is this a problem? It is because performance may be influenced by a large number of factors only one of which is learning and memory. For example, changes in motor function, perception, and motivation can all alter learning performance. Thus, these so-called performance characteristics must be thoroughly investigated before a conclusion as to the ability of a subject to learn and remember can be drawn (Gerlai, 2001b). Most often performance factors are tested in separate test paradigms after or before the learning task. However, some learning paradigms allow inferences to be made about performance factors as their design is such that "built-in" control tests or behavioral measures are available. The context and cue-dependent learning task is a prime example of this, and this is one of the reasons why the task has become popular among a large number of scientists.

Context and cue-dependent fear conditioning has been one of the most frequently employed learning paradigms in behavioral neuroscience and behavioral genetic research. The paradigm was first introduced by two independent laboratories in the early 1990s (Kim and Fanselow, 1992; Phillips and LeDoux, 1992) using laboratory rats, but quickly became popular among mouse researchers too. Because of its simplicity, speed, and elegance it is utilized in numerous behavior genetic studies ranging from transgenic (Mayford et al., 1996) and knockout approaches (Abeliovich et al., 1993) to large-scale high-throughput mutagenesis screens (Reijmers et al., 2006). Natural genetic variation in performance in this task has also been analyzed by comparing mouse strains. One of the first of these studies was by Paylor et al. (1994) who investigated the differences between two inbred strains of mice, DBA/2J and C57BL/6J. What does this paradigm measure and why is it so popular?

Method of fear conditioning: dissociation of hippocampal learning from non-hippocampal learning performance

Fear conditioning utilizes the natural tendency of rodents to remain still in response to noxious stimuli including pain. As it has been argued, this behavior has probably evolved as an adaptive response to predators (Blanchard and Blanchard, 1969). Given that it occurs as a natural response, no lengthy shaping or pre-training procedure is required to make the mouse perform it appropriately: when pain or other fear-inducing stimuli are presented the mouse usually freezes. Most often pain is delivered in the form of a mild electric shock. The shock (the unconditioned stimulus, or US) can be paired with a number of stimuli (the conditioned stimulus or CS) and the association between US and CS can later be tested by presenting CS alone and measuring the freezing response. The US employed is usually a tone cue, and indeed mice can learn the association between tone and electric shock and respond to the tone alone with significant amount of freezing days or even weeks after the training (e.g., Gerlai and McNamara, 2000). Also important is the fact that this robust response can be elicited by a short, e.g., 6 minutes long (Gerlai, 1998), training during which the tone and shock are paired only a few times (most often three times or less). That is, the paradigm is extremely fast and still produces a robust memory. Another appealing and perhaps even more important feature of the paradigm is its ability to dissociate hippocampal learning processes from other characteristics. In addition to the tone cue and shock association, rats and mice are capable of learning the place where they received the shock. When probed by placing them into the shock chamber subsequent to the training session (usually 24 h later), they will remember the place, usually referred to as the "context," and will respond to it by elevated freezing. Thus, even without the salient tone cue that was made contiguous with the tone, rodents can exhibit signs of fear in the original shock chamber. This context response was shown by Kim and Fanselow (1992) and

Behavioral Genetics of the Mouse: Volume I. *Genetics of Behavioral Phenotypes*, eds. Wim E. Crusio, Frans Sluyter, Robert T. Gerlai, and Susanna Pietropaolo. Published by Cambridge University Press.

Phillips and LeDoux (1992) to be dependent upon the hippocampal formation in rats and later by others, including our laboratory (Gerlai, 1998), in mice. According to current theories, the hippocampus supports relational learning of which contextual learning is a subtype (e.g., Dusek and Eichenbaum, 1997). By binding loosely related pieces of memories (information) together (e.g., Moscovitch and Nadel, 1998), the hippocampus enables the organism to remember complex spatial or episodic-like memories, a function crucial for the appropriate identification of the context in which a shock was delivered. Thus the experimenter can test whether the subject learned and remembered the context (context test, hippocampal function), whether it learned and remembered the cue (cue test, non-hippocampal function including performance factors such as motor function, perception, and motivation), and whether the subject could respond to the shock (training, performance factors including pain perception and motor function). In sum, the context and cue-dependent fear conditioning paradigm is fast, elicits robust memory, and allows the dissociation of numerous factors involved in the learning process. In the following pages we will discuss the power and utility of this paradigm and how it has been employed in the comparison of mouse strains.

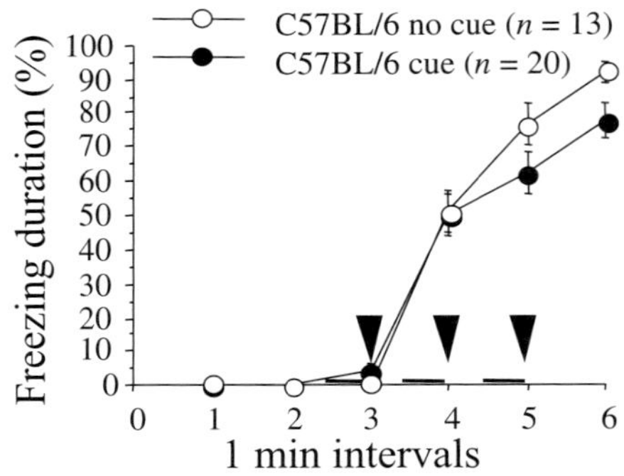

Figure 31.1 Repeated presentation of mild electric foot shocks leads to increased amount of freezing in C57BL/6J mice. Means ± SEM are shown. Arrowheads indicate the presentation of electrical shock (1 s, 0.7 mA). Short solid lines indicate the presentation of the tone cue (80 dB, 2900 Hz, 20 s long). Solid circles show the performance of mice that received the tone cue and the open circles show those that did not receive the tone during training. Sample sizes (*n*) are also indicated. Note the robust increase of freezing in both groups of mice. For statistical analysis and details see Gerlai (1998).

Differences between C57BL/6J and DBA/2J: the question of background and foreground cues

As stated above, Paylor et al. (1994) were perhaps the first to discover the robust difference between the above two inbred mouse strains in fear conditioning. Their results indicated that while DBA/2J mice are capable of learning the associative tone cue just like C57BL/6J, their performance in response to the context alone was significantly impaired. The impairment was unlikely to be due to reduced pain or inability to perform freezing as DBA/2J mice showed robust freezing responses to the shock during training and they also responded well to the tone cue. The impaired context response was also unlikely to be due to motivational problems, e.g., lack of fear, because DBA/2J mice responded well, i.e., showed strong freezing, to the tone cue. The latter finding also suggested that these mice do not suffer from a general learning or memory deficit. Thus, because of the context-specific aspect of their impairment, the results suggested that DBA/2J mice suffered from hippocampal dysfunction (Paylor et al., 1994). Numerous hippocampal abnormalities have been identified in DBA/2J (for examples see Nguyen and Gerlai, 2002) and thus any number of these may represent the underlying causes, including abnormal structure (e.g., smaller infra- and intrahippocampal mossy fiber projections; Schwegler et al., 1988), abnormal enzyme activity (e.g., protein kinase C; Wehner et al., 1990), and abnormal synaptic plasticity (e.g., CA1 long-term potentiation; Nguyen et al., 2000).

Gerlai (1998) also investigated the differences between DBA/2J and C57BL/6J using the context and cue-dependent fear conditioning paradigm and discovered that under certain circumstances DBA/2J mice were able to respond to the context with robust freezing. I will discuss these results in more detail as they illuminate an important general problem inherent in most studies investigating complex hippocampal learning and they also allow us to theorize about the basis of differences between the mnemonic capabilities of these two strains of mice.

First let us examine freezing performance during training. Figure 31.1 shows how much C57BL/6J mice froze during the 6 minutes of the training session. For the first 3 minutes of the session mice were allowed to roam around undisturbed. A short (20 second) tone cue terminated together with a 1-second-long 0.7-mA electric shock at the end of the 3rd, 4th, and 5th minute and mice started to exhibit an increasing amount of freezing, a clear indication of their elevated fear. The reader may notice that Figure 31.1 shows two line diagrams. The solid circles show the results for C57BL/6J mice that received the tone–shock pairing and the open circles represent the values for those mice that received only the shock but no tone cue. Importantly, both treatments elicited a robust fear response demonstrating that it is the pain-inducing shock stimulus that leads to the freezing response during training and not the tone cue. Twenty-four hours later mice were tested either in the cue or the context probe test. It is notable that in most studies mice are usually tested in both instead of one or the other probe tests. The reason why most investigators opt for this is that this way they have larger *N* values and thus more statistical power. The potential problem with this is that the exposure to the first probe test may interfere with, or affect performance in, the second probe test. Thus, we decided to test our mice in only one or the other probe test.

The results of these tests are shown for C57BL/6J mice in Figures 31.2 and 31.3. When presented with a tone cue C57BL/6J mice responded with a robust increase of freezing (Figure 31.2) but only if they received the tone–shock pairing the day before. If they received the shock treatment alone and no tone cue during training, they did not respond to the

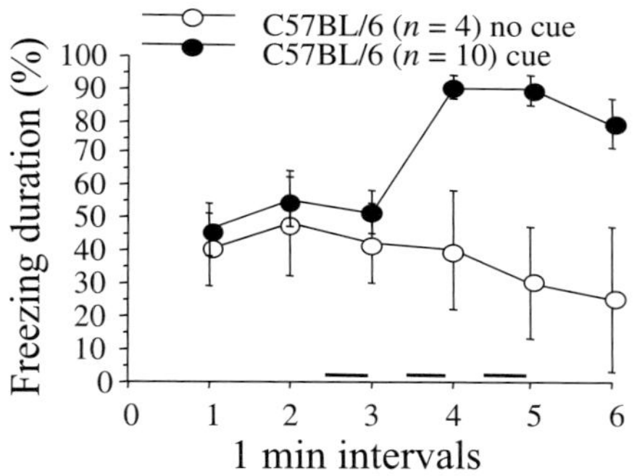

Figure 31.2 Presentation of the associative tone cue leads to increased amount of freezing in C57BL/6J mice 24 hours after training. Means ± SEM are shown. Short solid lines indicate the presentation of tone cue (80 dB, 2900 Hz, 20 s long). Solid circles show the performance of mice that received the tone cue during training (24 h before) and the open circles show those mice that did not receive the tone during training. Sample sizes (*n*) are indicated. Note the robust increase of freezing of mice that received the tone–shock pairing indicating the acquisition of the association between tone and shock. Also note that both groups of mice show a somewhat elevated amount of freezing even before the tone presentation despite the fact that the test chamber represents a context different from that of the training chamber. This suggests some level of generalized fear or the presence of a residual context. For more discussion on this topic please see text. For statistical analysis and details see Gerlai (1998).

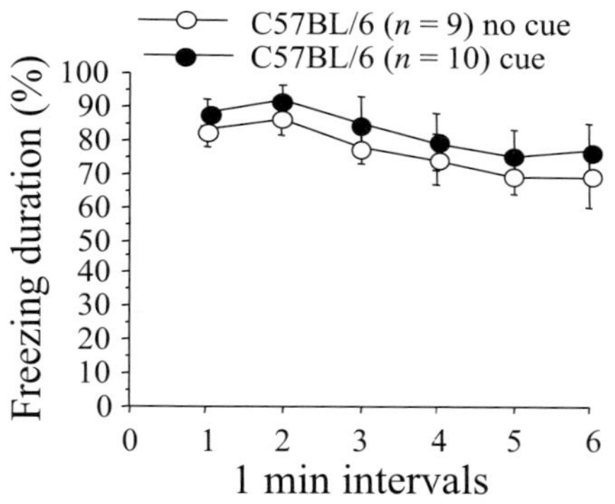

Figure 31.3 Exposure to the context where shocks were delivered 24 hours before leads to dramatic elevation of freezing in C57BL/6J mice. Means ± SEM are shown. Solid circles show the performance of mice that received the tone cue during training and the open circles show those that did not receive the tone during training. Sample sizes (*n*) are also indicated. Note that no shocks or tone cues are presented during the context probe task. For statistical analysis and details see Gerlai (1998).

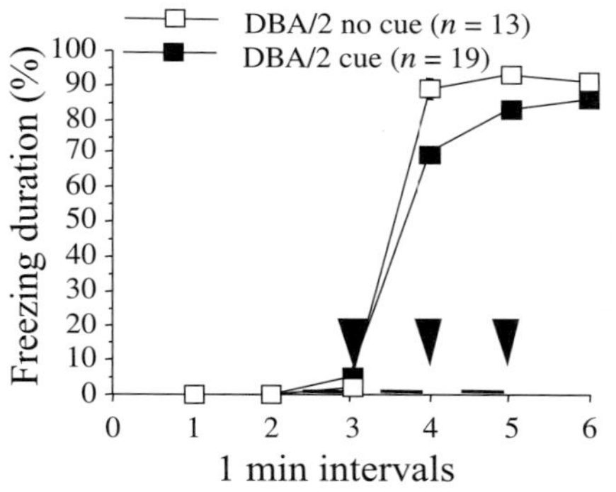

Figure 31.4 Repeated presentation of mild electric foot shocks leads to increased amount of freezing in DBA/2J mice. Means ± SEM are shown. Arrowheads indicate the presentation of shock (1 s, 0.7 mA). Short solid lines indicate the presentation of tone cue (80 dB, 2900 Hz, 20 s long). Solid squares show the performance of mice that received the tone cue and the open squares show those that did not receive the tone during training. Sample sizes (*n*) are also indicated. Note the robust increase of freezing in both groups of mice. For statistical analysis and details see Gerlai (1998).

tone cue the following day. This is not unexpected but clearly shows that the tone response does not represent a generally increased sensitization to *any* stimuli but rather it is the result of CS and US association, i.e., it is a result of associative learning. Another important observation may be made from Figure 31.2. All mice showed some freezing even before tone cue presentation, a freezing response that is significantly above the level of what was recorded before the first shock was presented during training (Figure 31.1). This is an interesting observation given that the test chamber in which the tone cue probe test was conducted was made different from the shock chamber in terms of visual, olfactory, and tactile cue characteristics. Most often investigators refer to context A (the shock chamber) and context B (the tone cue probe test chamber) to emphasize the fact that the context during tone test is different. Although the two chambers are indeed different, it appears that at least some aspects of the shock context are still apparent to the mice and, as their freezing levels indicate, they identify the tone cue probe test situation as somewhat dangerous even before the tone cue is presented. In other words, there is some overlap between context A and B, an assumption that is easy to explain if one realizes that both contexts were in the same test room with external visual, olfactory, and auditory stimuli identical, and also in both contexts the behavioral recording is preceded by handling of the animals by the same experimenter. Another potential (but less likely) explanation is that C57BL/6J experienced a generalized increase in fear.

When placed in the shock chamber where they previously received the shock treatment (context A), C57BL/6J mice responded with a high level of freezing (Figure 31.3), indicating that they remembered the place where they were shocked before. This memory appears independent of whether the mice received the shock paired with the tone or received the shock not paired with a tone cue, an important point to which we will return shortly when we examine the performance of DBA/2J mice. Briefly, C57BL/6J mice are usually found excellent in hippocampal tasks and thus their good freezing performance to the context was expected (e.g., Crawley et al., 1997).

What was unexpected, however, was the way DBA/2J mice responded to the context, a finding we will examine shortly; but, first, let me briefly discuss their performance during training. DBA/2J mice responded with robust freezing to repeated exposure to the electric shock (Figure 31.4). In fact they appeared to respond somewhat faster and more robustly compared to C57BL/6J mice (Figure 31.1 vs. Figure 31.4), a small difference that has been observed by others too (e.g., Fitch et al., 2002; Paylor et al., 1994). Note that again we had two groups of mice, one of which received the tone cue and another that did not (both received the shock, of course). The following day, when

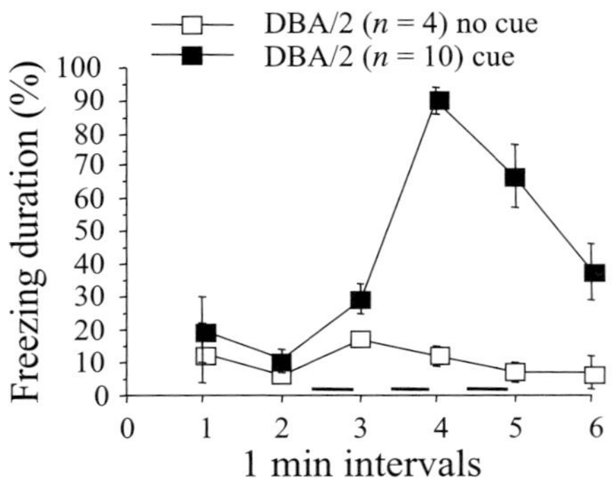

Figure 31.5 Presentation of the associative tone cue leads to increased amount of freezing in DBA/2J mice 24 hours after training. Means ± SEM are shown. Short solid lines indicate the presentation of tone cue (80 dB, 2900 Hz, 20 s long). Solid squares show the performance of mice that received the tone cue during training (24 h before) and the open squares show those mice that did not receive the tone during training. Sample sizes (*n*) are indicated. Note the robust increase of freezing of mice that received the tone–shock pairing, indicating the acquisition of the association between tone and shock. Also note the different freezing trajectories of this graph as compared to that of the C57BL/6J results (Figure 31.2). For example, unlike C57BL/6J, DBA/2J mice do not show elevated amount of freezing before the tone presentation and those that received the tone–shock pairing habituate their tone response faster than C57BL/6J. For more discussion on this topic please see text. For statistical analysis and details see Gerlai (1998).

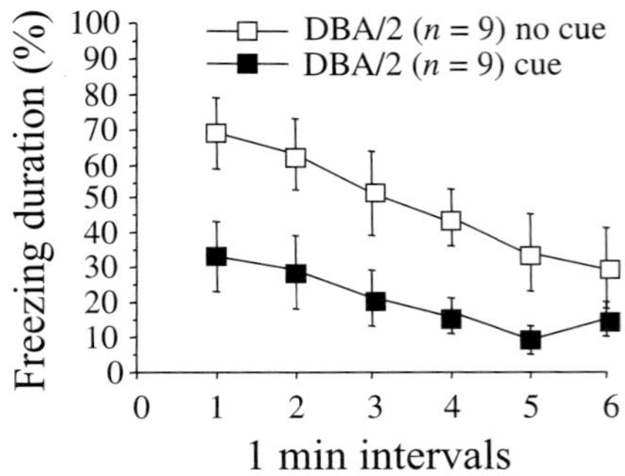

Figure 31.6 Exposure to the context where shocks were delivered 24 hours before leads to dramatic elevation of freezing in DBA/2J mice. Means ± SEM are shown. Solid squares show the performance of mice that received the tone cue during training and the open squares show those that did not receive the tone during training. Sample sizes (*n*) are also indicated. Note that no shocks or tone cues are presented during the context probe task. Note the robust difference between the two groups of mice indicating that presentation of tone cue during training significantly reduced the freezing response of DBA2J mice when tested 24 hours later in the shock context. For statistical analysis and details see Gerlai (1998).

some of these mice were tested for the tone response, those that received the tone–shock pairing responded to the tone cue with a robust elevation of freezing and those that did not get the pairing pretty much ignored the tone cue as expected (Figure 31.5). Interestingly, however, the tone response habituated fast as the mice gradually reduced their freezing level despite the repeated presentation of the cue, a finding that may not be a random error but rather a characteristic response of DBA/2J shown by others as well (e.g., Fitch et al., 2002; Paylor et al., 1994).

Perhaps the most interesting difference between C57BL/6J and DBA/2J mice was the way they responded to the context, and particularly how the presence or absence of a tone CS during training affected their context response (Figure 31.6). As we have seen before, C57BL/6J mice were not affected by the presence of the tone during training: irrespective of the past tone presentation they responded strongly to the context alone, i.e., they appeared to remember the place where they were shocked before. However, DBA/2J mice were significantly impaired compared to C57BL/6J in their context response if they received the tone cue–shock pairing during training, but much less so if they did not receive the tone cue. It was known before that DBA/2J mice are impaired at contextual fear conditioning and other relational learning tasks requiring the acquisition of complex relationships of diffuse cues, such as those characterizing a particular place or spatial location (e.g., Crawley et al., 1997). But why did the presentation of a tone cue during training made them more impaired in their ability to respond to the context?

Phillips and LeDoux (1994) put forth an interesting hypothesis with regard to the differences between "foreground" and "background" cues. According to their reasoning, foreground cues are very few, salient, and robust stimuli whereas background cues can be highly numerous, more diffuse, and less salient. In their paper they show that hippocampal lesions primarily affect the ability of rats to process background cues whereas the learning of foreground cues remains largely unaffected. This reasoning is in line with what we know about the function of the hippocampal formation, i.e., its ability to "tie" loosely connected pieces of information together. Association between a pair of salient cues does not require the hippocampus, whereas learning several cues characterizing the context that are always present, i.e., both before and after as well as during shock presentation, and none of which stand out as a particularly distinguishing feature of the chamber, should require the relational processing power of the hippocampus. This is exactly what Gerlai (1998) proposed was behind the differences between DBA/2J and C57BL/6J mice. When DBA/2J mice were presented with a salient associative cue together with the shock, they could learn this cue well but could not associate the shock with the less salient and more diffuse contextual background cues. However, when they were not provided with a salient tone cue with which to associate the shock, they could presumably focus on a single aspect of the context and treat it as a foreground cue. It is unclear what exactly this cue might have been but it could be a tactile cue (the metal bars delivering the shock), a visual cue (the walls of the apparatus), or an olfactory cue (the smell of the cleaning agent used). Once learned, this single cue could be used by DBA/2J to help them respond to the shock chamber even without "identifying" whether it is the same context or not. In other words, DBA/2J mice may have converted the contextual task into a single cue association task. But again, they could do it only if no salient single associative cue was experimentally provided for them during training. Unlike DBA/2J mice, C57BL/6J mice, however, were able to learn the context irrespective of whether or not they were given the salient tone cue. Their normally functioning hippocampus was able to perform the operation termed "automated recording of attended experience," as proposed by Morris and Frey (1997), and thus could learn both the salient associative cue and the contextual background cues at the same time.

The conclusion one may draw from the above results is several fold. First and foremost, it is clear that when the experimenter tries to teach his or her experimental subjects they may learn things differently than what the experimenter expected. Two, the learning task may be highly genotype-dependent. Most often it is assumed that the learning task is dependent upon the procedural and technical aspects exactly as designed by the experimenter. However, from the above it is clear that the experimental subject plays just as important a role with regard to what the learning task will do and what it will quantify. This argument is particularly important for the large number of mouse knockout studies in which "standard" tests are used without considering the possibility that these tests may be solved differently in a mutation-dependent manner. Similarly, when investigating naturally occurring variation, for example, when comparing inbred strains of mice, this possibility must also be considered, as the above example clearly demonstrated.

Other strains, other complications: impaired perception and different coping strategies

An important issue with which I started this chapter is that one cannot measure learning and memory directly, and thus any factor that influences the performance of mice in a learning task may lead one to believe that learning and memory is altered. Perception is an important performance factor that may influence behavioral performance. For example, blind mice will not be able to solve a task in which visual stimuli must be perceived, such as the water maze. This may seem a trivial problem but apparently it is still not well appreciated (for an experimental example and discussion on this topic see Gerlai et al., 2002). In the latter paper, mice carrying a null mutation at the *mGluR8* locus were analyzed: mGluR8 is a subtype of metabotropic, i.e., G-protein coupled, glutamate neurotransmitter receptor that was expected to be involved in hippocampal learning and memory. These mice were originally generated by others, and in the hope of making a robust and easy to breed mutant line, the host strain for the mutation was chosen to be ICR. ICR is an outbred albino strain that originates from CD1, another outbred albino strain (see e.g., Adams et al., 2002). Both are indeed quite healthy, grow quickly, and breed well. Unfortunately, however, in addition to being albino, ICR mice also carry the retinal degeneration (rd) allele and become virtually blind by their age of 3 months. Thus the original request by the molecular biologists that generated the mice to test these animals in the water maze, a visuo-spatial learning task, turned out to be a complete failure (Gerlai et al., 2002). However, fear conditioning was still an option because in this latter task modalities other than visual can also be used to acquire information about both foreground (the tone) and background (the contextual) cues. Indeed, using this task, subtle deficits were detected in mGluR8 null mutants as compared to age and background genotype-matched control wild type (Gerlai et al., 2002). However, it was notable that neither the mutant nor the wildtype animals performed much freezing at all in any phase of the task. Even after repeated electric shocks the mice appeared to freeze a lot less than mice from the DBA/2J and C57BL/6J strains (Gerlai et al., 2002). A possible explanation for this might be that a new behavioral quantification method was used and this may have detected immobility differently. The new method used was based on force transducer technology and allowed automated quantification of the acceleration forces the moving animal generated in the apparatus (Fitch et al., 2002) as opposed to observation of the crouching posture (Blanchard and Blanchard, 1969). However, comparison of the CD1 and C57BL/6J strains using the force transducer technology suggested otherwise (Figure 31.7). Clearly, C57BL/6J and CD1 differed dramatically. Although CD1 and ICR increased their freezing in response to the shock during training and in response to the context and to the tone cue during probe tests, it was clear that they could stay still for only short periods of time (Adams et al., 2002). There may be a number of potential explanations for the low freezing values seen in ICR and CD1. For example, motor characteristics (e.g., CD1 and ICR may not be able to stay still), or altered pain reaction (ICR and CD1 may be less sensitive to shock), or perhaps altered antipredatory strategies (active escape as opposed to passive avoidance). Activity monitoring of the mice in open fields and water mazes did not suggest increased levels of locomotion in the CD1 and ICR strains as compared to C57BL/6J, thus "hyperactivity" may be ruled out; in fact C57BL/6J appears to be slightly more active under some conditions (see e.g., Crawley et al., 1997; Gerlai et al., 2002). Notably, reduced pain sensitivity in CD1 as compared to C57BL/6J may also be ruled out. The latter conclusion is based on the results obtained during training in the fear conditioning paradigm: when CD1 mice received the electric shock their immediate flinching or jumping response as quantified by measuring the force their jump generated using the force transducer technology (Fitch et al., 2002) was not less than that of C57BL/6J mice (Gerlai et al., 2002). In fact, it was slightly higher and more immediate; i.e., it reached its peak at the very first shock presentation. Thus reduced pain sensitivity is not likely. It is therefore more plausible that the reason for less freezing in CD1 and ICR mice has to do with a more active avoidance strategy these mice perform in response to the occurrence of pain in fear conditioning.

The moral of this story is that not only perception but the manifestation of fear responses may also be different among genetic variants of the mouse and focusing on a single behavioral measure in a single test may be misleading. Fear responses and generally defensive behavioral responses may be complex and an animal under threat may exhibit numerous motor and posture patterns (e.g., Blanchard et al., 2003). Similarly,

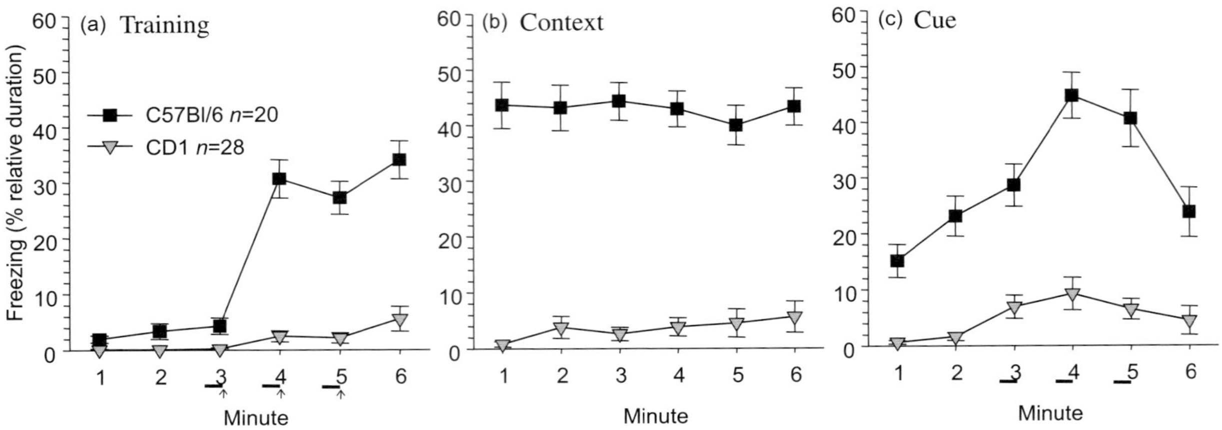

Figure 31.7 Mice from the CD1 outbred and C57BL/6J inbred strains show dramatically different amount of freezing in the cued and context–dependent fear conditioning tasks as measured using the force transducer technology. Means ± SEM are shown. Solid circles represent the CD1 and gray triangles the C57BL/6J mice. Sample sizes (*n*) are also indicated. Arrows indicate the presentation of shock (1 s, 0.7 mA). Short solid lines indicate the presentation of the tone cue (80 dB, 3000 Hz, 20 s long). Note that freezing is quantified automatically using force transducer technology. (a) Panel shows the freezing responses quantified during training. (b) Panel shows the freezing responses in the context probe test performed 24 hours after training. (c) Panel shows the freezing responses to the tone cue tested also 24 hours after training. For statistical analysis and details see Gerlai et al. (2002).

Gerlai et al. (1999), for example, measured multiple behavioral characteristics in their fear conditioning paradigm and found that increased fear was associated with increased long-body (stretch attend) posture and decreased grooming.

In sum, strain differences can manifest in multiple ways and a very few simple measures may not capture this complexity. Such strain differences may allow one to study how complex biological phenomena are organized at the mechanistic (genetic) level. But these strain differences also have practical relevance. They may allow one to pick and choose the strain most appropriate as a host for a particular null mutation, or they may allow one to select a genetic background best suited for the analysis of the effects of a pharmaceutical compound. In the following section, we detail one example for the latter.

Strain differences: a solution for ceiling and floor effects

Gerlai et al. (1999) wanted to analyze the *in vivo* function of a receptor tyrosine kinase family, the Eph tyrosine kinases. The family represents the largest number of receptor tyrosine kinases and their role in the adult brain was not known despite the clear indication that the kinases were expressed there. The investigators decided to use a somewhat unorthodox strategy they termed "protein targeting" (Gerlai, 2000; Gerlai et al., 1998) with the use of genetically engineered highly selective antibody-like proteins (EphA5-IgG immunoadhesin, an antagonist, and ephrinA5-IgG immunoadhesin, an agonist). These immunoadhesins allowed them to functionally shut down or activate an entire subclass of receptor tyrosine kinases, the Eph-A receptors. The question they posed was whether the deactivation and/or activation of the kinases would have any effect on behavior and brain function of the treated animals, particularly whether their hippocampal learning and synaptic plasticity (the manipulation of the kinases was mainly restricted to the hippocampus) could be altered. The investigators decided to use C57BL/6J and DBA/2J mice as their subjects. Although these mice showed no differences in their Eph tyrosine kinase receptor or ephrin (the ligand for the kinase receptor) expression pattern and activity, the argument was that these strains differ dramatically in their hippocampal learning and synaptic plasticity. The "bad" strain (DBA/2J) would have a large room to improve and the "good" strain (C57BL/6J) a large room to worsen as a result of the activation or inhibition of the EphA system. The implicit assumption behind this is the argument that manipulation of Eph kinases should be able not only to impair but also to improve neural plasticity even beyond "normal" function (Gerlai, 2001a).

The manipulation worked: the antagonist immunoadhesin EphA5-IgG disrupted fear conditioning in a context-specific manner leaving cued responses during the probe and shock responses during training intact in C57BL/6J mice, whereas the agonist immunoadhesin ephrinA5-IgG was able to improve fear conditioning performance in a context-specific manner in DBA/2J mice (Gerlai et al., 1999, also see Figure 31.8). These changes were confirmed using other hippocampal-dependent behavioral tasks and also using electrophysiological quantification of synaptoplastic changes in the hippocampus (Gerlai et al., 1999). In summary, the two mouse strains with characteristic learning performance and synaptic plasticity patterns allowed the investigators to avoid ceiling (inability to improve on an already excellent performance score) and floor (inability to impair an already suboptimal performance score) effects.

Fear conditioning and other mouse strains

Although C57BL/6J and DBA/2J are perhaps two of the most frequently used inbred strains in behavioral studies, and CD1

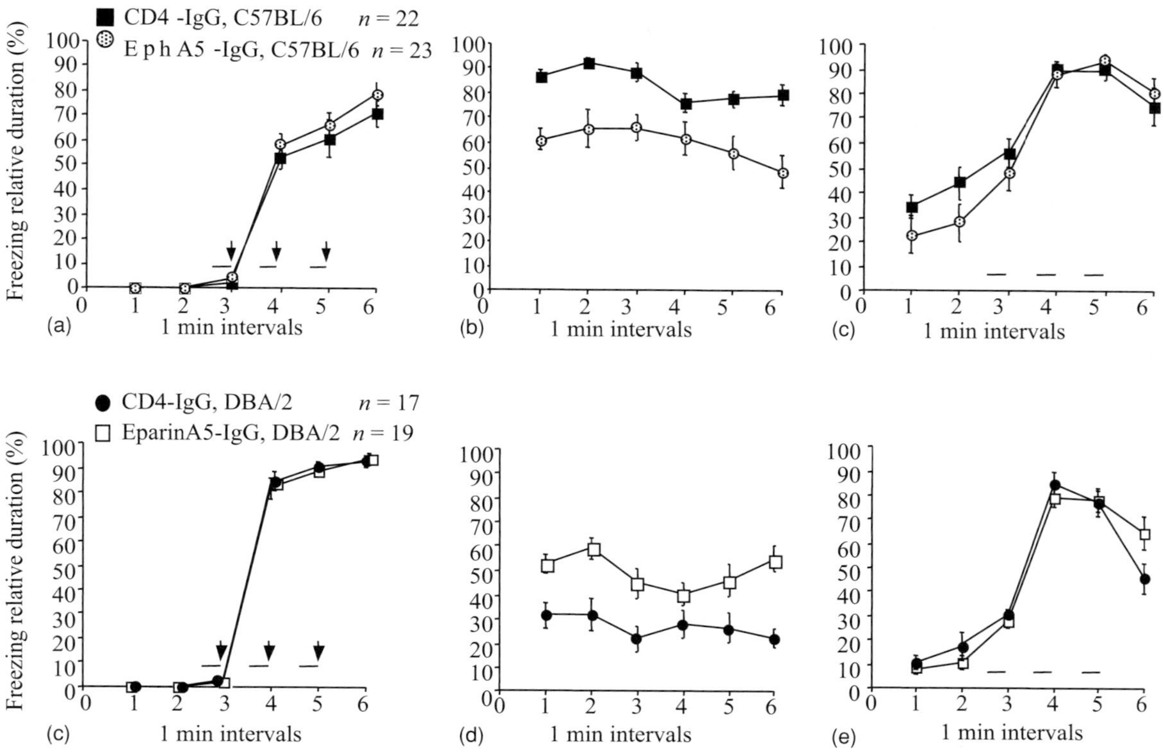

Figure 31.8 Context-specific impairment of learning performance in C57BL/6J and context-specific improvement of learning performance in DBA/2J mice after 7 days long localized bilateral hippocampal infusion of the antagonist EphA5-IgG and the agonist ephrinA5-IgG immunoadhesins. Means ± SEM are shown. Note that as a control a biophysically similar but functionally inert immunoadhesin CD4-IgG was infused (control mice). Solid square: control C57BL/6J, gray circle: EphA5-IgG infused C57BL/6, solid circle: control DBA/2J, open square: ephrinA5-IgG infused DBA/2J. Sample sizes (*n*) are also indicated. (a and d) Panels show the training responses, i.e., freezing measured using observation-based methods and the Observer (Noldus, Wageningen, The Netherlands) software. (b and e) Panels show the freezing responses observed in the context probe test, and (c and f) panels show the freezing responses observed in the tone cue probe test. Arrows indicate the shock and the short solid lines the tone cue presentation. For statistical analysis and details see Gerlai et al. (1999).

and its substrains, e.g., the ICR, are perhaps the most frequently used outbred strains, other mouse strains have also been thoroughly investigated using the context- and cue-dependent fear conditioning paradigm. Bolivar et al. (2001) compared several inbred strains including 129S1/SvImJ, A/J, BALB/cByJ, C3H/HeJ, C57BL/6J, CBA/J, DBA/2J, and FVB/NJ as well as C3.BLiA^{+Pde6b}, which is congenic with C3H but does not have the retinal degeneration allele, rd. Their systematic analysis of these strains is exemplary as far as activity levels are concerned. Their quantification technique involved analysis of photobeam breaks recorded automatically by infrared light sensitive photocells. Generally, levels of activity measured this way and amount of freezing negatively correlate and thus the Bolivar et al. (2001) experiment is useful when comparing the strains in terms of their fear conditioning performance. However, this activity measure does not differentiate grooming, sitting, rearing, leaning, sleeping, and other behaviors from freezing, behaviors that would be recorded as lack of locomotor activity (no photobeam breaks), and this represents an important limitation. It is also important to realize that even small locomotor movements may be missed by the photocell detectors that are usually placed a couple of centimeters apart from each other and often only in one dimension of the chamber (as must have been the case with the San Diego monitoring device used in the above study). Nevertheless, the general conclusion of Bolivar et al. (2001) is irrefutable: inbred strains differ from each other. Another important conclusion of this study is that gender differences should also be considered as they are often genotype-, i.e., strain-, dependent.

In a more recent study, Balogh and Wehner (2003) analyzed memory performance of multiple inbred mouse strains (A/Ibg, AKR/J, BALB/cByJ, CBA/J, C3H/HeIbg, C57BL/6J, DBA/2J, LP/J, SJL/J, and 129S6/SvEvTac) not only the usual 24 hours after fear conditioning training, but also 14 and 60 days later. The authors again found robust strain differences. Importantly, there was a very good correlation between 24 hours and 60 day post-training performance, indicating that the differences, i.e., the relative ranking of inbred strains in their cued and context responses remained even after this extended period of time. Based on these results, the authors concluded as follows:

In general, however, these data are consistent with the notion that those genes that regulate genetic variation in initial acquisition also have a major impact on long-term memory. Since the correlations are less than one, these data also suggest that there may be unique, but much smaller, subsets of genes regulating variation in retention across time. Thus, polymorphisms in additional genes that vary across strains, but are not involved in acquisition, might regulate retention. (Balogh and Wehner, 2003)

It must be noted, however, that the variation in "initial acquisition" always manifests in the way the animal performs later at a memory task because the information acquired, and thus the memory that is stored and can be recalled, depends on this initial phase. Thus genes that affect acquisition will indeed influence later memory performance, but this effect does not necessarily prove the direct involvement of these genes in the "regulation of retention" or other processes. In other words, the Balogh and Wehner (2003) study did not dissociate the process of acquisition from those of memory formation including consolidation, maintenance, or recall of memory. Nevertheless, the main conclusion as to potential additional genetic effects influencing memory formation in a genotype-dependent manner in inbred strains is probably correct.

It is also important to note that although a significant decay of freezing responses was evident after 60 days, even at this time point there was significant freezing present in mice of several strains. If one assumes that the amount of freezing reflects the strength of memory (i.e., the amount of fear remembered), a reasonable assumption, then these results indicate the presence of long-lasting memory of the aversive training event.

Last, and quite interestingly, instead of a linear time-dependent progressive decline of freezing values, Balogh and Wehner (2003) discovered that at 14 days post-training mice of many strains showed an elevated freezing response that appeared to be rather generalized, i.e., was shown both to the tone and to the context and even before the tone in a different context. This "incubation of fear" effect is noteworthy as it bares resemblance to what one may define in the human clinic as "anxiety" and thus may be an appropriate model for testing anxiolytic compounds for pharmacotherapy (Balogh and Wehner, 2003).

Fear conditioning: a collection of several paradigms

The last point we would like to consider in this chapter is the fact that fear conditioning is a theme and not a paradigm. There are many ways one can condition and test fear-related memories. Up till now we have focused on the context- and cue-dependent fear conditioning with the use of three electric shocks, a tone cue, and a short 6 minute training session, a set of procedures and parameters that is most frequently used in the analysis of the mouse and the rat. The fact that these parameters were used, i.e., the fact that the training and testing procedures were standardized, is important as it allows comparison of results across laboratories. However important standardization may be, a rigid test paradigm may not allow capturing the true complexity of fear responses and related mnemonic characteristics. Thus, numerous variations of several aspects of fear conditioning have been employed. A more systematic review of these methods will be presented in a separate volume of this book series. Here, only a single example will be given of a variant of the paradigm to illustrate how small changes in the procedure of fear conditioning may be utilized in the characterization of inbred mouse strains.

Holmes et al. (2002) compared C57BL/6J, 129S6, and DBA/2J strains of mice in cued and contextual fear conditioning, but instead of the more frequently used contiguity between tone cue and shock, which in most paradigms are made to terminate together, they employed a method called "trace" conditioning. In trace conditioning there is a small temporal gap between the presentation of the tone cue and the shock. As reviewed in Holmes et al. (2002) this trace conditioning method imposes a form of temporal complexity on learning. The hippocampus is known to be involved in relational memory and the sorts of relationships it processes and encodes may involve temporal factors (e.g., episodic memory). Therefore, when temporal complexity is imposed the hippocampus is expected to be involved. Indeed, trace conditioning has been shown to be sensitive to hippocampal dysfunction (e.g., reviewed in Holmes et al., 2002). Comparison of the three inbred strains showed that, as expected, C57BL/6J mice performed well in both the context and the cue probe test, whereas DBA/2J did poorly on both. Interestingly, the third strain, 129S6, also performed well in both the context and the cued probes, suggesting normal hippocampal function. This is notable because a considerable amount of discussion has been devoted to whether a group of strains all designated 129 are appropriate hosts for null mutations. Some of these strains have been used to generate embryonic stem (ES) cells and thus gene targeting is carried out on these genetic backgrounds (see, e.g., Gerlai, 1996). However, several of these strains possess serious abnormalities ranging from agenesis of corpus callosum, to impaired learning and memory performance, difficulty with breeding, etc. (see, e.g., Gerlai, 1996). As a result of these problems, most scientists opted for crossing the 129-type mouse carrying the null allele with a mouse from another strain, most often C57BL/6, and this hybridization led to a number of genetic problems (Gerlai, 1996). However, according to the results by Holmes et al. (2002) and also others (reviewed in Holmes et al., 2002), not all 129 strains are created equal and some of them perform well in learning tasks and thus may not need to be avoided, or to be hybridized with other inbred strains, in gene targeting studies in which impairments in learning and memory are expected as a result of the null mutation.

Conclusions

Perhaps the most important conclusion that is valid for all learning paradigms and maybe even for all behavioral tasks in general is that although the experimenter may have a clear idea what his or her subjects will need to do and what his or her test is designed to measure, whether the subjects actually conform to these ideas and do as they are expected to is another matter. In a learning task, mice may find strategies alternative to the learning processes the experimenter wants to investigate. The change in their behavior as a result of learning, i.e., the phenotypical manifestation of memory, may be different from what the experimenter expected. And the different learning strategies or the phenotypical expression of the acquired memories may be genotype-dependent. Another important, and again fairly general, point to remember is that often complex behaviors such as learning and memory cannot be directly measured and the inferences we make about them are all dependent upon numerous performance factors including motor function, perceptual function, and motivation. Thus a single task conducted in a standard manner is almost always not sufficient to really discover the true nature of the differences among mouse strains or the changes elicited by mutations introduced in an artificial manner. Last, comparison of mouse strains is important for numerous reasons. Here, we mainly focused on practical issues, including how one can measure fear-related memories and how one can utilize inbred strains for other genetic approaches including gene targeting. However, strain comparisons may be also used as a discovery tool, for example allowing the eventual identification of quantitative trait loci (QTL) or analysis of the potential selection forces that shaped the behavior, i.e., understanding of the evolutionary past, topics that are well covered in several other chapters of this Handbook.

References

Abeliovich, A., Paylor, R., Chen, C., Kim, J.J., Wehner, J.M., and Tonegawa, S. (1993) PKC gamma mutant mice exhibit mild deficits in spatial and contextual learning. *Cell* 75: 1263–1271.

Adams, B., Fitch, T., Chaney, S., and Gerlai, R. (2002) Altered performance characteristics in cognitive tasks: comparison of the albino ICR and CD1 mouse strains. *Behav Brain Res* 133: 351–361.

Balogh, S.A. and Wehner, J.M. (2003) Inbred mouse strain differences in the establishment of long-term fear memory. *Behav Brain Res* 140: 97–106.

Blanchard, D.C., Griebel, G., and Blanchard, R.J. (2003) the mouse defense test battery: pharmacological and behavioral assays for anxiety and panic. *Eur J Pharmacol* 463: 97–116.

Blanchard, R.J. and Blanchard, D.C. (1969) Crouching as an index of fear. *J Comp Physiol Psychol* 67: 370–375.

Bolivar, V.J., Pooler, O., and Flaherty, L. (2001) Inbred strain variation in contextual and cued fear conditioning behaviour. *Mamm Genome* 12: 651–656.

Crawley, J.N., Belknap, J.K., Collins, A., Crabbe, J.C., Frankel, W., Henderson, N., et al. (1997) Behavioral phenotypes of inbred mouse strains: implications and recommendations for molecular studies. *Psychopharmacology* 132: 107–124.

Dusek, J.A. and Eichenbaum, H. (1997) The hippocampus and memory for orderly stimulus relations. *Proc Natl Acad Sci USA* 94: 7109–7114.

Fitch, T., Adams, B, Chaney, S., and Gerlai, R. (2002) Force transducer based movement detection in fear conditioning in mice: a comparative analysis. *Hippocampus*, 12: 4–17.

Gerlai, R. (1996) Gene targeting studies of mammalian behavior: is it the mutation or the background genotype? *Trends Neurosci* 19: 177–181.

Gerlai, R. (1998) Contextual learning and cue association in fear conditioning in mice: a strain comparison and a lesion study. *Behav Brain Res* 95: 191–203.

Gerlai, R. (2000) Protein targeting: altering receptor kinase function in the brain. *Trends Neurosci* 23: 236–239.

Gerlai, R. (2001a) Eph tyrosine kinase receptors and neural plasticity. *Nat Rev Neurosci* 2: 205–209.

Gerlai, R. (2001b) Behavioral tests of hippocampal function: simple paradigms, complex problems. *Behav Brain Res* 125: 269–277.

Gerlai, R., Adams, B., Fitch, T., Chaney, S., and Baez, M. (2002) Performance deficits of mGluR8 knockout mice in learning tasks: the effects of null mutation and the background genotype. *Neuropharmacology* 43: 235–249.

Gerlai, R., Cairns B., Van Bruggen, N., Moran P., Shih A., Sauer, H., et al. (1998) Protein targeting in the analysis of learning and memory: a potential alternative approach to gene targeting. *Exp Brain Res* 123: 24–35.

Gerlai, R. and McNamara, A. (2000) Anesthesia induced retrograde amnesia is ameliorated by ephrinA5-IgG in mice: evidence for Eph receptor tyrosine kinase involvement in mammalian memory. *Behav Brain Res* 108: 133–143.

Gerlai, R., Shinsky, N., Shih, A., Williams, P., Winer, J., Armanini, M., et al. (1999) Regulation of learning by EphA receptors: a protein targeting study. *J Neurosci* 19: 9538–9549.

Holmes, A., Wrenn, C.C., Harris, A.P., Thayer, K.E., and Crawley, J.N. (2002) Behavioral profiles of inbred strains on novel olfactory, spatial and emotional tests for reference memory in mice. *Genes Brain Behav* 1: 55–69.

Kim, J.J. and Fanselow, M.S. (1992) Modality-specific retrograde amnesia of fear. *Science* 256: 675–677.

Mayford, M., Bach, M.E., Huang, Y.Y., Wang, L., Hawkins, R.D., and Kandel, E.R. (1996) Control of memory formation through regulated expression of a CaMKII transgene. *Science* 274: 1678–1683.

Morris, R.G. and Frey, U. (1997) Hippocampal synaptic plasticity: role in spatial learning or the automatic recording of attended experience? *Philos Trans R Soc Lond B Biol Sci* 352: 1489–1503.

Moscovitch, M. and Nadel, L. (1998) Consolidation and the hippocampal complex revisited: in defense of the multiple-trace model. *Curr Opin Neurobiol* 8: 297–300.

Nguyen, P.V., Abel, T., Kandel, E.R., and Bourtchouladze, R. (2000) Strain-dependent differences in LTP and hippocampus-dependent memory in inbred mice. *Learn Mem* 7: 170–179.

Nguyen, P.V. and Gerlai, R. (2002) Behavioural and physiological characterization of inbred mouse strains: prospects for elucidating the molecular mechanisms of mammalian learning and memory. *Genes Brain Behav* 1: 72–81.

Paylor, R., Tracy, R., Wehner, J., and Rudy, J.W. (1994) DBA/2J and C57BL/6J mice differ in contextual fear but not auditory fear conditioning. *Behav Neurosci* 108: 810–817.

Phillips, R.G. and LeDoux, J.E. (1992) Differential contribution of amygdala and hippocampus to cued and contextual fear conditioning. *Behav Neurosci* 106: 274–285.

Phillips, R.G. and LeDoux, J.E. (1994) Lesions of the dorsal hippocampal formation interfere with background but not foreground contextual fear conditioning. *Learn Mem* 1: 34–44.

Reijmers, L.G., Coats, J.K., Pletcher, M.T., Wiltshire, T., Tarantino, L.M., and Mayford, M. (2006) A mutant mouse with a highly specific contextual fear-conditioning deficit found in an N-ethyl-N-nitrosourea (ENU) mutagenesis screen. *Learn Mem* 13: 143–149.

Schwegler, H., Crusio, W.E., Lipp, H.-P., and Heinrich, B. (1988) Watermaze learning in the mouse correlates with variation in hippocampal morphology. *Behav Genet* 18: 153–165.

Wehner, J.M., Sleight, S., and Upchurch, M. (1990) Hippocampal PKC activity is reduced in poor spatial learners. *Brain Res* 523: 181–187.

Zola-Morgan, S. and Squire, L.R. (1993) Neuroanatomy of memory. *Annu Rev Neurosci* 16: 547–563.

Chapter 32

Taste and odor

Hans Welzl and David P. Wolfer

Taste and odor are essential for food selection and odor also for social interaction in mice. Several tastes and odors are innately linked to specific behaviors, reflecting their importance for survival. The significance of many gustatory and olfactory stimuli, however, has to be learned. Since associations of tastes and odors with specific consequences are biologically significant, they are quickly acquired and long lasting. For this reason, gustatory and olfactory learning is ideal for investigating mechanisms of learning and long-term memory formation in mice.

Similarly to what has been found with other learning and memory paradigms (see previous chapters in this section), genetic background modifies learning about the significance of tastes and odors (as will be discussed below). The background can directly affect mechanisms of synaptic plasticity underlying memory formation, or it can indirectly affect learning by modifying how task-specific sensory stimuli are perceived or by influencing task-associated motivational and emotional states.

Taste memory

Gustatory learning is learning about nutritional food qualities. Although basic tastes such as "sweet" or "bitter" are innately accepted or rejected depending on their concentration (e.g., Grill and Norgren, 1978; Ninomiya et al., 1984; Reilly and Pritchard, 1996), the nutritional consequences of many tastes have to be learned. Learning follows the rules of classical conditioning with a novel taste serving as the conditioned stimulus (CS). The unconditioned stimuli (USs) are food components that either have a desirable nutritional value or cause malaise. Learning is the association between the taste of a food and the appetitive or aversive consequences following its ingestion. As a result food with a specific taste will be approached or avoided in the future. For the investigation of memory formation the acquisition of a conditioned taste aversion (CTA) seems to be especially well suited due to its fast acquisition and easy experimental manipulation (Bures et al., 1998; Welzl et al., 2001).

The ability to acquire a CTA can vary from mouse strain to mouse strain (Belknap et al., 1977; Broadbent et al., 1996, 2002; Dudek and Fuller, 1978; Horowitz and Whitney, 1975; Risinger and Cunningham, 1992, 1995, 1998, 2000;). Strain differences have been detected with different tastes as CSs and different drugs as USs (but see also Ingram, 1982; Orsini et al., 2004). The difficulty is to determine whether the salience of the CS, the sensitivity to the aversive US, or the ability to associate the CS with the US is affected by the genetic background.

Sweet tasting fluids are frequently used CSs in CTA. Mouse strain differences in the perception of sweet stimuli have been investigated already more than 30 years ago. Pelz et al. (1973) suggested that the variability of saccharin preference among strains is based on their genetic variability. Fuller (1974) concluded from his research with C57BL/6J (B6) mice, DBA/2J (D2) mice, and derived crosses that a single locus (*Sac*) is the major determinant of saccharin preference. More recent research found the *Sac* locus located on chromosome 4 (Lush et al., 1995; Phillips et al., 1994) and corresponding to the *Tas1r3* gene that encodes the sweet taste receptor T1r3 (Bachmanov et al., 2001; Li et al., 2001; Max et al., 2001).

Whether mice preferred or avoided saccharin depended on the genotype of the mouse and the concentration of the solution (unpublished data by Cooper, 1966; cited in Fuller, 1974). B6 mice preferred saccharin already at lower concentrations than D2 mice, and preference for saccharin was much larger in B6 than D2 mice. 129P3/J mice seem to have a low saccharin preference similar to that of D2 mice (e.g., Lush, 1989).

In light of the strain differences in the salience of saccharin it is surprising that B6 and D2 mice acquired a CTA equally well. When a sweet solution (saccharin or sucrose) was the CS and lithium chloride the malaise inducing US, CTA was similar in B6 and D2 mice (Belknap et al., 1977; Ingram, 1982). It is possible that with the selected intensities of CS and US both strains reached optimal levels of learning, and consequently a ceiling effect covered any contribution the differential sensitivity to saccharin might have had.

Besides saccharin many other tastes served as CSs in CTA. The degree of preference or avoidance of these tastes might depend on the genetic background and influence the acquisition of an aversion. For example, mouse strains have been found to differ in their response to NaCl (e.g., Beauchamp and Fisher, 1993), umami-tasting solutions (e.g., Bachmanov et al., 2000; Murata et al., 2009), bitter substances (e.g., Blizard et al., 1999; Cagniard and Murphy, 2009; St. John and Boughter, 2004), ethanol (e.g., Belknap et al., 1977; Cagniard and

Behavioral Genetics of the Mouse: Volume I. *Genetics of Behavioral Phenotypes*, eds. Wim E. Crusio, Frans Sluyter, Robert T. Gerlai, and Susanna Pietropaolo. Published by Cambridge University Press.

Murphy, 2009), and lipids (Glendinning et al., 2008). For a general review on genetics and taste in mice, see Boughter and Bachmanov (2007). Some of these taste qualities have been used as the CS in CTA as will be described below.

Mouse studies reporting a strain difference in CTA predominantly attributed the difference to a genetic influence on perception of the US. With NaCl as the CS and lithium chloride as the US, mice from the D2 strain developed a stronger CTA than B6 mice (Risinger and Cunningham, 2000). D2 mice were also more sensitive to lithium chloride than B6 mice in a place aversion paradigm. Mainly for these reasons the authors concluded that strain differences in lithium chloride pharmacokinetics rather than differences in general learning abilities were responsible for their results. This conclusion was supported by the observation that after an injection of lithium chloride the accumulation of lithium in different body tissues and the elimination of lithium is strain dependent (El-Kassem and Singh, 1983; Smith, 1978).

With ethanol (or its metabolite acetaldehyde) as the US, D2 mice acquired a strong aversion against the associated CS whereas B6 mice developed no or only a much smaller aversion (Belknap et al., 1977; Broadbent et al., 1996, 2002; Dudek and Fuller, 1978; Horowitz and Whitney, 1975; Risinger and Cunningham, 1992, 1995, 1998). B6 mice, however, learned the task when the strength of the US was increased by injecting a higher dose of ethanol (Risinger and Cunningham, 1992). The combined observation of ethanol's effects on CTA and other behaviors suggested that D2 mice are more sensitive to the aversive properties of ethanol than B6 mice. The relatively continuous distribution of strain means in CTA in BXD recombinant inbred mice (Risinger and Cunningham, 1998) and in 15 different mouse strains (Broadbent et al., 2002) indicated an influence of multiple genes on this behavioral phenotype.

B6 and D2 mice also differ in other ethanol-sensitive behavioral and physiological measures such as locomotor activity, place preference (Cunningham et al., 1992), hyperglycemic response (Risinger and Cunningham, 1992), hypothermia (Broadbent et al., 2002; Risinger and Cunningham, 1998), ataxia, or acute withdrawal (Risinger and Cunningham, 1998), but not all differences correlated with differences in CTA. Further, the correlation between the effects of ethanol on CTA with that on other behaviors seemed to depend on the experimental design. Whereas the study with BXD mice found a correlation between CTA and hypothermia (Risinger and Cunningham, 1998), the study with 15 different mouse strains did not find such a correlation (Broadbent et al., 2002).

Injection of nicotine as the US after drinking NaCl as the CS led to a marked CTA in D2, but not in B6 mice (Risinger and Brown, 1996). The results resemble those obtained with ethanol as the US, mentioned above. The literature, however, provides no support for a correlation between the strain differences in nicotine-induced CTA and in nicotine metabolism or brain nicotine receptor densities (for discussion, see Risinger and Brown, 1996).

Amphetamine as US induced CTA of similar strength in B6 and D2 mice (Orsini et al., 2004). These data are surprising in light of the strain difference in amphetamine-induced place preference. B6 mice preferred the amphetamine-paired compartment whereas D2 mice avoided it. Thus, the strain differences possibly affected the rewarding but not the aversive properties of amphetamine. They also suggest that when a US elicits malaise of comparable strength in D2 and B6 mice, the two strains are equally able to learn the association.

Strain differences might also affect the extinction of a CTA. Crabbe and coworkers investigated CTA in inbred C57BL/6By and BALB/cBy mice and their crosses with a glucose solution as the CS and ethanol as the US (Crabbe et al., 1982). The speed of extinction of CTA varied among strains, and this was possibly due to a strain difference in the susceptibility to the aversive effects of ethanol. Mice perceiving the US as weak developed only a marginal CS–US association; as a consequence extinction of the association was accelerated. Extinction of CTA was also different in B6 and D2 mice (Ingram, 1982). After pairing sucrose with lithium chloride, extinction was faster in B6 than in D2 mice. Again a lower sensitivity to the aversive properties of the US might have led to a weaker association between sucrose and ethanol, and consequently to a faster extinction.

B6 and 129P3/J mice also differed in appetitive conditioning of taste preferences (Sclafani and Glendinning, 2005). B6 mice developed a stronger preference for a CS associated with intragastric infusion of a nutrient (sucrose or soybean oil) than D2 mice. Strain differences in the orosensory response led to differences in conditioned preference. Using a separate concentration of the CS for each strain eliminated strain differences in unconditioned CS preference. Pairing these adjusted CSs with intragastric injection of nutrients equally increased CS preference in B6 and D2 mice.

In conclusion, the majority of studies on strain differences in CTA used B6 and D2 mice as subjects with a few studies including other strains. Strain differences in the acquisition of CTA primarily depended on their different sensitivities to the aversive properties of the US. The strains also differ in their degree of preferences for tastes used as CSs, and this might also affect CTA acquisition. So far clear-cut evidence for an influence of genetic background on CTA via an effect on general learning abilities is lacking.

Experimental manipulation of several genes known to affect synaptic plasticity interfered with the formation of CTA (e.g., Balschun et al., 2003; Cui et al., 2005; Janus et al., 2004; Masugi et al., 1999; Pennanen et al., 2004; Plath et al., 2006). Thus, these genes are also critical for acquiring a taste aversion. However, this does not inevitably mean that allele variability of these genes is responsible for the variability in CTA strength between mouse strains.

Odor memory

Mice possess two separate olfactory systems that are involved in numerous forms of learning. The main olfactory system detects a vast variety of volatile odorants that can be associated with many different behavioral and physiological responses. The accessory olfactory system primarily detects pheromones and mediates predominantly fixed patterns of responses in the context of social interactions (for review, see e.g., Brennan, 2001; Brennan and Keverne, 1997; Wilson and Stevenson, 2003).

Olfactory sensitivity varies among mouse strains. Several studies investigating such variations used a protocol similar to that for conditioned taste aversion (Griff and Reed, 1995; Pourtier and Sicard, 1990; Wysocki et al., 1977). Mice drank from a bottle in the presence of a specific odor, and drinking was followed by an injection of lithium chloride. Animals that perceived the odor and learned the odor–malaise association avoided drinking from the bottle with the aversively conditioned odor during a choice test. B6 mice failed to detect low concentrations of isovaleric acid, a substance present in sweat, urine, and vaginal secretions (Griff and Reed, 1995; Pourtier and Sicard, 1990; Wysocki et al., 1977). At the same low concentrations several other strains (A/J, AKR/J, BALB/cJ, C3HeB/FeJ, D2, SJL/J, SWR/J) readily detected the presence of isovaleric acid. A linkage study with B6, D2, and BXD mice and their F1 hybrids and N2 backcrosses revealed candidate loci for the recessive anosmia for isovaleric acid on chromosomes 4 and 6 (Griff and Reed, 1995). Strains did not differ in their ability to detect other olfactory stimuli.

In an olfactory discrimination task, mice learned to dig for food morsels in a cup only when the covering material had a specific scent (cinnamon or nutmeg). B6 and D2 mice acquired this task equally well. However, D2 mice learned discrimination reversals faster than B6 mice (Mihalick et al., 2000). In a very similar task where again specific odors (e.g., rosemary, thyme, cinnamon) indicated the presence or absence of food, 129/SvEv mice outperformed B6 mice already during acquisition (Colacicco et al., 2002). The F1 hybrids had an intermediate success rate in detecting the scented cup that contained the food. With repeated testing both strains successfully reduced their errors to a similar low rate. In a more complex device, D2 mice failed to acquire an olfactory discrimination in contrast to several other strains (BALB/cByllco, CD-1, 129S2/SvPasCr1, B6; Restivo et al., 2006). The two studies indicate a role of genetic background on olfactory discrimination learning, but the exact outcome depends on procedural details and which strains are included in the comparison.

Another and more popular olfactory discrimination task is social transmission of food preference. A group-housed mouse is allowed to eat specifically flavored food in a separate cage. After transferring this demonstrator mouse back to its group cage the other mice immediately approach the demonstrator and sample the smell of the flavored food. After this social transmission of a food odor, the observer mice prefer to eat food with the now familiar odor instead of food with an unknown odor (Kogan et al., 1996). Surprisingly, three different strains (B6, 129S6, D2) acquired the task equally well (Holmes et al., 2002). It is important to keep in mind that normal social behavior is necessary to acquire this discrimination. This has been demonstrated with mutants displaying abnormal social interactions (Drew et al., 2007). These mice were unable to acquire a socially transmitted food preference but were unimpaired in other olfactory as well as non-olfactory learning tasks.

Olfactory learning also can be involved in tasks that primarily focus on non-olfactory learning. In the radial maze, a task designed to investigate spatial memory, mice used odor cues depending on genetic background and experimental protocol (Roullet et al., 1993). When the use of a radial strategy to solve the task was prevented, BALB/c mice exclusively used olfactory cues whereas B6 relied on spatial cues. CB6F1 hybrids were able to use either type of information to successfully navigate through the maze.

Strain differences in olfactory sensitivity and in control of behavior by olfactory stimuli are reflected in anatomical and physiological differences in the olfactory bulb. TH-immunoreactivity in the glomerular region of D2 and BALB/cBy mice was more intense in old than in young mice (Mirich et al., 2002). B6 mice did not show such an age-dependent increase. In contrast to the two other strains, B6 mice had noticeable larger astrocytes in all layers of the olfactory bulb. Further, kindling in the olfactory bulb developed significantly faster in D2 than in C3H inbred mice with the sensitivity of B6 mice lying in between but closer to the values of D2 than C3H mice (Green and Elizondo, 1989).

Staggerer mice, a mutant strain with reduced mitral cell number in the olfactory bulb, did not learn an associative olfactory task (Deiss and Baudoin, 1999). They habituated normally to two olfactory stimuli but were unable to use these stimuli to locate a reward. Mutant mice also displayed a reduced sensitivity for an aversive and an attractive odor (Deiss and Baudoin, 1997). Staggerer mice had several other motor, sensorimotor and cognitive deficits and altered cerebellar anatomy. The underlying genetic cause is very likely a recessive gene on chromosome 9 coding for RORα, a nuclear hormone receptor that seems to be crucial for axon extension, synaptogenesis, and myelination in the central nervous system (Sashihara et al., 1996).

In conclusion, mouse strains differ in their ability to learn and remember the significance of odor stimuli in one (Colacicco et al., 2002), but not in another learning paradigm (Holmes et al., 2002). The detected strain effects were small, and the data left it open whether these differences were due to differences in general learning abilities, or in other aspects, e.g., detecting and processing of olfactory stimuli.

As shown for taste memory, experimentally mutating genes known to be crucial for synaptic plasticity also affected olfactory learning and memory (e.g., Kogan et al., 1996; Mayeux-Portas et al., 2000; Rampon et al., 2000; Shimshek et al., 2005). Again, these findings neither support nor reject the possibility that allele variability of these genes is responsible for the variability in olfactory learning between mouse strains.

Summary

Several researchers found an influence of genetic background on gustatory and olfactory learning whereas others did not. How big a strain difference showed up depended on the type of procedure that was selected to measure learning. This is illustrated by the fact that strain differences observed in one type of olfactory discrimination task (Colacicco et al., 2002; Mihalick et al., 2000) were lacking in another one (Holmes et al., 2002).

Strains also differed in their gustatory and olfactory sensitivity, and rewarding or punishing stimuli used in taste and odor learning paradigms were perceived differently by different strains. As mentioned above, lithium chloride can be a more or less aversive stimulus, i.e., more or less effective in producing malaise, depending on how fast it is metabolized and eliminated. Finally, a stimulus of a specific intensity, such as a specific concentration of a saccharin solution, might be attractive for one strain, but less so, neutral, or even aversive for another one (Fuller, 1974).

In conclusion, the current data do not unrestrictedly support a strain difference in genes involved in synaptic plasticity. The extent of strain differences varied from task to task and was highly dependent on procedural details. In a number of cases learning differences could be traced back to differences in the perception of task-specific stimuli. However, one would expect that strain differences in synaptic plasticity genes affects performance to lesser or larger degree under different procedural conditions. That genes involved in synaptic plasticity are also important for taste and odor memory is supported by numerous experiments that found impairments after inactivation of synaptic plasticity genes – such as those encoding the transcription factor CREB or different glutamate receptors – by genetic engineering.

Chapter abbreviations

B6, C57BL/6J strain of mice
D2, DBA/2J strain of mice
CS, conditioned stimulus
CTA, conditioned taste aversion
US, unconditioned stimulus

Acknowledgments

This work was supported by the National Center of Competence in Research (NCCR) "Plasticity and Repair."

References

Bachmanov, A.A., Li, X., Reed, D.R., Ohmen, J.D., Li, S., Chen, Z., et al. (2001) Positional cloning of the mouse saccharin preference (*Sac*) locus. *Chem Senses* 26: 925–933.

Bachmanov, A.A., Tordoff, M.G., and Beauchamp, G.K. (2000) Intake of umami-tasting solutions by mice: a genetic analysis. *J Nutr* 130: 935S–941S.

Balschun, D., Wolfer, D.P., Gass, P., Mantamadiotis, T., Welzl, H., Schütz, G., et al. (2003) Does cAMP response element-binding protein have a pivotal role in hippocampal synaptic plasticity and hippocampus-dependent memory? *J Neurosci* 23: 6304–6314.

Beauchamp, G.K. and Fisher, A.S. (1993) Strain differences in consumption of saline solutions by mice. *Physiol Behav* 54: 179–184.

Belknap, J.K., Belknap, N.D., Berg, J.H., and Coleman, R. (1977) Preabsorptive vs. postabsorptive control of ethanol intake in C57BL/6J and DBA/2J mice. *Behav Genet* 7: 413–425.

Blizard, D.A., Kotlus, B., and Frank, M.E. (1999) Quantitative trait loci associated with short-term intake of sucrose, saccharin and quinine solutions in laboratory mice. *Chem Senses* 24: 373–385.

Boughter, J.D., Jr. and Bachmanov, A.A. (2007) Behavioral genetics and taste. *BMC Neurosci* 8 (Suppl 3): S3.

Brennan, P.A. (2001) The vomeronasal system. *Cell Mol Life Sci* 58: 546–555.

Brennan, P.A. and Keverne, E.B. (1997) Neural mechanisms of mammalian olfactory learning. *Progr Neurobiol* 51: 457–481.

Broadbent, J., Linder, H.V., and Cunningham, C.L. (1996) Genetic differences in naloxone enhancement of ethanol-induced conditioned taste aversion. *Psychopharmacol* 126: 147–155.

Broadbent, J., Muccino, K.J., and Cunningham, C.L. (2002) Ethanol-induced conditioned taste aversion in 15 inbred mouse strains. *Behav Neurosci* 116: 138–148.

Bures, J., Bermudez-Rattoni, F., and Yamamoto, T. (1998) *Conditioned Taste Aversion. Memory of a Special Kind.* Oxford University Press, New York, USA.

Cagniard, B. and Murphy, N.P. (2009) Endogenous nociception modulates diet preference independent of motivation and reward. *Physiol Behav* 96: 412–420.

Colacicco, G., Welzl, H., Lipp, H.-P., and Würbel, H. (2002) Attentional set-shifting in mice: modification of a rat paradigm, and evidence for strain-dependent variation. *Behav Brain Res* 132: 95–102.

Crabbe, J.C., Rigter, H., and Kerbusch, S. (1982) Analysis of behavioural responses to an ACTH analog in CXB/By recombinant inbred mice. *Behav Brain Res* 4: 289–314.

Cui, Z., Lindl, K.A., Mei, B., Zhang, S., and Tsien, J.Z. (2005) Requirement of NMDA receptor reactivation for consolidation and storage of nondeclarative taste memory revealed by inducible NR1 knockout. *Eur J Neurosci* 22: 755–763.

Cunningham, C.L., Niehus, D.R., Malott, D.H., and Prather, L.K. (1992) Genetic differences in the rewarding and activating effects of morphine and ethanol. *Psychopharmacol* 107: 385–393.

Deiss, V. and Baudoin, C. (1997) Hyposmia for butanol and vanillin in mutant staggerer male mice. *Physiol Behav* 61: 209–213.

Deiss, V. and Baudoin, C. (1999) Olfactory learning abilities in staggerer mutant mice. *C R Acad Sci III* 322: 467–471.

Drew, C.J., Kyd, R.J., and Morton, A.J. (2007) Complexin 1 knockout mice exhibit marked deficits in social behaviours but appear to be cognitively normal. *Hum Molec Genet* 16: 2288–2305.

Dudek, B.C. and Fuller, J.L. (1978) Task-dependent genetic influences on behavioral response of mice (*Mus musculus*) to acetaldehyde. *J Comp Physiol Psychol* 92: 749–758.

El-Kassem, M. and Singh, S.M. (1983) Strain dependent rate of Li+ elimination associated with toxic effects of lethal doses of lithium chloride in mice. *Pharmacol Biochem Behav* 19: 257–261.

Fuller, J.L. (1974) Single-locus control of saccharin preference in mice. *J Hered* 65: 33–36.

Glendinning, J.I., Feld, N., Goodman, L., and Bayor, R. (2008) Contribution of orosensory stimulation to strain differences in oil intake by mice. *Physiol Behav* 95: 476–483.

Green, R.C. and Elizondo, B.J. (1989) A non-stereotaxic method for olfactory bulb kindling reveals distinct kindling rates among inbred mouse strains. *J Neurosci Meth* 27: 109–113.

Griff, I.C. and Reed, R.R. (1995) The genetic basis for specific anosmia to isovaleric acid in the mouse. *Cell* 83: 407–414.

Grill, H.J. and Norgren, R. (1978) The taste reactivity test. I. Mimetic responses to gustatory stimuli in neurologically normal rats. *Brain Res* 143: 263–279.

Holmes, A., Wrenn, C.C., Harris, A.P., Thayer, K.E., and Crawley, J.N. (2002) Behavioral profiles of inbred strains on novel olfactory, spatial and emotional tests for reference memory in mice. *Genes Brain Behav* 1: 55–69.

Horowitz, G.P. and Whitney, G. (1975) Alcohol-induced conditioned aversion: genotypic specificity in mice (*Mus musculus*). *J Comp Physiol Psychol* 89: 340–346.

Ingram, D.K. (1982) Lithium chloride-induced taste aversion in C57BL/6J and DBA/2J mice. *J Gen Psychol* 106: 233–249.

Janus, C., Welzl, H., Hanna, A., Lovasic, L., Lane, N., St George-Hyslop, P., et al. (2004) Impaired conditioned taste aversion learning in APP transgenic mice. *Neurobiol Aging* 25: 1213–1219.

Kogan, J.H., Frankland, P.W., Blendy, J.A., Coblentz, J., Marowitz, Z., Schütz, G., et al. (1996) Spaced training induces normal long-term memory in CREB mutant mice. *Curr Biol* 7: 1–11.

Li, X., Inoue, M., Reed, D.R., Huque, T., Puchalski, R.B., Tordoff, M.G., et al. (2001) High-resolution genetic mapping of the saccharin preference locus (Sac) and the putative sweet taste receptor (T1R1) gene (Gpr70) to mouse distal Chromosome 4. *Mamm Genome* 12: 13–16.

Lush, I.E. (1989) The genetics of tasting in mice. VI. Saccharin, acesulfame, dulcin and sucrose. *Genet Res* 53: 95–99.

Lush, I.E., Hornigold, N., King, P., and Stoye, J.P. (1995) The genetics of tasting in mice. VII. Glycine revisited, and the chromosomal location of Sac and Soa. *Genet Res* 66: 167–174.

Masugi, M., Yokoi, M., Shigemoto, R., Muguruma, K., Watanabe, Y., Sansig, G., et al. (1999) Metabotropic glutamate receptor subtype 7 ablation causes deficit in fear response and conditioned taste aversion. *J Neurosci* 19: 995–963.

Max, M., Shanker, Y.G., Huang, L., Rong, M., Liu, Z., Campagne, F., et al. (2001) Tas1r3, encoding a new candidate taste receptor, is allelic to the sweet responsiveness locus Sac. *Nat Genet* 28: 58–63.

Mayeux-Portas, V., File, S.E., Stewart, C.L., and Morris, R.J. (2000) Mice lacking the cell adhesion molecule Thy-1 fail to use socially transmitted cues to direct their choice of food. *Curr Biol* 10: 68–75.

Mihalick, S.M, Langlois, J.C., and Krienke, J.D. (2000) Strain and sex differences on olfactory discrimination learning in C57BL/6J and DBA/2J inbred mice (*Mus musculus*). *J Comp Psychol* 114: 365–370.

Mirich, J.M., Williams, N.C., Berlau, D.J., and Brunjes, P.C. (2002) Comparative study of aging in the mouse olfactory bulb. *J Comp Neurol* 454: 361–372.

Murata, Y., Beauchamp, G.K., and Bachmanov, A.A. (2009) Taste perception of monosodium glutamate and inosine monophosphate by 129P3/J and C57BL/6ByJ mice. *Physiol Behav* 98: 481–488.

Ninomiya, Y., Higashi, T., Katsukawa, H., Mizukoshi, T., and Funakoshi, M. (1984) Qualitative discrimination of gustatory stimuli in three different strains of mice. *Brain Res* 322: 83–92.

Orsini, C., Buchini, F., Piazza, P.V., Puglisi-Allegra, S., and Cabib, S. (2004) Susceptibility to amphetamine-induced place preference is predicted by locomotor response to novelty and amphetamine in the mouse. *Psychopharmacol* 172: 264–270.

Pelz, W.E., Whitney, G., and Smith, J.C. (1973) Genetic influences on saccharin preference of mice. *Physiol Behav* 10: 263–265.

Pennanen, L., Welzl, H., D'Adamo, P., Nitsch, R.M., and Götz, J. (2004) Accelerated extinction of conditioned taste aversion in P301L tau transgenic mice. *Neurobiol Dis* 15: 500–509.

Phillips, T.J., Crabbe, J.C., Metten, P., and Belknap, J.K. (1994) Localization of genes affecting alcohol drinking in mice *Alcohol Clin Exp Res* 18: 931–941.

Plath, N., Ohana, O., Dammermann, B., Errington, M.L., Schmitz, D., Gross, C., et al. (2006) Arc/Arg3.1 is essential for the consolidation of synaptic plasticity and memories. *Neuron* 52: 437–444.

Pourtier, L. and Sicard, G. (1990) Comparison of the sensitivity of C57BL/6J and AKR/J mice to airborne molecules of isovaleric acid and amyl acetate. *Behav Genet* 20: 499–509.

Rampon, C., Tang, Y.P., Goodhouse, J., Shimizu, E., Kyin, M., and Tsien, J.Z. (2000) Enrichment induces structural changes and recovery from nonspatial memory deficits in CA1 NMDAR1-knockout mice. *Nat Neurosci* 3: 238–244.

Reilly, S. and Pritchard, T. (1996) Gustatory thalamus lesions in the rat: I. Innate taste preferences and aversions. *Behav Neurosci* 110: 737–745.

Restivo, L., Chaillan, F.A., Ammassari-Teule, M., Roman, F.S., and Marcheti, E. (2006) Strain differences in rewarded discrimination learning using the olfactory tubing maze. *Behav Genet* 36: 923–934.

Risinger, F.O. and Brown, M.M. (1996) Genetic differences in nicotine-induced conditioned taste aversion. *Life Sci* 58: PL223–PL229.

Risinger, F.O. and Cunningham, C.L. (1992) Genetic differences in ethanol-induced hyperglycemia and conditioned taste aversion. *Life Sci* 50: PL113–PL118.

Risinger, F.O. and Cunningham, C.L. (1995) Genetic differences in ethanol-induced conditioned taste aversion after ethanol preexposure. *Alcohol* 12: 535–539.

Risinger, F.O. and Cunningham, C.L. (1998) Ethanol-induced conditioned taste aversion in BXD recombinant inbred mice. *Alcohol Clin Exp Res* 22: 1234–1244.

Risinger, F.O. and Cunningham, C.L. (2000) DBA/2J mice develop stronger lithium chloride-induced conditioned taste and place aversions than C57BL/6J mice. *Pharmacol Biochem Behav* 67: 17–24.

Roullet, P., Lassalle, J.M., and Jegat, R. (1993) A study of behavioral and sensorial bases of radial maze learning in mice. *Behav Neural Biol* 59: 173–179.

Sashihara, S., Felts, P.A., Waxman, S.G., and Matsui, T. (1996) Orphan nuclear receptor ROR gene: isoform-specific spatiotemporal expression during postnatal development of brain. *Mol Brain Res* 42: 109–117.

Sclafani, A. and Glendinning, J.I. (2005) Sugar and fat conditioned flavor preferences in C57BL/6J and 129 mice: oral and postoral interactions. *Am J Physiol Regul Integr Comp Physiol* 289: R712–R720.

Shimshek, D.R., Bus, T., Kim, J., Mihaljevic, A., Mack, V., Seeburg, P.H., et al. (2005) Enhanced odor discrimination and impaired olfactory memory by spatially controlled switch of AMPA receptors. *PLoS Biol* 3: e354.

Smith, D.F. (1978) Lithium chloride toxicity and pharmacodynamics in inbred mice. *Acta Pharmacol Toxicol* 43: 51–54.

St. John, S.J. and Boughter, J.D., Jr. (2004) The contribution of taste bud populations to bitter avoidance in mouse strains differentially sensitive to sucrose octa-acetate and quinine. *Chem Senses* 29: 775–795.

Welzl, H., D'Adamo, P., and Lipp, H.-P. (2001) Conditioned taste aversion as a learning and memory paradigm. *Behav Brain Res* 125: 205–213.

Wilson, D.A. and Stevenson, R.J. (2003) Olfactory perceptual learning: the critical role of memory in odor discrimination. *Neurosci Biobehav Rev* 27: 307–328.

Wysocki, C.J., Whitney, G., and Tucker, D. (1977) Specific anosmia in the laboratory mouse. *Behav Genet* 7: 171–188.

Chapter 33

Object recognition in the mouse

Ekrem Dere, Armin Zlomuzica, Maria A. De Souza Silva, and Joseph P. Huston

Object memory

Mice readily approach and explore novel three-dimensional objects, such as coffee cups or bottles, which are assumed to have no natural significance for the animal and which have never been paired with reinforcement. They also show a spontaneous preference for novel over familiar objects. Typically, the mouse noses at the object and explores it with its whiskers and forepaws. The object recognition (OR) task measures spontaneous behavior, which does not require training or the maintenance of the behavior by positive reinforcement (e.g., by food delivery). Furthermore, the task is less afflicted with stress as compared to other popular learning tasks, such as the Morris water maze or fear conditioning tasks. Generally, the OR paradigm should be favored over other learning tasks when the experimental manipulation is likely to have an effect on food intake, reward, and reinforcement-related processes (which preclude food-rewarded tasks), pain perception, or stress (which preclude shock-motivated tasks), and thermoregulation (which preclude water navigation tasks). Since the OR task, in comparison to other animal models of learning and memory, does not require lengthy training and does not induce high levels of stress, it is more closely related to conditions under which human recognition memory is measured (Ennaceur and Delacour, 1988).

Different versions of the OR paradigm can be used to evaluate different memory domains, such as the recognition of a familiar object (Sik et al., 2003), object–place recognition (Massey et al., 2003), temporal order OR, and episodic-like memory in mice (Dere et al., 2005). Since the learning and test situations in these different versions of the OR paradigm are highly similar, it is possible to investigate the effects of experimental manipulations, such as the knockout of a gene, on these different forms of OR memory, excluding confounding influences of paradigm-specific demands on the animal's performance. For example, if a genetic manipulation disrupts motor or sensory systems required to perceive and explore objects, the animals should be equally impaired in the OR, spontaneous object–place recognition, temporal order OR, and episodic-like object memory versions. On the contrary, selective impairments in one, but not the other versions (experimental dissociation), would suggest a specific involvement of the gene in a specific type of object memory.

Object recognition

The OR task typically consists of a sample trial (of usually 3–15 min duration), during which animals explore two identical objects in a familiar arena, followed by a delayed test trial, in which a novel object is presented together with one of the familiar objects. In this situation, depending on the intertrial interval imposed between sample and test trials, the duration of the sample trial, as well as on the observation interval (usually 3–10 min), control animals spend more time exploring the novel object as compared to the familiar object (Sik et al., 2003). It is important to note that this task taxes the memory for unique episodes (one-trial learning), which makes it more sensitive to amnestic experimental interventions as compared to other tasks, in which incremental learning across multiple trials. Although there are other one-trial learning tasks available to tax the memory for an unique episode, such as the inhibitory avoidance task, they all rely on negative reinforcement and are therefore confounded by stress-effects on memory formation.

However, OR in the mouse is also measured after multiple sample trials usually given at a short, e.g., 3 minute intertrial interval. The sample trials are followed by tests of either object displacement and/or novel OR (Orsini et al., 2004; Roullet and Lassalle, 1990; Roullet et al., 1997). In contrast to the one-trial OR version where only one object (two copies) is presented during the sample trial, in the multiple sample trial version several different objects are presented in a stable spatial configuration during the sampling trials.

In the one-trial version, retention intervals ranging from seconds to hours (but sometimes up to days) are used. Since performance deteriorates as the delay between the sample and the test trial increases, one can distinguish between the effects of experimental manipulations on short/working and long-term memory. One can also assume an effect on the *de novo* protein synthesis-dependent consolidation stage of long-term object memory (given that other factors,

Behavioral Genetics of the Mouse: Volume I. *Genetics of Behavioral Phenotypes*, eds. Wim E. Crusio, Frans Sluyter, Robert T. Gerlai, and Susanna Pietropaolo. Published by Cambridge University Press.

such as state-dependent learning, are controlled), if an experimental intervention affects recognition memory after "long" (e.g., >2 h) but not "short" (e.g., <30 min intervals) retention. One advantage of the one-trial OR paradigm is that the effects of lesions or drugs on attention, sensory-motor functions, or the motivation to explore novel objects can be potentially distinguished from effects on memory, since the latter is inferred from delay-dependent differences in the time spent exploring a novel and familiar object. If a lesion or drug impairs, e.g., the motor performance required to explore objects, this impairment should have the same effect on the exploration of the novel and familiar object, independent of the delay interposed between the sample and test trial. Conversely, a specific effect on recognition memory is assumed if the experimental intervention impairs or improves recognition memory after long, but not short, delays.

Object–place recognition

A modification of the paradigm allows the measurement of the memory for spatial locations within an environment or context, where objects have been initially explored. In this version of the paradigm two identical and familiar objects are presented during the test trial, with one of the objects shifted to a novel location. Here, the animals spend more time exploring the object encountered in the novel location (Massey et al., 2003), suggesting that mice not only encode and maintain the features of an object, but also the spatial location in which it was encountered. After a sampling trial of only 3 minutes, adult male Swiss albino mice are able to recognize the displacement of an object to a novel location after both short- and long-retention delays of 30, 90, 180, or 360 minutes (Assini et al., 2009). In line with the extensive literature showing that hippocampal lesions impair spatial memory, the same study reported that lidocaine inactivations of the hippocampal CA1 region before the sample trial impaired object–place recognition in Swiss mice (Assini et al., 2009).

Temporal order memory

Another variant of the OR paradigm measures the memory for the temporal order in which two different objects were presented in the past. Unlike the spontaneous OR, object–place and object–context versions, this task is a three-trial procedure, composed of two sample trials, usually with an intertrial interval of about 1 hour, during which two copies of a novel object are presented. During the test trial one "old familiar" object, known from the first sample trial, and another "recent familiar" object from the second sample trial, are presented together. Here, one finds that the animals spend more time exploring the "old familiar" object relative to the "recent familiar" object, indicating that the previously explored objects are recognized and discriminated in terms of their relative recency (Dere et al., 2005; DeVito et al., 2009) and not in respect of their relative familiarity (Hannesson et al., 2004).

Episodic-like memory

The study of amnesic patients has led to the idea of multiple memory systems. Among these, episodic memory is especially important, because it permits the remembering of autobiographical events instead of the mere remembering of impersonal facts and rules. Episodic memory refers to unique experiences which have occurred only once and include the remembering of what happened and where and when it happened. This type of memory is extremely sensitive to aging and disease. Against this background it is desirable to have a valid animal model of episodic memory in order to evaluate possible therapeutic strategies. A recent attempt to model *what, where,* and *when* memory for unique experiences in mice made use of a three-trial object exploration task. The mice were placed into an open field containing four copies of a novel object. After a delay of 50 minutes, the mice received a second trial identical to the first, except that four novel objects were presented. After an additional delay of 50 minutes, the mice received a test trial, identical to the sample trials, except that two copies of the object from sample trial 1 ("old familiar" objects) and two copies of the object known from sample trial 2 ("recent familiar" objects) were present. Furthermore, one "old familiar" object was shifted to a novel location, whereas the "recent familiar" objects were presented at familiar locations. In this situation, the mice spent more time exploring two "old familiar" objects relative to two "recent familiar" objects, reflecting memory for what and when, and concomitantly directed more exploration at a spatially displaced "old familiar" object relative to a stationary "old familiar" object, reflecting memory for what and where (Dere et al., 2005). Thus, the OR paradigm can also be used to measure *what, where,* and *when* memory for unique events in mice (Dere et al., 2008) and rats (Kart-Teke et al., 2006)

Applications

Compared to the multiple sample trial version of the OR task and other paradigms based on incremental learning, the one-trial OR version allows the investigation of drug effects on different stages of memory formation (encoding, rehearsal, consolidation) and retrieval/ recollection (Pitsikas et al., 2003). To assess the effects of a drug on the encoding of object characteristics and/or early stages of memory consolidation, the drug can be administered prior to the sample trial. It can also be administered immediately after the sample trial to study drug effects on the consolidation of object information (Winters and Bussey, 2005). Finally, drug effects on object memory retrieval can be assessed by pre-test application. However, one has to consider that pre-sample or pre-test trial drug applications: (a) can lead to state-dependent learning; (b) might impair sensory-motor functions; (c) can alter motivational factors; or (d) can influence arousal and/or attention, all being confounding variables, which have to be excluded by appropriate control experiments and by considering the pharmacokinetic dynamics and metabolization rates of the drugs used (Spear and File, 1996). On the other hand, one has to consider that post-sample trial applications can have proactive effects on the test trial performance and other physiological consequences, which may preclude interpretation in terms of actions on the consolidation process (Carey, 1987).

Compared to the classical food rewarded delayed matching or non-matching to sample tasks, where the delays are usually as short as a few minutes, the OR task also allows the assessment of OR on a long-term memory scale (up to 24 h). Dependent on the duration of the sample trial and the retention interval applied, the OR paradigm is sensitive to both amnesic (Ennaceur and Meliani, 1992; Kart-Teke et al., 2006) and promnesic (Ennaceur et al., 1989; Kart-Teke et al., 2007; Zlomuzica et al., 2007, 2008) pharmacological manipulations. Finally, with at least 1 week of rest between tests, animals can be tested repeatedly in different versions of the paradigm, under different drug doses, or under different delays imposed between the sample and test trials in order to obtain within-subject dissociations between different types of object memory, dose response curves, or to show the delay dependency of an experimental manipulation.

After sufficient training (repeated sample object exposures) object memory can last for several weeks. Thus, the OR paradigm can be used to model human anterograde and (sometimes temporarily graded) retrograde amnesia, e.g., in rats with hippocampal or rhinal cortex lesions (Gaskin et al., 2003; Kornecook et al., 1999; Mumby et al., 1999, 2002, 2005).

The OR paradigm combined with the inactivation of targeted genes has been a powerful tool for elucidating the molecular and cellular mechanisms of learning and memory in mice. Since molecular genetics and transgenic technology have been improved to a level which allows the inactivation, overexpression, or replacement of targeted genes in a cell-type, region-specific, and/or temporally restricted manner, it is possible to investigate the effect of highly specific genetic manipulations on anterograde and retrograde amnesia in mouse mutants.

The OR task was initially developed for rats (Ennaceur and Delacour, 1988) and, later on, adapted for mice with only minor modifications (Dodart et al., 1997; Messier, 1997; Steckler et al., 1999). The OR task and its modifications are currently used in lesion and pharmacological studies, animal models of aging, studies comparing rodent strains, as well as for the cognitive phenotyping of genetically engineered mice. Critical factors that can influence the outcome in OR tasks are housing conditions, the rodent strains used (see following section), the age and sex of the animals (Jonasson, 2005 for review), whether rich spatial and contextual cues were present during testing (Murai et al., 2007), the duration of sample and test trials, and the length of the intertrial interval (Dere et al., 2007).

Natural genetic variation in spontaneous OR in the mouse

By the use of transgenic and gene targeting technologies mouse mutants have been generated that model human neurodegenerative and psychiatric diseases. Sometimes the behavioral phenotypes of these mouse mutants were rather unexpected and contradictory. Some of these results were certainly due to variations in the genetic background of those models. Therefore, it is important to have detailed information on natural genetic variation in OR performance among inbred mouse strains in order to be able to select the appropriate strain for molecular biological and pharmacological studies (Gerlai, 2001). Although there is an increasing number of pharmacological and gene-targeting studies using OR tasks to assess memory performance in mice (Dere et al., 2007), there are only a few studies available that directly compare the learning performance of different inbred mouse strains in such OR tasks. Brooks et al. (2005) assessed six inbred strains (129S2/Sv, BALB/c, C3H/He, C57BL/6J, CBA/Ca, DBA/2) commonly used in behavioral pharmacology and in the generation of genetic models of neuropsychiatric diseases. They first habituated the mice to the test apparatus and the presence of objects in order to control for strain differences in novel object neophobia, and thereafter presented the animals with two copies of a novel object during a singe sample trial. The sample trial was terminated after the animal had explored each object for a total of 30 seconds or after 5 minutes had elapsed, whichever occurred first. After this single sample trial three test trials were given after delays of 1, 4, and 24 hours. On test trials, the familiar object known from the sample trial was always presented together with a novel object. A significant effect of delay was observed which, however, was not strain-dependent. The mice could readily distinguish between novel and familiar objects after delays of 1 and 4 but not 24 hours after the sample trial. There were no significant strain differences in terms of novel object preference when the absolute difference between time spent exploring the novel and the familiar object was taken as the dependent variable. However, when the difference in the exploration of the novel and familiar object was expressed as percent of the total time spent exploring both objects, there was a significant strain × delay interaction. It further appeared that among the six strains tested, the 129S2/Sv strain performed worst, while BALB/c and DBA/2 strains performed best. In terms of learning and memory the C57BL/6 and DBA/2 strains, the latter being one of the first inbred strains produced, are the best-studied mouse strains. DBA/2 mice are impaired compared to C57BL/6 mice in spontaneous OR memory at a delay of 15 minutes (Podhorna and Brown, 2002) and 24 hours (Voikar et al., 2005). There is evidence that the gene transcription factor cAMP-response element-binding protein (CREB) is involved in *de novo* protein synthesis required for long-term memory (Pittenger et al., 2002) and that the hippocampus plays a role in OR in mice (Hammond et al., 2004). Interestingly, it has been shown that training in the hidden platform version of the water-maze task induced significantly higher levels of phosphorylated CREB in the hippocampus of C57BL/6 mice as compared to DBA/2 mice (Sung et al., 2008). It is therefore possible that the differences in the OR task seen after long retention intervals between these two strains are related to learning-induced gene expression, which is essential for long-term memory formation.

In contrast, Orsini et al. (2004) reported that C57BL/6 mice are impaired as compared to DBA mice in OR after a very short delay of 3 minutes. While DBA/2 mice showed a significant preference for the novel object, the C57BL/6 mice failed to do so. However, it is unlikely that C57BL/6 mice are unable to recognize a familiar object after a delay of only 3 minutes. It is more reasonable to assume that after a delay of only 3 minutes between the sample and the test trial, the C57BL/6 mice may be saturated, and not motivated to explore any objects again, regardless of whether they are familiar or novel. Indeed, compared to the DBA/2 mice, the C57BL/6 ones spent significantly less total time exploring both objects during the recognition trial. There is also evidence that mouse strains differ in terms of novel object-induced neophobia. Furthermore, there are interindividual differences in novel object-induced neophobia (Dere et al., 2004) that can affect performance in OR. Furthermore, for CF1 male adult mice, it has been reported that these interindividual differences are stable on retests at 1 week and 8 months after the initial assessment (Kazlauckas et al., 2005).

Podhorna and Brown (2002) showed that, compared to DBA/2 mice, the C57BL/6 mice showed significantly fewer total object approaches and significantly more stretch-attend postures during the sample phase of the OR task, suggesting higher levels of novel object-induced neophobia. Novel object-induced neophobia, as reflected by the frequency of stretch-attend postures, is generally negatively correlated with the time spent exploring novel objects. A low level of object exploration during the sample trial inevitably leads to a poorer object memory, because of insufficient encoding of the object features.

It has been reported that 129/SvPAS@Ico mice are impaired compared to C57BL/6, Swiss, and BALB/c mice in OR at retention delays of 1 and 24 hours, while the latter three strains perform at similar levels (Sik et al., 2003). The severe impairment of 129/SvPAS@Icomice is especially important in gene-targeting studies, since the embryonic stem cells used in these studies are mostly derived from this strain and most mutant mouse strains tested behaviorally still bear residual 129/SvPAS@Ico genes in their genome (see Gerlai 2001 for discussion). In general, the 129 inbred strains are hypoactive, show little exploratory behavior (Royle et al., 1999), and have some other neuroanatomical, electrophysiological, and behavioral peculiarities (Balogh et al. 1999, Bampton et al., 1999; Crawley et al., 1997; Gerlai, 1998; Magara et al., 1999; Montkowski et al., 1997; Wolfer et al., 1997).

Roullet et al. (1997) compared C57BL/6 and DBA/2 mice in a multiple-sample trial version of the OR paradigm. The mice were repeatedly exposed to a configuration of five different objects placed at fixed locations within an arena. After habituation to this spatial configuration of objects, the authors examined the animal's reactivity to object displacement or the presentation of a novel object. The authors report that C57BL/6 mice reacted to spatial novelty while DBA mice did not. In contrast, both strains responded to object novelty in a similar manner. It seems that the C57BL/6 mice exhibit superior performance to DBA/2 mice in OR tasks and display higher levels of novel-object-induced neophobia relative to DBA/2 mice. Interestingly, lesions of the nucleus accumbens, which has been implicated in behavioral correlates of neophobia (Green et al., 2006), had no effect on novel object or object–place recognition in C57BL/6 mice but restored the impaired object–place recognition in the DBA mice (Roullet et al., 1997). Therefore, it is possible that the differences between these two strains in OR tasks are partially due to differences in object-induced neophobia. It remains to be tested whether the above mouse strain differences in different versions of the OR task are due to neophobia, generally reflected by low amounts of object exploration and stretch-attend postures, rather than differences in learning and memory capacities. This could be done by introducing experimentally naïve mice of each strain into an unfamiliar open field with a novel object placed in the center. Here, neophobic animals show less exploration of the novel object compared to neophilic ones (Dere et al., 2004; Roy and Chapillon, 2004).

It has been proposed that certain sleep stages such as rapid eye movement (REM) and slow wave sleep play a role in memory consolidation (Axmacher et al., 2009). Palchykova et al. (2006) have shown that sleep deprivation immediately after the sample trials, but not after a delay of 6 hours, can impair OR after a 24 hour retention interval in OF1 mice, suggesting that sleep immediately after learning, where consolidation takes place, is important for the establishment of a long-term object memory. C57BL/6J mice and BALB/cJ mice have been examined for sleep behavior and architecture in blocks of 5 hours for 24 hours after placing a novel object into the home cage (Tang et al., 2005). The BALB/cJ mice showed greater reductions in total sleep time as compared to the C57BL/6J mice after introduction of the novel object. Compared to their respective controls, the C57BL/6J mice showed increased latency to enter REM and non-rapid eye movement (NREM) sleep stages after object placement, while in the BALB/cJ mice only the latency to enter NREM stage was increased. Furthermore, the BALB/cJ mice exhibited reductions in total REM and NREM sleep during the first 5 hours of object exposure. In C57BL/6J mice only the total amount of NREM sleep was reduced during the first 5 hour block. During the second 5 hour block only the C57BL/6J mice showed increased time of NREM and total sleep. There were no significant differences between C57BL/6J and BALB/cJ mice during the second and third 5 hour blocks (Tang et al., 2005). In conclusion, it is possible that strain differences in OR tasks might be at least in part also due to differences in memory consolidation during sleep.

There is a large natural genetic variation in visual abilities (measured in terms of visual detection, pattern discrimination, and visual acuity), among inbred mice strains including 129SI/Sv1mJ, A/J, AKR/J, BALB/cByJ, C3H/HeJ, C57BL/6J, CAST/EiJ, DBA/2J, FVB/NJ, MOLF/EiJ, SJL/J, SM/J, and SPRET/EiJ strains (Brown and Wong, 2007). It has been reported that these strain differences in visual acuity can account for a significant proportion of the variance between strains in measures of learning and memory in the Morris water maze (Brown and Wong, 2007). Fortunately, in the OR task the mouse does not depend on visual, but rather on tactile information to establish a memory representation of novel objects. As a nocturnal species mice investigate objects by physical contacts mainly with their whiskers and forepaws (Ferezou et al., 2006). Therefore, the OR task is more sensitive to detect changes in learning and memory performance in mice after experimental manipulations that might affect vision.

Hippocampal synaptic plasticity, e.g., long-term potentiation (LTP) and depression (LTD) have been proposed to play important roles in learning and memory. Although the evidence that suggests a role for synaptic changes such as those seen in LTP and LTD in learning and memory is mainly correlative, it is still interesting to see whether strain differences in OR performance might be paralleled by differences in hippocampal synaptic plasticity. Using *in vitro* slice stimulation and recording methods, Nguyen et al. (2000) compared the synapto-plastic properties of hippocampal CA1 neurons in C57BL/6J, CBA/J, DBA/2J, and 129/SvEms/J mice. It was found that hippocampus CA1-LTP induced by a combination of theta, and repeated multiburst 100 Hz stimulation led to a synaptic potentiation that was of longer persistence in C57BL/6J mice as compared to both CBA and DBA mice. In contrast, LTD was not significantly different between these four strains. Furthermore, forskolin/IBMX-induced activation of cAMP-dependent protein kinase in the hippocampus was deficient in area CA1 of DBA/2J and CBA/J slices, but not in BL/6J and 129/SvEms/J slices. These results suggest a genetic basis for some forms of synaptic plasticity. It is therefore possible that the strain-dependent variation in cAMP-induced synaptic facilitation may contribute to the reported strain differences in learning and memory performance.

Summary

The OR paradigm is a powerful tool for studying the neurobiology of learning and memory processes in the mouse. It measures spontaneous behavior which does not require food or water deprivation, the application of reinforcers (food delivery or electric shock application), or extensive rule learning. This task is also by far less stressful than, e.g., the water maze or inhibitory avoidance tasks, which have been widely used as models of learning and memory. Using the OR paradigm it is, for example, possible to investigate the effects of genetic interventions in the mouse genome on memory not confounded by concomitant effects on weight regulation, reinforcement, pain perception, and susceptibility to stress. Furthermore, variations of the OR paradigm can be used to measure different forms of recognition memory, including OR, object–place memory, and temporal order memory in mice.

Acknowledgments

This work was supported by the Deutsche Forschungsgemeinschaft (DFG) through grants no. DE 1149/5–1 to E. D. and DE 792/2–4 to MAdSS.

References

Assini, F.L., Duzzioni, M., and Takahashi, R.N. (2009) Object location memory in mice: pharmacological validation and further evidence of hippocampal CA1 participation. *Behav Brain Res* 204: 206–211.

Axmacher, N., Draguhn, A., Elger, C.E., and Fell, J. (2009) Memory processes during sleep: beyond the standard consolidation theory. *Cell Mol Life Sci* 66: 2285–2297.

Balogh, S.A., McDowell, C.S., Stavnezer, A.J., and Denenberg, V.H. (1999) Behavioral and neuroanatomical assessment of an inbred substrain of 129 mice with behavioral comparisons to C57BL/6J mice. *Brain Res* 836: 38–48.

Bampton, E.T., Gray, R.A., and Large, C.H. (1999) Electrophysiological characterisation of the dentate gyrus in five inbred strains of mouse. *Brain Res* 841: 123–134.

Brooks, S.P., Pask, T., Jones, L., and Dunnett, S.B. (2005) Behavioural profiles of inbred mouse strains used as transgenic backgrounds. II: cognitive tests. *Genes Brain Behav* 4: 307–317.

Brown, R.E., and Wong, A.A. (2007) The influence of visual ability on learning and memory performance in 13 strains of mice. *Learn Mem* 14: 134–144.

Carey, R.J. (1987) Post-trial hormonal treatment effects: memory modulation or perceptual distortion? *J Neurosci Methods* 22: 27–30.

Crawley, J.N., Belknap, J.K., Collins, A., Crabbe, J.C., Frankel, W., Henderson, N., et al. (1997) Behavioral phenotypes of inbred mouse strains: implications and recommendations for molecular studies. *Psychopharmacology (Berl)* 132: 107–124.

Dere, E., De Souza Silva, M.A., and Huston, J.P. (2004) Higher order memories for objects encountered in different spatio-temporal contexts in mice: evidence for episodic memory. *Rev Neurosci* 15: 231–240.

Dere, E., Huston, J.P., and De Souza Silva, M.A. (2005) Integrated memory for objects, places, and temporal order: evidence for episodic-like memory in mice. *Neurobiol Learn Mem* 84: 214–221.

Dere, E., Huston, J.P., and De Souza Silva, M.A. (2007) The pharmacology, neuroanatomy and neurogenetics of one-trial object recognition in rodents. *Neurosci Biobehav Rev* 31: 673–704.

Dere, E., Zlomuzica, A., Viggiano, D., Ruocco, L.A., Watanabe, T., Sadile, A.G., et al. (2008) Episodic-like and procedural memory impairments in histamine H1 receptor knockout mice coincide with changes in acetylcholine esterase activity in the hippocampus and dopamine turnover in the cerebellum. *Neuroscience* 157: 532–541.

DeVito, L.M., Konigsberg, R., Lykken, C., Sauvage, M., Young, W.S., and Eichenbaum, H. (2009) Vasopressin 1b receptor knock-out impairs memory for temporal order. *J Neurosci* 29: 2676–2683.

Dodart, J.C., Mathis, C., and Ungerer, A. (1997) Scopolamine-induced deficits in a two-trial object recognition task in mice. *Neuroreport* 8: 1173–1178.

Ennaceur, A., Cavoy, A., Costa, J.C., and Delacour, J. (1989) A new one-trial test for neurobiological studies of memory in rats. II: Effects of piracetam and pramiracetam. *Behav Brain Res* 33: 197–207.

Ennaceur, A. and Delacour, J. (1988) A new one-trial test for neurobiological studies of memory in rats. 1: Behavioral data. *Behav Brain Res* 31: 47–59.

Ennaceur, A. and Meliani, K. (1992) A new one-trial test for neurobiological studies of memory in rats. III. Spatial vs. non-spatial working memory. *Behav Brain Res* 51: 83–92.

Ferezou, I., Bolea, S., and Petersen, C.C. (2006) Visualizing the cortical representation of whisker touch: voltage-sensitive dye imaging in freely moving mice. *Neuron* 50: 617–629.

Gaskin, S., Tremblay, A., and Mumby, D.G. (2003) Retrograde and anterograde object recognition in rats with hippocampal lesions. *Hippocampus* 13: 962–969.

Gerlai, R. (1998) A new continuous alternation task in T-maze detects hippocampal dysfunction in mice. A strain comparison and lesion study. *Behav Brain Res* 95: 91–101.

Gerlai, R. (2001) Gene targeting: technical confounds and potential solutions in behavioral brain research. *Behav Brain Res* 125: 13–21.

Green, T.A., Alibhai, I.N., Hommel, J.D., DiLeone, R.J., Kumar, A., Theobald, D.E., et al. (2006) Induction of inducible cAMP early repressor expression in nucleus accumbens by stress or amphetamine increases behavioral responses to emotional stimuli. *J Neurosci* 26: 8235–8242.

Hammond, R.S., Tull, L.E., and Stackman, R.W. (2004) On the delay-dependent involvement of the hippocampus in object recognition memory. *Neurobiol Learn Mem* 82: 26–34.

Hannesson, D.K., Howland, J.G., and Phillips, A.G. (2004) Interaction between perirhinal and medial prefrontal cortex is required for temporal order but not recognition memory for objects in rats. *J Neurosci* 24: 4596–4604.

Jonasson, Z. (2005) Meta-analysis of sex differences in rodent models of learning and memory: a review of behavioral and biological data. *Neurosci Biobehav Rev* 28: 811–825.

Kart-Teke, E., Dere, E., Brandão, M.L., Huston, J.P., and De Souza Silva, M.A. (2007) Reinstatement of episodic-like memory in rats by neurokinin-1 receptor antagonism. *Neurobiol Learn Mem* 87: 324–331.

Kart-Teke, E., De Souza Silva, M.A., Huston, J.P., and Dere, E. (2006) Wistar rats show episodic-like memory for unique experiences. *Neurobiol Learn Mem* 85: 173–182.

Kazlauckas, V., Schuh, J., Dall'Igna, O.P., Pereira, G.S., Bonan, C.D., and Lara, D.R. (2005) Behavioral and cognitive profile of mice with high and low exploratory phenotypes. *Behav Brain Res* 162: 272–278.

Kornecook, T.J., Anzarut, A., and Pinel, J.P. (1999) Rhinal cortex, but not medial thalamic, lesions cause retrograde amnesia for objects in rats. *Neuroreport* 10: 2853–2858.

Magara, F., Muller, U., Li, Z.W., Lipp, H.P., Weissmann, C., Stagljar, M., et al. (1999) Genetic background changes the pattern of forebrain commissure defects in transgenic mice underexpressing the beta-amyloid-precursor protein. *Proc Natl Acad Sci USA* 96: 4656–4661.

Massey, P.V., Warburton, E.C., Wynick, D., Brown, M.W., and Bashir, Z.I. (2003) Galanin regulates spatial memory but not visual recognition memory or synaptic plasticity in perirhinal cortex. *Neuropharmacology* 44: 40–48.

Messier, C. (1997) Object recognition in mice: improvement of memory by glucose. *Neurobiol Learn Mem* 67: 172–175.

Montkowski, A., Poettig, M., Mederer, A., and Holsboer, F. (1997) Behavioural performance in three substrains of mouse strain 129. *Brain Res* 762: 12–18.

Mumby, D.G., Astur, R.S., Weisend, M.P., and Sutherland, R.J. (1999) Retrograde amnesia and selective damage to the hippocampal formation: memory for places and object discriminations. *Behav Brain Res* 106: 97–107.

Mumby, D.G., Glenn, M.J., Nesbitt, C., and Kyriazis, D.A. (2002) Dissociation in retrograde memory for object discriminations and object recognition in rats with perirhinal cortex damage. *Behav Brain Res* 132: 215–226.

Mumby, D.G., Tremblay, A., Lecluse, V., and Lehmann, H. (2005) Hippocampal damage and anterograde object-recognition in rats after long retention intervals. *Hippocampus* 15: 1050–1056.

Murai, T., Okuda, S., Tanaka, T., and Ohta, H. (2007) Characteristics of object location memory in mice: behavioral and pharmacological studies. *Physiol Behav* 90: 116–124.

Nguyen, P.V., Duffy, S.N., and Young, J.Z. (2000) Differential maintenance and frequency-dependent tuning of LTP at hippocampal synapses of specific strains of inbred mice. *J Neurophysiol* 84: 2484–2493.

Orsini, C., Buchini, F., Conversi, D., and Cabib, S. (2004) Selective improvement of strain-dependent performances of cognitive tasks by food restriction. *Neurobiol Learn Mem* 81, 96–99.

Palchykova, S., Winsky-Sommerer, R., Meerlo, P., Dürr, R., and Tobler, I. (2006) Sleep deprivation impairs object recognition in mice. *Neurobiol Learn Mem* 85: 263–271.

Pitsikas, N., Rigamonti, A.E., Cella, S.G., and Muller, E.E. (2003) The 5-HT1A receptor antagonist WAY 100635 improves rats performance in different models of amnesia evaluated by the object recognition task. *Brain Res* 983: 215–222.

Pittenger, C., Huang, Y.Y., Paletzki, R.F., Bourtchouladze, R., Scanlin, H., Vronskaya, S., et al. (2002) Reversible inhibition of CREB/ATF transcription factors in region CA1 of the dorsal hippocampus disrupts hippocampus-dependent spatial memory. *Neuron* 34: 447–462.

Podhorna, J. and Brown, R.E. (2002) Strain differences in activity and emotionality do not account for differences in learning and memory performance between C57BL/6 and DBA/2 mice. *Genes Brain Behav* 1: 96–110.

Roullet, P. and Lassalle, J.M. (1990) Genetic variation, hippocampal mossy fibres distribution, novelty reactions and spatial representation in mice. *Behav Brain Res* 41: 61–70.

Roullet, P., Mele, A., and Ammassari-Teule, M. (1997) Ibotenic lesions of the nucleus accumbens promote reactivity to spatial novelty in nonreactive DBA mice: implications for neural mechanisms subserving spatial information encoding. *Behav Neurosci* 111: 976–984.

Roy, V. and Chapillon, P. (2004) Further evidences that risk assessment and object exploration behaviours are useful to evaluate emotional reactivity in rodents. *Behav Brain Res* 154: 439–448.

Royle, S.J., Collins, F.C., Rupniak, H.T., Barnes, J.C., and Anderson, R. (1999)

Behavioural analysis and susceptibility to CNS injury of four inbred strains of mice. *Brain Res* 816: 337–349.

Sik, A., van Nieuwehuyzen, P., Prickaerts, J., and Blokland, A. (2003) Performance of different mouse strains in an object recognition task. *Behav Brain Res* 147: 49–54.

Spear, L.P. and File, S.E. (1996) Methodological considerations in neurobehavioral teratology. *Pharmacol Biochem Behav* 55: 455–457.

Steckler, T., Weis, C., Sauvage, M., Mederer, A., and Holsboer, F. (1999) Disrupted allocentric but preserved egocentric spatial learning in transgenic mice with impaired glucocorticoid receptor function. *Behav Brain Res* 100: 77–89.

Sung, J.Y., Goo, J.S., Lee, D.E., Jin, D.Q., Bizon, J.L., Gallagher, M., et al. (2008) Learning strategy selection in the water maze and hippocampal CREB phosphorylation differ in two inbred strains of mice. *Learn Mem* 15: 183–188.

Tang, X., Xiao, J., Parris, B.S., Fang, J., and Sanford, L.D. (2005) Differential effects of two types of environmental novelty on activity and sleep in BALB/cJ and C57BL/6J mice. *Physiol Behav* 85: 419–429.

Voikar, V., Polus, A., Vasar, E., and Rauvala, H. (2005) Long-term individual housing in C57BL/6J and DBA/2 mice: assessment of behavioral consequences. *Genes Brain Behav* 4: 240–252.

Winters, B.D. and Bussey, T.J. (2005) Glutamate receptors in perirhinal cortex mediate encoding, retrieval, and consolidation of object recognition memory. *J Neurosci* 25: 4243–4251.

Wolfer, D.P., Muller, U., Stagliar, M., and Lipp, H.P. (1997) Assessing the effects of the 129/Sv genetic background on swimming navigation learning in transgenic mutants: a study using mice with a modified beta-amyloid precursor protein gene. *Brain Res* 771: 1–13.

Zlomuzica, A., Dere, E., Huston, J.P., and De Souza Silva, M.A. (2008) NK3 receptor agonism promotes episodic-like memory in mice. *Neurobiol Learn Mem* 90: 420–425.

Zlomuzica, A., De Souza Silva, M.A., Huston, J.P., and Dere, E. (2007) NMDA receptor modulation by D-cycloserine promotes episodic-like memory in mice. *Psychopharmacology (Berl)* 193: 503–509.

Index

Bold page numbers indicate illustrations, *italic* page numbers indicate tables/figures